BUILDING CONSTRUCTION

A Practical Guide in the technique of building construction for students preparing for A.M.I.E., U.P.S.C., Degree, Diploma and other examinations conducted by Indian Universities

By

SUSHIL KUMAR

Author of :

Treasure of R.C.C. Designs

and

Engineering Materials

Standard Publishers Distributors

1705-B, Nai Sarak, DELHI-110006

P.B. No. 1066 Phone : 326 2700, 328 5798

Published by:
A.K. Jain (Prop.)
For Standard Publishers Distributors
1705-B, Nai Sarak, Delhi- 110006
Ph: 23262700, 23285798 Fax: 23243180

Edition: 1st 1965, 2nd 1966, 3rd 1967, 4th 1968, 5th 1969, 6th 1973, 7th 1975, 8th 1976, 9th 1978, 10th 1979, 11th 1981, 12th 1982, 13th 1984, 14th 1985, 15th 1986, 16th 1988, 17th 1991, 18th 1994, 19th 1997; Reprints 1999, 2001, 2003, 2005, 2008

20th Edition : 2010
Reprint : 2012
Reprint : 2014
Reprint : 2015
Reprint : 2016
Reprint : 2017
Reprint : 2018
Reprint : 2019
Reprint : 2021
Reprint : 2023

Price: Rs. 425-00

ISBN : 978-81-8014-168-3

Laser Typesetting by:
Bhargave Laser Printers, Delhi.

Printed at :
Lomus Offset Press, Delhi

PREFACE TO THE NINETEENTH EDITION

In this edition of the book two new chapters have been added. The first chapter pertains to 'Introduction to Buildings' and the second chapter pertains to 'Water Supply and Drainage Systems'. During the process of up-dating the text throughly it was considered necessary to carry out certain additions and deletion in the subject matters and to rearrange some of the chapters.

I express my thanks to all those who have assisted me in bringing out this edition of the book so expeditiously.

New Delhi SUSHIL KUMAR

PREFACE TO FIRST EDITION

This book has been written for the architect, builder, building surveyor and civil, structural and municipal engineer and is intended not only for the professional examinations but also for the degree and diploma examination of the various Indian institutions and universities, Technical and scientific knowledge of building construction takes new strides from year to year so much so that it is almost impossible for the modern engineer to keep abreast of the new development in the field. This book will be of real assistance in this context as it incorporates the results of modern construction in a selective manner, pointing out its pitfalls and bottle necks and shows ways of avoiding such failures.

The preparation of self-explanatory drawing is an art as much can be accomplished and the full story can be told without making the details cumbersome with a little thought and mature skill. This book gives an accurate insight into the preparation of working drawings for building construction as they are used in actual construction work. To achieve the object there has been a generous augmentation of more than thirteen hundred illustrations in this book.

The book has been divided into twenty three chapters. The chapters included in the book strike a high water mark for their significance and importance. A sincere attempt has been made to cover an extensive field and disburse the benefits as wide as possible. Knotty questions bearing on respective chapters set in A.M.I.E. and degree/diploma examinations have been solved in the text to provide all assistance a technical and aspiring professional needs and finds difficult to lay his hands upon. I have availed of every useful information and statistical data from great engineering works, reliable Government publications and eminent authors and if some of the chapters of an outstanding quality, scarcely any credit is mine.

Materials are essential elements, constituent parts or substances which are used to raise a building. But materials could not be turned into structures without a method of construction. Also the stability of a structure requires that its components be so designed and constructed as to render maximum benefit with minimum cost. A successful engineer needless to say, must be an adept in structural designs to provide new attributes to introduce new innovations and to serve new purposes in familiar designs. To enable the reader to have a thorough grasp of the subject, self-explanatory structural designs have been added to the various chapters.

The author owes special thanks to Shri Prabhu Dayal Bindoria but for whose unstinted labour in preparing the drawings the book could not have been published. It is indeed very pleasant to express my thanks to all those who found time to lighten my work by advice and assistance. It is particularly fortunate in having Mrs. Trishla Jain as my wife whose meticulous proof correction has, no doubt, proved to be of utmost value.

New Delhi
March 1965

Sushil Kumar

CONTENTS

Pages

Pages

Pages

1

Introduction

1·1. GENERAL

A building can be defined as a structure broadly consisting of walls, floors and roofs, erected to provide covered space for different uses such as residence, education, business, manufacturing, storage, hospitalization, entertainment, worship etc. The method adopted for construction and choice of material to be used in the building depends upon a number of factors like character of occupancy, location of site, climate, local materials and funds available.

Normally all buildings are constructed according to drawings and specifications prepared by architects. Each city has prescribed building byelaws to which buildings must conform. The building byelaws lay down norms like minimum front, side and rear set backs, minimum height and area of habitable rooms, kitchen, bath, minimum area of windows, width of stair case etc. which are required to be followed by the architect in evolving the design of the building. Apart from respecting the byelaws, the building design should ensure optimum utilization of built-up space (*i.e.* area under circulation should be minimum) thermal comfort, proper ventilation, desirable illumination and acoustical characteristics and it should satisfy the functional requirements of people who live and work in the building.

1·2. TYPES OF BUILDINGS

Depending upon the character of occupancy or the type of use, different types of buildings have been classified in following groups as per national building code :-

(1) *Residential Buildings*

(2) *Educational Buildings*

(3) *Institutional Buildings*

(4) *Assembly Buildings*

(5) *Business Buildings*

(6) *Mercantile Buildings*
(7) *Industrial Buildings*
(8) *Storage Buildings*
(9) *Hazardous Buildings*

1. Residential Buildings. These shall include one or two private dwellings, apartment houses (flats), dormitories, hotels etc.

2. Educational Buildings. These shall include any building used for school, college or day-care purposes involving assembly for instruction, education or recreation.

3. Institutional Buildings. These shall include any building or part thereof which is used for purposes such as medical or other treatment or care of persons suffering from physical or mental illness or disease, care of infants, convalescents of aged persons and for penal or detention in which the liberty of the immates is restricted. Institutional buildings ordinarily provide sleeping accommodation for the occupants. It includes hospitals, sanatoria, nursing homes, orphanages, jails, prisons, mental hospitals, reformatories etc.

4. Assembly Buildings. These shall include any building or part of a building where group of people congregate or gather for amusement, recreation, social, religious, periodic, civil, travel and similar purposes, for example, theatres, motion picture houses, assembly halls, auditoria, exhibition halls, museums, skating rings, gymnasiums, restaurants, places of worship, dance halls, club rooms, passenger stations, and terminals of air, surface, marine and public transportation services.

5. Business Buildings. These shall include any building or part of a building which is used for transaction of business, for the keeping of accounts and records for similar purposes, city halls, town halls, court houses, libraries shall be classified in this group in so far as principal function of these is transaction of public business and the keeping of books and records.

6. Mercantile Buildings. These shall include any building or part of a building which is used as shops, stores, market, for display and sale of merchandise either wholesale or retail.

7. Industrial Buildings. These shall include any building or part of a building or structure, in which products or materials of all kinds and properties are fabricated, assembled or processed, for example, refineries, gas plants, mills, dairies, Industries etc.

8. Storage Buildings. These shall include any building or part of a building primarily for the storage or sheltering of goods, wares or merchandise, like warehouses, cold storages, freight deptt, transit sheds, garages, hangers, grain elevators, stables etc.

9. Hazardous Buildings. These shall include any building or part of building which is used for the storage, handling, manufacture or processing of highly combustible explosive materials or products which are liable to burn with extreme rapidity and/or which may produce poisonous fumes or explosions; for storage, handling, manufacturing or processing which involve highly corrosive, toxic or noxious alkalies, acid or other liquids or chemicals producing flame, fumes and explosive, etc.

1·3. COMPONENTS OF A BUILDING

A building can be broadly divided in two parts *viz.* (*i*) *Sub-structure* and (*ii*) *Super-structure.* The portion of the building below the surrounding ground is known as sub-structure and the portion above the ground is termed as super-structure. The components of a building can be broadly summarised as under:

1. *Foundations*
2. *Plinth*
3. *Walls*
4. *Columns*
5. *Floors*
6. *Doors, windows and ventilators*
7. *Stairs*
8. *Roof*
9. *Building finishes*
10. *Building services*

1. Foundations. Foundation is the lowest part of a structure below the ground level which is in direct contact with the ground and transmits all the dead, live and other loads to the soil on which the structure rests. The provision of foundation is made in such a way that the soil below the foundation is not stressed beyond its safe allowable bearing capacity. Depending upon the type of soil existing at site, its safe bearing capacity and the type of building which is required to be constructed, a structure may need shallow or deep foundations. In case of load bearing walls, the foundation could be in the form of spread footings. For framed structures, the foundation could be in the form of independent column footings, combined footing, rafts or piles. The different types of foundation have been described in details in Chapter 2 and Chapter 3.

2. Plinth. The portion of the building between the ground surrounding the building and the top of the floor immediately above the ground is known as plinth. The level of the surrounding ground is known as *formation level* or simply *ground level* and the level of the ground floor of the building is known as *plinth level.* The plinth height should be such that after proper levelling and grading of the ground adjoining the building (for proper drainage) there is no possibility of the rain water entering the ground floor. The built up covered area measured at the floor level is termed as plinth area.

3. Walls. Walls are provided to enclose or divide the floor space in desired pattern. In addition, walls provide privacy, security and give protection against sun, rain, cold and other adverse effects of weather. The division of floor space varies according to the functions required to be performed in the building. In a well planned layout, the walls divide the space in such a manner so as to achieve maximum carpet area and minimum area of circulation. Walls are constructed by use of building units like bricks, stone, concrete blocks (hollow or solid) etc. The building units are bonded together with mortar in horizontal and vertical joints and the construction is termed as masonry. When bricks are used as building units it is known as brick masonry and when stones are used as building units it is termed as stone masonry. Walls can be broadly divided in two categories viz (*i*) *Load bearing walls* and (*ii*) *Non-load bearing walls.*

A load bearing wall supports its own weight as well as the super-imposed loads transferred to it through floors/roofs.

A non-load bearing wall on the other hand carries its own weight and is not designed to carry any super-imposed load from the structure. They are normally provided as partition walls.

4. Column. A column may be defined as an isolated vertical load bearing member the width of which is neither less than its thickness nor more than four times its thickness. *Pier* is a vertical load bearing member similar to a column except that it is bonded into load bearing wall at the sides to form an integral part and extends to full height of the wall. A pier is introduced to increase the stiffness of the wall to carry additional load or to carry vertical concentrated load. Pier also strengthens the wall to resist lateral pressure without buckling.

5 Floors. Floors are flat supporting elements of a building. They divide a building into different levels thereby creating more accommodation on a given plot of land. The basic purpose of a floor is to provide a firm and dry platform for people and other items like furniture, stores, equipment etc. Floor is generally referred to by its location.

A floor provided for accommodation below the natural ground level is termed as basement floor. A floor immediately above the ground is termed as ground floor and all other floors such as 1st floor, 2nd floor etc. are termed as upper floors. A floor basically consist of two parts namely (*i*) *sub-floor* (*ii*) *Flooring.*

The sub-floor is the structural component of the floor which supports all the loads (dead and super-imposed) and flooring is covering layer of desired specification (cement concrete, terrazzo, tiles etc.) provided over the sub-floor to serve as a finishing layer.

6. Doors, Windows and Ventilators. A door may be defined as a barrier secured in an opening left in a wall to provide usual means of access to a building, room or passage. This can be termed as the most constantly used moving component in a building. A door normally consists of two components namely (*i*) *Door frame*, and (*ii*) *door shutter*. The door frame is permanently held in position and fixed to the masonry of the opening with the help of hold-fasts or rawl plugs. Shutter is the moving part of the door. Doors are made out of material like wood, steel, aluminium, plastic, flexible rubber etc. They can be side hung, sliding, folding, revolving or rolling type depending upon the functional requirement. A window may be defined as an opening left in a wall for the purpose of providing day light, vision and ventilation. Similar to door, a window has a frame and one or more shutters. The shutters are normally fitted with glass or similar transparent material. The windows can be side-hung, top or bottom-hung, louvered type and the shutters can be fully glazed, panelled and glazed or fully panelled type.

7. Stair. A stair may be defined as a structure comprising of a number of steps connecting one floor to another. The stair must be constructed in such a manner that it is safe and comfortable to use and it should be so located as to permit easy communication. Stairs may be made from material like timber, stone, bricks, steel, reinforced concrete etc. The selection of the type of material to be used depends upon the aesthetical importance, funds available, durability and fire resisting qualities desired.

8. Roof. It is the uppermost component of a building and its main function is to cover the space below and protect it from rain, snow, sun, wind etc. A roof basically consist of two component namely (*i*) *The roof decking* and (*ii*) *the roof covering*. The roof decking is the structural component which supports the roof covering. A roof can be either flat, pitched or curved in shape. The choice of the type of roof is made keeping in view the location of the building, weather conditions, funds available and functional and aesthetics requirement. The structural components or roof decking in case of pitched roof is generally a truss, in case of curved roof it is a shell or dome and in case of flat roof it is a flat slab. The roof covering or roofing which is provided over pitched roof could be in the form of tiles, slates, A.C. sheets, G.I. sheets, etc. In case of flat roofs, the roof covering is termed as terracing, which could comprise of a layer of varying thickness of material like lime concrete, mud phuska etc. The terracing serves dual purpose *i.e.* (*i*) of providing suitable slopes on the roof top for draining of rain water and (*ii*) of acting as insulation layer for providing thermal comfort to the users of the space below.

9. Building Finishes. A building is considered incomplete till such time the surface of its components is given appropriate treatment. Building finishes include items like plastering, pointing, white/colour washing, painting, varnishing, distempering etc. The building finishes not only protect the surface from adverse effect of weather but also provide decorative effect.

10. Building Services. Building services include services like water supply, drainage, sanitation, lighting, electricity, accoustics, heating, ventilation, air conditioning, fire detection and fire control etc.

The services like water supply, drainage and sanitation are normally clubbed under the term 'plumbing services'. From consideration of safety of the users, the planning, designing and detailing of all services should be done based on provisions in national building code and also based on norms prescribed by various statutory municipal bodies.

1·4. TECHNICAL TERMS AND DEFINITIONS

Alteration. A change in usage of building or a structural change, such as an addition to the area or height, or the removal of a part of a building, or any change to the structure, such as the construction of, cutting into or removal of any wall, partition, column, beam, floor joist or other support, or a change to or closing of any required means of ingress or egress or a change to the fixtures or equipment.

Balcony. A horizontal cantilevered projection including a hand-rail or balustrade to serve as passage or sitting out place.

Barsati. Habitable room/rooms on the roof of building with or without toilets/kitchen.

Basement or Cellar. The lower storey of a building below or partly below ground level.

Building Height :

(*i*) *In case of flat roofs*—the vertical distance measured from the average level of the centre line of the adjoining street to the highest point of the building adjacent to the street wall.

(*ii*) *In case of pitched roofs*—the vertical distance measured from the average level of the centre line of the adjoining street to the point where the external surface of the outer wall intersect the finished surface of the sloping roof.

(*iii*) *In case of gable end of sloped roof facing the road*—the vertical distance measured from the average level of the centre line of the adjoining street to the mid-point between the eaves level and the ridge.

Note :

(*a*) If the building does not abut on a street, the height shall be measured above the average level of the ground around and contiguous to the building.

(*b*) Architectural features serving no other functions except that of decoration shall be excluded for the purpose of taking heights.

Building Line. The line upto which the plinth of a building adjoining a street or extension of a street or on a future street may lawfully extend.

Cabin. A room constructed with non-load bearing partition/ partitions with adequate provision of light and ventilation.

Chhajja or Sunshade. A sloping or horizontal structural overhang usually provided over openings on external walls for protection from sun and rain.

Courtyard. A space permanently open to the sky, enclosed fully or partially by building and may be at ground level or any other level within or adjacent to a building.

Covered Area. Ground area covered by the building immediately above the plinth level. It does not include the spaces covered by, canopy, chhajja, canopy or alike projections; uncovered staircase, compound wall, gate, well, tank, fountain, uncovered swimming pool, garden etc.

Damp Proof Course. A course consisting of some appropriate water proofing material provided to prevent penetration of dampness or moisture.

Drain. A line of pipes including all fittings and equipment such as manholes, inspection chambers, traps, gullies and floor traps used for the drainage of a building, or a number of buildings. Drain shall also include open channels used for carriage of surface water.

Drainage. The removal of any liquid (storm water, waste water, sewage etc.) by a system constructed for this purpose.

Exit. A passage, channel or means of egress from any building, storey or floor area to a street or other open space of safety.

External Wall. An outer wall of a building not being a party wall even though adjoining to a wall of another building and also means a wall abutting on an interior open space of any building.

Fire Lift. One of the lifts specially designed for use by fire service personnel in the event of fire.

Fire Proof Door. A door or shutter fitted to a wall opening, and constructed and erected with the requirement to check the transmission of heat and fire for a specified period.

Floor. The lower surface in a storey on which one normally walks in a building. The general term, floor, unless otherwise specifically mentioned, shall not refer to a mezzanine floor.

Floor Area Ratio (FAR). The quotient obtained by dividing the total covered area (plinth area) on all floors multiplied by 100 by the area of the plot.

$$FAR = \frac{\text{Total covered area of all floors} \times 100}{\text{Plot Area}}$$

Footing. A foundation unit constructed in brick work, masonry or concrete under the base of a wall or column for the purpose of distributing the load over a large area.

Foundation. That part of the structure which is in direct contact with and transmitting loads to the ground.

Front. As applied to a building means the portion facing the street from which it has access.

Gallery. An intermediate floor or platform projecting from a wall of an auditorium or a hall providing extra floor area, additional seating accommodation etc.,

Garage. A building or portion thereof used for shelter, storage or parking of a wheeled vehicle.

Ground Floor. Means the storey of the building which has its floor surface nearest to the ground around the building.

Habitable Room. A room occupied or designed for occupancy by one or more persons for study, living, sleeping, eating, kitchen (if it is used as living room), but not including bathroom, water closet, store, pantry, corridor etc.

Loft. An intermediary floor space created by introduction of a slab between floor and ceiling of a room, passage or wherever it is provided with maximum clear height of 1·5 metre for storage purposes only.

Mumti or stair cover. A structure with a covering roof over a staircase and its landing, built to enclose only the stairs for the purpose of providing protection from, weather and not used for human habitation.

Mezzanine Floor. An intermediate floor between two floor levels above ground floor and at least one side of it should form an integral part of space floor/below.

Open Space. An area, forming an integral part of the plot, left open to the sky.

Parapet. A low wall or railing built along the edge of a roof or a floor.

Partition. An interior non-load bearing wall, one storey or part storey in height.

Plinth. The portion of a structure between the surface of the surrounding ground and surface of the floor, immediately above the ground.

Plinth Area. The built up covered area measured at the floor level of the basement or of any storey.

Porch. A covered surface supported on pillars or otherwise for the purpose of pedestrian or vehicular approach to a building.

Room Height. The vertical distance measured from the finished floor surface to the finished ceiling surface.

Service Road. A road/lane provided at the rear or side of a plot for service purposes.

Set-Back Line. A line usually parallel to the plot boundaries and laid down in each case by the local municipal bodies beyond which nothing can be constructed towards the plot boundary.

Site or (Plot). A piece of land enclosed by definite boundaries.

Storey. The portion of a building included between the surface of any floor and the surface of the floor next above it, or if there is no floor above it, then the space between any floor and ceiling next above it.

Street. Any means or access *i.e.* highway road, street, lane, pathway etc. over which public have a right of passage.

Structural wall. Means a load bearing wall or a wall that carries load in addition to its own weight.

Travel Distance. The distance from the remotest point on a floor of a building to a place of safety be it a vertical exit, horizontal exit or an outside exit measured along the line of travel.

Water Closet (WC). A privy with arrangement for flushing the pan with water. It does not include a bathroom.

1·5. REQUIREMENT OF PARTS OF BUILDINGS

1. Plinth

Main Building. The plinth or any part of a building shall be so located with respect to surrounding ground level that adequate drainage of the site is assured. The height of plinth shall not be less than 45 cm from surrounding ground level.

Interior Court Yard. Every interior courtyard shall be raised at least 15cm above the level of the centre of the nearest street and shall be satisfactorily drained.

2. Habitable Rooms

(*i*) *Size.* The area of habitable room shall not be less than 9·5 sq.m. in cases where there is only one room. The minimum width of a habitable room shall be 2·4 m. Where there are two rooms, area of one out of these shall not be less than 9·5 sq. m. and area of the other not less than 7·5 sq.m.

(*ii*) *Height.*

(*a*) The height of the habitable room shall not be less than 2·75 m measured from the surface of the floor to lowest point of the ceiling (bottom of slab).

(*b*) In case of pitched roof, the average height of rooms shall not be less than 2·75 m.

(*c*) The minimum clear head room under a beam, folded plates or eaves shall be 2·4 m.

(*d*) In case of air-conditioned rooms, a height of not less than 2·4m, measured from the surface of the floor to the lowest point of air conditioning duct or the false ceiling shall be provided.

(*e*) In case of educational buildings, ceiling height shall be 3·6 m for all regions. It could be 3 m in cold climate.

(*f*) In case of industrial buildings, ceiling height shall be 3·6 m and 3 m if air-conditioned.

3. Kitchen

Size. The area of kitchen shall not be less than 5 sq. m. with a minimum width of 1·8 m. Where there is a separate store, the area of the kitchen may be reduced to 4·5 sq.m . A kitchen which is also intended to be used as a dining room shall have a floor area not less than 9·5 sq.m. with a minimum width of 2·4 m.

Height. The height of a kitchen measured from the surface of the floor to the lowest point in the ceiling (bottom of slab) shall not be less than 2·75 m. except for the portion to accommodate floor trap of the upper floor.

Other Requirements. Every kitchen should have a window of not less than 1 Sq.m . in area, opening directly on to an interior or exterior open space but not into a shaft.

4. Store Room

Size. The area of a store room shall not be less than 3 sq.m. In case, the area of the store room is more than 3 sq.m, the light and ventilation requirement to the extent of 10% of the floor area shall be provided.

Height. The height of a store room shall not be less than 2·2m.

5. Garage

Garage Private. The size of a private garage in residential buildings shall be not less than 2·75 m x 5·4.m.

Height. The height of a garage shall not be less than 2·4 m.

Other Requirements. The plinth of the garage located at ground level shall not be less than 15 cm above the surrounding ground level.

6. Mezzanine Floor

Size. The minimum size of the mezzanine floor (if it is to be used as a living room) shall not be less than 9·5 sq. m. The mezzanine floor area could be upto a maximum of 25% of the actual covered area on the ground floor. Area of mezzanine floor is normally not counted towards FAR calculations.

Height. The height of Mezzanine floor shall not be less than 2·20 m, and not more than 2·70 m.

7. Bathrooms and Water closets

Size. The size of a bathroom shall be not less than 1·8 sq.m. with a minimum width of 1·2 m The minimum size of water closet shall be 1·1 sq.m. with a minimum width of 0·9 m. If it is a combined bathroom and water closet, the minimum area shall be 2·8 sq.m. with a minimum side of 1·2 m.

Height. The height of a bathroom or water closet measured from the surface of the floor to the lowest point in the ceiling (bottom of slab) shall be not less than 2·2 m.

Other Requirements. Both the bathroom as well as the water closet should be so located that at least one of its walls open to external air and is provided with ventilation in the form of window or ventilator of minimum size to the extent of 0·37 sq.m.

8. Roof

The roof of building shall be so constructed as to permit satisfactory drainage of rain water by means of sufficient rain water pipes of adequate size.

Parapet. Parapet walls and handrails provided on the edges of roof terrace, balcony, varandah etc. shall not be less than 1 m and more than 1·5 m in height.

9. Basement

The basement shall have following requirements :-

(*i*) Every basement shall be in every part at least 2·4 m in height from the floor to the underside of the roof slab or ceiling.

(*ii*) The minimum height of the ceiling of any basement shall be 0·9 m and the maximum of 1·2 m above the average surrounding ground level.

(*iii*) The basement should be properly ventilated.

(*iv*) Adequate arrangement should be made so that surface drainage does not enter the basement.

(*v*) In case of basements for office and commercial use, sufficient number of exit ways and access ways shall be provided with a travel distance of not more than 15 m.

(*vi*) Basement shall not be used for residential purposes.

10. Stairway

The following minimum width shall be provided for each stairway :-

(*a*) Residential building (dwelling) upto 3 storey = 0.9 m

(*b*) Other residential building *e.g.* flats, hostels, group housing etc. = 1.25 m

(*c*) Residential hotel buildings = 1.5 m

(*d*) Assembly buildings like auditorium, theatres and cinemas = 1.5 m

(*e*) Institutional buildings like hospitals = 2 m

(*f*) Educational buildings like schools, colleges = 1.5 m

(*g*) All other buildings = 1.5 m

Other Requirement s:

(*i*) The minimum width of treads without nosing shall be 25 cm for staircase for residential buildings. In the case of other buildings, the minimum width of tread shall be 30 cm.

(*ii*) The maximum height of riser shall be 19 cm in the case of residential buildings and 15 cm in the case of other buildings. There should not be more than 12 risers per flight.

(*iii*) Handrails shall be provided with a minimum height of 90 cm from the centre of the tread.

(*iv*) The minimum headroom in a passage under the landing of a staircase and under the flights of the staircase shall be 2·2 m.

11. Ramps

(*i*) Ramps with a slope of not more than 1 in 10 may be substituted for stairways. Larger slopes shall be provided for special uses but in no case greater than 1 in 8. For all slopes exceeding 1 in 10 and where the use is such as to involve danger of slipping, the ramp shall be surfaced with approved non-slipping material.

(*ii*) The minimum width of the ramps in hospitals shall be 2·25m.

(*iii*) Handrail shall be provided on both sides of the ramp.

(*iv*) Ramps shall lead directly to outside open space at ground level.

Questions for Review

1. Describe the classification of different types of buildings as per National Building Code.
2. What are the different components of a building? Explain the same with the help of sketches wherever possible.
3. Elaborate different requirements as per building byelaws which are to be kept in view while making provision of following items in a building.

   (*a*) Basement
   (*b*) Stairways
   (*c*) Mezzanine floor
   (*d*) Kitchen
   (*e*) Bathroom and W.C
   (*f*) Size and height of habitable rooms.

2

Foundations

2·1. DEFINITION

Foundation is the lowest part of a structure which provides a base for the super-structure proper. This term includes the portion of the structure below the ground level as well as the artificial arrangement of concrete block, piles, raft, grillage, etc. provided to transmit the loads on the structure including the dead weight of the structure itself to the soil below.

If a steel girder provided in a structure is found to be weak and flexible, it can usually be strengthened by suitable means; if a column acting as a load bearing member in a building is found to be inadequate, it can be made safe by providing additional reinforcement or by introducing intermediate columns, but if the foundation of a building yields and ruptures the structure, then little can be done to improve the situation.

2·2. PURPOSE OF FOUNDATION

It is often misunderstood that the foundation is provided to support the load of the structure. In fact it is a device to transmit the load of the structure to the soil below. Foundation is provided for the following four main purposes:

(*i*) To distribute the weight of the structure over large area so as to avoid over-loading of the soil beneath.

(*ii*) To load the sub-stratum evenly and thus prevent unequal settlement.

(*iii*) To provide a level surface for building operations.

(*iv*) To take the structure deep into the ground and thus increase its stability, preventing overturning.

2·3. SITE EXPLORATION FOR FOUNDATIONS

The designer should have adequate information regarding the type and nature of soil available at different depths at the site for designing safe, sound and economical foundation for a structure. The aim of site exploration is to get as much information as possible about the physical properties and characteristics of the underlying material at site as well as details of other geological features of the area. The study of site exploration can be broadly sub-divided in following groups :

(*i*) *Purpose of site exploration.*

(*ii*) *Site inspection and preliminary investigation.*

(*iii*) *Methods of site exploration.*

(*iv*) *Depth and spacing of trial pits and bore holes.*

(*v*) *Choice of method of exploration.*

2·4. PURPOSE OF SITE EXPLORATION

The purpose of site exploration is to collect complete details of the site to enable the designer to take following decisions :

(*i*) To fix the value of the safe bearing capacity of soil.

(*ii*) To select an economical yet safe type of foundation.

(*iii*) To fix the depth upto which the foundation must be taken inside the ground.

(*iv*) To predict the likely settlement of the selected foundation and to make allowance for the same in the design.

(*v*) To know the underground water level and if needed, to decide upon the method to be adopted to solve the ground water problem.

(*vi*) To forecast the difficulties which are likely to be encountered due to nature of the sub-soil during construction and to take advance actions in this regard.

2·5. SITE INSPECTION AND PRELIMINARY INVESTIGATION

Inspection of the site is the first step which should be taken prior to the design and construction of a structure. This includes the study of all neighbouring quarries or cuts made in the nearby areas for construction purposes, the study of existing structures and, if possible, of their foundations. Information gathered in respect of other's experience in making excavation in that locality may prove to be of great help. The study of the strata exposed to view in the old pits, unlined wells or banks or nallahas in the near vicinity may reveal the nature of the soil at different depths.

The other important information which can be gathered during the inspection is summarised below :

(*i*) Whether the ground is soft, hard, marshy, water logged or made-up type.

(*ii*) Classification of soil by visual examination.

(*iii*) Behaviour of the ground during changes in ground water-level. Whether the rain water gets drained off or causes flooding of site.

(*iv*) Whether there is likelihood of excessive movement of ground (This may be due to unstable site slopes, subsidence due to mining or the site being very near some old drain, pits well, underground working etc.).

(*v*) Whether the sub-soil water contains sulphates or other chemicals in quantities sufficient to cause damage to foundations.

2·6. METHODS OF SITE EXPLORATION

The methods commonly adopted for site exploration are :

(*i*) *Test pits* (*ii*) *Probing*

(*iii*) *Sub-surface sounding* (*iv*) *Boring*

(*v*) *Geo-physical method.*

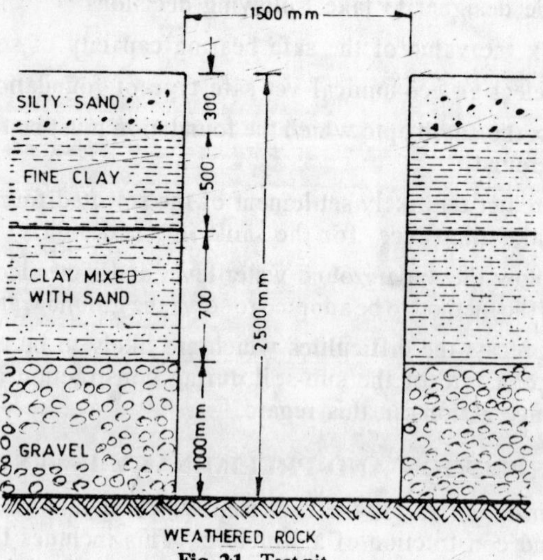

Fig. 2·1. Test pit

(*i*) **Test Pits.** The best way to ascertain the nature of the sub-soil strata is to dig a hole and see. The holes which are large enough to permit the entry of persons for inspection are called 'Test Pits'. The test pits are normally square in plane and are dug by hand or by excavating equipment. The dimensions of the pit depend largely upon the depths upto which the excavation is to be made. In cohesionless soils, the sides of the test pit are sharply sloped. In cohesive soils, at depth below 3 m, bracing is required to keep the sides of the pit vertical. Test Pits are comparatively expensive and hence they are only used for structure having shallow foundations (upto 3 meters).

(*ii*) **Probing:** In this method of exploration, a steel bar of 25 to 40 mm. in diameter with a pointed end is driven in the ground until a hard sub-stratum is met with. The bar is driven by allowing it to fall vertically under its own weight or by drop hammer. At intervals the bar is drawn out and some idea of the nature of the soil is obtained by examining the soil sticking to the sides of the bar. This method is suitable to be performed in places where the soil is soft like clay, gravel or sand. A rough idea of the nature of strata is obtained from the number of blows required to drive the rod inside the ground. Experienced workmen engaged in the execution of the driving operation may be able to judge the soil at various depths by reading the way the rod penetrates into the ground. This method is not always to be relied upon completely. More definite analysis should be done to ascertain the results arrived at by this method.

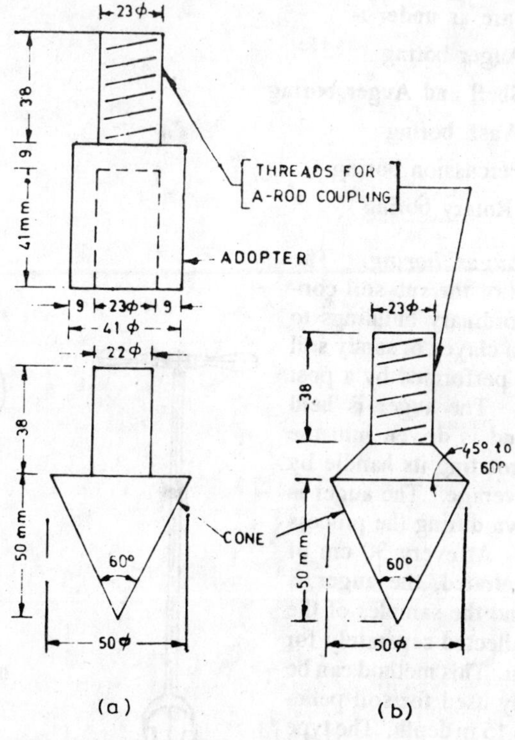

Fig. 2·2 Different types of standard cones

(*iii*) **Sub-surface soundings.** The method of sub-surface soundings consists of measuring the variation in the resistance offered by the soil with the depth by means of a tool known as *penetrometer*. The penetrometer may consist of a 50 mm diameter mild steel cone fitted loosely to a steel rod (Refer Fig. 2·2) or it may be a tool known as *Standard Split Spoon*

Sampler (refer 2·13). The sampler can be split into two parts longitudinally for removal of the soil sample which get filled up in its tube when it is driven in the ground. The penetrometer is driven in the ground with the help of blows from a 650 *N* weight falling from a height of 750 mm. The number of blows required to drive the penetrometer into the ground through a distance of 300 mm is termed as *Standard Penetration Resistance* or *SPT* of the soil and it is designated as *N*. The '*N*' values of soil are thus determined at different depths. The resistance offered by the soil to penetration is co-related with the engineering properties of soil like density, consistency (strength), permeability and bearing capacity. Curves co-relating the *N* values with bearing capacity of certain types of soil have been drawn and are available for ready reference.

(*iv*) **Boring.** The type of boring methods commonly adopted for soil exploration are as under :-

(*a*) Auger boring
(*b*) Shell and Auger boring
(*c*) Wash boring
(*d*) Percussion boring
(*e*) Rotary boring

(*a*) *Auger boring.* The examination of the sub-soil conditions for ordinary buildings to be erected in clayey or sandy soil can be best performed by a post hole auger. The auger is held vertically and is driven into the ground by rotating its handle by applying leverage. The auger is pressed down during the process of rotation. At every 30 cm of depth penetrated, the auger is taken out and the samples of the soils are collected separately for examination. This method can be conveniently used for soil penetration upto 15 m depth. The type of augers commonly used are shown in Figs. (2·3 and 2·4). For deeper holes or in grounds where gravel, boulders or compact material is present, this method is not adopted.

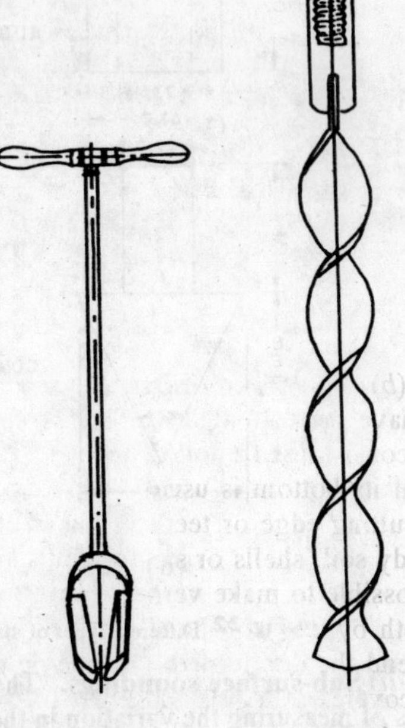

Fig. 2·3 Fig. 2·4

Different types of augers

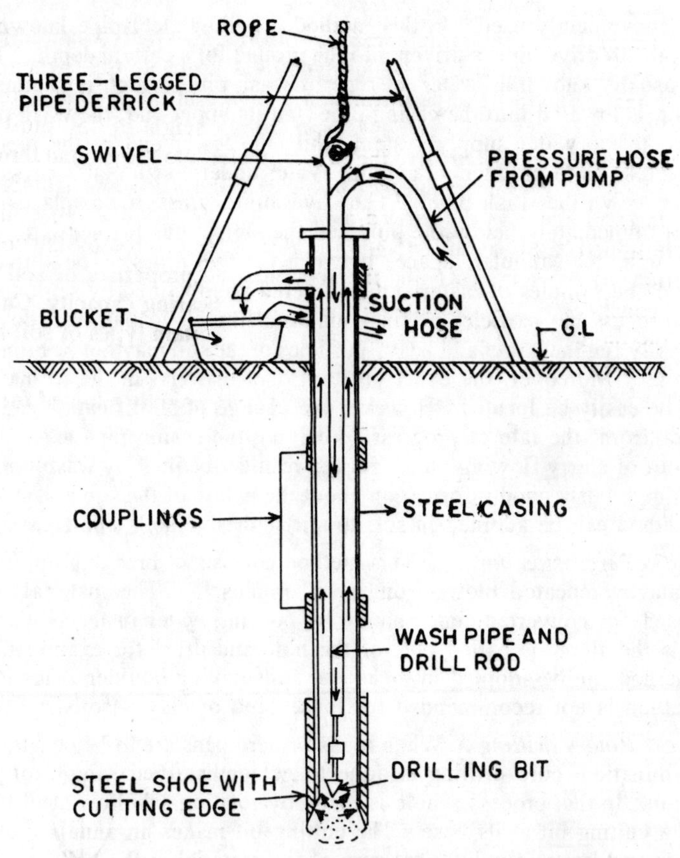

Fig. 2·5

Installation for Wash boring

(*b*) *Shell and auger boring*: In this method different type of tools have to be adopted for boring. In case of soft to stiff clay, cylindrical auger consisting of a hollow tube of 75 to 200 mm in diameter with a cutting edge at its bottom is used. In case of various stiff and hard clay, shells with cutting edge or teeth at lower end are to be adopted while in case of sandy soil, shells or sand, pumps are used for boring. By this method it is possible to make vertical boring upto 200 mm in diameter and 25 m in depth by use of a hand rig. By use of mechanical rig it is possible to extend the depth of the bore hole upto 50 m. The samples of the soil are recovered at regular intervals (or whenever there is a change in strata) for conducting tests in laboratory for identification of soils and establishing properties of the sub-soil strata at various depths.

(*c*) *Wash boring*. For test boring over 3 m. in depth, this method can be conveniently used. In this method a hollow steel pipe known as casing pipe or drive pipe is driven into the ground for a certain depth. Then a pipe usually known as water jet pipe or wash pipe, which is shorter in diameter, is lowered into the casing pipe. At its upper end, the wash pipe is connected to water supply system while the lower end of the pipe is contracted so as to produce jet action. Water under considerable pressure is forced down the wash pipe. The hydraulic pressure displaces the material immediately below the pipe and the slurry thus formed is forced up through the annular space between the two pipes. The slurry is collected and samples of material encountered are obtained by settlement. In this process the particles of finer material like clay, loam etc. do not settle easily and the larger and heavy particles of the soil may not be brought up at all. Moreover, the exact position of a material in the formation cannot be easily be located. However the change of stratification can be guessed from the rate of progress of driving the casing pipe as well as the colour of slurry flowing out. Yet the results obtained by wash boring process give fairly good information about the nature of the sub-soil strata. This method can be adopted in soft to stiff cohesive soils and fine sand.

(*d*) *Percussion boring* . This method consists of breaking up of the sub-strata by repeated blows from a bit or chisel. The material thus pulverised is converted into slurry by pouring water in the bore. At intervals the slurry is bailed out of the hole and dried for examination. This method can be adopted in rocks and soils having boulders. However this method is not recommended for loose sand or clayey soils.

(*e*) *Rotary drilling* . When rocks or hard pans are to be penetrated for examination, core drilling is done to get undisturbed samplesof the formation. In this process a hole is made by rotating a hollow steel tube having a cutting bit at its base. The cutting bit makes an annular cut in the strata and leaves a cylindrical core of the material in the hollow tube. Two types of cutting bits are generally used, namely, diamond bit and shot bit. Diamond bit consists of industrial diamonds set in the face of the bit and in shot bit, chilled shot is used as an abrasive to cut the hard pan. When core samples of small diameter are needed, diamond bit is preferred.

(*v*) *Geo-physical methods.* Geo-physical methods are used when soil exploration is to be carried out over large area and where speed is of prime importance. These methods are based on principle that physical properties like electrical conductivity, elasticity or seismicity, magnetic susceptibility. density etc. vary for different types of soils. There are 4 methods of geo-physical survey, namely, (*i*) Seismic refraction method (*ii*) Electrical resistivity method, (*iii*) Magnetic method and (*iv*) Gravitational method. However, out of these only two methods *viz* (*i*) Seismic refraction method and (*ii*) Electrical resistivity method are widely used.

(*i*) *Seismic method* . This method is based on the principle that sound waves travel faster in rock than in soil. This is on account of the fact that

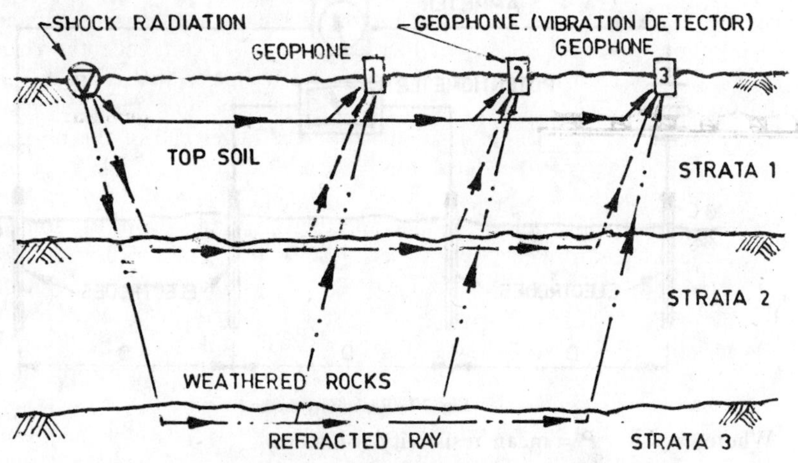

Fig. 2·6

Seismic refraction method

velocity of sound waves is different in different media. In this method shock waves (or sound waves of vibration) are created into the soil at ground level or at a certain depth below it, either by striking a plate on the soil with the hammer or by exploding small charge in the soil. The shock waves so produced travel down in the sub-soil strata and get refracted after striking a hard rock surface below. The refracted or radiated shock waves are picked up by the vibration detector (also known as geophone) where the time of travel of the shock waves gets recorded. Knowing the time of travel of the primary and refracted waves at various geophones, time and distance graphs are drawn based on which it is possible to evaluate the depth of various strata in the sub-soil. Different materials such as clay, gravel, soft rock, hard rock etc. have characteristics seismic velocities and hence it is possible to establish their identity in the sub-soil based on time distance graph.

(*ii*) *Electrical resistivity method.* This method is based on the principle that each soil has different electrical resistivity, depending upon the type of soil, its water content, compaction and composition. Thus saturated soil has lower electrical resistivity as compared to loose dry gravel or solid rock. In this method 4 electrodes are driven in the ground at equal distance apart and in a straight line. The distance between two electrodes being the depth of exploration or depth upto which the ground resistance is to be measured. A current is passed between the two outer electrodes and the potential drop between the inner electrodes is measured by use of potentiometer.

The mean resistivity is calculated by the following formula :-

$$P = 2\,\pi D\ \frac{E}{I}$$

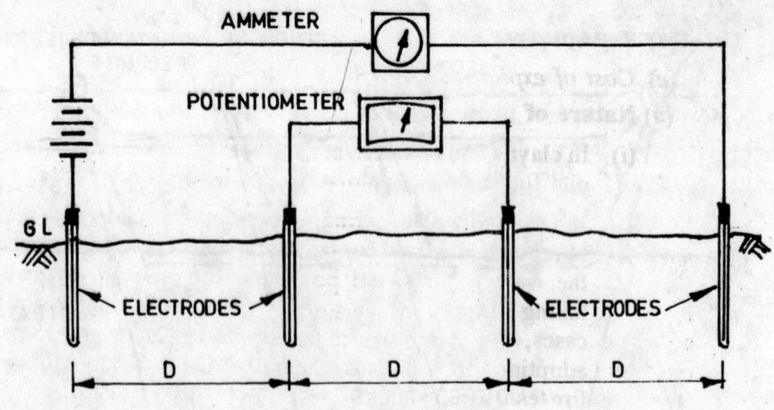

Fig. 2·7. Resistivity method

Where P = mean resistivity (ohm.cm)

D = distance between electrodes (cm)

E = potential drop between inner electrodes (*volts*)

I = current flowing between outer electrodes (*amperes*)

Average value of resistivity for various types of soils have already been established based on tests. Thus on knowing the values of change in mean resistivity of sub-soil strata at site, it is possible to establish the nature and distribution of different type of soils in the formation.

2·7. DEPTH AND SPACING OF TRIAL PITS OR BORE HOLES

The location and the spacing of trial pits or bore holes for a particular site require special consideration. The pits or the bore holes should be so located so as to give adequate information in respect of changes in properties of the underlying strata with depth. The number and spacing of the trial pits or bore holes to be adopted for a site will depend upon the area of the plot as well as the type of structure to be built. For a plot of 0·4 hectare we may have one trial pit or bore hole near each corner and one in the middle. For smaller size of plot and for ordinary structure, one trial pit or bore hole near the centre should be sufficient. The depth of the pit or bore hole will depend upon the characteristics of the soil as well as the type of structure, its shape, size and loading condition. As a thumb rule, its depth should be one and a half times the probable width of the footing or 1·5 m whichever is more. In case of weak soils, however, the pits or the bore holes should be taken to a depth at which the loads can be carried by the soil without undesirable settlement.

2·8. CHOICE OF METHOD OF SITE EXPLORATION

The choice of method to be adopted for site exploration for a particular site depends upon the following factors :-

(*a*) *Nature of ground*

(*b*) *Topography*

(*c*) *Cost of exploration.*

(*a*) Nature of ground

(*i*) In clayey soils borings are suitable for deep exploration and pits for shallow exploration.

(*ii*) In sandy soils boring is easy but special equipment is required to be used for taking representative samples below the water table. Trial pits, however, present difficulty in taking samples, when ground water table is high. In such cases, it is necessary to lower the ground water table (by adopting suitable methods) to facilitate taking soil sample for testing.

(*iii*) In case of rocks, boring is considered suitable in hard rocks and pits in soft rock. Core borings are suitable for identification of types of rocks but they do not give adequate information in respect of joints and fissures in rock formation. Such information can only be obtained from trial pits or boring of large diameters.

(*b*) Topography. **In case the site is located in hilly terrane, the choice between vertical openings (*i.e.* bores and trial pits) and horizontal openings (*i.e.* exploratory drifts or headings) may depend upon the topography and geological structure of the sub-strata. In case of steeply inclined strata and steep slopes, inclined boring or drifts are considered suitable, whereas for horizontal or gentle slopes, trial pits or vertical borings are considered suitable.**

(*c*) Cost of exploration. **From consideration of cost, deep exploration borings are preferred to deep shafts. In case of shallow exploration in soil, the choice between pits and borings will depend on the nature of ground and the extent of information required. In case of shallow exploration in rock, use of core drilling can be justified, if number of bore holes required to be drilled are large. Alternatively, trial pits prove to be more economical.**

2·9. BEARING CAPACITY OF SOILS

The topic of bearing capacity is perhaps the most important of all the topics in soil engineering. Soils behave in a complex manner when loaded. Soil when stressed due to loading, tend to deform. The resistance to deformation of the soil depends upon factors like water content, bulk density, angle of internal friction and the manner in which load is applied on the soil. The maximum load per unit area which the soil or rock can carry without yielding or displacement is termed as the *bearing capacity of soil.*

When excessive load is transmitted to the soil by a structural

foundation, the settlement of the foundation takes place which can endanger the stability of the structure. The settlement due to load is caused basically on account of two factors, namely, (*i*) the soil below footing gets compressed by certain amount and (*ii*) since the foundations cover only a limited area there is a possibility that the concentrated stresses developed are so high as to cause actual rupture (shear failure) and displacement of soil below. The loading intensity which causes soil rupture and lateral displacement resulting in rapid sinking of the loaded area into the ground is termed as *ultimate bearing capacity of soil.*

Past experience shows that very often a structure fails due to unequal settlement or differential settlement. This happens when a part of building is founded on compressible *stratum* and the remaining part rests on firm soil strata. Thus the part of the building on compressible soil settles at a rate well in excess of the part of building on firm soil leading to the differential settlement. Differential settlement can also occur when one part of the building is loaded much more than other or intensity of load is varying and is more than the bearing capacity of soil. In case, however, the settlement is uniform and small in magnitude it does not endanger the structure in any way.

Based on the survey of data concerning the damage due to differential settlement, it has been observed that no structural damage or damage to interior finish or partition walls etc. occurred in buildings investigated at angular distortion smaller than 1:300. For any given structure there is certain amount of settlement (either uniform or differential or both) which can be tolerated without creating unsafe conditions. This is termed as *permissible settlement.* It is on this account that almost all design codes permit certain maximum value of permissible settlement in the design of structures. IS 1904-1978 gives details of maximum permissible settlement which can be allowed in the design of structures. This value varies from 50 mm to 100 mm for different type of structures, foundations (isolated or raft) in different types of soils.

The various terms which are used in connection with the bearing capacity are summarised under :-

1. *Ultimate bearing capacity of soil*. The intensity of loading, at the base of foundation, at which soil support fails in shear.

2. *Safe bearing capacity.* The maximum intensity of loading that the soil will safely carry without risk of shear failure is called safe bearing capacity of soil. This is obtained by dividing the ultimate bearing capacity by a certain factor of safety, and it is the value which is used in the design of foundation. The factor of safety normally varies from 2 to 3.

3. *Net Pressure intensity.* To understand this term, let us consider one metre length of a continuous footing of a wall shown in Fig. 2·8.

Let B = width of the footing

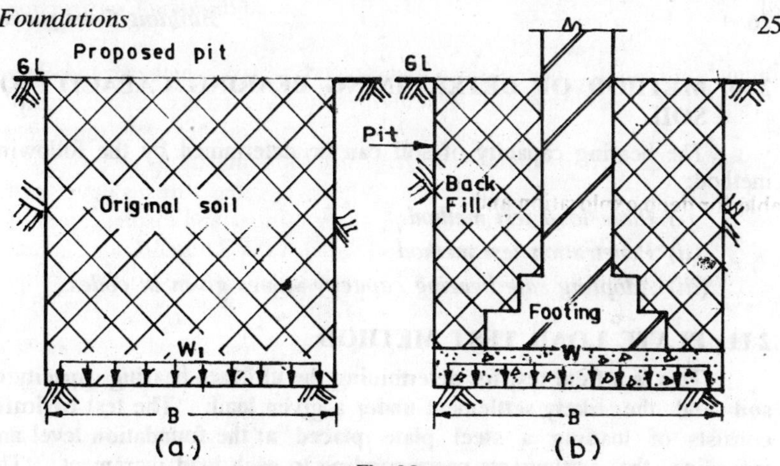

Fig. 2·8

W_1 = weight of soil that was existing above the base of foundation prior to excavation.

W = total load on the base of the foundation *[This being sum of (i) total dead load upto footing top + (ii) Self weight of footing + (iii) Weight of backfill soil located above the footing + (iv) Live load on the foundation due to floors above, and due to snow, wind, seismic forces etc. (wherever applicable)]*

q_n = net pressure intensity

$$\text{Net Pressure intensity} = \frac{\text{Net load on base of foundation}}{\text{Area of footing}}$$

Net load on base of foundation = $W - W_1$

$$\therefore \quad q_n = \frac{W - W}{B \times 1}$$

Hence net pressure intensity refers to loading acting on the bottom of the foundation trench, in excess of the weight of soil removed from the trench. Net pressure intensity is normally considered in the design of foundation, thereby utilising the advantages of reduction of loading brought about on account of removal of soil from the foundation trench.

4. *Allowable bearing pressure.* It is the maximum allowable net loading intensity which can be applied to the soil taking into account the ultimate bearing capacity, the amount and kind of settlement expected and the ability of the given structure to withstand the settlement. It is, therefore, dependent upon both the sub-soil and the type of building proposed to be erected thereon. The allowable bearing pressure adopted in the design of foundation is lesser of the following two values :

(*a*) The safe bearing capacity of soil, or

(*b*) The maximum allowable bearing pressure that the soil can take without exceeding the specified limits of permissible settlement.

2·10. METHOD OF DETERMINING BEARING CAPACITY OF SOIL

The bearing capacity of soil can be determined by the following method:

(*i*) *Plate load test method*

(*ii*) *Penetration test method*

(*iii*) *Adopting safe bearing capacity values given in codes.*

2·11. PLATE LOAD TEST METHOD

This is a field test for determining the ultimate bearing capacity of soil and the likely settlement under a given load. The test basically consists of loading a steel plate placed at the foundation level and recording the settlements corresponding to each load increment. The test load is gradually increased till the plate starts to sink at a rapid rate. The total value of load on the plate in such a stage divided by the area of the steel plate gives the value of the ultimate bearing capacity of soil. The ultimate bearing capacity is divided by suitable factor of safety (which varies from 2 to 3) to arrive at the value of safe bearing capacity of soil. For proper understanding, this test can be sub-divided into the following heads :

1. *Test set-up*
2. *Testing procedure*
3. *Interpretation of results*
4. *Limitations of the test.*

1. **Test set up.** A pit is dug at site upto the depth at which the foundation is proposed to be laid. The width of the pit should be at least 5 times the width of the test plate. At the centre of the pit a small square depression or hole is made whose size is equal to the size of the test plate and bottom level of which corresponds to the level of actual foundation. The depth of the hole should be such that the ratio of depth to width of the loaded area is approximately the same as the ratio of the actual depth to width of the foundation.

The mild steel plate (also known as bearing plate) used in the test should not be less than 25 mm in thickness and its size may vary from 300 to 750 mm. The plate could be square or circular in shape. Circular plate is adopted in case of circular footing and square plate is used in all other types of footings. The plate is machined on side and edges.

2. **Testing procedure.** The load is applied to the test plate through a centrally placed column. The test load is transmitted to the column by one of the following two methods :

(*i*) *By gravity loading or reaction loading method*

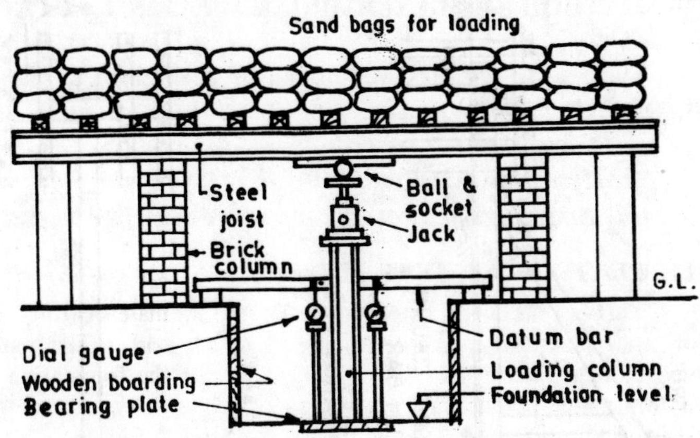

Section A - A

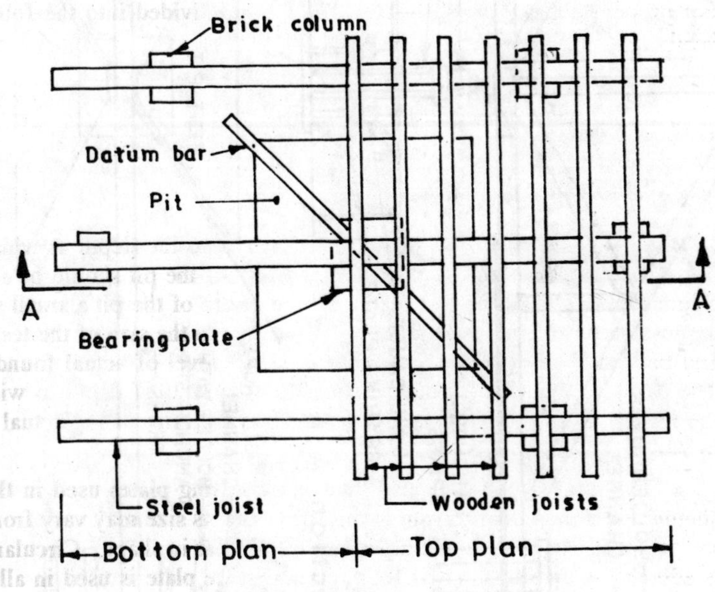

Fig. 2·9

Plate Load Test (Reaction by gravity loading)

(*ii*) *By loading truss method.*

(*i*) In case of gravity loading method, a loading platform is constructed over the column placed on the test plate and test load is

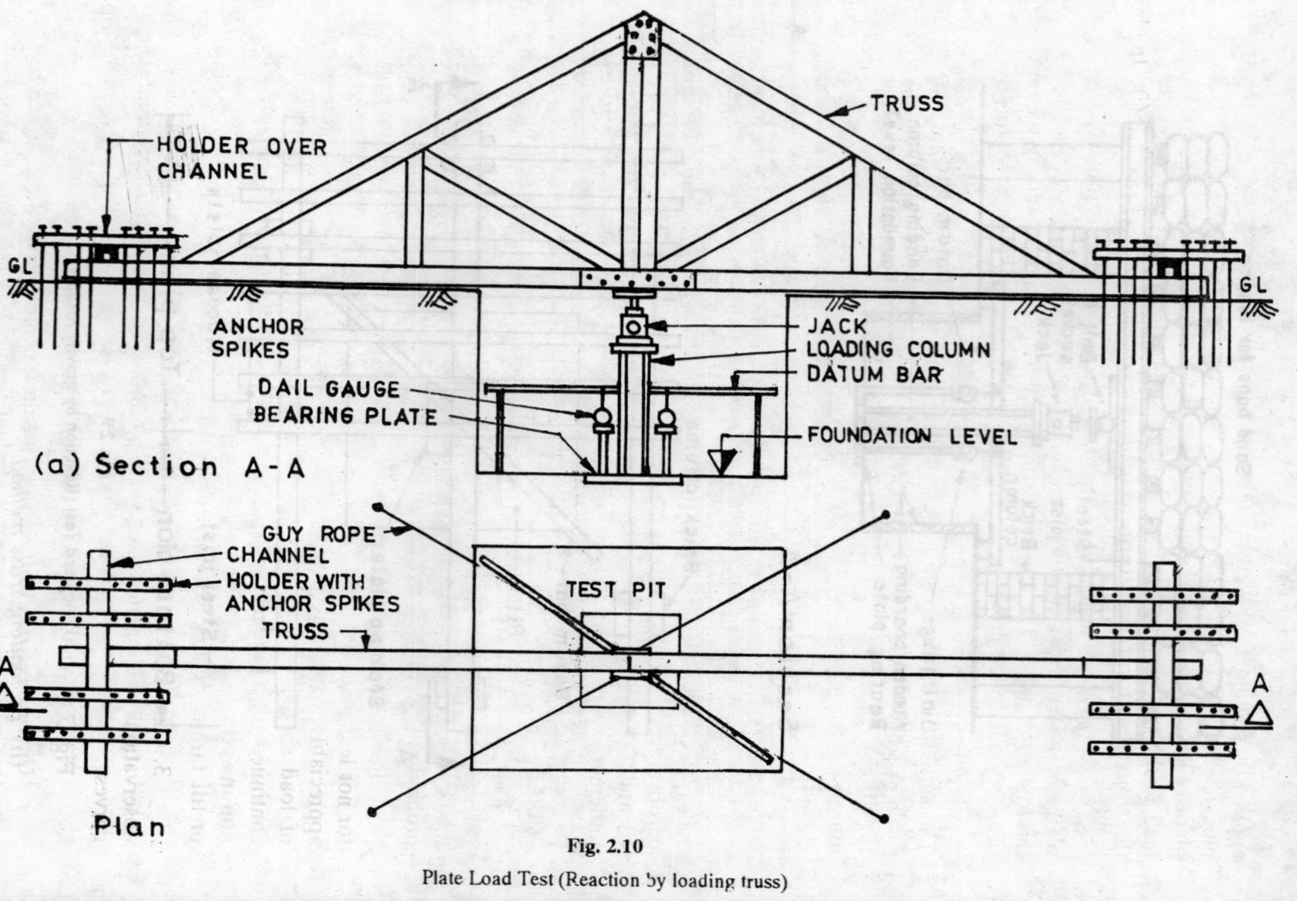

Fig. 2.10

Plate Load Test (Reaction by loading truss)

applied by placing dead weight in the form of sand bags, pig iron, concrete blocks, lead bars etc. on the platform. Many a times a hydraulic jack is placed between the loading platform and the column top for applying the load to the test plate - the reaction of the hydraulic jack being borne by the loaded platform. This form of loading is termed as *reaction loading.*

(*ii*) In case of reaction truss method, instead of constructing a loading platform, a steel truss of suitable size is provided to bear the reaction of the hydraulic jack. The truss is firmly anchored to the ground by means of steel anchors and guy ropes are provided for ensuring its lateral stability.

When the load is applied to the test plate, it starts sinking slowly. The settlement of the plate is recorded to an accuracy of 0·02 mm with the help of sensitive dial gauges. At least two dial gauges are used to account for differential settlement. The dial gauges are placed at diametrically opposite ends of the plate and one dial gauge is mounted on independently supported references beam or datum rod. As the plate sinks, the ram of the dial gauge moves down and the settlement is recorded. The magnitude of load is indicated on the load-gauge of the hydraulic jack.

The load is applied in regular increment of about 2KN or 1/5th of the expected ultimate bearing capacity, whichever is less. Settlement should be observed for each increment of load after an interval of 1, 4, 10, 20, 40 and 60 minutes and thereafter at hourly intervals until the rate of settlement becomes less than 0·02 mm per hour. The maximum load to be applied for the test should be about 1·5 times the expected ultimate bearing capacity of the soil.

In case of clayey soils the time settlement curve should be plotted at each load stage and load should be increased to next stage either when the curve indicates that the settlement has exceeded 70 to 80% of the probable ultimate settlement at that stage or at the end of 24 hour period.

For soils other than clayey soils, each load increment should be kept for not less than one hour or upto a time when the rate of settlement gets appreciably reduced (to a value of 0·02 mm per min.). The next increment of load should then be applied and observations repeated. The test is continued till a settlement of 25 mm under normal circumstances or 50 mm in special cases (such as dense gravel, and sand mixture) is obtained or till failure occurs, whichever is earlier.

3. **Interpretation of results.** The load intensity and settlement observations of the plate load test are plotted in the form of load settlement curves.

Fig. 2·11 shows four typical curves applied to different soils.

Curve I is typical for loose to medium non-cohesive soils. It can be seen that initially this curve is a straight line, but as the load increases it flattens out. There is no clear point of shear failure.

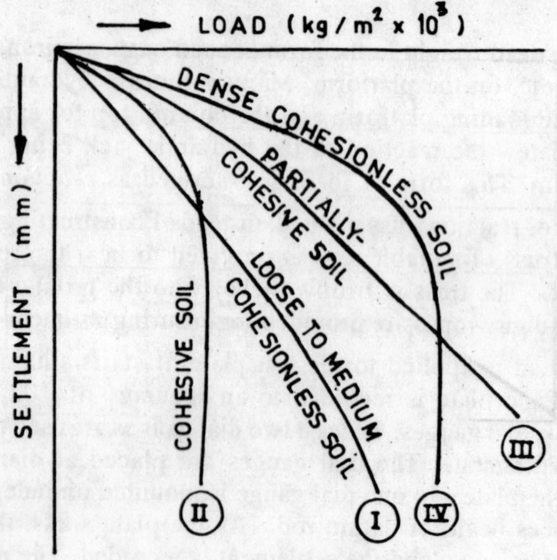

Fig. 2·11

Load settlement curve

Curve II is typical for cohesive soils. This may not be quite straight in the initial stages and leans towards settlement axis as the settlement increases.

Curve III is typical for partially cohesive soils.

Curve IV is typical for purely dense non-cohesive soil.

Curves II and IV have well defined points of failure (yield points) and as such there is no difficulty in arriving at the value of ultimate bearing capacity of soil in case of dense non-cohesive soils or cohesive soils.

In case of Curve I and III, where yield point is not well defined, settlement observations during test should be plotted as abcissa against corresponding load intensity as ordinates, both to logarithmatic scales. Refer Fig. 2·12, which will give two straight lines. The inter-section of the two lines is taken as yield point or yield value. If the load settlement curve does not indicate any marked breaking point, shear failure may be assumed corresponding to a settlement equal to 1/5th of the width of test plate. The value of load on the horizontal axis corresponding to yield point (or change of load settlement plot from straight line to curve) is taken as the ultimate bearing capacity of soil.

The safe bearing capacity is obtained by dividing the ultimate bearing capacity by a factor of safety varying from 2 to 3. The value of safe bearing capacity thus arrived at, is considered to be based on criterion of *shear failure.*

Safe bearing capacity (SBC) based on permissible settlement. As indicated earlier the settlement of footing is also related to the SBC

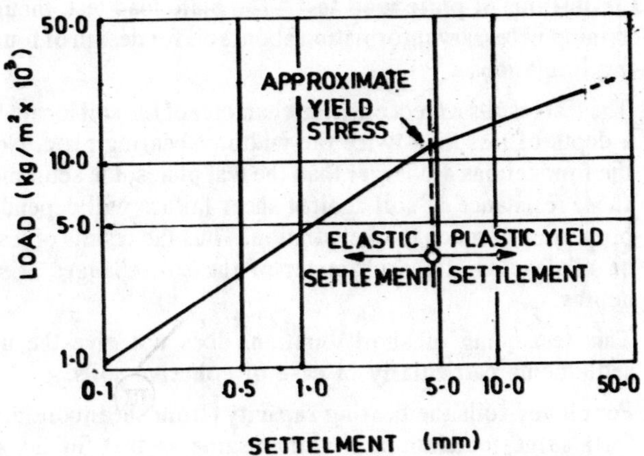

Fig. 2·12

Load settlement curve (Log – log plot)

of the soil. The value of ultimate bearing capacity and hence the SBC, in this case, can be obtained from the load settlement curves by reading the value of load intensity corresponding to the desired settlement of test plate. The value of permissible settlement (S_f) for different types of footings (isolated or raft) for different types structures are specified in the I.S. code. The corresponding settlement of test plate (S_p) can be calculated from the following formula :

$$S_f = S_p \left[\frac{B\,(B_p + 0{\cdot}3)}{B_p\,(B + 0{\cdot}3)} \right]^2$$

where

B = the width of footing in mm.

B_p = width of test plate in mm.

S_p = settlement of test plate in mm.

S_f = settlement of footing in mm.

After knowing the value of settlement of test plate (S_p) from the above formula, the value of load intensity corresponding to this settlement (S_p) is taken from the curve in Fig. 2·12. The value of load intensity or ultimate bearing capacity, divided by factor of safety gives the value of SBC.

The final value of safe bearing capacity is taken as lesser of the following two values

(*i*) SBC based on shear failure and

(*ii*) SBC based on settlement (S_p) of test plate.

4. **Limitations of plate load test.** The plate load test, though very useful in obtaining necessary information about soil for design of foundation has following limitations :

(*i*) The test results reflect only the character of the soil located within a depth of less than twice the width of bearing plate. Normally the foundations are larger than the test plates, the settlement and shear resistance of soil against shear failure will depend on the properties of much thicker stratum. Thus the results of test could be misleading if the character of the soil changes at shallow depths.

(*ii*) This test being of short duration, does not give the ultimate settlements particularly in case of cohesive soils.

(*iii*) For clayey soils the bearing capacity (from shear consideration) for a large foundation, is almost same as that for the smaller test plate. But in dense sandy soils the bearing capacity increases with the size of the foundation and hence the test with smaller size test plate tends to give conservative values in dense sandy soils.

In view of the above limitations, the plate load test method of determining SBC of soil may be considered adequate for light or less important structures under normal condition. However, in case of unusual type of soil stratum and for all heavy and important structures, relevant laboratory tests or field test are essential to establish the SBC of soil.

2·12. PENETRATION TEST METHOD

The equipment used for this test may be a standard split spoon sampler, a cone or other specially shaped tool. The test consists of measuring the resistance offered by the soil to the penetration of the test-tool under dynamic or static loading. The values of resistance, also known as N values, are co-related with properties of soil, such as density, bearing capacity etc. Thus, based on information about N values, it is possible to determine the bearing capacity of soil by use of standard graphs or tables. In this method, a hole 55 to 150 mm in diameter is made in the ground with the help of suitable equipment for conducting the test. It may sometimes be necessary to use casing or drilling mud where the soil at site is sandy, soft clay or other such type where the sides of the hole are likely to cave in. At sites where casing is used to keep the walls of the test hole stable, it should not be driven below the level at which the test is to be made or soil sample is to be taken. The distance between end of casing and bottom of bore hole should not be more than 150 mm.The casing should preferably be sunk in ground by turning it slowly instead of driving it so as to ensure that it does not disturb the soil and hence the test is performed on undisturbed soil.

A standard split spoon sampler consists of a thick wall tube having outer diameter of 50·8 mm and internal diameter of 35 mm and a length of 600 mm. The tube has a drive shoe attached to its bottom and coupling head at top to accommodate the drill rod, used for testing. The drill rod is coupled to sampler head and the sampler is lowered into the clear hole, made in the ground in advance. The sampler is driven into the undisturbed soil at the bottom of the hole, with the help of a driving weight assembly consisting of a driving head and a 650 N weight with 750 mm free fall. The blows from the driving weight fall on the drill rod which drives the sampler into the soil. The sampler is first driven through 150 mm in the hole. This is known as seating drive. The sampler is then driven further through 300 mm and the number of blows required for 300 mm penetration are recorded. This number of blows is termed as penetration resistance of soil and is represented by symbol *N*.

The tests are made and samples are taken out at every change in stratum or at intervals of 1·5 m whichever is less.

In case of very fine or silty saturated sand, it is seen that although they have less value of bearing capacity, yet, due to apparent increase in resistance that such type of soil offer to penetration of the sampler, the *N*

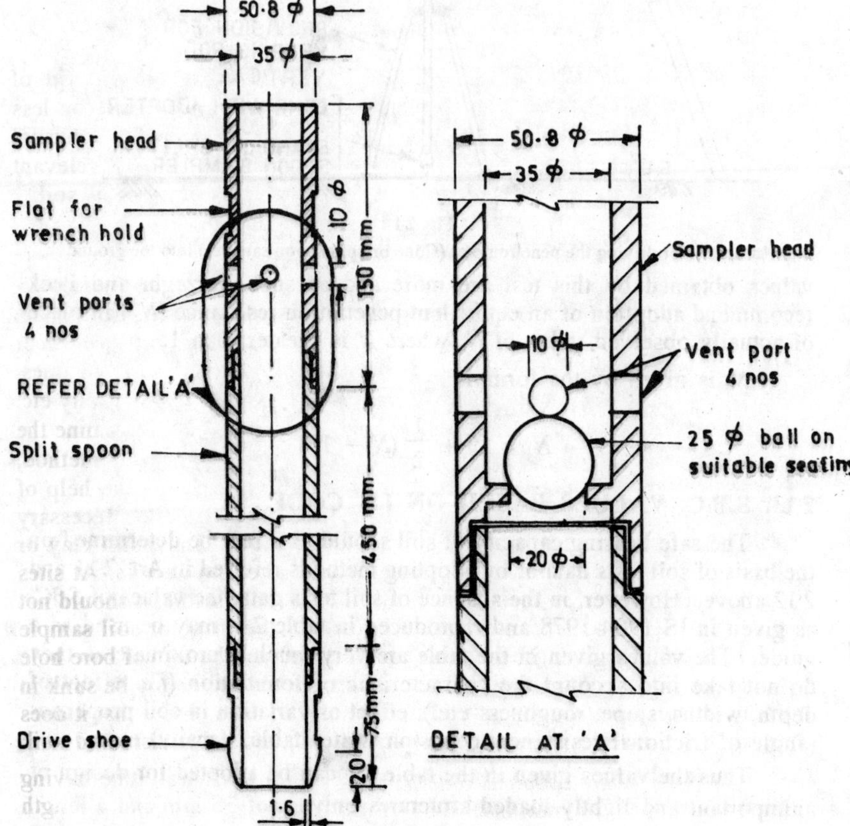

Fig. 2·13. Standard split spoon sampler

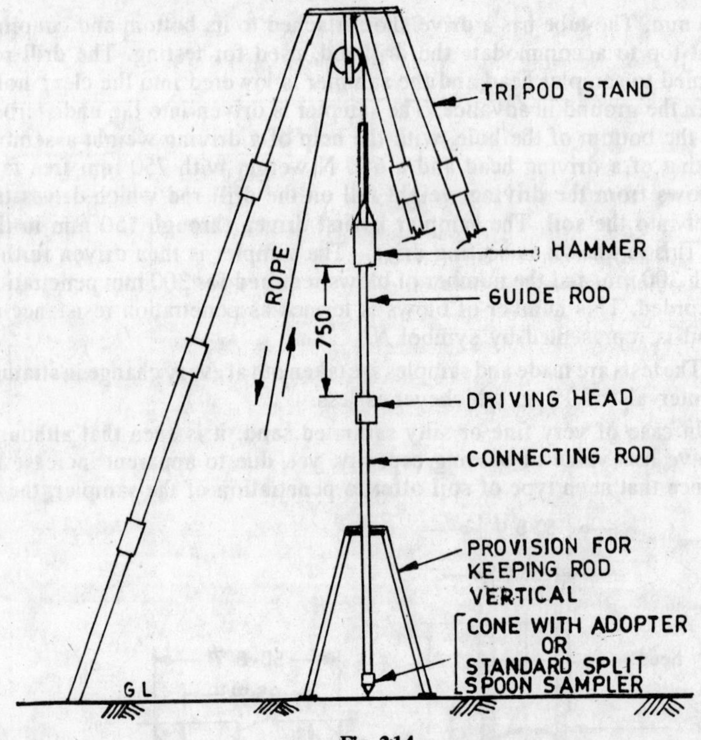

Fig. 2·14

Assembly for driving the penetrometer (Cone or split spoon sampler) into the ground

values obtained by this test are more and as such Terzaghi and Peck recommend adoption of an equivalent penetration resistance (N_s) in place of actually observed value of N, where N is greater than 15.

N_s is given by the formula

$$N_s = 15 + \frac{1}{2}(N - 15)$$

2·13. S.B.C. VALUES BASED ON I.S. CODE

The safe bearing capacity of soil should as a rule be determined on the basis of soil tests data or by adopting methods referred in Art. 2·11 and 2·12 above. However, in the absence of soil tests data, the values of *SBC*, as given in IS 1904-1978 and reproduced in table 2·1, may be used as a guide. The values given in the table are very much approximate as they do not take into account the characteristic of foundation (*i.e.* effect of depth, widths, slope, roughness etc.), effect of variation in soil properties (angle of frictional resistance, cohesion, water table, density) etc.

Thus the values given in the table 2·1 can be adopted for design of unimportant and lightly loaded structures only.

TABLE 2·1

Safe Bearing Capacity of Soil Based on I.S. Code

Sl. No.	*Types of Rocks/Soils*	*Safe Bearing Capacity*	
		KN/m²	*Kg/cm²*
	(*a*) Rocks		
1.	Rocks (hard) without lamination and defects, for example, granite, trap and diorite	3300	33
2.	Laminated rocks, for example, sand-stone and lime stone in sound condition	1650	16.5
3.	Residual deposits of shattered and broken bed rock and hard shale, cemented material	900	9
4.	Soft rock	450	4.5
	(*b*) *Non-cohesive Soils*		
5.	Gravel, sand and gravel, compact and offering high resistance to penetration when excavated by tools. (Refer Note 4)	450	4.5
6.	Coarse sand, compact and dry (with ground water level at a depth greater than width of foundation below the base of footing)	450	4.5
7.	Medium sand, compact and dry	250	2.5
8.	Fine sand, silt (dry lumps easily pulverised by the fingers)	150	1.5
9.	Loose gravel or sand gravel mixture; loose coarse to medium sand, dry (Refer Note 4)	250	2.5
10.	Fine sand, loose and dry.	100	1
	(*c*) *Cohesive Soils*		
11.	Soft shale, hard or stiff clay in deep bed, dry	450	4.5
12.	Medium clay, readily indented with a thumb nail.	250	2.5
13.	Moist clay and sand clay mixture which can be indented with strong thumb pressure	150	1.5
14.	Soft clay indented with moderate thumb pressure.	100	1
15.	Very soft clay which can be penetrated several centimeters with the thumb	50	0.5
16.	Black cotton soil or other shrinkable or expansive clay in dry condition (50 percent saturation) (Refer Note 1)	To be determined after site investigations	
17.	Peat (Refer Note 1 and 2)		
18.	Fills or made up ground. (Refer Note 3 and 4)		

Notes:

(1) No generalized values for safe bearing capacities can be given for these type of soils. In such area, adequate site investigation is required to be carried out and expert advice should be sought.

(2) Peat may occur in a very soft spongy condition or may be quite firm and compact. While ultimate bearing capacity may be high in the compact cases, very large consolidation settlements occur even under small pressures and the movements continue for decades.

(3) The strength of made up ground depends on the nature of the material, its depth and age, and the method used for consolidating it.

(4) Compactness or looseness of non-cohesive material may be determined by driving a wooden picket of dimension 5 cm x 5 cm x 70 cm with a sharp point. The picket is pushed vertically into the soil by full weight of a person and if the penetration of the picket exceeds 20 cms, the loose state is assumed to exist.

(5) The safe bearing capacity value may be increased by an amount equal to weight of the material (virgin oil) removed from above the bearing level, that is, the base of the foundation.

2·14. METHODS OF IMPROVING THE BEARING CAPACITY OF SOILS

Following methods are generally adopted to improve the bearing capacity of soils:

(1) Increasing the depth of the footing is the simplest method of increasing the bearing capacity, This method is restricted to sites where the sub-soil water level is much below and deep excavations do not increase the cost of foundations disproportionately.

(2) Drainage is a well known method to improve the bearing capacity of certain soils. Drains (with open joints) are laid in trenches just at the footing base. The sub-soil water thus collected is drained out through a system of pipe drains provided outside the external walls of the building.

(3) By blending granular material, like sand, gravel or crushed stone into the natural soil by ramming. The layer of soil thus formed is much stronger and is of improved bearing capacity.

(4) By confining the soil in an enclosed area with the help of sheet piles. This method is used with advantage in shallow foundations in sandy soils.

(5) By driving sand piles. This method is based on the principle of reducing the void volume of the natural soil. Holes are made in the soft soil with the help of wooden piles or other means and then sand is filled

in the holes and rammed. These are called sand piles. Bearing capacity of soft soil can be appreciably improved by driving sand piles at close spacing.

2·15. TYPES OF FOUNDATIONS

Foundations can be broadly classified into two types *i.e.* (*i*) *deep foundations* and (*ii*) *shallow foundations.* Pile, cofferdams and caisson fall under the category of deep foundations. Deep foundations are separately described in chapter 3.

2·16. SHALLOW FOUNDATIONS

When the foundation is placed immediately beneath the lowest part of the super structure, it is termed as shallow foundation. The object of this type of foundation is to distribute the structural loads over a wide horizontal area at shallow depth below the ground level. The various types of foundations which can be included under shallow foundation are :

1. *Spread footings*
2. *Grillage foundation*
3. *Eccentrically loaded footings*
4. *Combined footings*
5. *Mat on raft foundation.*

2·17. SPREAD FOOTINGS

As the name suggests, in case of spread footings, the base of the member transmitting load to the soil is made wider so as to distribute the load over wider area. Broadly speaking, all types of foundation mentioned above can be covered under the term *spread*. However, from design and construction point of view they have been treated separately. The various footings described under this classification are :

(*a*) *Wall footings*
(*b*) *Reinforced concrete footings*
(*c*) *Inverted arch footings*
(*d*) *Column footings.*

(*a*) **Wall Footings.** It consists of several courses of bricks, the lowest course being usually twice the breadth of the wall above. The increased base width of the wall is achieved by providing 5 cm (½ brick length) offsets on either side of the wall. The depth of each course is usually 10 cm (thickness of one brick). In some cases, however,the bottom courses are made 20 cm. (depth of two bricks) deep. The footing of various wall thicknesses have been shown in Figs. 2·15 and 2·16. In case of footings for *stone walls,* the size of offsets is slightly more than that of the brick wall footing. A bed of lean concrete of uniform thickness is first spread over the entire length of the wall. The depth of lean concrete bed is normally not less than 15 cm. and its projection on either side of the wall base varies between 10 to 15 cm. In any case, the depth of the concrete bed should never be less than its projection beyond the wall base. The concrete bed provides a plain surface upon which the masonry work for wall footing can

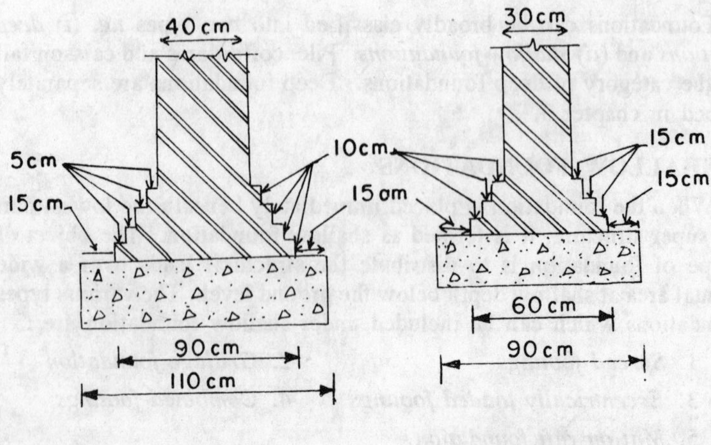

Fig. 2·15 **Fig. 2·16**

Two and $1\frac{1}{2}$ brick (modular bricks) thick wall footing using standard bricks in metric units.

be started directly. It not only rectifies the inequalities of excavation but also bridges over soft patches in the soil below. The design of the brick wall footings and the concrete bed block is fully described in Chapter 4.

(*b*) **Reinforced concrete footings.** In places where the walls are subjected to relatively heavy loading and the bearing capacity of the soil on which the wall footing is to rest is low, the type of footing described in (*a*) above will work out to be massive in size. In such cases it is desirable to provide reinforced concrete footing below the wall. This appreciably reduces the volume of masonry work in footing and depth of bed concrete and as such proves to be economical. A 7 to 8 cm. thick bed of lean concrete is usually provided below the reinforced concrete footing to

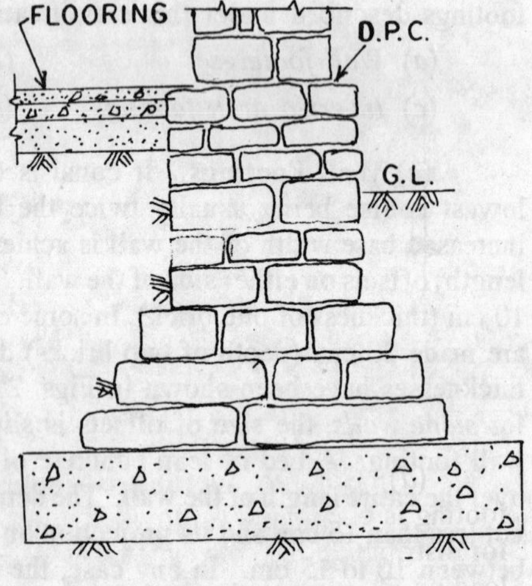

Fig. 2·17

Stone-wall footing

perform the function of concrete bed block. The reinforced concrete footing for a wall is shown in Fig. 2·18.

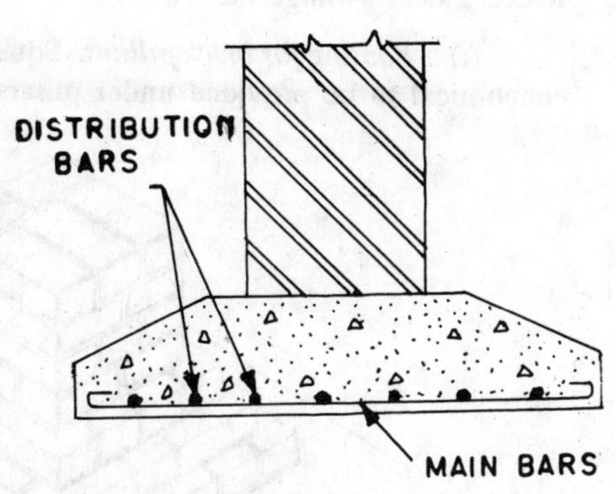

Fig. 2·18

Reinforced concrete footing for a wall

(*c*) **Inverted arch footings.** This type of foundation used to be provided for multistoried buildings in olden times. However, with the advent of reinforced cement concrete construction practice, inverted arch construction is rarely done these days. One of the drawbacks in this type of construction is that the end piers have to be specially strengthened by buttresses to avoid the arch thrust tending to rapture the pier junction. However, the advantage of inverted arch construction is that in soft soils the depth of foundation is greatly reduced. Inverted arch footing has been shown in Fig. 2·19.

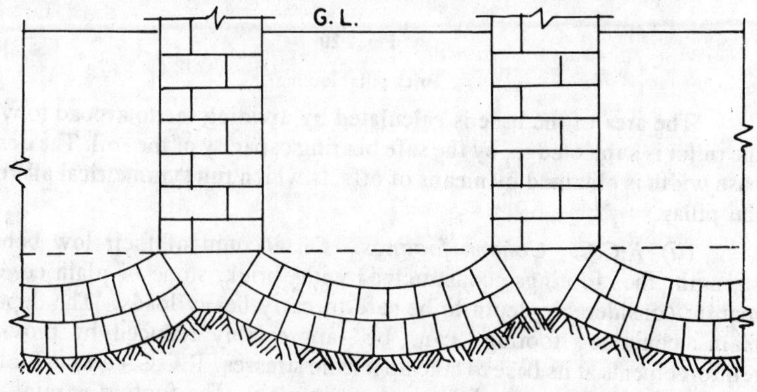

Fig. 2·19

Inverted arch foundation

(*d*) **Column footings (Independent footings).** An independent footing is one which is provided under a column or other similar member for distributing the concentrated loads in the form of uniformly distributed load on the soil below. The footing may be square, rectangular or circular in plan. Depending upon the load to be carried and the bearing capacity of the soil, independent footing may be of brick masonry, stone masonry, R.C.C., steel grillage etc.

(*i*) *Footing for brick pillars.* Square footing is the simplest and most economical to be provided under pillars, columns or stanchions.

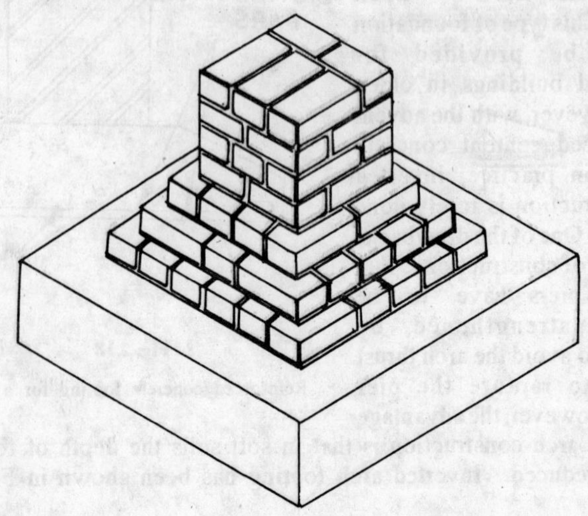

Fig. 2·20

Brick pillar footing

The area of the base is calculated by dividing the total load to which the pillar is subjected to, by the safe bearing capacity of the soil. The desired base width is obtained by means of offsets which run symmetrical all-round the pillar.

(*ii*) *R.C.C. Column footings.* On account of their low bending strength the footings constructed with brick, stone or plain concrete require considerable depth to be safe to carry heavy loads. The depth of plain concrete footing can be appreciably reduced by providing reinforcements at its base to take up tensile stresses. R.C.C. column footings may be circular, rectangular or square in plan. The footing is reinforced both-ways by means of mild steel ribbed bars placed at right angles to one another at equal distances apart.

(*iii*) *Stone pillar footings.* This type of footing is similar in construction to the footing for brick pillar. In this case, the regular offsets on all the four sides of pillar are slightly bigger in width and depth.

2·18. GRILLAGE FOUNDATIONS

When heavy structural loads from columns, piers or stanchions are required to be transferred to a soil of low bearing capacity, grillage

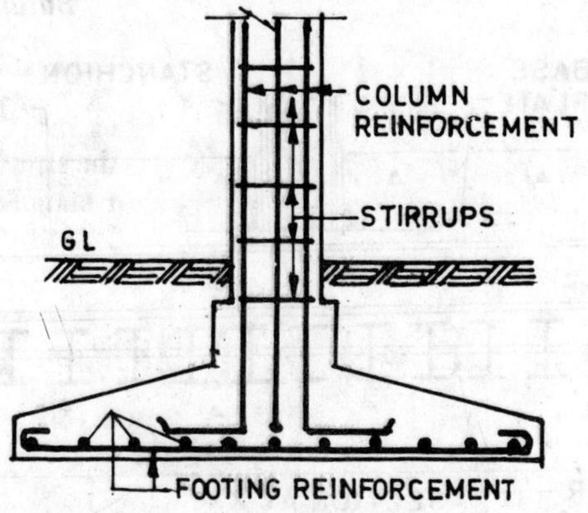

Fig. 2·21

R.C.C. column footing

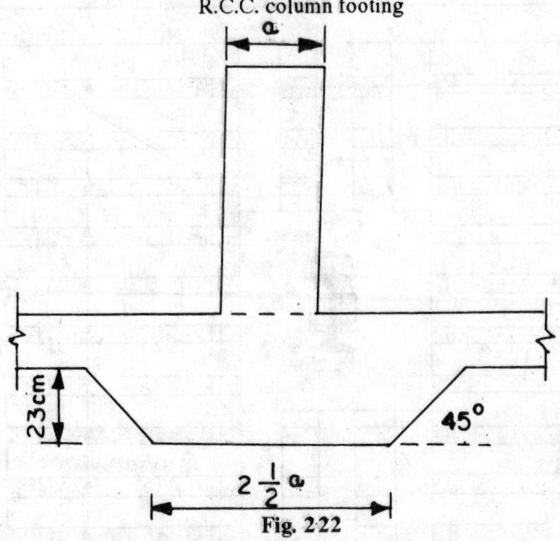

Fig. 2·22

Shape and proportion of a column foundation monolithic with floor slab.

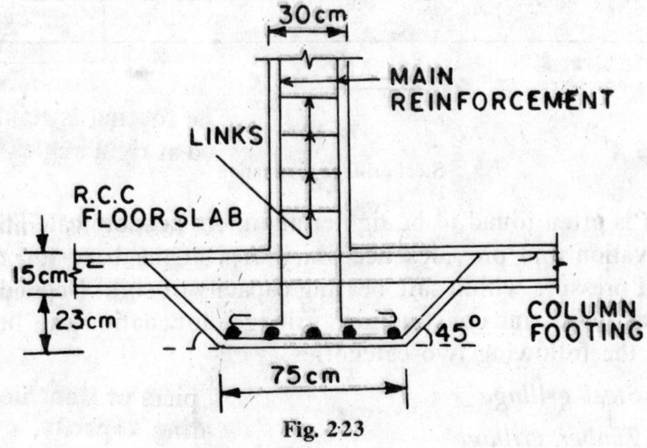

Fig. 2·23

Position of reinforcements in column in Fig. 2.22

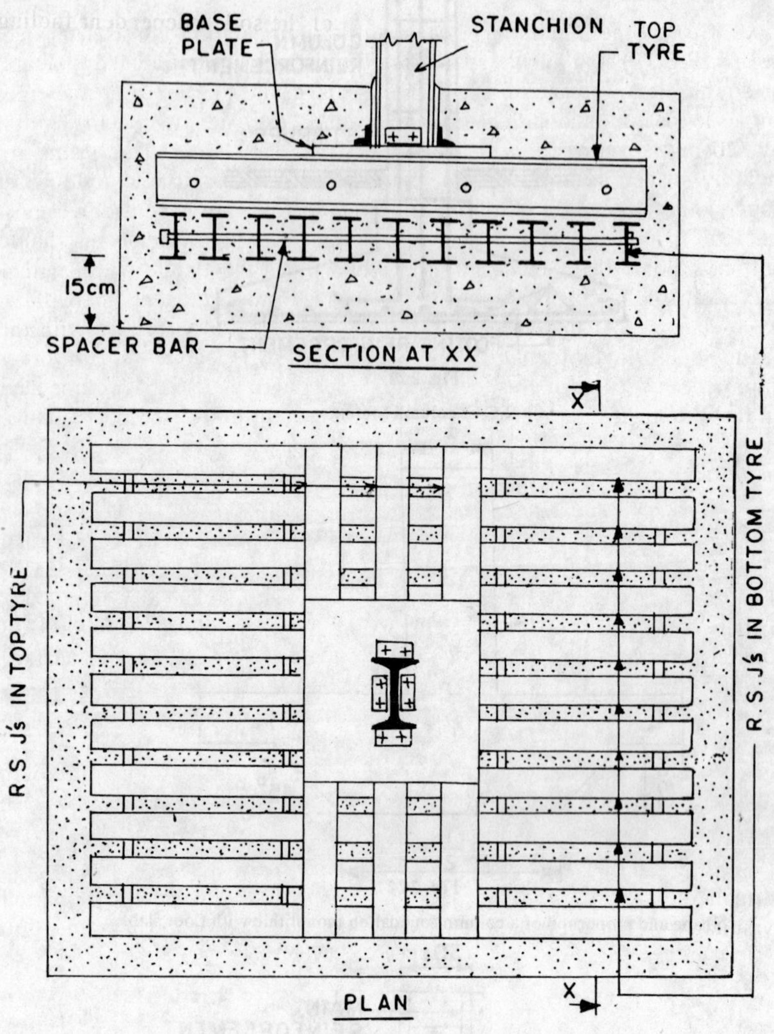

Fig. 2·24

Steel grillage foundation

foundation is often found to be lighter and more economical. This avoids deep excavation and provides necessary area at the base to reduce the intensity of pressure within safe bearing capacity of soil. Depending upon the material used in construction grillage, foundation can be broadly divided in the following two categories.

(*a*) *Steel grillage*

(*b*) *Timber grillage*

(*a*) **Steel grillage.** Steel grillage foundation consists of steel beams (R.S.J's) also known as grillage beams which are provided in single or double tiers. In case of double tier grillage, the top tier is laid at right angles to the bottom one. The grillage beams of each tier are held in position by 20 mm spacer bars with 25 mm dia pipe separators. The beams are suitably spaced so as to provide facility for the placing and rodding of concrete between them. A minimum clearance of 8 cm is considered most suitable. In any case, the distance between the flanges of the beams should not be more than one and half to two times the flange width with a maximum of 30 cm. If the beams are spaced more distance apart, there is a danger of the concrete filling not acting monolithically with the beams, and as such, may result in the failure of the foundation. In order to protect the beams against corrosion, a minimum cover of 10 cm is kept on the outer sides of the external beams as well as above the upper flange of the top tier. Cover of concrete under the lower beam should not be less than 15 cm.

Method of construction. Excavations are carried to the designed depth and the bed is well levelled. This foundation bed is covered with a layer of rich mixture of concrete not less than 15 cm in thickness. This is well compacted so as to make the layer of concrete an impervious bed. Grillage beams of the designed dimensions are then placed on this bed of concrete at specified distance apart using separators. The upper surface of the grillage beam flanges is brought in a horizontal plane and rich cement grout is poured all round the lower flanges of the beams so as to secure the beams to the concrete bed. The concrete is then placed between and around the beams. If another tier is required to be provided, it is kept at right angles to the already laid tier and the entire space is filled with concrete.

(*b*) **Timber grillage.** Where the soil encountered is soft and is permanently water-logged building walls can be economically supported by suitably designed timber grillage foundations. This type of foundation

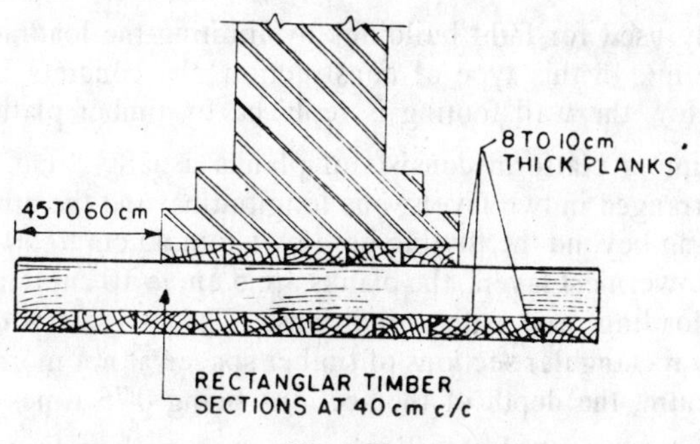

Fig. 2·25

Timber grillage foundation for a wall

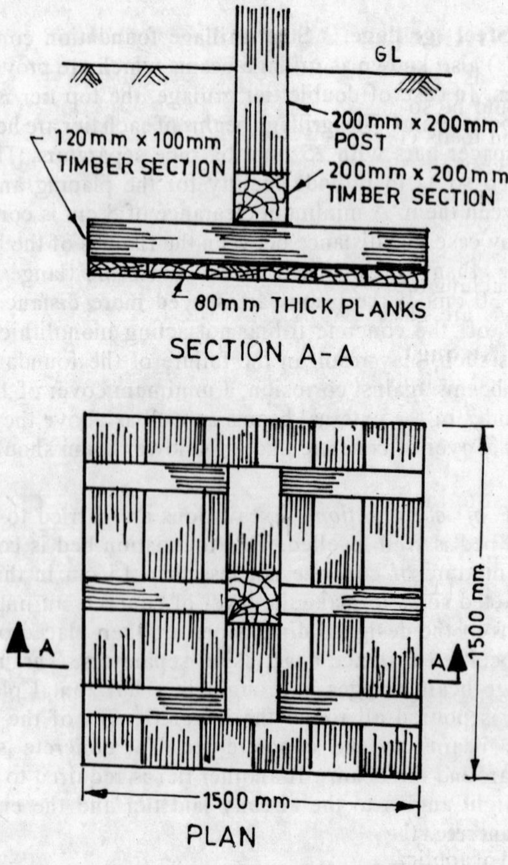

Fig. 2·26 Grillage foundation for a timber post

can be safely used for light buildings by limiting the loading on the soil to 5·5 tonne/m². In this type of construction, the concrete block usually provided below the wall footing is replaced by timber platform.

The timber platform consists of planks usually 8 cm to 10 cm. in thickness, arranged in two layers, one longitudinal and the other across the wall extending beyond the footing base by about 45 cm to 60 cm on either side. In the lowermost layers, the planks are 5 cm to 10 cm thick depending upon the loading and site conditions. The two layers of planks are separated by rectangular sections of timber spaced at not more than 38 cm. centre to centre, the depth of the sections being 0·75 times the width.

The arrangement of scantlings in case of grillage foundations for a column or a timber post is shown in Fig. 2·26.

2·19. ECCENTRICALLY LOADED FOOTINGS

As far as practicable, the foundation, (it may be for a wall or column), should be so shaped and proportioned that the centre of gravity of the imposed loads coincide with the *c.g.* of the supporting area of base. However, when walls or columns are to be placed close to property lines, the required supporting areas of the base cannot be placed concentrically with the imposed loads without overlapping the property lines. Hence, different methods are adopted to ensure the stability of the wall or column without encroaching the area outside the property line of the building. Some of the methods are given below :

(*i*) **Offsetting the footings.** The footing of a boundary load bearing wall is so shaped so as to have a considerable wider base with regular offsets on the inside while the outside wall face is kept flush with the boundary line. As is obvious, the imposed load acting on one side of the centre of the footing loads the supporting area unequally. The pressure at the edge of wall nearest to the point of application of imposed load will be found to be

$$= p_1 = \frac{W_1}{b}\left(1 + \frac{6e}{b}\right) \qquad ...(i)$$

where

W_1 = total load per unit length coming on the supporting area.

b = width of the supporting area at base of footing.

e = eccentricity of load or the distance between the *c.g.* of the load and *c.g.* of supporting area of base.

The pressure at the edge of footing farthest from the point of application of imposed load will be found to be

$$= p_2 = \frac{W_1}{b}\left(1 - \frac{6e}{b}\right) \qquad ...(ii)$$

Referring equation (*ii*), we find that if the eccentricity $= \frac{b}{6}$, p_2 becomes zero. However, if if e is greater than $\frac{b}{6}$, the value of p_2 will be negative and as such the footing will have tension at B. In masonry construction excessive tension is not permissible and hence in order that the supporting area is fully in compression, the width of footing is so adopted that the *c.g.* of the load falls within the middle third of the base.

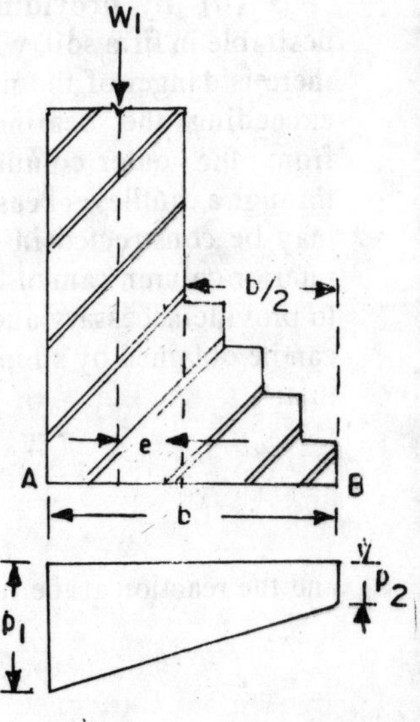

Fig. 2·27

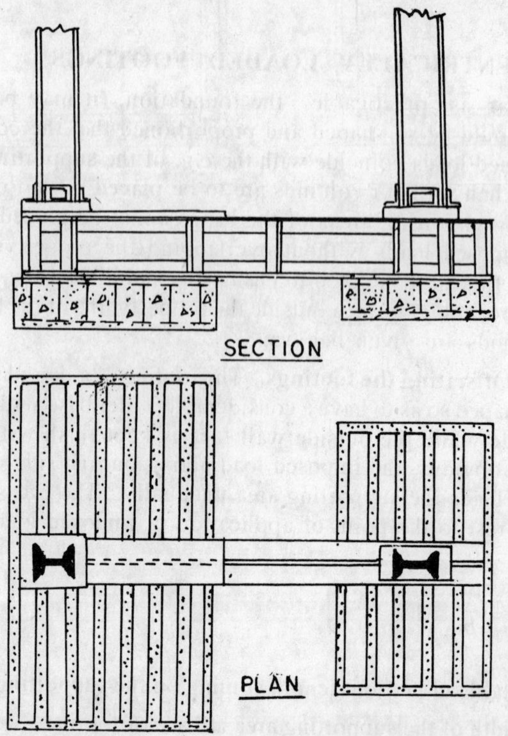

Fig. 2·28 Sectional elevation and plan of steel grillage strap footing

(*ii*) **By providing strap footing.** Offsetting of the footing is desirable in firm soil, while in soft soil and under heavy loading conditions, there is danger of the maximum pressure at the outer edge of the footing exceeding the bearing capacity of soil. In such circumstances, the load from the outer column is balanced by the load from the inner column through a cantilever beam, acting about a fulcrum. Cantilever or strap footing may be constructed in reinforced concrete, or steel grillage. In case, the interior column cannot be provided, a suitable concrete block may be built to provide necessary anchorage. The value of the weight of concrete block can be obtained by simple principle of lever. Referring to Fig. 2·30 we find that:

$$W_1 \times l_1 = W_2 l_2$$

$$W_2 = \frac{W_1 l_1}{l_2}$$

and the reaction at the fulcrum is given by :

$$R = W_1 \left(\frac{l_1}{l_2} + 1 \right)$$

2·20. COMBINED FOOTINGS

A combined footing is so proportioned that the centre of gravity of

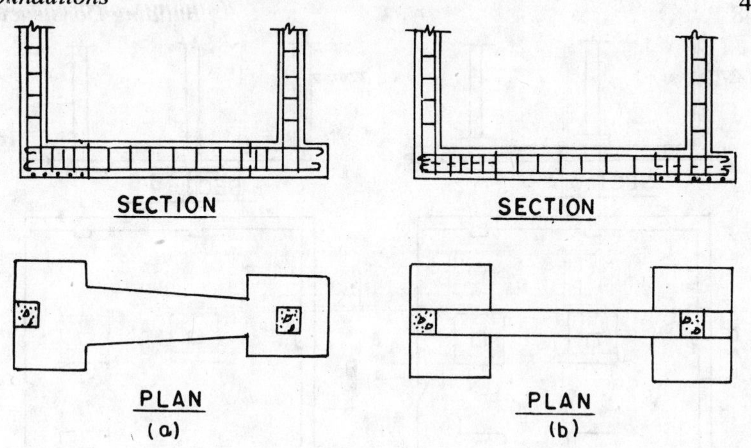

Fig. 2·29

R.C.C. Strap footing

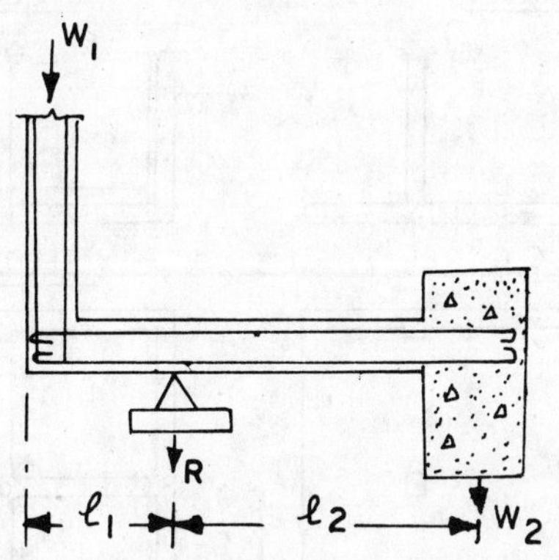

Fig. 2·30

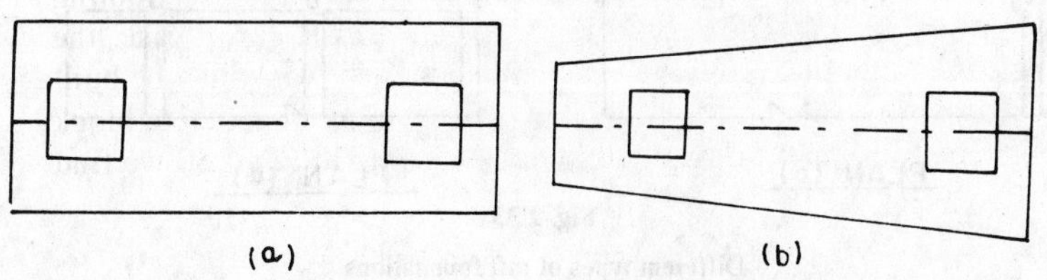

Fig. 2·31

Combined rectangular footing

Fig. 2·32

Combined trapezoidal footing

the supporting area is in line with the *c.g* of the two column loads. A combined footing may be rectangular or trapezodial in shape. Rectangular shape is only possible where loading condition is such that either the two columns

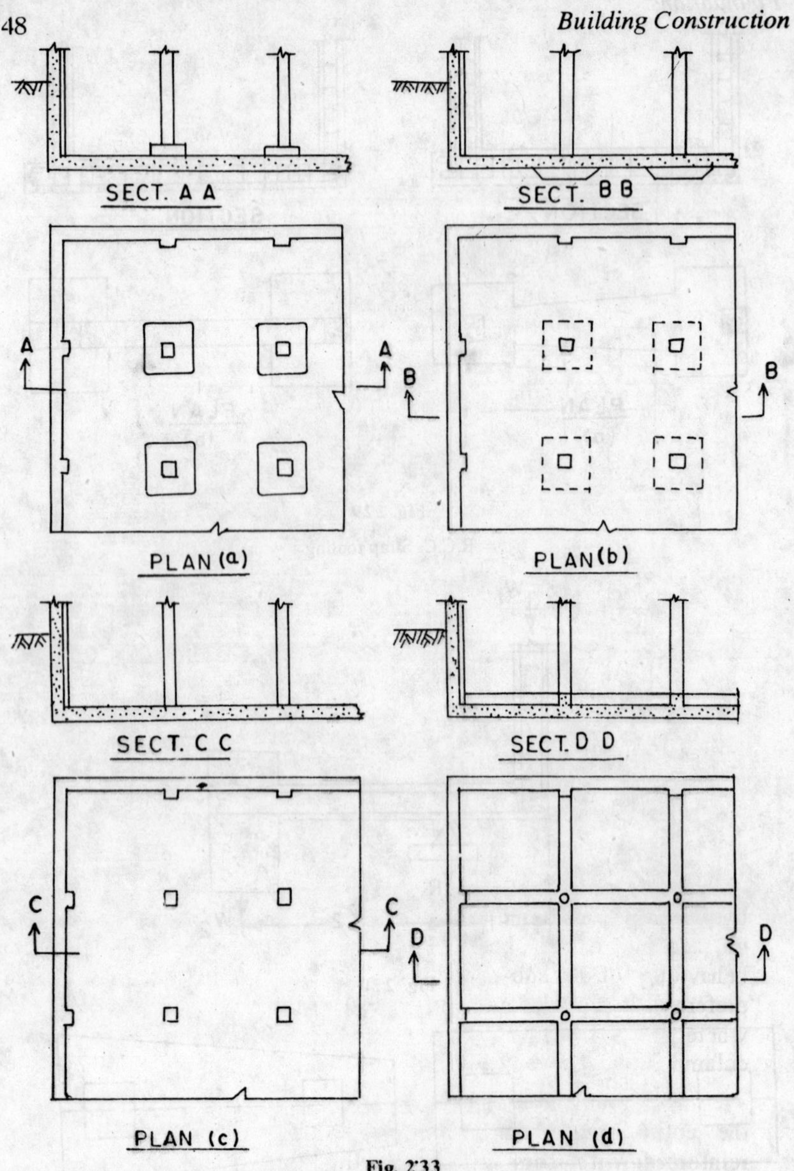

Fig. 2·33
Different types of raft foundations

are equally loaded or the interior column carries greater load. On the other hand, in case of trapezodial footing, no such condition is applicable.

2·21. RAFT FOUNDATION

In made-up ground, soft clay or marshy site having low value of bearing capacity, heavy concentrated structural loads are generally supported

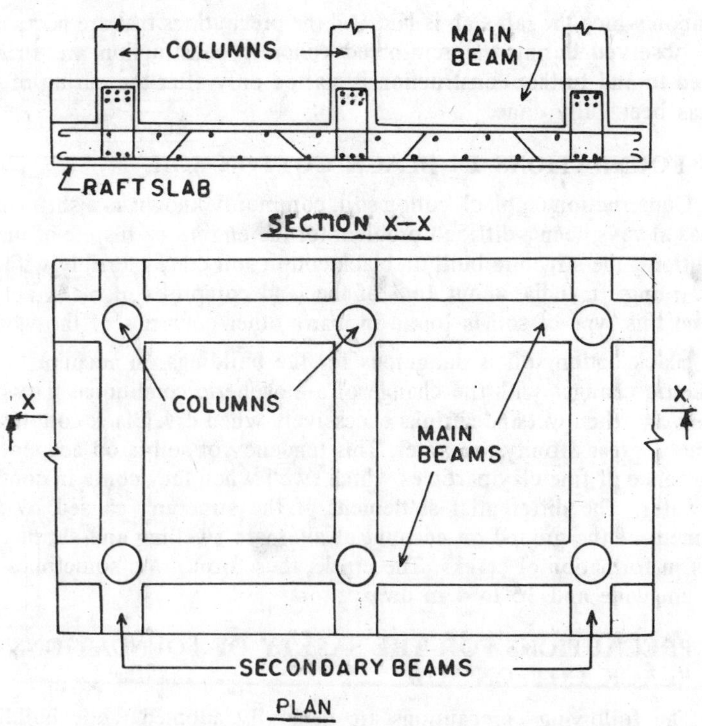

Fig. 2'34
Raft foundation

by providing raft foundation. Also if the structure is liable to subsidence on account of its being located in mining area or due to uncertain behaviour of its sub-soil water condition, raft foundation should be preferred. It provides an economical solution to difficult site conditions, where pile foundation cannot be used advantageously and independent column footing becomes impracticable.

Raft foundation consists of thick reinforced concrete slab covering the entire area of the bottom of the structure like a floor. The slab is reinforced with bars running at right angles to each other both near bottom and top face of the slab. Sometimes it is necessary to carry the excessive column load by an arrangement of inverted main beams and secondary beams, cast monolithically with the raft slab.

Method of construction. The raft slab generally projects for a distance of 30 cm. to 45 cm. on all the sides of the outer walls of the structure and as such the area of excavation is slightly more than the area of the structure itself. The excavation is made to the required depth and the entire excavated area is well consolidated. This surface, when dry, provides the base upon which the raft slab is laid. All the precautions that are necessary to be observed during the reinforced concrete construction are strictly adhered to and further construction is started only after the curing of the raft has been fully done.

2·22. FOUNDATIONS IN BLACK COTTON SOIL

Construction on black cotton soil, commonly known as a shrinkable soil has always been a difficult problem for the engineers. Inspite of many precautions, the structure built on black cotton soil cracks terribly without any warning. In India, about 16% of the land comprises of black cotton soil and this type of soil is found in many other countries of the world.

Black cotton soil is dangerous for the buildings on account of its volumetric changes with the change of atmospheric conditions. It swells excessively when wet and shrinks excessively when dry. Black cotton soil has thus a great affinity for water. This tendency of soil is on account of the presence of fine clay particles which swell when they come in contact with water. The differential settlement of the structure, caused by the movement of the ground on account of alternate swelling and shrinkage, results in formation of cracks. The cracks thus formed are sometimes 15 to 20 cm wide and 2·5 to 4 m deep.

2·23. PRECAUTIONS FOR THE SAFETY OF FOUNDATIONS IN BLACK COTTON SOIL

The following precautions are generally adopted while building structures on black cotton soil :

(1) To limit the loads on the soil to 5·5 tonnes/m². If water is liable to find an access to the foundations, the limit of loading should be restricted to 4900 kg/m².

(2) To take the foundation to such depths where the cracks cease to extend. The minimum depth of foundation should be at least 1·5 m

(3) To provide reinforced concrete ties or bands all round the main walls of the building. The R.C.C. ties or bands which may be 10 cm to 15 cm deep should be placed at plinth level, lintel level and eaves level. In case of flat roof, the R.C.C. slab itself acts as a tie and as such no extra band need be provided near the roof in such cases.

(4) If the depth of black cotton soil at a given site is only 1 to 1·5 m, the entire black cotton soil above the hard bed may be completely removed and the foundation laid on the hard bed below.

(5) The swelling of soil in direct contact with the foundation material causes maximum damage. Hence it is necessary to prevent direct contact of black cotton soil with masonry work below ground level. This can be achieved by making wider trenches for foundation and filling spaces on the either side of foundation masonry with sand or moorum.

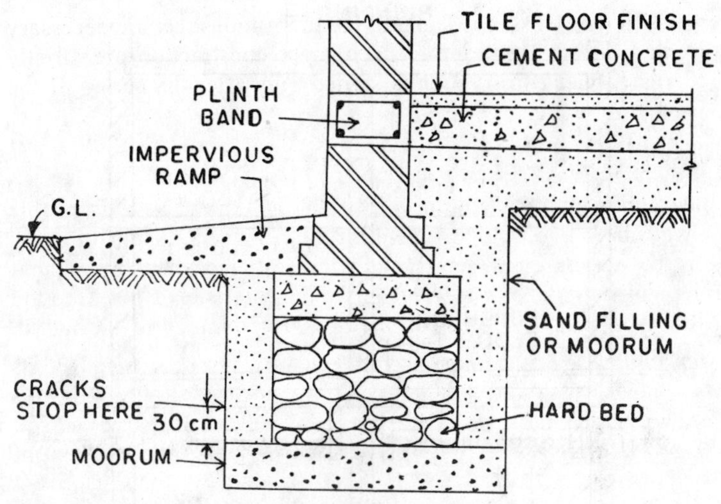

Fig. 2·35

Wall footing in black cotton soil

(6) The bed of foundation trench should be made firm or hard by ramming it well. On the rammed bed a 30 cm layer of good hard moorum should be spread in layers of 15 cm, each layer being well watered and rammed before laying the next layer. On this layer either stone or sand bed should be provided to the desired height to place the foundation concrete bed block upon it.

(7) In case of ordinary buildings, the foundation should be taken at least 30 cm deeper than the depth where the cracks stop.

(8) In important structures raft foundation should be provided so as to float the building on the bed below, quite independent of the surrounding soil.

(9) For less important structures like compound walls etc., the foundations should preferably be taken at least 15 cm below the depth at which cracks in soil cease to occur.

(10) Construction in black cotton soil should be undertaken during dry season.

(11) The masonry for the walls should start at least 15 cm below the general ground level.

(12) The width of trench for main walls or load bearing walls of a building should be dug 40 cm wider than the width of foundation. This is necessary to ensure provision of at least 20 cm wide layer of coarse sand on either side of foundation masonry thereby separating the sub-structure from having direct contact with black cotton soil.

In case of compound wall, width of sand layer on either side of foundation masonry could be 15 cm.

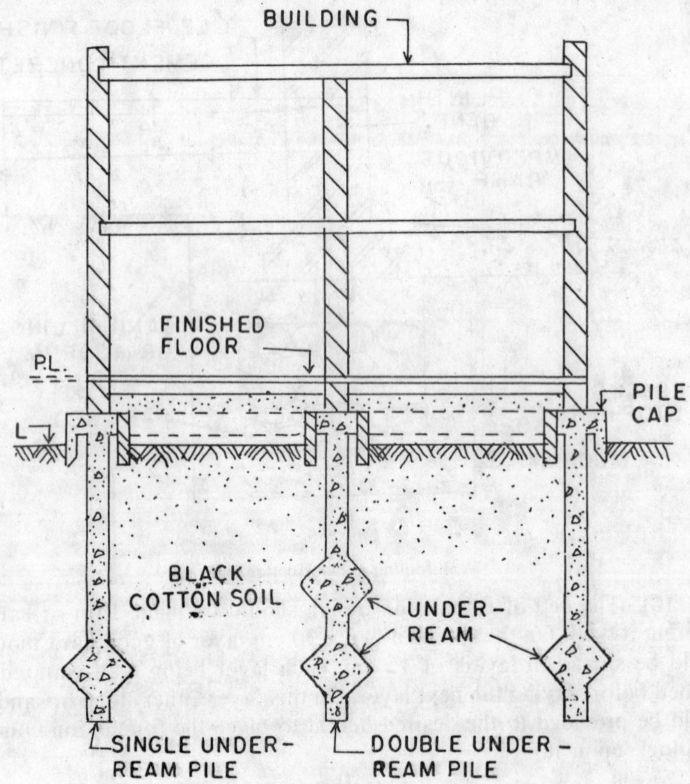

Fig. 2·36

Under-reamed pile foundations

Based on research carried out at Central Building Research Institute Roorkee (INDIA) and elsewhere it is found that under-reamed piles provide an ideal solution to foundations in black cotton soil or other similar types of expensive soils. The construction of under-reamed pile is briefly discussed below :

Under-reamed pile foundations. Under-reamed piles are bored cast-in-situ concrete piles having bulb shaped enlargement near base. These piles are commonly recommended for providing safe and economical foundations in expensive soils such as black cotton soil, filled up ground and other types of soils having poor bearing capacity. In this type of foundation the structure is anchored to the ground at a depth where ground movement due to changes in moisture content are negligible. A pile having one bulb is known as single under-reamed pile. It is seen that the load bearing capacity of the pile can be increased by increasing the number of bulbs at the base. In such a case the pile is named as multi-

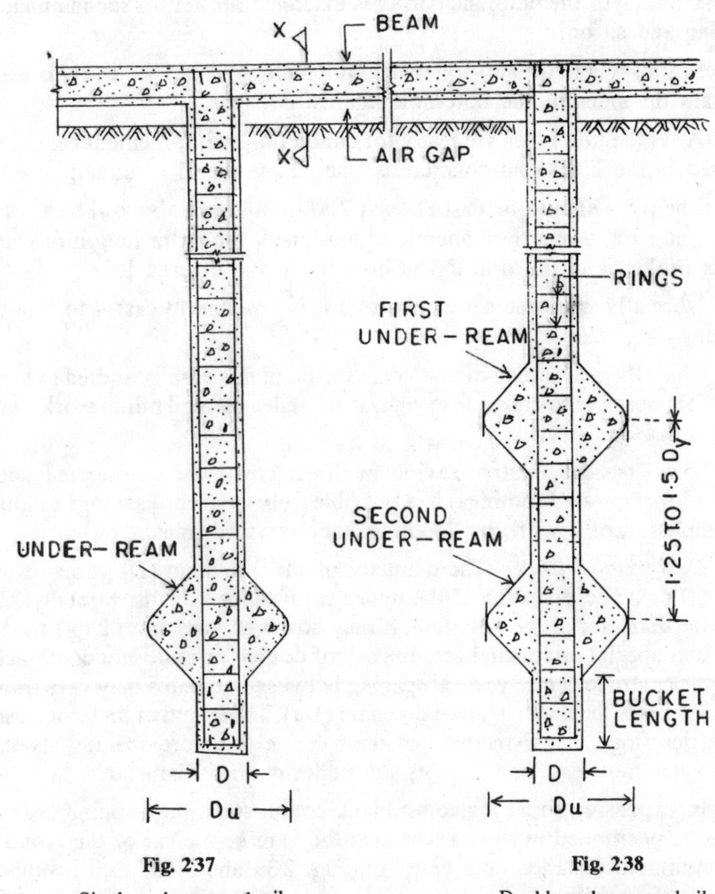

Fig. 2·37
Single under-reamed pile

Fig. 2·38
Double under-reamed pile

under reamed pile. The increase in the bearing capacity of the pile can also be achieved by increasing the diameter and the length of the pile.

The method of construction of under-reamed pile is very simple. The holes for casting piles in the ground may be bored by using hand augers. The auger should be rotated slowly with a constant downward pressure and should be taken out when it is just full. For deeper borings suitable extension rods are attached to the auger. In projects where the magnitude of piling work is more, electric power winch may be used for expediting boring operation.

After boring is carried out to the required depth, the base of the bore hole is enlarged in the form of a bulb near its base by use of a tool, known as *under-reamer*.

In case of double or multi-under-reamed piles, after the boring and under-reaming of the bulb, the boring is extended further for second under-reaming and so on.

In sites where the sub-soil water table in high, bentonite slurry is used to retain the sides of the bore-hole against collapse.

After the pile holes are ready for concreting, reinforcement cages are lowered in the holes and concrete is poured by use of a funnel.

The piles should be cast at least 200 to 300 mm above the cut-off-level. Later on, when the concrete is hardened, the extra length of each pile is broken and the pile top is brought to the desired level.

Normally with one set of two rigs it is possible to cast 4 to 5 piles in a day.

Since the magnitude of open excavation in this case is limited to bore holes, it is possible to continue execution of under-reamed piling work even in rainy season.

Thus, besides relative saving in direct cost (when compared with conventional isolated footings) it is possible to have overall savings in time of completion of a work by adopting under-reamed piles.

Design guidelines : The diameter of the pile stem (D) varies from 20 to 50 cm. The diameter of the under-ream bulbs (Du) is normally 2.5 times the diameter of the pile stem. It may however, vary from 2 to 3 times (D) under special circumstances. In case of double or multi-under-reamed piles, the centre to centre vertical spacing between two bulbs may vary from 1¼ to 1½ times the under-reamed diameter (Du). The length of under-reamed pile varies from 3 to 8 metres and their centre to centre spacing should normally be not less than 2 times the under-reamed diameter.

In expensive soils, including black cotton soils, the capping beams should be positioned in such a way that they are kept clear of the groun'! by a minimum distance of 8 cm (Ref. Fig. 2·39 and 2·40)

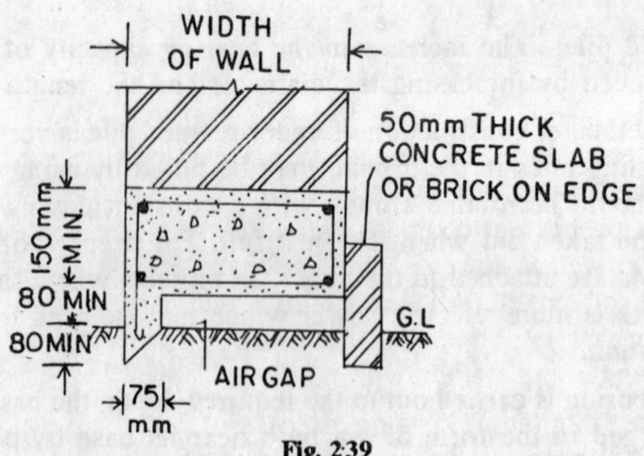

Fig. 2·39

Detail applicable to interior beam

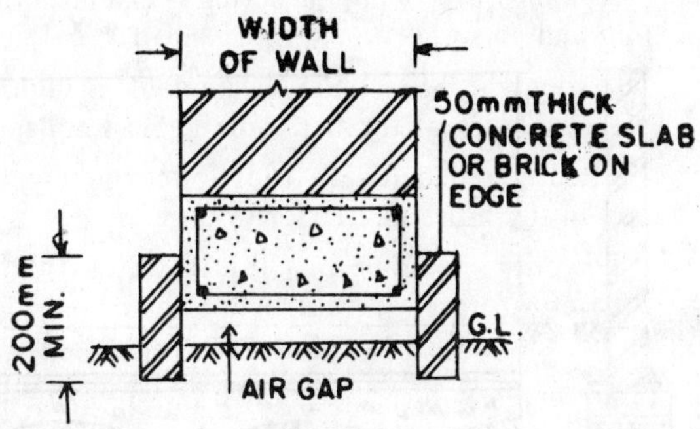

Fig. 2·40

Detail applicable to exterior beam

By this arrangement, it is ensured that the soil from below does not heave against the beam and the full load of the structure is borne by the piles.

Exact size of the piles, including reinforcements can be worked out by use of design table given in IS : 2911 (Part III–1973)

Bored Compaction Pile : Bored Compaction pile is a modified form of under-reamed pile, wherein the compaction of concrete and its surrounding ground is achieved by driving a suitable core through the freshly filled concrete in the pile bore hole. The process of compaction results in increasing the bearing capacity of the pile by 50 to 100%. These piles serve as suitable foundation for heavy structures in loose to medium sandy and silty stratas even with high sub-soil water table.

In case of ordinary under-reamed pile, the reinforcement cage is placed in position first in the bore hole and concrete is poured subsequently whereas in bored compaction pile the bore hole is first filled with concrete and the reinforcement assembly (which serves as a core) is driven into it immediately afterwards. The assembly normally consists of reinforcement cage in which the vertical bars are bent to form a funnel at lower end on which a hollow sheet iron core is fitted. This holds a detachable cast iron shoe and a 10 cm diameter heavy steel driving pipe. The driving pipe is fitted with a driving head. As the assembly is driven, it pushes out concrete on all sides resulting in compaction of the concrete as well as the soil simultaneously. During the process of driving, the level of concrete is constantly maintained upto the ground level by adding extra concrete from time to time. Finally, the driving pipe is filled with concrete and withdrawn.

The minimum spacing in case of bored compaction pile may be kept as 1·5 times the under-reamed diameter.

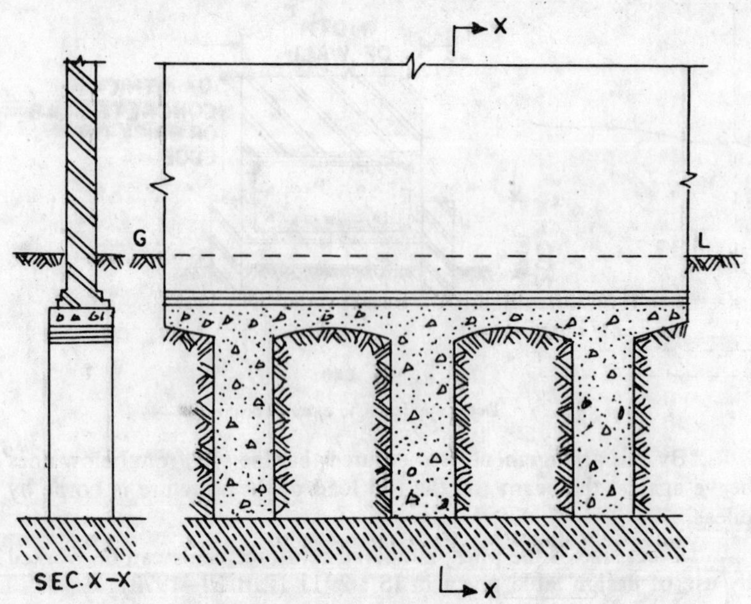

Fig. 2·41

Wall footing resting on pier foundation

2·24. PIER FOUNDATION

When a heavily loaded building is to be situated in sandy soil or soft soil, overlying hard bed at reasonable depth, pier foundations are sometimes used to transfer the load of the building to the hard bed below. This method consists in sinking vertical shafts up to hard bed and filling them with concrete.

The diameter of the shaft and their centre to centre spacing depend upon the loading condition, nature of soil and depth at which the hard bed is situated. However, in no case the diameter or the least horizontal dimension (in case of square or rectangular pier) of plain concrete pier should be less than 1/12th its height. To prevent the side earth from falling in, the inside of the shaft is sometimes lined with timber. The timber lining is removed during the filling up of the shaft with concrete. The shafts are connected to each other by a reinforced cement concrete or steel grillage cap. Concrete cap is sometimes built in the form of arch.

2·25. FOUNDATIONS ON SLOPING GROUND

To avoid sloping foundation bed or excessive depth of excavation at the top end, stepped foundation is necessary to be provided in a considerably sloping ground. This is achieved by cutting the portion of the

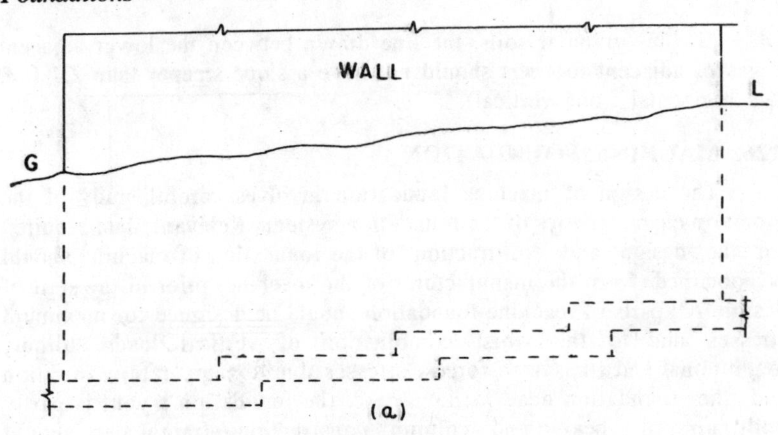

Fig. 2·42

Elevation of stepped foundation

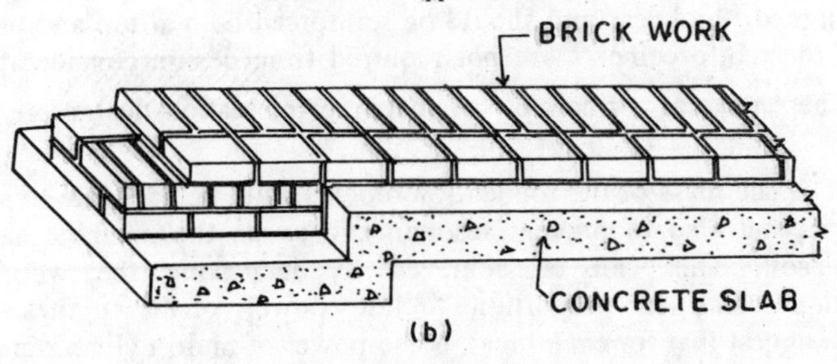

Fig. 2·43

Isometric view of stepped foundation

foundation trench in steps. The steps should not preferably be more than the depth of the concrete bed and each step should be a multiple of the depth of one brick so as to fit in with the brick courses. The lap of concrete at each step should never be less than the vertical thickness of the concrete.

In some cases, the bottom of the footings of different walls of the same structure may be at different levels. The following limitations (as given by I.S.I.) are necessary to be observed in deciding the depth of footings in such circumstances.

(1) The distance between the lower edge of the footing to the sloping surface should not be less than 1 m for soils and 60 cm for rocks.

(2) In clayey soils, the line drawn between the lower adjacent edge of the upper footing and the upper adjacent edge of the lower footing should not have a steeper slope than 2:1 (*i.e* two horizontal : one vertical).

(3) In granular soils, the line drawn between the lower adjacent edges of adjacent footings should not have a slope steeper than 2:1 (*i.e.* two horizontal : one vertical).

2·26. MACHINE FOUNDATION

The design of machine foundation involves careful study of the vibration characteristics of the foundation system. Relevant data required for the design and construction of the foundation of machine should be obtained from the manufacturer of the machine, prior to the start of design. All parts of machine foundation should be designed for maximum stresses due to the worst combination of vertical loads, torque, longitudinal and transverse forces, stresses due to temperature variation and the foundation dead load. In case, the foundation layout is partly built up of beam and column construction, straight bars should be provided both at top and bottom of the beams and the spacing of the stirrups should be close. The main foundation block should have the designed thickness and should be reinforced both at top and bottom, even if the reinforcements are not required from design considerations.

The general principles of machine foundation design are given below:

(1) The mass of the foundation block should be adequate to absorb vibrations and also to prevent resonance between the machine and the adjacent soil. This can be achieved by increasing the weight of foundation block in proportion to the power of the engines. Some authors suggest that for each break horse power of multi-cylinder engines, a minimum of 725 kg. weight of foundation should be provided for gas engines, 565 kg. for diesel engines and 225 kg. for steam engines. For single cylinder engines, the above value should be increased by 40 to 60%. As a thumb rule, the weight of the foundation should be at least 2½ times the weight of the whole machine.

(2) To avoid the possibility of differential settlement, the foundation should be so dimensioned that the resultant force due to the weight of the machine and the weight of the foundation passes through the centre of gravity of the base contact area.

(3) The foundation should be stiff enough to have necessary rigidity, since the slighest deflection of foundation can cause considerable bearing troubles.

(4) To avoid transmission of vibration from the machine to the adjoining parts of the building, a gap should be left alround the machine foundation to isolate it from the adjoining parts of the building.

(5) As far as possible, overhanging cantilevers should be avoided. However, in situation where it is not possible to avoid cantilever projections, they should be designed for strength and rigidity against vibrations.

(6) All units of machine foundation should be provided with reinforcement running both ways along the surface of the concrete block. The concrete cover to the reinforcement should not be less than 75 mm at the bottom, 50 mm on sides and 40 mm at the top. In case of foundation for steam turbo-generators, cover for the reinforcement at bottom, side and top of base slab should not be less than 100 mm.

(7) The amount of reinforcement in foundation units should not be less than 25 kg/m^3 of concrete for impact type or reciprocating type of machines, 50 kg/m^3 of concrete for rotary type of machines and 100 kg/m^3 of concrete for steam turbo generators.

(8) M 150 to M 200 grade of concrete can be used in the foundations and as far as possible, the entire block should be concreted in one operation without construction joints.

2·27. CAUSES OF FAILURE OF FOUNDATIONS

The causes of failure of foundations may be summarised under the following heads :

(1) *Unequal settlement of the sub-soil*

(2) *Unequal settlement of masonry*

(3) *Horizontal movement of the soil adjoining the structure.*

(4) *Shrinkage due to withdrawal of moisture from the soil below the foundation.*

(5) *Lateral pressure tending to over-turn the structure.*

(6) *Action of atmosphere.*

(7) *Lateral escape of the soil below the foundation.*

(1) **Unequal settlement of the sub-soil.** Unequal settlement of foundation results in dangerous cracks which ultimately lead to the failure of the structure. If the nature of soil and the loading conditions are uniform over the entire site of the building, the distribution of pressure can be safely assumed to the uniform and the danger of unequal settlement is minimised. In practice, however, it is frequently found that different conditions exist under different parts of the building. If some portion of a building is higher, the soil below the higher portion will be subjected to greater pressure than the soil below the lower portion of the building. Or one section of a building may rest on rock and the adjacent section on compressible clayey soil. In either of the cases, the soil is subjected to unequal settlement, and as such in designing the foundation the pressure on soil should be reduced to such a value that the difference in settlement between the two portions of structure is reduced to minimum. In no case, the pressure on the soil should exceed the safe-bearing capacity of soil. Materials used for construction should be durable so as to avoid danger of the disintegration of the foundations. As far as possible foundation should

be so proportioned that the *c.g.* of load from structure coincides with the centre of gravity of the foundation. By adopting these precautions, the danger to structures on account of unequal settlement of sub-soil can be minimised.

(2) **Unequal settlement of masonry.** Mortar used as binding material in the masonry construction shrinks and gets compressed when loaded excessively before it has fully set. This defect may lead to the unequal settlement of masonry. The following measures should be adopted to avoid unequal settlement of masonry :

(*i*) The consistency of mortar should be such that it provides easy workability. The mortar should neither be very lean nor very stiff.

(*ii*) As far as possible the execution of work should be so planned that the masonry is raised to the same uniform level throughout.

(*iii*) The progress of construction should be such that not more than 1.5m high masonry is constructed in a day.

(*iv*) Masonry should be properly cured for a period of at least 10 days to ensure development of adequate strength of the mortar joints.

(3) **Horizontal movement of the soil adjoining the structure.** This defect is very common in clayey and black cotton soils. Such types of shrinkable soils undergo volumetric changes with the changes in atmospheric conditions. They swell excessively when wet and shrink excessively when dry. The differential settlement caused by the movement of the ground on account of alternate swelling and shrinkage of soil result in formation of cracks in the structure. Following precautions are necessary to be observed to minimise the effect of soil movement on foundation :

(*i*) To limit the loads on the soil to 5·5 tonnes/m². If water is liable to find an access to the foundation, the limit of load on the soil should be restricted to 4900 kg/m²

(*ii*) To take the foundation to such depths where the cracks cease to extend.

(*iii*) To prevent intimate contact of such soils with the part of the structure below ground level. This is achieved by digging trenches on either side of the foundation and filling them with granular material like sand or moorum. Similarly, a layer of sand or moorum is provided below the concrete block, or bottom face of masonry foundation.

For important structures, R.C.C. raft foundation or pile foundation should be used.

(4) **Shrinkage due to withdrawal of moisture from the soil below the foundations.** The roots of an isolated tree generally acquires land equal to the circumference of a circle whose radius is more than

the height of the tree. In some soils it has been found to cause significant dying effect. The suction of sub-soil moisture by the roots, specially in abnormal spells of dry weather, affects the foundation adversely. On account of the differential shrinkage below the foundation, cracks are produced in the foundations, which can ultimately lead to the failure of structure. To avoid the damage due to the trees, certain safe distance should always be maintained between the trees and the foundations. It is considered reasonable to keep the foundations at least the height (*H*) of the tree away from a mature single tree and about 0.5 H from rows of trees. For mature trees the minimum distance should preferably be 15 m and for young trees, the distance should be suitably increased. In addition, the foundation of the structure should be taken to depth of at least 90 cm. to minimise the cracking action.

This defect is also liable to occur in structure founded on damp soils overlying a band of gravel or sand or some other porous material. An abnormally dry weather may cause the sub-soil water level to descend to such a depth that the soil immediately below the foundations may loose its moisture. This may result in formation of cracks in the building on account of differential settlement. Pile foundations or raft foundations are found to be safe for such places.

(5) **Lateral pressure tending to overturn the structure.** This defect may be caused by a force which acts on a wall in such a way that the wall develops a tendency to tilt or overturn. The force may be the thrust due to sloped roof or an arch pressure, or due to a violent storm or the effect of a wide cantilever projection. This defect can cause the foundation of the wall to yield due to the entire load being concentrated near the foundation edge. In such cases, the foundation of the wall should be so shaped and proportioned that the necessary conditions of stability are achieved.

(6) **Action of atmosphere.** A part of rain water finds its way inside the ground. The rain water during its passage downwards brings down salts from the surface which react chemically with the material of foundation and cause it to disintegrate. Also if the foundation is not taken deep inside the ground, rain water may scour the soil above the foundation and expose it. The changes in the sub-soil water table, as a result of seasonal change, result in expansion or shrinkage of the soil, which causes cracks in foundations. Following precautions are necessary to be observed to avoid the danger on this account :

(*i*) The foundations should be taken to such a depth that the adverse effects of atmosphere are nullified.

(*ii*) There should be adequate provision for the drainage of the sub- soil water when it rises up and causes danger to the foundation.

(*iii*) If necessary, dense cement concrete, M150 or stone masonry should be used in the foundations at sites where the ground water and soil contain excessive sulphates or other injurious compounds.

The sides of the foundation trenches should be well-filled and consolidated. At ground level, the outer surface adjacent to the wall should be given a good slope, so as to allow the rain water to flow away from the wall. A 60 cm. to 90 cm. wide plinth protection serves the purpose more efficiently.

(7) **Lateral escape of the soil below the foundation.** This defect is liable to occur when the building is situated near a river bank, or deep cutting where there is danger of the soft or loose soil below the foundation getting disturbed. In case of such sites, sheet piles of timber or steel, should be driven to confine the soil and avoid the danger.

2·28. SETTING OUT

Before commencement of the excavation of trenches for foundation, a setting out plan is prepared on paper. The setting out plan is a dimensioned ground floor plan, usually drawn to scale of 1 : 50. The plan is fully dimensioned at all breaks and openings. One of the methods of laying out

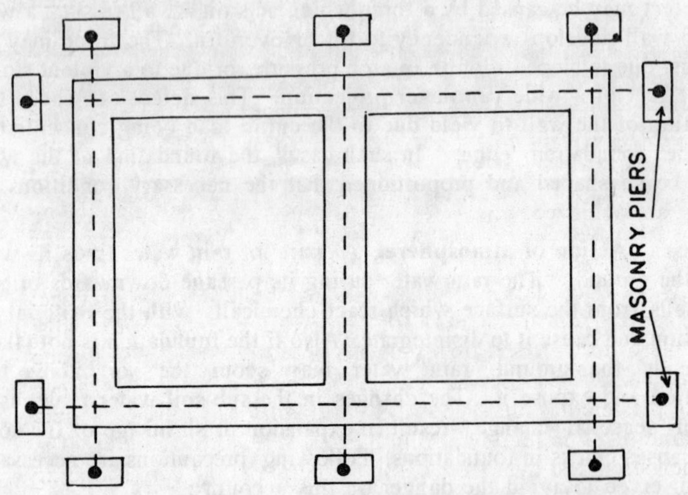

Fig. 2·44

Setting out of foundations with the help of masonry piers

is to first mark the centre line of the longest outer wall of building by stretching a string between wooden pegs driven at its ends. This serves as the reference line for marking the centre line of all the walls of the building. The centre line of the wall, which is perpendicular to the long wall, is marked

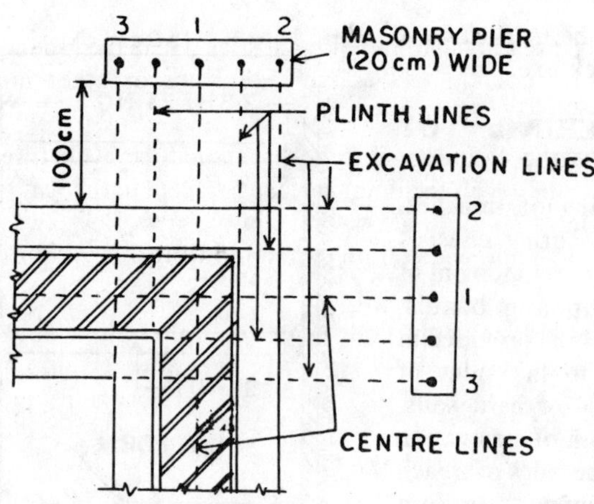

Fig. 2·45

Plan showing a corner of a building to illustrate how the marks are made on the top of masonry pier in the process of setting out of foundations

by setting up a right angle. Right angle is set up by forming triangles with sides 3, 4 and 5 units long. If we fix the two sides of the right angles triangle to be 3 m, and 4 m, then the third side *i.e.* the hypotenuse should be taken a 5 m. The dimensions should be set out with a steel tape. The alternative method of setting out right angle is by the use of theodolite. This instrument is also helpful in setting out acute or obtuse angles. Small right-angled projections are usually set out with mason's square.

The method of setting out described above is not so reliable for important works as there is likelihood of the wooden pegs being pulled up or displaced. In an accurate method, the centre lines of the building walls are carefully laid by means of small nails fixed into the head of the wooden pegs driven at the quoins. In case of rectangular buildings, the diagonal from the opposite corners are checked for their equality. Small brick walls, pillars or platforms are constructed 90 cm clear of the proposed foundation trench. The platforms are about 15 cm wider than the trench width and are plastered at top. The tops of all platforms or pillars should be at the same level preferably at plinth or floor level of building. The strings are then strenched over the nails in the pegs and the corresponding lines are marked on the wet plastered platforms top by pressing the stretched string on the plastered surface by a trowel. The outside lines of the foundation trench and the plinth lines are marked on the wet plastered platform top in the similar manner.

Before starting excavation, the strings are stretched between the outside lines of the foundation trench marked over the platform top and the cutting lines are marked on the ground by lime powder. If necessary, the lines may be marked by a daghbel or pick-axe.

2·29. TIMBERING OF TRENCHES

Timbering of trenches, also known as shoring, consists of a temporary arrangement of boardings or poling boards, wailings and struts provided to give support to the sides of trench. In case of hard soils, where the depth of excavation is moderate, the sides of trench require no support. In firm soils, if the depth of trench is small, shoring may not be necessary, but in case the depth of excavation is more, it is safe to provide light support to check any possibility of the yielding of trench sides. As a rule, all trenches in soil other than rock or hard compact soil more than 1.5 m deep, should be properly shored and timbered. Similarly all trenches in soft or fissured rock or hard soil and more than 2 m deep should be properly shored and timbered. A pair of poling boards of size 20 to 25 cm x 4 to 5 cm and length varying with the depth of excavation are struted apart by 10 cm bullies or struts and are spaced at a minimum distance of 1·8 m along the length of trench (Refer Fig. 2·47). Although one strut between two poling boards is sufficient, but sometimes, the site condition may require the use of two struts between boards. (Refer Fig.2·48).

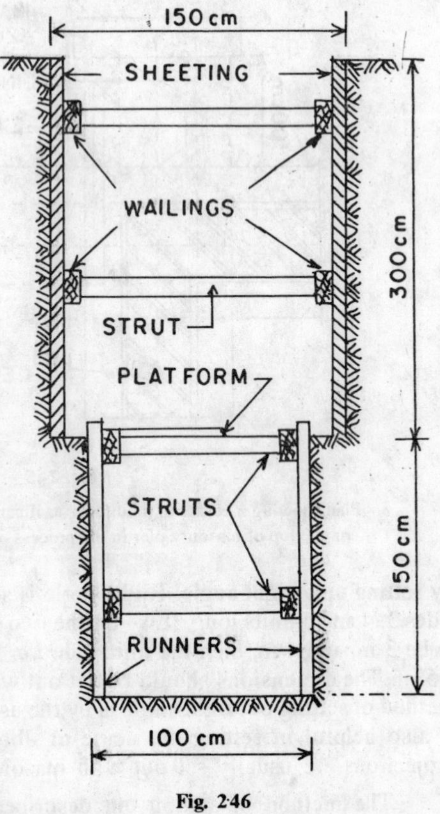

Fig. 2·46

Timbering of deep trench in loose soil

In stiff clay or moderately firm soils, the support to sides of trench can be given by the use of horizontal boarding or wailing.

The wailing of size 20 to 25 cm x 5 to 7.5 cm is run continuous along the length of trench on either sides and the pair of wailings is strutted apart as shown in Fig. 2·50. In case of softer soil, the poling boards are placed closer, say 90 cm apart and are held by wailings. The wailing on either sides is strutted apart as shown in Fig. 2·52. In this arrangement, it is assumed that the poling boards resist the soil pressure and transfer the

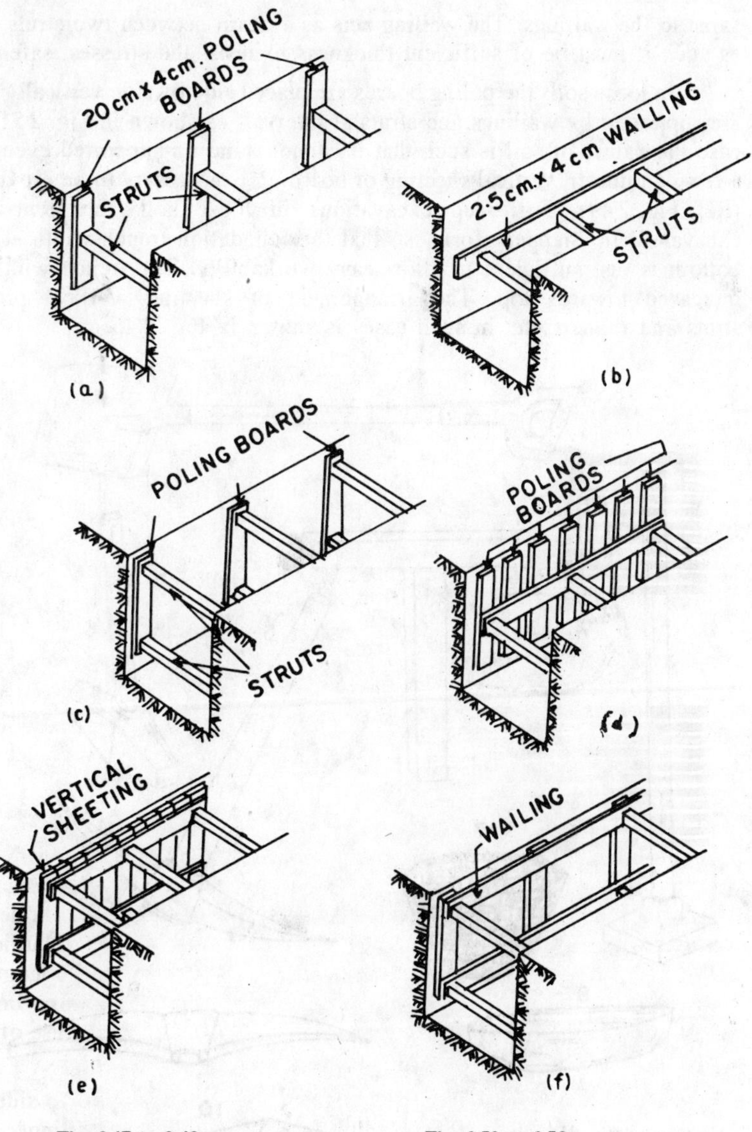

Fig. 2·47 to 2.49 **Fig. 2·50 to 2·52**

Fig. 2·47 and 2·50	Figures showing various methods of the timbering of trenches in firm soil.
Figs. 2·48 and 2·52 and 2·50	Figures showing various methods of the timbering of trenches in mod erately firm soils.
Figs. 2·49 and 2·51	Figures showing timbering of trenches in loose soil.

same to the wailings. The wailing acts as a beam between two struts and as such it must be of sufficient thickness to resist the stresses, safely.

In loose soils the poling boards are placed side by side vertically and are supported by wailings and struts at intervals as shown in Fig. 2·51. In case the nature of soil is such that it cannot stand un-supported even for a few centimeter, vertical sheeting or boarding is necessary to be provided (Ref. Fig. 2·49). For deep excavations in loose soil, the trench is excavated in stepped form so that the foundation trench width at the bottom is just sufficient to allow easy workability. The trench width is increased towards top. The arrangement of sheeting, vertical props, struts and runners etc. in such cases is shown in Fig. 2·46.

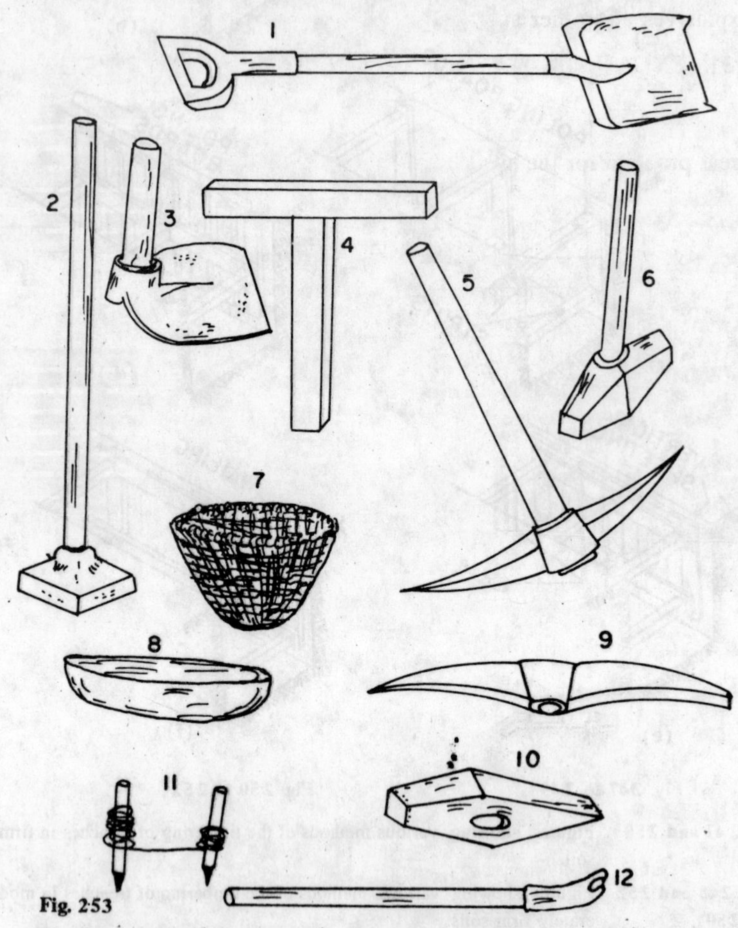

Fig. 2·53

(1) Spade (2) Rammer (3) Kassi (4) Boning rod (5) Pick-axe (6) Sledge hammer (7) Cane basket (8) Iron-pan (9) Pick-axe with flat end (10) Wedge (11) Line and pins (12) Crow-bar.

Complete information regarding the under ground services like sewers, electrical conduits, gas mains, water pipe lines etc. passing through the site should be available before excavation work is taken up. These informations are essentially needed to enable the workmen engaged in excavation work to take sufficient precautions to prevent accident and damage to these services.

2·30. IMPLEMENTS FOR EXCAVATION

The common implements used for excavation of foundation trenches are shown in Fig. 2·53. These tools are useful for loose, moderately firm or firm soils when the depth of foundation is not much. In case of deeper excavation or cutting in hard bed like rock etc., explosives and other such materials become necessary.

2·31. EXCAVATION FOR FOUNDATIONS IN WATER-LOGGED SITES

Digging of foundation trenches in water-logged grounds poses a great problem for the site engineer. There are various methods of dealing

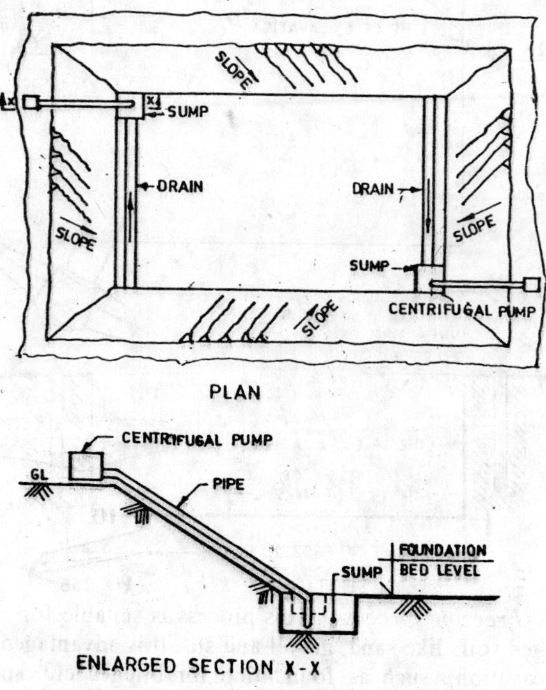

Fig. 2·54. Dewatering of foundations by constructing drains

with the situation which depend upon the depth of excavation, depth of water table and many other factors. Following methods are generally adopted while digging foundation trenches in water-logged sites.

(1) *By constructing drains.* This method is generally adopted in shallow foundations in water-logged ground. In this method, drains of suitable size are constructed by the sides of the foundation trench. The drains collect sub-soil water from the sides and the enclosed area and convey it into a shallow pit or sump well. From the sump, the water is continuously bailed or pumped out. This is the cheapest method of draining excavated area and can be easily adopted by deploying unskilled labour and by using simple equipment.

(2) *By constructing deep wells.* In coarse soils, porous rock or in sites where large quantity of sub-soil water is required to be drained out, 30 to 60 cm diameter wells are sometimes constructed at 6 to 15 m centres all round the site for temporary drainage of the ground. The water collected in the wells is pumped out continuously . This method can be adopted for depths of excavation upto 24 m.

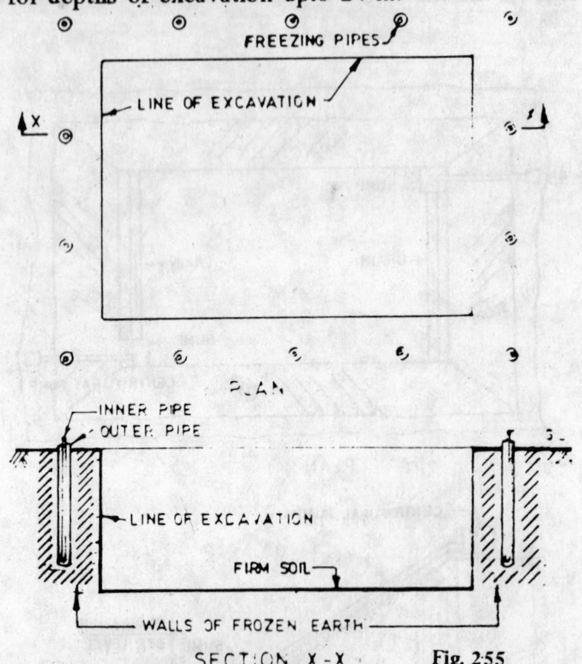

Fig. 2·55

(3) *Freezing process.* This process is suitable for excavations in water-logged soils like sand, gravel and silt. It is advantageously used for deep excavation such as foundation for bridges etc. specially when excavation is to be made adjacent to an existing structure or near some waterways. The process consists in forming a sort of cofferdam by

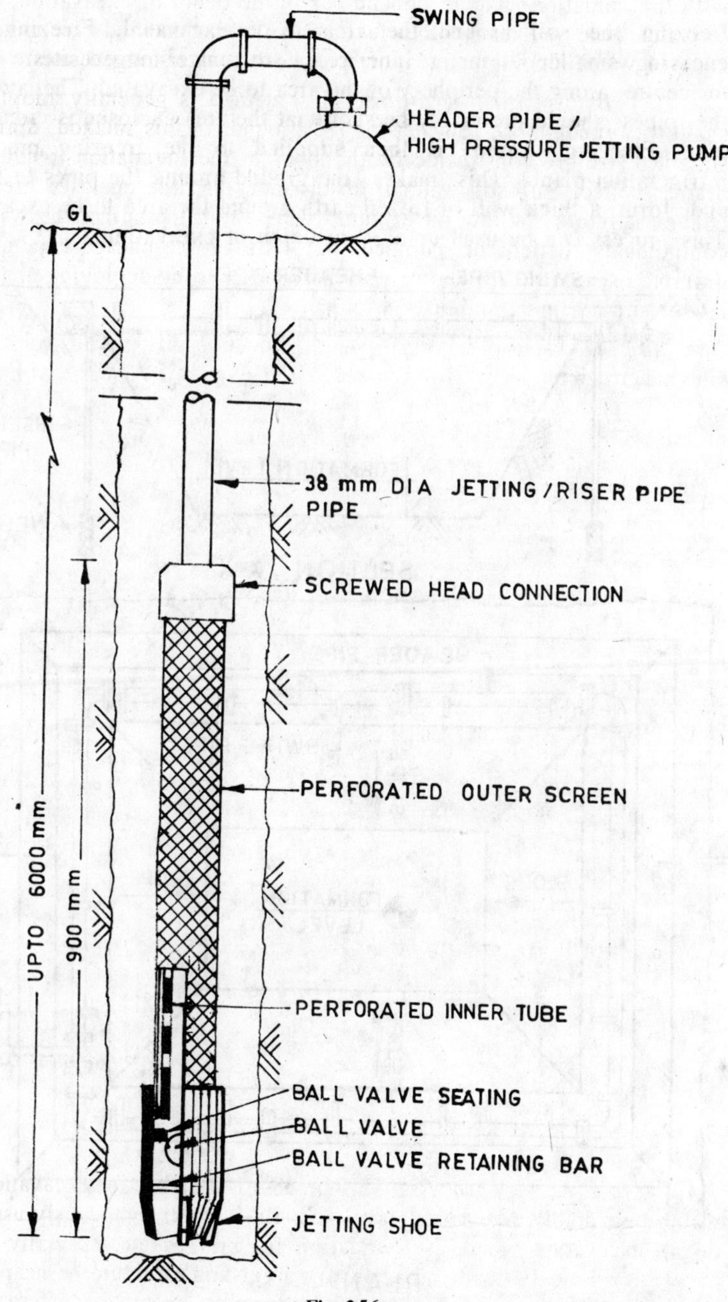

Fig. 2·56

Details of well point

freezing the soil around the area to be excavated. Freezing pipes encasing smaller diameter inner pipes are sunk about one metre centre to centre along the periphery of the area to be excavated. The layout of the pipes should preferably be such that the area enclosed is circular in plan. Freezing liquid is then supplied to the freezing pipes by refrigeration plant. This makes the ground around the pipes to freeze and form a thick wall of frozen earth around the area to be excavated. This process can be used upto 30 m depth of excavation.

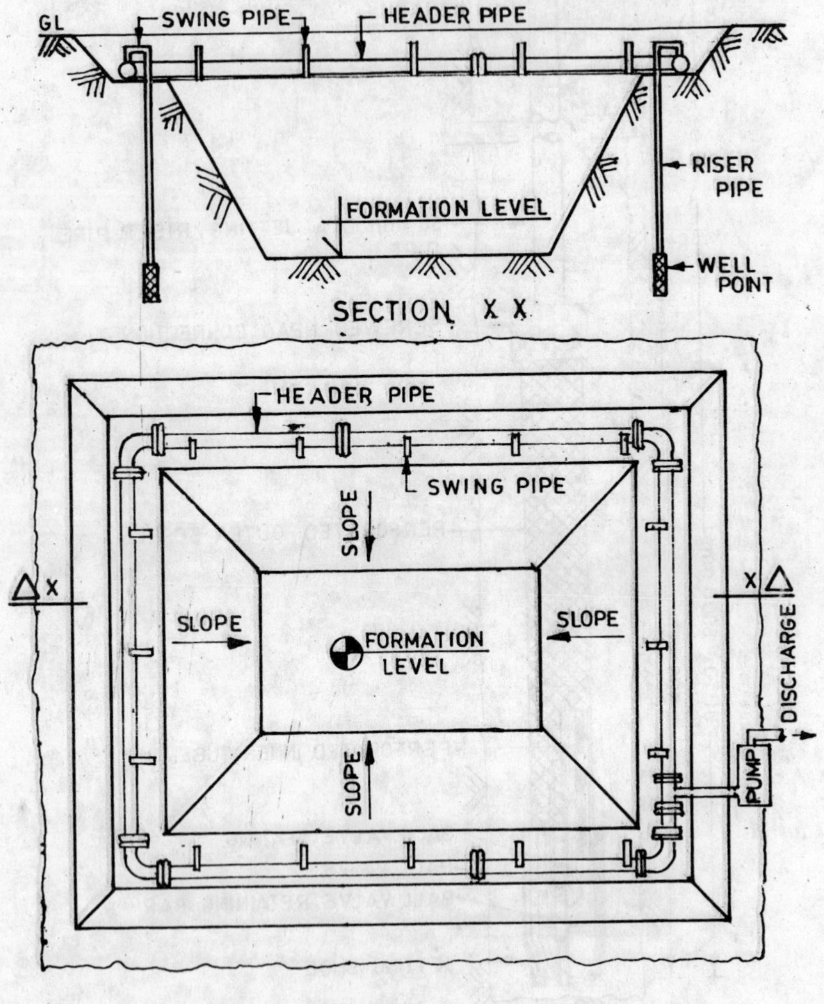

Fig. 2·57

Well point system

(4) *By chemical consolidation of soil.* In this method, the soft water-logged soil is converted into a semi-solid mass by forcing chemicals like *silicates of soda and calcium chloride* into the soil. This method is used for small works.

(5) *Well point system.* This is a method of keeping an excavated area dry by intercepting the flow of ground water with pipe wells

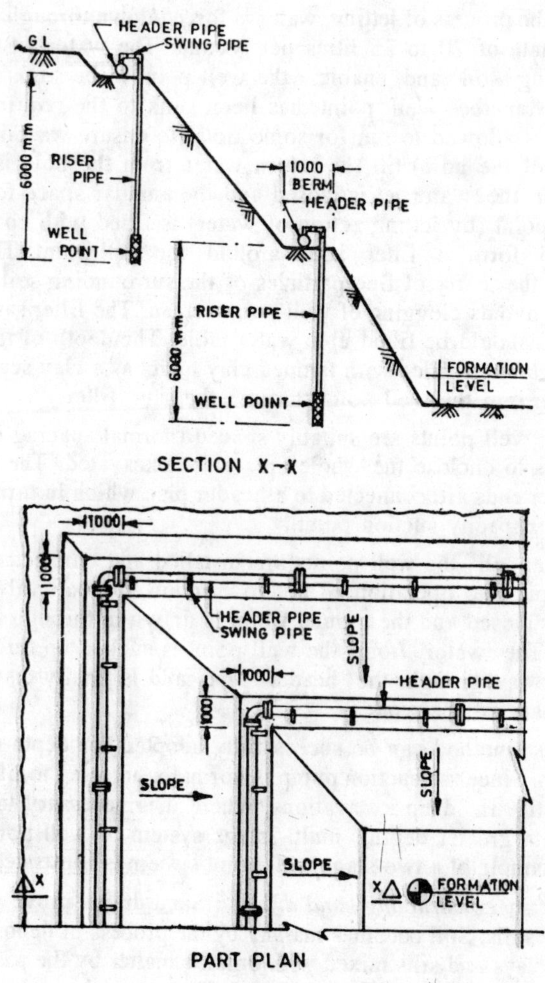

Fig. 2·58

Two-stage well point system

driven deep into the ground. The main components of a well point system are : (*i*) the well points, (*ii*) the riser pipe, (*iii*) the header pipe and (*iv*) the pumps.

The well point consists of a perforated pipe about 120 cm long and 4 cm in diameter. This pipe has a ball-valve to regulate the flow of water and a screen to prevent the mud from entering into the pipe. The well point tube is connected to 5 to 7·5 cm diameter pipe known as riser pipe and is sunk into the ground by jetting.

In the process of jetting, water is forced down through the well point at the rate of 20 to 25 litres per second. The water jet dislodges the surrounding soil and enables the well point to be sunk to the desired depth. After the well point has been sunk to the required depth, the water jet is allowed to run for some time (to ensure washing all sand or silt out of the hole) till the return water from the hole is quite clean. Thereafter the water jet is closed and the annular space formed around the well point (by jetting action of water) is filled with coarse sand and gravel to form a filter zone around the well point. The filter zone prevents the entry of fine particles of the surrounding soil into the well point and avoids clogging of well point screen. The filter sand around the well point should be filled upto water table. The depth of the hole above the water table is filled with tamped clay to act as a clay seal to minimise air getting into the well point through the sand filter.

The well points are suitably spaced (normal spacing being 100 cm c/c), so as to enclose the whole area to be excavated. The riser pipes at their upper ends are connected to a header pipe which in turn is connected to a high capacity suction pump.

After all the well points are installed and connected, the suction pump is put into operation. Due to suction, the ball valve in the well point gets closed and the ground water is drawn in through the well point screen. The water from the well point is sucked up through the riser pipes, flows through the header pipe and is finally discharged away from the site of the work.

This method can be successfully adopted for depth of excavation upto 18 m. Since the suction pump is normally not used to lift water above 6 m depth, in deep excavations, where it is necessary to lower water table to a greater depth, multi- stage system of well point is used. A typical example of a two-stage well point system is illustrated in Fig. 2·58.

(6) *By constructing sand drains* : Sand drains prove very effective in marshy soils. Soil becomes marshy by the process of deposition of thick layers of clays and silts mixed with organic matter by the passage of time. Marshy soil is thus subjected to capillarity and has a high pore water pressure. When this type of soil is subjected to load, its wet soils contents are gradually pushed out on either side and this results in subsidence of the ground. To avoid this, sand drains are made in the ground. The diameter

of the sand drains normally varies between 300 mm to 450 mm and their centre to centre spacing may vary from 3 to 6 metre. The hole for making the sand drain can be made by driving steel pipe casting into the ground. The drain holes are driven deeper than the marshy layer possibly upto an underlying rock or firm base. The marsh in the pipes is removed by means of jets. Selected type of sand is then filled into the pipes and the pipes are withdrawn leaving vertical sand piles in the ground. A thick layer of sand (sand blanket) is spread over the entire area to be consolidated. When the sand layer is subjected to load, the water from the muck of the marshy soil gets squeezed into the vertical sand drains.

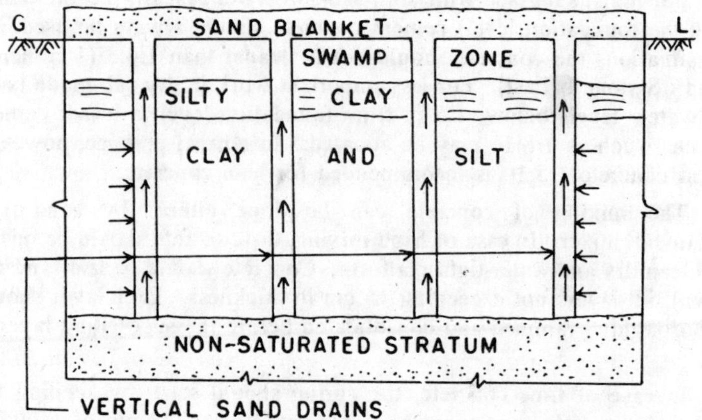

Fig. 2·59

Sand drains

By capillary action, the water from the sand drains rises up and is fed into the sand blanket from where it can be drained out. The objective of consolidation of soil by this method is to develop increased soil resistance to superimposed loads usually consisting of earth fills in highway or airport construction.

(7) *Electro-Osmosis.* Well point system is rendered ineffective in very fine sands, silts or clay, because such soils tend to hold the water by capillary action and offer great resistance to percolation. It has been established that if a direct current is passed through a soil of low permeability, its rate of drainage is greatly increased. In the process of Electro-Osmosis, steel rods forming the positive electrodes are driven into the soil midway between the well-points, which are made to act as negative electrodes. When electric current is passed, the ground water flows towards the negative electrode (well-points) and is pumped out. This process requires very expensive equipment and hence it is rarely used.

2·32. FOUNDATION CONCRETE

The type of concrete and the proportion of the ingredients used in its making depend upon the nature of the structure, quality of the materials used and the site conditions. Since lime is fairly cheap, lime concrete is generally used for foundations in dry-subgrade. Lime concrete is produced by mixing one cu. metre of wet ground lime mortar with 2·5 cu. meter of ballast. The proportion of ingredients in lime mortar may be 1 lime : 2 sand or 1 lime : 1 surkhi : 1 sand or 1 lime : 2 surkhi. The ballast may be of brick, stone or shingle. The size of the ballast is generally restricted to 40 mm.

For moist subgrade with high sub-soil water (usually 1·5 m. or less below the foundation level) cement concrete should always be used. In such situations the concrete should not be leaner than 1:4:8 (1 cement : 4 sand : 8 stone ballast). For less important work in dry sub-grade (sub-soil water level below 1.5 m from foundation level) a leaner cement concrete such as 1:8:16 may be adopted. In normal practice, however, cement concrete 1:5:10 is recommended for lean concrete.

The mixing of concrete can be done either by hand or in mechanical mixer. In case of hand mixing, the concrete should be mixed on a clean dry and water-tight platform. Concrete should be laid (and not thrown) in layers not exceeding 15 cm in thickness. Each layer should be thoroughly rammed and consolidated before the succeeding layer is laid.

In case of lime concrete, the curing should start (by keeping the concrete damp with moist gunny bags, sand etc.) after 24 hours of its laying and should be continued for a minimum period of 7 days. The masonry work over the foundation lime concrete should be started only after 7 days. In case of cement concrete, however, the masonry work over the foundation concrete may be started after 48 hours of its laying. The curing of the cement concrete, which starts 24 hours after its laying, is continued along with the masonry for at least 10 days.

2·33. SPECIFICATION NO. 1

Earth work in excavation for foundation trenches :

1. All foundation pits and trenches shall be dug out to exact dimensions as shown on the drawings or as directed by competent authority. In firm soil, the sides of the foundation trench shall be kept vertical upto a depth of 1.8 m, from the bottom and for greater depth of excavation, the trench shall be widened by providing 45 cm wide steps on either side for every 1.8 m depth measured from the bottom. In soft or loose soil, the sides of the trenches must be sloped down or shored up carefully to prevent the earth from the sides falling in.

2. The bed of all foundation trenches shall be dressed perfectly

level and made firm by watering and ramming. Pot holes and soft spots shall be dug out square and filled with the concrete of the same mixture as the foundation concrete or any other material as approved by the competent authority.

3. If the foundation trenches are dug to depth greater than that shown on the drawing or as required by the competent authority, no earth filling shall be allowed in bringing the foundation bed to proper level. The excess depth shall be filled with the concrete of the same mixture as the foundation concrete or any other material as approved by the competent authority at the cost of the contractor.

4. Before laying foundation concrete the trenches must be inspected and passed by the competent authority.

5. No material excavated from foundation trenches shall be placed within 1.5 m or half the depth of the trench, whichever is more, from the outer edge of the trench.

6. Soon after the concrete and masonry work in the foundations has been completed and measured, the spaces around the masonry in trenches shall be cleared of all debris, brick bats, mortar dropping etc., and refilled with earth in layers not exceeding 15 cm to the original surface of the ground. Each layer of refilled earth shall be watered, rammed and consolidated before the succeeding one is laid.

7. The excavation shall be fenced and marked by red lights at night to guard against risk of accidents.

8. If the work is done by contract, the contractor's rate shall include the following :

(*a*) Excavation, including lifting and depositing the earth as specified.

(*b*) Shoring, pumping or bailing out water keeping the excavation free of water during the foundation masonry work is in progress.

(*c*) Site clearance.

(*d*) Setting out works, profiles etc.

(*e*) Protection and supporting of existing services, *i.e* pipes, water mains, cables etc. coming across during excavation.

(*f*) Fencing to guard against accidents.

(*g*) Initial lead of 50 m and a lift of 1·5 m.

Questions for Review

1. What do you understand by the term 'Foundation' ? What is the purpose of providing foundations ? What are the causes of failure of foundation and what methods would you adopt to ensure the stability of the structure ?

2. Why is surface exploration necessary ? State the different methods of exploration of the sub-soil conditions and describe one of the methods in details?
3. Describe one method of determination of the safe bearing capacity of soil at site. What will be the difference in the applicability of the result if the soil is (*a*) clay (*b*) sand ?
4. (*a*) Describe with the help of sketches one method of ascertaining 'Bearing capacity of soils'. Describe the various methods of improving 'Bearing capacity of soils'.

   (*b*) What are the characteristics of 'Black cotton soil' ? What precautions would you take in designing and laying foundations in Black cotton soil ?
5. (*a*) What is meant by 'safe bearing power' of a soil ? Explain how it is determined.

   (*b*) Discuss the various methods of constructing foundations in Black cotton soils.

   (*c*) Sketch the foundation for a brick pillar 45 cm, square carrying 50 tonne of load on a soil where bearing capacity is 11 tonne/m^2. Weight of soil is 1800 kg/m^3 and the angle of internal friction is 25°.
6. What are the characteristics of 'Black cotton soil' ? What precautions and measures will you take in designing the foundations of a three-storeyed hospital in such a soil? Give a typical cross-section of the type of foundation you consider best.
7. Why do building constructed on shrinkable clays get damaged ? Describe briefly any two methods adopted to safeguard against this damage in new buildings to be constructed on such soil.
8. (*a*) Describe different method of testing the capacity of soil.

   (*b*) What are the minimum and maximum bearing capacities of the following soils ?

   (*i*) Black cotton soil

   (*ii*) Hard moorum

   (*iii*) Compact gravel

   (*iv*) Soft rock.
9. State the various investigations you would make and the different methods you would adopt for the examination of soils to obtain data for the design of foundations.

   What value of safe bearing pressure would you adopt in the following cases?

   (*i*) Compact clay nearly dry

   (*ii*) Compact sand (prevented for spreading)

   (*iii*) Hard rock.
10. Name the different types of foundations you recommend under different situations and soils. Explain them briefly.

11. Sketch and explain briefly the various types of footings you consider suitable for :
    (*a*) Isolated R.C.C. column.
    (*b*) Isolated steel column.
    (*c*) Isolated timber column.
    (*d*) R.C.C. column monolithically founded with the footing.
12. What type of foundation would you propose for a multi-storeyed office building in ?
    (*i*) Very soft soil
    (*ii*) Black cotton soil.
    Describe briefly.
13. Trace the principal causes of unequal settlement of structures and mention the precautions to be taken to prevent uneven settlement of foundations.
14. Draw neat sketches to explain the following types of foundations :
    (*i*) Grillage foundations.
    (*ii*) Raft foundations.
    (*iii*) Combined footing for steel stanchion.
    (*iv*) Combined footing for concrete column.
    (*v*) Treatment necessary for columns or walls carrying unequal loading.
15. Describe the process of 'setting out' for excavation of the foundation of a three storeyed building. The foundations are required to be taken 1.8 m deep inside the ground which may be considered firm. Draw suitable sketches to show the necessary procedure.
16. What are the usual permissible loads on the following soils :
    (*a*) Alluvial clay
    (*b*) Black cotton soil
    (*c*) Compact sand
    (*d*) Moorum
    If you were to construct a eight-storeyed building on one of these soils what type of foundation would you adopt ? Give reasons.
17. Write short notes on any three of the flowing :
    (*a*) Raft foundation.
    (*b*) Typical footing for a brick wall and an R.C.C. column giving suitable dimensions.
18. (*a*) What is meant by bearing capacity of a soil ? What are the soil properties on which the bearing capacity of soil depends ?
    (*b*) Describe the various methods generally adopted to improve the bearing capacity of soils.
19. What type of foundation would you adopt under a heavy steel column over soft soil ? Explain with reasons and make a neat sketch.

20. (*a*) A five storeyed building is to be constructed on a site with black cotton soil. What investigations will you carry out before deciding the suitability and the choice of the type of foundation to be used ?

    (*b*) Explain the method by which you will ascertain the bearing capacity of the soil after the excavation for the foundation is complete.

21. Describe different types of foundations used for buildings, load bearing masonry walls, abutments and piers of bridges. Sketch a suitable grillage foundation for a compound column consisting of ISMB 400 mm x 250mm with flange plates 300 mm x 18 mm on each flange and carrying a load of 230 tonne. The allowable bearing capacity of soil is 15 tonne/m^2.

22. Briefly describe the methods for dealing with foundations on expensive or made up type of soil. Illustrate with neat sketches.

23. Why are foundations necessary ? Name the various types of foundations used for buildings and give brief description of each and its suitability.

24. Give in brief the tests you would carry out for finding the bearing capacity of soil at a site for constructing a multi storeyed building.

25. (*a*) Enumerate and explain with sketches the various methods used for improving the bearing capacity of soils.

    (*b*) Write short notes with sketches on Shallow foundations.

3

Deep Foundations

3·1. GENERAL

In case, the strata of good bearing capacity is not available near the ground, the foundation of the structure has to be taken deep with the purpose of attaining a bearing stratum which is suitable in all respects. In addition there may be many other conditions which may require deep foundations for ensuring stability and durability of a structure. For example, the foundation for a bridge pier must be placed below the scour depth, although suitable bearing stratum may exist at a higher level. The most common forms of construction pertaining to deep foundations are :

(*a*) Piles
(*b*) Cofferdams
(*c*) Caissons

3·2. PILE FOUNDATIONS

Pile foundation is generally used when simple spread foundation at a suitable depth is not possible either because the stratum of required bearing capacity is at a greater depth or steep slopes are encountered. In compressible soil or water-logged soil or soil of made-up type, piles are used with advantage for providing safe foundation for any type of structure. Piles are used for foundation for buildings, trestles, bridges and water front installations (piers, docks, etc). In general, pile foundation provides a common solution to all difficult foundation site problems.

3·3. TYPES OF PILES

Depending upon their function or use piles may be classified into the following types :

(*i*) ***Bearing piles***
(*ii*) ***Friction piles***
(*iii*) ***Sheet piles***
(*iv*) ***Anchor piles***
(*v*) ***Batter piles***
(*vi*) ***Fender piles***
(*vii*) ***Compaction piles.***

(*i*) *Bearing piles.* Bearing piles are those which are driven into the ground until a hard stratum is reached. Such piles act as pillars supporting the super-structure and transmitting the load down to the level at which it can be safely borne by the ground. Thus bearing piles, by themselves do not support the load, rather they act as a medium to transmit the load from the foundation to the resisting sub-stratum.

(*ii*) *Friction piles.* When piles are required to be driven at a site where the soil is weak or soft to a considerable depth, the load carried by a pile is borne by the friction developed between the sides of the pile and the surrounding ground (skin friction). In such cases the pile is named as friction or floating pile. Thus friction piles are driven in the type of soil whose strength does not increase with depth or, where rate of increase in strength with depth is very slow.

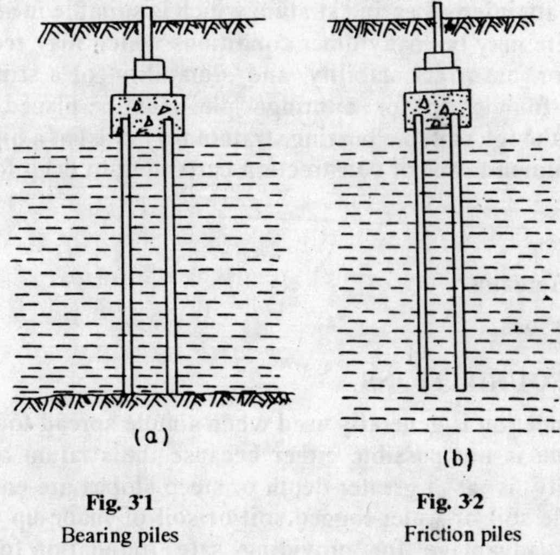

Fig. 3·1
Bearing piles

Fig. 3·2
Friction piles

(*iii*) *Sheet piles.* Sheet piles differ from bearing or friction piles in that they are rarely used to furnish vertical support but are used to function as retaining wall. Sheet piles are used for retaining soil that is liable to escape laterally when subjected to pressure or to enclose the area required for some foundation and protect it from the action of running water or leakage.

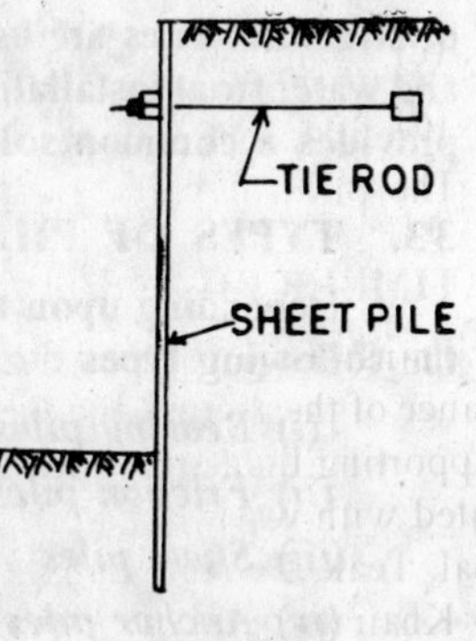

Fig. 3·3

(*iv*) *Anchor piles.* When piles are used to provide anchorage against horizontal pull from sheet piling walls or other pulling forces, they are termed as anchor piles.

(*v*) *Batter piles.* When piles are driven at

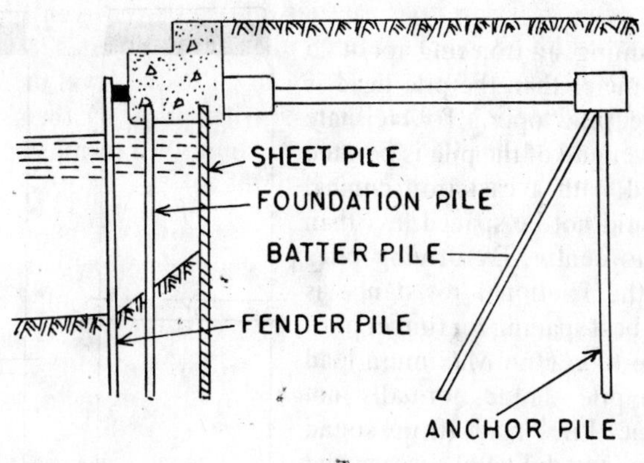

Fig. 3·4

an inclination to resist large horizontal or inclined forces, the piles are termed as batter piles.

(*vi*) *Fender piles.* When the piles are used to protect concrete deck or other water front structures from the abrasion or impact that may be caused from the ships or barges (when they are tied up at the deck) they are called fender piles. The fender piles are ordinarily made up of timber.

(*vii*) *Compaction piles.* When piles are driven in granular soil with the aim of increasing the bearing capacity of the soil, the piles are termed as compaction piles.

3·4. CLASSIFICATION OF PILES

Depending upon the material used in their manufacture, piles can be broadly classified as :

(*a*) *Timber piles*

(*b*) *Concrete piles*

(*c*) *Composite piles*

(*d*) *Steel piles.*

The piles of various categories are described below in details.

3·5. TIMBER PILES

Transmission of load through timber piles takes place by the frictional resistance of the ground and the pile surface. Timber piles prove economical for supporting light structures to be located in compressive soils constantly saturated with water. The piles are made from timber obtained from trees like Sal, Teak, Deodar, Babul, Khair etc. It has been found that piles made from Khair wood can stand action of sea water better and are thus commonly used for marine works. Timber piles may be circular or square in cross-section. Piles are driven with the help of pile-driving machine in which a drop hammer delivers blows on the pile head. To prevent the pile head from brooming, an iron ring about 25 mm less in diameter than the pile head is provided at the pile top. To facilitate driving, the lower end of the pile is pointed and is provided with a cast iron conical shoe. Piles should not be spaced less than 60 cm centre to centre. By driving piles much closer, the frictional resistance is destroyed. The best spacing for timber piles is 90 cm centre to centre. Maximum load on a wooden pile should normally not exceed 20 tonne. Piles made from sound timber free from any defect or disease and driven in the soils which are either permanently wet or permanently dry, will remain in good condition for centuries. However, when subjected to alternate dry and wet conditions (on account of variations in ground water level) they get decayed. It is on this account that timber piles are cut a little below the lowest water-mark and capped with concrete, steel grillage, stone or timber. If timber capping is used, the cap should be permanently under water.

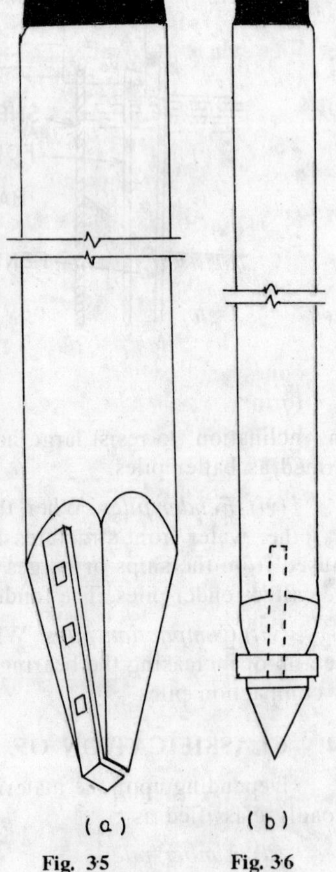

Fig. 3·5 **Fig. 3·6**

Timber piles

Advantages of timber piles :

(1) They are economical.

(2) They can be driven rapidly and as such there is great saving in time in execution of piling work.

(3) On account of their elasticity, timber piles can be recommended for sites where piles are likely to the subjected to unusual lateral forces.

(4) They do not need heavy machinery and elaborate technical supervision.

Disadvantages of timber piles :

(1) Timber piles must be cut off below the permanent ground water level to prevent them from decay and thus if the water table at a site is at a greater depth, extra cost of excavation needed to provide the pile cap may render the choice uneconomical.

(2) Timber piles can not be driven in filled up ground without injury and as such they can not be recommended for such sites.

(3) They are liable to decay or deteriorate by salt water or insects.

(4) On account of their restricted length, they cannot be used for jobs where long piles are needed.

(5) They have low bearing capacity.

3·6. CONCRETE PILES

Concrete piles may be classified as :

(*a*) *Pre-cast piles*

(*b*) *Cast in place or cast-in-situ piles*

(*c*) *Prestressed concrete piles.*

(*a*) *Pre-cast piles.* Pre-cast pile may be defined as a reinforced concrete pile which is moulded in circular, square, rectangular or octagonal form. The piles are cast and cured in a casting yard and then transported to the site for driving. In case space is available, pile can also be cast and cured near the site of works. They are driven in a similar manner as timber piles with the help of pile drivers. The diameter of the pile normally varies from 35 cm to 65 cm and their length varies from 4·5 m to 30 m.

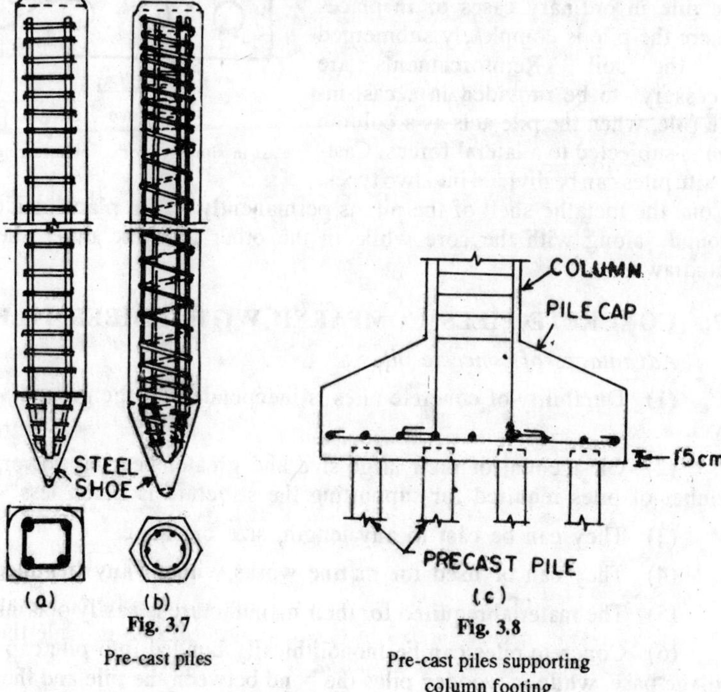

Fig. 3.7
Pre-cast piles

Fig. 3.8
Pre-cast piles supporting column footing

The function of reinforcement in a pre-cast pile is to resist the stresses produced on account of its handling, driving and the loading which the pile is finally expected to receive. Longitudinal reinforcement usually consists of one bar 20 mm to 50 mm in diameter at each angle of the section of the pile. The vertical rods are tied horizontally by bars 6 mm to 10 mm in diameter. The horizontal bars may be provided in the form of stirrups wound around the verticals. For lengths of approximately 90 cm at head and toes, the spacing of the stirrups should be 8 cm c/c. Circular piles are seldom tapered but when tapering of the piles becomes necessary due to site conditions, their length is restricted to 12 m.

(*b*) *Cast-in-situ piles.* Cast-in-situ piles are those piles which are cast in position inside the ground. Since the pile is not subjected to handling or driving stresses, it is not necessary to reinforce the pile in ordinary cases or in places where the pile is completely submerged n the soil. Reinforcements are cessary to be provided in a cast-in-u pile, when the pile acts as a column and is subjected to a lateral forces. Cast-in-situ piles can be divided into two types. In one the metallic shell of the pile is permanently left in place inside the ground along with the core while in the other type the outer shell is withdrawn.

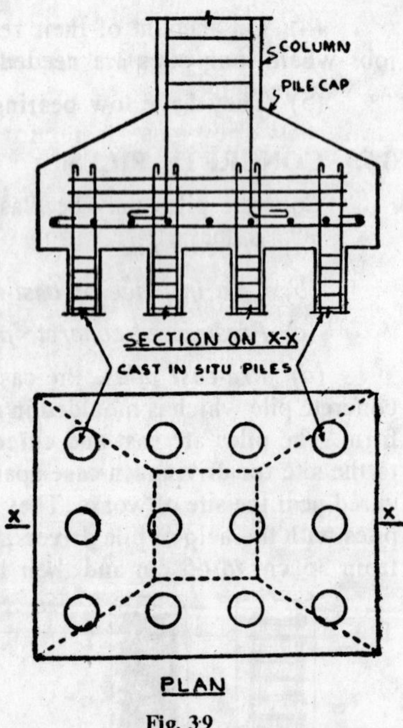

Fig. 3·9
Cast-in-situ piles for column forting.

3·7. CONCRETE PILES COMPARED WITH TIMBER PILES

Advantages of concrete piles :

(1) Durability of concrete piles is independent of the ground water level.

(2) On account of their large size and greater bearing power, the number of piles required for supporting the structure is much less.

(3) They can be cast to any length, size or shape.

(4) They can be used for marine works without any treatment.

(5) The material required for their manufacture is easily obtainable.

(6) Concrete piles can be monolithically bonded into pile cap or a grillage base, while in wooden piles the bond between the pile and the pile cap is not perfect.

Disadvantages of concrete piles :

(1) They cost much more than timber piles.

(2) They cannot be driven rapidly.

(3) Concrete piles require elaborate technical supervision and heavy driving machines.

(4) They must be reinforced to withstand handling stresses.

3·8. TYPES OF CAST-IN-SITU PILES

The names of the various types of commonly used cast-in-situ piles are:

(*a*) *Simplex pile*
(*b*) *Franki pile*
(*c*) *Vibro pile*
(*d*) *Vibro-expanded pile*
(*e*) *Raymond pile*
(*f*) *Mac Arthur pile*

(*a*) *Simplex pile.* This type of pile may be driven through soft or hard soil. A steel tube having an internal diameter equal to the dia of the pile and 20 mm in thickness, is driven into the ground. To facilitate driving

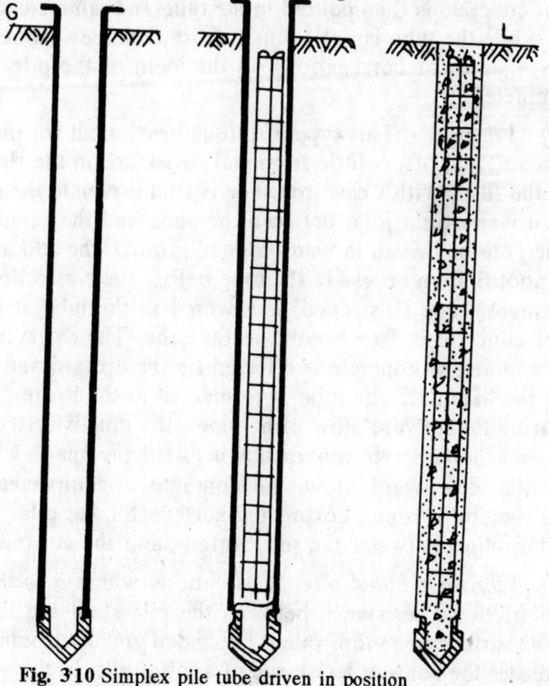

Fig. 3·10 Simplex pile tube driven in position
Fig. 3·11 Reinforcement cage lowered in the tube
Fig. 3·12 Completed simplex pile

of the pile, the steel tube is fitted with a detachable steel shoe that completely closes the bottom of the tube. When the tube has been driven to the required depth, a charge of concrete is poured into the tube and the tube is gradually withdrawn leaving the charge of concrete below. Thus by alternately pouring the concrete and withdrawing the tube, the pile is constructed to its full length. The metallic shoe remains in place and hence a new one is needed for each pile. In case the pile is required to be reinforced, the reinforcement cage is lowered into the steel tube prior to the pouring of concrete.

(*b*) *Franki pile.* This type of pile has an enlarged base and a corrugated stem. A steel tube, having its internal diameter equal to the diameter of the pile required, is held vertical at the ground level with the help of leads. A charge of concrete is poured at its base filling the bottom 60 cm to 90 cm of the tube. The charge of concrete is consolidated into a solid plug by the blows of the drop hammer working inside the tube. Further blows of the hammer on the plug pull the tube down on account of the friction developed between the concrete and the inside surface of the tube. When the tube has been driven to the required depth, it is slightly raised and the plug is forced out of the tube by hammering. The reinforcement cage (if needed) is then lowered inside the tube. A fresh charge of concrete is then poured in the tube and rammed well by the drop hammer while the tube is pulled up a short distance. The repeated process produces a series of corrugations on the stem of the pile and the pile is thus completed.

(*c*) *Vibro pile.* This type of pile is best suited for places where the ground is soft and offers little frictional resistance to the flow of concrete. A steel tube fitted with a cast iron shoe is first driven to the required depth. There is a water-tight joint between the shoe and the casing so that even if the pile is to be driven in water-logged ground, the soil and the sub-soil water cannot find an access in the tube before the concreting is done. The reinforcement cage (if needed) is lowered in the tube at this stage. The charge of concrete is then poured in the tube. The extraction of the tube and the ramming of concrete is effected by the upward and the downward blows of the hammer. The tube is connected to the hammer by extracting links. During the upward blow of hammer, the tube is raised up by a short distance and the concrete moves down to fill the space left by the tube. During the downward blow, the concrete is compacted and rammed outwards thereby forming corrugated surface for the pile. This results in increased friction between the pile surface and the surrounding ground.

(*d*) *Vibro-expanded pile.* In situations where it is desired to have increased frictional resistance between the pile stem and the surrounding ground, the surface of a vibro-pile is expanded greatly to achieve the object. This increases the bearing resistance of a vibro pile. In this process, a steel tube of the required diameter of the pile having a detachable cast iron conical shoe at its base, is driven to the required depth. A charge of concrete

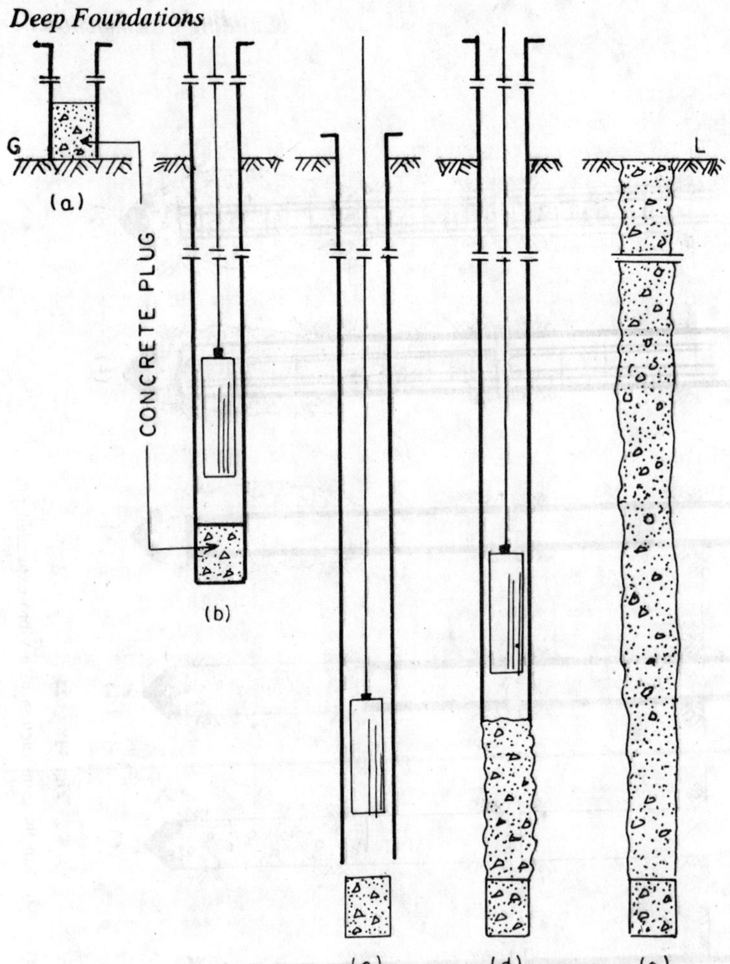

Fig. 3·13 Concrete plug at pile tube base

Fig. 3·14 Concrete plug dragging the pile tube down on hammering

Fig. 3·15 Concrete plug forced out

Fig. 3·16 Concrete filled in the tube and tube partially raised up

Fig. 3·17 Completed Franki pile

(filling a good length of tube) is poured and the tube is completely withdrawn leaving the cast iron shoe and the charge of concrete down in the pile hole. The withdrawn tube is fitted with a special flat iron shoe and once again driven in the same hole. The charge of concrete down below gets expanded to nearly double its area by this process. If required, another charge of concrete is poured and the process repeated. The reinforcement cage is thereafter lowered in the tube (if needed) and the pile is completed as usual. The different stages in the construction of a vibro-expanded pile are shown in Figs. 3·18-3·24.

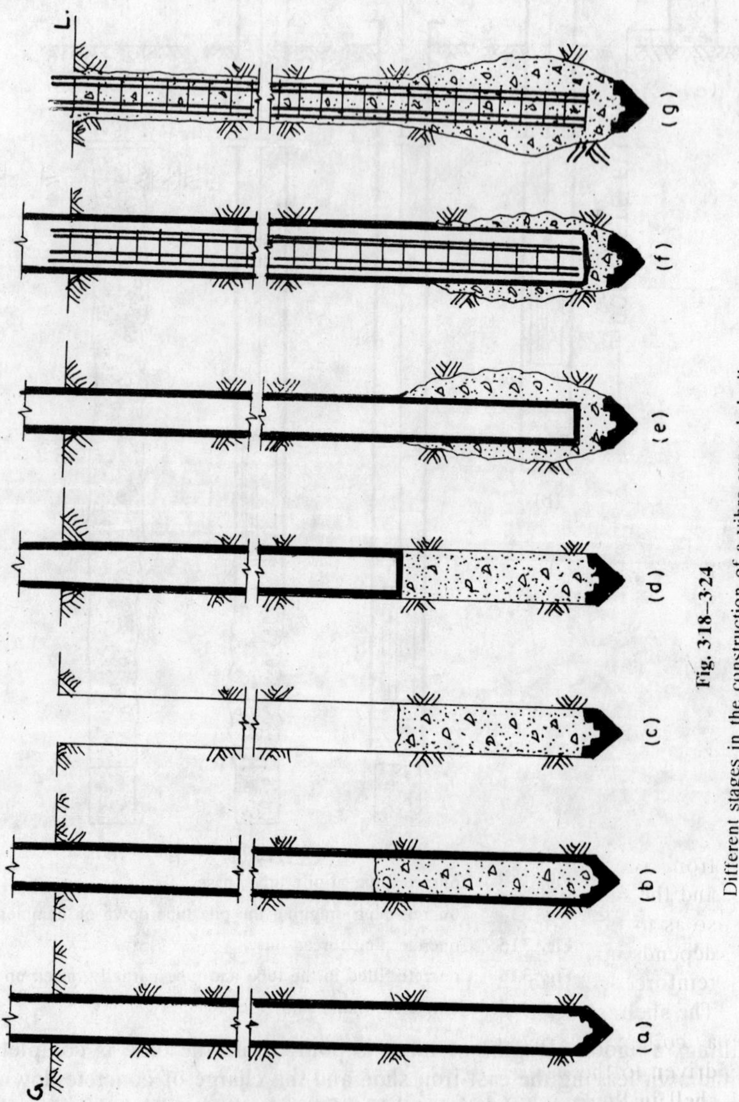

Fig. 3·18—3·24 Different stages in the construction of a vibro-expanded pile

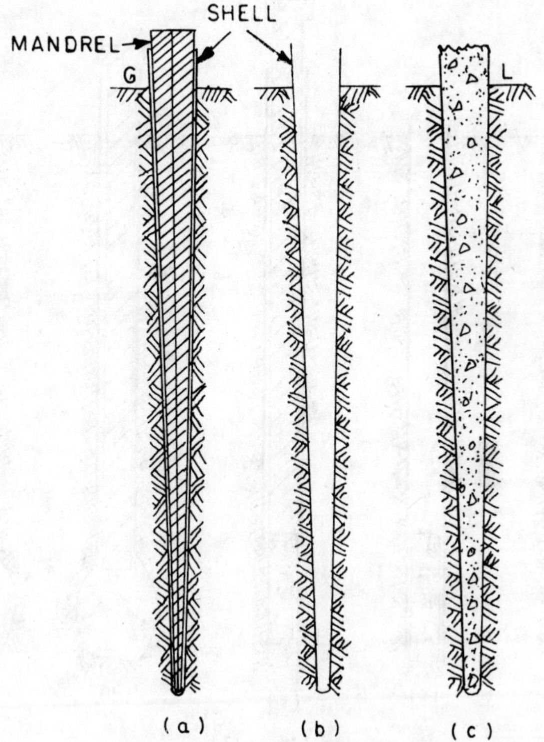

Fig. 3·25 Mandrel and shell driven in position

Fig. 3·26 Mandrel taken out

Fig. 3·27 Completed Raymond pile

(*e*) *Raymond pile.* This type of pile is constructed in lengths varying from 6 to 12 m. The diameter of the pile varies from 40 to 60 cm at top and the diameter at its base is slightly smaller, varying from 20 to 28 cm so as to give a uniform taper to the pile. The thickness of the outer shell depends upon the pile diameter and site conditions. The thin steel shell is reinforced with hard drawn wire spiral spaced at 8 cm centre to centre. The shell is closed at the bottom with a steel boot. The shell is placed over a collapsible mandrel having the same taper as the pile and both are driven to the desired depth. The mandrel is then withdrawn leaving the shell in the ground. The shell is gradually filled with concrete up to the top. This forms a Raymond pile. The function of shell outside the concrete core is to prevent the adjoining soil and the sub-soil water coming in contact with fresh concrete.

(*f*) *Mac Arthur pedestal pile.* In this type of pile the apparatus consists of an outer casting (a hollow steel pipe) and an inner core. The bottom of the core is of such a size that it completely closes the open base

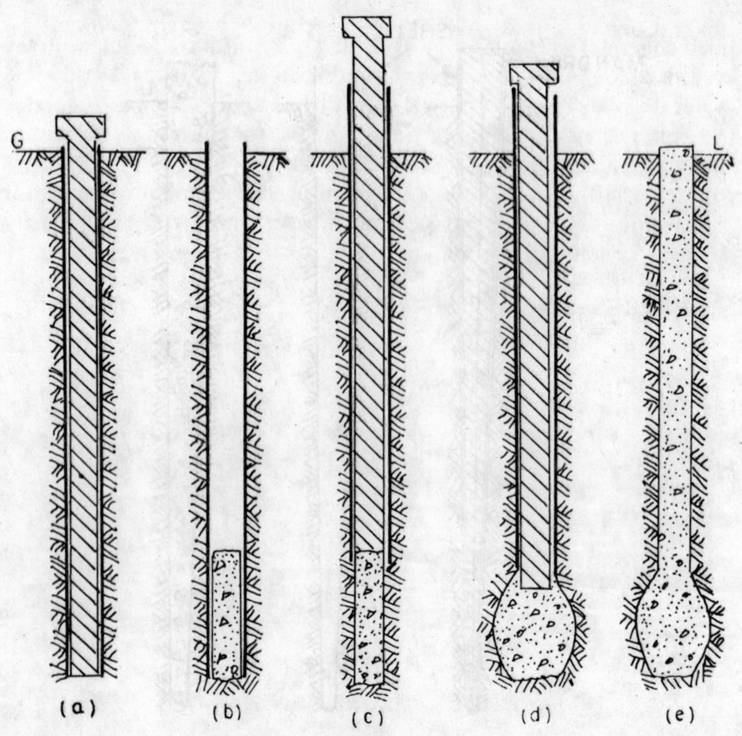

Fig. 3·28 – 3·32
Mac Arthur pedestal pile in di fferent stages of construction

of the casting when inserted inside. The core and the casting are together driven into the ground to the required depth. The core is removed and a charge of concrete is poured in the casting. The core is replaced in the casting, so that it is in contact with the concrete. The casting is then pulled up a short distance of 60 cm to 90 cm and on account of the pressure of the core and the hammer, the concrete is rammed out. Again the core is removed, another charge of concrete poured and the operation repeated. When the bulb at the base has been formed to the required area, the casting is gradually withdrawn after filling with concrete so as to form a stem of constant diameter. If desired, the pile can also be reinforced.

3·9. DRILLED PILES

This is a recent development in foundation engineering. This development owes its origin to the manufacture of large mechanical augers, capable of drilling holes upto about 160 cm in diameter. In normal practice, however, the diameter of the drilled piles does not exceed 60 cm. In this method holes are drilled deep enough so as to penetrate a hard strata of required bearing capacity. The holes are then filled with plain or reinforced concrete. The holes may be drilled to a uniform diameter or they may have an enlarged diameter at the base. In places where it is not possible to de-water the piles bore, the piles may be drilled under water and concrete may be placed in position with the help of suitable tremic pipes. In cohesive soils drilling is carried out with the help of auger or bucket type drills. Steel shaft casings are used (where needed) to shut off water or keep the soil out of the pile holes. In such cases, the casing is gradually withdrawn as concreting proceeds.

3·10. PRESTRESSED CONCRETE PILES

The biggest disadvantage of pre-cast piles is their large weight. The precast pile is reinforced to take care of the handling and driving stresses rather than to strengthen the pile to act as a column. In some cases the handling stresses are anticipated to be so severe that the pile has to be lifted at a number of points (In normal practice the pile has to be lifted at two points only) to be able to withstand the handling stresses.

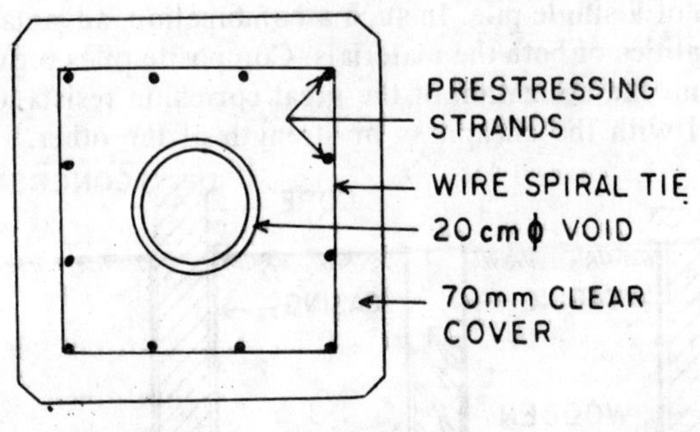

Fig. 3·33

Cross-section of a typical prestressed concrete pile

With the introduction of prestressed concrete piles, it has been possible to make pre-cast piles which are lesser in weight and which can be lifted from one or at the most two points. The weight of very large pile is reduced by casting 200 mm to 300 mm diameter voids in its stem. The voids are formed by casting 200 mm to 300 mm diameter fibre tubes inside the piles at the time of concreting. In casting prestressed concrete piles, the pre-tensioning cables required for each pile are subjected to the required pull (tension) in the casting bed. The fibre tube for forming void in the pile (if needed) is securely held in position inside the form work and the piles reinforced with the pre-tensioned cables are concreted in a row. The casting bed can be made to accommodate 2 to 5 lines of piles 120 to 150 metre long. Prestressed piles are provided with lifting hooks at 1/5th of the pile length from each end.

In general, square prestressed concrete piles of length 50 times the thickness can be handled with a single point pick-up and upto 60 times the thickness with two point pick-up. Piles 500 mm square and smaller are usually cast solid, whereas pile above 500 mm square may be cast with 200 mm to 300 mm diameter cored hole (void).

On account of their reduced weight, prestressed concrete piles have definite handling advantage over precast piles and as such they are getting increasingly popular these days. When compared with reinforced concrete piles, prestressed piles have proved to be extremely durable in tidal zones and for harbour installations, where the unsupported lengths of the piles becomes considerable. Their load carrying capacity is high and they are capable of resisting uplift forces and combined moments. In addition, they can withstand extremely hard driving stresses which may be far more than that which a conventional type of reinforced concrete pile can bear.

3·11. COMPOSITE PILES

This is a type of construction in which piles of two different materials are driven one over the other, so as to enable them to act together to perform the function of a single pile. In such a combination, advantage is taken of the good qualities of both the materials. Composite piles prove economical as they permit the utilization of the great corrosion resistance property of one material with the cheapness or strength of the other.

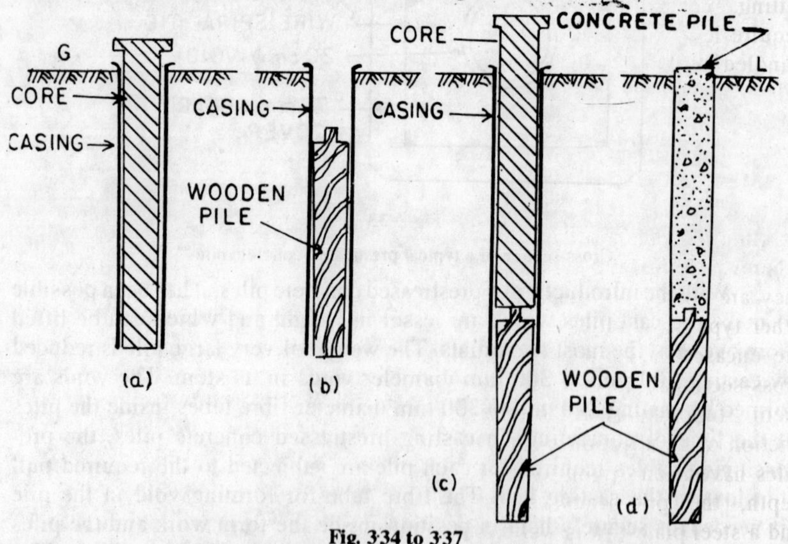

Fig. 3·34 to 3·37

Different stages in the construction of composite pile

The different stages in the construction of a composite pile having a timber pile at its lower part and precast concrete pile above are shown in Figs. 3·34 to 3·37. This type of composite pile is used with the object of achieving economy in the cost of piling work.

Another type of composite pile commonly used consists of a steel pipe or H-pile at the bottom and cast-in-situ concrete pile at the top. This type of composite pile is recommended in cases where the designed length of the pile works out to be greater than that available for the cast-in-situ type of pile.

3·12. STEEL PILES

The types of steel piles commonly used are :

(*a*) *H-Piles*

(*b*) *Pipe-piles*

(*c*) *Screw piles*

(*d*) *Disc piles*

(*a*) *H-piles.* The use of rolled steel *H*-beams to function as bearing piles is a comparatively recent development in piling industry. *H*-piles can withstand large impact stresses developed during hard driving. This type of pile has proved to be especially useful when the pile is expected to penetrate a rock or through hard substratum. On account of the smaller cross-sectional area of the pile, it can be driven to the desired depth without jetting, coring or adopting other methods of soil displacement. They require less storage space and can be handled without much difficulty. *H*-piles can be advantageously driven close to an existing structure as they produce very small soil displacement. *H*-piles can be securely spliced. Spliced *H*-piles have been driven to a maximum depth of 100 m. In situations where the piles are liable to corrode, they are coated with coal tar or some other type of suitable coating or they are encased in concrete. The smaller cross-sectional area of *H*-pile makes them function less effectively as friction or compaction piles. After the piles have been driven to the required depth, the pile heads are cut square and a steel plate cap is welded to each pile and finally the pile heads are embedded in reinforced concrete pile cap. *H*-piles are commonly used in the construction of bulkheads, trestles, retaining walls, bridges and cofferdams.

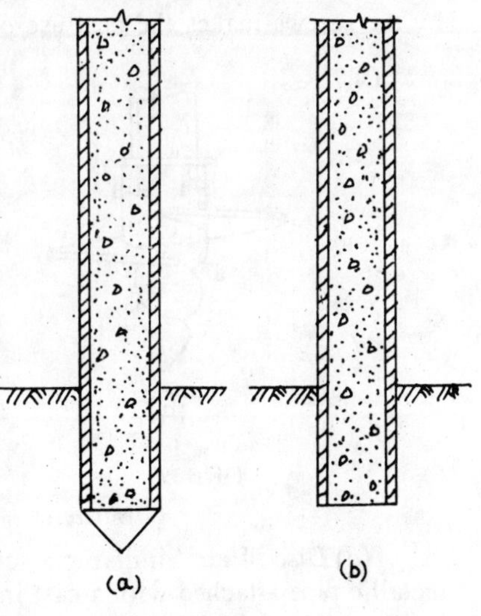

Fig. 3·38

(*a*) Closed end pipe pile

(*b*) Open end pipe pile

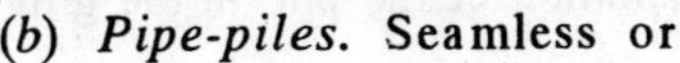

(*b*) *Pipe-piles.* Seamless or

welded steel pipes are often driven to function as end bearing or friction piles. The pipe piles may be driven either open ended or close ended. When the driving end of the pipe is left open, (without any driving point) it is called open end pile. Open ended piles are usually driven to penetrate rock or hard pan. As the open ended piles are sunk in the ground, the soil inside the pipe is cleaned out (by means of compressed air, water jet etc) simultaneously. When the pipes have been driven to the required depth, they are cleaned from inside and filled with concrete.

In case of closed end piles, the driving end of each pipe is closed by welding a conical steel or cast iron shoe to the pipe end. In this case also, after driving, the hollow space inside the pipe is normally filled with concrete. The diameter of the pipes used for piling varies from 25 cm to 120 cm and their shell thickness varies from 8 mm to 12 mm. The use of this type of pile for depths of 30 m or more is quite common.

(*c*) *Screw piles.* A screw pile consists of a cast iron or steel shaft of external diameter normally varying from 15 to 30 cm and terminating into a helix or screw base. The pile shaft may be hollow or solid. The diameter of the screw at its base varies from 45 cm to 150 cm. The pile is sunk by screwing the pile down inside the ground by use of an electric motor. Screw piles function most efficiently in soft clay or loose sand. In such a ground it is easy to install the piles and also the large bearing area provided by the screws makes the best use of the low bearing capacity of the soil.

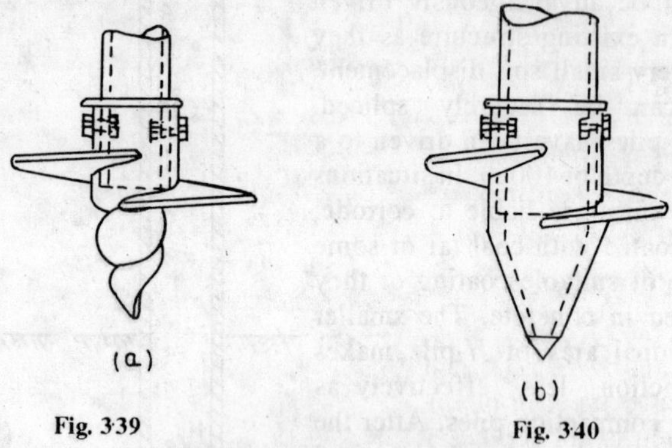

Fig. 3·39 Fig. 3·40

Different forms of screw piles

(*d*) *Disc piles.* Similar to a screw pile, a disc pile consists of hollow metallic pipe attached with a cast iron disc to its foot so as to enlarge the bearing area of the pile. There is a hole at the bottom of the pile to permit the water jet pipe to pass through during the sinking of the pile by jetting. Disc pile can be used only in sandy or soft soils which may permit sinking of the pile by water jets. Disc piles are used mostly in marine installations, where the total penetration of the pile in the ground is required to be large.

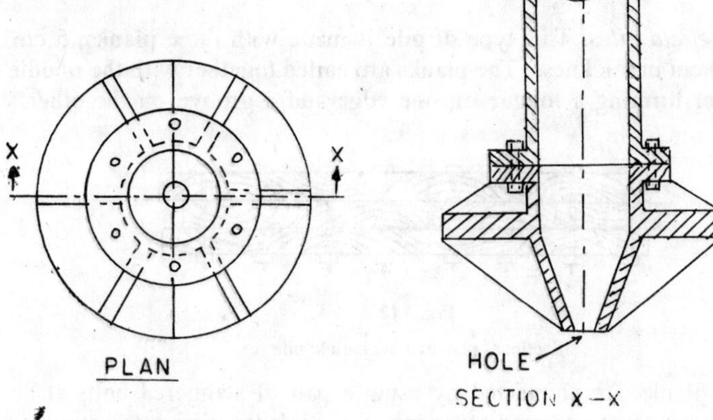

Fig. 3·41

Disc pile

3·13. SHEET PILES

Sheet piles may be made up of wood concrete or steel. Steel piles are driven side by side into the ground to form a continuous vertical wall for retaining soil. The alignment and resistance or thrusts are normally provided by horizontal wallers, braces or tiebacks. Factors affecting the choice of a particular type of pile include nature of ground, cost, ease of installation, availability of material, ability to withstand driving, lateral strength and ease of making connections. Depending upon the material used in their manufacture, sheet piles can be divided into the following categories:

(*a*) *Wooden sheet piles*

(*b*) *Precast concrete sheet piles*

(*c*) *Prestressed concrete sheet piles*

(*d*) *Steel sheet piles.*

(*a*) *Wooden sheet piles.* Wooden sheet piles are made in various sizes and forms. The nature of site conditions determine, the choice of a particular type. In places where excavation is small and the ground water problem is not serious, 5 cm x 30 cm to 10 cm x 30 cm wooden planks arranged in a simple row will serve the purpose. If the water-tightness is required to a great extent, lapped sheet piling is used. In this case, each pile is made up of two planks, either spiked or bolted to one another. Thus if only earthen banks of small height are to be supported, a single or double row of planks properly erected will perform the function of sheet piling. If complete water tightness is desired or pressure of the retained material is large, *wakefield* or tongue and grooved sheeting is generally used. To facilitate the driving of the piles, they are usually bevelled at foot. This not only assists in driving but also prevents bruising, if the piles encounter hard stratum.

Wakefield piles. This type of pile is made with three planks, 5 cm, 8 cm or 10 cm in thickness. The planks are nailed together with the middle plank offset forming a tongue on one edge and a groove on the other.

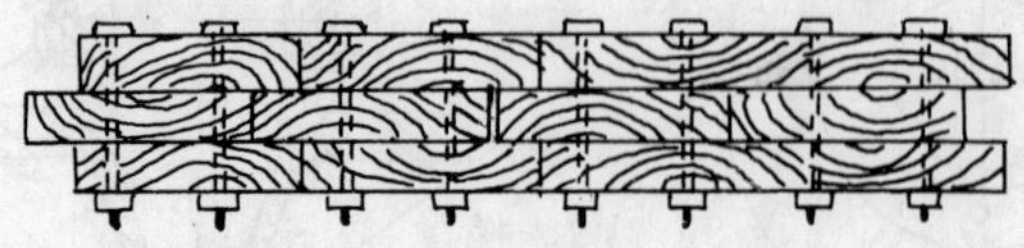

Fig. 3·42

Sectional plan of a Wakefield pile

The planks are connected by using a pair of staggered bolts at 80 cm centre to centre at intermediate points. The triple lap piles prove stronger in driving. There is no wastage in forming the tongue and groove joints and the piles have less tendency to warp. Timber sheet piles have light weight and as such the equipment required for pile driving is also light. This is considered to be an important advantage timber piles have over piles of other materials.

(*b*) *Precast concrete sheet piles.* Precast concrete piles are made in square or rectangular cross-section and are driven similar to wooden piles to form a continuous wall. The interlock between two piles is normally provided with the help of tongue and groove joint. The tongue and groove extend to the full length of the piles in most of the cases.

An alternative method of providing joint between two piles is shown in Fig. 3·44. In this method, after the piles are driven to the required depth, the joint is grouted with cement mortar 1 : 2 (1 cement : 2 sand).

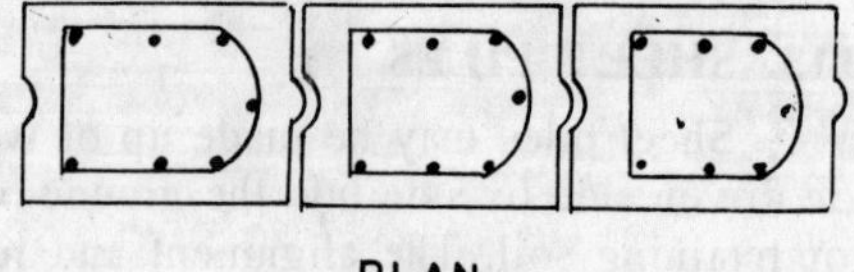

PLAN

ELEVATION

Fig. 3·43

Precast Concrete piles

The piles are reinforced to avoid formation of cracks due to rough handling or shrinkage stresses. In order to reduce the possibility of damage due to driving impact, the stirrups should be spaced closely near the top and bottom of the piles. The piles are normally bevelled at their feet to facilitate tightly close driving of a pile against the already driven one.

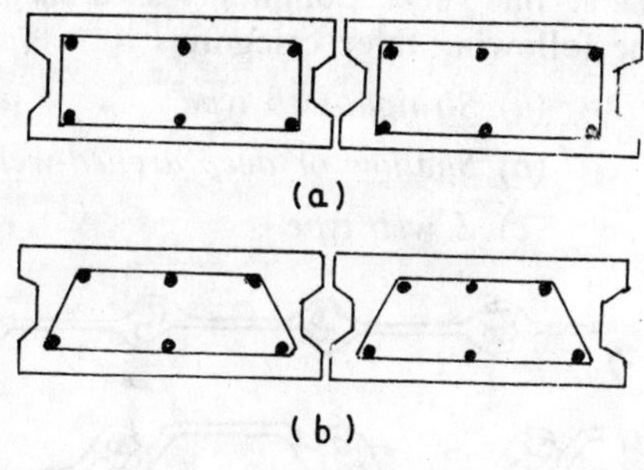

Fig. 3·44
Sectional plans of different types of precast concrete piles

Reinforced concrete sheet piles are bulky and heavy and as such they are gradually being superseded by prestressed concrete piles.

(*c*) *Prestressed concrete sheet piles.* On account of the numerous advantages the prestressed concrete members have over the conventional type of reinforced concrete members, prestressed concrete sheet piles are commonly used for sheet piling jobs. Similar to concrete sheet piles, they are reinforced on both the faces so that they could be handled from either side. They are comparatively lighter in weight, more durable and economical in the long run. They are advantageously used in sea water, since the danger of cracking of concrete is negligible and also the corresponding danger of corrosion of pile reinforcement is reduced.

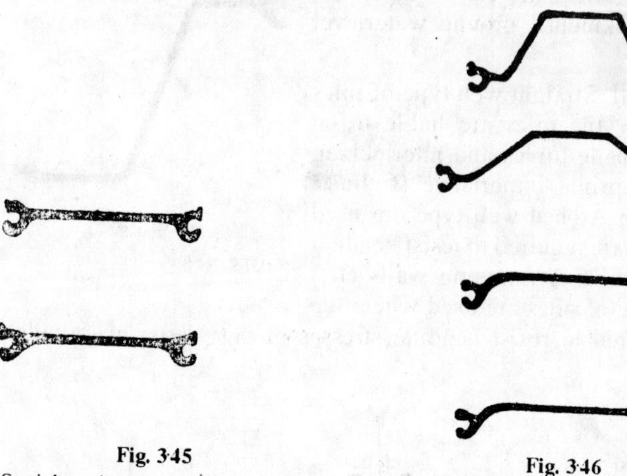

Fig. 3·45
Straight web type of steel sheet pile

Fig. 3·46
Arched web type of steel sheet pile

(*d*) *Steel sheet piles.* Steel sheet pile is a rolled steel section consisting of a plate called the web with integral interlocks on each edge. The interlocks consist of a groove, one of whose legs has been suitably flattened. This flattening forms the tongue which fits into the groove of a the second sheet. Commonly used sheet piles can be broadly divided into the following three categories :

(a) *Straight-web type*

(b) *Shallow or deep arched-web type*

(c) *Z web type.*

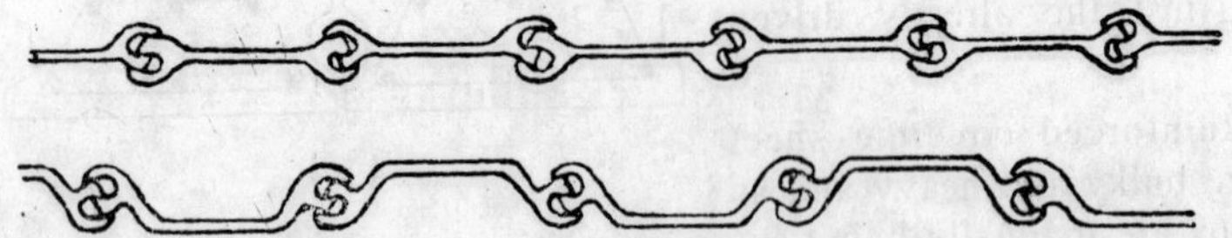

Fig. 3·47 – Fig. 3·48

Locking arrangement of different forms of steel sheet piles

Special shapes and sizes of steel sheet piles are manufactured for meeting the requirement of junctions and other similar situations. Each of the above mentioned type of piles is manufactured in varying widths and lengths. The selection of the type of pile and the section to be adopted depend upon the depths upto which the pile is to be driven, the nature of soil to be penetrated, the elevation of the earthen embankment, ground waterlevel etc.

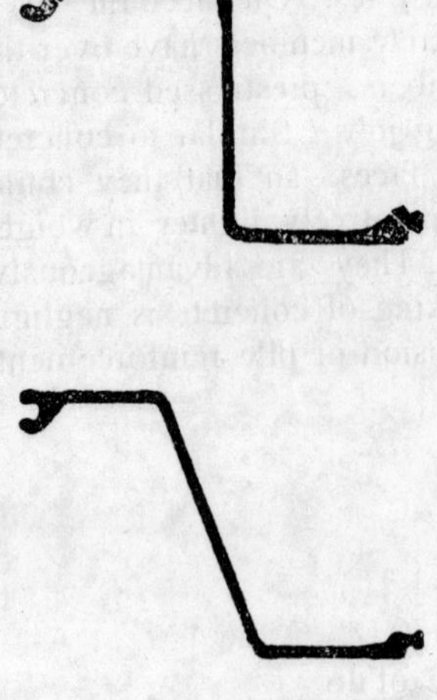

Fig. 3·49

Z-type of steel sheet pile

In general, Straight web type of piles are used where the piles are liable to be subjected to tensile forces and interlocking strength is of prime importance (Cellular cofferdam etc); Arched-web type are used where the piles are required to resist bending stresses (in cantilever retaining walls etc,) and Z-web type of piles are used where the piles are required to resist bending stresses of very large magnitude.

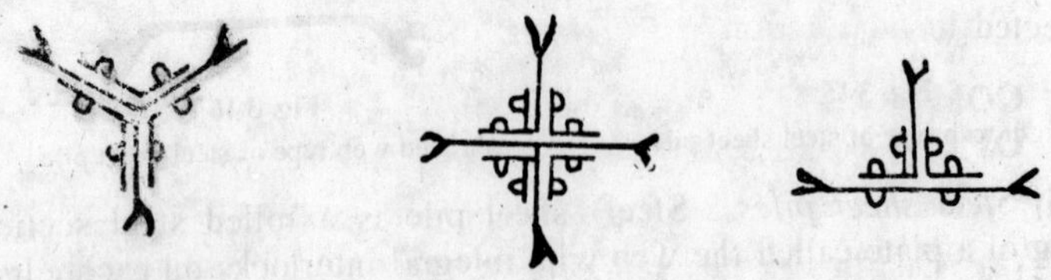

Fig. 3·50

Junction details

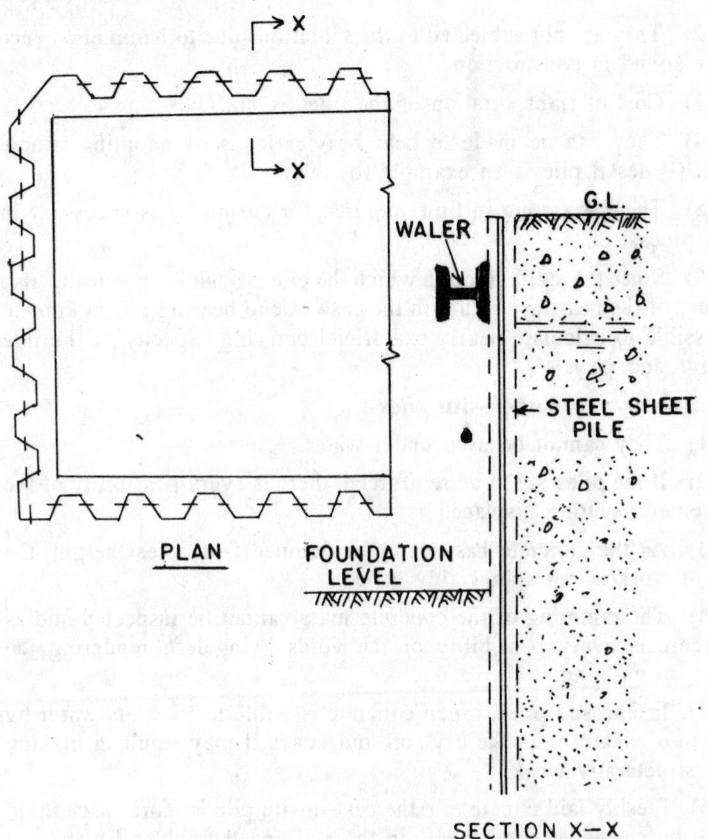

Fig. 3·51

Steel sheet piles are driven with the help of pile drivers which may be of drop hammer type or single or double acting hammer driven by steam or compressed air. The outstanding feature of steel sheet piles is that they can be used for greater depths. The continuous interlocking arrangement of the piles gives strength and rigidity to the supported structure. A wall made from properly driven sheet piles leaks very little, hence steel sheet piling is used with advantage in the construction of deep cofferdams. They are commonly used in coastal defence works which are likely to be subjected to tidal action.

3·14. COMPARATIVE ADVANTAGES AND DISADVANTAGES OF CAST-IN-SITU AND PRECAST PILES

(*a*) *Advantages of* **cast-in-situ** *piles :*

(1) Cast-in-situ piles are cast in exact length and as such there is no wastage of time, money and material, as is common in precast piles.

(2) They are not subjected to the vibrations due to hammers, hence they are sound in construction.

(3) Cost of transportation of the piles is nil.

(4) They can be made to bear heavier loads by adopting simple method. (Pedestal pile is an example for this).

(5) There is saving in time required for curing, as is necessary in case of precast piles.

(6) Since the strata through which the pile is sunk is known and also the nature of the bearing stratum in the case of end bearing pile is known, it is possible to calculate nearly exact load carrying capacity of the pile for design accuracy.

Disadvantages of **cast-in-situ** *piles* :

(1) They cannot be used under water.

(2) If the piles are to be reinforced, there is every possibility of the reinforcements getting displaced.

(3) As the concrete has got to be dumped from great height, the quality of work is not appreciably good.

(4) The ramming of the concrete mass cannot be inspected and as such there is every possibility of the voids being left, rendering the construction unsound.

(5) In uncased piles, green concrete is liable to loose its water by coming into contact with the dry soil and hence it may result in making the pile structurally weak.

(6) Freshly laid concrete in the cast-in-situ pile is more susceptible to attack by corrosive constituents in the soil and the sub-soil water.

(7) The driving of adjacent piles may rupture shell-less cast-in-situ pile on account of the tensile stresses set up in the ground during driving.

(*b*) *Advantages of* **precast** *piles* :

(1) They can be cast well before the commencement of the work resulting in rapid execution of work.

(2) Their construction can be well supervised and any defect detected can be rectified before use.

(3) The reinforcements remain in their proper position and do not get displaced.

(4) They can be driven under water.

(5) The driving of adjacent pile does not produce adverse effect upon the already driven pile.

(6) They can be loaded soon after they have been driven to the desired depth.

Disadvantages of **precast** *piles* :

(1) They are heavy and hence great difficulty is experienced for their handling and transportation.

(2) The exact length of a pile can rarely be pre-determined and as such it has to be lengthened sometimes, rendering the pile weak at the joint.

(3) If a pile is found to be too long after driving, it has got to be cut down which involves extra labour, time and expense.

(4) They are subjected to the driving shocks after the concrete has fully set. This may result in unsound construction.

3·15. PILE DRIVING

Piles are driven in the ground by the following methods :

(1) *By drop hammer*

(2) *By steam hammer*

(3) *By water jets*

(4) *By boring.*

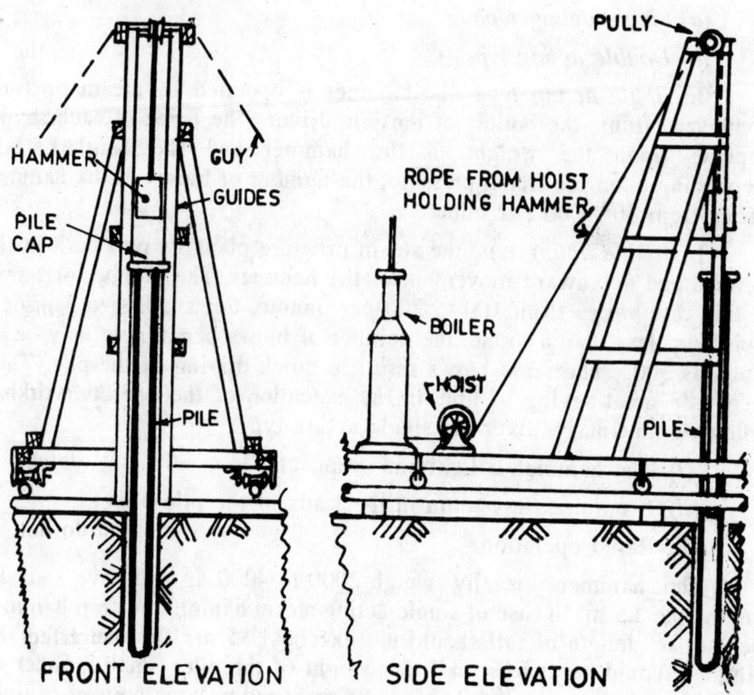

Fig. 3·52

Drop hammer type of pile driving machine

(1) *By Drop Hammer.* It is a pile driver in which the hammer is raised by a rope or steel cable and then allowed to drop on the pile cap. The drop hammer type of pile driver essentially consists of :

(*i*) A solid casting with jaws on each side that fit into the guides of the pile driver leads. (*ii*) A pin near the top for attaching the hoisting rope. (*iii*) A broad base on which it strikes the pile.

The driving of pile by drop hammer takes a lot of time and thus it is being replaced by steam hammer these days. The weight of drop hammer varies from 230 to 1800 kg. The exact weight of the drop hammer to be used is governed by the shape and length of pile as well as the nature of ground to be penetrated.

(2) *By Steam Hammer.* In this type of pile driver the hammer is automatically raised and then dropped a short distance by means of a steam cylinder and piston. The frame-work supporting the hammer also has pile lifting device. On account of uniform steam pressure, the rate of hammer blows on the pile is also uniform. Thus the short, uniform and quick blows keep the pile in continuous motion.

There are two types of steam hammer pile drivers :

(*a*) *Single acting type*

(*b*) *Double acting type*

In *single acting type*, the hammer is operated by steam pressure conveyed from the boiler of the pile driver. The force of each stroke depends upon the weight of the hammer and the height of fall. Depending upon the steam pressure, the number of blows of the hammer varies from 50 to 60 per minute.

In *double acting type* the steam pressure plays its part both in the upward and downward movement of the hammer. The number of blows in this case varies from 100 to 200 per minute. For small size hammers weighing less than a tonne, the number of blows per minute may be as much as 300. The rapid blows result in quick driving of the pile. Thus there is a great saving of time in the execution of the operation. It has following advantages over the single acting type.

(*i*) The hammer is light and compact.

(*ii*) It requires less room in the leads of the pile driver.

(*iii*) Rapid operation.

The hammers usually weigh 2000 to 4000 kg and have a stroke length upto 1·5 m. In case of single-acting steam hammer or drop-hammer the stroke length of fall should not exceed 1·35 m. The weight of the hammer should be at least half the weight of the pile. Small number of heavy blows with a small fall should be preferred to large number of light blows with a large fall. During the process of driving, the pile head should be protected by a cast steel helmet. The helmet is fitted with a timber stub dolly at its top and a cushion or packing saw dust, coir mat, coiled hemp rope, asbestos fibre or other similar material is placed between the timber dolly and the pile head. The hammer blow is taken on the timber dolly. (Refer Fig. 3·53)

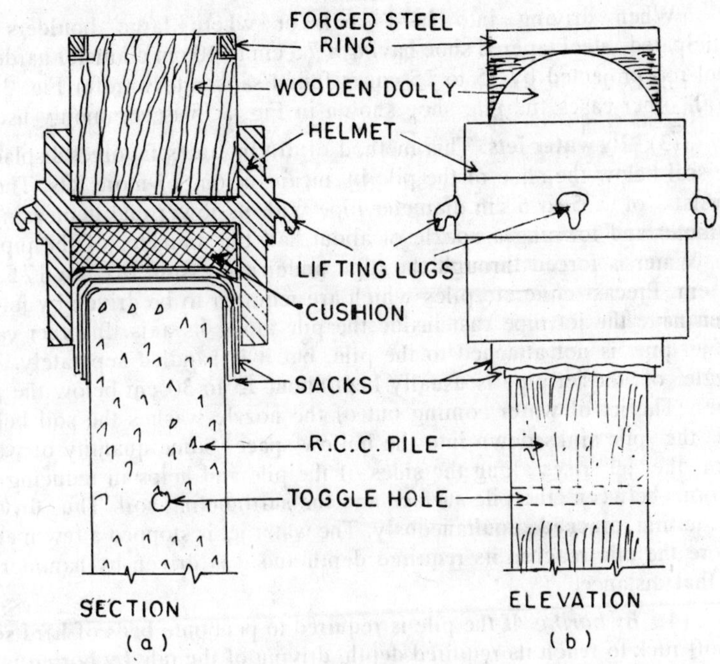

Fig. 3·53

Arrangements made for protecting pile head during pile driving operation

Pile Shoes. To facilitate easy penetration in the ground the pre-cast pile may be tapered or cast blunt at its lower end or it may be provided with pile shoes made up of cast iron, steel or steel plate. The most usual type being a cast iron conical shoe attached to the concrete by wrought iron straps. (Fig. 3·55) For deciding the type of shoe, nature of the soil to be penetrated is the guiding factor. For driving into soft soil

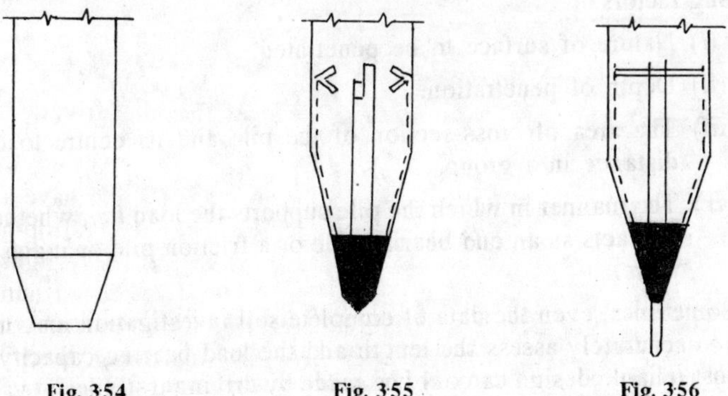

Fig. 3·54 Fig. 3·55 Fig. 3·56

like loose filling vegetable earth, river silt etc., the metallic shoe may not be used and the pile cast with a blunt end will meet the requirements. This type of pile is shown in Fig. 3·54.

When driving into hard rock or where large boulders are anticipated, steel tapered shoe having a 7·5 cm to 10 cm diameter hardened steel rod projected by 15 to 25 cm is found satisfactory (refer Fig. 3·56). In all other cases the pile shoe shown in Fig. 3·55 is commonly used.

(3) **By water jets.** This method of driving piles involves displacing the soil below the shoe of the pile by means of one or more jets. The jet consists of a 5 to 6 cm diameter pipe with its lower end decreasing in diameter and forming a nozzle of about half the diameter of the pipe at top. Water is forced through the pipe under a pressure of 7 to 17·5 kg/sq. cm. Precast concrete piles which are required to be driven by jetting often have the jet pipe cast inside the pile along its axis. In other cases the jet pipe is not attached to the pile, but it is handled separately. The nozzle of the jet pipe is usually kept about 15 to 30 cm below the pile shoe. The jet of water coming out of the nozzle washes the soil below and the pile sinks down into the hollow space. Some quantity of water from the jet rises along the sides of the pile and helps in reducing the friction between the pile surface and the surrounding soil. Thus driving and jetting proceed simultaneously. The water jet is stopped a few meters before the pile reaches its required depth and it is driven by hammering for that distance.

(4) *By boring.* If the pile is required to penetrate beds of hard soil, or soft rock to reach its required depth, driving of the pile by boring is an economical solution. In dry soils, boring may be done by auger, while in soft rocks, rotary well drills may be used. The bore is made to the required depth of the pile. Precast piles is then put in the bore hole. In case of continuously stiff soil, in-situ piles may be made in the bore holes.

3·16. LENGTH AND SAFE BEARING CAPACITY OF PRE-CAST PILES

The load bearing capacity of pre-cast pile depends upon the following factors.

(*i*) Nature of surface to be penetrated.

(*ii*) Depth of penetration.

(*iii*) The area of cross-section of the pile and its centre to centre distance in a group.

(*iv*) The manner in which the pile supports the load *i.e.*, whether the pile acts as an end bearing pile or a friction pile or it acts both ways.

Sometimes, even the data of complete soil investigation may not be helpful to accurately assess the length and the load bearing capacity of a pile. Most reliable design can only be made by driving test piles at various locations over the construction site. However, in the absence of loading test, or test piles informations, a large number of penetration formulae are now available to make a reliable estimation of the length and the safe loads bearing capacity of a pile.

3·17. FORMULA TO FIND THE LENGTH OF A PILE

Let the properties of the soil penetrated be represented by the symbols given below :

Weight of the soil $= w$ kg/cu.m

Angle of repose of soil $= \phi$

Co-efficient of friction between concrete and soil $= \mu$

Bearing capacity of the soil at the foot of the pile $= p$

If P be the load bearing capacity of the pile and 'h' be its depth of penetration, by Rankine's theory, we have

$P = p \times$ area of the cross section of pile + frictional resistance of pile

$$= (p \times A) + \left(C_p \times \frac{wh^2}{2} \times \mu \times \text{perimeter of the pile}\right) \quad ...(i)$$

where A = area of cross-section of the pile and

C_p = coefficient of pressure, which is

$$= \frac{1 + \sin\phi}{1 - \sin\phi}$$

For wet soil, ϕ may be assumed to vary between 15 to 22° and μ between 0·3 to 0·35.

For known values of ϕ, μ, p, w, A and P, h can be determined by equation (i).

3·18. ALLOWABLE LOAD ON SINGLE PILE

It has been observed that a pile exerting greater resistance against driving can sustain greater load. A number of formulae have been evolved to determine the load carrying capacity of a pile on the principle that the energy supplied to the pile is utilised in useful work done in driving the pile and in other loses. Among these formulae, also known as 'Dynamic pile formulae', Engineering News formula is considered to be the simplest and is thus commonly used.

The Engineering News formula is represented by the following equation

$$Q_a = \frac{WH}{F(S + C)} \quad ...(i)$$

where Q_a = allowable load in kN

W = weight of hammer in kN

H = height fall of hammer in cm

F = factor of safety (usually taken = 6)

S = average settlement of pile in cm per blow

C = empirical constant. Its value is taken

= 2·5 cm for drop hammer and = 0·25 cm for steam hammer (single or double acting type)

Substituting the value of S and C in equation (*i*) above, the formula for different types of hammers can be written as under :

(*i*) For pile driven by drop hammer

$$Q_a = \frac{WH}{6(S + 2{\cdot}5)}$$

(*ii*) For pile driven by single or double acting steam hammer

$$Q_a = \frac{WH}{6(S + 0{\cdot}25)}$$

3·19. GROUPING OF PILES

It has been observed both theoretically and practically that the total bearing value of a group of friction piles is less than the product of the load bearing capacity of an individual pile multiplied by the number of piles in that group. Some reduction takes place on account of the overlapping of the zone of stresses in soil supporting the piles. The magnitude of reduction per pile depends on the size and shape of the pile group and the size, spacing and length of the individual pile.

However, in case of end bearing piles where assistance for supporting the load by friction is not made use of, no such reduction takes place. Hence the load bearing capacity of a group of piles in cohesionless soils should be taken as the product of the load bearing capacity of an individual pile multiplied by the number of piles in that group.

3·20. BASIC RULES FOR THE DESIGN OF PRE-CAST R.C. PILES

1. **Main reinforcement**. The area of main longitudinal reinforcement in a pile should not be less than the following values :

(*i*) 1·25 percent of the gross cross-sectional area of the pile for piles of length upto 30 times their least width.

(*ii*) 1·5 percent of the cross-sectional area of the pile, for piles of the length between 30 and 40 times their least width.

(*iii*) 2 percent of the gross cross-sectional area of the pile for piles of length upto 40 times their least width.

2. **Lateral reinforcement**. Similar to columns, the lateral reinforcement in the piles may be provided in the form of links made out of *m.s.* bars of diameter not less than 5 mm. The volume of the lateral reinforcement in the body of the pile should not be less than 0·2% of the gross volume of the pile and their c/c spacing should not exceed distance equal to half the least width of the pile. The volume of lateral reinforcement at either end of the pile for a distance equal to 3 times the least width of the pile should not be less than 0·6% of the gross volume.

The transition between the close spacing of the laterals at the end of the pile and the wider spacing in the body of the pile should be made gradually. If the pile is required to penetrate through hard strata, additional reinforcement in the form of helix, should be provided in the top of the pile for a distance equal to three times the least width.

The longitudinal bars should be held apart by pairs of 12 mm to 16 mm mild steel bars known as forks or spreaders, spaced at centre to centre distance not exceeding 1·5 m. To prevent the outward displacement of the bars, 6 mm ϕ links should also be provided with the spreaders as shown in Fig.3·57.

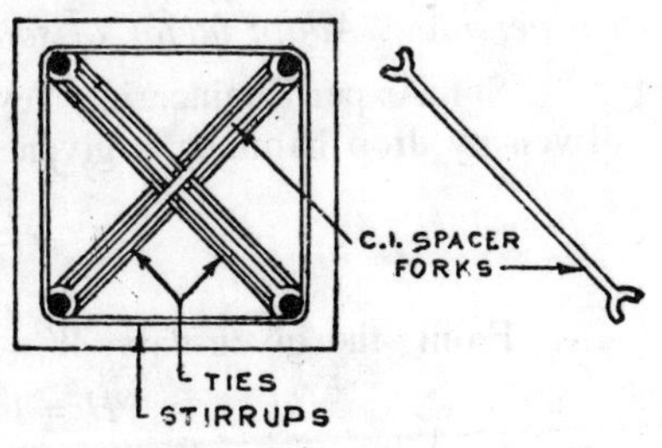

Fig. 3·57

3. **Concrete mix**. Concrete used should preferably be of higher grade than M 15. Concrete mix of M 20 with well graded aggregate is normally recommended.

4. **Cover**. The thickness of concrete covering the main bars should not be less than 40 mm. In places where the steel is liable to corrode, say in sea water, the cover should be increased to 55 mm.

5. **Stripping pile heads**. In order to ensure monolithic behaviour of the piles and the pile cap, the concrete from the pile heads of the driven piles is stripped and the longitudinal reinforcement exposed for a length, sufficient for bonding in the pile cap.

6. **Arrangement of piles.** The arrangement of piles and their spacing in a group needs a careful consideration. It mainly depends upon the sub-soil properties, types of loading and the requirement of the site. Close spacing of piles may result in reduced bearing capacity of the soil, uneven distribution of loads,heaving or uplifting of the piles, which may finally lead to the settlement of foundations. In order that each pile in a group may carry an equal share of load, the centre of gravity of the group should coincide with the c.g. of the super-imposed load. The c/c distance between any two piles should generally be not less than 1000 mm.

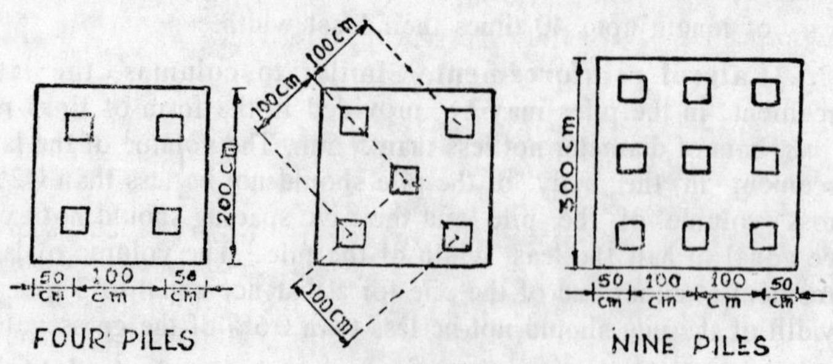

Fig. 3·58

Example 3·1. *Determine the safe allowable load on a precast pile driven by drop hammer weighing 50 kN. The free fall of the hammer = 1·20 m and the average penetration recorded in the last few blows is 0·5 cm per blow. Adopt factor of safety = 6.*

Sol. As per Engineering News formula, safe allowable load on a pile driven by drop hammer is given by

$$Q_a = \frac{WH}{F(S + 2 \cdot 5)} \quad \ldots(i)$$

From the given data, $W = 50$ kN

$H = 1 \cdot 2 \text{ m} = 120 \text{ cm}$

$F = 6$

$S = 0 \cdot 5 \text{ cm}$

Substituting the values in the formula in (*i*) above we have

$$Q_a = \frac{50 \times 120}{6(0 \cdot 5 + 2 \cdot 5)}$$

$$= 333 \cdot 33 \text{ kN}$$

Example 3·2. *A pre-cast R.C.C. pile is driven by a steam hammer. The striking part of the hammer weighing 25 kN has a stroke length of 0·75 m which results in average penetration of 0·3 cm in the last few blows of the hammer. Find safe allowable load on the pile.*

Sol. As per Engineering News formula, safe allowable load on a pile driven by steam hammer is given by

$$Q_a = \frac{WH}{F(S + 0 \cdot 25)}$$

From the given data,

$W = 25$ kN

$H = 0 \cdot 75 \text{ m} = 75 \text{ cm}$

$S = 0 \cdot 3 \text{ cm}$

Adopting factor of safety $F = 6$, we get

$$Q_a = \frac{25 \times 75}{6(0{\cdot}3 + 0{\cdot}25)}$$

$$= 568{\cdot}18 \text{ kN}$$

3·21. SELECTION OF TYPE OF PILE

Broadly speaking the different types of piling have been devised to suit different ground conditions. Hence the nature of the ground, where piling operation is to be carried out, determines to a large extent the choice of type of pile to be used. In addition, the other important factors which must be considered in this regard are :

(*i*) The nature of the structure
(*ii*) Loading conditions
(*iii*) Elevation of ground water level with respect to the pile cap.
(*iv*) Probable length of pile required.
(*v*) Availability of materials and equipment.
(*vi*) Factors which may cause deterioration of pile.
(*vii*) Probable cost of pile.

Unless the information about the identities, strength and compressibilities of the sub-soil stratum is known, it is difficult to determine as to how the pile will transmit the load to the ground, *i.e.* whether by end bearing or by skin friction.

If the sub-soil consists of silt on alluvium for appreciable depth, the pile will act as a friction pile. Depending upon the load, suitable pile, such as vibro pile, vibro-expanded pile or a tapered pile may be used in such situations.

In case the ground is very weak or loose and it is intended to use cast-in-situ pile, casing will be required to prevent the inflow of soil into the pile hole. In adverse conditions, where the ground water flow is strong, pile with permanent casing should be used. Alternatively, precast piles may be used.

In case the strata is comparatively firm uncased cast-in-situ pile or precast pile may be selected. Pedestal pile will be more effective where the pedestal can help in reducing the length of a pile in the firm strata. In situations where it is necessary that the pile should penetrate into rock for some depth, H-piles may work out to be a good selection. In case the soil is firm clay, drilled piles with or without enlarged base, may work out to be an economical choice.

In case piling is to be done very near to some existing structure, open ended tube piles or H-pile may work out to be good choice. Piles used to support marine structures or structures above open water should be precast type. In situations where there is possibility of the soil being chemically aggressive, high quality precast concrete piles should be selected. The piles should, however, be suitably treated before driving. Timber piles may work out to be economical, if they are used for the foundation of timber trestles structures with moderate loads or if the piles are to be used for protecting concrete docks etc.

3·22. COFFERDAMS

A cofferdam may be defined as a temporary structure constructed in a river or a lake or any other water-bearing surface for excluding water from a given site to enable the building operation to be performed on dry surface. The walls of the temporary structure should be practically water tight or at least they should be able to exclude water to such an extent that the quantity of water that leaks inside the enclosed area, can be easily pumped out. Cofferdams are classified according to the type of construction. The type of construction is dependent upon the depth, soil conditions,, fluctuations in the water level, availability of material etc. Cofferdams are advantageously constructed where a large area of site is to be enclosed and the hard bed is at reasonable depth.

Considering the material used in their construction, cofferdams may be divided into the following categories.

(*a*) *Earthen cofferdam*

(*b*) *Rock-fill cofferdam*

(*c*) *Single-walled cofferdam*

(*d*) *Double-walled cofferdam*

(*e*) *Crib cofferdam*

(*f*) *Cellular cofferdam (Circular or diaphragm type)*

(*a*) *Earthen cofferdam.* It essentially consists of an earthen embankment built around the area to be enclosed. It is constructed in places where the depth of water is not much, say 1·3 to 1·8 m. and the velocity of the current is very low. As a precautionary measure, the earth bank is carried about one metre above the water level. The top width of the bank should not be less than 1 m. and the side slopes may vary from 1 : 1½ to 1 : 2. The earth embankment should be built from a mixture of clay and sand or clay and gravel. If the estimated quantity of clay is not easily obtainable, the banks may be constructed with a central clay wall with slopes of sand on either side. In order to prevent the embankment from scouring due to the action of water, side slopes of the bank on water side should be pitched with rubble boulders. If the current of water is such that there is a danger of the earthen embankment getting washed away, canvas bags half filled with material

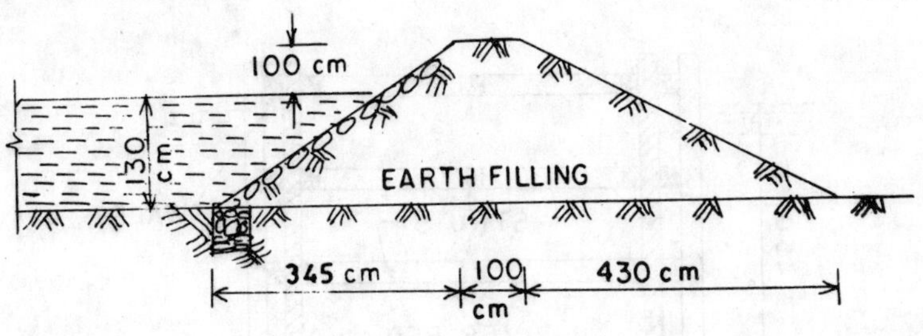

Fig. 3·59

Section of an earthen cofferdam

of embankment (mixture of clay, sand or gravel) are stacked one over the other to form the embankment. After the work of construction of cofferdam is over, the water from the enclosed area is pumped out so as to leave a dry surface inside. Excavations can then be performed to the required depth and the work of construction of foundations carried out.

(*b*) *Rockfill cofferdam.* If the depth of water to be retained by the embankment of cofferdam is of order of 1·8 to 3 m., stone or rubble is used for the embankment. This construction is adopted only if the stone is easily available in the nearby areas. The stones are assembled in the required shape of the embankment and the voids are partially filled with earth and stone-chips. The side slope on the water side is protected by pitching.

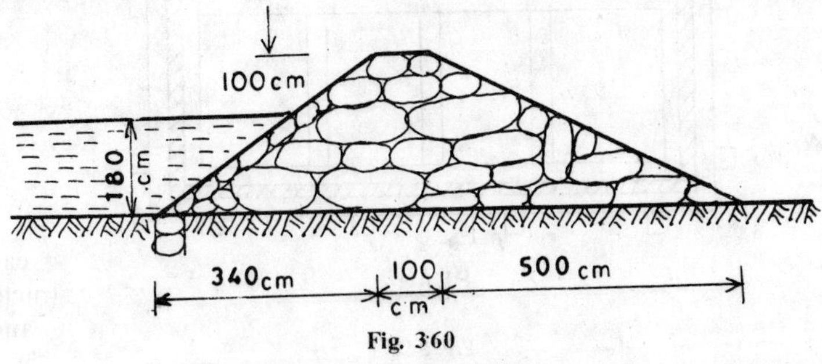

Fig. 3·60

Section of rockfill cofferdam

(*c*) *Single-walled cofferdam.* This type of cofferdam is used in places where the area to be enclosed is very small and the depth of water is more, say 4·5 to 6 m. Timber piles known as guide piles are first driven deep into the firm ground below the river bed. Depending upon the velocity of the current of the water in the river, the centre to centre spacing of the piles may vary between 1·8 to 4 m. Longitudinal runners called wales are then bolted to the guide piles at suitable distance apart. Steel or wooden sheet piles are then driven into the river bed along the wales and are secured to the wales by bolts. The sheets on the two faces are braced by trussed arrangement of struts. This helps in increasing the stability of walls

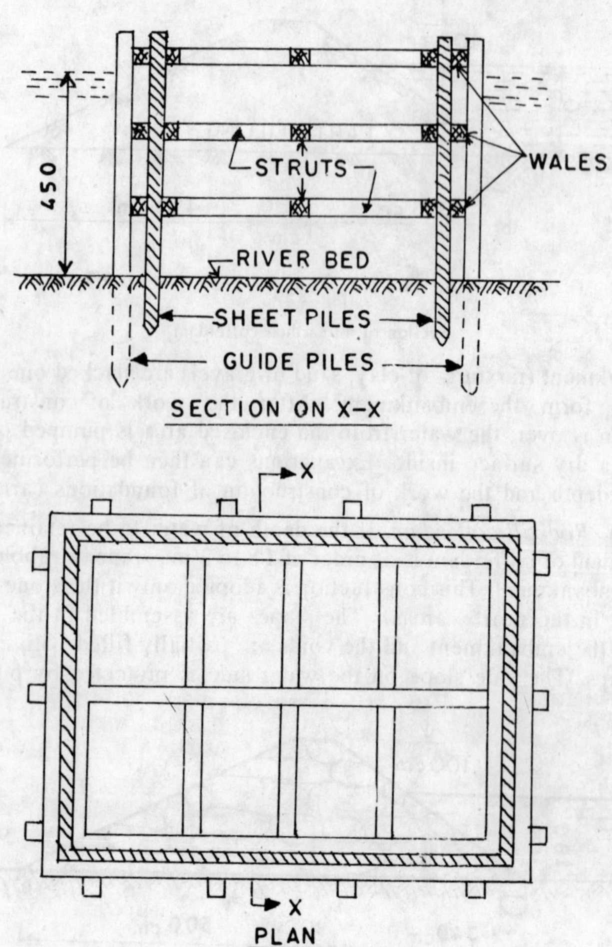

Fig. 3·61

Single-walled cofferdam

against the water pressure. Half-filled bags of sand stacked on the inside and the outside faces of the sheets help in increasing the stability of cofferdam. After the cofferdam is constructed, the water in the enclosed area is pumped out and the construction work is taken up.

(*d*) *Double-walled cofferdam.* For cofferdams required to enclose larger areas in deep water, single wall type becomes uneconomical as larger sections of trussed struts would be necessary to resist the water pressure. Double-walled cofferdam is provided in such situations. Its construction is essentially the same as that of a single-walled cofferdam

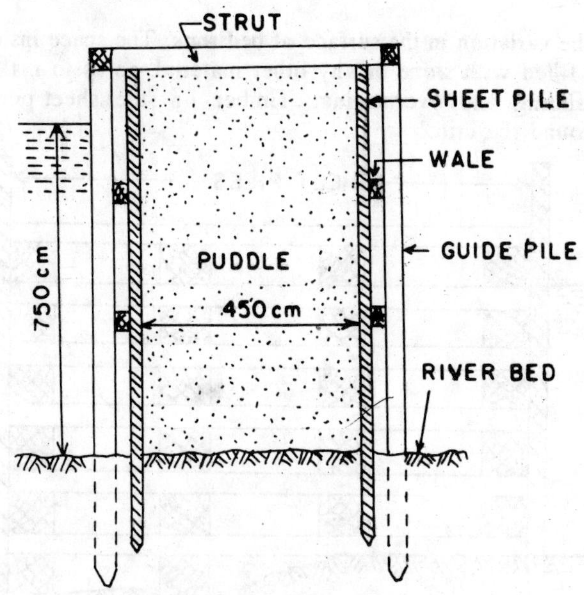

Fig. 3·62

Double-walled cofferdam

except that in place of one wall, a pair of walls with a gap in between is used all along the boundary of the space to be enclosed. This type of cofferdam can be used in depth of water upto 12 m. As the depth of water increases, the wall should be made wider in order to make it stable against over-turning and sliding. The distance between the two walls depends upon the depth of water. The thickness of wall should be equal to the depth of water upto 3 m. For greater depths of water, the thickness of wall should be 3 m. plus ½ the depth of water in excess of 3 m. At their top, the two faces of the walls are connected by steel rods spaced at close intervals. To prevent the leakage from the ground below, the sheet piles are driven to a good depth in the bed.

(*e*) **Crib cofferdam.** In deep waters where it is difficult to penetrate the guide piles or sheet piles into the hard bed below, crib cofferdam is used. In this type of construction, the sheet piles are supported by a series of wooden cribs. A crib is a framework of horizontal timbers installed in alternate courses to form pockets which can be filled with earth or stones. The length and breadth of each crib depend upon the depth of water and the current of flow. The framework of the cofferdam (made from logs of wood) is prepared on ground and then floated to the site where the cofferdam is to be constructed. The layers of sand and the other loose material overlying the impervious hard bed is dredged out. Crib is then sunk to the position, the bottom of each crib is given a shape to fit in the variation in the surface of bed rock. The space inside the crib is then filled with stone or any other material, so as to make it stable against sliding and overturning. Timber or steel sheet piles are then driven around the crib.

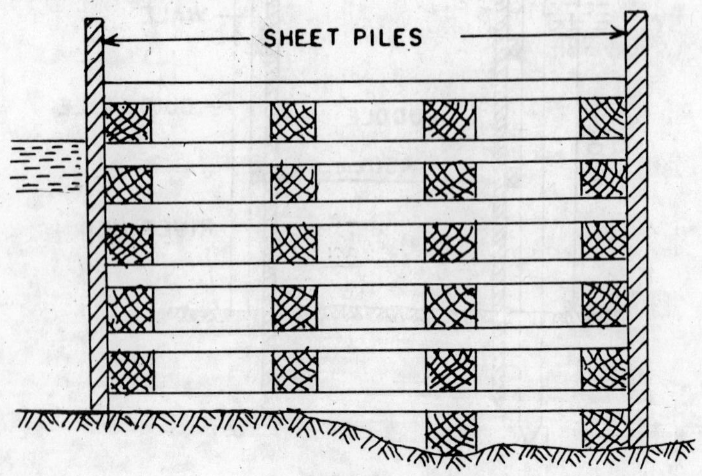

Fig. 3·63
Crib cofferdam

(*f*) *Cellular cofferdam.* This type of cofferdam is mostly used for de-watering large areas in places where the depth of water may be of the order of 18 to 21 m. Cellular cofferdams are mostly used during the construction of marine structures like dams, locks, whares etc. Cellular cofferdam is made by driving straight web steel sheet piles, arranged to from a series of inter-connected cells. The cells are constructed in various shapes and styles to suit the requirements of site. Finally the cells are filled with clay, sand or gravel to make them stable against the various forces to which they are likely to be subjected to. The two common shapes of the cellular cofferdam are :

(*i*) *Circular type cellular cofferdam.*

(*ii*) *Diaphragm type cellular cofferdam.*

(*i*) **Circular type cellular cofferdam.** The circular type of cellular cofferdam has the advantage that each cell may be filled completely to the top before starting the construction of the next cell without causing any distortion to the shell of the cofferdam. Thus, when one cell is completely filled up it can be used for placing crane or other equipment required for the construction of other cells. In addition, each cell acts as a self-supporting independent unit and in case one of the cells collapses due to scour or interlock damage or some other reason, it does not produce any adverse effect on the neighbouring cells. It is found that the interlock stresses reach their maximum permissible value when

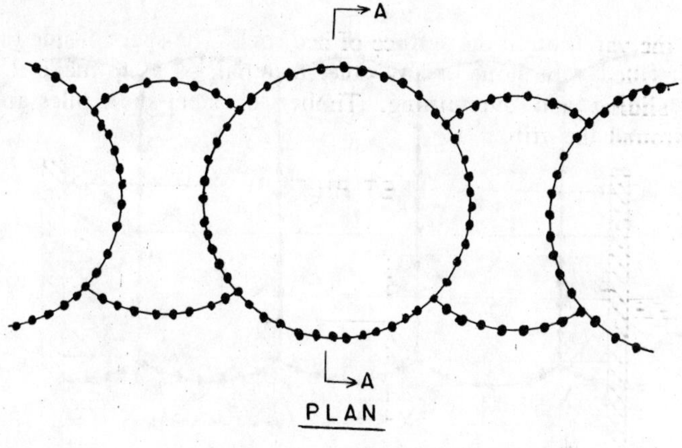

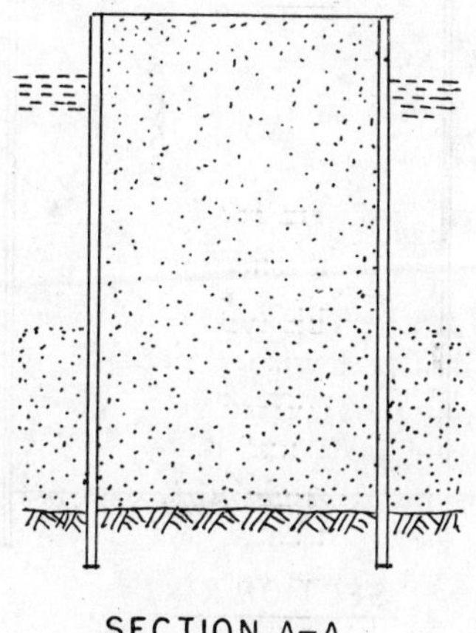

Fig. 3·64
Circular type cellular cofferdam

the diameter of cell is about 21 meter. Hence in case, from design consideration it is necessary to have effective width of the cofferdam more than 21 meter, diaphragm type of cofferdam must be used.

(*ii*) **Diaphragm type cellular cofferdam**. This consists of a series of diaphragm of steel sheet piles connected as shown in Fig. 3·65. The straight diaphragm walls are connected to each other by steel piles arranged in the form of arches on either sides. The radius of the connecting arcs is generally made equal to the distance between the straight diaphragm walls. With this arrangement, the tension in the arcs and cross walls remain equal. After the cells are driven to the required

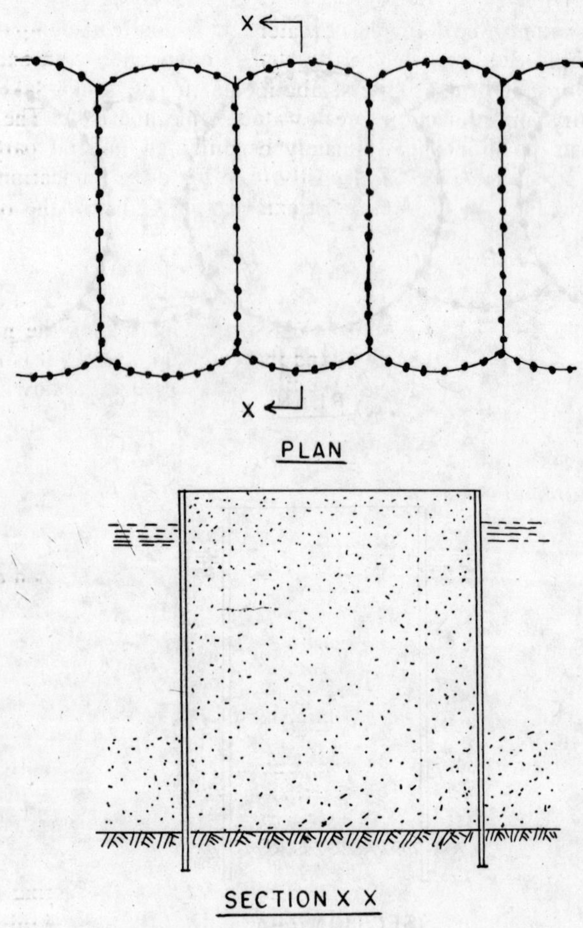

Fig. 3·65
Diaphragm type cellular cofferdam

depth, they are filled with earth, sand, gravel or other filling material. In this type of cofferdam, as the diaphragm which separates the two cells is a straight wall, it is necessary to fill adjacent cells at approximately the same rate. If this is not done, the unbalanced pressure from the fill will distort the diaphragm (cross-walls) which may result in the failure of the interlocks. In this respect, the circular type cofferdam has the advantage over the diaphragm type cofferdam because in the former, it is not necessary to fill the adjacent cells at the same time. This type of cofferdam has the advantage that the effective width of the cofferdam can be increased to desirable limits without increasing the interlock stresses.

3·23. CAISSONS

A caisson may be defined as a watertight structure made up of wood, steel or reinforced concrete, constructed in connection with excavation for foundations of bridges, piers, abutments in rivers and lakes, dock structures for shore protection, break waters, lamp house etc. The caisson remains in its position and ultimately becomes an integral part of the permanent structure. They are mostly used for deep foundations under water where the foundations must extend upto or below the river bed from stability considerations.

3·24. TYPES OF CAISSONS

The shape and size of a caisson depends upon the nature of structure for which it is to be built and the depth upto which it is required to be sunk. Caissons can be broadly classified in the following three different types.

(1) *Open caisson*

(2) *Box caisson*

(3) *Pneumatic caisson.*

(1) **Open Caissons**. Depending upon their shape, open caissons can be further classified as :

(*i*) *Single wall open caisson*

(*ii*) *Cylindrical open caisson*

(*iii*) *Open caisson with dredging wells.*

(*i*) *Single wall open caisson.* This is a box type structure having no top or bottom(during construction) and mainly consists of vertical walls. The method of construction of a typical single wall open caisson is describe below.

Wherever necessary the site is made dry by dewatering the area by adopting suitable method (By constructing cofferdam, sheet piling etc.). A pit of suitable size and about 3 to 4 meter deep is excavated at the site where the caisson is to be sunk. The cutting edge is then placed at the bottom of the pit. Above the cutting edge, the wall reinforcements are suitably tied up and the shuttering for casting the concrete wall is erected. The caisson walls are then concreted in lifts to a suitable height. When the freshly laid concrete has gained sufficient strength, sinking operation is started. The caisson sinks due to its own weight when the soil from the space inside is removed by use of clam shell or any other method. As sinking proceeds, additional sections of the caisson walls are successively cast. Sinking is stopped while each section is concreted and it is resumed only after the concreted section has gained sufficient strength. When the caisson is sunk to the required depth, its base is plugged by providing 1·5 to 4·5 m thick layer of concrete (concrete seal).

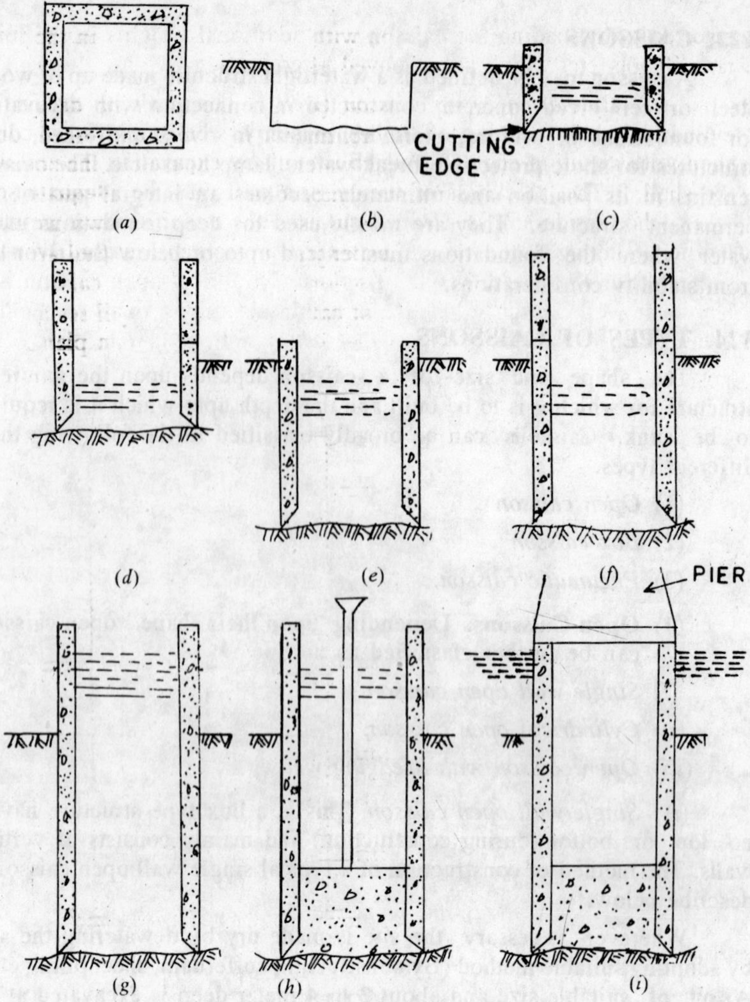

Fig. 3·66 (*a*) Shows sectional plan of the single wall open caisson required to be constructed.
(*b*) to (*i*) Show the different stages in the construction of the caisson.

The concrete for the seal is placed by use of a tremie pipe or by a bottom dump bucket. After the concrete seal has set, the water inside the caisson is pumped out and the empty space is filled with sand, gravel or concrete.

The rate of sinking of the caisson is always slow because the downward moment of the caisson is resisted by the skin friction of the ground on its walls. At times the skin friction becomes so great that the caisson does not sink even after all the earth has been dredged out from the inside clean down to the cutting edge. In such situations, sinking is resorted to by loading the caisson with additional weights in the form of rails, ingots etc. which are removed afterwards.

(*ii*) *Cylindrical open caisson* (*well*). This may be defined as a cylindrical shell made up of timber, masonry, steel or reinforced concrete shod with a cutting edge and which is sunk by excavating the soil within the shell. The thickness of the caisson wall must be adequate so that when the inside soil is dredged out, it sinks under its own weight. To facilitate sinking of the caisson water jets are sometimes used around the sides which decrease the skin friction. Cylindrical open caisson is also known as well caisson. This type of caisson is similar in all respect to the single wall open caisson except that its wall is circular in plan.

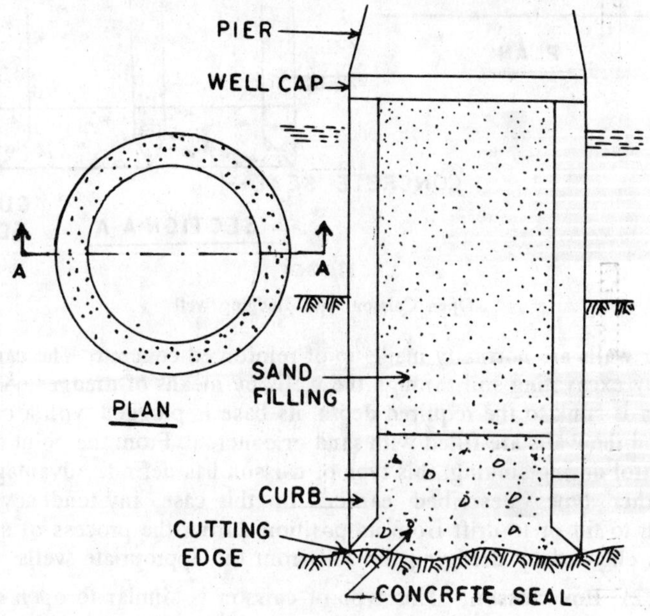

Fig. 3·67 Cylindrical open caisson (well foundation)

The method of construction of well caisson is exactly similar to that of a single wall open caisson described earlier. After the well is sunk to the desired depth its bottom is sealed with concrete. This type of caisson is commonly adopted for providing foundation for bridges and other structures to be built in rivers and waterways.

(*iii*) *Open caisson with dredging wells.* This type of caisson has the distinction of being employed for the deepest foundation for bridge piers, abutments and other similar structures. The caisson in this case is rectangular or square in plan and is further sub-divided into smaller sections from inside forming open walls. The outside walls as well as the inside

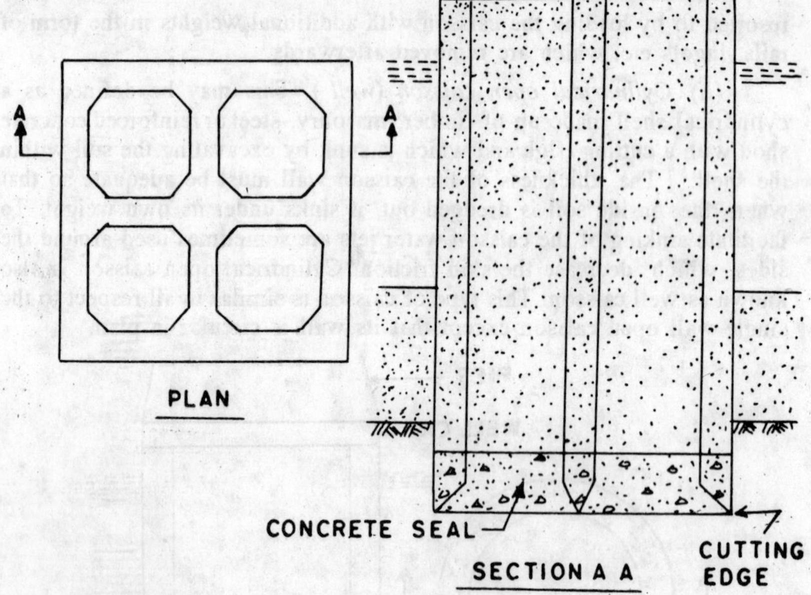

Fig. 3·68

Open Caisson with dredging wells

divider walls are normally made up of reinforced concrete. The caisson is sunk by excavating soil through the wells by means of dredges. After the caisson is sunk to the required depth, its base is plugged with a concrete seal and the walls are filled with sand or concrete. From the point of view of control during sinking, this type of caisson has definite advantage over the other types described earlier. In this case, any tendency of the caisson to tilt or to drift from its position during the process of sinking, can be checked by dredging the soil from the appropriate wells.

(2) **Box caisson.** This type of caisson is similar to open caisson except that it is closed at bottom. The caisson is cast and cured on land and when required, it is launched in water and towed to the site for sinking. The caisson is sunk by filling sand, gravel, or concrete in the empty space inside. The place where the caisson base is to rest must be levelled and as such box caissons are used in places where the strata of sufficient bearing capacity is available near the ground. In normal practice, the soft natural bottom soil of the river bed is dredged out to some depth and the trench thus formed is filled with sand to have a levelled base. The function of the sand layer is to uniformly distribute the superimposed loads over the soil below and thus avoid tilting of the caisson.

(3) **Pneumatic caisson.** This type of caisson is closed at top and open (during construction) at the bottom. The water is excluded from the

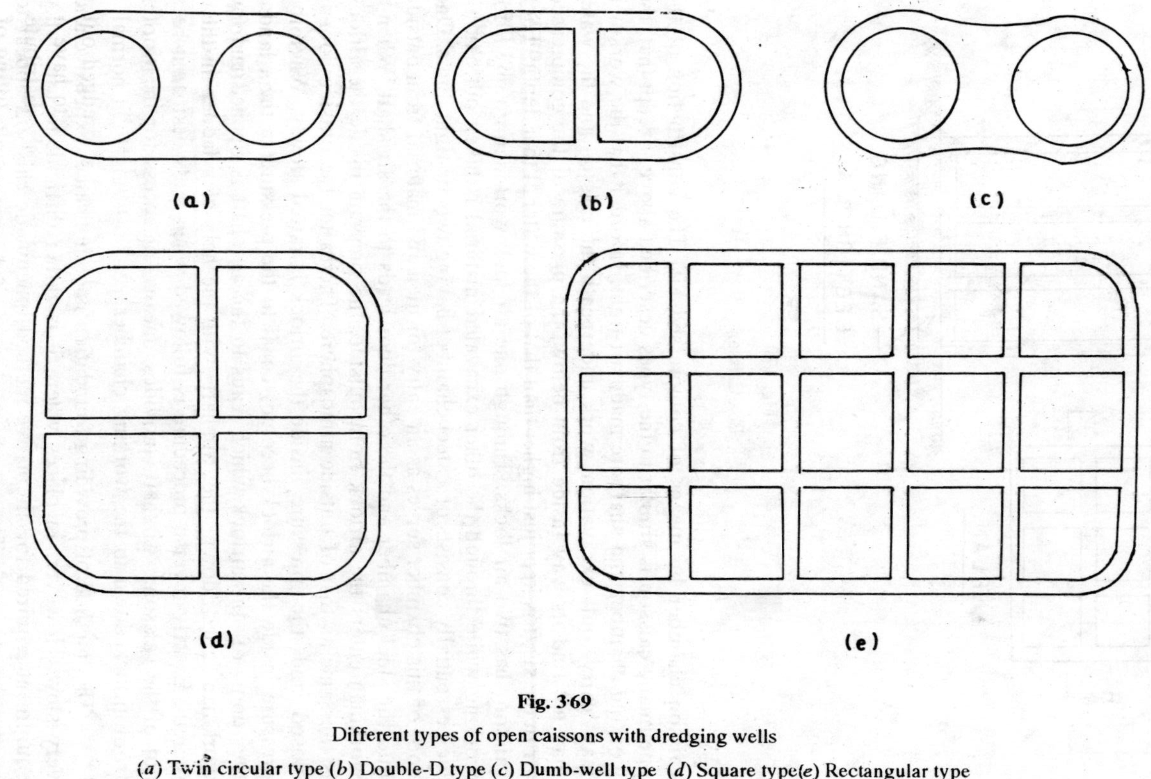

Fig. 3·69
Different types of open caissons with dredging wells
(*a*) Twin circular type (*b*) Double-D type (*c*) Dumb-well type (*d*) Square type(*e*) Rectangular type

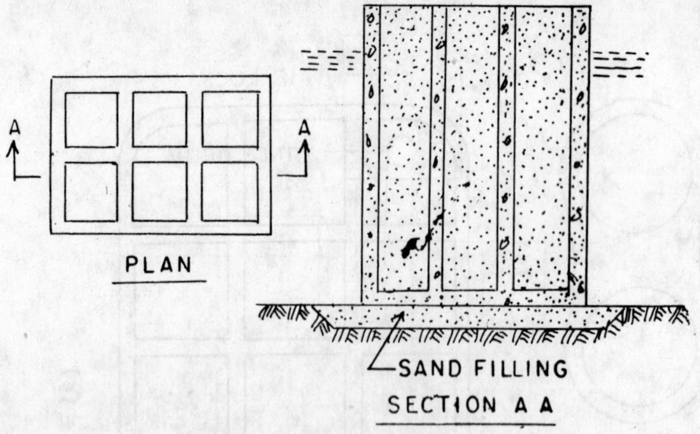

Fig. 3·70

Box Caisson

caisson chamber by means of compressed air. The construction of the pneumatic caisson is similar to the types described above, except that, the working chamber and shaft are made air-tight. In order that the workmen may carry out excavation work underneath the caisson and the water may not find its way inside from below, the pressure of the compressed air in the shaft is kept just higher than that of the water at that depth. Each caisson has two air locks. Through one air lock workmen go down for working while through the other excavated material is taken out. An air lock essentially consists of a steel chamber having two air-tight doors. One door of this chamber serves as an entry for men and material from outside into the steel chamber and the other door leads to the air shaft. When a workman enters the airlock from outside, the pressure inside the airlock is the same as that of outside atmosphere. Thereafter the outside door is closed and the pressure inside the airlock is raised slowly. When the pressure inside the airlock becomes equal to the pressure in the caisson, the door of the airlock which leads to the air shaft is opened and the workman goes down the airshaft with the help of a ladder installed therein. Exactly reverse procedure is followed when the workman comes out of the caisson. Airshaft provides means of access for the worker from airlock down to the working chamber.

This method of providing foundation is complicated, expensive and very slow. However, in places where it is rather difficult to use bulky equipment required for alternative method, pneumatic caissons appear to be the only choice. The chief advantage of this method is that entire operation of sinking the caisson can be carried out under controlled conditions. This method provides better facilities for removing obstructions, inspection of work and concreting of foundations bed.

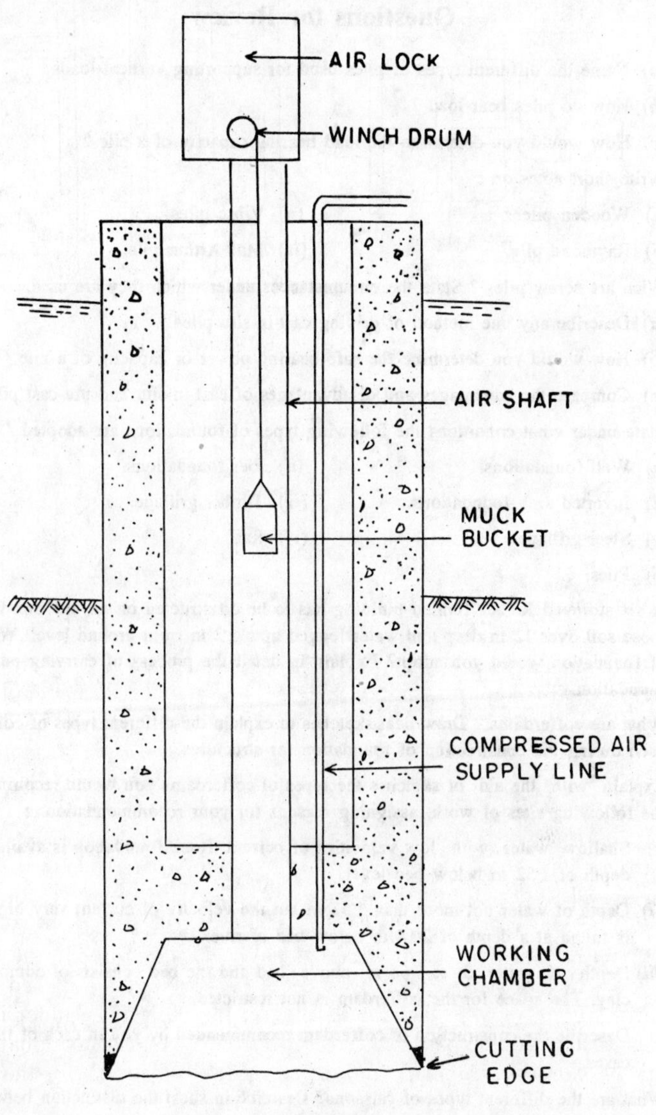

Fig. 3.71

Pneumatic caisson

This type of caisson is suitable for depths ranging from 25 m to 40 m. At higher depths, the persons working inside the caisson for sinking operation are liable to get caisson disease (resulting from the expansion of bubbles of air trapped on joints, muscles etc.).

Questions for Review

1. (*a*) Name the different types of piles used for supporting vertical loads

   (*b*) How do piles bear load ?

   (*c*) How would you determine the load bearing capacity of a pile ?

2. Write short notes on :

   (*i*) Wooden piles. (*ii*) Vibro piles.

   (*iii*) Raymond pile. (*iv*) Mac Arthur pile.

3. What are screw piles ? State the circumstances under which they are used.

4. (*a*) Describe any one method of driving cast-in-situ piles.

   (*b*) How would you determine the safe bearing power or capacity of a pile ?

   (*c*) Compare the advantages and disadvantages of cast-in-situ and pre-cast piles.

5. State under what conditions the following types of foundations are adopted ?

   (*i*) Well foundations. (*ii*) Pier foundations.

   (*iii*) Inverted arch foundations. (*iv*) Timber grillage.

   (*v*) Steel grillage. (*vi*) Raft

   (*vii*) Piles.

6. A six-storeyed R.C.C. framed building has to be constructed on an old tank bed with loose soil over 12 m deep and water logged upto 1·2 m from ground level. What type of foundation would you adopt? Explain in detail the process of carrying out such a foundation.

7. What are cofferdams ? Draw neat sketches to explain the different types of cofferdams used during the construction of foundation for structures.

8 Explain with the aid of sketches the types of cofferdams you would recommend for the following sites of work, assigning reasons for your recommendations :

   (*i*) Shallow water with low velocities of current. Rock foundation is available at a depth of 1·52 m below bed level.

   (*ii*) Depth of water not more than 1·52 m but the velocity of current very high. Rock is found at a depth of 2·14 m below bed of river.

   (*iii*) Depth of water 3 to 3·66 m above bed and the bed consists of compact blue clay. The space for the cofferdam is not restricted.

   Describe the construction of cofferdam recommended by you in each of the above cases.

9. What are the different types of caissons? Describe in short the distinction between the three different types of caissons commonly adopted. Sketch the types selected to justify your answer.

10. Write short notes on any five of the following :

    (*i*) Boundary wall footings. (*ii*) Steel sheet piles.

    (*iii*) Core drilling (*iv*) Spread footings.

    (*v*) Shoring of trenches. (*vi*) Foundations on sloping ground.

    (*vii*) Raft foundations.

11. Enumerate the different types of piles and describe each type briefly. Give advantages and disadvantages of each type.

12. (*a*) What are the advantages of concrete piles over other types of piles ?

    (*b*) What are the factors governing the choice of the most appropriate type of foundations for a given structure ?

    (*c*) What are the factors governing the minimum depth of shallow foundations ?

13. What are cofferdams ?

    Explain how they could be constructed and used for the construction of bridge pier in a river with a minimum depth of water (3m), where the river bed is rocky.

14. What type of foundation would you propose for a multistoreyed office building in :

    (*i*) Soft soil (*ii*) Black cotton soil.

    (*iii*) Made-up ground. (*iv*) Over soft rock.

15. What do you mean by the term Bearing Capacity of a soil? Give the bearing capacity of various types of soils.

    Distinguish between spread and pile foundations. Describe with sketches the method of constructing any one of them.

16. (*a*) What is bearing capacity of soils?

    Describe the various methods of improving the bearing capacity.

    (*b*) What are the advantages and disadvantages of

    (*i*) Pre-cast and (*ii*) Cast-in-situ, concrete piles.

17. While excavating the foundation, 90 cm wide x 90 cm deep for the main wall of a building, an abandoned 2·44 m diameter well, with 23 cm thick walls and filled up with earth was encountered. Explain what measures you should take to build the foundation of the wall in this portion

18. Describe with sketches the method of timbering in a trench of size 180 cm deep x 120 cm wide for laying underground sewer in moderately firm ground.

19. Explain with diagram grillage foundation under a steel stanchion. What type of foundation is replacing grillage foundation in modern construction?

20. Calculate the safe load which an R.C.C. pile will support if a hammer weighing 500 kg. falling from a height of 2 m drives the pile by 6 mm per blow.

21. What are the different types of foundations used in the construction of buildings and bridges? What is a well foundation? Describe in details the well foundation and its method of construction. Under what circumstances it is preferred to the other types of foundations?

22. What are the different types of piles that are used in high rise buildings? Describe in details the construction of a typical foundation used for a tall building. What do you understand by sheet piling? When and where are they used in practice?

23. State in detail the difference between the pre-cast and the cast-in-situ piles. Explain the Simplex pile in detail. What do you understand by uneven settlement of foundation and how it could be avoided?

24. State the different kinds of piles used for building purposes. Discuss the method of construction and suitability of each under varying conditions.

4

Structural Design of Foundations

4·1. DESIGN LOADS

The basic requirement to be satisfied in structural design is that the member must be strong enough to support all the possible types of loads to which it is liable to be subjected. Hence assessment of the design load is essentially the first step that should be taken before proceeding for the design proper. The loads coming on the foundation of a structure can be classified in three different categories :

(1) *Dead loads*

(2) *Live loads*

(3) *Wind loads, seismic forces, etc.*

4·2. DEAD LOADS

It is the sum of loads comprising of the self-weight of the structure (weight of all walls, partitions, floors, roofs etc), weight of its footings foundation and loads of all other permanent construction in the building. These loads can be calculated by finding the weights of the cubical contents of the different materials used in the structure. Table 4·1 gives the weight of the different materials for facility of load estimation.

TABLE 4·1

WEIGHT OF STRUCTURAL ITEMS

Description of material	Weight in	
	Kg/m^3	(N/m^3)
Concrete		
Reinforced cement concrete ...	2500	25000
Plain concrete M15(1: 2 : 4) ...	2400	24000
Concrete with brick aggregate ...	1920	19200
Breeze concrete ...	1120 to 1180	11200 to 11800

Description of material	*Weight in*	
	Kg/m³	***(N/m³)***
Lime concrete ...	1920	19200
Pumic concrete ...	800 to 1100	8000 to 11000
Saw-dust concrete ...	1120	11200
Aerated concrete ...	260	2600
Masonry		
Granite ashlar masonry ...	6400	64000
Granite rubble masonry ...	2400	24000
Dry rubble masonry ...	2080	20800
Lime stone ashlar masonry ...	2560	25600
Sand stone masonry ...	2240	22400
Brick masonry in cement mortar ...	1920	19200
Brick masonry in lime mortar ...	1760	17600
Mortars		
Lime mortar ...	1760	17600
Cement mortar ...	2080	20800
Gypsum mortar ...	1200	12000
	Weight in	
Flooring	***Kg/m²***	***N/m²***
25 mm. thick terrazzo flooring ...	60	600
25 mm. thick granolithic flooring ...	60	600
25 mm. thick tile flooring (excluding screeding)	20	200
28 mm. thick hard wood flooring laid in mastic	22	220
28 mm thick soft wood flooring laid in mastic ...	15	150
Partitions		
10 cm. thick brick partition ...	195	1950
6 cm. thick hollow concrete block partition ...	61.5	615
6 cm. thick hollow clay block partition ...	61.5	615
10 cm. thick hollow glass block partition ...	90	900
Roofing		
Thatch roofing on battens ...	35	350
Mangalore tiles with battens in single layer ...	65	650
Country tiles with battens in single layer ...	70	700
Country tiles with battens in double layer ...	120	1200
Allahabad tiles with battens in single layer ...	85	850
Allahabad tiles with battens in double layer ...	160	1600
Complete roofing with A.C. sheets ...	49	490
Complete roofing with G.I. sheets ...	20	200
First class mud roofing with 12 mm thick boarding and a single layer of tiles ...	342	3420

Description of material	*Weight in*	
	Kg/m²	*(N/m²)*
First class mud roofing with a double layer of tiles ...	390	3900
Jack arch roofing with 85 mm concrete filling over crown and 85 mm mud plaster on level ...	730	7300
Timber trusses and purlins ...	12	120
Asbestos cement sheet ...	12 to 75	120 to 750
G.I. sheets and bolts ...	15	150
5 mm. thick Bitumen macadam roof finishing ...	11	110

TABLE 4·2

WEIGHT OF STORED MATERIALS

Description of material	*Weight in*	
	Kg/m³	*(N/m³)*
Ashes ...	640	6400
Bitumen ...	1040	10400
Cement ...	1440	14400
Coal solid ...	1300	13000
Coal crushed ...	850	8500
Charcoal ...	300	3000
Coke ...	520	5200
Hay and straw ...	290	2900
Paper in rolls ...	700	7000
Lime stone ...	2650	26500
Granite stone ...	1640	16400
Slaked lime ...	800 to 1020	8000 to 10200
Surkhi ...	1010	10100
Brick ballast ...	1010 to 1450	10100 to 14500
Stone ballast (38 mm size) ...	1440	14400
Stone ballast (13 mm size) ...	1920	19200
Fire wood (chopped) ...	1400	14000
Dry sand ...	1600	16000
Dry earth filling (moderately rammed) ...	1520	15200
Compacted earth filling ...	2200	22000
Puddled clay (stiff) ...	2160	21600
Slag wool ...	270	2700
Gravel (loose) ...	1600	16000
Gravel (rammed) ...	1920	19200
Stocks of books and files ...	850	8500
Building rubbish ...	1440	14400
Glass in sheets ...	2600	26000
Timber ...	640 to 960	6400 to 9600
Wheat ...	830	8300
Water ...	1000	10000

4·3. LIVE LOADS

Also called as super-imposed load consists of moving or variable loads like that of people using the structure, temporary stores, snow load etc. The live loads to be used for the design of floors and roofs in buildings are shown in Table 4·3.

TABLE 4·3

SCHEDULE OF LIVE LOADS AS GIVEN BY I.S.I
(LIVE LOADS ON FLOORS)

Loading class No.	*Types of floors*	*Minimum live loads* *kg/m²* *of floor area*	*Alternative minimum live loads*
(1)	(2)	(3)	(4)
200	Floor for residential purpose including dwelling houses. Floors of tenements, hospitals, wards, bedrooms in hotels and dormitories.	200	Subject to a minimum total load of 2·5 times the values in column (3) for any given slab panel and 6 times the value in column (3) for any given beam.
250	Office floors other than entrance halls, floor of light work-rooms.	250—400	
	Note : The lower value of 250 kg/m² should be taken where separate storage facilities are provided and the higher value of 400 kg/m² should be taken where such provisions are lacking.		This total load shall be assumed to be uniformly distributed on the entire area of the slab panel or the entire length of the beam.
300	Floors of banking halls, office entrance halls and reading rooms.	300	
400	Shop floors used for the display and sale of merchandise ; floors of work-rooms generally ; floors of class-rooms in schools ; floor of places of assembly with fixed seating, churches, chapels, restaurants ; circulation space in machinery halls, power stations etc., where not occupied by plant or equipment.	400	Subject to a minimum total load of 2·5 times the value in column (3) for any given panel and 6 times the value in column (3) for any given beam.

(1)	(2)	(3)	(4)
500	Floor of ware house, work-shops, factories and other buildings or parts of buildings of similar category for light weight loads ; office floors for storage and filling purposes; floors of places of assembly without fixed seating, public rooms in hotels, dance halls, waiting halls, etc.	500	This total load shall be assumed to be uniformly distributed on the entire area of the slab panel or the entire length of the beam.
750	Floors of warehouses, workshop, factories and other buildings or parts of buildings of similar category for medium weight loads.	750	
1000	Floors of warehouses, workshops, factories, and other buildings of similar category for heavy weight load, floors of book stores and libraries and roof and pavement light over basements projecting under the public footpath.	1000	
	Stairs, corridors, landings and balconies not liable to/liable to overcrowding:		
Stairs	Stair landings and corridors for class 200 loading but not liable to overcrowding.	300	Subject to a minimum of 130 kg. concentrated load at the unsupported end of each step for stairs constructed out of structurally independent cantilever steps.
	Stairs landings and corridors for class 200 loading but liable to over-crowding and for all other classes.	500	
Balconies	*Balconies not liable to overcrowding :*		
	(i) For class 200 loading.	300	
	(ii) For all other classes.	500	
	Balconies liable to overcrowding :	500	

(1)	(2)	(3)	(4)
Garage (light)	Floors used for garages for vehicles not exceeding 2·5 metric tonnes gross weight.	500	
	(*i*) Slabs	400	Or the combination of actual wheel load whichever is greater
	(*ii*) Beams	250	Or the combination of actual wheel load whichever is greater
Garage (heavy)	Floors used for garages for vehicles not exceeding 4 metric tonnes gross weight.	750	Subject to a min. of 1½ x max. wheel load but not less than 900 kg. considered to be distributed over 75 cm. square.

TABLE 4·4

LIVE LOADS ON ROOFS

Type of roof	*Live loads as measured on plan*	*Minimum live load as measured on plan*
1. Flat, slopping or curved with slope upto and including 10°.		375 kg uniformly distributed over any span of one metre width of the roof slab and 900 kg uniformly distributed over the span in case of beams.
(*a*) Access provided.	150 kg./m²	190 kg uniformly distributed over any span of one metre width of the roof slab and 450 kg uniformly distributed over the span in case of beam.
(*b*) Roof to which access is not provided except for repairs or maintenance etc.	75 kg./m²	
2. Sloping roof with slope greater than 10°.	75 kg/m² minus one kg./m² for every degree increase in slope over 10° upto and including 20° and 2kg./m² for every degree increase in slope over 20°.	Subject to a minimum of 40 kg./m²

Note : (1) Multiply by 10 to convert loads in kg/m² to N/m².

Note : (2) Roofs of buildings used for accidental assembly purpose should be designed for a minimum load of 4000 N/m² or more.

These loads are considered as uniformly distributed dead load on the plan area of the floor or roof and also include the normal effect of impact. These loads do not take into account snow loads, special concentrated loads due to heavy machine, vibrational effect of moving machinary, crane etc.

The following reduction in assumed total live load on floors may be made in designing columns, walls, piers, their support and foundation (This reduction is not applicable in case of warehouses, garages and other buildings used for storage purposes and for factories and workshops designed for super-imposed load of more than 5000 N/m²).

Number of Floors carried by member under consideration.	*% Reduction of total live load on all floors above the member under consideration.*
1.	0
2.	10%
3.	20%
4.	30%
5. or more.	40%

4·4. PROVISION OF LOADS FOR PARTITION WALLS

In places where it is possible to fix the position of the partition walls which may be of brick or wood or other desired material, the actual load due to partition walls should be calculated separately and then included in the dead load for the design of the floors and the supporting structures. If it is not possible to decide the exact position of the partition wall beforehand, the floors and the supporting structures should be designed to carry, in addition to other loads, a uniformly distributed load per square metre of not less than 33⅓% of the weight per metre run of the finished partition over the entire floor area subject to a minimum uniformly distributed load of 100 Kg/m² in the case of floors used for office purposes.

4·5. WIND LOADS

Wind acts horizontally on the exposed vertical surfaces of walls, columns and inclined roof of the structure. It is assumed to exert uniform pressure on the component of the structure on which it acts and it tends to disturb the stability of the structure. Wind pressure is expressed in terms of a basic pressure (p) which is an equivalent static pressure in the direction of flow of wind. The basic wind pressure is decided paying due regard to the meteorological data and local conditions such as characteristics and location of structure and characteristics and duration of wind flow. In the absence of meteorological data the basic wind pressure or wind load can be calculated by the following formula :

$$P = k.V^2$$

Where P = the wing pressure in Kg/m²

and V = the wind velocity in Km per hour.

and k = the coefficient, the value of which depends on a number of factors, such as the wind speed, the type, proportion and shape of structure and temperature of air. The National Building Code contains wind pressure maps and charts based on value of $k = 0.006$ from which the value of wind pressures for different height of the building located in different parts of the country can be easily read.

As per National Building Code, for structures of various plan shapes other than rectangular plan shape, the external pressure acting on the projected area in the plane perpendicular to the wind, should be the product of basic pressure (p) (worked out from maps/charts or formula given above) and the shape factors given in Table 4·5 below :

TABLE 4·5

FACTORS FOR VARIOUS SHAPES (IN PLAN) OF STRUCTURE

Shape of structure in plan	*Factor*		
	Ratio of height of structure to base width 0–4	*Ratio of height of structure to base width 4–8*	*Ratio of height of structure to base width 8 or above*
1. Circular	0·7	0·7	0·7
2. Octagonal	0·8	0·9	1·0
3. Square (Wind perpendicular to diagonal)	0·8	0·9	1·0
4. Square (Wind perpendicular to face)	1·0	1·15	1·3

Note : In the case of projections above the general roof level, the base width shall be taken as the width of the construction where it rises from the surface of the roof ; and the height governing the ratio of height to base width shall be that from the roof surface to the top of the construction.

The stability calculations of the building as a whole should be carried out by considering the combined effect as well as separate effect of the imposed load and wind load on vertical surfaces, roofs and any other part of the building above the general roof level.

Calculation should be made for the effect of wind on the design of individual components of the building. When, however, adequate stiffening is provided by walls or floors and walls, calculations for the effect of wind, except in regard to wall panel, roofs and foundations, need not be made on:

(*i*) a building or a part of a building of which the height does not exceed twice the effective width.

(*ii*) a section adjoining two parts of an adequately stiffened building, if height of the section exceeds twice its width but the length of the section does not exceed four times its width.

(*iii*) a wing of such a building if it does not project more than twice its own width.

4·6. WIND PRESSURE ON SLOPED ROOFS

In case of sloped roofs, the wind pressures varies with the degree of slope. The wind is assumed to be acting on one side of the roof and that side is termed as *windward side* or *windward slope.* The opposite side of the roof is termed as *leeward side* or *leeward slope.* For flat and pitched roofs, the wind pressures normal to the surface, due to the wind blowing at right angles to the ridge, in terms of basic wind pressures (p) as given in National Building Code are reproduced below.

TABLE 4·6

WIND PRESSURE ON ROOFS (WIND NORMAL TO RIDGE)

S.No.	*Slope of roof on windward side (α)*	*External wind pressure*	
		Windward Slope	*Leeward slope*
1	0°	− 1.00 p	− 0.50 p
2	10°	− 0.70 p	− 0.50 p
3	20°	− 0.40 p	− 0.50 p
4	30°	− 0.10 p	− 0.50 p
5	40°	+ 0.10 p	− 0.50 p
6	50°	+ 0.30 p	− 0.50 p
7	60°	+ 0.40 p	− 0.50 p
8	70°	+ 0.50 p	− 0.50 p
9	80°	+ 0.50 p	− 0.50 p
10	90°	+0.50 p	− 0.50 p

Note 1. In the above table, plus (+) means an external positive pressure on the roof normal to the surface of the roof and minus (−) means an external suction on the roof normal to the surface of the roof.

2. In the case of a flat roof, windward slope means windward half of the roof and leeward slope means leeward half of the roof.

47. DESIGN OF MASONRY WALL FOUNDATION

This consists in calculating (1) the depth upto which the foundation of the wall of the building should be taken below the ground level, (2) depth of the concrete bed block and (3) the width of foundation.

(1) **Depth of foundation.** The depth at which the foundation should be located depends on the character of the sub-soil and the magnitude of load on the structure. However the foundation must be carried below:

(*i*) The depth of frost penetration in regions of temperate climate.

(*ii*) The depth at which high volume changes in soil due to moisture fluctuation do not cause any adverse effect. In case of black cotton soil or clayey soils, under-reamed piles are recommended for providing safe and economical foundations.

(*iii*) Below the depth of unconsolidated materials like mud, garbage dumps and other similar type of made up ground. In case this type of soil strata exists at site for appreciable depths, under-reamed pile or strip or raft foundation works out to be safe and economical.

(*iv*) The depth upto which the soil is weakened by root holes, or cavities produced by animals, worms etc. In addition the foundation should be taken deep enough to eliminate the possibility of rain water scouring the ground and exposing the foundation.

(*v*) Irrespective of the above, all foundations should extend to a minimum depth of 500 mm below natural ground level. However, if good rock is available at smaller depth it is sufficient to remove top soil and construct the foundation over a cleaned and prepared bearing surface.

(*vi*) For all important buildings it is necessary to get the soil investigation of the site carried out by some specialist agency. The test report should contain details regarding the type of sub-soil strata at various depths, depth of water table, and recommendations regarding the bearing capacity of soil at different depths.

For normal buildings, the depth of foundation below ground level is commonly calculated by the Rankine's formula. According to this formula the minimum depth of foundation is given by

$$D_f = \frac{p_0}{\gamma}\left(\frac{1 - \sin\phi}{1 + \sin\phi}\right)^2$$

Where

D_f = minimum depth of foundation in metres.

p_o = bearing capacity of soil in kN/m^2

γ = density of soil or the unit weight of soil in kN/m^3

ϕ = the angle of repose of the soil.

(For values of angle of repose for different materials refer appendix II)

(2) **Depth of concrete bed-block.** The depth of foundation concrete bed-block depends upon the type of concrete used (whether cement or lime concrete), the projection of the concrete block beyond the bottom-most course of wall footing and the upward pressure of the soil below. The concrete block may fail by shearing, crushing or by bending. It is usual to design the depth from considerations of maximum bending moment and the depth thus obtained is found to render the concrete block safe against shear or crushing.

Thus, if d = depth of concrete block in cm.

x = projection of concrete block on either side of the lowermost course of wall footing in metre.

m_r = safe modulus of rupture of concrete mix used in N/mm^2

p = net upward soil pressure in kN/m^2

the depth 'd' of the concrete block can easily be worked out from consideration of maximum bending moment.

Refer Fig. 4·1. The cantilever projection of the concrete block = $AB = x$

Bending moment (M) in kNm about a vertical plane through B

$$= \frac{px^2}{2}$$

The moment of resistance of the concrete block

$$= \frac{m_r d^2}{6} \text{ kNm}$$

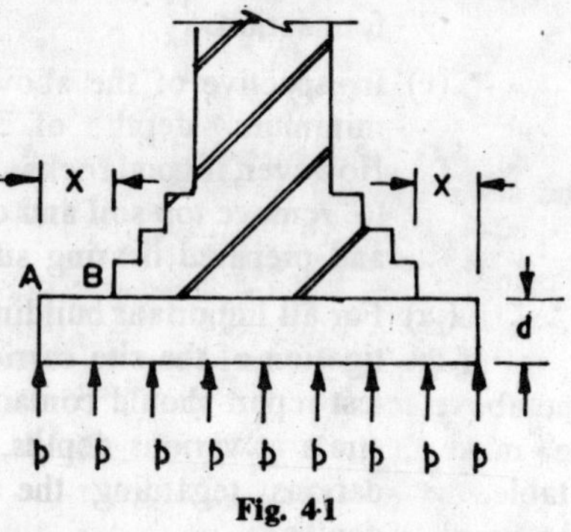

Fig. 4·1

Equating the moment of resistance to the maximum bending moment, we have

$$\frac{m_r d^2}{6} = \frac{p\,x^2}{2}$$

or

$$d^2 = \frac{6 \times px^2}{m_r \times 2} = \frac{3px^2}{m_r}$$

$$\therefore \quad d = \sqrt{\frac{3px^2}{m_r}}$$

The values of m_r for various grades of concrete mix are given in the Table 4·7.

TABLE 4·7

MODULUS OF RUPTURE OF VARIOUS GRADES OF CONCRETE MIX

S.No	Type of concrete	(m_r) in N/mm²
1	Pure lime and surkhi concrete (1 mortar : 3 stone ballast)	.155
2	Moderately hydraulic lime and sand concrete (1 mortar : 3 stone ballast)	.155
3	1 : 4 : 8 Plain cement concrete	.246
4	1 : 3 : 6 Plain cement concrete (M 10)	.352
5	1 : 2 : 4 Plain cement concrete (M 15)	.527

Since the resistance of concrete to tensile and shear stresses is very low, the value of x must be limited. The least value of x should vary between 100 to 150 mm. The plane of greatest shear in concrete block occurs at an angle of 45°. The depth of concrete can thus be fixed by using the above property.

3. **Width of footing.** The width of footing is obtained by dividing the total load per unit length on the footing bed by the safe bearing capacity of the soil. Thus if W be the total load including self weight of footing in kN per metre and p_0 be the safe bearing capacity of the soil in kN/m², then,

$$\text{the width of footing} = \frac{W}{p_0} \text{ metre.} \qquad \text{...}(i)$$

If the wall is to rest on a hard substratum, the width obtained from the above equation will be very small and as such it may not be sufficient to make the wall safe against tilting. In order that the wall may be stable, the lowermost course of the wall footing is normally made twice the width of the wall (b) and the concrete bedblock is made to extend at least 100 to 150 mm on either side of the lowermost course, this extension being denoted by x.

Hence the width of footing

$$= 2b + 2x \qquad \text{...}(ii)$$

The greater of the two results obtained from equations (*i*) and (*ii*) should be adopted as the final width of footing.

Example 4·1. *Design the foundation of the end walls of a* *residential building three storey high. The foundation is to rest on sandy soil having angle of repose of 30° and safe bearing capacity of 160kN per sq. m. The soil weighs 16·8 kN per cubic metre. The walls are 300 mm thick and 13·5 m high above ground level (this includes the height of parapet wall). The roof consists of 150 mm thick R.C. slab finished with 100 mm thick (average) lime concrete terracing. The floors consists of 120 mm thick R.C. slabs finished with 25 mm thick terrazzo flooring. The maximum span of the R.C. slab for the roof or floor between the end walls and the wall next to the end wall is 4 metres. The roof is accessible. The live load over the roof slab and floor slab may be taken @ 1500 N/m² and 2000 N/m² respectively.*

Sol. *Depth of foundation.* The depth of foundation as given by Rankine's formula is

$$D_f = \frac{p_o}{\gamma}\left(\frac{1-\sin\phi}{1+\sin\phi}\right)^2$$

$$\frac{160}{16{\cdot}8} \times \left(\frac{1-\sin 30^\circ}{1+\sin 30^\circ}\right)^2$$

$$= \frac{160}{16{\cdot}8} \times \frac{1}{9} = 1{\cdot}06 \text{ m}$$

Adopt depth of foundation = 1·1 m

Load calculations :

To calculate load per metre carried by wall. The total load carried by the wall consists of sum of the following :

(1) Self wt. of wall
(2) Load due to roofing
(3) Load due to floors
(4) Self wt. of foundations

(1) *Self weight of wall per metre :*

$= L \times B \times H \times$ wt. per m³ of masonry

$= 1 \times 0{\cdot}3 \times 13{\cdot}5 \times 19200 = 77760$ N

(2) *Load due to roofing per metre :* Load per sq. m due to roofing consists of the following :

(*i*) Self wt. of 150 mm thick slab

$= 0{\cdot}150 \times 25000 = 3750$ N/m²

(*ii*) wt. of 100 mm thick lime concrete terracing

$= 100 \times 19200 = 1920$ N/m²

(*iii*) Live load $= 1500$ N/m²

Total 7170 N/m²

As the maximum span between walls is 4 m,

load per metre on each wall

$$= 7170 \times \frac{4}{2} = 14340 \text{ N}$$

(3) *Load due to floors :* As the building is three storied, the live load on the floors will be reduced as specified on page no. 132. Hence the calculations for dead load and live load are made separately.

(*a*) Dead load per sq. m due to floor slab consists of the following

(*i*) Self wt. of 120 mm thick slab

$$= \cdot 120 \times 25000 = 3000 \text{ N/m}^2$$

(*ii*) wt. of 25 mm terrazzo flooring

$$= 0{\cdot}025 \times 24000 = 600 \text{ N/m}^2$$

$$\text{Total} \quad 3600 \text{ N/m}^2$$

Dead load due to each floor on wall per metre

$$= 3600 \times \frac{4}{2} = 7200 \text{ N/m}$$

Since there are two floors, total dead load per metre

$$= 7200 \times 2 = 14400 \text{ N/m}$$

(*b*) Live load per metre : As on page no. 132 reduction in live load due to each floor shall be under :

(*i*) Live load per metre due to first floor

$$= 2000 \times \frac{4}{2} \times \frac{90}{100} = 3600 \text{ N/m}$$

(*ii*) Live load per metre due to 2nd floor

$$= 2000 \frac{4}{2} \times \frac{80}{100} = 3200 \text{ N/m}$$

Total live load per metre length of wall

$$3600 + 3200 = 6800 \text{ N/m}$$

Hence total dead + live load due to floors

$$= 14400 + 6800 = 21200 \text{ N/m}$$

Total load (dead + live) per metre length of wall

$$= W = (1) + (2) + (3)$$

or $$W = 77760 + 14340 + 21200 = 113300 \text{ N/m}$$

Design of footing :

Total load on footing :

$$= W = 113300 \text{ N/m}$$

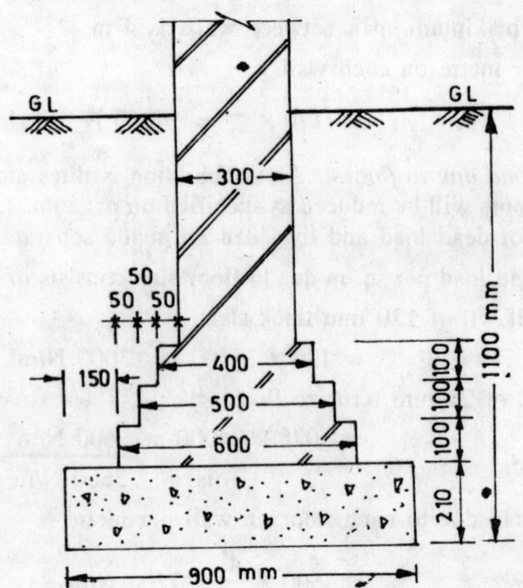

Fig. 4·2

Self wt. of foundation :

(Assumed @ 10% of *W*)

$$= 113300 \times \frac{10}{100} = 11330 \text{ N/m}$$

∴ Total design load per metre

$$= 113300 + 11330 = 124630 \text{ N/m}$$

$$= 124{\cdot}63 \text{ kN/m}$$

∴ Width of foundation

$$= B = \frac{124{\cdot}63}{p_o} = \frac{124{\cdot}63}{160} = 0{\cdot}78 \text{ m}$$

$$= 780 \text{ mm}$$

Width of foundation from practical consideration : Let the projection of concrete bed block from the lower most course of footing be $= x = 110$ mm.

$$\therefore \quad B = 2 \times b + 2x$$

$$= 2 \times 300 + 2 \times 150 = 900 \text{ mm}$$

Hence adopt width of wall footing = 900 mm

Depth of the concrete bed block :

Let the mix of concrete for the bed block be 1 :4 :8 for which the value of m_r (modulus of rupture) from Table 4·7 = 0·246 N/mm²

$$\text{Depth of conc. block} = \sqrt{\frac{3.p_o.x^2}{m_r}}$$

$$= \sqrt{\frac{3 \times 160 \times 1000 \times \cdot 150 \times \cdot 150}{\cdot 246}}$$

$$= 209{\cdot}5 \text{ mm say } 210 \text{ mm.}$$

4·8. DESIGN OF GRILLAGE FOUNDATION

Fig. 4·3 shows a simple arrangement of one tier grillage for a steel stanchion. The formulae for the bending movement and shear force required for the design of grillage beams can be arrived at by the simple method described below.

Let W = load supported by one beam in Newton (N)

L = length of the beam in metre (m)

l = length of the base plate in metre (m)

The load distribution is shown in Fig. 4·3. It is evident that maximum bending moment in the beam

$$= \frac{W}{2} \times \frac{L}{4} - \frac{W}{2} \times \frac{l}{4}$$

$$= \frac{W}{8}(L - l) \text{ Nm}$$

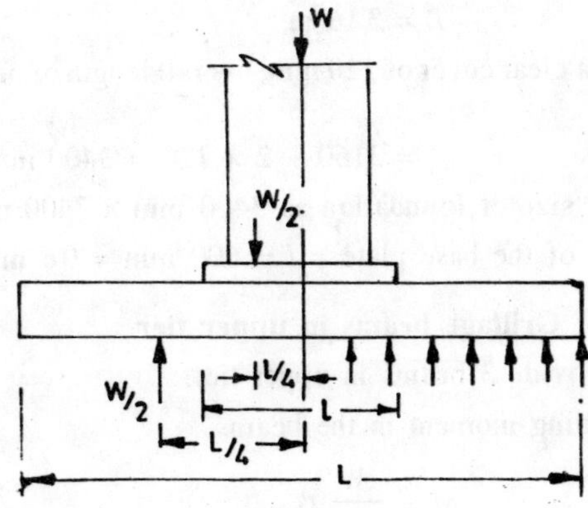

Fig. 4·3

Maximum shear in beam occurs at the edge of the base plate.

Upward soil pressure on the beam

$$= \frac{W}{L} \text{ N/m}$$

Cantilever projection of the beam beyond the edge of the base plate

$$= \left(\frac{L-l}{2}\right)$$

∴ Max S.F. in the beam

$$= \frac{W}{L}\left(\frac{L-l}{2}\right)$$

Example 4·2. *Design a suitable grillage foundation for a steel stanchion carrying load of 1000 kN. Adopt safe bearing capacity of soil = 110 kN/sqm. The size of base plate of the stanchion is 600 mm x 600 mm.*

Sol. Load on stanchion (W) = 1000 kN

Self wt. of foundation and overlying earth fill

(assumed @10% of W) $= 1000 \times \frac{10}{100} = 100$ kN

Total load = 1100 kN

∴ Area of foundation

$$= \frac{1100}{p_o} = \frac{1100}{110} = 10 \text{ m}^2$$

Let the foundation be square in plan

∴ One side of square base $= \sqrt{10} = 3{\cdot}16$ m

or $L = 3{\cdot}16$ m

Keeping a clear cover of 120 mm, overall length of one side of square base

$$= 3160 + 2 \times 120 = 3400 \text{ mm}$$

∴ Adopt size of foundation as 3400 mm x 3400 mm

One side of the base plate = l = 600 mm = 0·6 m

Design of Grillage beams in upper tier :

Let us provide 3 beams in upper tier.

Max bending moment in the beams

$$M = \frac{W}{8}(L-l)$$

$$= \frac{1000}{8}(3{\cdot}160 - {\cdot}6) = 320 \text{ kNm}$$

B.M per beam $= \frac{M}{3}$

$$= \frac{320}{3} = 106{\cdot}67 \text{ kNm}$$

$$= 106{\cdot}67 \times 10^6 \text{ N mm}$$

Permissible bending stress

$$f = 165 \text{ N/mm}^2 + 33\% \text{ of } 165 \text{ N/mm}^2$$

$$= 219{\cdot}45 \text{ N/mm}^2$$

Since $M = f\,z$

$$\therefore \quad z = \frac{M}{f}$$

$$= \frac{106{\cdot}67 \times 10^6}{219{\cdot}45} = 48606 \text{ mm}^2$$

Provide ISLB 300 having following properties

(*i*) Overall depth = d = 300 mm

(*ii*) Flange width = b = 150 mm

(*iii*) $t_f = 9{\cdot}4$ mm

(*iv*) $t_w = 6{\cdot}7$ mm

(*v*) $z_{xx} = 488{\cdot}9 \times 10^3 \text{ mm}^3$

(*vi*) Wt. per meter = 377 N

Check for shear :

$$\text{S.F.} = \frac{W}{L}\left(\frac{L-l}{2}\right)$$

$$= \frac{1000}{3{\cdot}16}\left(\frac{3{\cdot}16 - 0{\cdot}6}{2}\right) = 405 \text{ kN}$$

$$\text{S.F. per beam} = \frac{405}{3} = 135 \text{ kN}$$

$$\therefore \quad \text{Shear stress} = \frac{135 \times 1000}{300 \times 6{\cdot}7} = 67{\cdot}2 \text{ N/mm}^2$$

which is less than permissible value of shear stress

$= 94{\cdot}5 \text{ N/mm}^2$, hence safe.

Check for vertical buckling :

Actual web height $= h = 300 - (2 \times 9{\cdot}4) = 281{\cdot}2$ mm

$$t_w = 6{\cdot}7 \text{ mm}$$

$$\text{Slenderness ratio} = \frac{h}{t_w}\sqrt{3}$$

$$= \frac{281{\cdot}2}{6{\cdot}7} \times \sqrt{3} = 7{\cdot}3$$

Permissible compressive stress corresponding to the above slenderness ratio = 109·54 N/mm^2

Length on which load is assumed to be spread

$$= B = \text{plate length} + \frac{1}{2}\text{beam depth}$$

$$= 600 + \frac{1}{2} \times 300 = 750 \text{ mm}$$

Corresponding bearing area

$$= B \times t_w = 750 \times 6{\cdot}7 = 5025 \text{ mm}^2$$

Total bearing area provided by the webs of 3 beams

$$= 3 \times 5025 = 15075 \text{ mm}^2$$

$$\therefore \quad \textit{Bearing stress} = \frac{1000 \times 1000}{15075}$$

$$= 66{\cdot}34 \text{ N/mm}^2 < \text{permissible value, hence safe}$$

Clear space between the flanges

$$= \frac{(600 - 3 \times 150)}{2} = 75 \text{ mm} - \text{O.K.}$$

Design of beams in the lower tier :

$$\text{Max. B.M.} = 320 \text{ kNm}$$

Let us provide 10 beams in the lower tier

$$\therefore \quad \text{B.M./beams} = \frac{320}{10} = 32 \text{ kNm}$$

$$z = \frac{M}{f} = \frac{32 \times 10^6}{219{\cdot}45} = 145819 \text{ mm}^3$$

Provide ISLB 200 having following properties

(*i*) Overall depth = d = 200 mm

(*ii*) Flange width = b = 100 mm

(*iii*) t_f = 7·3 mm

(*iv*) t_w = 5·4 mm

(*v*) z_x = 169·7 x 1000 mm^3

(*vi*) Wt. per meter = 377 N

Check for shear :

Shear force per joist

$$= \frac{405}{10} = 40{\cdot}5 \text{ kN}$$

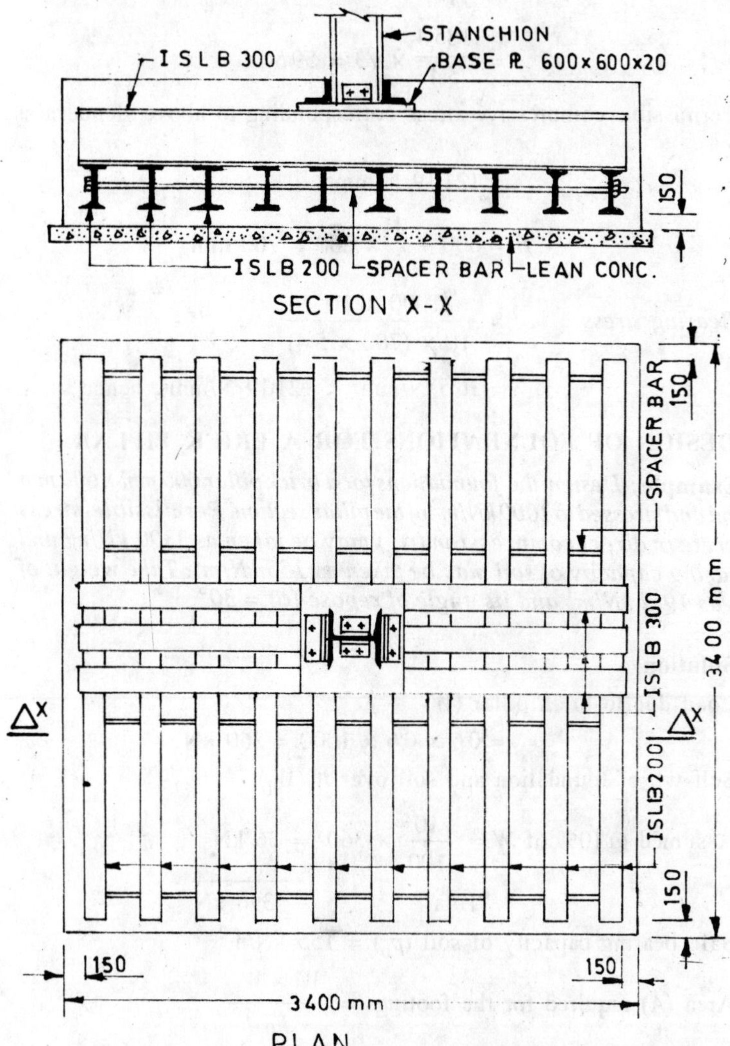

Fig. 4·4

$$\text{Shear stress} = \frac{40{\cdot}5 \times 10^3}{200 \times 5{\cdot}4} = 37{\cdot}5 \text{ N/mm}^2, \text{ safe}$$

Check for vertical buckling :

Actual web height = h = 200 − (2 × 7·3) = 185·4 mm

$$\therefore \text{ Slenderness ratio} = \frac{h}{t_w}\sqrt{3}$$

$$= \frac{185 \cdot 4}{5 \cdot 4} \times \sqrt{3} = 59 \cdot 5$$

Permissible compressive stress corresponding to above slenderness ratio

$$= 121 \cdot 19 \text{ N/mm}^2$$

$$B = 600 + \frac{1}{2} \times 200 = 700 \text{ mm}$$

Bearing stress $= \dfrac{1000 \times 10^3}{10 \times (700 \times 5 \cdot 4)}$

$= 26 \cdot 5 \text{ N/mm}^2 < 121 \cdot 19 \text{ N/mm}^2$, hence safe.

4·3. DESIGN OF FOUNDATIONS FOR A BRICK PILLAR

Example : *Design the foundations for a brick pillar 600 mm x 600 mm in section and stressed to 1000 kN/m² of the pillar section. Permissible stress in concrete in direct compression (σ_{sc}) may be taken as 3300 kN/m² and safe bearing capacity of soil may be taken as 155 kN/m². Take weight of soil (γ) as 19·2 kN/m³ and its angle of repose (ϕ) = 30°*

Solution

Load due to brick pillar (W)

$$= 0 \cdot 6 \times 0 \cdot 6 \times 1000 = 360 \text{ kN}$$

Self wt. of foundation and soil over it (W_1)

Assumed @10% of $W = \dfrac{10}{100} \times 360 = 36$ kN

Total 396 kN

Safe bearing capacity of soil (p_0) = 155 kN/m²

Area (A) required for the footings $= \dfrac{W + W_1}{p_0}$

$$= \frac{396}{155} = 2 \cdot 55 \text{ sq.m}$$

Let B be one side of the square footing

$\therefore$ $B \times B = A = 2 \cdot 55$

or $B = \sqrt{A} = \sqrt{2 \cdot 55} = 1 \cdot 598 \text{ m} \approx 1 \cdot 60 \text{ m}$

ˈnce footing of brick pillar rests on concrete block, minimum area ›nry required at the base of pillar

$$= \frac{W}{\sigma_{sc}}$$

or $$\frac{360}{3300} = 0{\cdot}109 \text{ m}^2$$

One side of the bottom most course of pillar masonry

$$= \sqrt{0{\cdot}109} = 0{\cdot}33 \text{ m}$$

which is less than the size of the pillar.

From considerations of economy the width of the pillar at its bottom most course should be

$$= 2 \times 600 = 1200 \text{ mm} = 1{\cdot}2 \text{ m}$$

∴ Projection of concrete block on either side of the lower most course of pillar masonry

$$= x = \frac{1}{2}(1{\cdot}6 - 1{\cdot}2) = 0{\cdot}2 \text{ m}$$

Depth of concrete block :

Let the mix of concrete for the bed block be 1 : 3 : 6 for which the value of modulus of rupture as per Table 4·7.

$$m_r = 0{\cdot}352 \text{ N/mm}^2 = 350 \text{ kN/mm}^2$$

The depth of concrete block is given by

$$d = \sqrt{\frac{3px^2}{m_r}}$$

$$= \sqrt{\frac{3 \times 155 \times 0{\cdot}2 \times 0{\cdot}2}{350}}$$

$$= 0{\cdot}230 \text{ m}$$

$$= 230 \text{ mm}$$

Depth of foundation : The depth of foundation as given by Rankine's formation is

$$D_f = \frac{p_0}{\gamma}\left(\frac{1 - \sin\phi}{1 + \sin\phi}\right)^2$$

$$= \frac{155}{192}\left(\frac{1 - \sin 30°}{1 + \sin 30°}\right)^2$$

$$= \frac{155}{192} \times \frac{1}{9}$$

$$= 0{\cdot}897 \text{ m}$$

However provide a depth of foundation as 1·1 m so that there is sufficient depth of soil available above the top-most course of footing

and the chances of the footing top getting exposed due to soil erosion are minimised.

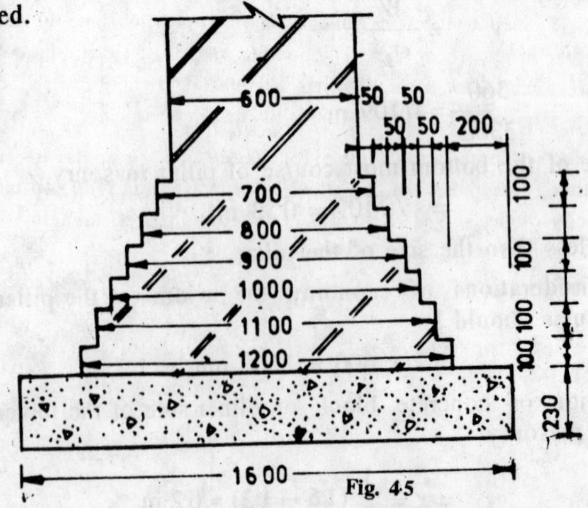

Fig. 45

Questions for Review

1. A room 3 m. by 3·66 m. internally has 30 cm. thick (1½ modular brick) wall all round. Load transferred to foundation per running metre is 10 tonne and the bearing capacity of soil at 90 cm. depth is 14 tonne/sq. m. Plinth is to be kept 60 cm. above the ground level. Design a suitable section for the footing and wall upto plinth level and estimate the quantities involved, Weight of earth is 1760 kg/m^3 and its angle of repose is 30°.

2. A shed is to be constructed 12.9 m long and 7·4 m wide internally having 30 cm. thick walls on the shorter sides. Three pillars 30 cm. x 30 cm. are provided on each longer side with clear spacing of 3 m. Design the foundations with the following data :

   (*i*) Angle of repose : 30°

   (*ii*) Wt. of earth = 1760 kg/m^3.

   (*iii*) Safe bearing capacity of soil = 11 tonne/sq. m

   (*iv*) Plinth 45 cm. above ground

   (*v*) Load on wall at plinth level = 10 tonne/running meter.

   (*vi*) Load on each pillar at plinth level = 10 tonne.

3. A steel column carrying an axial load of 255 tonne is to be founded on a soil having its safe bearing capacity of 16 tonne/sq. m. The column base plate is to be 60 cm. square. Design a suitable steel grillage foundation and make a detailed sketch of the same.

4. A weight of 27 kg. is dropped vertically on the bottom of foundation trench from a height of 3 metres and makes an impression of 7 mm. The base of the weight is 18 cm x 18 cm in size. Calculate the safe bearing pressure per sq. metre assuming a factor of safety of 8.

5. Calculate the width and depth of a plain cement concrete foundation supporting a wall carrying a load of 33 tonne/meter run at the foundation level. The safe allowable pressure on the soil is to be limited to 11 tonne/sq. m. The soil weighs 1760 kg/m^3 and has an angle of repose of 30°.

6. **Design the foundation of a four-storeyed office building to be constructed on a soil having its safe bearing capacity of 22 tonne/m^2. The soil weighs 1920 kg/m^3 and has an angle of repose of 30°. The main walls of the building upto first two floors are 40 cm. thick and for the rest of the height 30 cm thick. The flooring consists of 13 cm. thick R.C.C. slab finished with 25 mm. thick mosaic flooring. The maximum span between the two main walls is 3·7 m. The roof slab is finished with 10 cm. thick terracing.**

7. **Determine the size of footings and depth of foundation for a brick pillar 40 cm. square. The concentrated load carried by the pier at the footing top is 15 tonne. The bearing capacity of the soil at the depth of foundation is 16 tonne/m^2. The soil weighs 1920 kg/m^3 and has an angle of repose of 30°.**

5

Damp-Proofing and Water-Proofing

5·1. GENERAL

One of the essential requirements of a building is that it should be dry. Dampness in building may occur due to bad design, faulty construction and use of poor quality of materials. Dampness not only affects the life of the building adversely, but also creates unhygenic conditions for the occupants. Hence, treatment to check dampness is considered to be one of the important items of work in the construction of a building. The treatment given to prevent leakage of water from roof is generally termed as *water-proofing*, whereas the treatment given to keep the walls, floors and basement dry is termed as *damp-proofing*.

5·2. DEFECTS CAUSED BY DAMPNESS

The various defects caused by dampness in buildings may be summarised as under :

1. It causes efflorescence which may ultimately result in disintegration of bricks, stones, tiles etc.
2. It may result in softening and crumbling of plaster.
3. It may cause bleaching and flaking of paint with the formation of coloured patches.
4. It may result in the warping, buckling and rotting of timber.
5. It may lead to the corrosion of metals.
6. It may deteriorate electrical fittings.
7. It promotes growth of termites.
8. It breeds mosquitoes and creates unhealthy living conditions for the occupants.

5·3. CAUSES OF DAMPNESS

Absorption of moisture by the building materials is one of the chief cause: of dampness. On account of granular nature of materials, moisture finds an easy access through the voids and this aided by capillary action assists the moisture to travel in different directions. Thus, either on account of faulty design of structure or bad workmanship or by use of defective materials, moisture may find its way to interior of the building either through the foundations and plinth, walls, chhajjas, floors or the roof.

5·4. SOURCES OF DAMPNESS

The important sources of dampness may be summarised as below :

1. Dampness rising through the foundation walling. Moisture from wet ground may rise well above ground level on account of capillary action.
2. Splashing rain water which rebounds after hitting the wall surface may also cause dampness.
3. Penetration of rain-water through unprotected tops of walls, parapets, compound walls etc. may cause dampness.
4. In case of sloped roofs rain-water may percolate through defective roof covering. In addition, faulty eaves course and eaves or valley gutters may allow the rain-water to descend through the top of supporting wall and cause dampness.
5. On condensation, the moisture present in the atmosphere gets deposited on the components of the building which may cause dampness.
6. In case of flat roofs, inadequate roof slopes, improper rain-water pipe connections and defective junction between roof slab and parapet wall may prove to be the sourcesof dampness.
7. Dampness caused due to leakages in wet areas of a building which include kitchen, bathroom, water closet etc. The leakages could be on account of use of sub-standard plumbing fittings and fixtures as well as faulty execution of work and bad workmanship.

5·5. METHODS OF DAMP-PROOFING

Following methods are generally adopted to prevent the defect of dampness in a structure.

1. *Membrane damp-proofing*
2. *Integral damp-proofing*
3. *Surface treatment*

4. *Guniting*
5. *Cavity wall construction*

1. Membrane Damp-proofing

This consists in providing layer or membrane of water repellent material between the source of dampness and the part of the structure adjacent to it. This type of layer is commonly known as damp proof course (or D.P.C.) and it may comprise of materials like bitumenous felts, mastic asphalt, silicon, epoxy, polymers,plastic or polythene sheets, cement concrete etc. Depending upon the source of dampness, D.P.C. may be provided horizontally or vertically in floors, walls etc. Provision of D.P.C in basement is normally termed as'*tanking*'.

General principles to be observed while laying D.P.C. are as under.

(*i*) The D.P.C. should cover full thickness of walls excluding rendering.

(*ii*) The mortar bed upon which the D.P.C. is to be laid should be made levelled, even and free from projections. Uneven base is likely to cause damage to D.P.C.

(*iii*) When a horizontal D.P.C. is to be continued to a vertical face, a cement concrete fillet 75 mm in radius should be provided at the junction, prior to the treatment. (Ref. Fig.5.5).

(*iv*) Each D.P.C. should be placed in correct relation to other D.P.C. so as to ensure a complete and continuous barrier to the passage of water from floors, walls or roof.

2. Integral Damp-proofing

This consists in adding certain water-proofing compounds with the concrete mix to increase its impermeability. Such compounds are available in market in powdered as well as liquid form. The compounds made from clay, sand or lime (chalk, talc, fuller's earth etc.) help to fill the voids in concrete and make it water proof.

Another form of compound like alkaline silicates, aluminium sulphates, calcium chlorides etc. react chemically when mixed in concrete to produce water proof concrete.

Pudlo, Permo, Impermo, etc. are some of the many commercially made preparations of water-proofing compounds commonly used. The quantity of water proofing compound to be added to cement depends upon the manufacturers recommendations. In general, one kg. of water proofing compound is added with one bag of cement to render the mortar or concrete water proof.

3. Surface Treatment

As described earlier the moisture finds its way through the pores of materials used in finishing. In order to check the entry of the moisture into the pores, they must be filled up. Surface treatment consists in filling up the pores of the surfaces subjected to dampness. The use of water repellent metallic soaps such as calcium and aluminium oleates and stearates is much effective in protecting the building against the ravages of heavy rain. Bituminous solution, cement coating, transparent coatings, paints and varnishes fall under this category. In addition to other surface treatments given to walls, the one commonly used is lime cement plaster. The walls plastered with cement, lime and sand mixed in proportions of 1 : 1 : 6 is found to serve the purpose of preventing dampness in wall due to rain effectively.

4. Guniting

This consists in depositing an impervious layer of rich cement mortar over the surface to be water-proofed. The operation is carried out by use of a machine known as cementgun. The assembly broadly consists of a machine having arrangements for mixing materials and a compressor for forcing the mixture under pressure through a 50 mm dia flexible hose pipe. The hose pipe has nozzle at its free end to which water is supplied under pressure through a separate connection.

The surface to be treated is first thoroughly cleaned of dirt, dust, grease or loose particles and wetted properly. Cement and sand (or fine aggregate) usually taken in proportion of 1 : 3 to 1 : 4 are then fed into the machine. This mixture is finally shot on the prepared surface under a pressure of 2 to 3 kg/cm^2 by holding the nozzle of the cement gun at a distance of 75 to 90 cm from the working face. The quantity of water in the mix can be controlled by means of regulating value provided in the water supply hose attachment. Since the material is applied under pressure it ensures dense compaction and better adhesion of the rich cement mortar and hence the treated surface becomes water proof.

5. Cavity wall Construction

This consists in shielding the main wall of the building by an outer skin wall leaving a cavity in between the two. The cavity prevents the moisture from travelling from the outer to the inner wall. The cavity wall has been described in detail in Chapter 6.

5·6. MATERIALS USED FOR DAMP-PROOFING

An effective damp proofing material should have the following properties:

(*i*) It should be impervious.

(*ii*) It should be strong and durable and should be capable of withstanding both dead as well as live loads without damage.

(*iii*) It should be dimensionally stable.

(*iv*) It should be free from deliquescent salts like sulphates, chlorides and nitrates.

The materials commonly used to check dampness can be divided into the following four categories :

(1) *Flexible material.* Material like bitumen felts (which may be hessian-based or fibre/glassfibre-based), plastic sheeting (Polythene sheet) etc.

(2) *Semi-rigid materials.* Materials like mastic asphalt or combination of materials or layers.

(3) *Rigid materials.* Materials like first class bricks, stones, slates, cement concrete etc.

(4) *Grout* consisting of cement slurry and acrylic based chemicals/polymers.

5·7. SELECTION OF MATERIAL FOR D.P.C.

The choice of material to function as an effective damp-proof course requires a judicious selection. It depends upon the climatic and atmospheric conditions, nature of structure and the situation where D.P.C. is to be provided. The points to be kept in view while making selection of D.P.C. materials are briefly discussed below.

(*i*) *D.P.C. above ground level.* For D.P.C. above ground level, with wall thickness generally not exceeding 40 cm, any one of the type of material mentioned above may be used. Cement concrete is however, commonly adopted material for D.P.C. at plinth level. A 25 to 50 mm thick layer of cement concrete M15 (1: 2: 4 mix) serves the purpose under normal conditions. In case of damp and humid atmosphere richer mix of concrete should be used. The concrete is further made dense by adding water-proofing materials like Pudlo, Impermo, Waterlock etc. in its ingredients during the process of mixing. It is usual to apply two coats of hot bitumen over the dried surface of concrete D.P.C.

(*ii*) *D.P.C. material for floors, roofs etc. :* For greater wall thickness or where D.P.C. is to be laid over large areas such as doors, roofs etc. the choice is limited to flexible materials which provide lesser number of joints like mastic asphalt, bitumen felts, plastic sheets etc. The felts, when used should be properly bonded to the surface with bitumen and laid with joints properly lapped and sealed.

(*iii*) *D.P.C. material for situations where differential thermal movements occur.* In parapet walls and other such situations materials like mastic asphalt, bitumen felts and metal (copper or lead) are recommended. It is important to ensure that the D.P.C. material is flexible so as to avoid any damage or puncture of the material due to differential thermal movements between the material of the roof and the parapet wall.

(*iv*) *D.P.C. material for cavity walls.* In cavity wall construction the cavity over the door or window should be bridged by the flexible material like bitumen felt, strips of lead etc.

(*v*) In case of expansion and construction joints, in R.C.C. slab and retaining walls in basements it is neecessary to provide water bar made out of P.V.C or G.I. sheet to seal the joint against passage of sub-soil water into the building.

5·8. DAMP-PROOFING TREATMENT IN BUILDINGS

Damp-proofing treatment in buildings can be broadly divided into the following categories :

(*i*) Treatment to foundations
(*ii*) Treatment to floors
(*iii*) Treatment to walls
(*iv*) Treatment to flat roof
(*v*) Treatment to parapet wall
(*vi*) Treatment to pitched roof.

5·9. DAMP-PROOFING TREATMENT TO FOUNDATIONS

Before dealing with the problem of damp-proofing treatment to be given to foundations, it is necessary to have an idea about the ground water level.

When it rains, the rain water seeps through the ground until it is stopped by an impervious layer in the sub-soil strata.

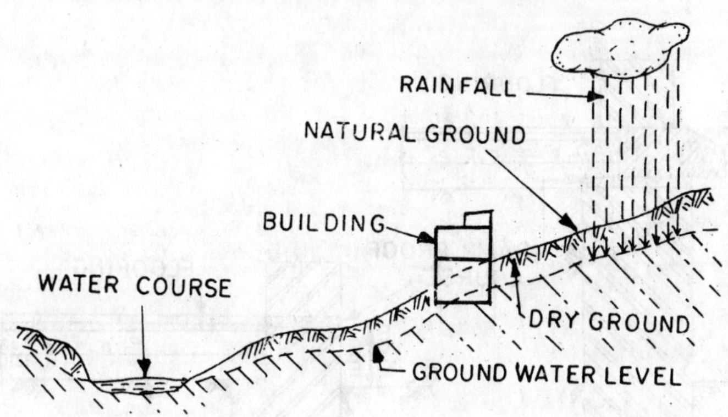

Fig. 5·1

Figure showing relationship between the ground water level and the general contour of the ground

The water in the pores of soil is known as ground water. The top level of the ground water is known as *sub soil water level or* ground water level. The ground water level does not remain constant. It varies with season, magnitude of rainfall and the nature of sub-soil strata. The ground water generally follows the contours of the ground and flows from higher level to lower level underneath the ground. Therefore it appears at greater depth in high ground (ridge) and at shallower depths in low ground (valley). In addition, irregular distribution of soft and hard strata, in the composition of the ground may also cause the ground water level to rise or drop appreciably within small areas.

It is observed that the soil above the water table contains certain amount of moisture in their voids. Near the ground level the soil voids are only partly filled with water and they gradually get fully saturated with water by the time the depth of ground water level is reached. The strata of ground above the water table attracts water by capillary action. The height of capillary rise depends on the size of voids in the soil. It is noticed that in case of fine grained soils like clay, silty clay etc., ground water can rise more than 6 m by capillary action. In case of granular soils like gravel, coarse sand etc. the capillary rise of water is almost negligible.

Depending upon the depth of the ground level, the treatment to be given to the foundation can be subdivided into the following four categories.

(1) Treatment to foundations on ordinary soil

(2) Treatment to foundations on damp soil

(3) Treatment to basements in ordinary soil

(4) Treatment to basement in damp soil.

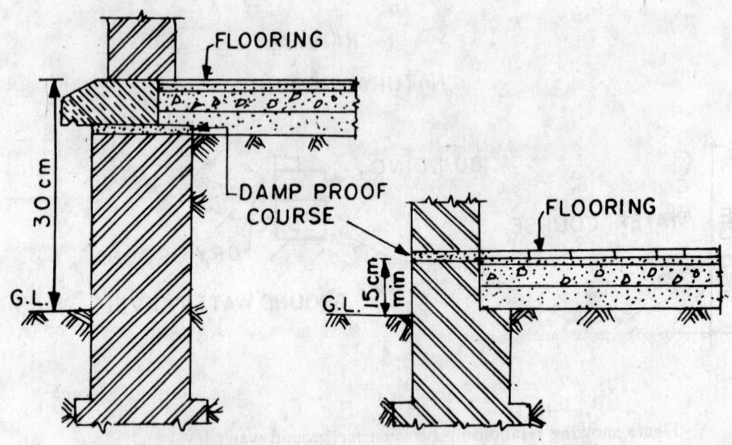

Fig. 5·2 Fig. 5·3

Alternative methods of providing D.P.C. for buildings on ordinary soil.

1. Treatment to foundations on ordinary soil

Building foundations on ordinary soil where the sub-soil water table is not high are also liable to get damp. Bricks being porous, brick masonry below ground level can absorb moisture from adjacent ground. This moisture travels up from one course to another by capillary action and can make the wall damp for a considerable height. This can be checked by providing D.P.C. at appropriate place.

In case of buildings without basement, the best portion for damp-proof course lies at plinth level. In case of structure without plinth, D.P.C. should be provided at least 150 mm above ground level. If the damp-proof course is laid just at the ground level, earth, dust or leaves might accumulate outside the wall and by the passage of time the level of outside earth may be raised above the D.P.C. level. In such a case, moisture can travel from outside ground level to brickwork above D.P.C. and hence the purpose of providing D.P.C. will not be served.

2. Treatment to Foundations on Damp Soil

In case of building constructed on damp soil in wet areas, both the walls as well as the ground floor are liable to become damp due to capillary rise of moisture from ground. In such cases, the D.P.C. is laid over the entire area of ground floor including wall thickness. Bitumen felts can be used for damp-proofing treatment. The sequence of operations for laying D.P.C. can be divided in the following steps :

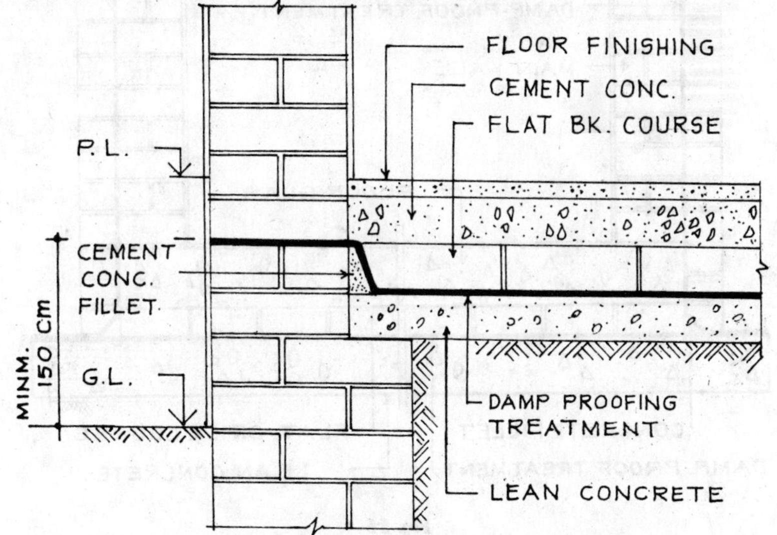

Fig. 5·4 Figure showing provision of D.P.C. for buildings in damp soil

(*i*) Apply hot bitumen at the rate of 1·5 kg/m² over the prepared surface to serve as primer coat.

(ii) Lay bitumen felt in single layer over the primer coat.

(iii) Apply hot bitumen at the rate of 1·5 kg/m² over the bitumen felt to serve as finishing coat.

Immediately after laying, the D.P.C. is protected with a course of bricks laid flat on a cushion of fine sand. This prevents damage to D.P.C. specification on account of droppage of sharp edged implement or other materials during construction.

3. Treatment to basements in ordinary soil

In sites where subsoil water table is low or, where the hydrostatic pressure is not much, the treatment consists in providing a horizontal D.P.C. over the entire area of basement floor and then extending it in the form of vertical D.P.C. on the external faces of the basement walls. The D.P.C. material thus functions like a water-proof tank on the external faces of the basement and keeps it dry.

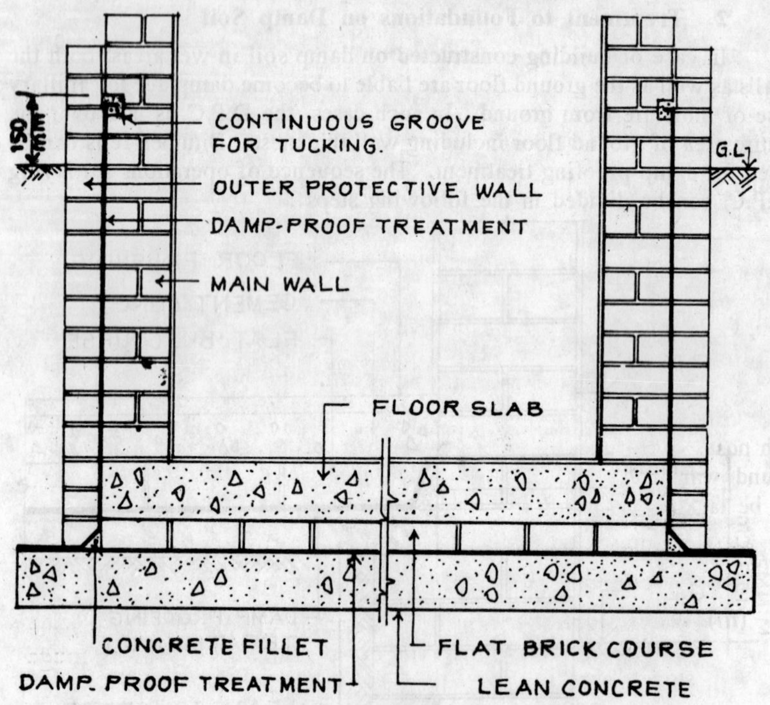

Fig. 5·5

Provision of D.P.C. for basement in ground where the sub-soil water table is not very high.

It is common to use bitumen felts in multiple layers for damp-proofing treatment to basements. For normal duty treatment or in places where moisture ingress is not considered excessive, two layers of bitumen felts are used. In case of heavy duty treatments or in places where heavy moisture ingress is encountered, three layers of bitumen felts are used. The sequence of operations for laying D.P.C. in basement for normal duty treatment can be divided in the following steps.

(*i*) Apply hot bitumen at the rate of 1·5 kg/m² over the prepared surface to serve as primer coat.

(*ii*) Lay bitumen felt in single layer over the primer coat.

(*iii*) Apply hot bitumen at the rate of 1·5 kg/m² over the bitumen felt.

(*iv*) Lay another layer of bitumen felt in single layer over the hot bitumen layer in step (*iii*) above.

(*v*) Apply hot bitumen at the rate of 1·5 kg/m² over the bitumen felt laid in step (*iv*) above.

The horizontal D.P.C. is laid on the smoothened top of the lean concrete bed. The lean concrete bed should be thick or strong enough to withstand the construction traffic. As explained earlier, immediately after laying, the D.P.C. is protected with a course of bricks laid flat on a cushion of fine sand to prevent damage to D.P.C. specification on account of droppage of sharp edged implement or other material during construction.

The vertical D.P.C. is laid continuous with the horizontal one on the external face of the basement wall and it is continued 150 mm above the ground level where it is tucked into a 65 mm deep groove made in the wall. The groove is subsequently filled with cement mortar 1:4. The vertical D.P.C., unless protected, is likely to get punctured by roots of trees or get damaged by the salts/acids in the soil. Necessary protection in this regard is given by constructing half brick outer skin wall as shown in Fig. 5·5.

4. Treatment to Basements in Damp Soil

Ground water always produces hydrostatic pressure and as such poses great problem in the design of basement. In sites where the ground water table is high, the problem of damp-proofing of basement can be tackled by one of the following three methods.

(*i*) By providing foundation drains and D.P.C.

(*ii*) By providing R.C.C. floor and wall slab and D.P.C.

(*iii*) Water proofing treatment by using grout consisting of cement mortar admixed with acrylic based chemicals alongwith rough stone slabs.

(*i*) **By providing foundation drains and D.P.C.** The foundation drains help in lowering the ground water level of the site and avoid building up of hydrostatic pressure against floor or walls of the basement. This in turn helps in preventing leakage and in keeping the basement dry.

In this method, the area of the basement is disconnected from the adjacent ground by making a 70 to 100 cm wide trench at basement

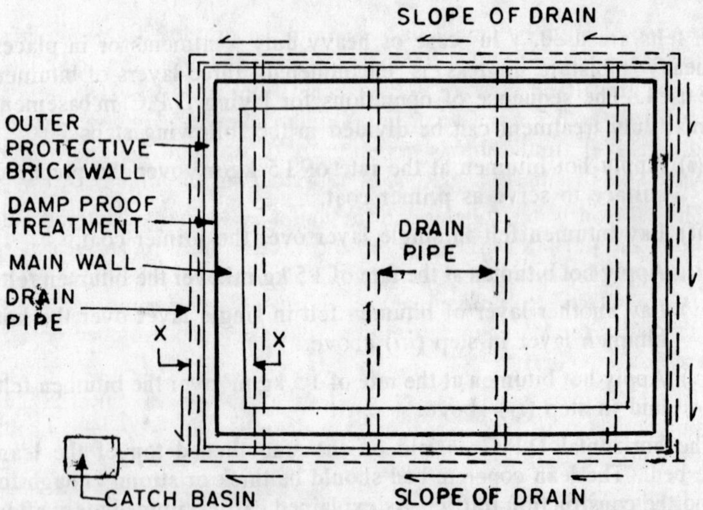

Fig. 5·6. Plan showing layout of Foundation drains and catch basin.

level along the periphery of the basement. Open jointed drainage pipes are then placed in the trench for drainage of sub-soil water, the bed of the trench being given adequate slope to ensure easy flow of the water.

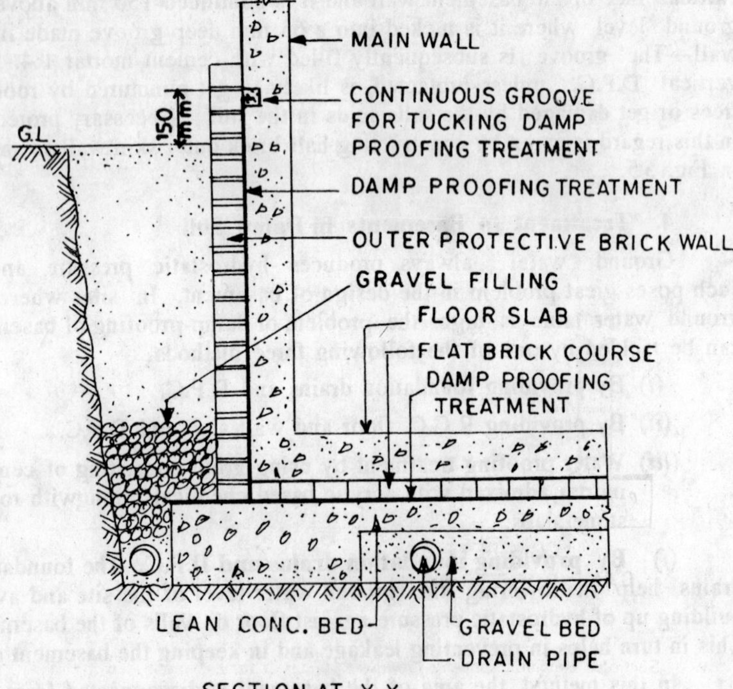

Fig. 5·7 Provision of D.P.C. in damp soils using foundation drains

In case of large basement, it is desirable to provide drainage pipes below the concrete floor of the basement as well. The pipes should be laid on a bed of gravel and covered with selected filter material. The trench is also filled upto ground level in a similar manner. The open joints and bed of gravel around drains permit entry of water in the drains, but prevent the soil or silt from getting in and clogging the drain.

The sub-soil water collected in the drains is discharged into an out fall point. The out fall point may be storm water drain or an area at a level lower than the footings of building. In case it is not possible to have an out fall point lower than the footing, the subsoil water from the drains is collected in a catch basin or sump and thereafter, it is pumped out to the nearest surface drain.

(*ii*) **By providing R.C.C. floor and wall slab and D.P.C.** In some situations where the underground water pressure is severe, mere provision of drainage pipes or pumping out the sub-soil water may not solve the problem. In such cases both the basement walls as well as the floor slab have to be designed and suitably reinforced to substain the hydrostatic pressure safely. The construction of such basements can be treated to be almost similar to the construction of under ground water tanks provided with D.P.C. treatment. The other details remain more or less same as applicable to treatment of basements in ordinary soil, except that at least three layers of bitumen felts are used for D.P.C.

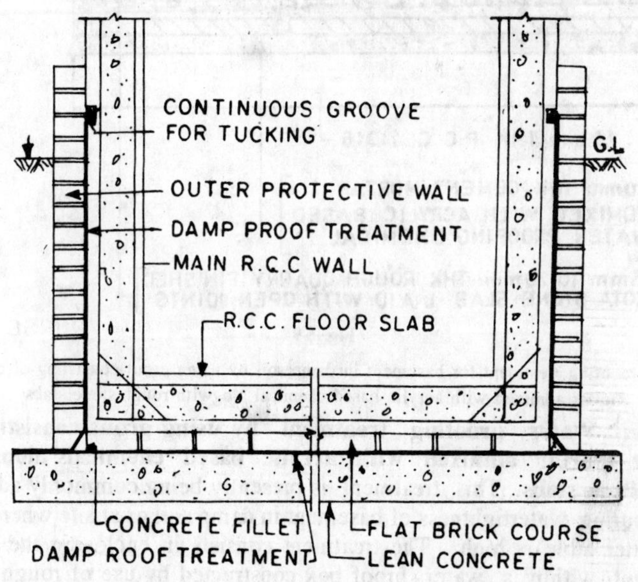

Fig. 58

Sectional elevation showing provision of D.P.C. for basement in damp soil

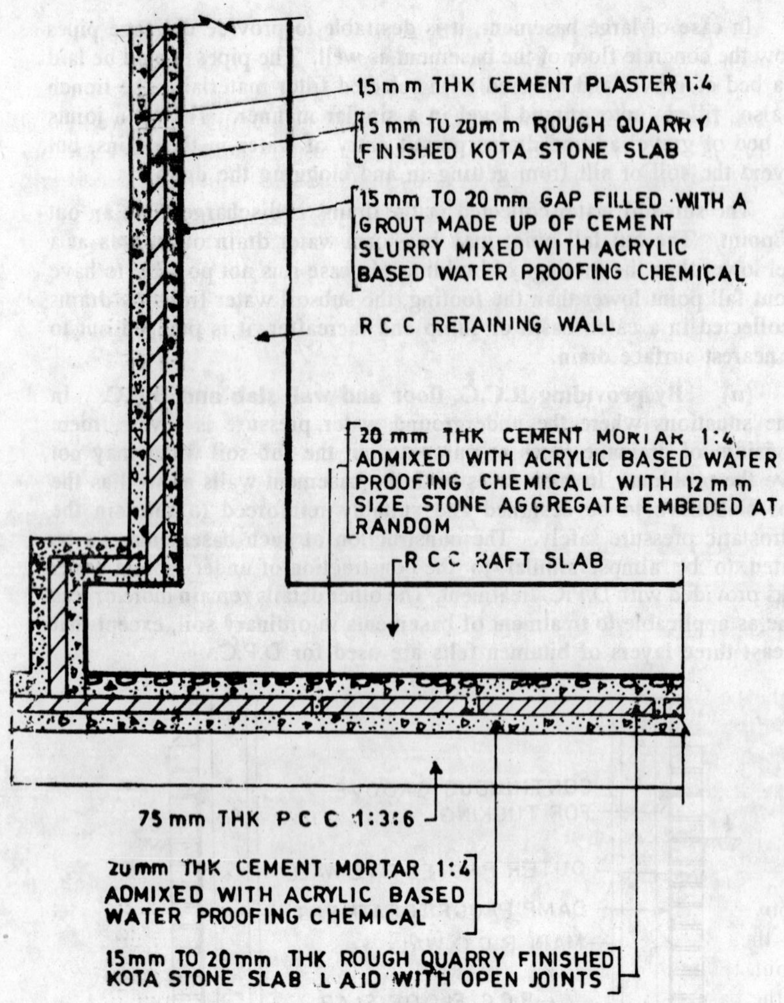

Fig. 5·9

Water proofing treatment for basement in damp soil by using grout consisting of cement mortar admixed with acrylic based chemical alogwith rough stone slabs

(*iii*) **Water proofing treatment by using grout consisting of cement mortar admixed with acrylic based chemicals alongwith rough stone slabs.** This treatment is presently being commonly adopted for ensuring watertightness of basement in damp soil or at site where sub-soil water table is high. The treatment consists in enclosing the entire basement within a water-proof box constructed by use of rough stone slabs and cement mortar admixed with acrylic based chemical. The various steps involved in applying this treatment are described on the next page.

Step 1. The excavation of the ground for the basement is carried out in such a manner that a working space of at least 750 mm is available around the periphery of the external walls of the proposed basement. The excavated area is kept dry by continuous pumping using well-point equipment or by adopting any other suitable method. Over the dry levelled ground a 75 to 100 mm thick layer of lean concrete 1:3:6 is laid to serve as mud mat or levelling course for the raft slab. This layer of lean concrete should project by 150 mm beyond the external vertical surface of the proposed raft slab or the external wall. Over the layer of lean concrete a layer of 15 to 20 mm thick rough quarry-finished stone slabs is laid over 20 mm thick cement mortar (1:4) admixed with acrylic based chemical. The stone slabs are laid side by side leaving a gap of about 15 to 20 mm between them. The joints thus left are raked open and grouted with cement slurry admixed with acrylic based chemical. A 20 mm thick layer of cement mortar (1:4) admixed with acrylic based compound is laid over the stone slabs and stone chips 12 mm & down are embedded at random in this layer of mortar.

The total maximum thickness of the above water proofing layer should be around 60 mm. This layer should cover the entire area of lean concrete *i.e.* it should be extended 150 mm beyond the external face of raft slab/walls.

Step 2. The R.C.C. raft slab suitably designed to withstand the highest applicable water pressure during monsoon is laid over the water proofing layer in step (1). The R.C.C. walls of the basement are constructed monolithically with the R.C.C. raft. After the curing of the R.C.C. wall is over, the water proofing treatment is continued unbroken on the external sides of the walls, so as to form an external box of the water proofing layer.

Step 3. To continue the treatment along the vertical plane, a groove about 30 to 40 mm deep is made in top mortar layer or bottom layer of the water proofing treatment [laid in step (1)] at a clear distance of about 18 mm from the external face of the wall all along the periphery of the basement. Rough quarry-finished stone slabs 15 mm to 20 mm in thickness are fixed vertically in the above groove. The stone slabs are fixed side by side without leaving any gap between the edges. Maximum of two to three horizontal layers of stone are laid at a time one over the other vertically up. A coat of very rough cement plaster (1:4) is applied to the external face of the rough stones. After the layer is set, the gap between the wall and stone layer is filled with a grout made up of cement slurry admixed with acrylic based chemical, which on gellation forms an impermeable layer, monolithic with the external face of the wall.

A final coat of 15 mm cement plaster (1:4) is applied over the rough external cement plaster and finished smoothly. This treatment to external face of walls is continued upto 300 mm above ground level.

The acrylic based chemical is normally added @ 1% *i.e.* 1 kg of chemical to be mixed with 100 kg of cement.

5·10. TREATMENT TO FLOORS

In places where the soil water table is low and rainfall is not much, a 75 to 100 mm thick layer of coarse sand is first spread over the entire area of flooring on the prepared bed of rammed earth. Alternatively, this layer can comprise of stone soling with voids filled with smaller stones. This layer is known as base course and its material is well rammed. A 75 to 100 mm thick layer of lean cement concrete (1 : 3 : 6 or 1 : 4 : 8) mix or lime concrete is thereafter laid over the base course. This forms the base for the floor topping, which may comprise of tiles, stone or cement concrete etc.

In places where the sub soil water table is high, or in damp or humid areas, where there is possibility of moisture rising up in the floors, it is necessary to provide membrane D.P.C. of flexible material like bitumen felt etc. over the entire area of flooring. For details refer Fig. 5·4.

5·11. TREATMENT TO WALLS

As explained earlier wall can get damp due to penetration of moisture from its external face to internal one, due to porosity of bricks and mortar joints. Various treatments given to exposed surface of walls to prevent dampness include pointing, plastering, painting etc. It is observed that plaster made out of cement, lime and sand mixed in proportion of 1:1:6 serves as very effective rendering to protect the wall against dampness in normal weather conditions. In areas of heavy rainfall, cement plaster 1:4 (1 cement : 4 sand) mixed with water proofing compounds like Pudlo, Permo etc. serves the purpose effectively. In exposed brick work, dampness can be prevented by painting the surface with water proof cement paint or with colourless liquid waterproofing compound.

5·12. TREATMENT TO FLAT ROOFS

Flat roofs require relatively heavier and costlier water-proofing treatment as compared with pitched or sloped roofs. The specification of material used for the purpose should be such that it performs the function of water-proofing as well as provides adequate thermal insulation. Stagnation of water on the roof is considered to be the root cause of leakage and dampness in flat roofs. This can be avoided by providing adequate roof slope and rain water pipes. In case of R.C.C. or R.B.C. slab roofing with proper grading above, a slope of 1 in 40 to 1 in 60 is considered desirable. This may be achieved either by varying the thickness of the terracing material or by constructing the roof slab with a slope, or by providing part slope in the roof slab and part in the terracing material.

In addition to the slope, the size and the spacing of the rain water pipes or the outlets require due consideration for the proper drainage of the roof. In general practice, one 10 cm diameter pipe is considered suitable for every 30 sq.m. of the roof area to be drained.

In cases, where slopes for the drainage of the roof are given in the roof slab itself or in situations where thermal insulation is not important and the problem of slopes in flat roof is tackled suitably, the waterproofing treatment for the roof may consist in laying bitumen felt directly over the surface of roof slab after painting the roof top with hot bitumen. The bitumen felt may be hessian based or fibre based. Depending upon the type of building, climate and atmospheric conditions of the site, the treatment with felts may be with four courses, six courses or eight courses. The four course treatment is recommended for moderate conditions, whereas the six and eight course treatments are recommended for severe and very severe conditions respectively. The (1) four course treatment and (2) six course treatment are briefly given below:

(1) *Four Course Treatment.* The method of laying a four course treatment may be broadly divided into the following steps.

(*a*) Apply hot bitumen at the rate of 1·2 kg/m^2 on the roof surface.

(*b*) Lay bitumen felt in single layer over the hot bonding material laid in (*a*), the end and side laps for the felt being not less than 100 and 75 mm. respectively.

(*c*) Apply hot bitumen at the rate of 1·2 kg/m^2 over bitumen felt laid in (*b*)

(*d*) Spread pea-sized gravel or grit at the rate of 0.008 m^3 per square metre over the layer of hot bitumen in (*c*).

(2) *Six course treatment.* In case of six course treatment, the first three courses are laid in a similar manner as explained in steps (*a*),(*b*) and (*c*) above. The fourth courses consists of laying another layer of bitumen felt over the course of hot bitumen explained in step (*c*) above.

The fifth course consists of applying hot bitumen at the rate of 1·2 kg/m^2 over fourth course.

The sixth course consists of spreading pea-sizes gravel or grit at the rate of 0·008 m^3 per square metre over the layer of hot bitumen laid in fifth course.

The other commonly adopted water-proofing treatment for flat roofs in the various regions of this country consists in providing a grading (terracing) of selected materials over the roof slab. In general, one of the following methods of grading is adopted to meet the requirements of water-proofing:

(1) *Grading of lime concrete*

(2) *Grading of lime concrete with tiles.*

(3) *Grading of mud phuska with tiles.*

(4) *Grading of brick coba laid with grout consisting of cement mortar admixed with acrylic based chemical.*

1. **Grading of lime concrete.** This method of water proofing roof is adopted in hot and humid regions in India. The method consists in laying lime concrete grading (terracing) over the roof slab. The process of laying can be divided into the following steps:

(*i*) Laying lime concrete of 10 cm. average thickness directly over the roof slab. The slope for proper drainage of roof being given to the lime concrete layer.

(*ii*) Applying a priming coat of hot bitumen solution over the dried layer of lime concrete.

(*iii*) Laying bitumen sheets over the primed surface.

In some places, the laying of bitumen sheets over the grading is not considered necessary and only the lime concrete is finished to an impervious surface. This practice is common in some parts of Bombay and Calcutta.

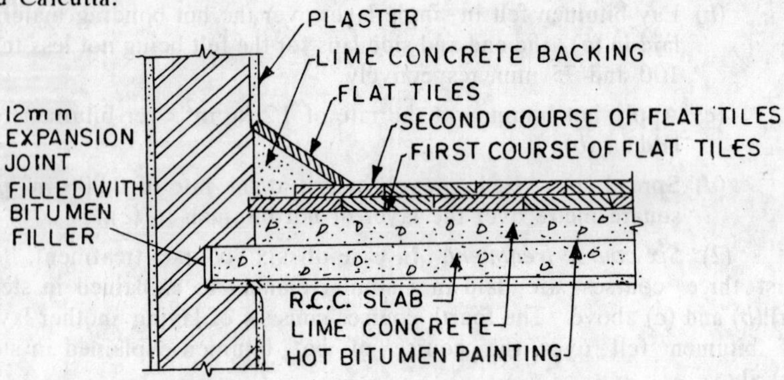

Fig 5·10

Water-proofing treatment of flat roof using Lime concrete with tiles.

2. **Grading of lime concrete with tiles.** In case the roof is likely to be used for sleeping or other such purpose, the grading provided must be strong enough to withstand the wear and tear due to the traffic of the users. The lime concrete layer in such places is covered with two courses of brick tiles set in cement mortar. The following operations are generally involved in laying such grading:

(*i*) Painting the top of roof slab uniformly with a layer of hot bitumen spread at the rate of 1·70 kg. of bitumen per sq. m. of roof surface.

(*ii*) Spreading immediately, coarse sand at the rate of 0.6 cu.m. of sand per 100 sq.m. of roof surface when the bitumen is still hot.

(*iii*) Laying lime concrete in an average thickness of 10cm. The slope for proper drainage of roof being given to the lime concrete layer.

(*iv*) Laying two courses of flat brick tiles (with breaking joints) over the compacted later of lime concrete in cement mortar 1:3. Thickness of each course of tile varies from 13 to 20 mm. The joints of the top course of tiles being pointed with cement mortar 1:3. Instead of flat tiles, precast concrete tiles or 25 mm thick Shahabad stone can also be used depending upon availability and cost.

Water-proofing of flat roof, by adopting this method, is very common in the southern region of this country. This type of treatment is best suited for building in hot and humid regions like Calcutta, Bombay and Madras. For details refer Fig. 5·10.

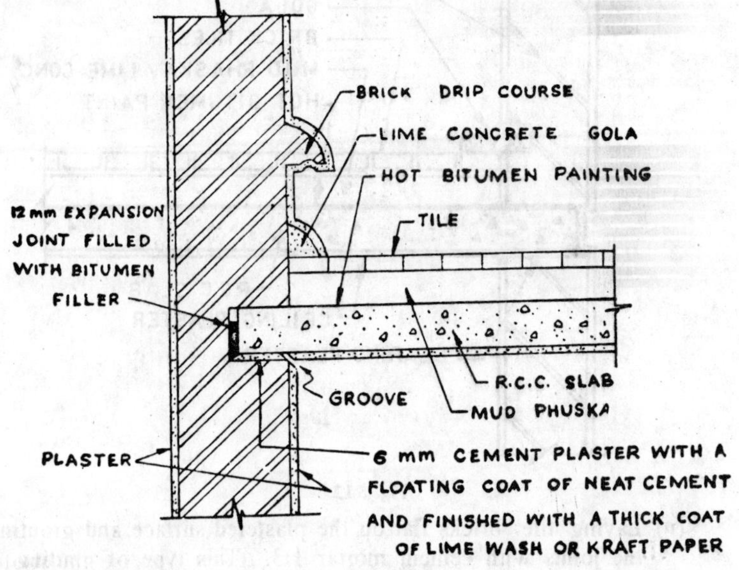

Fig 5·11

Water proofing treatment of flat roof using mud phuska with tiles.

3. **Grading of mud phuska with tiles.** This method of water proofing flat roofs is best suited for hot and arid regions. It is commonly adopted in Delhi, Punjab, Rajasthan and U.P. and in other regions where the rainfall does not exceed 1000 mm. The various operations involved in this type of grading may be divided in the following steps:

(*i*) Painting the top of roof slab uniformly with a layer of hot bitumen spread at the rate of 1·70 kg. of bitumen per sq. m. of roof surface.

(*ii*) Spreading immediately coarse sand at the rate of 0.6 cu. m. of sand per 100 sq. m. of roof surface, when the bitumen is still hot.

(*iii*) Laying 10 cm. thick (average) mud phuska prepared from puddled clay mixed with bhusa at the rate of 8·1 kg. of bhusa per cu. m. of clay, the slope for the proper drainage of roof being given to this layer. The slope should not be less than 1 in 40. If, however, it is not possible to give the entire slope by the mud phuska layer, a part of the slope required for the roof drainage may be given in the roof slab itself.

(*iv*) Plastering the consolidated layer of mud phuska with a 13 mm. thick coat of mud gobri mortar 3:1 (3 mud : 1 cowdung).

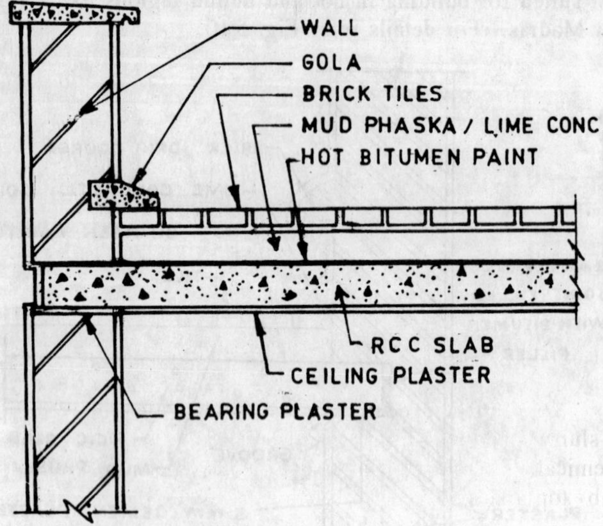

Fig. 5·12

(*v*) Laying tile bricks flat on the plastered surface and grouting the joints with cement mortar 1:3. This type of grading is illustrated in Fig. 5·11 & 5·12.

4. **Grading of brick coba laid with grout consisting of cement mortar admixed with acrylic based chemical.** An effective modern water-proofing treatment to flat roofs which is commonly adopted in buildings in regions of heavy to moderate rainfall consists in using grout consisting of cement mortar admixed with acrylic based chemical. The method of applying this treatment is as under:

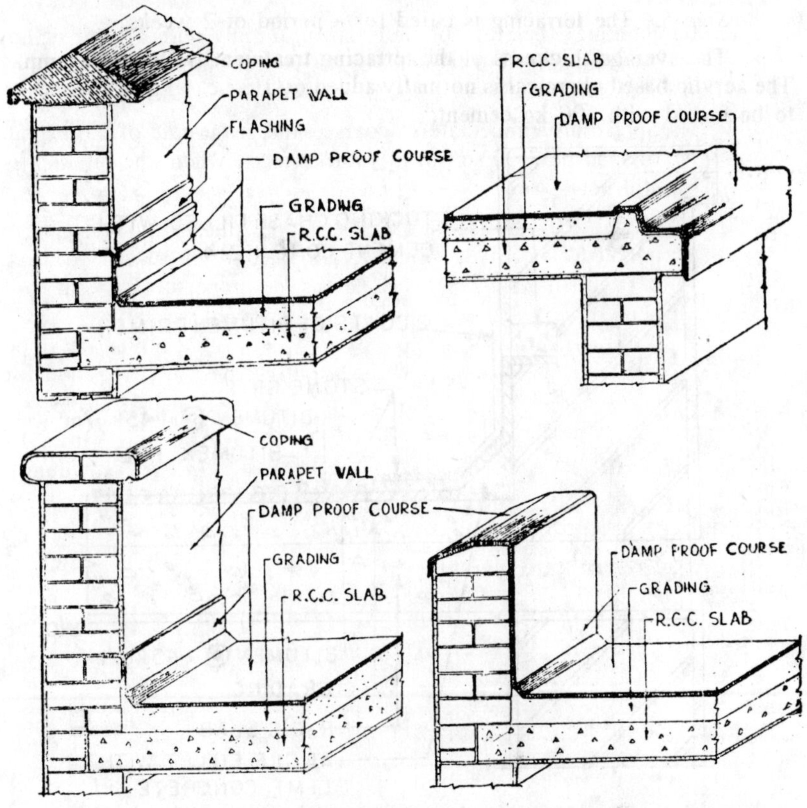

Fig 5·13–5·16

Figures showing water-proofing treatment of flat roofs with or without parapet walls.

Step 1. Clean the surface of the roof slab and lay a 15 mm thick coat of slurry made up of neat cement mortar (1:4) mixed with acrylic based chemical. The slurry fills up all the crevices and fine cracks in the roof slab top and also acts as binding layer for the treatment in next step.

Step 2. Provide a layer of broken brick bats in cement mortar (1:4) admixed with acrylic based chemical. This layer is laid in slope for proper drainage of rain water.

Step 3. The layer of terracing in step (2) is cured for about 2 days and again a thick coat of cement mortar of same composition as in step (1) is applied to fill up the joints or voids.

Step 4. A 20 mm thick layer of cement mortar (1:4) admixed with acrylic based chemical is applied on top and finished smooth. The top surface of this layer is marked with 300 mm x 300 mm false squares. The joint between the slab and wall is rounded off.

Step 5. The terracing is cured for a period of 2 weeks.

The average thickness of the terracing treatment is about 100 mm. The acrylic based chemical is normally added @ 1% *i.e.* 1 kg of chemical to be mixed with 100 kg cement.

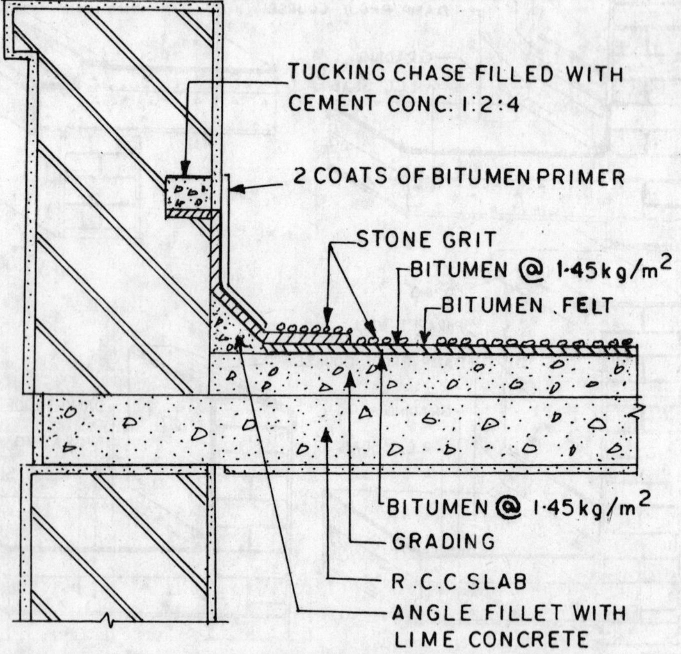

Fig. 5·17

Alternative detail showing treatment to parapet

5·13. TREATMENT TO PARAPET WALL

If the flat roof has a parapet and there are cracks in it or its plaster is very porous or defective, rain water may find and easy access to the wall below and make the wall and some portion of the ceiling damp. Rain water may also leak through cracks at the junction of the parapet and the roof slab. In cases where asphalt layer is provided over the grading material for the water-proofing treatment to roof slab, the asphalt layer covering the roof is turned up against the parapet for a height of at least 15 cm. The parapet wall is further protected by providing coping of brick, concrete or stone on its top (Refer Fig. 5·13 to 5·18).

In case, the waterproofing of roof is carried out by providing lime terracing, the layer of terracing is taken right upto the edge of the roof and a 15 cm high rounded triangulated fillet is built up there. This is necessarily done to prevent the soakage of water through the parapet in

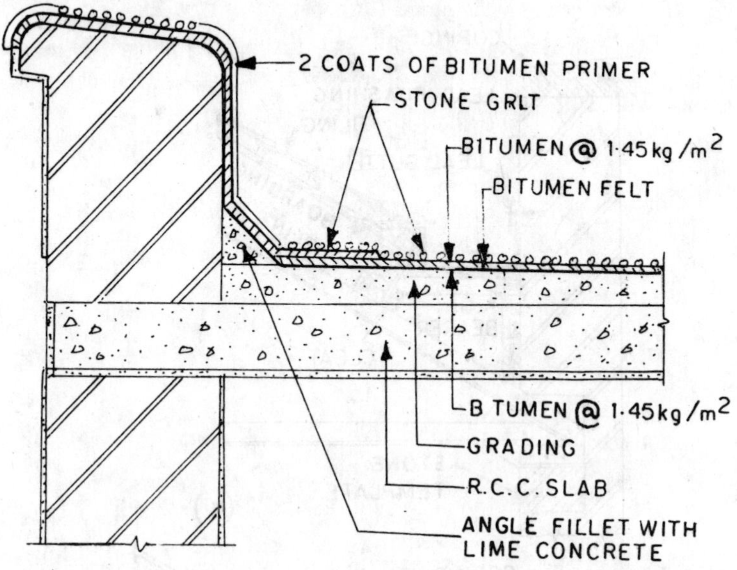

Fig. 5·18

Figure showing alternative method of water-proofing roof and parapet

case the rain water pipes get choked up and the water stands on the roof. In addition, a drip course is also some times provided over the lime gola to prevent water from parapet to fall directly over its junction with roof (Refer Fig. 5·11).

5.14. TREATMENT FOR THE ROOFS OF MULTISTOREYED BUILDINGS

In case of multi-storey buildings, besides water-proofing, thermal insulation is an equally important factor to be kept in view. This is achieved by increasing the thickness of the grading by using economical and effective materials. The waterproofing treatment generally given to the roof of multi-storey buildings and other important structures may be divided in the following steps:

1. Painting the top of roof slab uniformly with a layer of hot bitumen spread at the rate of 1·70 kg. of bitumen per sq. m. of roof surface.

2. Spreading immediately coarse sand at the rate of 0·6 cu.m. of sand per 100 sq.m. of roof surface when the bitumen is still hot.

3. Laying cinder concrete 1:15. (1 cement : 15 cinder of 13 mm and down gauge) in an average thickness of 15 cm, the slope for the proper drainage of roof being given to this layer.

4. Laying 7·6 cm. thick layer of lime concrete over the consolidated

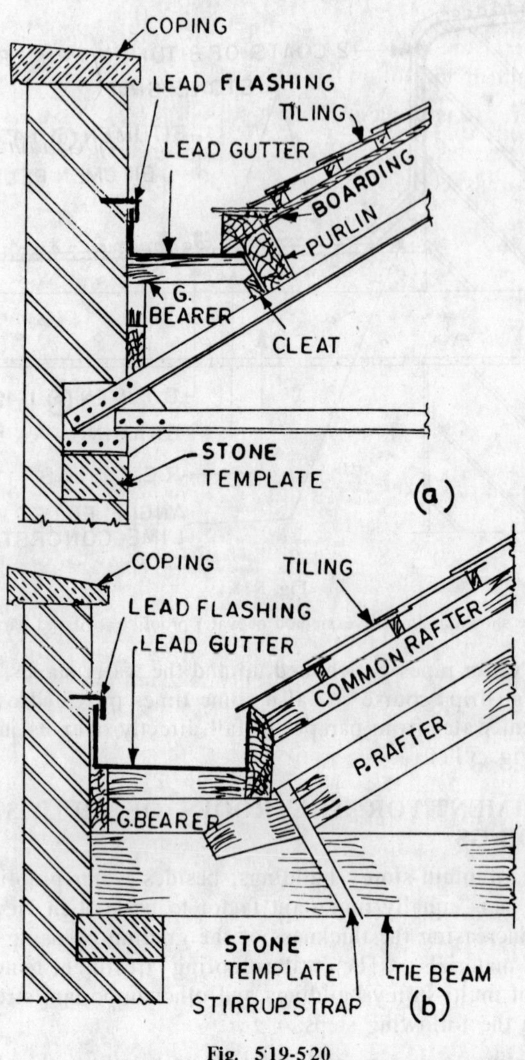

Fig. 5·19-5·20

Damp-proof treatment for sloped roof of steel and timber

layer of cinder concrete, the lime concrete being prepared by mixing 50% of mortar consisting of lime and sand mixed in the ratio of 1:2 with brick ballast 25 mm. and down gauge.

5. Spreading 13 mm. thick layer of cement mortar 1:3 and laying tile bricks flat and open jointed over the mortar. Finally grouting the joints in the brickswith cement mortar 1:3.

5·15. TREATMENT TO PITCHED ROOFS

In general, rainwater is liable to leak through a pitched or sloped roof on account of the following reasons:

(*i*) Due to insufficient roof slope.

(*ii*) Due to insufficient lap in roof covering material *i.e.*, A.C. or C.G.I. sheets, tiles, slates etc.

(*iii*) Due to inadequate treatment to rainwater gutter.

All the above referred causes can be taken care of by proper design and construction. However, design and installation of rainwater gutter or valley gutter needs extra special attention with respect to its capacity, position, fixing, water tightness and freedom to accommodate expansion or contraction in any direction. The slope of the gutter should not be flatter than 1:100 in straight lengths and it should be made steeper in portions where the gutter is not straight. The gutter should be leak-proof and all the joints in the gutter should be made tight. The work of laying roofing should be carefully supervised and it should be ascertained that the sheets or tiles project beyond the edge of the gutter. Lead flashing should be continued upto the vertical face of the parapet wall and should stop inside the body of the wall. As in case of flat roof, the parapet wall should itself be protected by stone or brick coping. The provision of damp-proof course layer in lead in a valley gutter is shown in figs 5·19-5·20.

If the architect wants to provide a projection in the face wall to decorate the structure, a sloped junction should be stressed. The junction should further be protected by a layer of damp-proof course, as shown in Fig. 5·21.

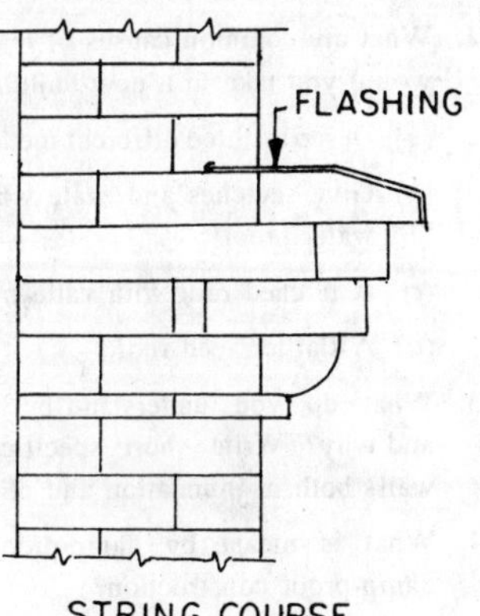

STRING COURSE

Fig. 5·21

5·16. SPECIFICATIONS FOR LAYING 38 MM. THICK DAMP-PROOF COURSE WITH CEMENT CONCRETE 1 : 2 : 4 AT PLINTH LEVEL.

1. The damp-proof course shall cover the full thickness of the wall.

2. The base of the damp-proof course shall be clear, even and free from projections liable to cause damage to the D.P.C.

3. The side shuttering shall be strong and so fixed that it does not get disturbed during compaction and the concrete slurry does not leak out.

4. The concrete prepared by mixing ingredients in the proportion of 1:2:4 (1 cement : 2 sand : 4 stone ballast 12 mm. and down gauge) shall be of workable consistency.

5. The concrete shall be laid and tempered roughly to make a dense mass.

6. After 24 hours of its laying, the concrete layer shall be cured for at least seven days.

7. After curing is complete, the surface shall be left to dry out to receive the coat of hot bitumen.

8. The dried surface of concrete shall be properly cleaned with brushes and finally with a piece of cloth soaked in kerosine oil. Hot bitumen in specified quantity shall be applied uniformly all over the treated surface of concrete.

Questions for Review

1. What are common causes of leakage and dampness in flat roofs and what precautions would you take in a new building to avoid possibility of leakage?
2. (*a*) Describe three different methods you know for water-proofing 'Roofs and Terraces.'

   (*b*) Give sketches and state what precautions you would take to make the following water-proof :

   (*i*) A pitched roof with valleys and gutters.

   (*ii*) A flat-terraced roof.
3. What do you understand by Damp-proof course? Where is it placed in a building and why? Write short specification on damp-proof course laying to brick masonry walls both in foundation and plinth and in super-structure.
4. What is meant by damp-proof construction and what are the materials used for the damp-proof construction?

   Describe, with sketches, how you would make a cellar floor of building which has a floor 2·4 m below ground level, damp-proof. The sub-soil water level is 90 cm below ground level, *i.e.*. 1.5 m above the cellar floor.
5. Discuss the problem of drainage of foundations of residential and public buildings.

   How will you ensure water-proofing and damp-proofing of such buildings constructed in localities where water table rises quite high during the monsoon season ?
6. What are the visible signs of the action of dampness in a building and what precautions are necessary to avoid the same.
7. An underground R.C.C. drinking water reservoir is proposed to be constructed at a site having high sub-soil water level. What specification would you suggest for water-proofing treatment ? Explain in brief the specification of materials and also methods of execution of the entire work.
8. What is dampness ? What are its ill effects ? Explain with sketches how damp-proofing is done in building.

9. (*a*) Indicate the type of waterproofing treatment you would recommend for flat R.C.C. roof slabs in :

   (*i*) Hot and dry regions like Rajasthan.

   (*ii*) Hot and humid regions like U.P. and Andhra Pradesh

   (*iii*) Hot and very humid regions like West Coast and Assam.

   (*b*) What steps will you follow for laying the water-proofing treatment with bitumen felt for a large building in a coastal region with annual rainfall exceeding 250 cm.

0. (*a*) What is damp-proof course ? Where and now is it used ?

   (*b*) Explain how you will provide damp-proof treatment to a basement floor 2·40 m below plinth level with R.C. walls and R.C. floor slab over 7·6 cm lean concrete. Height of plinth is 60 cm above the ground level.

11. The floor of a basement is 3 m below the ground level. The permanent water table is 2 m above the basement floor level. Draw a neat sketch showing the damp-proofing treatment for the external wall of the basement.

12. Describe in details the construction of basement in a building to be used as a store. How will you ensure water-proofing and proper ventilation? Draw a cross-section showing all the details.

6

Cavity Walls

6·1. GENERAL

Cavity wall consists of two walls with a 5 cm. to 8 cm. cavity between them. The outer wall also known as outer leaf, consists of a 10 cm. (half brick) thick wall and the inner wall is sufficiently thick and strong to carry the imposed load safely The minimum thickness of the inner wall is restricted to 10 cm. (half brick). The provision of a continuous cavity in the wall efficiently prevents the transmission of dampness to the inner wall.

6·2. ADVANTAGES

The advantages of cavity walls may be summariseed as below:

(*i*) As there is no intimate contact between the two leaves except at the wall ties (which are of impervious material), there is no possibility of the moisture travelling from the outer leaf to the inner.

(*ii*) The layer of air in the cavity being non-conductor of heat reduces the transmission of heat from the external face to the internal one and as such cavity walls are best suitable for a tropical country like India. Tests have revealed that cavity walls have 25% greater insulating value than soild walls.

(*iii*) They have good sound insulation property.

(*iv*) They are economical.

6·3. POSITION OF CAVITY

The cavity should start near ground level and terminate near eaves level in case of sloping roof or near coping in case of flat roof with parapet wall. The cavity should preferably start 15 cm. below the damp-proof course level. This has the advantage of draining any condensed moisture below the level of damp-proof course. Similarly, to prevent infiltration of moisture, the bottom of cavity should lie at least

15 cm. above the out side ground level. The damp-proof course for the two leaves is laid separately, although at the same level. This is necessary to continue the cavity below damp-proof course. The cavity is kept fully ventilated by providing air bricks in the external wall immediately above damp-proof course. The air bricks should be kept 90 cm. apart. Similarly, air bricks are provided near the top of the cavity.

6·4. WALL TIES

The two walls are tied together with metallic or terra-cotta ties to obtain structural stability. They are spaced 90 cm. apart horizontally and 30 to 45 cm. apart vertically in staggered positions.

6·5. CONSTRUCTIONAL DETAILS

The constructional treatment given to the cavity wall at the door or window openings requires special considerations.

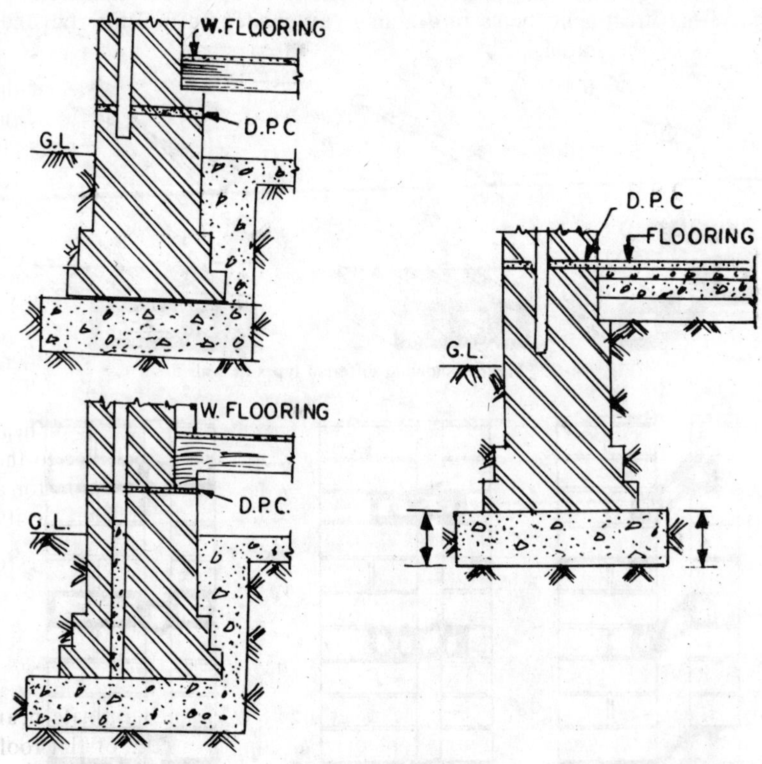

Fig 6·1–6·3

Figures showing the various positions of cavity and the relative positions of damp-proof course layer.

Since the object of cavity is to separate the two leaves, a vertical damp-proof coarse is essential to be provided when sealing the cavity at window and door reveals, as the openings can provide an easy access to the dampness to pass into the cavity. At the sill level it is always advantageous to keep the back of the sill flush with the back face of the external wall. In case of wooden sills a horizontal damp-proof course is introduced under the sill. The damp-proof course used should be of flexible material.

At the top of openings of doors and windows, stone or brick arch or concrete lintel is provided to carry the loads from structure above the opening. In order that the cavity may be continuous, the two walls should have separate lintels or arches. Sometimes a common lintel is provided for both the leaves as shown in Fig. 6·13. In such a case, an inclined strip of lead or any other flexible damp-proof course membrane is provided so as to bridge over the cavity. The damp-proof course should extend lengthwise beyond the frame for a distance of about 15 cm on either side. This treatment helps in draining any condensed moisture beyond the sides of the frame.

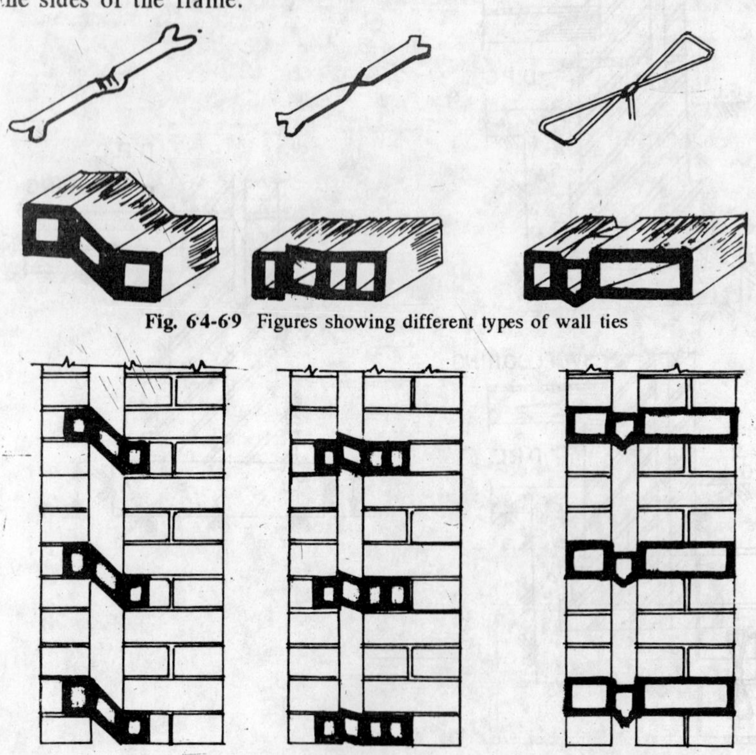

Fig. 6·4-6·9 Figures showing different types of wall ties

Fig 6·10-6·12

Sectional elevations of brick cavity walls showing different types of wall ties

In case of flat roof, the cavity is continuous for the full height of parapet wall and is sealed by the coping. In case of inclined roofs, cavity is closed at eaves level. It is important to take special precautions so that no mortar should fall inside the cavity during the process of construction. For this purpose a wooden batten of adequate length and having width equal to that of the cavity is used. The batten is supported over the wall ties and the construction of brick work is carried out (on either side of the batten) to the height where next row of wall ties have to be provided. There after the wooden batten is lifted up, cleaned of mortar droppings, and replaced over the next row of wall ties. By this method, the cavity remains clean of mortar droppings.

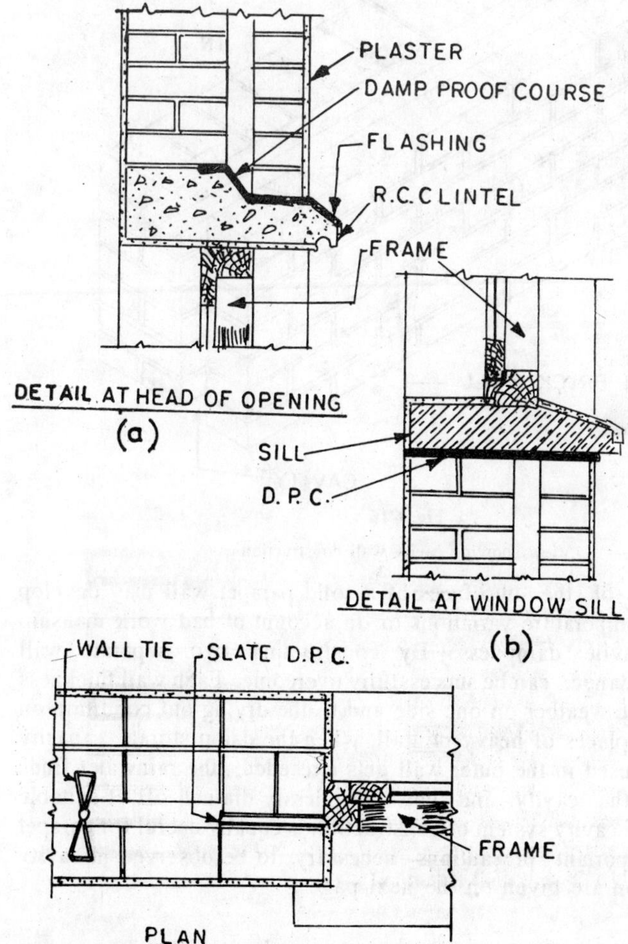

Fig. 6·13-6·15 Figures showing necessary treatment for cavity walls at door or window openings

The treatment of cavity to parapet walls is shown in Figs. 6·17-6·19. If the cavity is to be continued to the coping level, the layer of damp-proof course should be laid continuous covering the full thickness of wall immediately below the coping. In case, it is desired to stop the cavity at the mid-height of the parapet wall, damp-proof course layer is laid to the full thickness of the wall at level where the cavity joins the soild wall.

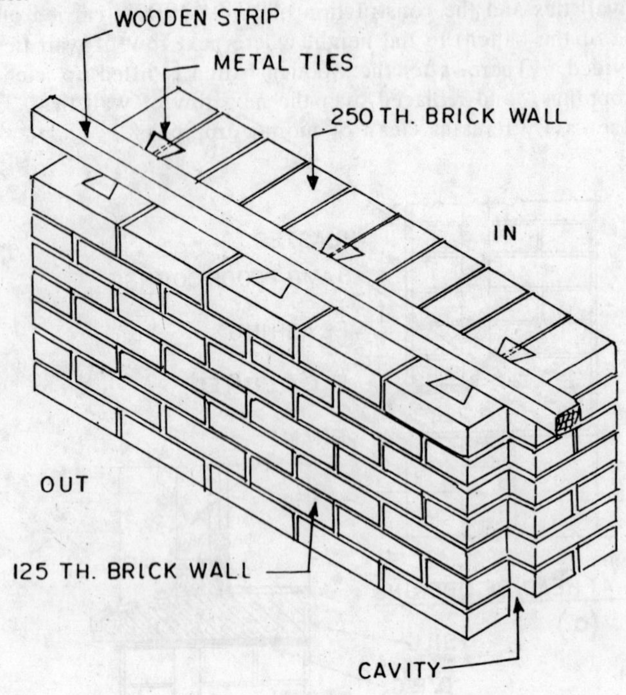

Fig. 6·16

View showing cavity wall construction

Exposure of the two faces of a solid parapet wall may develop cracks due to temperature variations or on account of bad work manship which may invite dampness. By constructing two separate wall thickness, this danger can be successfully overcome. Each wall thickness is exposed to the weather on one side and to the drying out condition on the other. In places of heavy rainfall, when the damp storage capacity of the material used in the outer wall gets exceeded, the rainwater finds its way into the cavity and gets harmlessly drained off at suitable points. Thus the cavity system of construction is equally useful for parapet walls. The important precautions necessary to be observed in cavity wall construction are given on the next page :

6·6. PRECAUTIONS

(1) As far as possible there should be no intimate contact between the two leaves of the cavity wall.

(2) The half brick thick leaf of the cavity wall should be constructed with stretcher bond.

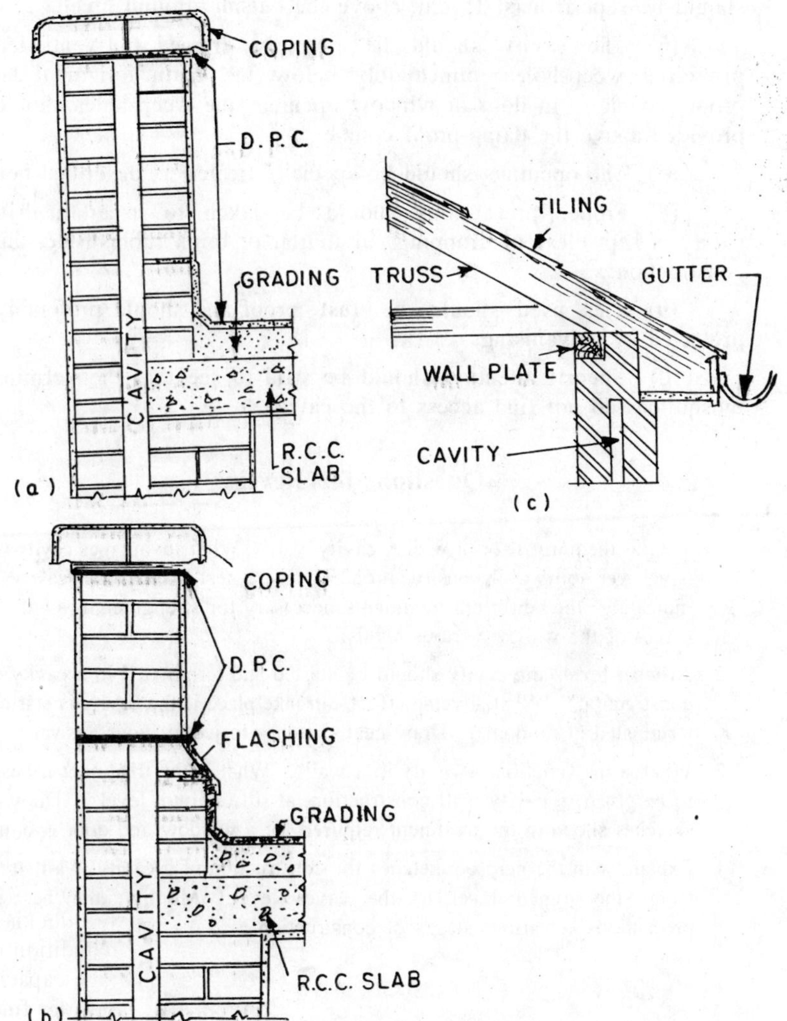

Fig. 6·17-6·19

Figures showing necessary treatment for a cavity wall at eaves level and parapet level

(3) The damp-proof course layer should be laid separately for the two leaves.

(4) The horizontal damp-proof course layer should be laid at least 15 cm. above the bottom of the cavity.

(5) To prevent infiltration of moisture, the bottom of cavity should be kept at least 15 cm. above the outside ground level.

(6) The cavity should be properly drained and ventilated by providing weep-holes immediately below the main horizontal damp-proof course. In door or window openings the weep-holes should be provided above the damp-proof course.

(7) The openings should be specially treated as described before.

(8) Proper precautions should be taken to ascertain that the cavity is kept clear of droppings of mortar or brick rubbish etc. during construction.

(9) Ties used should be rust proof and should preferably be protected by galvanising.

(10) Special attention should be paid to see that the vermins or mosquitoes do not find access to the cavity.

Questions for Reveiw

1. What is the purpose of providing cavity walls? What advantages cavity walls have over solid wall construcution? Draw a neat sketch of a cavity wall indicating the different treatments necessary for strengthening of the two leaves of the wall at various levels.
2. At what levels the cavity should be started and terminated in a cavity wall construction ? What adverse effect can take place if the cavity is started or terminated at random ? Draw neat sketches to justify your answer.
3. What is the function of cavity in a wall ? What is the treatment necessary to be given in cavity wall construction, at sill or lintel level ? Draw neat sketches showing the treatment required for a window and door opening.
4. Explain with the help of sketches the construction of a cavity wall starting from the trench level to the eaves level. Mention the necessary precautions at various stages of construction.

7

Shoring, Underpinning and Scaffolding

7·1. SHORING

The term 'shoring' is applied to construction of the temporary structure required to support an unsafe structure.

The main objects of shoring may be summarised as below :

1. When the walls of a building develop signs of bulging or leaning outwards, shoring is necessary to prevent further development of the defects.

2. When defective walls of a building are to be dismantled and rebuilt, shoring is resorted to for supporting the floors or roofs connected to that wall.

3. Shoring is necessary to support the super-structure when large openings are required to be made in the main walls. Shoring may be essential to give support to the walls of two adjacent buildings when the intermediate building is to be pulled down and rebuilt.

7·2. TYPES OF SHORING

There are three different types of shores :

1. *Raking shores*
2. *Flying shores*
3. *Dead shores*

(1) **Raking shores.** This is a system of giving temporary support to an unsafe wall. The construction of raking shore, also known as inclined shore, varies with the conditions of site. In all cases wall-plate 23 cm. x 5 cm. to 23 cm. x 7·6 cm. in size is fixed against the unsafe wall with hooks. The wall-plate is further secured to the wall by means of needles. The needles which are 10 cm. x 7·6 cm. in section penetrate inside the wall

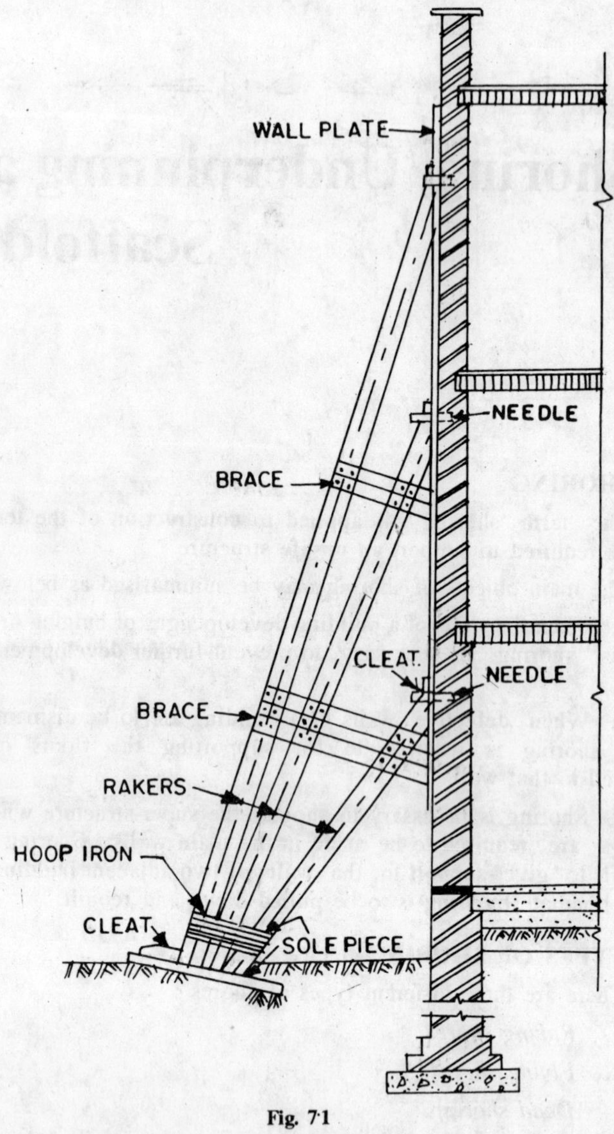

Fig. 7·1

Raking shores

for a distance of about 10 cm . In turn, the needles are strengthened by providing wooden cleats. The top end of the inclined rakers rest against

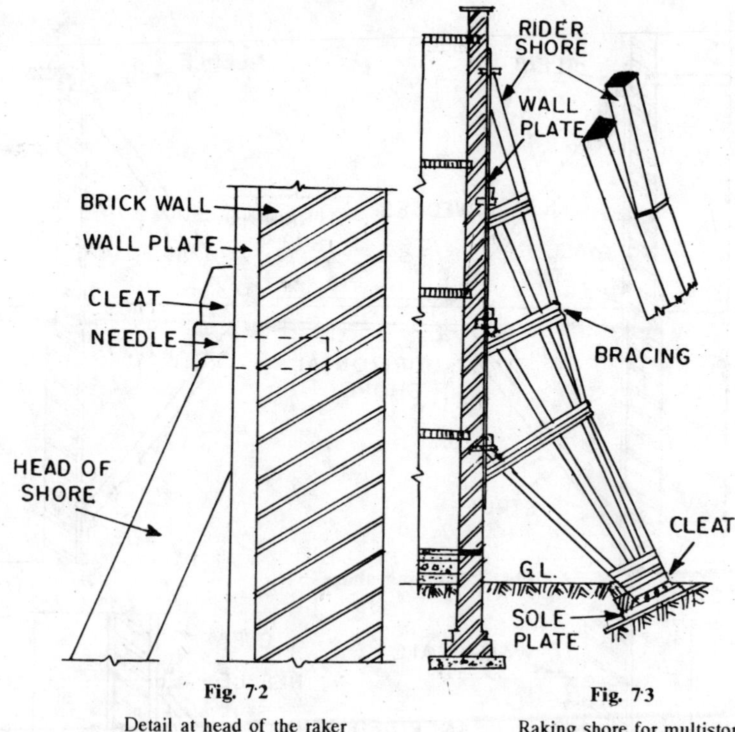

Fig. 7·2
Detail at head of the raker

Fig. 7·3
Raking shore for multistoreyed building where inclination of the rakers has to be limited due to short land width available

the needles. At their base the rakers are supported by a sole piece bedded in an inclined position in the ground. The rakers are secured to the sole piece by cleats and dogs. In soft ground the area of the sole piece is increased so as to distribute the pressure over large area. In places where more rakers are provided, they are bound together by means of hoop iron or braces 2·5 cm. thick and 15 cm. wide. The inclination of the outer raker to the ground should vary between 60° to 75°. The sets of shores should be usually placed at 3 to 4·6 m. centre to centre along the wall length.

(2) **Flying shores.** It is a system of providing temporary support to the party walls of two buildings where the intermediate building is to be pulled down and rebuilt. All types of arrangements of supporting the unsafe structure in which the shores do not reach the ground come under this category. The flying shore consists of wall plates, needles, cleats, horizontal struts (commonly known as horizontal shores) and inclined struts arranged in different forms which varies with the situation. In this system also, the wall plates are placed against the wall and secured to it. A horizontal strut is placed between the wall plates and is supported by

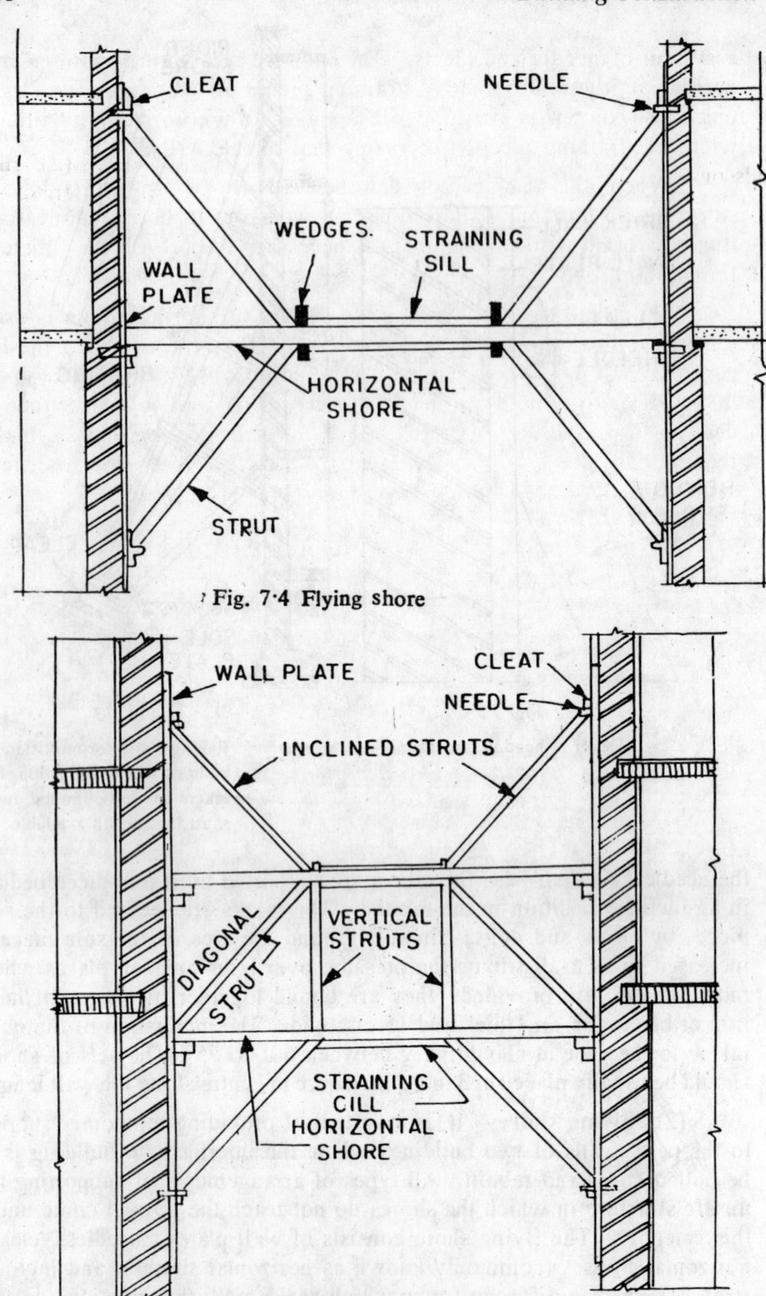

Fig. 7·4 Flying shore

Fig.7·5

Flying shore: When the distance between two wall is considerable

a system of needle and cleats. The inclined struts are supported by the needle at their top and by straining pieces at their feet. The straining piece also known as straining sill is spiked to the horizontal shore. The width of straining piece is the same that of the struts.

When the distance between the walls (to be struted apart) is considerable, a horizontal shore does not work out to be economical and a trussed frame work of members is necessary to perform the function of flying shore.

(3) **Dead shores.** This is the system of shoring which is used to render vertical support to walls and roofs, floors, etc., When the lower part of a wall has to be removed for the purposes of providing an opening in a wall or to rebuild a defective load bearing wall in a structure. The dead shore consists of an arrangement of beams and posts which are required to support the weight of the structure above and transfer the same to the ground on firm foundations below.

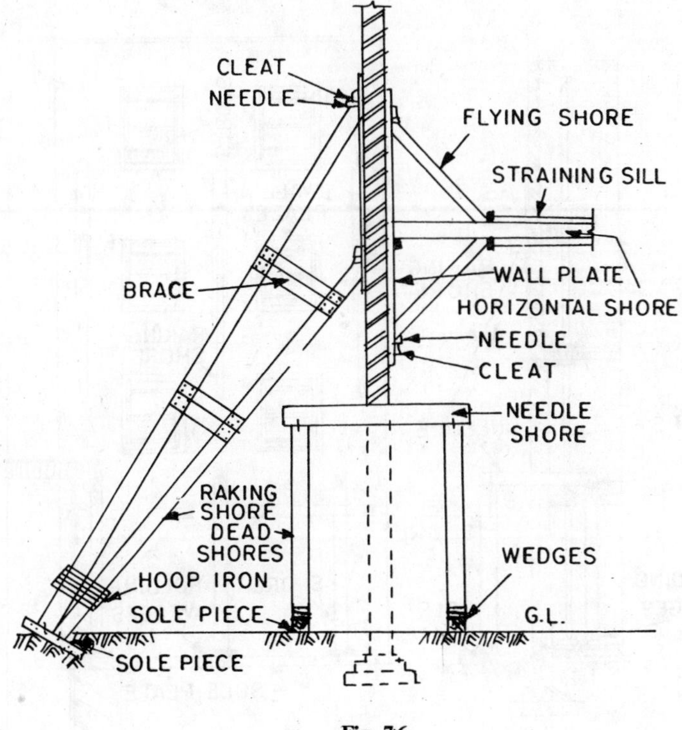

Fig. 7·6

Dead shore

When openings in the wall are to be made, holes are cut in the wall at such a height so as to allow sufficient space for insertion of the beam or girder that will be provided permanently to carry the weight of the structure above. Distance at which the holes are cut depend upon the type of masonry and it varies from 1·2 m. centre to 1·8 m. centre. Beams called needles are placed in the holes and are supported by vertical props called dead shores at their ends on either side of the wall. The needles may be of timber or steel and are of sufficient section to carry the load from above. The dead shores stand away from wall on either side so as to allow for working space when the needle and props are in position. The props are tightened up by folding wedges provided at their bases while the junction between the prop and the needle is secured with the help of dogs. Before the dismantling work is started, all the doors, windows or other openings are well struted. In order to relieve the wall of load of floors and roof above, they are independently supported. Vibrations and shocks are bound to occur when wall cutting is done and as such as a measure of safety, raking shores are sometimes erected before the commencement of wall-cutting operation.

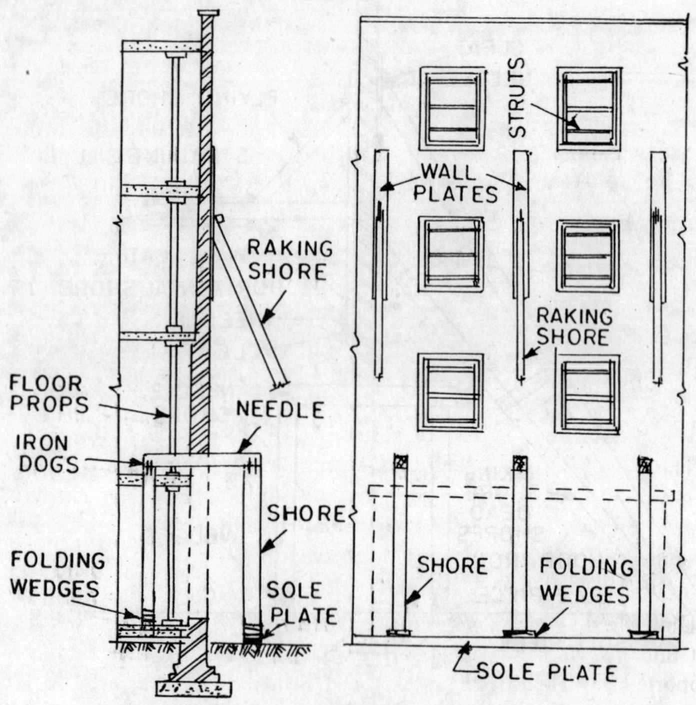

Fig. 7·7

Section and elevation showing arrangement of dead shores for making an opening in an existing wall

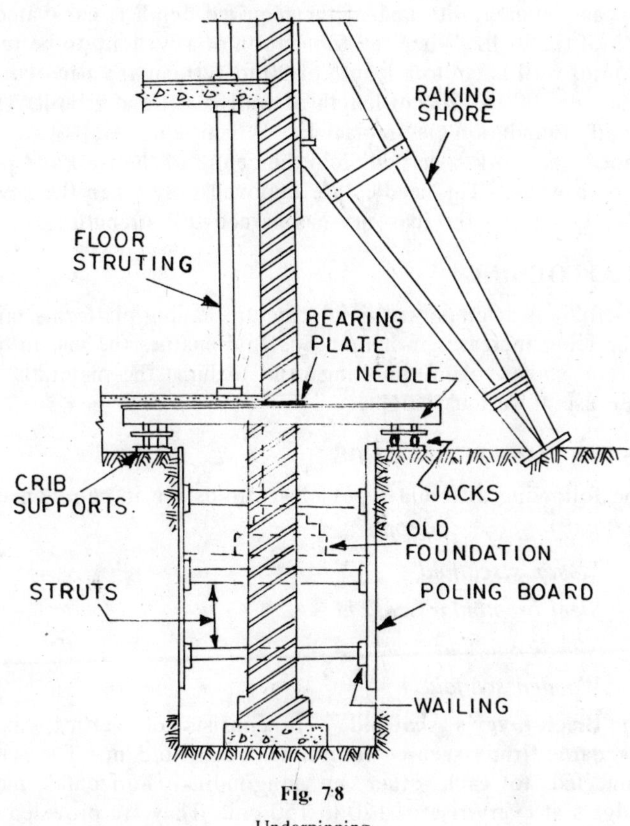

Fig. 7·8
Underpinning

7·3. UNDERPINNING

This term is applied to the building of new work underneath an existing structure without disturbing its stability. Underpinning may be necessary when defective foundation of a wall is to be replaced with new foundation or when the existing foundation of a wall is required to be strengthened to enable it to carry more loads.

Underpinning is done in short lengths of 1·2 m to 1·8 m. Holes are cut into the walls at regular intervals and the entire load of the wall is supported by the needles which rest on sole pieces at ground level. The section of the needle depends upon the load to be supported. The needle may be a timber beam or a rolled steel joist. Thus the structure is held well supported in a secured manner and the footings are relieved of the loads from the wall above. When the R.S.J. beams are used as needles, wooden bearing plates are inserted between the beams and the supported wall to avoid crushing of masonry. A pit of sufficient width to provide easy workability and of the required depth is excavated along the length of the wall. When the foundation of a wall are to be replaced, the section of wall is cut to a length of 90 to 120 cm and new work built. Again, the next 90 to 120 cm length of wall is cut and rebuilt. Thus the entire wall foundation is replaced. When long walls are to be underpinned, the work is started from the centre of the wall and progress is made both ways. The needles are removed only when the new work is completely set and the masonry has gained full strength.

7·4. SCAFFOLDING

Scaffold is a temporary rigid structure having platforms raised up as the building increases in height. Scaffold enables the mason to work at different stages of a building and to hoist the materials for the immediate use at various heights.

7·5. TYPES OF SCAFFOLDS

The following different types of scaffolds are in common use :

1. *Brick-layer's scaffold*
2. *Mason's scaffold*
3. *Steel or tubular scaffold*
4. *Needle scaffold*
5. *Wooden scaffold.*

(1) **Brick-layer's scaffold.** It consists of vertical members called standards firmly secured in grounds at 2·4 to 3 m. The standards are connected to each other by longitudinal horizontal members called ledgers at every rise of 120 to 150 cm. They are provided on the building side of the standards and are secured in position by rope-lashings. Transverse horizontal members called putlogs are lashed on the ledgers at one end and into the holes in the wall at the other end. Putlogs are 90 cm. in length and are generally spaced 120 cm. apart. They are provided to support the working platform. In high scaffolding, cross braces are invariably used to stiffen the temporary structure. Braces consist of poles tied by rope-lashing on the standards.

(2) **Mason's scaffold.** Since it is difficult to leave holes in the stone masonry to provide bearing for the putlogs, in mason's scaffold two frames of standards, ledgers and braces are used. One is placed close to the wall and the other at a distance of 1.5 m from the first one. Thus the mason's scaffold is entirely independent of the stone wall.

Close mason's scaffold. **They are supposed to carry the loads of heavy blocks of stones and the stresses produced on account of the lifting device and hence they are made stronger by placing the standards at fairly close distance apart.**

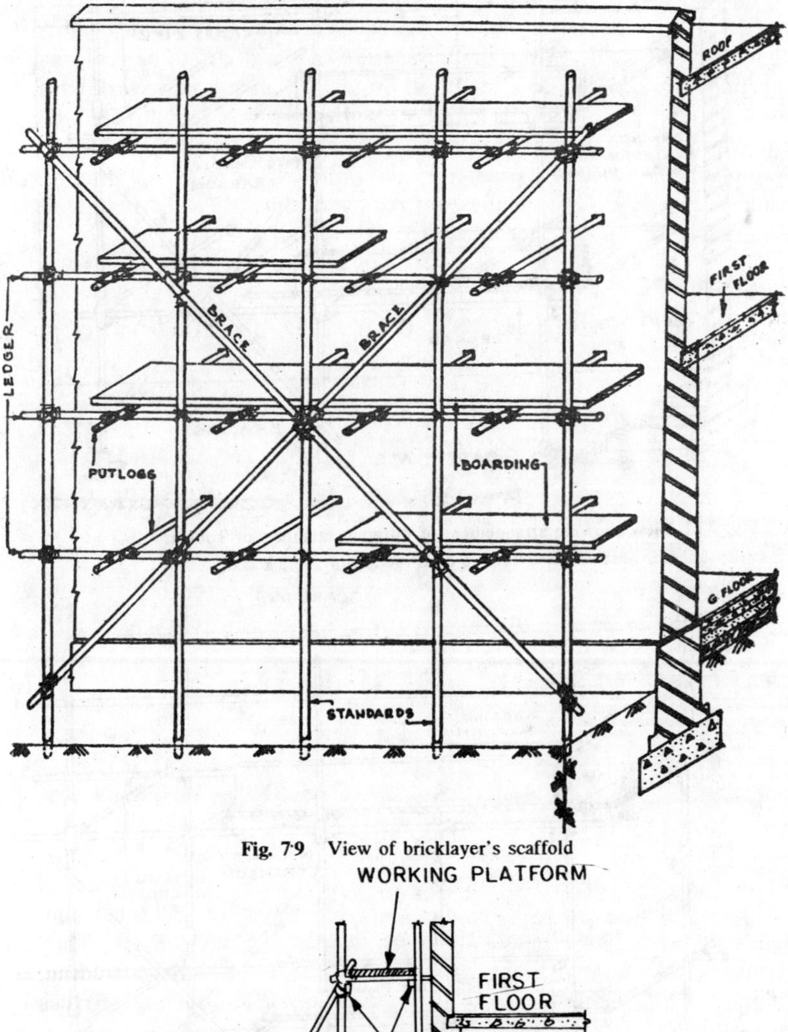

Fig. 7·9 View of bricklayer's scaffold

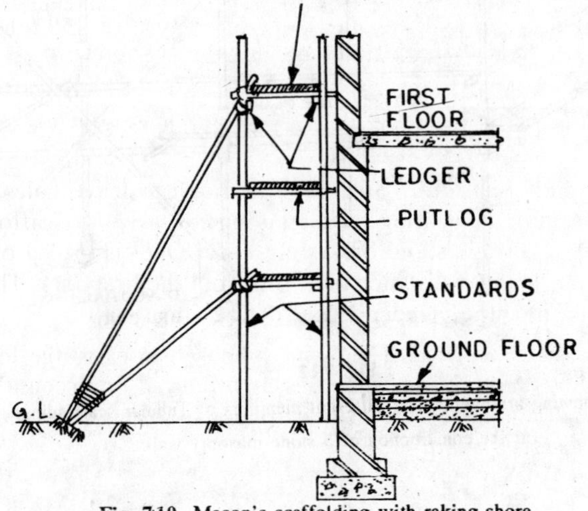

Fig. 7·10 Mason's scaffolding with raking shore

VERTICAL TOE PLANK
GUARD RAIL
MAIN CROSS PIPE
WORKING PLATFORM OF WOODEN PLANKS
INTERMEDIATE TRANSOM
COUPLING
LEDGER
DIAGONAL BRACING
BASE PLATE

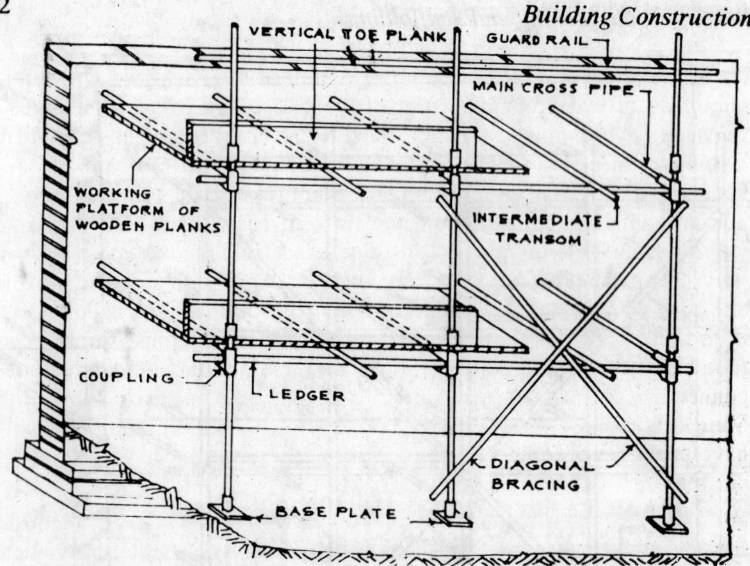

Fig. 7·11 View showing arrangement of different members of Tubular scaffolding for the construction of a brick wall

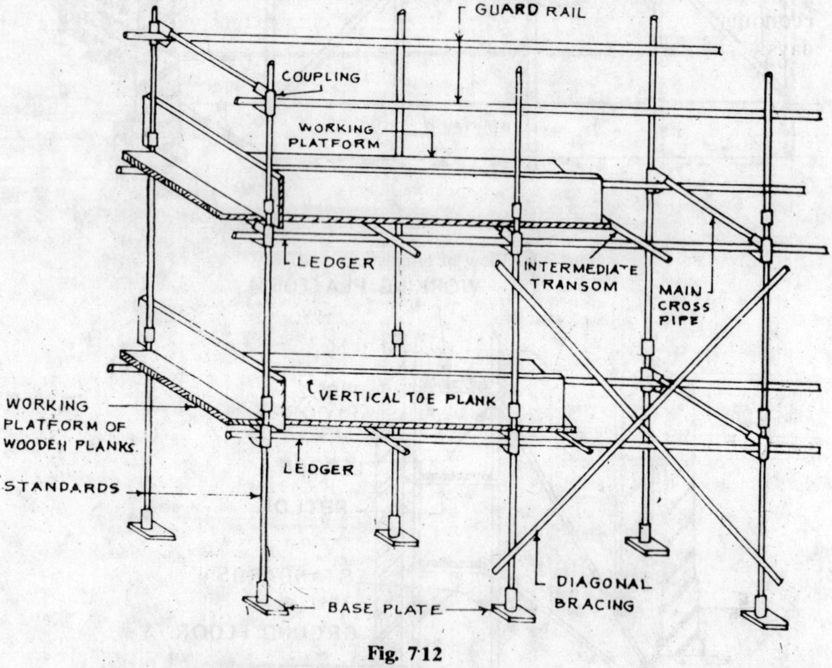

GUARD RAIL
COUPLING
WORKING PLATFORM
LEDGER
INTERMEDIATE TRANSOM
MAIN CROSS PIPE
VERTICAL TOE PLANK
WORKING PLATFORM OF WOODEN PLANKS
LEDGER
STANDARDS
DIAGONAL BRACING
BASE PLATE

Fig. 7·12

View showing arrangement of different members of Tubular scaffolding for the construction of a stone masonry wall.

(3) **Steel scaffold.** The construction of steel scaffold is essentially similar to that of timber scaffold. In this case, the wooden members are replaced by 38 mm to 64 mm diameter steel tubes and instead of rope lashings, special types of steel couplets or fittings are used for connecting different members. The steel tubes used for scaffolding for normal building construction work are of heavy class and of diameter varying from 40 to 60 mm. In this type of scaffold the vertical tubes called uprights or standards are spaced 2·5 to 3 m apart. Each standard is welded to a base plate, square or circular in plan. The base plate has holes so that it can be spiked to a timber or concrete base, thus forming rigid foundation for the scaffolding. Ledgers or the longitudinal tubes connecting the standard and are spaced at 1.8 m vertically apart. Short tubes or putlogs are normally 1·2 m to 1·8 m in length. The putlogs have special ends to fit into the joints in the wall and thus no big holes are required to be left in the wall for their bearing.

Tubular scaffolding has several advantages over the timber scaffolding such as rapid erection and dismantling, greater strength and durability and higher fire resisting qualities and salvage value. Although it works out to be costlier initially, but considering its several advantages and the increased number of reuses, it proves to be economical in the long run and hence it is being extensively used these days.

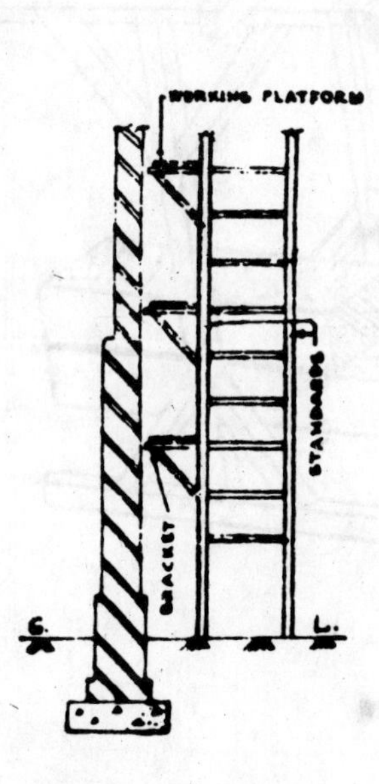

Fig. 7·13 Steel scaffold

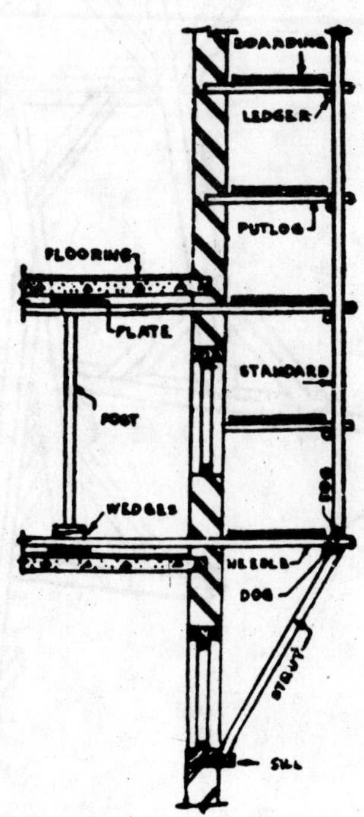

Fig. 7·14 Needle scaffold

Tubular scaffolding, mounted on wheels can be used for carrying out the work of painting the walls, under side of roofs etc.

(4) **Needle scaffold.** When scaffolding is to be provided for a building on the side of a busy street where the construction of ordinary scaffolding will obstruct the traffic on road, needle scaffold is use. The scaffold is erected from window corners or string courses by means of projecting needles. A needle is a timber beam which cantilevers out through the holes cut in wall. From inside the needles are supported on sole pieces and are prevented from lifting up by vertical struts wedged between the needles and the head pieces. The projected end of needle is supported by an inclined strut which rests on the window sill. The joint between the raking or inclined strut and the needle is clamped by means of dogs. This arrangement provides the staging for erection of ordinary type of scaffold.

(5) **Gantries.** There are two different types of wooden gantries:

(*a*) Gantry with traveller or crane

(*b*) Platform gantry.

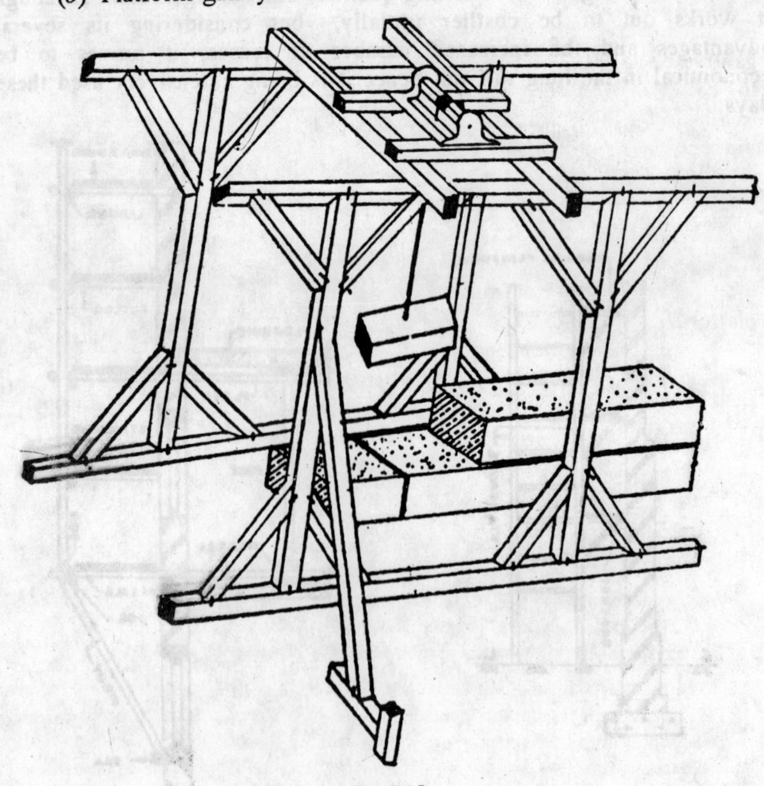

Fig. 7·15

Gantry with a traveller or crane

(*a*) *Gantry With Traveller or Crane.* When the stone blacks used in the wall construction are too heavy, crane travellers have to be used. The framework of gantry consists of timber standards raised on either side of wall. Longitudinal members called runners are fixed on the top of the standards, while at the bottom, the standards are fixed to common sole piece. Rails are fixed on the runners to carry the travelling platform. The travelling platform consists of trussed beams carrying lifting tackle, which can run on wheels on rails fixed on runners. This enables the travelling platform to move parallel to the wall along its length.

In addition, the beams of traveling platform have rails, so that the lifting tackle can move in a direction perpendicular to the wall. Thus the blocks of stones can be placed in any desired direction while building the wall.

(*b*) *Platform Gantry.* This is needed for providing a working platform above ground level and leaving the space below free from obstruction. Platform gantry is adopted in stone masonry construction as it can be used independent of wall. The gantry consists of vertical posts fixed to common sole piece at its base and to a head piece at the top. The headpieces support the joists on which working platform is laid. The entire framework is thoroughly braced and dogged. A 90 cm. high boarding is fixed on the outside of the gantry to prevent the material on the platform from falling down.

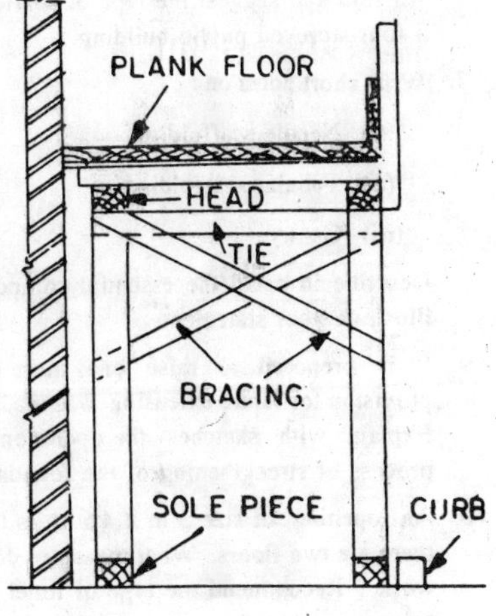

Fig. 7·16

Platform gantry

Questions for Reveiw

1. (*a*) Explain with sketches what you understand by the terms 'underpinning and shoring.'

   (*b*) It is proposed to form an opening 7·6 x 3·7 m. high in front wall of an existing two-storey building with the help of a steel beam supported by stanchions at both ends to support the load from above. The existing building is 24 m x 9 m. on plan. The floor to floor heights for ground and first floor are 5 m. and 3·7 m. respectively. Elaborate the method of executing the above work.

2. The roadside wall of a big hotel in 'Calcutta', situated in a very busy street, cracked dangerously right from the foundation to the third storey. It was ordered to demolish the said wall and construct a new brick wall in its place. Show by sketches the various operations involved in carrying out the work without causing much dislocation to the

traffic and ensuring the safety of the occupants at the same time.

3. **Write short notes on :**
   (*i*) Ranking shore
   (*ii*) Flying shore
   (*iii*) Dead shore

   Draw neat sketches showing the position of the various shores in an existing building.

4. What are the types of scaffolding commonly used? Draw a neat sketch to illustrate the difference between the various types giving short notes in favour of the sketches shown.

5. What is the difference between a gantry and scaffolding? Explain with sketches the different types of gantries used in building operation.

6. How do the steel scaffold differ from timber scaffold ? Discuss the advantage of steel scaffold and suggest the type of scaffold you would recommend in the construction of a four-storeyed public building.

7. **Write short notes on :**
   (*i*) Needle scaffolding
   (*ii*) Tubular scaffolding
   (*iii*) Gantry.

8. Describe in detail the essential components of raking shore showing neat sketches to illustrate your statement.

9. It is proposed to raise one more storey on an existing three-storey building. No provision for future extension was made in the design of foundations of the said building. Explain with sketches the operation of carrying out new construction, including the process of strengthening of the foundation.

10. An opening of size 3 m x 4·6 m. is to be provided in a solid brick wall above which there are two floors. What measures do you propose and how do you go ahead with the work ? Recommend the type of lintel you purpose to provide above the opening.

11. In a brick masonry wall building five-storey high, outer walls on the ground floor in- between shop-doors have developed cracks below the floor beams at the first floor level. Show, by neatly drawn sketches, how you would rectify them and strengthen the structure so as to make it safe ?

12. What do you understand by the terms 'under-pinning' and 'shoring'? Explain where these operations are made use of and why ? Give sketches.

13. Describe how will you provide a 1.2 m. wide door opening in a 38 cm. thick, load carrying wall of an old building.

14. (a) Describe with sketch the method of timbering in a trench of size 1·8 deep x 1·2 m wide for laying underground sewer in moderately firm ground,
    (b) State how will you provide a steel joist in place of an old rotten timber beam supporting tiled roofing. Sketch the details and mention all precautions to be taken.

15. Describe briefly with necessary sketches method of under-pinning used in foundation work of a building.

8

Masonry—Stone Masonry

8·1. GENERAL

The construction of stone, brick or tile which in hardened by heat is termed as masonry, It may also be defined as construction of building units bonded together with mortar. The selection of the type of material (*i.e.* brick or stone etc) for the masonry is made keeping in view the requirements of strength, water proofing, thermal insulation, fire resistance, durability and economy.

Masonry is basically a wall material. Masonry walls can be divided into following three categories.

(*i*) *Load bearing walls*

(*ii*) *Non-load bearing walls*

(*iii*) *Retaining walls*

(*i*) **Load Bearing Wall.** A wall designed to carry superimposed loads from floors and roof is termed as load bearing wall. Such walls have continuous foundation to carry the entire super imposed load including their self weight.

(*ii*) **Non-load Bearing Wall.** Also known as panel wall, curtain wall or filler wall, is a type of wall which carries no superimposed load. It is provided to serve as screen for privacy and to keep out wind and weather. The load from floors and roof in this case is borne by either brick piers or by a system of R.C.C. or steel beam and column frames.

(*iii*) **Retaining Wall.** This is a type of wall built to resist the pressure of earth, granular material or liquid filled behind it after it is built.

Depending upon the type of material used, masonry can be broadly divided in the following categories.

(*i*) *Stone masonry*

(*ii*) *Brick masonry*

(*iii*) *Reinforced brick masonrv*

(*iv*) *Composite masonry*

(*v*) *Hollow concrete block masonry*

(*vi*) *Load bearing wall tile masonry*

Stone masonry has been covered within the scope of this chapter whereas the remaining types of masonry have been detailed in Chapter 9 (Brick masonry) and Chapter 10 (Composite masonry).

8·2. COMPARISON BETWEEN STONE MASONRY AND BRICK MASONRY

(1) From times immemorial, stone has been used both for residential as well as public buildings. Historical buildings that stand today are the living examples of the strength, durability and the excellent weather-resisting qualities of stone masonry. Brick on the other hand has much less strength, durability and weather-resisting qualities.

(2) On account of its high crushing strength stone is used in the construction of piers, docks, dams and other marine structures. Brick, on the other hand, is not considered suitable in all such places.

(3) Shining texture of good class of stone masonry requires no treatment to enhance its appearance. On the other hand, plastering is necessary to conceal the defects in brick masonry.

(4) In buildings of monumental nature where architecture requires heavy mouldings with large projections, stone is best suited, brick being suitable for light ornamental work.

(5) Bricks when exposed are liable to get damp. Dampness may ultimately lead to the disintegration of the masonry. Stonework on the other hand suffers from no such danger.

(6) Certain salts present in the sewage react chemically with exposed brick and as such when brick-work happens to come in contact with sewage, it is always plastered. Stone on the other hand can be used in such places without providing any protective coat.

(7) On account of the high cost of stone masonry, its usage is generally restricted to hilly areas or stone districts; bricks on the other hand are easily available in allmost at all places and the masonry constructed with bricks costs much less.

(8) First class bricks possess all such qualities which are required for a good construction and hence brick masonry has now practically replaced stone masonry.

(9) On account of their regular shape and size, bricks afford great facility in maintaining proper bond in the masonry. It also results in quick construction. On the other hand, in stone masonry, the process of dressing and placing stones requires a great deal of time and extra labour.

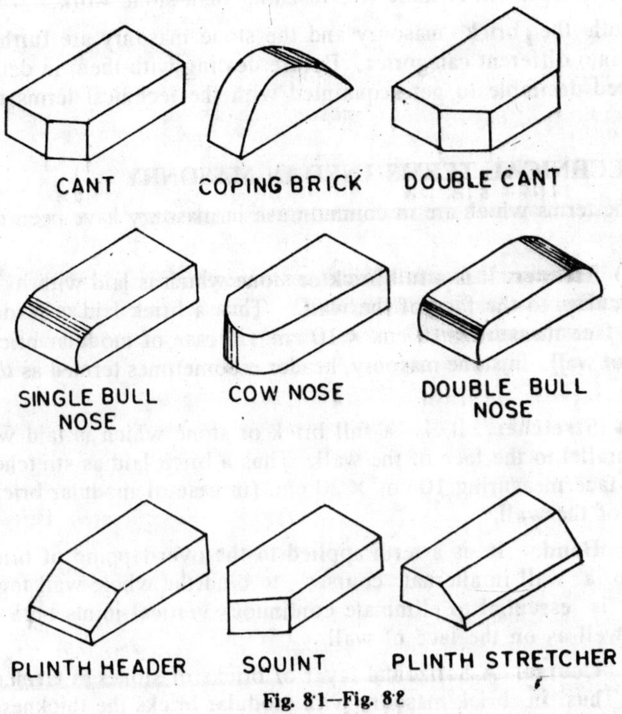

Fig. 8·1—Fig. 8·2

Bricks moulded in different shapes

(10) For the construction of jambs of doors and windows and for the walls meeting at obtuse or acute angle, bricks offer greater facility than stone.

(11) Bricks can be conveniently moulded into any desired shape at reasonable cost while the expense of the moulding of stone work is far more than that of brick.

(12) Skilled labour is required for good stone masonry construction because the defects left by a careless mason cannot be rectified easily, while in brick masonry the defects of a careless bricklayer can be concealed by plastering.

(13) On account of their convenient size and light weight, bricks require no lifting tackle while in stone masonry the large blocks of stone have to be kept in position with the aid of some lifting device

(14) Brick walls can be easily built in small thickness of 10 cm. and 20 cm. while the minimum thickness of a stone wall is generally restricted to 30 cm.

(15) Brickwork is more fire resisting than stone work.

Both the brick masonry and the stone masonry are further subdivided into different categories. Before dealing with them in detail it is considered desirable to get acquainted with the technical terms used in masonry.

8·3. TECHNICAL TERMS USED IN MASONRY

The terms which are in common use in masonry have been defined below :

(1) **Header.** It is a full brick or stone which is laid with its length perpendicular to the face of the wall. Thus a brick laid as header will show its face measuring 10 cm. × 10 cm. (in case of modular brick) on the face of wall. In stone masonry, header is sometimes termed as *through stone.*

(2) **Stretcher.** It is a full brick or stone which is laid with its length parallel to the face of the wall. Thus a brick laid as stretcher will show its face measuring 10 cm. × 20 cm. (in case of modular brick) on the face of the wall.

(3) **Bond.** It is a term applied to the over-lapping of bricks or stones in a wall in alternate courses, to bind the whole wall together. Bonding is essential to eliminate continuous vertical joints both in the body as well as on the face of wall.

(4) **Course.** A horizontal layer of bricks or stones is termed as a course. Thus in brick masonry with modular bricks the thickness of a course will generally be equal to 10 cm. plus thickness of one mortar joint.

(5) **Header course.** It is a course of brickwork entirely composed of headers.

(6) **Stretcher course.** It is a course of brickwork in which all the bricks are laid as stretchers.

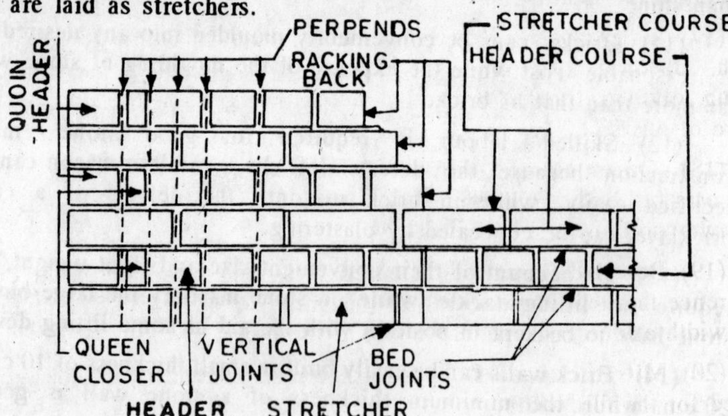

Fig. 8·9 Elevation of a masonry wall

(7) **Bed.** It is a term used to indicate the lower surface of bricks or stones in each course. It may also be termed as the surface of bricks on which it rests.

(8) **Face.** The surface of a wall exposed to weather is termed as face.

(9) **Facing.** The material used in the face of the wall is known as facing.

(10) **Back.** The inner surface of wall which is not exposed to weather is termed as back.

(11) **Backing.** The material used in forming the back of the wall is known as backing.

(12) **Hearting.** The portion of a wall between facing and backing is termed as hearting.

(13) **Joint.** The junction of two or more bricks or stones is called joint. Joints may be of cement mortar or of lime mortar. Joints parallel to the bed of bricks or stone in a course are termed as bed joints. Bed joint can also be defined as the horizontal mortar joint upon which masonry units are laid. Thus the pressure always acts normal to the bed joints. The joints (other than bed joint) which are perpendicular to face of wall are termed as cross joints or simple joints. Joints which are parallel to the face of wall are termed as wall joints.

(14) **Racking back.** The process of stopping the unfinished end of a wall in stepped fashion.

(15) **Bat.** It is the portion of a brick cut across the width or a brick cut by some fraction of its length.

(16) **Closer.** It is a portion of a brick cut in such a manner that its one long face remains uncut.

(17) **King closer.** It is a brick which is cut in such a way that the width of one of its end is half that of a full brick. It is formed by cutting off the triangular piece, between the centre of one end and the centre of one side.

(18) **Queen closer.** It is a term applied to a brick which is half as wide as a full brick. Queen closer is made by cutting a brick lengthwise into two portions.

(19) **Bevelled closer.** It is similar to king closer with the only difference that the whole length of the brick is bevelled for maintaining half width at one end and full width at the other.

(20) **Mitred closer.** It is a brick whose one end is cut, splayed or mitred for the full width.

(21) **Perpend.** It is a vertical joint on the face of a wall directly over vertical joints in alternate course.

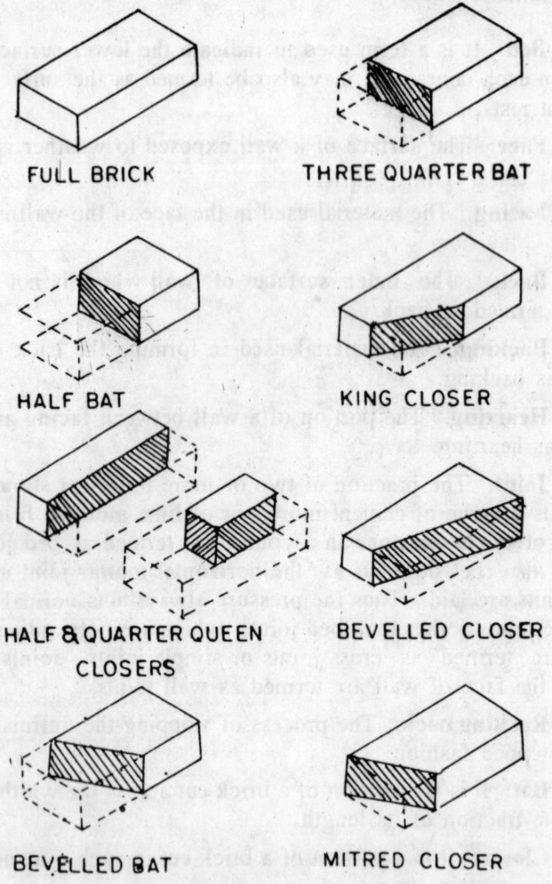

Fig. 8·10-8·17

(22) **Frog**. It is a depression on the top face of a brick. Frog provides a recess for the mortar which on setting forms a key and prevents the displacement of the brick above.

(23) **Quoin**. The exterior angle or corner of a wall is termed as quoin. The brick or wedge shaped stone used for the corner of walls is also known as quoin.

(24) **Plinth**. The horizontal projecting or flush course of stone or brick provided at the base of the wall above ground level is known as plinth. It raises the level of the ground floor of the building above the natural ground level with the aim of protecting the ground floor from flooding due to rain water and other weather effects.

(25) **Plinth Course**. It is the topmost course of plinth and is finished flush with the ground floor surface.

(26) **Sill.** It is a horizontal member of brick, stone, concrete or wood provided to give support for the vertical members of a window. It is also employed for the purpose of shedding off rain water from the face of wall immediately below the window opening.

(27) **Jambs.** The vertical sides of a finished opening for door, window or fire place etc. are termed as jambs. Jambs may be plain or rebated or splayed.

(28) **Reveals.** Reveals are the exposed vertical surfaces left on the sides of an opening after the door or window frame has been fitted in position.

(29) **Lintel.** A horizontal member of stone, brick, wood, steel or R.C.C., used to support the masonry or load above an opening.

(30) **Arch.** A mechanical arrangement of wedge-shaped blocks of stone or brick arranged in the form of a curve supporting the masonry or load above an opening.

(31) **String course.** It is a horizontal course of masonry projecting from the face of the wall and is generally at every floor or sill level throughout the length of the wall. It is intended to improve the elevation of the structure.

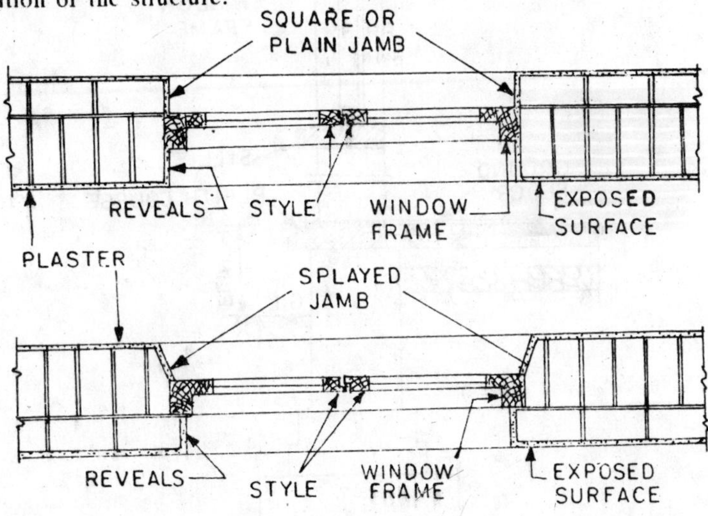

Fig. 8·18-8·19 Plain and splayed jambs

(32) **Cornice.** It is a horizontal moulded projection provided near the top of a building or at the junction of a wall and ceiling. It not only increases the architectural beauty of the structure, but also serves as a barrier for shedding the rain water off the face of the wall.

(33) **Frieze.** It is a course of stone masonry provided imme-

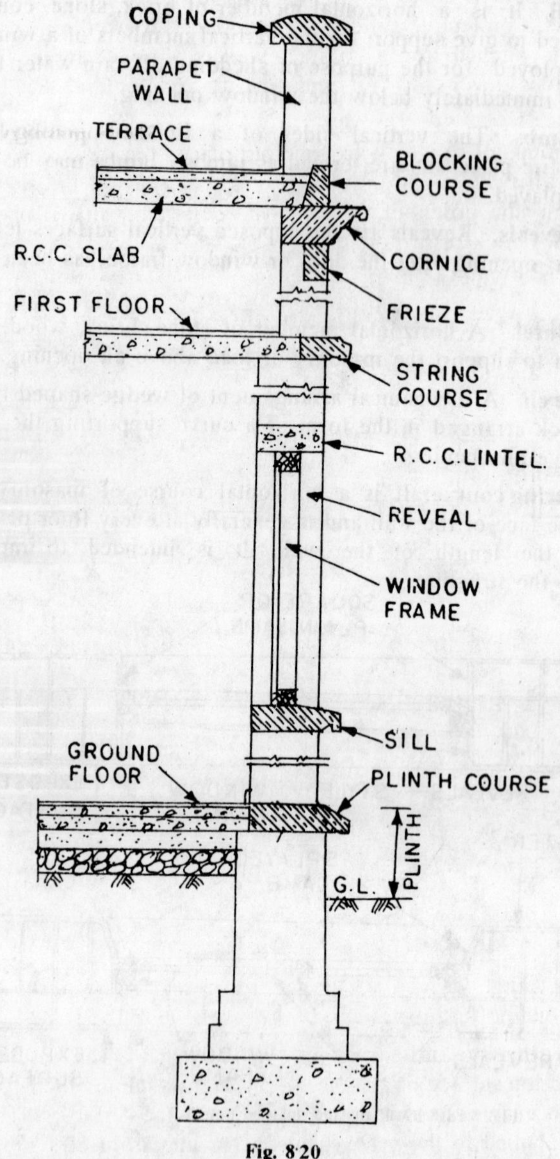

Fig. 8·20

diately below a cornice. This may be flushed with the wall or may be moulded. It is provided to enhance the appearance of wall.

(34) **Blocking course.** It is a course of stone masonry provided immediately above the cornice to check the tendency of the cornice to overturn and incidentally it adds to its appearance.

(35) **Parapet.** It is a term applied to a low wall built around a flat roof to act as a protective solid balustrades for the users of the terrace (flat roof). In pitched roof, parapet wall is provided to conceal the gutter at eaves level and for imparting an architectural effect to the structure.

(36) **Coping.** Coping is a covering placed on the exposed top of an external wall. It is essentially provided to prevent the seepage of water through the joints of the topmost course of wall. It may be of concrete, stone, brick or terra-cotta.

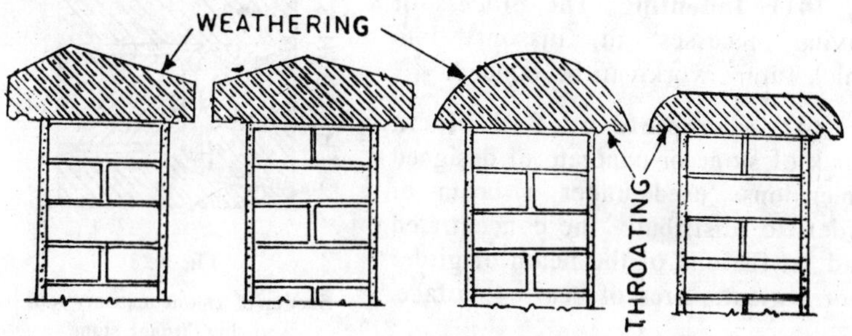

Fig. 8·21
Different forms of stone copings

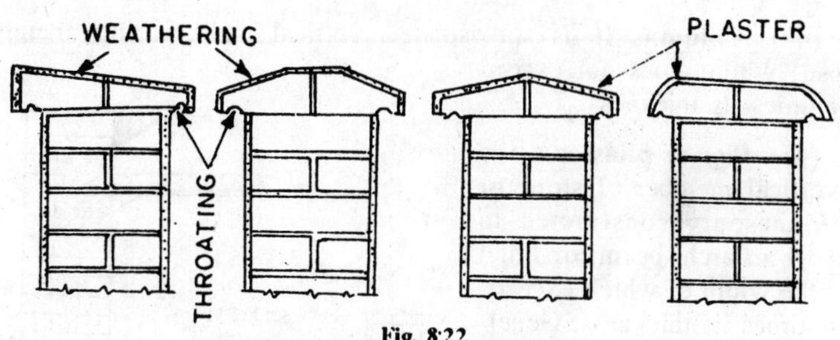

Fig. 8·22
Different forms of brick copings

(37) **Toothing.** Bricks left projecting in alternate courses for the purposes of bonding future masonry work.

(38) **Weathering and throating.** Weathering is the term applied to the bevelled top surface of a stone. Bevelling or sloping of the surface is necessary to enable the rain water to flow off the surface. Throating is the term applied to the groove cut on the underside of a projecting course of masonry in order to check the creepage of rain water from the underside of projected portion. If the rain water is not intercepted by such groove, it will spread across the bottom of the projected surface and will finally start flowing along the vertical surface of the wall, making it damp.

(39) **Gable.** It is a triangular shaped portion of masonry at the end of a sloped roof.

(40) **Through Stone.** A stone passing through a wall from front to back face and acting as a binder for the two faces of the wall is termed as through stone.

(41) **Indenting.** The process of leaving recesses in masonry into which future work can be bonded.

(42) **Template or Bed block.** Block of stone or concrete of designed dimensions used under a beam or girder to distribute the concentrated load at the end of the beam or girder over a greater area of bearing surface.

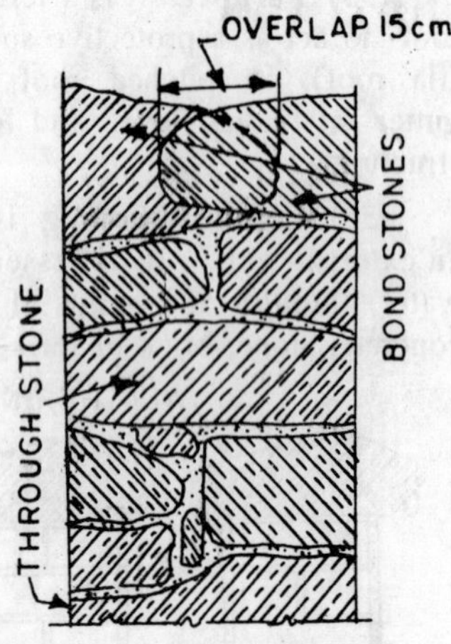

Fig. 8·23

Section of a stone masonry wall showing through stone

(43) **Spalls.** Chips or small pieces of stone broken off a large block are termed as spalls. They are used in filling the interstices in stone masonry.

(44) **Column.** It is an isolated vertical load bearing member whose width does not exceed four times its thickness.

(45) **Pier or pilaster.** It is a vertical member of stone or brick masonry constructed to support an arch, beam, or lintel etc., the width of which exceeds four times its thickness. Generally piers are made monolithic with the wall for the purpose of increasing the stability and stiffness of the wall to carry more concentrated loads.

(46) **Buttress.** It is similar to a pier built on the exterior of a wall and properly bonded to it. Buttresses are placed at intervals along the wall to make it stable for resisting outward thrust.

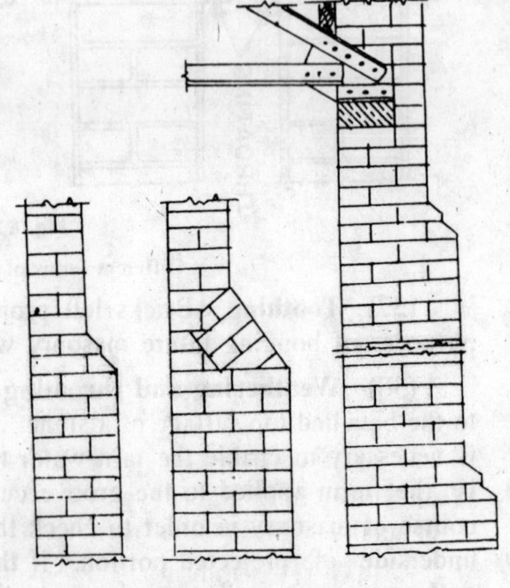

Fig. 8·24

Different forms of buttresses

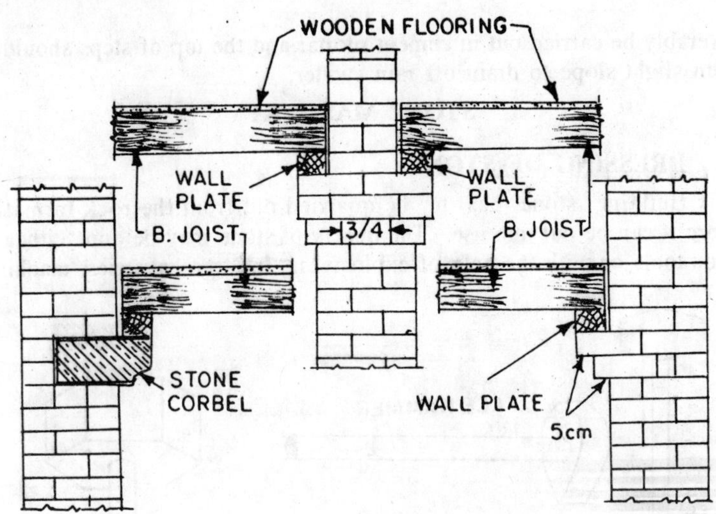

Fig. 8·25

Different types of Corbels

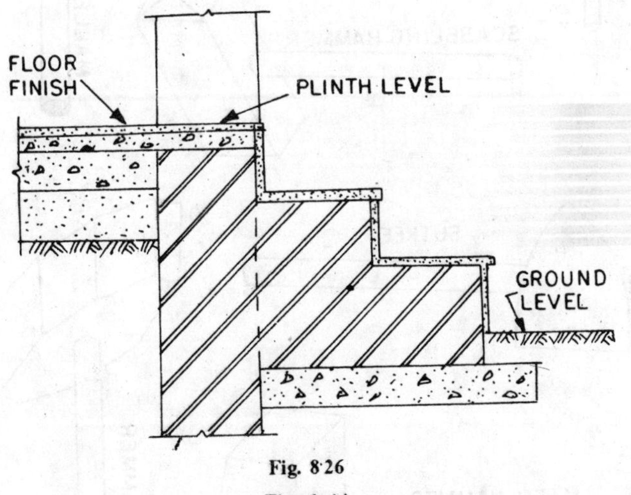

Fig. 8·26

Threshold

(47) **Corbel**. It is the extension of one or more courses of brick or stone from the face of a wall. The projection of courses, which is generally ornamental in shape, serves as a support for wall plate etc.

(48) **Thresholds**. The arrangement of steps provided from ground level to reach plinth level on external doors and verandah is termed as threshold. Thresholds are constructed at the last stage when other construction activities have almost come to an end. This work should

preferably be carried out in cement mortar and the top of steps should be given slight slope to drain off rain water.

STONE MASONRY

8·4. DRESSING OF STONE

Building stone has to be quarried out from the rock formation before it can be put into use. The quarry of stone may be done either by hand tools or with the help of explosives. In large quarries, machines

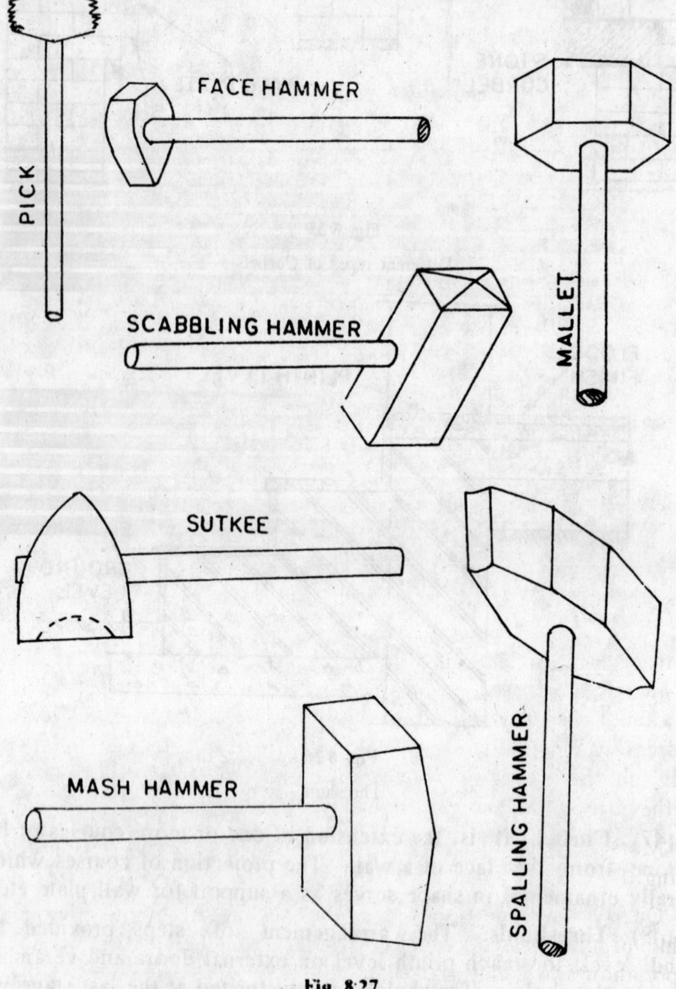

Fig. 8·27
Tools and implements used in dressing of stone

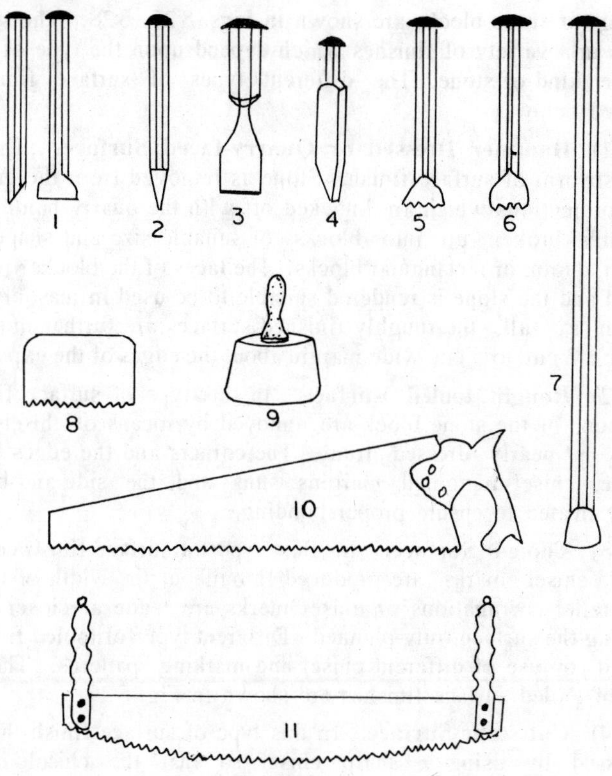

Fig. 8·28

(1) Pitching tool ; (2) Punch ; (3) Chisel ; (4) Chisel ; (5) Toothed chisel; (6) Claw chisel ; (7) Jumper ; (8) Drag ; (9) Mallet ; (10) Saw for soft stone ; (11) Crosscut saw.

have to be used for the purpose. Rough blocks of stone, as obtained from quarry, are irregular in shape and non-uniform in size and as such they cannot be used in masonry without proper dressing and cutting. The dressing of stone is done at the quarry itself, because stones, when freshly quarried, contain some moisture called "quarry sap" and in this state they are softer and can be easily worked upon.

Moreover, the local workmen are well experienced in the art of dressing.

Quarry dressed stones are thus economical and the reduced weight of the dressed blocks results in still further economy in transportation and handling costs.

The commonly used tools and implements for the cutting and

dressing of stone blocks are shown in Figs. 8·27—8·28. The stones are dressed in a variety of finishes which depend upon the type of masonry and the kind of stone. The different types of surface finishes are described below.

(1) **Hammer Dressed or Quarry-faced Surface.** This is the roughest form of surface finish. Stone as removed from the quarry has large projections which are knocked off with the quarry hammer and it is finally broken up into blocks of suitable size and shape such as khandki, quoin, or rectangular blocks. The faces of the blocks are roughly planned and the stone is rendered suitable to be used in masonry. When used in a wall, the roughly finished surfaces are further modified by forming a 2 cm. to 5 cm. wide margin about the edges of the exposed face.

(2) **Rough tooled surface**. In this type of surface finish, the projection of the stone block are removed by means of chisels and the surface is nearly dressed true. The corners and the edges are made accurate, chisel-draughted margins sunk and the side and bed joints roughly treated to ensure proper bonding.

(3) **Tooled Surface.** In this type of surface finish continuous parallel chisel marks are produced throughout the width of the stone. The parallel corrugations or chisel marks are made at closer intervals rendering the surface truly planned. Different types of tooled finishes are obtained by use of different chisel and marking patterns. The various forms of tooled surface finishes are shown in Fig. 8·29.

(4) **Cut stone Surface**. In this type of surface finish the surface is dressed by using a sharp chisel so that the chisel marks are practically imperceptible. It is considered superior to tooled surface.

(5) **Rubbed Surface**. This type of surface finish is obtained by grinding or rubbing a cut stone surface by hand or machine until it gets perfectly smooth.

(6) **Polished Surface.** The rubbed surfaces of granite, marble of lime stones are polished to enhance their texture. Polishing may be done by manual labour using sand and water, pumice stone etc. or by rubbing machine.

8·5. SELECTION OF SURFACE FINISH

The important factors influencing the selection of a particular type of surface finish are : the position of the stone and its function in the structure ; the architectural effects; the type of stone ; the atmospheric consideration and the fundsavailable. Hammer-dressed and quarry faced surface finish is best suitable for random rubble and coursed rubble masonry. Rough tooled surface and tooled surface are suitable for exposed surface. Polished finishes are used where absolute cleanliness is desired *i.e.,* for hospitals, bath-rooms, etc.

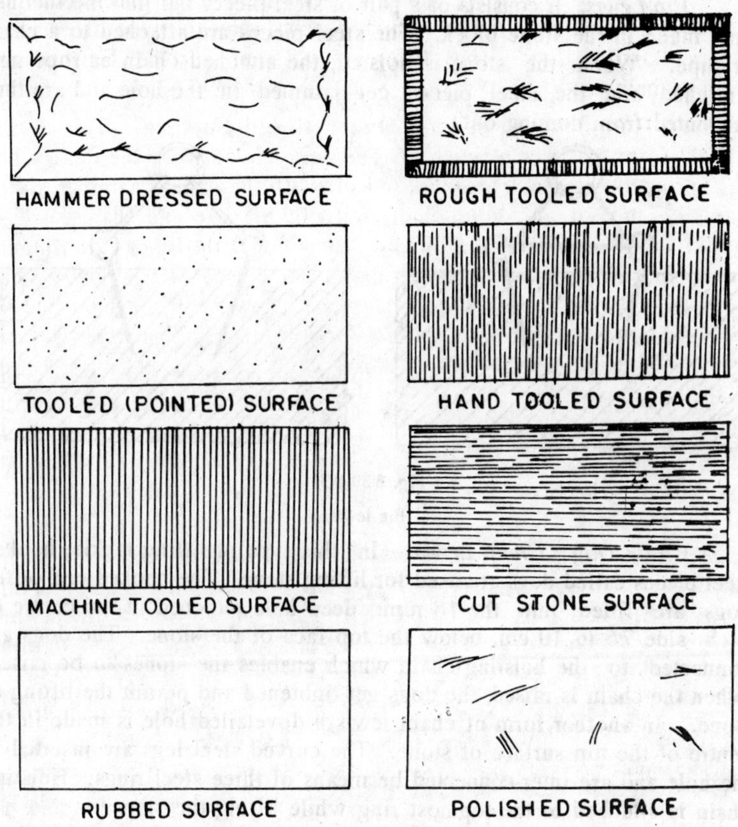

Fig. 8·29

Different types of surface finishes

8·6. SETTING OF STONE WORK (LIFTING APPLIANCES)

The process of placing a stone in its position in masonry construction is termed as "Setting". On account of its weight and size, lifting appliances are necessary for the setting of stone. The stone block is held by suitable means and is hoisted in position. For bigger blocks of stones, a derrick or a gantry is used for its lifting and placing purposes. The various appliances used for lifting stone blocks are described below.

1. Lewis

There are may kinds of Lewis, the most commonly used are pin lewis, chain lewis and three legged -lewis.

Pin Lewis. It consists of a pair of steel pieces put into the inclined hole made in the stone block. The steel pieces are attached to a chain or rope. When the stone is hoisted, the attached chain or rope gets stretched and the steel pieces get jammed in the hole and are thus prevented from coming out.

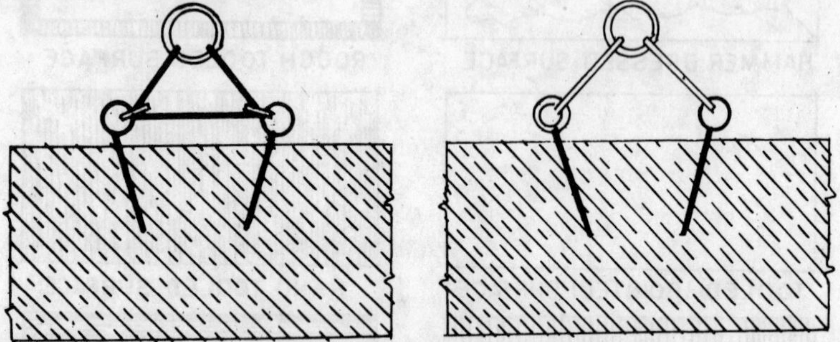

Fig. 8·30-8·31

Pin lewis

Chain Dog ; Chain Lewis. In this type of lifting tackle, hooked steel pieces called dogs are used for lifting stone. The hooked end of the dogs are fitted into the 18 mm. deep hole punched in the centre of each side 7·6 to 10 cm. below the top face of the stone. The dogs are connected to the hoisting chain which enables the stones to be raised. When the chain is raised, the dogs get tightened and permit the lifting of stone. In another form of chain lewis, a dovetailed hole is made in the centre of the top surface of stone. The curved steel legs are inserted in the hole and are inter-connected be means of three steel rings. Hoisting chain is attached to the topmost ring while lifting the stone.

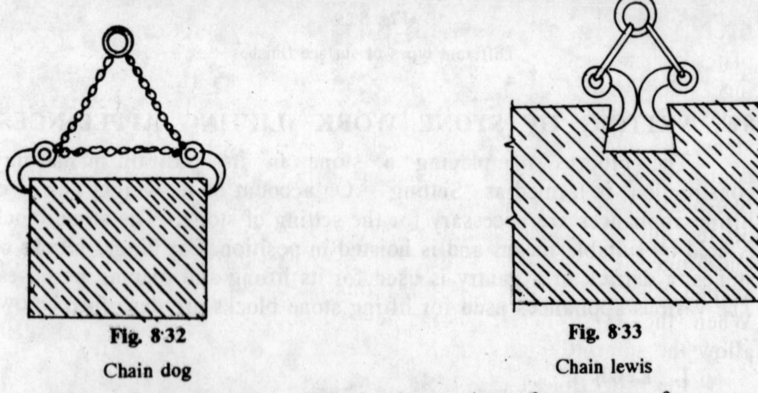

Fig. 8·32

Chain dog

Fig. 8·33

Chain lewis

Three-legged Lewis. This lewis consists of a system of a rectangular piece of steel with two dovetailed steel pieces on either side of it, a shackle at the top and a round steel pin connecting these components together. A dovetailed hole is made in the stone first. The dovetail steel pieces are inserted in the hole and the rectangular steel piece is then driven between them. The shackle is held in its position and it is secured to the three pieces by means of the round steel pin. When the stone is lifted, the lewis gets tightly fitted in the stone permitting the stone to be raised.

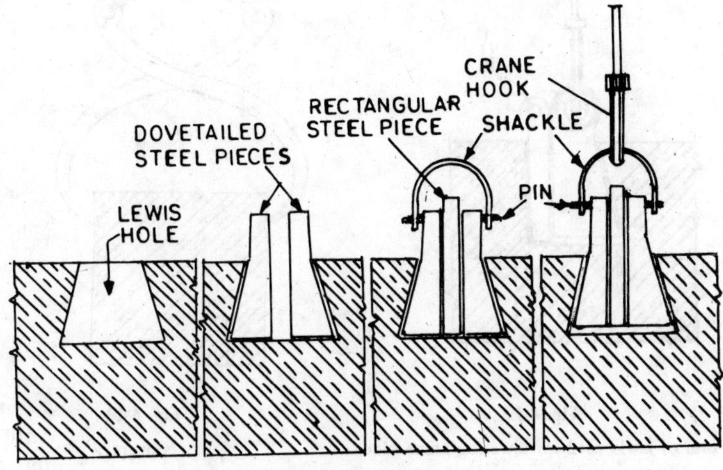

Fig. 8·34-8·37

Figures showing different stages in fixing the three-legged lewis in a stone block

Modified form of Three Legged Lewis. A modified form of the above lewis consists of a wedge-shaped block which can freely slide between two steel plates placed against the tapering sides of the wedge shaped hole made in the centre of the stone top. The steel plates are connected to each other through two pieces placed on either side of the plates on the stone top and hinged to them. The wedge shaped block has an eye for fixing the hook of the hoisting chain. When the stone is raised, the lewis on account of its shape does not allow the stone to slide down.

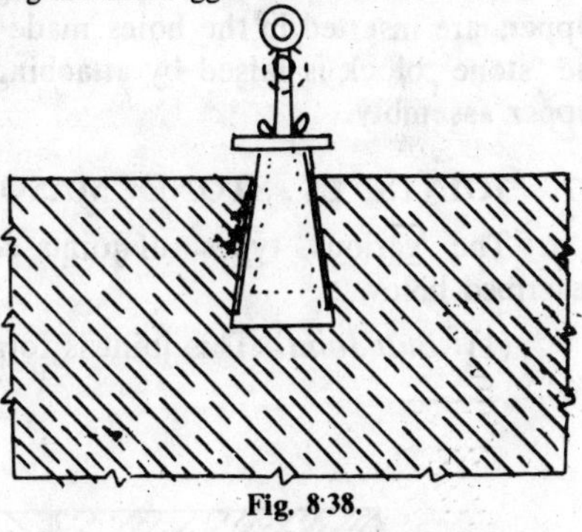

Fig. 8·38.

Modified form of three-legged lewis

Fig 8·39 illustrates a lewis, which is commonly used for lowering and setting stone under water. The wedge-shaped steel piece is inserted first and the rectangular straight piece is placed next in the hole made in the centre of the stone top. It is obvious from the shape of the lewis that when it is lowered down, the slant side of the dovetailed piece will prevent the slant side of the hole from permitting the lewis to get loosened.

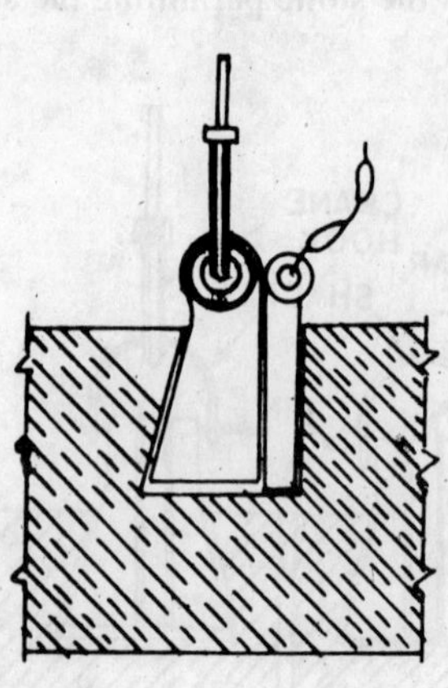

Fig. 8·39

Form of lewis used for lowering stone block

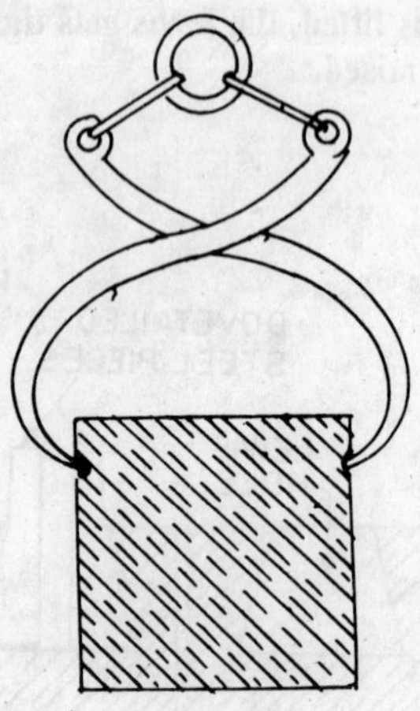

Fig. 8·40

Nippers

2. Nippers

This is another device for lifting stones. The pointed ends of the nippers are inserted in the holes made little below the stone top face and the stone block is raised by attaching a chain to the central ring of the nipper assembly.

8·7. JOINTS IN STONE MASONRY

The various types of joints commonly used in stone work are described below :

(1) *Butt Joint.* This joint is formed by placing two stones side by side.

Fig. 8·41

Butt joint

(2) *Rebated or Lapped Joint.* This joint is usually provided for stones laid on slopes to prevent the movement of one stone over the other by sliding.

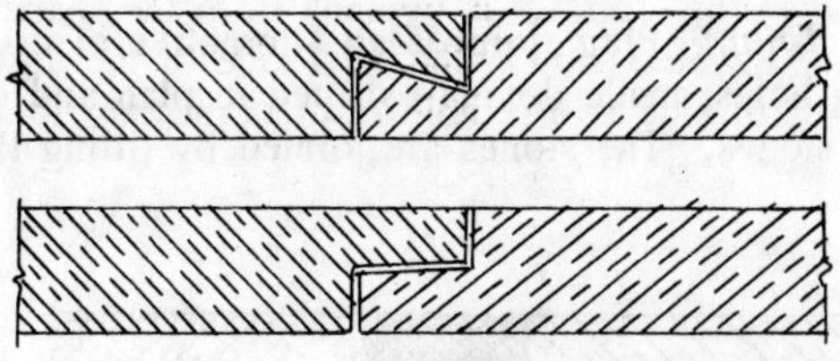

Fig 8·42

Different forms of rebated joints

(3) *Table Joints.* Its function is similar to that of rebated or lapped joint.

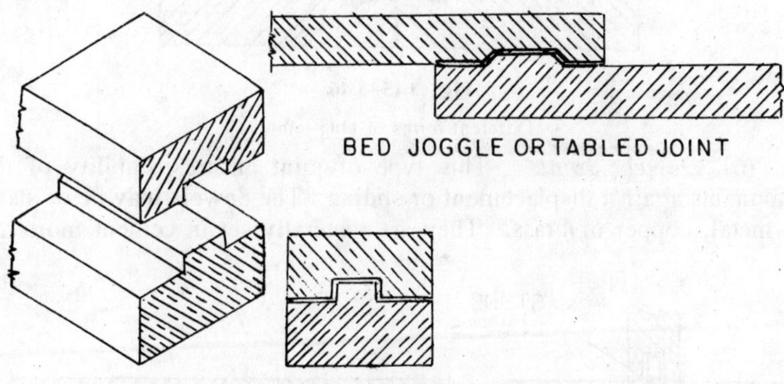

Fig. 8·43

Bed joint or tabled joint

(4) *Cramped joint.* The function of cramp which may be of slate or any metal, is to prevent the tendency of the joint to be pulled apart.

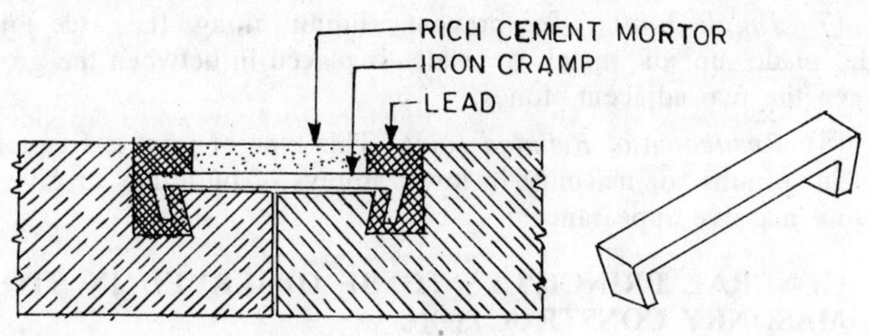

Fig. 8·44

Cramp joint

Wrought iron cramps may also be used, but they must be dipped in oil while hot or galvanized to prevent them from corrosion. The cramp may be of any metal and it should be completely encased with lead, cement or asphalt.

(5) *Plug Joint.* Plug joint is an alternative of cramp joint and consists of plug holes, made dovetail-shaped in plan and cut below the top surface of stones. The stones are jointed by filling the holes with cement or lead.

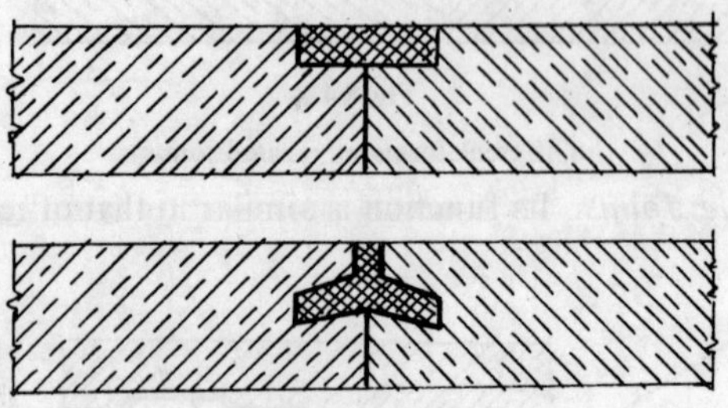

Fig. 8·45–8·46

Different forms of plug joint

(6) *Dowel Joint.* This type of joint ensures stability of the components against displacement or sliding. The dowels may be of slate, gun metal, copper or brass. They are generally set in cement mortar.

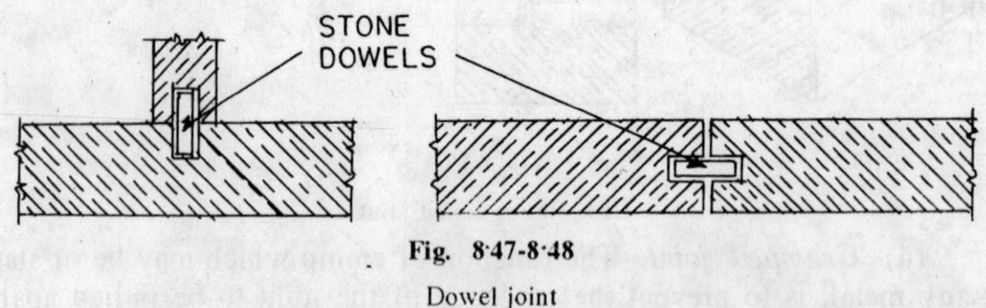

Fig. 8·47-8·48

Dowel joint

(7) *Joggle Joint.* To prevent sliding along the side joints, joggle made up of metal or slate is placed in between the groove between the two adjacent stones.

(8) *Rusticated or Rebated Joint.* This type of joint is frequently used in plinths or masonry in lower storeys of buildings to give the structure massive appearance.

8·8. GENERAL PRINCIPLES TO BE OBSERVED IN STONE MASONRY CONSTRUCTION.

(1) Stone used in a good masonry should be well seasoned, hard, tough, compact grained and uniform in texture and should be free from defects like cracks, flaws, cavities, veins and patches of soft or loose materials.

(2) Stone masonry may be constructed in lime or cement mortar above plinth level, but in damp-proof construction, or water-logged site

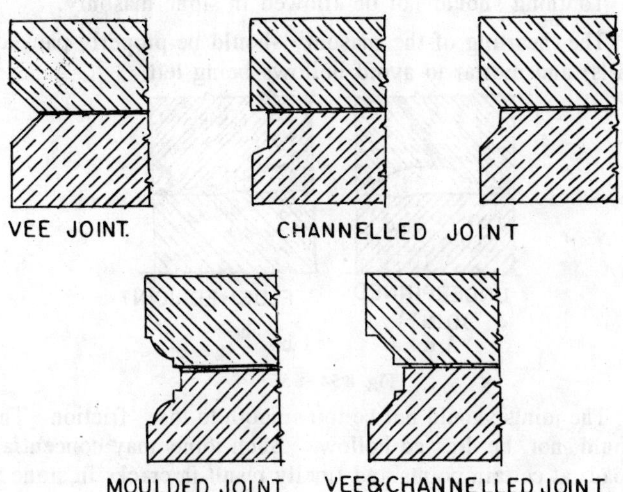

Fig. 8·49-8·53
Rusticated or rebated joints

or in basements, hydraulic lime or cement mortar with surkhi should be used in foundations and upto plinth level.

(3) In stone masonry construction, all stones should be laid upon their natural bed (construction of corbels, and other such projecting features are exceptions), so that pressure on each stone is at right angles to its laminations. Natural bed of stone is the term applied to the original position occupied by the stone during its formation. Corbels, cornice, string courses and other projecting courses are built by so placing the stones that their natural bed is parallel to the face of the supporting wall.

(4) Proper bond should be maintained throughout the masonry. The construction should be well supervised.

(5) The facing and the backing of the wall should be well bound by through stones. They should be laid staggered in the successive courses and kept apart with their centre to centre distance not exceeding 1.5 m.

(6) Quoins used to form the jambs for door or window openings should be of the full height of the course. The breadth of the quoin should be at least 1½ times the depth of the course and its length should be at least twice its depth.

(7) As far as possible all connected walls in a structure should be raised up uniformly and regularly. If breaks are unavoidable and one part of the wall has to be kept behind, the wall should be raked back at an angle of 45° or less. This is necessary to allow proper connection between the old work and the new work.

(8) Toothing should not be allowed in stone masonry.

(9) The hearting of the masonry should be properly packed with chips (spalls) and mortar to avoid hollows being left.

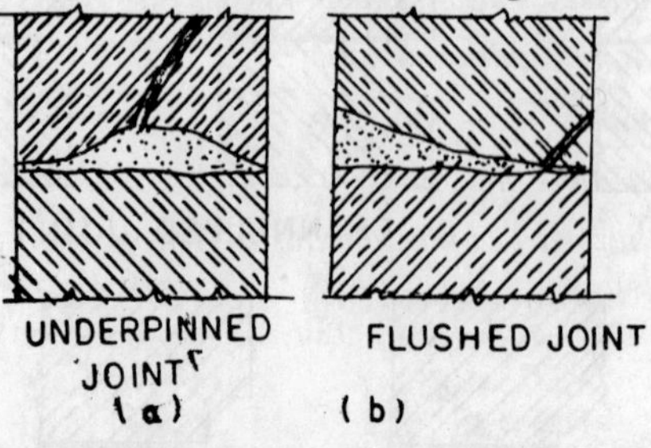

Fig. 8·54—8·55

(10) The joints should not be too smooth to lose friction. The bed joints should not be dressed hollow. Such joints may concentrate the uniform load at certain points and finally result in cracks in stone work. Refer Figs. 8·54-8·55 for explanation.

(11) The vertical joints should be staggered as far as possible.

(12) The vertical surfaces should be constructed truly vertical and checked with plumb rule. Similarly, the battered surfaces should be checked with a wooden template and plumb for the correctness of the slope.

(13) Necessary chases for dowels and cramps etc should be formed in stone before use in masonry.

(14) Stones to be used in the work should be well-wetted before use.

(15) The masonry should not be subjected to tensile stresses.

(16) When it is desired to construct over old or dry surface, it should be well cleaned and wetted before starting new construction.

(17) The exposed joints in stone masonry should be properly pointed. This is done by raking the exposed joints to a depth of 2 cm. and filling them with cement or lime mortar.

(18) The entire work should be cured well, *i.e.* it should be kept well-wetted for at least two weeks.

8·9. CLASSIFICATION OF STONE MASONRY

Depending upon the degree of refinement used in shaping the stone, the finishing adopted and the arrangement of the stone in the construction, stone masonry can be broadly classified in the following two types:

(1) *Rubble masonry*

(2) *Ashlar masonry*

8·10. RUBBLE MASONRY

It is further sub-divided in the following categories :

(*a*) *Uncoursed rubble masonry*

(*b*) *Random Rubble masonry*

(*c*) *Coursed rubble masonry*

(*d*) *Dry rubble masonry.*

(*a*) *Uncoursed Rubble Masonry.* This is the poorest form of stone masonry. The stones to be used for the work are directly obtained from the quarry, after merely knocking off weak corner and edges with the mason's hammer. The stones are carefully laid so as to break joints as much as possible. To avoid thick joints, chips of stone or spalls are wedged into the hearting. The face stones selected from the heap should have uniform colour, uniform bed and greater size. One-fourth of the face stones should tail back in the hearting in the form of headers. Bond stones provided to interlock the two faces should extend up to the full thickness of wall if the wall is less than 60 cm. in thickness. For wall thicker than 60 cm., a line of headers overlapping each other for a length of at least 15 cm. is laid right through the wall. The quoins are chisel or hammer dressed and are laid as header and stretcher alternately. In this work the thickness of joints should not exceed 13 mm.

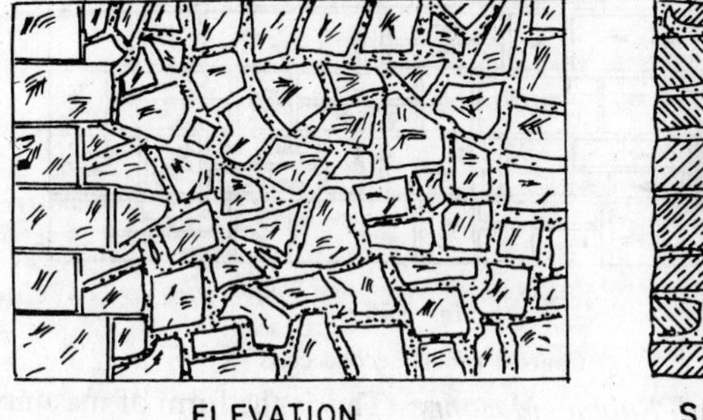

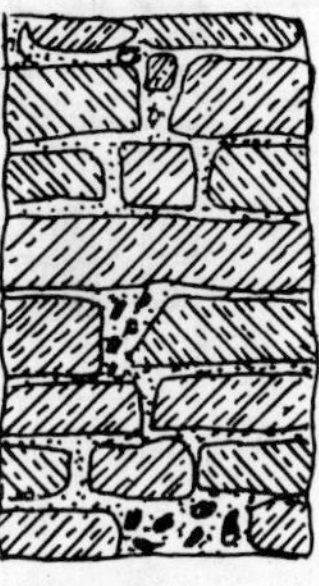

Fig. 8·56

Random rubble masonry (uncoursed)

(*b*) *Random Rubble Masonry.* This form is slightly superior to uncoursed rubble masonry. In this form the stones used in the work are hammer or chisel-dressed. The stones are not suitably shaped or finished and as such the elevation of this type of masonry shows irregular shaped stones with non-uniform joints. In a good work the face stones are of uniform colour and approximately equal in size. The height of stones should be greater than their breadth or length of tail into the work.

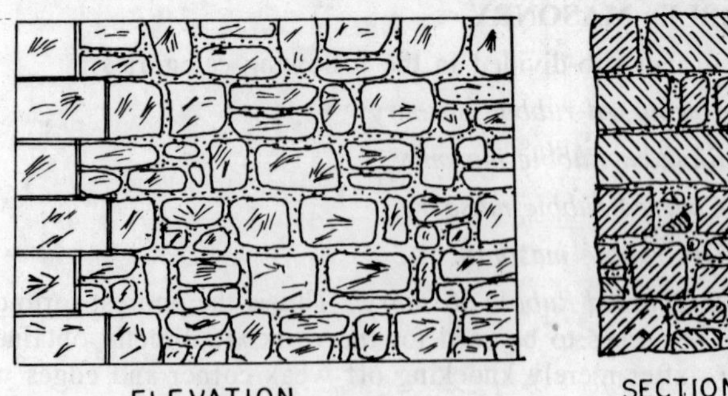

Fig. 8·57

Random rubble masonry

At least one fourth of the face stone should tail back into the hearting for ensuring proper strength to the work. The quoins and the through stones are provided in a similar manner as described earlier. The thickness of joints should not exceed 13 mm.

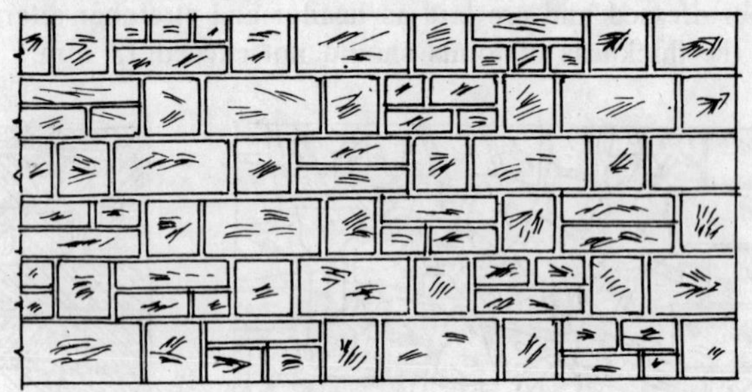

Fig. 8·58

Coursed rubble second class

(*c*) *Coursed Rubble Masonry.* This is the form of masonry which is commonly adopted in the construction of residential buildings public buildings, piers and abutments for ordinary bridges. Considering the dressing and finishing of the stones, it is further sub-divided into first class, second class and third class masonry. In first class masonry, generally all the courses are of the same height and the minimum height of the course is limited to 15 cm. The face stones are hammer dressed and the projection of the rock-faced surface does not exceed 38 mm. beyond the side or bed joints. The beds of the face stones are hammer or chisel dressed and rendered true and square. In good work, at least one third of the face stones tail back into the hearting for a distance equal to two times their height normally, and three times their height for thicker

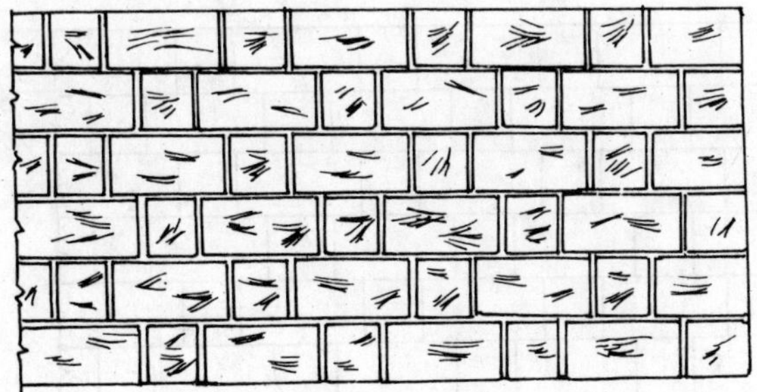

Fig. 8·59

Coursed rubble first class

walls. The through stones provided to bind the two faces together are spaced at 1·8 m. apart and the quoins are of the same height as the height of the course. The length of the quoin is generally kept 45 cm. The thickness of joints in this type should not exceed 10 mm.

(*d*) *Dry Rubble Masonry.* The construction of this form is similar to that of ordinary rubble masonry without mortar.

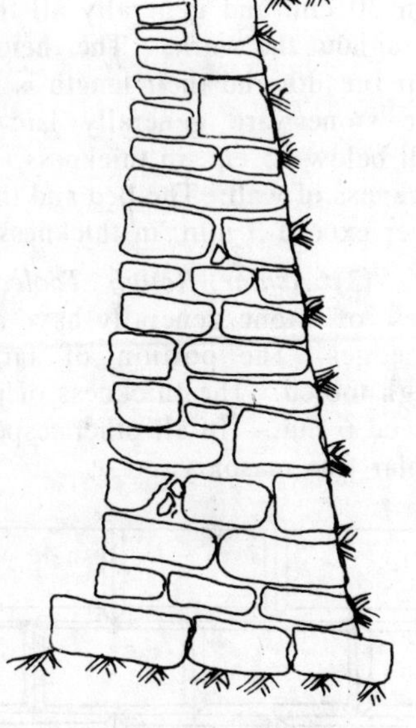

Fig. 8·60

Dry stone retaining wall

8·11. ASHLAR MASONRY

The work built from carefully dressed stones with accurate bedding and jointing is termed as ashlar masonry. This construction is further subdivided into various types. The different types of ashlar masonry are given below :

(1) *Ashlar fine*

(2) *Ashlar rough tooled*

(3) *Ashlar rock, rustic or quarry faced*

(4) *Ashlar chamfered*

(5) *Ashlar facing*

(1) *Ashlar Fine.* In this type of masonry, all the stones are fine tooled, on all bed and side joints, and the faces are rendered perfectly true to the pattern desired. The height of the courses is never less

Fig. 8·61

Ashlar fine

than 30 cm. and generally all the courses are kept of the same height throughout the work. The height of the stones used is never less than their breadth and their length is never less than twice their height. The face stones are generally laid as header and stretcher alternately. For wall below 75 cm. in thickness, the through stones extend to the full thickness of wall. The bed and the side joints in this type of work should never exceed 3 mm. in thickness.

(2) *Ashlar Rough Tooled.* In this type of masonry, the exposed faces of stone generally have a fine dressed chisel drafting all round the edges. The portion of face stone enclosed by the chisel draft is rough tooled. The thickness of joints in this type of work should never exceed 6 mm. In all other respects it conforms to the specifications of Ashlar fine masonry.

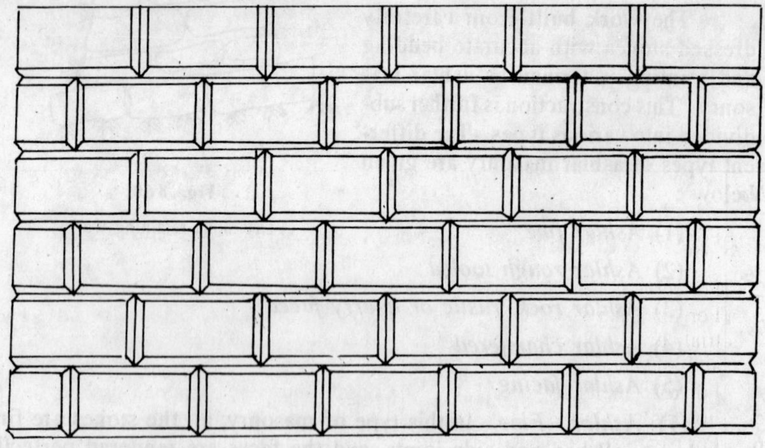

Fig. 8·62

Ashlar Chamfered

(3) *Ashlar Rock, Rustic or Quarry Faced.* This type of construction is similar to ashlar rough tooled except that the exposed faces of face stones between the chisel drafting are left rough. The projections in the space enclosed by chisel drafts are restricted to 75 mm only.

(4) *Ashlar Chamfered.* This type of masonry is similar to the one described above with the only difference that the edges around the exposed faces of stone are beveled off at an angle of 45° for depth of 25 mm. or more.

(5) *Ashlar Facing.* In this type of masonry the faces of stone are rough tooled, and chamfered and the stones are provided in face work only. The backing may be made in brick, concrete, or rubble as desired. The composite construction reduces the cost of work appreciably. The height of the course is never kept less than 20 cm. and the width of each stone is about 1½ times its height. The bed joints of all the stones are dressed perfectly true and square. For walls upto 75 cm. in thickness, the bond stones should extend for the full thickness of the wall and for thicker walls, the bond stones should overlap each other by 15 cm.

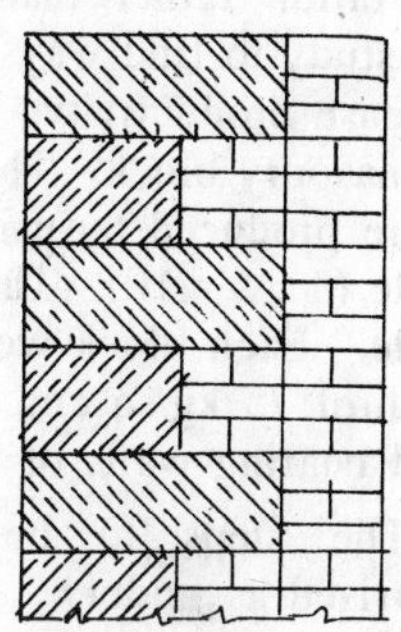

Fig. 8·63
Ashlar facing (stone facing with brick backing)

8·12. SELECTION OF STONE FOR MASONRY

In selecting a stone the most important factor is the situation in which it is to be used. Table (8·1) gives the recommended use of common types of stones for different situations.

TABLE 8·1
Recommended Use of Common Types of Stone

S.No.	*Situation*	*Types of Stone Used*
1.	For face work of building.	Marble, granite and close-grained sand stone.
2.	For curved or ornamental works.	Soft sand stone and marble.
3.	For masonry work in industrial town exposed to smoke and chemical fumes.	Granite, compact sand stone and quartzite.
4.	For road metal	Basalt and coarse-grained granite.
5.	For railway ballast.	Sand stone, granite and quartzite.
6.	For pavings, door sills and steps.	Marble, slates and sand stones.
7.	For bridges, piers, docks and other marine structures	Fine-grained granite and gneisses.
8.	For fire resistant masonry	Compact sand stone.

8·13. STONE MASONRY BLOCK CONSTRUCTION

The minimum thickness of random rubble masonry wall varies from 30 to 38 cm. in different parts of the country. Stone masonry wall construction is not only massive but it is time consuming and also requires skilled labour. The thickness of the wall is adopted mainly from practical consideration rather than strength requirements. CBRI Roorkee undertook a study to find ways and means to reduce the thickness of wall and to increase productivity. This has resulted in the production of pre-cast stone masonry block. The stone blocks of 30 cm. x 20 cm. x 15 cm. in size are produced by using stones upto 12 cm. in size and lean cement concrete (1 : 5 : 10). These blocks have facing of stone and backing of concrete. Each block weighs about 21 kg and has an average crushing strength of 75 kg/sq cm. These blocks can be used in the construction of load bearing walls of buildings upto four storey high.

The steps to be followed in casting the stone blocks are summarised as under :

(1) Place the mould (of timber or steel) on the casting platform and apply a coat of oil on the inner surfaces.

(2) Arrange two to three stones of nearly 12 cm. in size inside the mould and fill the gap between stone pieces with cement concrete 1 : 5 : 10 upto a height of about 5 cm.

(3) Fill the balance portion of the mould by placing 75 mm to 60 mm size pieces of stone in gaps and pouring same mix of cement concrete upto top of mould.

(4) The contents of the mould are thereafter compacted either by use of a plate vibrator or by roding and tamping.

(5) The cast stone blocks are cured for 7 days and subsequently dried in air for about 21 days before use.

The advantages of stone blocks are :

(1) The appearance of coursed rubble masonry is achieved by use of natural stone textured face of the stone block at much lesser cost.

(2) The blocks can be manufactured by ordinary labour either at factory or at the site itself.

(3) On account of regular shape and size of blocks, the speed of construction is faster.

(4) It is possible to construct 15 or 20 cm. thick load bearing walls thereby effecting considerable saving in material and achieving larger usable floor area.

(5) The internal plastering of wall can be avoided.

8·14. SAFE PERMISSIBLE LOADS ON STONE MASONRY.

Stone masonry work can be carried out in cement mortar or in lime mortar. The strength of stone masonry work depends upon the quality of stone, type of masonry (Ashlar, coursed rubble or random rubble) and the type of mortar used.

Safe allowable loads on various types of stone masonry (using good quality stone) with different types of mortar is given in table 8·2.

TABLE 8·2

Safe Permissible Load on Stone Masonry

S.No.	*Type of Stone Masonry*	*Safe Permissible Stress in*	
		Tonne/m²	*kN/m²*
1.	Ashlar masonry in 1:3 Cement Mortar		
	(*a*) Using Granite Stone	160	1600
	(*b*) Using sand stone	110	1100
2.	Ashlar masonry in 1:6 cement mortar		
	(*a*) Using Granite stone	130	1300
	(*b*) Using Sand stone	90	900
3.	Ashlar masonry in lime mortar		
	(*a*) Using Granite stone	110	1100
	(*b*) Using Sand stone	80	800
4.	Coursed rubble masonry in 1 : 3 cement mortar		
	(*a*) Using Granite Stone	120	1200
	(*b*) Using Sand Stone	100	1000
5.	Coursed rubble masonry in 1 : 6 cement mortar		
	(*a*) Using Granite Stone	100	1000
	(*b*) Using Sand Stone	80	800
6.	Coursed rubble masonry in lime mortar		
	(*a*) Using Granite Stone	90	900
	(*b*) Using Sand Stone	70	700
7.	**Random rubble masonry in 1 :3 cement mortar**	**70 to 100**	**700 to 1000**
8.	**Random rubble masonry in 1:6 cement mortar**	**50 to 60**	**500 to 600**
9.	**Random rubble masonry in lime mortar**	**40 to 50**	**400 to 500**

SPECIFICATION—8·1

Random Rubble Masonry

1. *Quality of Stone.* The stone used shall be hard, sound and free from defects of any kind. It should be of specified type and of approved quality and colour.

2. *Dressing.* The stone shall be hammer dressed on the sides and bed. The *'Bushing'* in the exposed face of the stone shall not project more than 40 mm. In case of stone face to be plastered, the *'Bushing'* shall not exceed 10 mm.

3. *Laying.* As far as possible the stones selected for the face work shall be nearly equal in size. The stone shall be laid so as to tail back and bond well into the backing. The height of face stones shall not be greater than either the breadth of face or the length of tail into the work. The stones used for hearting shall not be less than 150 mm in size and shall be solidly bedded in the mortar. Chips and spawls of stones shall be wedged into the work wherever necessary to avoid thick mortar beds on joints. The chips used shall not be more than 20% by volume of the masonry. No hollow space shall be left in the masonry.

4. *Joints.* Every stone shall be carefully laid so as to break joints as much as possible and to avoid long vertical line of joints. The face joints shall be neat and close and shall not exceed 25 mm. in thickness. In case pointing or plastering is required, the joints shall be raked to a minimum depth of 20 mm by a raking tool during progress of work when mortar is still green.

5. *Bond Stones.* Headers, bond stones or through stones running right through the full thickness of the wall, shall be provided in walls upto 60 cm. in thickness. For walls more than 60 cm. in thickness, two or more headers (bond stones) overlapping each other by about 15 cm shall be laid in a line from face to back. At least one bond stone or a set of bond stones (for thicker walls shall be inserted in every 0.5 sq. m. of the wall face.

6. *Quoins.* The quoins shall be neatly dressed to form the required angle at the corner. The quoins shall be laid as header and stretcher alternately. The length of the quoins shall be at least 30 cm. (25% of them being not less than 50 cm in length) and its volume shall not be less than 0.025 cu. m.

7. *Curing.* In case the masonry is in cement mortar, the curing shall start just after 24 hours of the laying of masonry and shall be continued for at least seven days. In case of masonry in fat lime mortar, curing shall start two days after the laying of masonry and shall be continued for at least seven days.

SPECIFICATION 8·2

Coursed Rubble Masonry

1. *Quality of Stone.* Same as for random rubble masonry.

2. *Dressing.* The face stones shall be hammer or chisel dressed to approximate rectangular block shape on all joints and beds. The bed joints and the side joints shall be dressed true and square for a depth of at least 50 mm. and 40 mm. respectively from the face. The *'Brushing'* on the exposed face of stones shall not project more than 40 mm. In case of the stone face to be plastered the *'Brushing'* shall not exceed 10 mm.

3. *Height of Courses.* The stones shall be laid in horizontal courses not less than 15 cm. nor more than 30 cm. in height. The stones in each course shall be of equal height so as to get all the courses truly horizontal and all joints truly vertical.

4. *Size of stones.* The height of the faces stones shall not be less than their breadth and they shall tail back into the work for a length not less than their height.

5. *Laying.* The face stones shall be laid alternate headers and stretchers. Stones shall break joint at least half the height of course. At least $\frac{1}{3}$rd of the stones shall tail back into the work for a distance not less than twice their height. The interior filling of the wall shall consist of flat bedded stones carefully laid on their proper beds and solidly bedded in mortar. Chips and spawls shall be wedged into the work, wherever necessary, to avoid thick joints. The chips used shall not be more than 10% by volume of the masonry. No hollow space shall be left in the masonry.

6. *Joints.* All side joints shall be vertical and bed joints horizontal. The mortar joints in the face work shall not exceed 10 mm. in thickness.

7. *Bond Stones.* Bond stones running right through the full thickness of the wall shall be provided in walls upto 60 cm. in thickness. For walls more than 60 cm. in thickness, two or more headers overlapping each other by at least 15 cm. shall be laid in a line from face to back. Bond stones or a set of bond stones shall be inserted 1·5 to 1·8 m. apart clear in every course.

8. *Quoins.* The height of quoins shall be the same as that of the course in which they occur. the quoins shall be laid as stretchers and headers alternately and shall be formed of stones at least 40 cm. in length. They shall be laid square on their bed, which shall be fair dressed to a depth of at least 10 cm.

9. *Curing.* Same as for random rubble masonry.

Questions for Review

1 What general principles would you keep in mind while supervising a (*i*) Stone work and (*ii*) Brick work under construction ? What guiding principles would you apply to ensure a structurally sound, aesthetically pleasant and a comfort-cum-service type construction ? Give only a general view.

2 (*a*) What is the most common type of building construction that you see being practised in Bombay and what are its advantages over any other type ?

(*b*) Explain with neat sketches the following :

(*i*) Coursed rubble masonry construction indicating clearly the position of through stone. What is the quality of mortar used ?

(*ii*) The different devices used for lifting heavy and moulded stone.

(*iii*) Method of finishing pointing in stone masonry construction.

3. Give detailed specification for the following items of work :

(*a*) First classs brick work in cement mortar.

(*b*) Ashlar stone masonry in cement mortar.

4. Sketch a section of a wall showing all the details starting from the foundation concrete, bed block upto the coping. The section should indicate two storied building with a flat roof. The section line should pass through a door at ground floor and a window at the first floor.

5. Describe with sketches the difference between uncoursed rubble masonry and random rubble masonry.

6. What are the types of joints used in stone masonry ? Describe each joint with neat sketch.

7. What is the purpose of dressing stone ? Discuss the different types of surface finishes employed in stone masonry construction.

8. Write short notes on :

(*i*) Lewis

(*ii*) Cramp joint

(*iii*) Joggle joint

(*iv*) String course

(*v*) Chain dog

(*vi*) Rusticated joints

9. Define the following terms with sketches where necessary.

(*i*) Perpend

(*ii*) Plinth course

(*iii*) String course

(*iv*) Blocking course

(*v*) Copings

(*vi*) Weathering and throating

(*vii*) Through stone.

10. Compare between stone masonry and brick masonry.

11. Explain with neat sketches and proper reasoning where you would adopt the following:

(*i*) Fine ashlar construction

(*ii*) Coursed random rubble

(*iii*) Dry stone revetment

(*iv*) Artificial stones

(*v*) Headers quoins ; Copings.

12. Explain the following with sketches wherever necessary :

(*i*) Corbelling

(*ii*) Template

(*iii*) Composite masonry

(*iv*) Buttress.

13. Give detailed specifications for coursed rubble masonry in cement mortar for a wall 45 cm. thick. Illustrate the same with sketches.

14. Name the different types of rubble masonry. What are the important points which have to be kept in view in the construction of rubble masonry ?

15. What do you mean by masonry work ? Name different kinds of masonry. Give general specifications for masonry work.

16. State the different, classifications of stone masonry. State also the general principles to be observed and the precautions to be taken in constructing stone masonry.

17. (*a*) What do you understand by dressing of stones ? Describe various methods of dressing. Explain with sketches.

(*b*) What are the general principles to be observed in the construction of masonry ?

18. (*a*) State the comparative merits and demerits of stone masonry and brick masonry.

(*b*) Give a section of a wall from ground level to roof and show clearly the plinth course, string course, cornice, blocking course and coping. There are two floors supported by the wall.

19. (*a*) What are the objects of dressing of stones ?

(*b*) Write a short note on block-in-course masonry and draw a neat sketch of the same.

(*c*) What types of stone masonry will you use for

(*i*) boundary wall

(*ii*) parapet wall

(*iii*) pier of a bridge

(*iv*) monument and

(*v*) market-place

9

Brick Masonry

9·1. DEFINITION

Construction of brick units bonded together with mortar is termed as brick masonry. The strength of the brick-work primarily depends upon quality and strength of the brick, the type of mortar and the method of bonding, adopted in construction. In addition, the strength of a brick wall is also dependent upon its slenderness ratio, lateral pressure due to wind and the degree of soundness in construction. Mortar not only acts as a cementing material but also imparts strength to the work by holding the individual bricks together to act as a homogeneous mass. Mortar is usually a mixture of cement and sand or lime and sand, or a mixture of three. For works of temporary nature, mud is generally used as a mortar and the construction is termed as kuccha pukka masonry. Cement mortar is used for works of permanent character, where strength of work is of vital importance. Lime mortar is used for all types of constructions.

9·2. TYPES OF BRICKS

Bricks can be divided into two types (*i*) Traditional bricks and (*ii*) Modular bricks.

The dimensions of traditional bricks vary from 21 to 25 cm in length, 10 to 13 cm in width and 7·5 cm in height in different parts of the country. The commonly adopted nominal size of a traditional brick is 23 x 11·4 x 7·6 cm.

With a view to achieve manufacture of uniform size of bricks all over the country, Indian Standard Institution has established a standard size for the brick. Such a brick is known as modular brick. The nominal size of a modular brick is taken as 20 x 10 x 10 cm whereas its actual dimensions are 19 x 9 x 9 cm.

Masonry with modular bricks work out to be cheaper since there is saving in the consumption of bricks, mortar and labour as compared with masonry with traditional bricks.

Nominal thickness of wall. The thickness of wall which is to be considered in evaluating the quantity of brickwork in estimates is called nominal thickness of wall. In fact it is a fictitious dimension which represents neither the actual thickness of wall (after excluding surface finishes like plaster renderings etc.) nor the overall thickness of wall including plaster rendering etc. Thus the nominal thickness of one brick wall using traditional brick of 23 cm. length (with allowance of 12 mm mortar joint) is taken as 23 cm. whereas its actual thickness varies from 21·8 cm to 22·4 cm. Similarly the nominal thickness of one brick wall using modular brick (with allowance of 10 mm. for mortar joint) is taken as 20 cm. where as its actual thickness is of 19 cm.

9·3. GENERAL PRINCIPLES TO BE OBSERVED IN BRICK MASONRY CONSTRUCTION

Brick masonry is classified as first class or second class according to the type of bricks used (whether first class or second class) and according to the method of laying. The strength of masonry work, however, depends very much upon the type of material used, nature of workmanship and supervision. Bad workmanship assisted by lack of supervision may lead to unsound construction inspite of the materials used being of the best possible type. The general principles which should be observed for a sound brick masonry construction are given below. (The technical terms used in this chapter have already been described in the chapter on Stone Masonry.)

(1) The bricks used in a good work should be sound, hard and well burnt with uniform size, shape and colour. They should have no cracks or flaws and a fractured surface should be free from holes, grit or lumps of lime etc. The surface of the bricks should be such that no mark should be made when scratched by the finger nail. The bricks when struck together should produce metallic ringing sound and they should not break when dropped from a height of 90 to 120 cm. on the ground. The bricks should not as a rule absorb water more than 20% of their weight when dry, when immersed in water for twenty-four hours.

(2) The bricks should be thoroughly soaked in clear water before use for suitable period so that the water just penetrates the full depth of bricks. This not only helps in removing the dirt, dust and other soluble salts (which cause efflorescence) from the bricks, but also reduces their tendency of suction of water from wet mortar. *In case, however, masonry is to be carried out in mud or fat lime mortar, the bricks should not be soaked in water before use.*

(3) The bricks should be laid on a full bed of mortar. They should be slightly pressed into the bed mortar while laying so as to ensure proper adhesion. All the courses should be laid truly horizontal and all the vertical joints should be truly vertical.

(4) All the joints should be properly flushed and filled with mortar so that no cavity is left in between.

(5) In case of walls two brick or more in thickness, the joints should be grouted at every course in addition to bedding and flushing with mortar.

(6) No brick-bats should be used in the work except when it is absolutely necessary for obtaining the specified bond.

(7) Brick work is generally laid in the English bond. In all cases, it should be ensured that a proper bond is maintained throughout the work.

(8) Only specified mortar of a good quality should be used in the work, taking great care that uniform mortar joint is obtained throughout the construction. Thickness of joints should not exceed 13 mm in any case.

Fig. 9·1

Raking back of a brick wall

(9) Unless brick -on-edge is specified, the bricks must be laid on their proper beds with their frogs pointing upwards.

(10) The courses of bricks at the plinth, window sill, floor/roof level and at the top of parapet wall should invariably be laid with brick on edge.

(11) The progress of work in raising masonry should be such that all the connected brick work should be carried up in uniform layers at one level. It should be ensured that no part of the wall should preferably rise more than 100 cm. than the rest of it. When it is not possible, the work should be raked back according to the bond. The walls should be raised up truly vertical.

(12) The buttresses, counterforts etc. should be built up course by course, maintaining proper bond with the main wall and should not be added or joggled afterwards.

(13) Iron fixtures like pipes, hold-fasts of doors and windows etc., which are to be fixed in the brick wall should be embedded in cement mortar or in cement concrete.

(14) The face joints are generally raked to a minimum depth of 10 to 15 mm. when the mortar in the joints is still green. This is done to provide proper key for the plastering or pointing.

(15) Plastering should be done after about 28 days of completion of brick masonry. This permits adequate time for the shrinkage in masonry and concrete to take place before plastering operation is carried out.

(16) In works where plastering or pointing is not desired, the mortar joints should be struck flush and finished at the time of laying.

(17) All the finished masonry work should be kept wet for at least seven days.

(18) When an existing wall has to be strengthened by making it thicker, 20 cm x 20 cm x 10 cm. (dimensions of one brick) recesses are cut in the existing wall. The recesses are staggered taking care that one recess is made in every sq. metre of the wall. This enables the new work to be properly bonded to the old one. This operation is termed as *back-bonding*.

(19) When it is desired to increase the length of the wall under construction at a future date, the wall is stopped with a toothed end. This is necessary to ensure continuous bonding between the old and the new work.

(20) It is found that with an average winter-summer temperature differences of about 100°F, a brick masonry wall expands by about 10 mm. in every 30 m. length of wall. Hence it is desirable to provide minimum 18 mm. wide expansion joints after every 30 to 45 m. length of wall.

(21) When timber floor or roof is required to be supported on masonry walls, the ends of timber joists (supporting the floor or roof) should rest on corbels or brackets as far as possible. In cases where the ends of timber joists have to be built into the wall itself, it is necessary to apply suitable preservative treatment to the embedded portion of the joists and in addition some space should be left around them to minimise chances of attack by termite and to ensure free circulation of air.

(22) Bed blocks of stone, concrete or reinforced concrete should be provided at the ends of beam (it may be of timber, steel or R.C.C.) carrying heavy loads.

(23) The bearing of R.C.C. floor or roof slab resting on brick masonry walls should not be less than the effective depth of slab or 10 cm., which ever is more.

(24) As a general practice, quoin brick should be headers and stretchers laid in alternate courses and the bond is developed by placing a quoin closer next to the queen header.

(25) Half brick partition walls should be reinforced with suitable

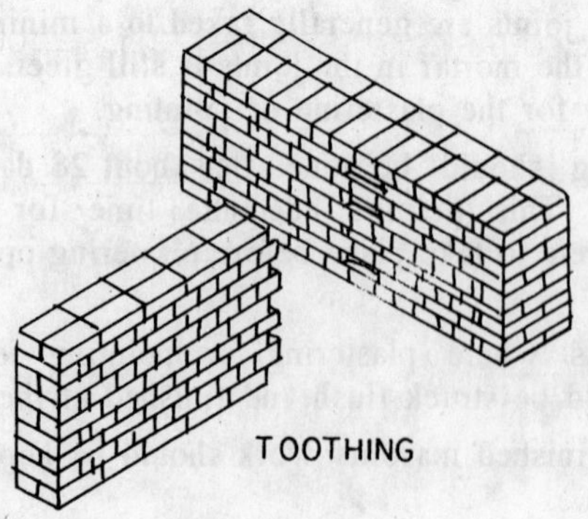

Fig. 9·2

reinforcement placed at every third or fourth course of the brick work. The reinforcement may be in form of mild steel bars, flat bars, hoops iron, expanded mesh, or steel fabric etc.

9·4. BONDS IN BRICK-WORK

On account of their uniform size and shape, the bricks can be arranged in a variety of patterns giving rise to different types of bonds. Bonding is essential to eliminate continuous vertical joints both in the body as well as in the face of wall thereby imparting strength to masonry. A wall having defective arrangement of bricks, reduces the strength and stability of the structure. A wall having continuous vertical joint does not act as a homogeneous mass to distribute the superimposed load. On the other hand, it may be assumed to be consisting of small columns and when a particular column of this wall comes under the load, it fails on account of its inability to distribute the load to the portion of wall on either side of it. The different types of bonds commonly adopted are given below :

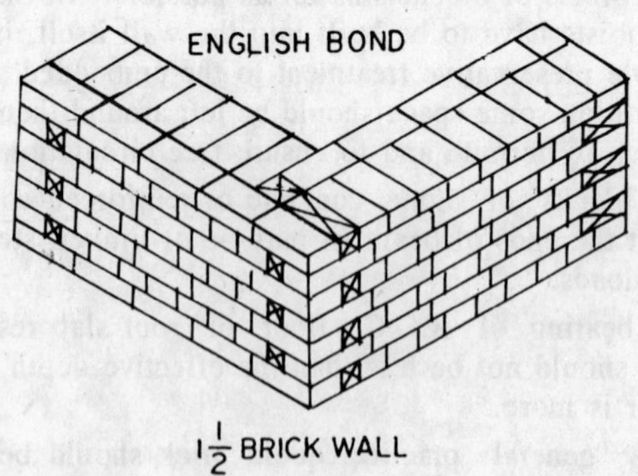

Fig. 9·3

Isometric view of 1½ brick thick wall in English bond.

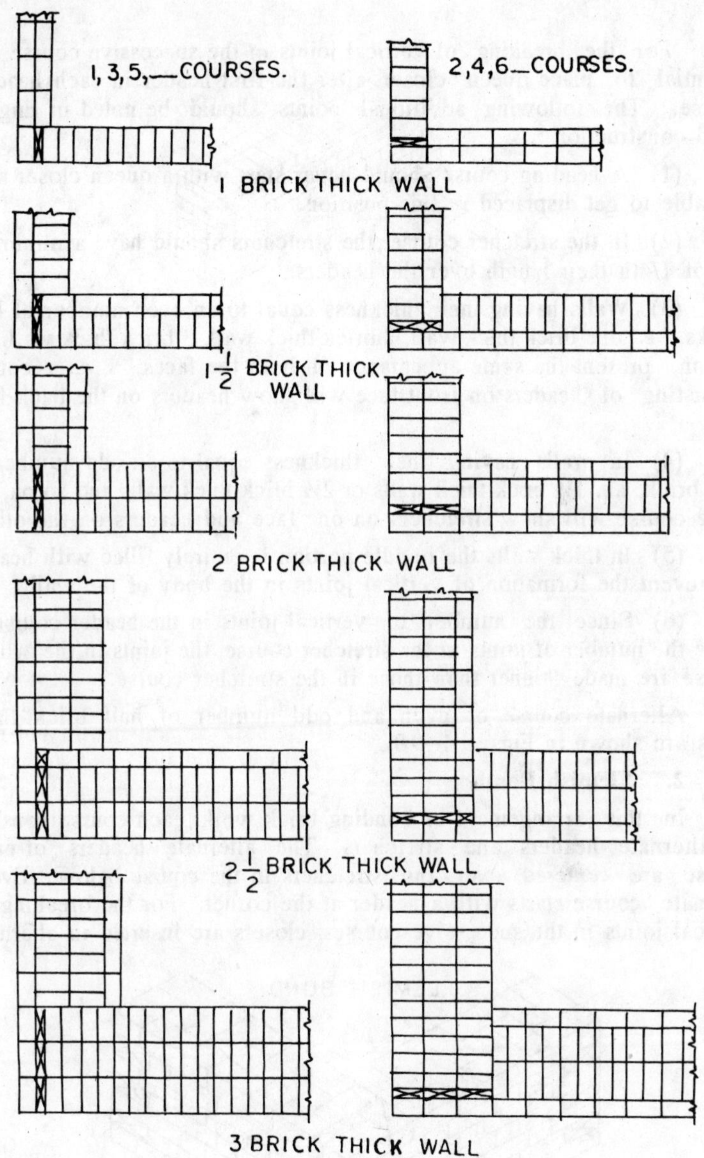

Fig. 9·4 — 9·8

Plans showing alternate courses of brick walls in different thickness in English Bond

1. English Bond

This bond consists of alternate course of heaáers and stretches. In this arrangement, vertical joints in the header courses come over each other and the vertical joints in the stretcher course are also in the same line. For the breaking of vertical joints in the successive course it is essential to place queen closer, after the first header in each heading course. The following additional points should be noted in English bond construction :

(1) A heading course should never start with a queen closer as it is liable to get displaced in this position.

(2) In the stretcher course, the stretchers should have a minimum lap of 1/4th their length over the headers.

(3) Walls having their thickness equal to an even number of half bricks, *i.e.*, one brick thick wall, 2 brick thick wall, 3 brick thick wall and so on, present the same appearance on both the faces, *i .e.* a course consisting of headers on front face will show headers on the back face also.

(4) In walls having their thickness equal to an odd number of half brick, *i.e.* 1½ brick thick walls or 2½ brick thick walls and so on, the same course will show stretchers on one face and headers on the other.

(5) In thick walls the middle portion is entirely filled with header to prevent the formation of vertical joints in the body of the wall.

(6) Since the number of vertical joints in the header course is twice the number of joints in the stretcher course, the joints in the header course are made thinner than those in the stretcher course.

Alternate course of even and odd number of half brick thick walls are shown in Figs. 9·4—9·8.

2. Flemish Bond

In this arrangement of bonding brick work, each course consists of alternate headers and stretchers. The alternate headers of each course are centered over the stretchers in the course below. Every alternate course starts with a header at the corner. For the breaking of vertical joints in the successive courses, closers are inserted in alternate

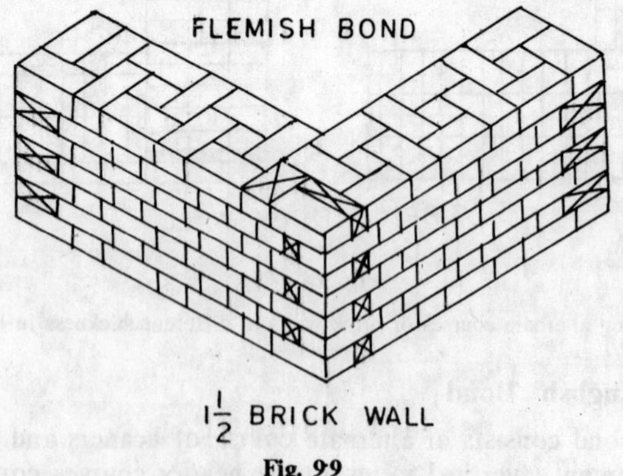

Fig. 9·9

Isometric view showing 1½ brick thick wall in Flemish Bond.

courses next to the quoin header. In walls having their thickness equal to odd number of half bricks, bats are essentially used to achieve the bond.

Flemish bond is further divided into two different types *viz*. Single Flemish bond and Double Flemish bond.

Single Flemish Bond. This bond is a combination of English and Flemish bond. In this work the facing of the the wall consists of Flemish bond and the backing consists of English bond in each course. This type of bonding cannot be adopted in walls less than one and a half brick in thickness. This bond is adopted to present the attractive appearance of Flemish bond with an effort to ensure full strength in the brick work. The alternate courses of brick walls of various thickness in Single Flemish bond are shown in Figs. 9·10-9·12.

Double Flemish bond. In this system of bonding brick work, each course presents the same appearance both in the front and back elevations. Every course consists of headers and stretchers laid alternately. This type of bond is best suited from considerations of economy and appearance. It enables the one brick wall to have flush

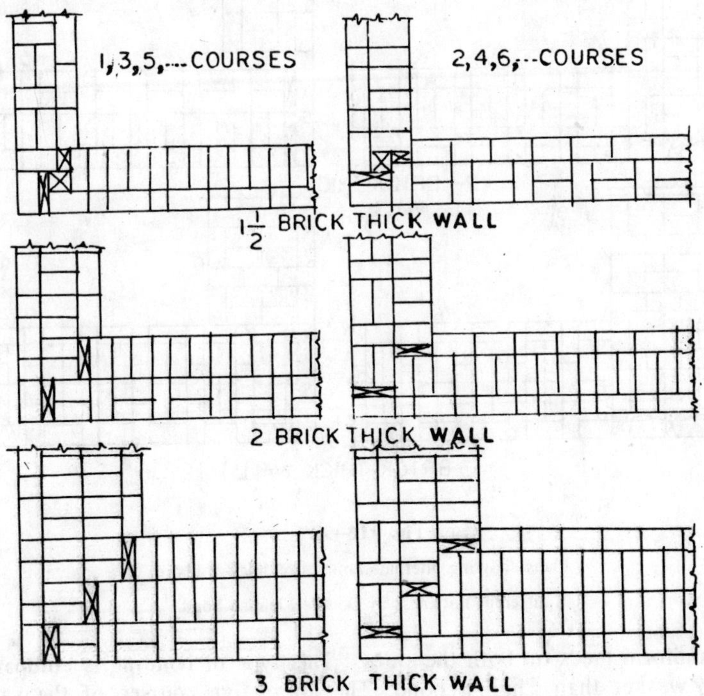

Fig. 9·10-9·12

Plans showing alternate course of brick wall in different thickness in single Flemish bond

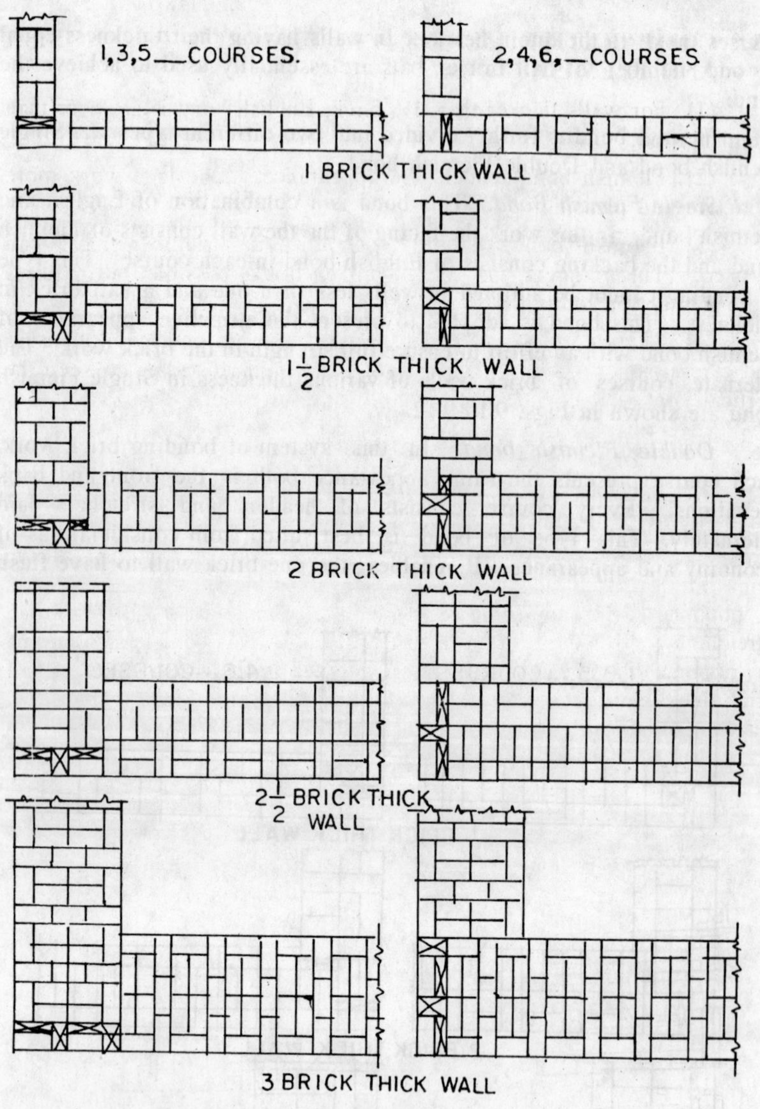

Fig. 9·13–9·17

Plans showing alternate courses of brick walls in different thickness in Double Flemish bond.

and uniform faces on both the sides. This type of bonding is comparatively weaker than English bond. The alternative courses of the walls of various thickness in Double Flemish bond are shown in Figs. 9·13-9·17.

9·5. COMPARATIVE MERITS AND DEMERITS OF ENGLISH AND FLEMISH BOND.

(1) For walls thicker than 1½ brick, English bond is stronger than Flemish bond.

(2) Flemish bond renders the appearance of the face work more attractive and pleasing.

(3) Flemish bond is slightly economical as a number of bats can be used. This renders the use of broken bricks possible, but requires more mortar for additional joints.

(4) The adoption of Flemish bond requires good workmanship and careful supervision. Thus extra attention is necessary to keep the vertical joints in alternate course one above the other.

9·6. OTHER TYPES OF BONDS

The bonds described below are not very common in use. Their use is restricted to special types of construction.

(*i*) **Stretching bond**. In this arrangement of bonding, all the bricks are laid as stretchers. The overlap, which is usually of half brick, is obtained by commencing each alternate course with a half brick bat. Stretching bond is used for half brick wall only. This bond is also termed as running bond and is commonly adopted in the construction of half brick thick leaves of cavity walls, partition walls, etc. Since there are no headers, suitable reinforcement should be used for structural bond.

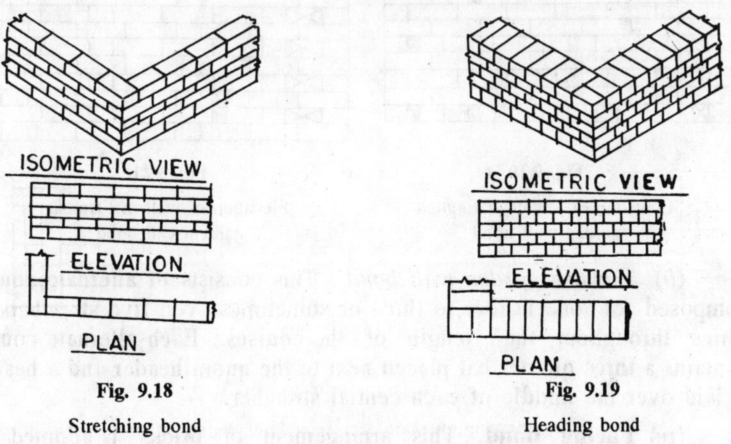

Fig. 9.18

Stretching bond

Fig. 9.19

Heading bond

(*ii*) **Heading bond.** In this type of bonding all the bricks are laid as headers on the faces. The overlap, which is usually of half the width of the brick is obtained by introducing a three-quarter bat in each alternate course at quoins. This bond permits better alignment and as such it is used for walls curved on plan. This bond is chiefly used for footings in foundations for better transverse distribution of load.

(*iii*) **Garden wall bond**. This type of bond is suitably adopted for one brick thick wall which may act as a garden wall or a boundary wall. In garden wall bond, it is possible to build uniform faces for a wall without much labour or expense. This type of bond is not so strong as English bond and its use is restricted to the construction of dwarf walls or other similar types of walls which are not subjected to large stresses. On accounts of its good appearance, this bond is sometimes used for the construction of the outer leaves of cavity walls.

There are two types of garden wall bond.

(*a*) *English garden wall bond*

(*b*) *Flemish garden wall bond*

(*a*) *English garden wall bond.* The general arrangement of bricks in this type of bonding is similar to that of English bond except that the heading courses are only inserted at every fourth or sixth course. Usually the arrangement consists of one course of headers to three courses of stretchers. A queen closer is placed next to the quoin header of the heading course to give the necessary lap.

Fig. 9·20

Elevation of a wall in English garden wall bond

Fig. 9·21

Elevation of wall in Flemish garden wall bond

(*b*) *Flemish garden wall bond.* This consists of alternate course composed of one header to three or sometimes even five stretchers in series throughout the length of the courses. Each alternate course contains a three quarter bat placed next to the quoin header and a header is laid over the middle of each central stretcher.

(*iv*) **Facing bond.** This arrangement of bricks is adopted for thick walls, where the facing and backing are desired to be constructed with bricks of different thickness. This bond consists of heading and stretching courses so arranged that one heading course comes after several stretching courses. Since the number of joints in the backing and the facing differ greatly, the load distribution is not uniform. This may sometimes lead to unequal settlement of the two thickness of the wall.

(*v*) **Raking bond.** This is a bond in brick work in which he bonding bricks are laid at any angle other than zero or ninety degrees. This arrangement helps to increase the longitudinal stability of thick walls built in English bond. In this arrangement of bonding, the space between the external stretchers of a wall is filled with bricks inclined to the face of the wall. This bond is introduced at certain intervals along the height of a wall. There are two common forms of raking bond ; (*a*) Herring bone bond (*b*) Diagonal bond.

(*a*) *Herring-bone bond.* This type of bond is best suited for very thick walls usually not less than four bricks thick. In this arrangement of brick work, bricks are laid in course inclined at 45° in two directions from the centre. This bond is also commonly used for brick pavings.

Fig. 9·22

Plan showing arrangement of bricks in Herring-bone bond.

Fig. 9·23

Plan showing arrangement of bricks in Diagonal bond

(*b*) *Diagonal bond.* This bond is best suited for walls which are 2 to 4 brick thick. This bond is usually introduced at every fifth or seventh course along the height of the wall. In this bond, the bricks are placed end to end in such a way that extreme corners of the series remain in contact with the stretchers.

(*vi*) **Dutch bond.** This bond is a modification of the old English cross bond and consists of alternate courses of headers and stretchers.

Fig. 9·24

Elevation of a wall in Dutch bond

In this arrangement of brick work, each stretching course starts at the quoin with a three-quarter bat and every alternate stretching course has a header placed next to the three-quarter brick bat provided at the quoin.

(*vii*) **English cross-bond.** This is similar to English bond and consists of alternate course of headers and stretchers. However, in this bond, queen closer are introduced next to quoin headers and each alternate stretching course has header placed next to quoin stretcher. This bond is sufficiently strong and bears a good elevation.

Fig. 9·25

Elevation of wall in English cross-bond.

(*viii*) **Zig-Zag bond.** This is similar to herring-bone bond with the only difference that in this case the bricks are laid in a zig-zag fashion. This is commonly adopted in brick paved flooring.

Fig. 9·26

Plan showing the arrangement of bricks in Zig-Zag bond.

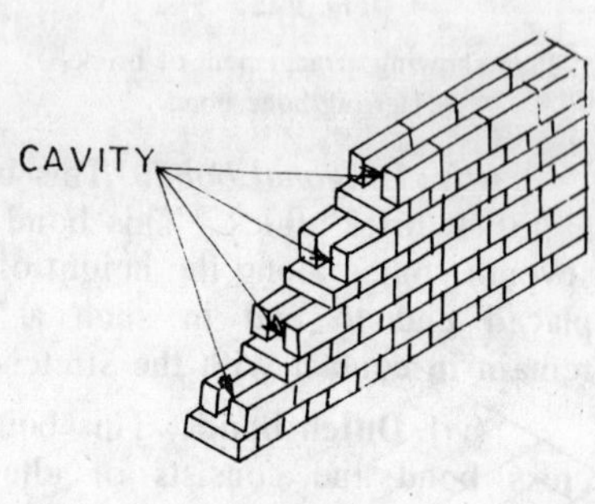

Fig. 9·27

View showing a wall in Silverlock's bond

(*ix*) **Silverlock's bond.** This is a form of bonding brick-work in which bricks are laid on edge. It is economical but weak in strength and hence it is only recommended for garden walls or partition walls. In this bond, the bricks are laid as headers and stretchers in alternate courses in such a way that headers are laid on bed and the stretchers are laid on edge forming a continuous cavity as shown in Fig. 9·27.

9·7 TEE-JUNCTION

A Tee junction is formed when two walls meet each other at right angles forming the letter 'T' in plan. T-junction may be formed in a number of

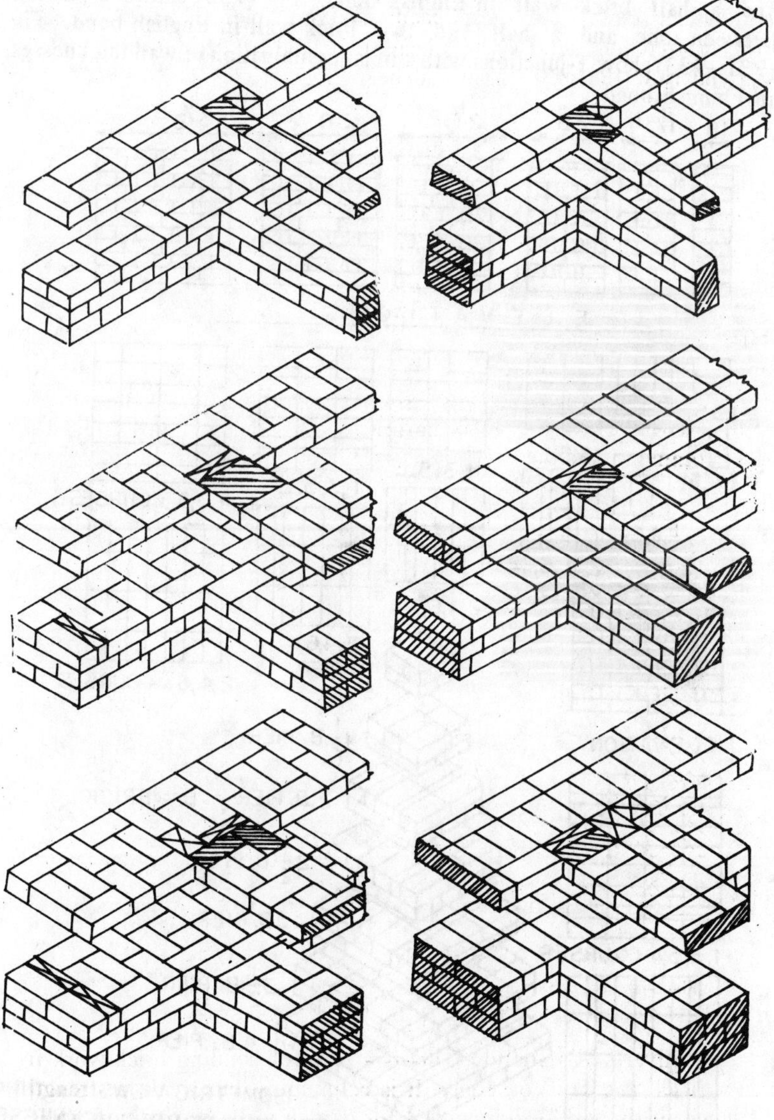

Fig. 9·28—9·30
Isometric view of T-junctions in English bond

9·31—9·33
Isometric view of T-junctions in Flemish bond

different bonds described earlier. Figs. 9·28—9·33 show the isometric views of T-junction between walls of various thickness. Fig. 9·28 shows T-junction between half brick and one and a half brick thick wall in English bond. Fig. (9·29) shows T-junction between one brick and one and a half brick wall in English bond. Fig (9·30) shows T-junction between one and a half and two brick wall in English bond. Fig. (9·31—9·33) show T-junctions with similar combination of wall thicknesses, in Flemish bond.

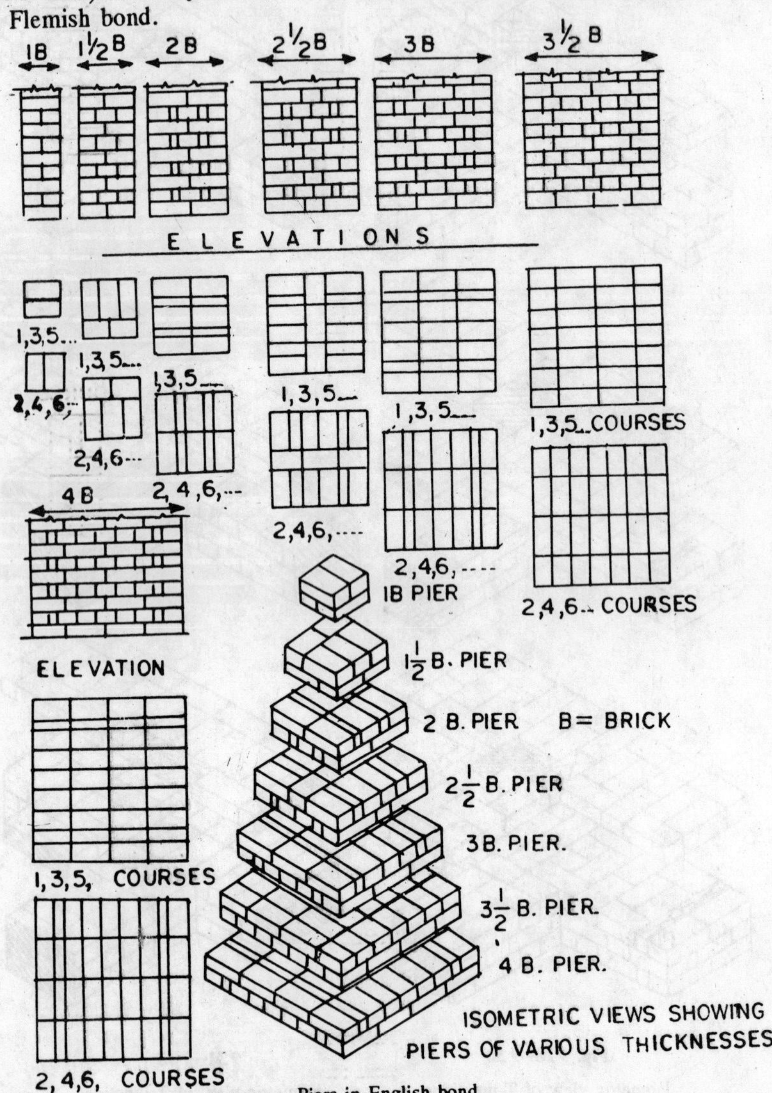

Fig. 9·34 Piers in English bond

9·8. BONDS IN PIERS

Piers may be constructed in a variety of bonds described earlier. The selection of the type of bond depends much upon shape of the pier as well as choice of the individual. In common practice, brick piers are built either in English bond of in Double Flemish bond.

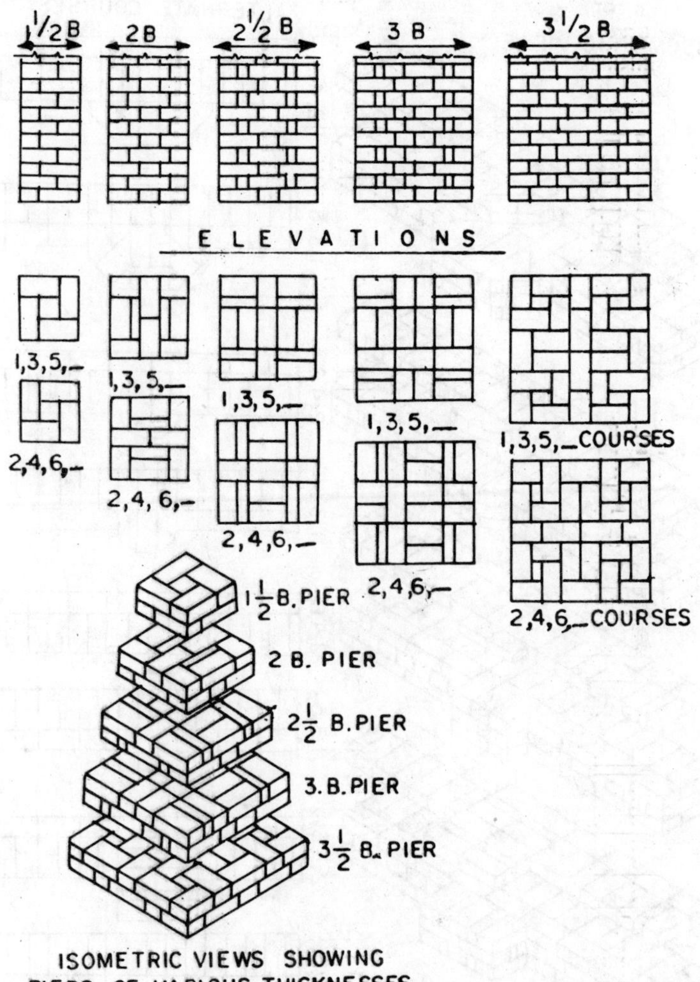

Fig. 9·35
Piers in Double Flemish bond

9·9. PIERS ATTACHED TO MAIN WALLS

Fig. 9·36 shows plans of alternate course of various shapes of piers attached to 1½ brick thick wall. On the left hand side of the plans is given the isometric view of the described shapes of attached pier. The function of the piers is to provide a large bearing area for giving

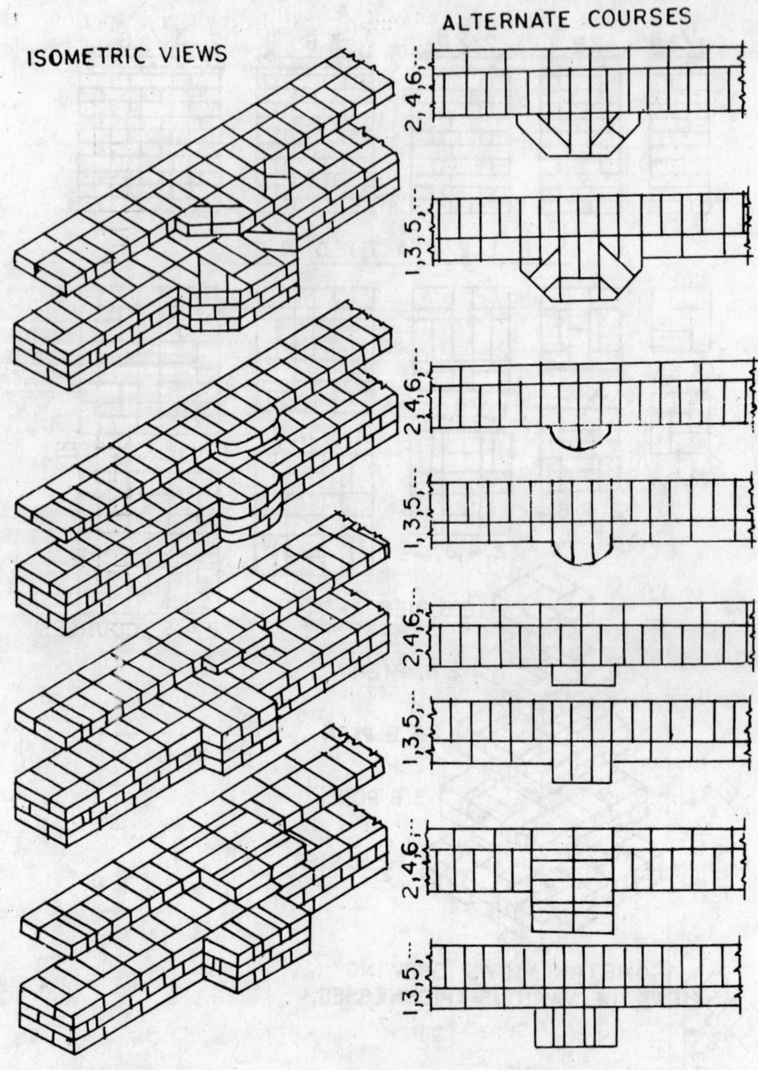

Fig. 9·36
Attached piers

support to the roof. It also helps in increasing the stability of the wall by stiffening it at intermediate points along length of the wall.

9·10. SQUINT JUNCTION

A squint junction is formed when two walls meet each other at an angle other than a right angle without forming a quoin. The junctions may be in English or Flemish bond. Great difficulty is experienced in forming the junction accurately and that is why they are rarely adopted.

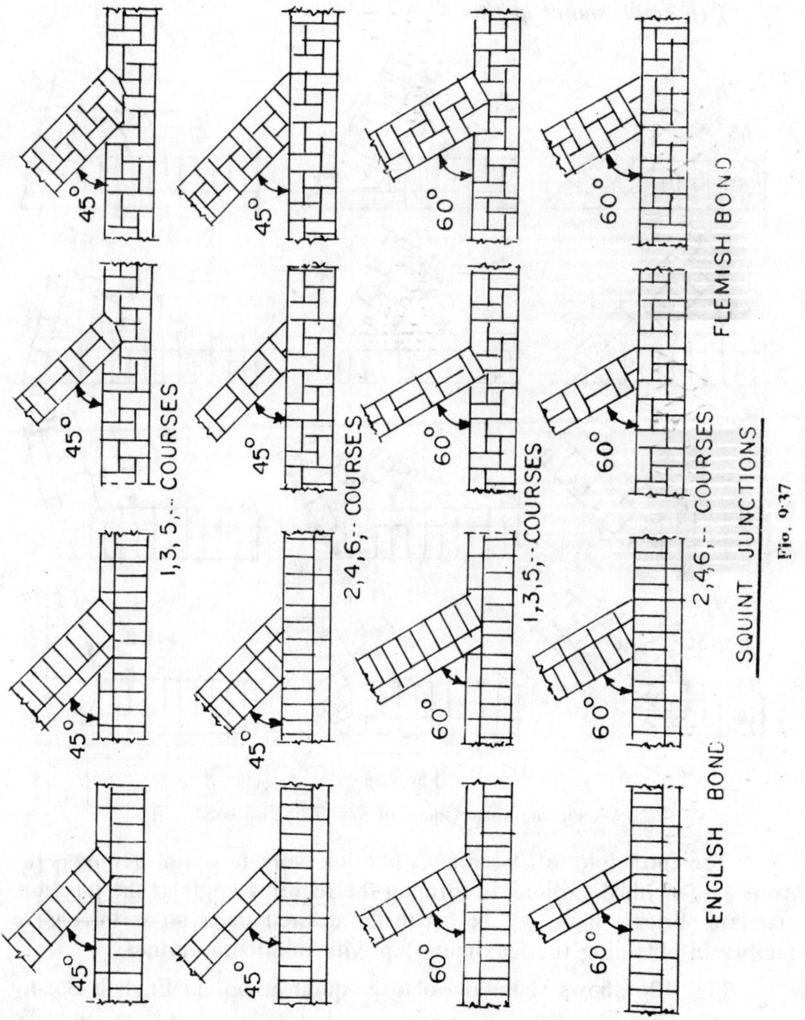

Fig. 9·37

Squint junctions between walls of various thickness have been shown in Fig. 9·37.

9·11. SQUINT QUOINS

When two walls meet at an angle other than a right angle, the junction or quoin thus formed is termed as squint. The squint quoins may be divided into the following two categories :

(*i*) *Obtuse squint quoins*

(*ii*) *Acute squint quoins*

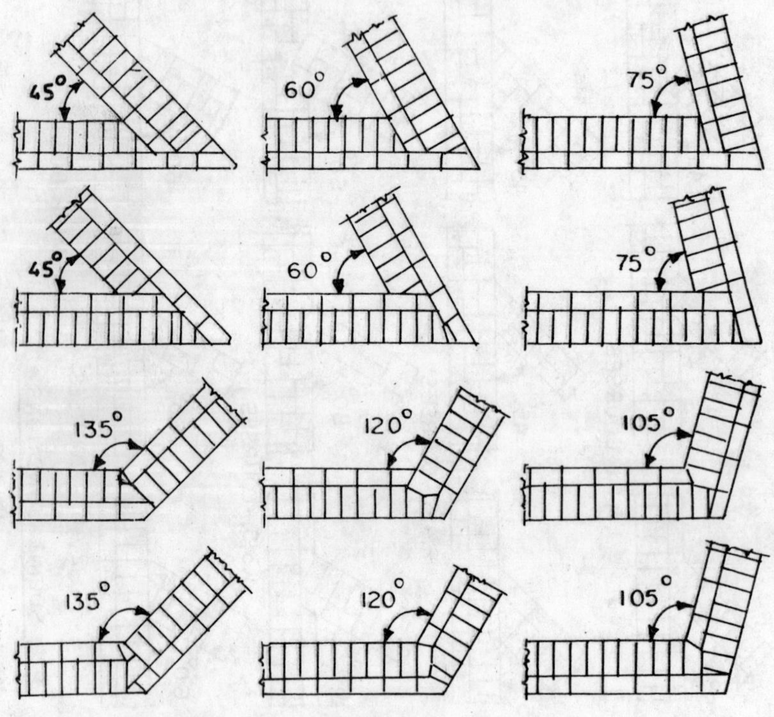

Fig. 9.38

Acute and obtuse squint quoins in English bond

The principle of bonding of brick work in squint quoins is the same as described earlier. In forming the required angle at the junction, standard bricks have to be cut to the desired shape so as to achieve facility in obtaining the maximum lap with minimum cutting.

Fig. 9·38 shows acute and obtuse squint quoins in English Bond.

Fig. 9·39 shows acute and obtuse quoins squint in Double Flemish bond.

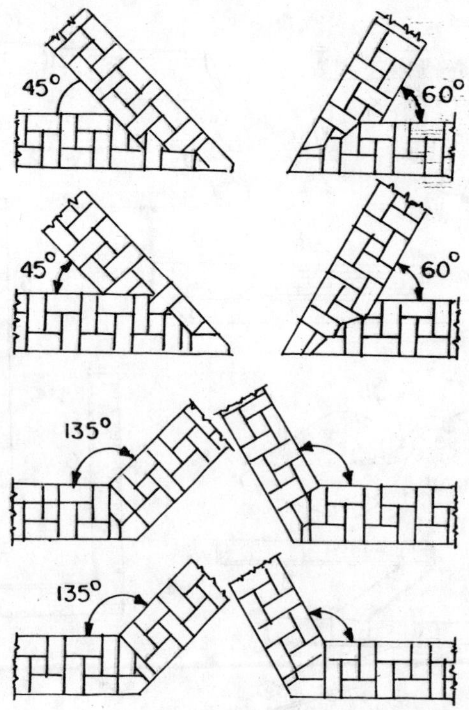

Fig. 9·39

Acute and obtuse squint quoins in Flemish bond.

9·12. TOOLS OF BRICK-LAYER

Tools used for brick masonry construction have been shown in Fig. 9·40.

(1) Hammer ; (2) Line and pins ; (3) Brick axe ; (4) Spirit level; (5) Plumb rule ; (6) Mason's square ; (7) Trowel ; (8) Two foot four fold rule.

9·13. RETAINING WALL

It may be defined as a wall built to resist the pressure of liquid, earth filling, sand, or other granular material filled behind it after it is built. It is commonly required in the construction of hill roads, masonry dams, abutments and wings walls of bridges and so on. Depending upon the site conditions, type of material to be retained and the height of the wall to be constructed, retaining wall may be built in dry stone masonry, stone masonry, brick masonry, plain cement concrete and reinforced cement concrete.

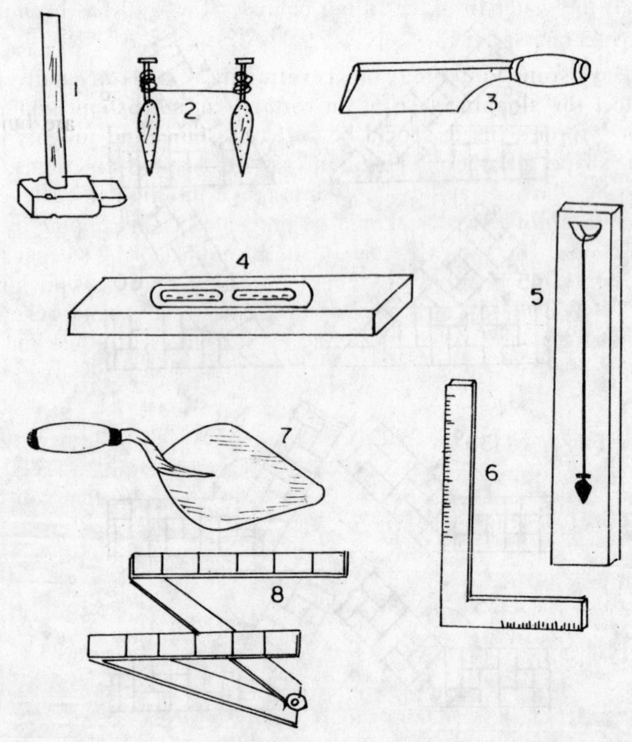

Fig. 9·40
Tools of Brick-layer

Dry stone retaining wall. This is the simplest form of retaining wall. The stability of such walls depends upon the arrangement of stones in the wall and the friction between the individual stones. The stones used in the wall should be of large size and roughly hammer-dressed so as to ensure maximum bedding area. The wall should have a minimum top width of 60 cm. and the front face should have a batter varying from 1 in 4 to 1 in 3. The batter of 1 in 4 is adopted for walls lesser than 4·5 m. in height. In principle, the height of dry stone masonry wall should be restricted to 6 m. For walls above 4·5 m. in height, the upper 4·5 m. of the walls is usually built of dry rubble stone masonry and the portion below this height is built with mortar. The stones used in the wall construction are laid at right angle to the face batter. A proper bond is maintained and the front and the rear faces of the wall are nicely bonded with the hearting. The filling immediately behind the wall should consist of stone chips gravel or similar granular material and not earth. 75 to 100 mm. square weep-holes should be provided in the wall at 2 m. c/c vertically and horizontally to drain off the water from the filling behind. The wall has been shown in figure 9·43 on page (252).

Dry stone pitching or revetment. It is generally provided to protect the slopping face of an earthen cutting or embankment from erosion. Stones used, should be perfectly sound and roughly cut to fit in the shape of the pitching. In case of channels and dams, pitching should be carried at least 90 cm. above the high flood level and to ensure its stability, the toe should be prevented from slipping by suitable construction. The slopes of embankment should not be steeper than 1:1, a slope of 1½ : 1 being usually adopted. The thickness of pitching varies from 30 cm. to 75 cm. Selected stones are tightly handpacked and all the interstices are filled up with smaller pieces of stone and wedged up tight. Every stone in pitching is laid flat and no projecting stones are allowed.

Breast walls. They are stone walls provided to protect the slopes of cutting in natural ground from the action of weather. The section of wall to be adopted depends upon the height of wall, the nature of the backing and the slope of cutting. The front and back batters of the wall vary from 1 in 4 to 1 in 2 (1 in horizontal : 4 or 2 vertical), with the minimum top width of 60 cm.

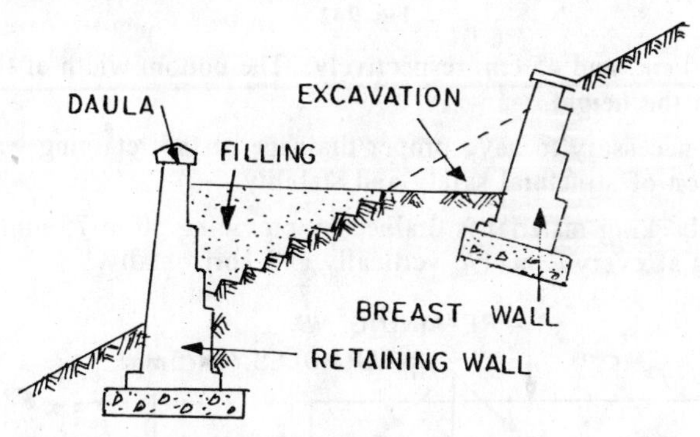

Fig. 9·41

Breast wall and retaining wall in respective positions

Brick masonry, stone masonry or plain concrete retaining walls: These walls are also provided to support earth, loose stone, coal etc. The wall acts as one mass to resist the thrust from the backing and is much more stronger than dry stone masonry wall. The stability of the wall depends entirely upon its dead weight. They are designed on the assumption that masonry or concrete is not subjected to any tensile stresses. In order that the walls may be stable they have to be very thick in section and as such they are seldom constructed for heights beyond 6 m. The top width of masonry walls and concrete walls should not be

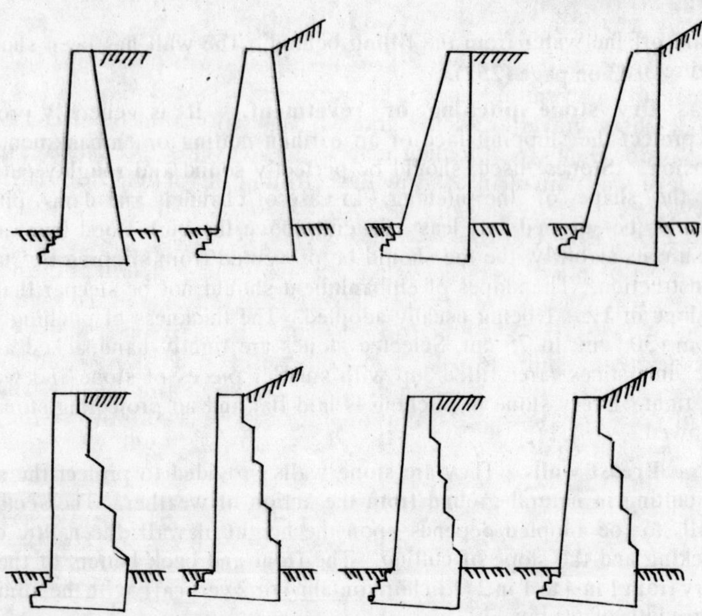

Fig. 9·42

less than 60 cm. and 45 cm. respectively. The bottom width of the walls varies with the height.

It is necessary to have proper drainage of the retaining wall from consideration of structural safety and stability.

The backing material is drained by providing 50 to 75 mm. square weep holes at every 2 m. c/c vertically and horizontally.

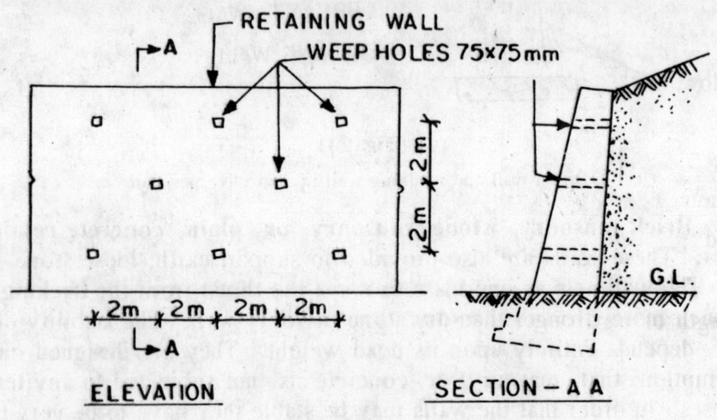

Fig. 9·43

Figure showing arrangement of weep holes in masonry retaining wall

The lowest weephole is keep 300 mm. above the ground level. In order to prevent blockage of the weep holes, a 450 mm. thick layer of stone chips, gravel or similar granular material should be placed behind the wall right from footing upto its top (covering the full area of the back of wall) simultaneously with the filling of backing material.

9·14. DESIGN OF RETAINING WALLS

The thrust from the backing which tends to overturn the wall or causes it to slide, is the deciding factor in the selection of the section and type of the wall. There are many conditions upon which thrust exerted by the backing depends, such as cohesion of the soil, dryness of the backing material, the manner in which the material is filled against the wall and so on. There are various theories by the help of which the value of thrust under different conditions can be worked out. Having known the thrust, the section of the wall is so designed that the selfweight is sufficient to resist the tendency of the thrust to slide the wall and the bottom width of the wall is such that the resultant force (resultant of the weight of wall and pressure of filling behind) lies within the middle third of the base. This condition is necessary to prevent the tendency of thrust to overturn the wall and to ensure that there is no tension at the wall base. It is equally essential to ascertain that the maximum stress at the toe of the wall does not exceed the safe bearing capacity of the soil.

Calculation of earth pressure. The thrust due to the back filling, which may be assumed to be earth, is generally calculated by Rankine's Theory. The theory is based on the assumption that the backing material or earth consists of cohesionless granular particles. The formulae derived from this theory under different conditions of back filling are given below:

Case 1 : *Walls with earth levelled with the top of wall.*

(*a*) Horizontal pressure per sq. m. (*ph*) at a depth of (*h*) metre below the levelled top is given by the formula :

$$ph. = w \times h \times \frac{1 - \sin \phi}{1 + \sin \phi}$$

where w = weight of filling in kg/m^3

and ϕ = angle of repose of the soil

(*b*) Total horizontal pressure (*P*) at a depth of (*h*) metre per metre length of wall is given by the formula :

$$P = \frac{wh^2}{2} \times \frac{1 - \sin \phi}{1 + \sin \phi}$$

Acting at $\frac{h}{3}$ metre from the base.

Case 2 : *In case of surcharged retaining wall* or wall retaining earth filled at slope of $a°$ to the horizontal, the formula giving lateral earth pressure (ph) is given by :

$$ph = wh \cos a \frac{\cos a - \sqrt{\cos^2 a - \cos^2 \phi}}{\cos a + \sqrt{\cos^2 a - \cos^2 \phi}}$$

acting parallel to the surcharge slope of the filling.

Total pressure (P) at depth of h metre per metre length of wall is given by the formula :

$$P = \frac{wh^2}{2} . \cos a . \frac{\cos a - \sqrt{\cos^2 a - \cos^2 \phi}}{\cos a + \sqrt{\cos^2 a - \cos^2 \phi}}$$

9·15. CONDITIONS OF STABILITY OF RETAINING WALLS

A satisfactory retaining wall must meet the following requirements for ensuring its stability :

(1) The wall should be structurally capable of resisting the pressure applied to it.

(2) The section of the wall should be so proportioned that it will not overturn by the lateral pressure.

(3) The wall should be safe from consideration of sliding, *i.e.*, the wall should not be pushed out by the lateral pressure.

(4) The weight of wall together with the force resulting from the earth pressure acting on it, should not stress its foundation to a value greater than safe bearing capacity of the soil on which it is founded.

(5) It is important to prevent accumulation of water behind a retaining wall. The backing material should be suitably drained by providing weep holes as detailed earlier.

(6) As far as possible, long masonry retaining walls should be provided with expansion joints located at 6 to 9 metre apart.

9·16. STRENGTH OF BRICK MASONRY

The permissible compressive stress in brick masonry depends upon the following factors :

(1) *Type and strength of bricks.*

(2) *Mix of mortar*

(3) *Size and shape of masonry construction.*

The strength of bricks masonry depends mainly upon the strength of bricks used in the masonry construction. The strength of bricks depends upon the nature of soil used for brick making and the method adopted for moulding and burning of bricks. Since the nature of soil varies from region to region, the average strength of bricks varies from

TABLE NO. 9·1

BASIC COMPRESSIVE STRESSES FOR MASONRY

S.No.	Description of mortar	Mix proportion (parts by volume)					Hardening	Basic compressive stress in Kg/cm² corresponding to masonry units having crushing strength in kg/cm²					
		Cement	Lime	Lime Pozzolana	Pozzolana	Sand	Time (after completion of work)	35	70	105	140	175	210
1.	Cement mortar	1	0 — ¼ (C)*	-	-	3	7	3.5	7.0	10.5	12.5	14.5	16.5
2.	Cement mortar	1	½ (C)*	-	-	4 ½	14	3.5	7.0	10.0	11.5	13.0	14.5
3.	Cement lime mortar	1	1 (C)	-	-	6	14	3.5	7.0	10.0	11.0	12.0	13.0
4.	Cement lime mortar	1	2 (B)	-	-	9							
5.	Cement mortar	1	-	-	-	6	14	3.5	5.5	8.5	10.0	11.0	12.0
6.	Lime-Pozzolana mortar	-	-	1	-	1.5							
7.	Cement lime	1	3(B) or (C)	-	-	12	14	2.5	5.0	7.0	8.0	9.0	10.0
8.	Hydraulic lime mortar	-	1 (A)	-	-	2	14	2.5	5.0	7.0	8.0	9.0	10.0
9.	Lime Pozzolana	-	1 (C)	-	1	2							
10.	Lime	-	1 (B)	-	-	3	28	2.5	4.0	5.5	6.0	6.5	7.0

Notes.

1. A, B and C are the types of lime conforming to relevant Indian Standards.
2. * shows that inclusion of lime in mortar is optional.
3. This table is valid for slenderness ratio upto 6.

as low as 30 kg/cm² (3N/mm²) in Andhra Pradesh to as high as 100 to 150 kg/cm² (10 to 15 N/mm²) in U.P.

The basic compressive stress in masonry for bricks/masonry units of different crushing strengths and for different mix and type of mortars is given in Table 9·1.

9·17. PERMISSIBLE COMPRESSIVE STRESS IN BRICK MASONRY

The permissible compressive stress in masonry shall be based on the value of basic compressive stress (f_b) as given in Table 9·1 and multiplying this value by (*i*) factor known as *Stress reduction factor* (K_f) and (*ii*) *Shape modification factor* (K_p).

(*i*) *Stress reduction factor* : Table 9·2 gives the stress reduction factor for various slenderness ratios.

TABLE 9·2

STRESS REDUCTION FACTOR FOR SLENDERNESS RATIO

S.No.	*Slenderness ratio*	*Stress Reduction factor* (K_f)
1	6	1.000
2	8	0.920
3	10	0.830
4	12	0.750
5	14	0.660
6	16	0.580
7	18	0.490
8	20	0.410
9	22	0.320
10	24	0.240

(*ii*) *Modification factor for shape of brick :*

Table 9·3 gives the value of modification factors for different ratios of height to width of bricks or masonry units

TABLE 9·3

MODIFICATION FACTOR FOR SHAPE OF BRICK

Ratio of height to width of brick	*Shape modification factor (K_p)*
upto 0.75	1.0
1.0	1.2
1.5	1.5
2 to 4	1.8

9·18. DEFECTS IN BRICK MASONRY

Brick work carried out by using good materials and by observing the specified principles of construction should require no maintenance. Defects in brick work may however, occur due to the following :

(*i*) Sulphate attack

(*ii*) Crystallization of salts from bricks

(*iii*) Corrosion of embedded iron or steel

(*iv*) Shrinkage on drying.

(*i*) *Defect due to sulphate attack :* Sulphate salts present in brick work react with alumina content of cement and with hydraulic lime in the lime mortar causing considerable increase in the volume of mortar. This in turn results in chipping and spalling of bricks and formation of cracks in joints and rendering. This defect occurs in situations where the brick work is exposed like boundary walls, parapets etc. or where it is likely to be in contact with moisture like manholes, retaining walls etc. This defect can be remedied by adopting suitable construction details and choosing materials in such a way that entry of moisture into the body of brick work can be checked.

(*ii*) *Defect due to crystallization of salts from bricks :* This defect commonly occurs in masonry made out of bricks which contains excessive soluble salts. When such bricks come in contact with water, the soluble salts get dissolved and appear in the form of fine whitish crystals on the surface of brick work. The phenomenon is also known as *efflorescence.* In this defect, the surface of brick work gets disfigured and presents an ugly look. Efflorescence can be remedied by brushing and washing the effected surface repeatedly.

(*iii*) *Defect due to corrosion of embedded iron or steel :* Iron or steel embedded in brickwork gets corroded in the presence of dampness. On corrosion the metal expands in volume and tends to crack the brick work. This defect can be prevented by encasing the reinforcement or iron member in dense cement mortar and providing a cover of 15 to 25 mm. around the embedded member.

(*iv*) *Defect due to shrinkage on drying :* Brick work normally swells with the absorption of water and subsequently shrinks when the water evaporates. In the process of shrinkage it develops cracks in the masonry joints. If the mortar used in the masonry is lean, the cracks are distributed over large number of joints and in case of rich mortar the cracks are few but wider. Such cracks do not effect the structural strength of the brick work and can be easily rectified. This defect can be prevented by using good quality bricks and by protecting masonry from moisture penetration.

9·19. REINFORCED BRICK WORK

Brick work strengthed by introduction of mild steel flats, hoop iron, expanded mesh or bars is termed as reinforced brickwork. This type of masonry is capable of resisting both compressive as well as tensile and shear stress. On account of its ability to resist lateral forces, reinforced brick masonry is extensively used in siesmic areas. It is essential to use first class bricks (having crushing strength of 140 kg/sq. cm or more) and rich and dense cement mortar in the reinforced brick work. The reinforcement should be effectively bedded and surrounded with mortar cover of 15 to 25 cm. This is necessary to protect the reinforcement against corrosion.

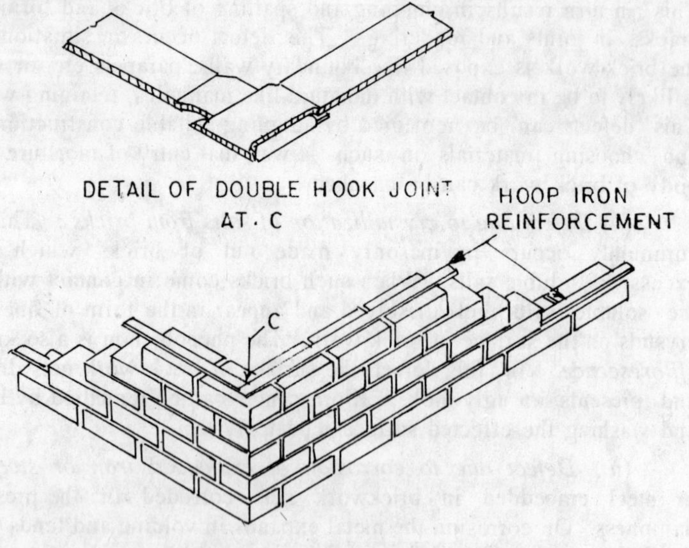

Fig. 9·44

Sketch showing arrangement of reinforcement in half brick wall

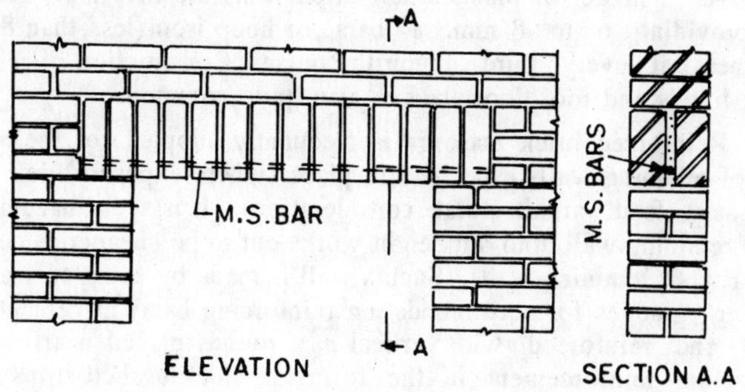

Fig. 9·45
Reinforced brick lintel

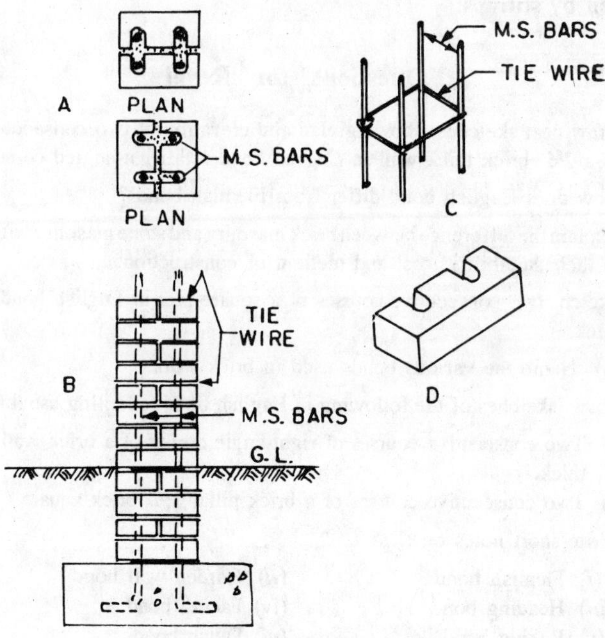

Fig. 9·46
Reinforced brick column

A— Plan showing arrangement of bricks in alternate courses.

B— Part elevation of column.

C— View showing arrangement of reinforcement.

D— Special brick having groove for accommodating reinforcement.

As a matter of practice half brick walls are invariably reinforced by providing 6 to 8 mm. ϕ bars or hoop iron (less than 8 mm in thickness) at every third or fourth course. Construction of reinforced brick lintels and roof/floor slabs is also quite common.

Reinforced brick masonry is frequently adopted for the construction of retaining walls especially in places where exposed brick work is necessary from architectural considerations. It is seen that reinforced brick retaining wall upto 4m height works out to be cheaper as compared with R.C.C. retaining walls. Such a wall is made by using special bricks (having grooves for accommodating reinforcing bars) in cement mortar 1 : 3 and reinforced with vertical *m.s.* rounds placed near each face. Horizontal reinforcement in the form of steel meshed strips are also provided at every third or fourth course.

Construction of reinforced brick column is more or less similar to that of reinforced brick wall except that the horizontal steel mesh is replaced by stirrups.

Questions for Reveiw

1. Draw neat sketches, showing plan and elevation of two consecutive courses of a 2½ brick thick wall in English bond at a right-angled corner.

   How does English bond differ from Flemish bond ?

2. Explain the difference between brick masonry and stone masonry with reference to their suitability, cost and method of construction.

   Sketch two consecutive courses of a square pier in English bond 2½ brick thick.

3. (*a*) Name the various bonds used in brick-work.

   Draw sketches of the following in English bond (adcpting usual brick size).

   (*i*) Two consecutive courses of right-angle corner of a brick wall 1½ brick thick.

   (*ii*) Two consecutive courses of a brick pillar 1½ brick square.

4. Write short notes on :

   (*i*) Flemish bond (*ii*) Garden wall bond
   (*iii*) Heading bond (*iv*) Facing bond
   (*v*) Raking bond (*vi*) Dutch bond
   (*vii*) Silverlock's bond

5. Supposing you are supervising the construction of a residential building in a good locality, write the general principles you would expect the brick-layer to observe during the construction of the brick masonry.

6. Define bond. Distinguish with the help of sketches between a wall in English bond and a wall in Flemish bond.

7. Sketch free-hand to a suitable scale the following :

   (*a*) Un-coursed and coursed rubble masonry in elevation.

   (*b*) Corner of 2 brick thick wall.

8. Sketch free hand to a suitable scale the following :

   (*a*) Ashlar random and coursed rubble masonry in elevation.

   (*b*) Isometric view of corner of one and half brick wall.

9. Answer briefly the following.

   (*a*) Of what and how are brick made ?

   (*b*) Mention all the equipment required by the brick layer and his assistant for erecting a brick wall in lime mortar.

10. What are retaining walls ? Underline the design procedure adopted in deciding the section of a retaining wall.

11. Define the following :

    (*i*) Header (*ii*) Stretcher
    (*iii*) King closer (*iv*) Queen closer
    (*v*) Prepend (*vi*) Bat.

12. Write short notes on any four of the following :

    (*a*) Damp-proof courses.

    (*b*) Tests conducted on bricks.

    (*c*) The English and Flemish bond in brick work.

    (*d*) Joints in stone work.

    (*e*) Composite masonry construction.

13. Give detailed specifications for the following item of work :

    First class brick work in cement mortar.

14. How many kinds of brick bonds are there ? Compare the two main kinds. In what respect is one superior to the other ? Show two courses of a small length of wall of one and a half brick in thickness in English bond.

15. Draw neat sketches of the following in English bond (adopting usual brick size).

    (*i*) Two consecutive course of the right-angled corner of a 1 ½ brick wall.

    (*ii*) Two consecutive courses of a 1½ x 2 brick pillar.

    (*iii*) Two consecutive courses of a junction of 1½ brick wall with that of a 2 brick wall.

16. Draw plan and elevation of two consecutive courses in English bond for :

    (*a*) 1 ½ brick wall meeting a 2 brick wall at a right-angled corner of a building.

(*b*) Single brick-cross wall meeting a 1 ½ brick cross-wall at any intermediate point.

17. Describe with neat sketches the different types of bonds used in brick work.

Give sketches for any two of the following showing alternate course in English bond.

(*i*) 2 brick wall at a corner of a rectangular building.

(*ii*) 2 x 3 ½ brick pillar.

(*iii*) 2 brick wall crossing with 3 ½ brick wall.

18. What is the difference between English bond and Flemish bond in brick masonry construction?

19. (*a*) What are the general principals observed in brick masonry ?

(*b*) Name the different bonds used in brick masonry. Show by sketches any two.

(*c*) Describe with sketches a dry stone retaining wall.

20. In which cases would you prefer brick masonry to stone masonry ?

State the general principles to be observed and the precautions taken in constructing brick masonry as well as stone masonry.

21. What do you mean by bond in brick work ? Distinguish between English and Flemish bond. What are the various precautions to be taken in brick masonry construction ?

22. (*a*) State the general principles to be observed during the construction of brick masonry work.

(*b*) Draw plans of

(*i*) Two consecutive courses of a right angled quoin between two walls each 1½ brick, in English bond and

(*ii*) Two consecutive courses of a junction of two brick walls in English bond.

23. (*a*) What is meant by efflorescence in brick work ? List the sources which give rise to dampness and efflorescence in building constructed near the sea coast.

(*b*) Show by free hand (well-proportioned sketches) plan and elevations of two successive courses of a brick pier 2½ brick thick x 3 brick wide.

10

Composite Masonry

10·1. DEFINITION

Walls constructed from two or more building materials are termed as composite walls. The combination may be stone slab in the face work and brick masonry at the backing, or superior stones in the facing with rubble stone masonry at the backing and so on. Composite masonry is adopted to improve the appearance of a structure by concealing the inferior work by use of superior quality of material in the desired positions. This results in great saving in cost and, if properly constructed, the structure remains equally strong and durable. Stones, bricks, concrete and tiles are the materials commonly used in such constructions.

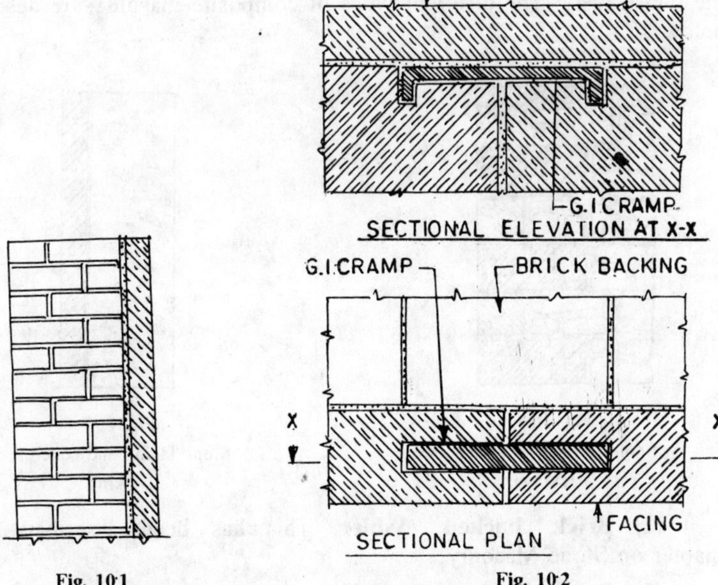

Fig. 10·1
Stone facing with brick backing

Fig. 10·2
Method of securing stone slabs to the backing by means of G.I. cramp

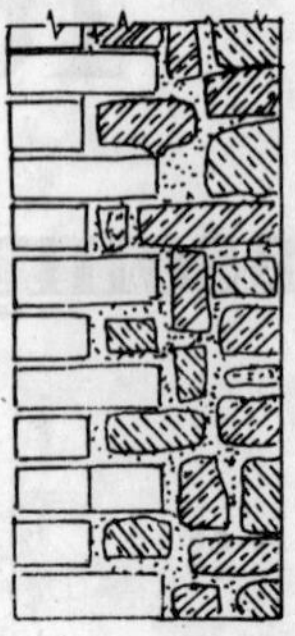

Fig. 10·3
stone facing and
Brick backing

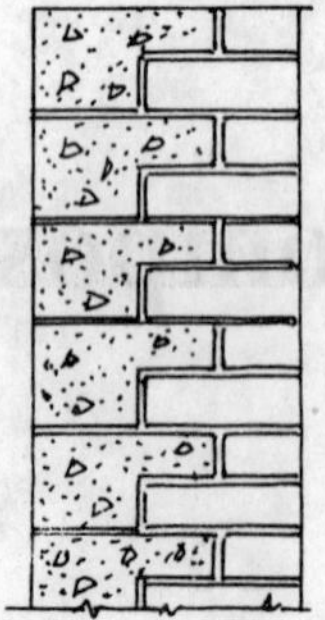

Fig. 10·4
Brick facing and
concrete backing

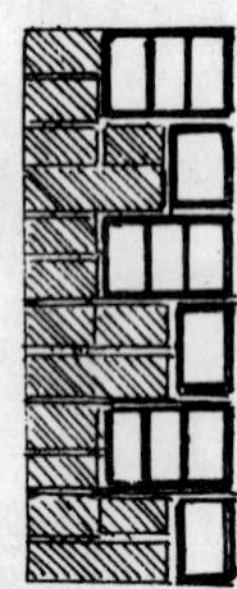

Fig. 10·5
Tile facing and
brick backing

10·2. COMPOSITE MASONRY IN STONE AND BRICK

In this type of construction stone slabs are used in the face work and brick masonry is used as backing (Ref. Fig. 10·1). Stones used may be natural or artificial. The stones are secured to the backing either by means of metal cramps or by projecting the headers in the face work well inside the backing. It is obvious that the number of joints in the backing will be much more than the joints in the facing and as such the bonding between the two faces should be perfect to avoid unequal settlement of the two leaves. The different types of composite masonry are described below :

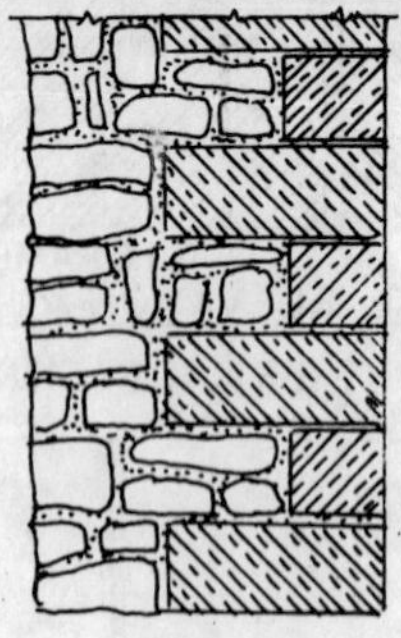

Fig. 10·6
Stone facing and rubble
backing

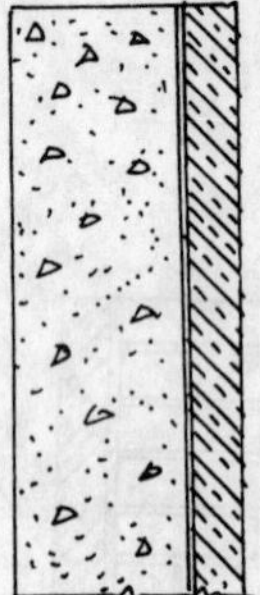

Fig. 10·7
Stone facing and concrete
backing

(1) **Brick backed Ashler.** This has been described in the chapter on Stone Masonry.

(2) **Brick facing with rubble or concrete backing.** In this type of construction, special types of glazed tiles or bricks are used as

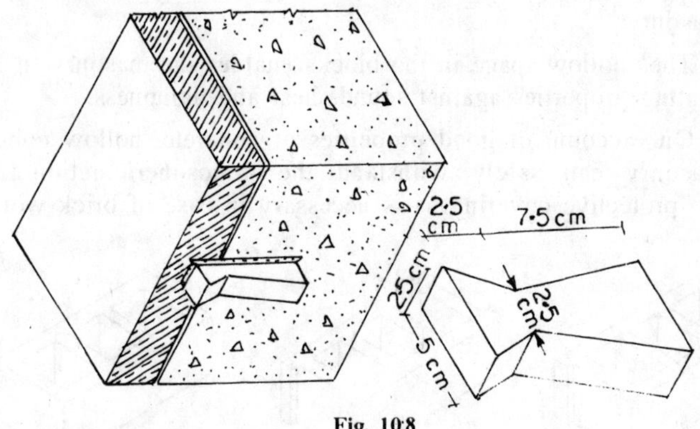

Fig. 10·8

Method of fixing stone facing with concrete backing

facing material and rubble or concrete or brick is used as backing. Terra-cotta tiles and other such glazed tiles are commonly used in lavatory blocks and other similar situations.

(3) **Stone facing with brick, rubble or concrete backing.** The purpose of this type of masonry is to make a wall constructed in brick to appear as if constructed in stone. It is adopted to improve the appearance of the structure at a lesser cost. The stone slabs seldom exceed 5 to 10 cm. in thickness and are properly secured to the backing.

10·3. CEMENT CONCRETE MASONRY

Cement concrete is largely used for construction purposes on account of its numerous advantages over the other building materials. Cement concrete may be used in the form of pre-cast concrete blocks, plain concrete slabs or reinforced concrete slabs. Hollow units may be defined as those units having core-void areas greater than 25% of their gross area. The blocks may have one or multi-core design to suit the choice of manufacturer. The various advantages of hollow concrete block masonry are described below:

(1) On account of their regular and uniform size and less weight, concrete blocks afford great facility in masonry construction which ultimately results in rapid execution of work.

(2) Hollow blocks of concrete permit adoption of thinner walls thereby achieving increased floor area and hence saving of space.

(3) There is a great saving of material on account of the blocks being hollow from inside. The air space (hollow space) in concrete blocks varies from 20 to 40 per cent of their volume.

(4) On account of the large size of concrete blocks, the number

of joints in a work are lesser and hence there is saving of mortar in this type of masonry.

(5) The hollow space in the blocks enables the masonry to have good insulating properties against sound, heat and dampness.

(6) On account of good properties of concrete, hollow concrete blocks masonry can safely withstand the atmospheric action and it requires no protective covering as is necessary in case of brick work.

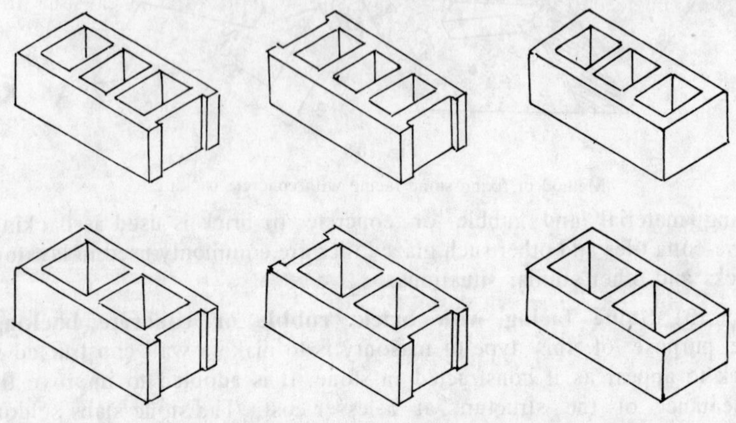

Fig. 10·9--10·14

Different forms of hollow concrete blocks

(7) Unskilled labour can be employed in this type of work and on account of the light weight of the blocks, the construction is speedy and it results in economy in labour cost and in time.

(8) The rough surface of the concrete blocks makes plastering (if needed) easier.

The hollow concrete blocks are manufactured in various shapes and sizes. The blocks are moulded in machine and can be obtained in the following standard sizes :

(1) Standard size hollow concrete blocks ...39 cm. x 19 cm. x 30 cm.

(2) Hollow building tiles ... 39 cm. x 19 cm. x 20 cm.

(3) Hollow concrete blocks for partition ... 39 cm. x 19 cm. x 10 cm.

The face thickness of blocks is kept 5 cm. or more.

The various shapes of hollow concrete blocks are shown in Figs. (10·9--10·14).

Specially shaped blocks are manufactured for corners, sills, lintels and jambs, wall closures, piers between doors and windows etc. to fit in the job suitably.

In order to ensure maximum economy in construction, the overall length and height of walls should be so fixed as to make maximum use of full and half length blocks possible. The width and height of door and window openings and the wall area between them should be planned keeping in view the size of blocks available. This is essential to minimise cutting of blocks and avoid wastage of material and time.

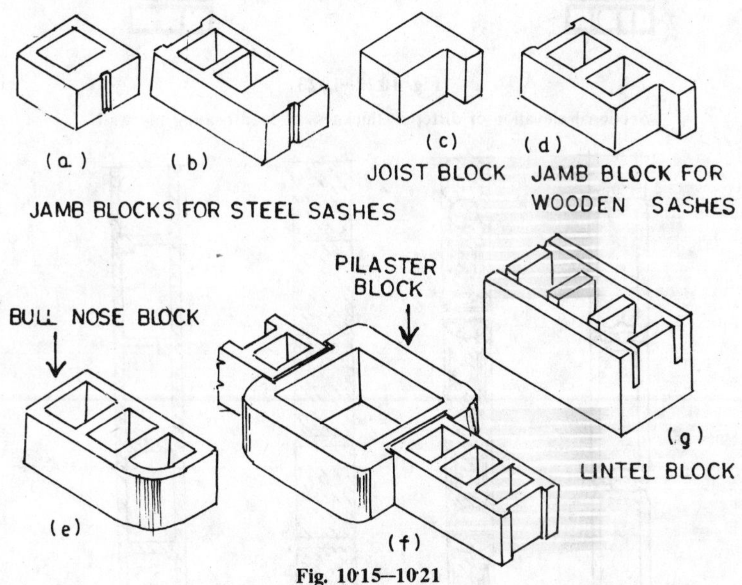

Fig. 10·15—10·21

Specially shaped hollow concrete blocks for jambs, lintels, corners, etc.

10·4. LOAD BEARING WALL TILES

There are a variety of shapes for load bearing clay tiles. The tiles may be used for either interior or exterior load bearing walls. As in the case of hollow concrete blocks, the surface of the tiles is rough and can be easily plastered if desired. The tiles may be used with or without any backing or facing. The tiles are designed to carry the superimposed load plus the weight of the facing material such as plaster etc.

10·5. GLASS BLOCK MASONRY

There are various shapes and sizes of glass blocks available these days. The blocks are made from structural glass annealed to withstand rigorous use of the material. In addition, the glass used is such that it is impervious to grease, chemicals, moisture and does not change its colour

Fig. 10·22--10·23

Sectional elevation of different thickness of load bearing tile walls

Figs. 10·24--10·27

Sections of different types of glass block masonry

with age. The blocks may be made hollow or solid. Glass block panels should never be made to carry loads other than their own weight. They are generally laid in cement-lime mortar (1 : 4). In glass block masonry with blocks more than 30 cm high, every horizontal joint should be reinforced with galvanized expanded metal strips. The construction is so executed that there is complete freedom of movement of the panel within the enframing members. Provision for expansion is made along the jambs and the head of each panel and there is proper anchorage of panel at head, sill and jambs.

Advantages of glass block masonry :

(1) It is attractive in appearance.

(2) On account of the dead air space within, the glass block masonry has great heat insulating value.

(3) It is non-porous, non-absorbent and impervious to moisture.

(4) It can be easily cleaned and as such it is considered more hygienic.

(5) It diffuses the light (which passes through) and with proper selection of suitable patterns, a wide range of light and decorative effects can be produced.

(6) It has good sound-proofing property and thus the room with glass block masonry walls has improved acoustical effects.

(7) Glass block masonry is best suited in places where window openings cannot be provided and clear glass in objectionable from considerations of privacy.

Questions for Review

1. Explain, what do you understand by the term 'composite masonry' Draw neat sketches wherever necessary to support your answer.
2. What do you understand by hollow block masonry ? What are the advantages of hollow concrete block masonry over the stone masonry construction ?
3. Write an essay on glass walls.
4. Write short notes on :

   (*i*) Method of securing stone slabs to concrete or brick backing.

   (*ii*) Composite masonry.

   (*iii*) Hollow concrete block masonry.

   (*iv*) Glass block masonry.

11

Partition Walls

11·1. DEFINITION

A partition wall may be defined as a wall or division made up of bricks, studding, glass or other such material and provided for the purpose of dividing one room or portion of a room from another. Generally, partition walls are designed as non-load bearing walls. It may be of folding, collapsible or fixed type. So far as the functional use of portion walls is concerned, they are employed to provide privacy from consideration of sight or sound or both. A good non-load bearing partition wall should be cheap, light, thin, fire-resistant and easy in construction.

11·2. TYPES OF PARTITION WALLS

Depending upon the material used, partition walls may be divided into the following different types :

(1) *Brick partitions*

(2) *Hollow brick partitions of clay, terra-cotta or concrete*

(3) *Glass partitions*

(4) *Concrete partitions-plain or reinforced*

(5) *Metal lath and plaster partitions*

(6) *A.C. sheet or G.I. sheet partitions*

(7) *Timber partitions.*

(1) **Brick Partitions.** They may be constructed with plain bricks, reinforced bricks or bricks nogged.

Plain brick partition : This type of wall is constructed by laying bricks as stretchers in cement mortar. Thus the wall is generally 10 cm. (half brick) thick and plastered on both the faces. If properly constructed, it is considerably strong and fire-resistant.

Reinforced brick partition wall : This type of wall is similar to plain brick partition wall except that at every third or fourth course, the

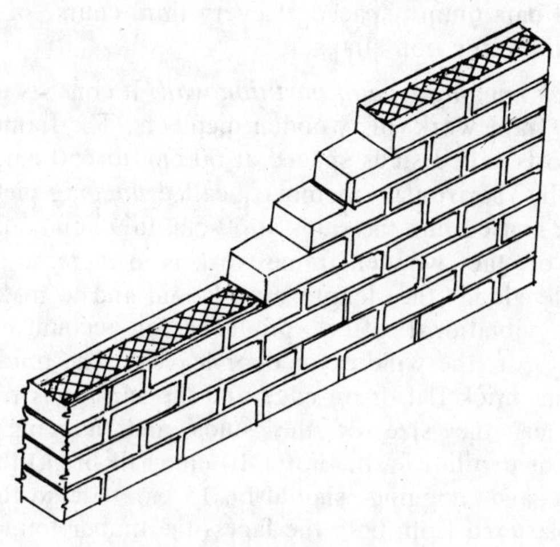

Fig. 11·1

Reinforced brick partition wall

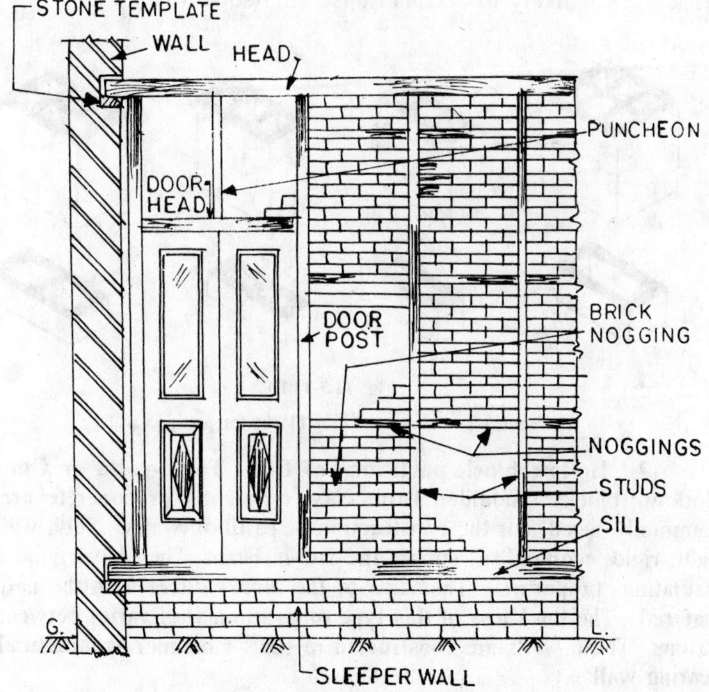

Fig. 11·2

Brick-nogged partition walls

bricks are reinforced with iron straps 25 to 28 mm. wide and 1·6 mm. thick. Mild steel bars 6mm.ϕ spaced at every third course of wall may be used as a substitute for iron straps.

Brick nogging type of partition wall: It consists of brick work built within a framework of wooden members. The frame work consists of vertical posts called studs spaced at 60 cm. to 150 cm. apart and held in position by horizontal members called *nogging* pieces. The nogging pieces are housed into the studs at 60 cm. to 90 cm. apart vertically. The function of the wooden frame work is to increase the stability of the wall both along the length and height and to make it more rigid to withstand vibrational effect produced on account of careless opening or closing of the window or door leaves . The brick work is built by laying the brick flat or on edge and the surface is plastered from both sides. Thus the size of the studs and nogging depends upon the thickness of partition wall. For a 10 cm (half brick) thick partition wall, the studs and nogging should be 15 cm wide so that after the brick-work is plastered from both the faces, the timber frame work may finish flush with the wall face. This type of partition wall suffers from the draw back of the timber getting decayed over a period of time. Moreover the mortar used may not stick well to the timber members and thus the brick work is likely to become loose after some time.

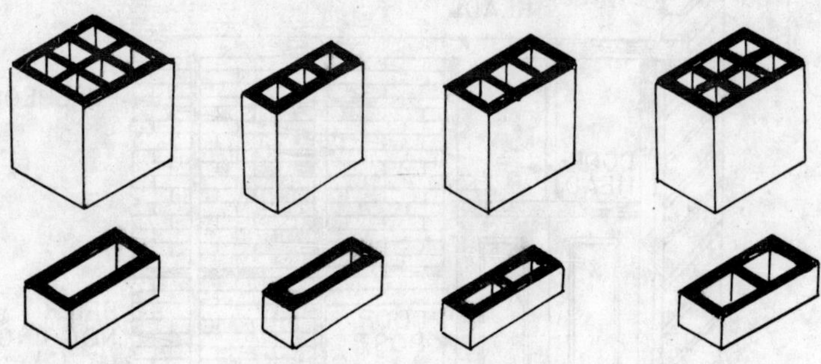

Fig. 11·3-11·10

Different types of hollow clay blocks for partition walls

(2) **Hollow block partitions of Clay, Terra-cotta or Concrete.** Hollow blocks moulded from clay, terra-cotta or concrete are now commonly used for the construction of partition walls. Such walls are light, rigid, economical, strong and fire-resistant. They have good sound insulating properties. The size of the blocks differ with the nature of material. The thickness of this type of partition wall varies between 6 to 15 cm. These walls are constructed in similar manner as structural load bearing walls.

(3) **Glass Partition.** These may be made from sheet glass or

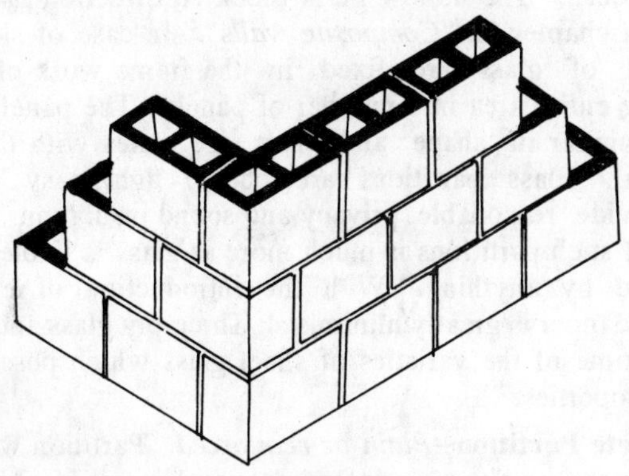

Fig. 11·11

View showing hollow block partition wall

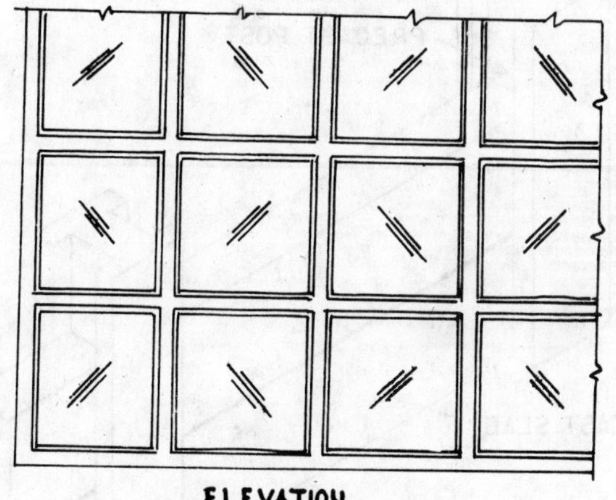

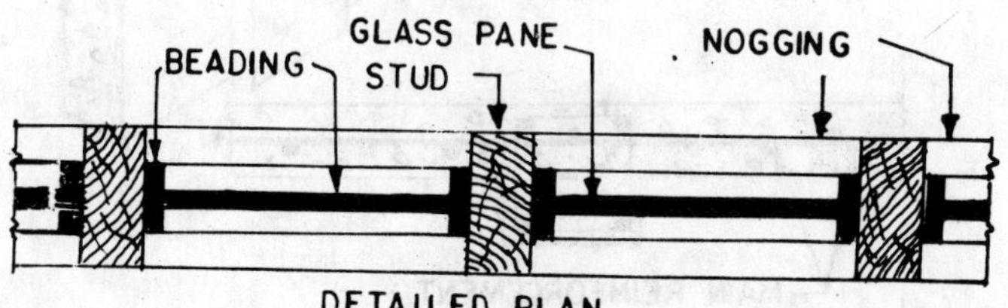

Fig. 11·12

Elevation and sectional plan of glass partition

hollow glass blocks. The hollow glass block construction has already been described in chapter on "*Composite walls*". In case of sheet glass partitions, sheets of glass are fixed in the frame work of wooden members dividing entire area in a number of panels. The panels may be square or rectangular in shape and their size varies with the choice of the individual. Glass partitions are cheap, light, easy in construction and provide reasonable privacy and sound insulation. The cost of maintenance of such partitions is much more as glass is liable to break when struck hard by anything. With the introduction of re-inforced glass sheets, this danger is greatly minimised. Three-ply glass and armour plate glass are some of the varieties of sheet glass which possess good shock-resisting properties.

(4) **Concrete Partition**—*Plain or reinforced.* Partition walls constructed in concrete, plain or reinforced, may be cast-in-situ or built from panels or blocks precast well in advance of the commencement of

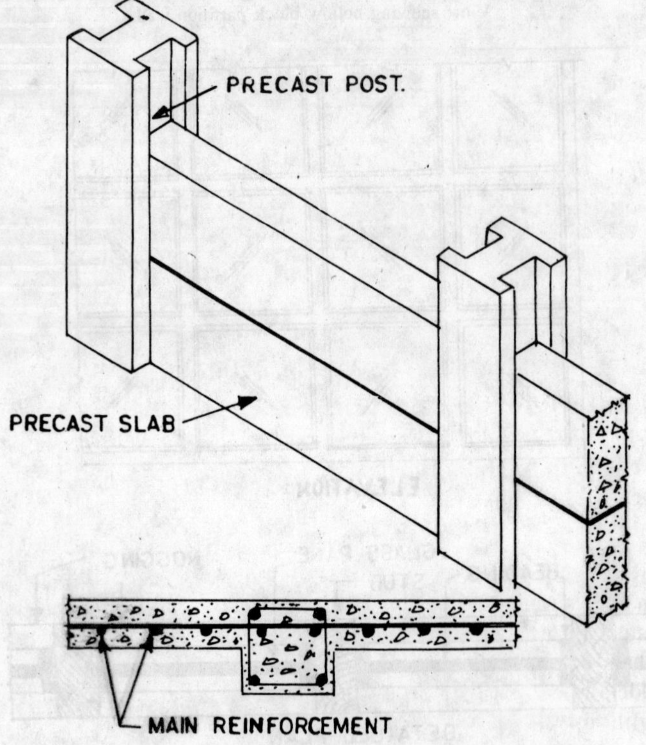

Fig. 11·13. View of a pre-cast concrete partition wall

Fig. 11·14. Sectional plan showing arrangement of reinforcement in a cast-in-situ partition wall.

work. Generally for cast-in-situ walls, 10 cm thick and below, the reinforcement consisting of mild steel bars or B.R.C. fabric is placed in the centre of the wall thickness. Concrete mix usually adopted in the work is M 15 (1:2:4). The wall is cast monolithically with the intermediate columns so as to be rigid and stable both along its length and height.

In case of precast concrete partition wall, pre-cast concrete slab panels and special shaped concrete posts are used. The slabs are generally 32 mm thick and are inserted in the grooves of the precast post and the joints are subsequently filled with cement mortar.

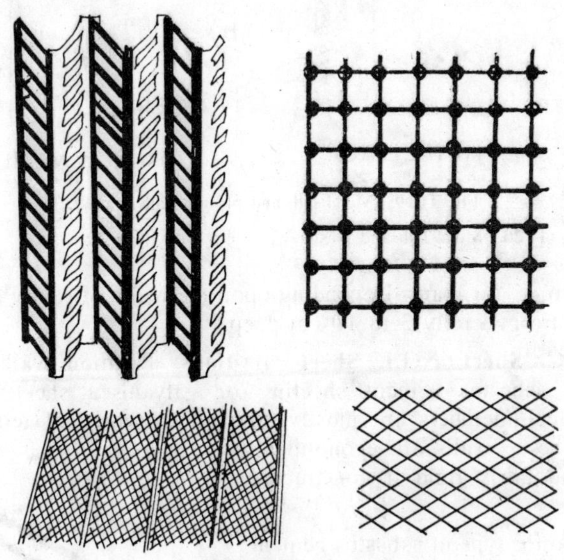

Fig. 11·15—Fig. 11·18

Different forms of Metal laths

(5) **Metal Lath and Plaster Partitions.** Metal lath and plaster when properly laid forms a reinforced partition wall which is thin, strong, durable and is considerably fire-resistant. The metal lath is available in a variety of patterns and it generally requires a frame-work of steel or timber for the purposes of fixing it in position. In case of partition walls with steel frames, lath is generally tied by galvanized iron wire to mild steel bars or channels spaced 15 to 30 cm apart. The latter is fixed on one side and the plaster is applied on both the sides of the lath. This type of partition wall may be 5 to 7·6 cm in thickness.

In order to achieve improved insulation against heat and sound, metal lath and plaster partition walls can also be made with a cavity between the wall thickness. This type of hollow partition wall is constructed by fixing metal lath on both sides of specially shaped steel channels

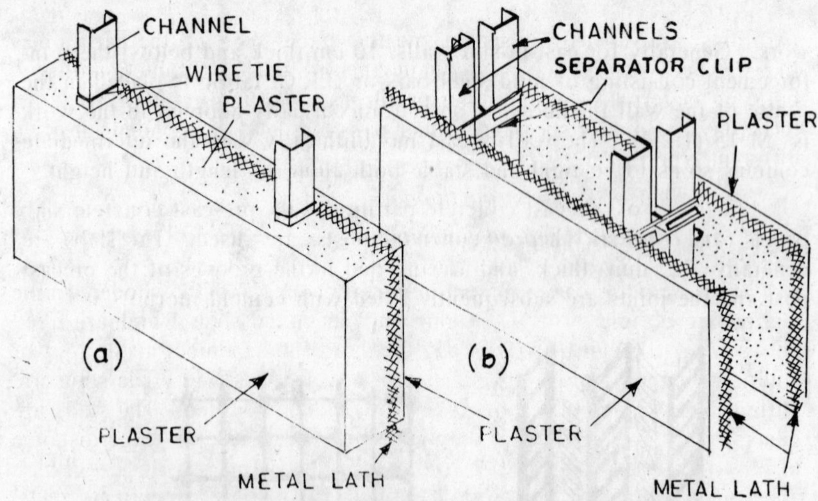

Fig. 11·19. Metal lath and plaster wall

Fig. 11·20. Metal lath and plaster wall with a cavity in between

spaced at 30 to 45 cm apart. Depending upon the width of cavity desired, the channels are generally 3 to 10 cm deep.

(6) **A.C. Sheet or G.I. Sheet Partitions.** Partition walls constructed from asbestos cement sheeting or galvanised sheet fixed to wooden or steel members are mostly adopted in works of temporary character. These walls are economical, light, thin and fairly rigid if constructed properly.

For superior type of asbestos cement sheet partition walls, specially manufactured slabs of the said material are used. Each slab consists of core of corrugated asbestos cement sheet with the plain asbestos cement sheet attached to it on either side. The use of such slabs renders the partition wall more fire-resistant and makes it have good heat and sound insulation properties.

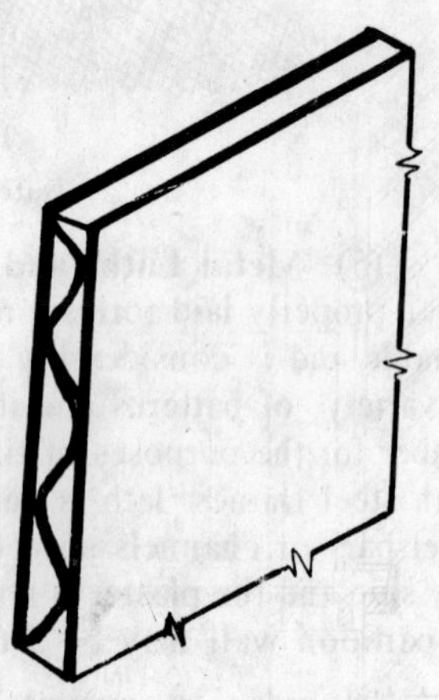

Fig. 11·21

(7) **Timber Partitions.** This type of partition consists of a wooden framework either supported on the floor below or by side walls. The frame work consists of a rigid arrangement of timber members which may be plastered or covered with boarding etc. from both the sides. Such

partitions are not fire resistant and the timber forming the partition is likely to decay or eaten away by white ants. With the introduction of new building materials, use of timber partitions is getting gradually reduced these days. Broadly speaking timber partitions may be divided into two different types :

(*a*) *Stud or common partition*

(*b*) *Trussed or braced partition*

(*a*) *Stud or common partition* . It consists of a wooden frame comprising of light vertical upright members called studs which are fixed between two horizontal members. The horizontal member at the foot of stud is called sill and the one at its top is called the head. The studs are stiffened by horizontal timber pieces called nogging, The studs are generally 10 cm x 5 cm in section and are spaced at an average distance of 30 to 45 cm apart. Head and sill are members of varying size which are rigidly secured to the ceiling joists and floor joists respectively. The door posts are sufficiently strong studs, capable of withstanding the impact due to the usage of the opening. The studs or short length verticals provided between the door head and the partition wall head

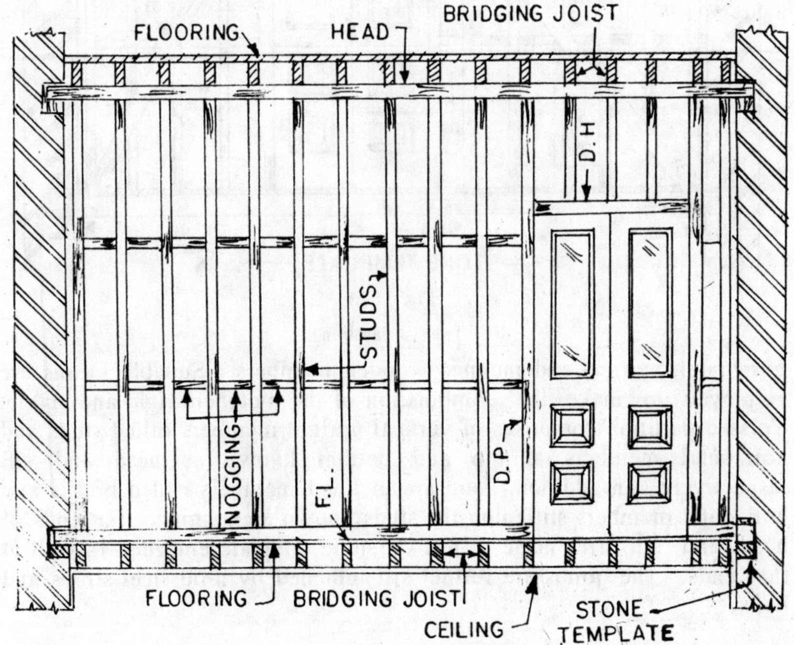

Fig. 11·22

Stud or common paritition

are termed as puncheons. To increase the stability of the partition, the ends of head and sill are usually embedded in the side walls for a short distance. The entire weight of the partition is borne by the floor and as such solid support for the sill member must be ensured. The support may be a solid wall or girder or beam below the floor.

Trussed partitions. In places where the provision of solid support below the sill is not possible, the wooden trussed partitions have to be employed. The weight from the trussed partition walls is borne by the side walls and thus it may be constructed quite independent of the floor. Trussed partition wall structure consists of a triangulated frame-work of

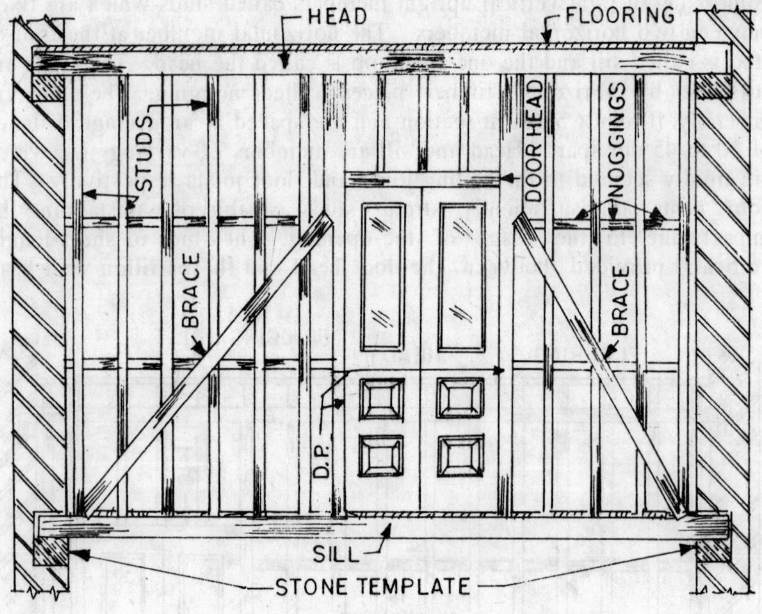

Fig. 11·23

Trussed partition

horizontal, vertical and inclined wooden members. Suitable means are employed to make the combination of the members rigid and stable. Trussed partition comprises of vertical upright members called studs and horizontal members at top and bottom known as head and sill respectively. In addition, there are inclined members called braces and horizontal members stiffening the studs known as nogging. The ends of head and sill are made to rest on stone template embeded in wall at the ends. The joints are further strengthened by mild steel straps and bolts.

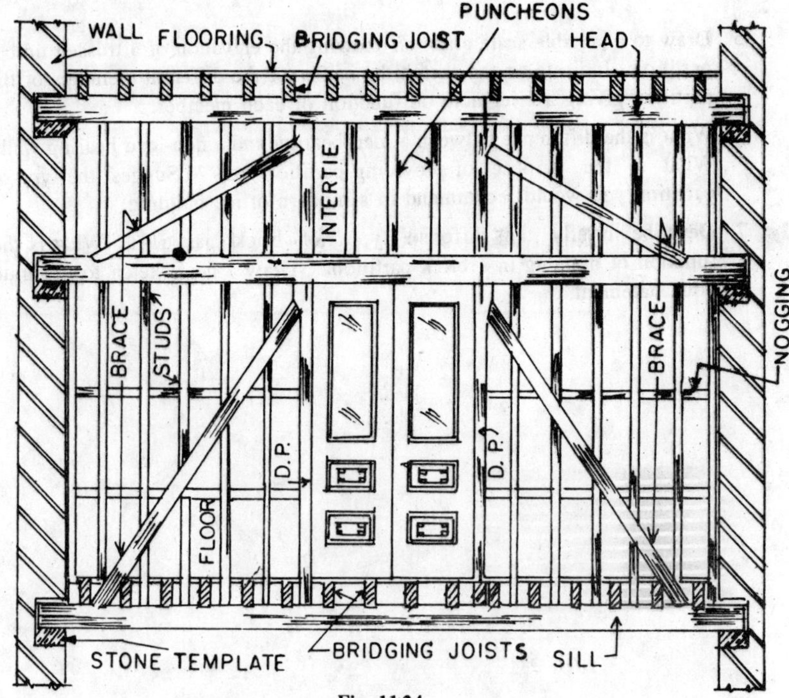

Fig. 11·24

Trussed partition supporting floor loads

Questions for Review

1. Sketch the details of wooden partition 2·44 m. high with a door 120 cm. x 214 cm. in the middle to be provided half way in a room of 12 m. x 6 m. in size.

2. Describe different types of partitions, giving the merits and demerits of each.

3. (*a*) Write a brief essay on the employment of different types of partition walls in modern construction. Illustrate your answer with sketches and give a critical note about each type in respect of efficiency, durability, economy and constructional aspect.

4. Write short notes on :

   (*i*) Metal lath and plaster partitions

   (*ii*) Pre-cast concrete partitions

   (*iii*) Asbestos cement partitions

   (*iv*) Glass partitions.

5. Draw to a suitable scale a sketch showing the elevation of a trussed timber partition. Indicate on the sketch the names of the different members of the partition giving a brief note on function of each member.
6. What is the difference between a load bearing and a non-load bearing wall? What is the purpose of providing partition wall ? Suggest the type of partition you would recommend in a modern office building.
7. Describe briefly the different types of brick partitions. What is the function of nogging in a brick partition.? Draw a neat sketch to illustrate your statement.

12

Ground Floors

12·1. GENERAL

The floor of a building immediately above ground is known as ground floor. In case, part of the building has basement, the floor is termed as basement floor.

There are various types of floors which are commonly adopted in residential or public buildings these days. Each type of floor has its own merits and there is not a single type that could be suitably provided under all circumstances. The different types of floors which are commonly used for ground floor construction are given below :

(1) *Mud flooring*
(2) *Muram flooring*
(3) *Brick flooring*
(4) *Flag stone flooring*
(5) *Tiled flooring*
(6) *Cement concrete flooring*
(7) *Granolithic flooring*
(8) *Terrazzo flooring*
(9) *Mosaic flooring*
(10) *Marble flooring*
(11) *Timber flooring*
(12) *Asphalt flooring*
(13) *Rubber flooring*
(14) *Linoleum flooring*
(15) *Acid proof flooring*

The selection of the type of floor often poses a problem and as such the following points are carefully considered before making final choice:

(*a*) It should be durable.

(*b*) It should be easy to clean.

(*c*) As far as possible it should be noiseless.

(*d*) It should have a good appearance.

(*e*) It should be free from dampness.

(*f*) It should be fire-resistant.

(*g*) It should have low maintenance cost.

(*f*) It should not be very costly.

The description of various types of floors is given below :

12·2. MUD FLOORING

In a tropical country like India, mud floors are commonly constructed in villages. They are cheap, hard, fairly impervious, easy in construction and easy in maintenance. They remain warm in winter and cold in summer and hence are most suitable for places where the temperature is extreme during these seasons. The method of its construction is very simple. Upon the prepared bed, a 25 cm thick layer of selected moist earth is evenly spread out and is rammed well so as to get a consolidated thickness of about 15 cm.

It is important to note that no water is used during the process of ramming. In order to prevent the formation of cracks after drying, chopped straw in small quantity is mixed with the moist earth before ramming. The floor is maintained by giving a thin cement cow-dung wash (1:2 to 1:3) once or twice a week.

12·3. MURAM FLOORING

Any disintegrated rock is called muram. Muram floors are constructed in villages in India and have the same advantages as that of mud-floors. Unlike mud floors, they are constructed with great care. Upon the prepared sub-grade a 15 cm. thick layer of muram is laid. A 25 mm. thick layer of powder muram is spread over the already laid muram layer and water is sprinkled over the entire surface. The surface is then rammed well. After ramming the surface is saturated with water, so that a thin layer is formed on the top of the rammed surface. Then the surface is trampled well till the cream of muram rises to the top. The surface is left in this state for about a day and then it is rammed again with wooden rammers called thapies for about three days. The dry hard surface thus formed is then smeared with a thick coat of cow dung and rammed once again for two days in the morning. Finally the surface is finished with a cement cow-dung plaster 1:4 (1 cement : 4 cow dung). To maintain the floor in good condition, it is given a wash of cement cow-dung plaster once a week and the surface is wiped clean immediately.

The merits and demerits of mud or muram flooring may be summarised below :

Merits :

(*i*) It is cheap

(*ii*) It has smooth, hard and fairly impervious surface.

(*iii*) It is easy in construction.

(*iv*) It has sufficiently long life if properly maintained.

(*v*) It maintains comfortable temperature in all seasons.

Demerits : For perfect maintenance of the surface it has to be given cement cow-dung wash once or twice a week. This may be objectionable from sanitary consideration.

12·4. BRICK FLOORING

This type of flooring is commonly provided in warehouses, stores and godowns or in places where heavy articles are stored. The flooring may be done with brick laid flat or on edge arranged in herring bone fashion or set at right angle to the walls.

The earth filling under floor is well compacted first and thereafter 10 to 15 cm. thick layer of cement or lime concrete is laid over the entire area of floor. This layer is known as subgrade. The floor is laid directly over the subgrade. The slope required to be given to the floor to take care of floor washing etc. is provided in subgrade itself. Prior to use, the bricks should be properly soaked and wetted in water.

The bricks are laid on edge on 12 mm. thick mortar bed in such a manner that all the joints are full of mortar. Where the floor has to be plastered or painted, all the face joints should be raked to a depth of about 15 mm. during the progress of work (when the mortar is green). This is necessary to ensure adequate bond between the flooring and the mortar of pointing or plastering. In case, however, pointing or plastering is not needed, the joints need not be raked and instead rendered flush and finished during laying of bricks.

The flooring should be cured for a minimum period of seven days before use.

Merits :

(*i*) It is durable and sufficiently hard.

(*ii*) It is cheaper than cement concrete, wooden or mosaic flooring.

(*iii*) It is non-slippery.

(*iv*) It is easily repairable.

Demerits :

It is absorbent.

12·5. FLAG-STONE FLOORING

Any laminiscated sandstone available in uniform thickness is called flag stone. The stone slab for flooring may be square or rectangular with width not less than 38 cm. and thickness varying from 20 to 40 mm. The sub-grade is prepared by laying a 10 to 15 cm. thick layer of lime concrete over a bed of well consolidated earth. On this sub-grade well-wetted flag stones are laid on 20 to 25 mm. thick layer of bed mortar. When the stone slabs are properly set, mortar in the joints is raked out to a depth of about 20 mm. and flush pointed with cement mortar (1:3). A slope of 1 to 40 is necessary to be given in flag stone flooring for proper drainage.

Merits :

(*i*) It is hard, durable and resistant to wear and tear and as such is used in workshops, garages and godowns.

(*ii*) It is easily repairable.

(*iii*) It is easy in construction.

(*iv*) In stone districts, it can be used with economy.

Demerits :

(*i*) It does not give a pleasing appearance and hence it is not suitable to be provided in places like residential building or important public buildings.

(*ii*) Its usage is not comfortable.

12·6. TILED FLOORING

Depending upon the type of material used tiled flooring can be broadly divided into the following categories.

(1) Terrazzo flooring

(2) Chequered tile flooring

(3) Glazed tile flooring

(4) PVC. tile flooring.

1. Terrazo Tile Flooring

Terrazo tiles manufactured under hydraulic pressure are available in following three standard sizes.

(*i*) 200 x 200 x 20 mm.

(*ii*) 250 x 250 x 22 mm.

(*iii*) 300 x 300 x 25 mm

The total thickness of the tile comprises of two layers *i.e.*, the upper layer and the backing. The thickness of the upper layer (also known as wearing layer) varies from 5 mm. to 6 mm. and it consists of mix of cement, marble chip aggregates and pigment (if required) taken in suitable proportions. The backing of the tile consists of leaner mix of cement and aggregates usually taken in proportion of 1:3.

The tiles are laid over reasonably hard sub-grade which may be of concrete or R.C.C. slab.

Prior to laying tiles, a 30 mm. thick layer of lime mortar 1:3 (1 lime putty : 3 surkhi or coarse sand) is spread over the sub-grade to serve as bedding. The bedding mortar is allowed to harden for a day. Immediately before laying, neat cement slurry is spread over the bedding mortar and the tiles are then fixed over the slurry grout. Each tile is gently tapped with a wooden mallet till it is properly bedded and levelled. The joints in the tile flooring should not exceed 15 mm. in thickness. Next day all the joints in the flooring are cleaned of loose mortar, dust etc. to the depth of about 5 mm. by use of wire brush. The joints are thereafter grouted with cement slurry of the same shade as that of tiles and the same slurry is applied over the flooring in the form of thin coat.

The flooring is cured for 7 days and thereafter the surface of the tiles is ground with grinding machine fitted with grinding stone No. 60. After grinding, the surface is washed and again covered with a thin coat of cement and cured second time. After curing, the surface is ground second time by using grinding stone No. 120. Day after the second grinding, the entire surface is finally ground using grinding stone No. 320.

The grinding of surface is also termed as polishing. Finally the surface of flooring is rendered clean by using oxalic acid water.

2. Chequered Tile Flooring

Chequered tiles are available in the following standard sizes.

(*i*) 200 x 200 x 22 mm.

(*ii*) 250 x 250 x 22 mm.

(*iii*) 300 x 300 x 22 mm.

The method of laying tile is identical to that of terrazzo tile flooring except that the polishing of the tile has to be done by hand.

3. Glazed Tile Flooring

Glazed tiles are available in the following standard sizes :

(*i*) 100 x 100 x 5 or 6 mm.

(*ii*) 150 x 150 x 5 or 6 mm.

The tiles have only top surface glazed. The tiles are laid over a reasonably hard sub-grade which may be of concrete or R.C.C. slab. Prior to laying, a 10 mm. thick layer of cement mortar 1: 3 is spread over the sub-grade to serve as bedding. The mortar is allowed to harden so that it may permit the mason, to work over the surface without damage to bedding mortar. Immediately before laying tiles, neat cement slurry is spread over the bedding mortar and the tiles are then fixed over the slurry grout. Each tile is gently tapped with a wooden mallet till it is properly bedded and levelled. The joints in the tiles should be as thin as possible and should be in straight lines. After laying, the joints are cleaned off all dirt and mortar to a depth of 2 to 3 mm. with the help of wire brush or trowel and then flush pointed with white or coloured cement to suit the colour of the tiles. The flooring is cured for seven days and thereafter washed before use.

4. P.V.C. Tile Flooring

P.V.C. tiles are now manufactured in variety of shades and design and are being commonly used in residential as well as non-residential buildings. It gives a decorative floor finish which is resilient, smooth and can be cleaned easily. It can be laid over a prepared base which can be of concrete, timber etc.. Prior to laying the tiles, it is necessary to see that the base is perfectly dry and brought to the temperature at which it will be while in use. The layout of the P.V.C. tiles (to the required design/pattern) is marked on the base with the help of guide lines. Adhesive of specified make is thereafter applied on the base and the back of the P.V.C. tiles with the help of a notched trowel. Laying of tile commence when the adhesive has set sufficiently (say within half an hour). After laying, the tiles are pressed suitably with wooden rollers (weighting 5 kg.) to ensure intimate contact with the base. Extra adhesive that oozes out is wiped off and the flooring is finally cleaned with warm soap water before use.

Merits and demerits of tiled flooring.

Merits :

(*i*) It is non-absorbent.

(*ii*) It is easily repairable in patches.

(*iii*) It offers pleasant appearance.

(*iv*) It is durable.

(*v*) It permits quick laying of floor.

(*vi*) It is resistant to wear and has fairly good strength.

Demerits :

(*i*) This type of construction is very costly both in initial cost as well as in maintenance.

(*ii*) Terrazo tile and glazed tile flooring becomes slippery when wet.

(*iii*) PVC tile flooring gets damaged when in contact with burning objects like paper, match stick etc.

12·7. CEMENT CONCRETE FLOORING

This type of flooring is commonly used both in residential as well as public buildings.

The method of laying cement concrete flooring on ground floor of a building can be broadly divided in the following steps.

(*i*) **Preparation of sub-base.** The earth filling in plinth is consolidated thoroughly so as to ensure that no loose pockets are left in the whole area. A 10 to 15 cm. thick layer of clean coarse sand is then spread over the whole area. The sand layer a consolidated and dressed to the required level and slope.

(*ii*) **Laying of base concrete.** The base concrete used under floors may be cement concrete or lime concrete laid to a thickness varying from 7·5 to 10 cm. In case of cement concrete, the mix commonly used is 1:5:10 (1 cement : 5 sand : 10 stone or over burnt brick aggregates 40 mm. nominal size). Lime concrete, if used, should be made up out of 40 mm. nominal size stone/over burnt brick aggregate and 40% mortar comprising of 1 lime : 2 sand/surkhi or 1 lime : 1 surkhi/ash : 1 sand.

The base concrete layer is deposited over the whole area, thoroughly tamped and levelled to a rough surface.

(*iii*) **Laying the topping.** When the base concrete layer has fully set and hardened, its surface is thoroughly cleaned and the entire area is divided into rectangular or square panels by use of 4 mm. thick glass strips or 5 mm. thick plain asbestos strips. The size of panel is basically governed by the location of floor (exposed or indoor), temperature and other climatic conditions, thickness of topping and proportions of ingredients in concrete mix. etc. In general, the area of one such panel should not preferably be more than 2 sq. m. The surface of base concrete should be made damp and applied with a coat of neat cement slurry prior to laying the topping., The rough finished surface of base concrete coated with cement slurry ensures adequate bond between the base and the topping. The cement concrete topping, normally consisting of 1:2:4 (1 cement :2 sand : 4 stone aggregate) is then laid in required thickness in one operation in the panels. The concrete is spread evenly by using a straight edge and the surface is thoroughly tamped and floated with wooden floates till the cream of the mortar comes at top. The surface is then smoothened and finally finished by means of steel trowels. In case glass or plain asbestos strips are not required, the panels are formed by use of wooden or angle iron battens. The battens should have depth equivalent to that of concrete topping. The surface of the battens which is to come in contact with concrete, is coated with raw linseed oil before concreting. Concrete for the topping is then laid in alternate panels and finished as described above. The battens used for forming panels are removed next day and the topping concrete laid in the alternate panels.

In case the cement concrete flooring is to be laid over R.C.C. slab, it is usual to allow the slab concrete to harden and then lay the flooring. Prior to laying, the slab top is cleaned of all dirt, dust, loose particles, mortar droppings and debris etc. and the flooring laid as described under the sub-head '*laying of topping*' above.

In order to prevent the tendency of separation of the cement concrete flooring from the R.C.C. slab, a 20 mm. thick cushioning layer of lime mortar (1 lime : 2 sand/surkhi or/lime : 1 surkhi/sand) or 50 to 75 mm. thick lime concrete is sometimes provided between the R.C.C. slab and the cement concrete flooring.

If the working conditions permit, the flooring can also be laid monolithically over the R.C.C. slab while the slab concrete is still green. In this case, the slab concrete is roughened with wire brushes so as to ensure a good bond between the base and the flooring layer. Prior to laying of flooring, the slab surface is cleaned and a coat of cement slurry is applied over it. In this case, any slope required for the floor is given in structural concrete itself.

The flooring should be cured for 10 days before use.

Merits :

(*i*) It is non-absorbent and thus it is very useful for water stores.

(*ii*) It is durable and hence it is commonly used in kitchens, bathrooms, schools, hospitals, drawing rooms etc.

(*iii*) It is smooth and pleasing in appearance.

(*iv*) It is economical and has the advantages of costlier types of floors.

(*v*) It possesses good wearing properties and can be easily maintained clean.

Demerits :

(*i*) Defects in carelessly made floor cannot be rectified, and as such, it requires proper attention while laying.

(*ii*) It cannot be satisfactorily repaired by patch work.

12·8. GRANOLITHIC FLOORING

It is a finishing coat provided over the concrete surface to form a hard, resistant to abrasion and durable flooring. Granolithic concrete is composed of cement, sand and specially selected aggregates. The grading of aggregates is very important. Coarse aggregates from basalt or limestone or quartzite are suitably graded from 13 mm. to I.S. Sieve No.

.40. The concrete mix is usually of 1:1:2 or 1:1:3. In order to get monolithic construction, the granolithic concrete should be laid before the base concrete has set. The minimum thickness of finishing should be 13 mm. After laying,the surface is tamped and floated with wooden floats and finally smoothened by means of steel trowel.

12·9. TERRAZO FLOORING

On account of its decorative and good wearing properties, this type of flooring is becoming very popular these days and is being commonly provided in offices, schools, hospitals, residential buildings, banks etc. Terrazzo is a concrete surface with special aggregate of marble chips (or other decorative material) mixed with white and/or coloured cement in proportion of 3:1 (three marble chips : one cement) The aggregates are exposed by grinding the surface mechanically or manually. The flooring may be laid in different thicknesses. The construction of 40 mm. thick flooring is described below.

The operations of the preparation of sub-base and laying of base concrete are carried out in similar manner as described earlier in case of cement concrete flooring. The 40 mm. thick flooring will consist of a 6 mm. thick layer of terrazzo topping and 34 mm. thick under layer of cement concrete 1:2:4. Prior to laying flooring, the whole area is divided into suitable panels by use of dividing strips (made from copper, brass, aluminium or glass) of thickness not less than 1.5 mm. and width not less than 25 mm. In order to reduce the risk of cracking, the area of one panel should not exceed two square metre.

The surface of base concrete should be cleaned of all dirt, mortar dropping debris etc. and then wetted with water. Prior to laying the cement concrete under layer, the wet surface of base concrete is smeared with a coat of neat cement slurry to ensure proper bond between the base concrete and flooring.

The concrete under layer is then spread, consolidated, levelled and finished to a slightly rough surface. When the under layer has hardened sufficiently, already prepared plastic terrazzo mixture is laid over the treated under-layer of concrete and compacted thoroughly by tamping or rolling. The terrazzo mixture consists of one part of cement (white or coloured) or a mixture of the two mixed with two to three parts of marble chips with water just sufficient to make it a workable mix. During tamping or rolling, it should be ensured that at least 80% of the finished surface should show exposed marble chip aggregates and if necessary additional marble chips may be added during the process of tamping or rolling to satisfy this condition.

After tamping or rolling, the surface is floated and troweled and left to dry out for 12 to 18 hours. After this period the surface is cured under standing water for 2 days.

Thereafter surface is subject to first grinding by machine by use of carborundum stones of coarse grade (No. 60). The whole surface is then scrabbed using plenty of water. All pores and holes (if any) are then filled with cement grout (using the same coloured cement as in the original mix for terrazzo topping) and the surface coated with neat cement grout of cream like consistency. The surface is again cured for five to seven days and second grinding is carried out by machine by using carborundum stones of fine grade (No. 120). In the similar manner, the surface is cleaned and repaired as before and cured again for a period of 4 to 5 days and subjected to third grinding by machine by using carborundum stones of 320 grit size.

The finished surface is finally worked with dilute oxalic acid solution. The floor is finally polished by floor polishing machine fitted with felt or hessian bobs, till it shines.

12·10. MOSAIC FLOORING

Over the hard base concrete bed, a 5 cm. to 6 cm. thick layer of lime surkhi mortar is spread in a small area. The area taken is such that the flooring can be easily completed in that portion without the mortar getting dry. On this a layer of paste consisting of two parts of sifted slaked lime, one part of powdered marble and one part of puzzolana is laid in thickness not exceeding 3 mm. This surface is left to get dry for four hours. Tiles or marble pieces, cut to the desired shapes, are then hammered on this surface in the desired pattern. The surface is gently rolled by light stone roller about 30 cm. in diameter and 45 to 60 cm in length. Rolling is done till the tile or marble pieces form an even surface and are cemented together properly. The surface is allowed to dry for a day and then polished by rubbing it with pumice stone fitted with long wooden handle. The polished surface is finally allowed to dry for some weeks before use.

12·11. MARBLE FLOORING

Marble flooring is commonly provided in places of worship (temples, churches, mosques, gurudwaras etc.) and in public buildings having rich specifications. In view of high cost of material, use of marble flooring in residential buildings in normally restricted to areas like kitchen, bath room, entrance etc.

The size of marble slabs to be used in flooring depends upon the pattern to be adopted. The slabs for normal works are square or rectangular in shape and their thickness varies fro. 20 mm to 40 mm. The flooring is laid on prepared sub-grade of concrete or on the R.C.C. floor slab. Prior to laying flooring, the sub-grade in cleaned, wetted and mopped properly. A layer of bedding mortar which can be 1:4 cement mortar (1 cement :4 coarse sand) or lime mortar (1 lime putty : 1 surkhi : 1 coarse sand) is spread in average thickness of about 20 mm. under the area of each slab. The marble slab is laid on top of the bedding mortar, pressed and tapped with wooden mallet and brought to the level with the already laid adjoining slabs. The slab is thereafter lifted up and placed aside. The top surface of the bedding mortar is then corrected by adding mortar at hollows. After the bedding mortar is hardened slightly, cement slurry is spread over the same. at the rate of 4.4 kg of cement per sq m. The edges of the slab already laid are then smeared with cement slurry of the same shade as that of the marble slab and the slab to be paved is again placed in position and tapped with wooden mallet till it is properly bedded in level with the adjoining slabs. The joint between two slabs has to be very fine (paper thick). The cement that oozes out of the joint is cleaned. Remaining slabs are also laid in the same manner and the flooring work completed. The paved area is cured for a minimum period of seven days.

12·12. TIMBER FLOORING

In hilly areas, where the climate is damp and wood is easily available, wooden floors prove economical and as such are commonly used. Its use is not extensive in other places. However, they are used for dancing halls, carpentry halls, auditoriums etc. In this type of floor, prevention of dampness is of great importance and hence every precaution is taken to check the dampness from rising above. The entire area of ground below the floor is covered with a 15 cm. layer of

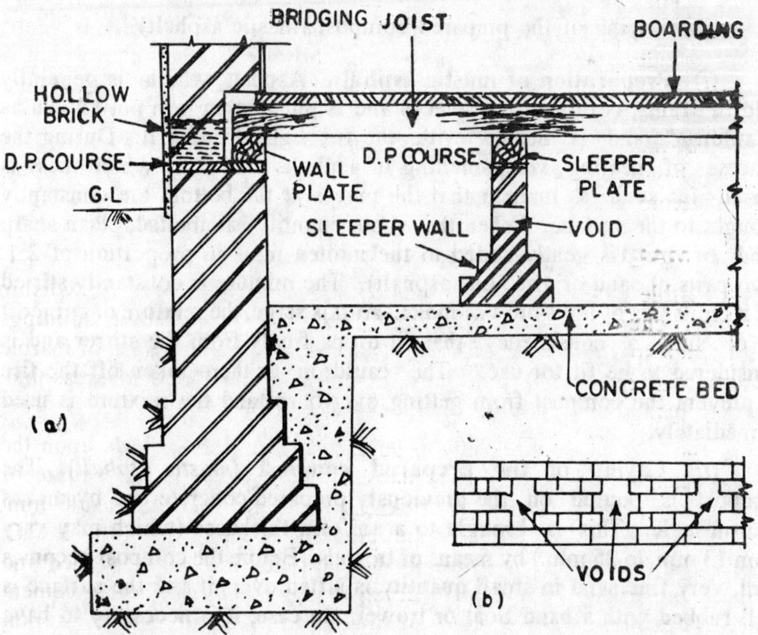

Fig. 12·1 (*a*). Section through a timber floor

Fig. 12·2 (*b*). Detail showing position of voids in the honey-combed sleeper wall

cement concrete. This layer is called "*site or oversite concrete*". Timber floors essentially consist of boarding supported on timber joists called bridging joists or floor joists which are nailed to wall plates at their ends and supported by intermediate walls called sleeper or dwarf walls, along their length. The sleeper walls are generally 10 cm. thick and are honey-combed. They are intended to reduce the span for the floor joists. The sleeper walls are seldom spaced more than 1·8 m. apart and are honey-combed to enable free circulation of air in the space below the floor. Longitudinal timber members called 'sleeper plates' are fixed on the top of sleeper walls. The floor joists are secured to the sleeper walls by being nailed to the sleeper wall plates. A damp-proof course layer is laid immediately below the wall plates to prevent the rising of dampness. The hollow space between the flooring and the oversite concrete is kept dry and fully ventilated.

12·13. ASPHALT FLOORING

On account of its being dustless, elastic, durable, waterproof, acidproof and attractive in appearance, mastic asphalt flooring is becoming very popular these days. This type of flooring is non-slippery and noiseless too, as such it is recommended for use in factories, loading platforms swimming pools, and terrace floors etc. The construction of floor involves the following operations:

(*i*) Preparation of mastic asphalt.

(*ii*) Laying of the prepared compost (mastic asphalt)

(*i*) **Preparation of mastic asphalt.** Asphalt which is generally sold in drums, is broken into pieces and is put into an iron pot known as "Cauldron" and is heated with the fire lighted under it. During the process of heating when melting of asphalts starts, the whole mass is stirred in such a manner that the pieces at the bottom are constantly brought to the surface. When the whole quantity has melted, clean sharp sand or grit is gently added to the molten mass in proportion of 2:1. (two parts of sand or grit : one asphalt). The mixture is constantly stirred till it emits jets of light-brown smoke. At this stage, the mixture or compost is of such a consistency that it drops freely from the stirrer and is considered to be fit for use. The cauldron is then taken off the fire to prevent the compost from getting over-burnt and the mixture is used immediately.

(*ii*) **Laying of the prepared compost** (*mastic asphalt*). The compost is poured on the previously prepared concrete bed by means of iron ladle. This is brought to a uniform thickness (which may vary from 13 mm. to 25 mm.) by means of trowels. Before the compost becomes hard, very fine sand in small quantity is sifted over it and the surface is well rubbed with a hand float or trowel. In case, it is necessary to have a second layer of asphalt, it should be laid before the bottom layer solidifies (no sand being sifted over the bottom layer in that case.). The top layer is then finished in the same manner as described above.

12·14. RUBBER FLOORING

This type of flooring is not common in India but in foreign countries, it is used in residential buildings as well as public buildings like offices, hospitals, school etc. It is noiseless, comfortable, clean and durable. Like linoleum, this type of flooring is manufactured in the form of sheets or tiles in a variety of colours. Its thickness seldom exceeds 10 mm. Rubber sheets or tiles are made by combining at very high temperature pure rubber with cotton fibre, granulated cork, asbestos fibre, other glues and colour pigments to get the designed colour. The rubber tiles are laid by glueing them to a smooth and thoroughly clean and dry base by use of a special adhesive. The base may be of concrete, R.C.C. or wood. It is expensive in its initial cost but it serves as a most durable wearing surface in long run. However, oil or grease renders the floor slippery and it becomes difficult to restore it in good condition.

12·15. LINOLEUM FLOORING

Linoleum is a covering laid over wooden or concrete floors in order to hide the defects of flooring or to enhance its appearance. It is a material manufactured by mixing oxidised linseed oil with powdered cork, wood floor, various type of gums and suitable colouring pigments. The plastic mass thus obtained is pressed in the designed form and dried in ovens. It is finally cured and seasoned. It is available in a variety of colours both in plain and printed form. It is mostly sold in rolls which are 1.8 to 3.66 m. in width and the thickness of the felt seldom exceeds 6 mm.

Linoleum can be laid in the following three ways :

(*i*) By spreading the covering loose on dry and smooth floor

(*ii* By pinning down the ends of the loosely laid linoleum covering to the floor below.

(*iii* By fixing the linoleum covering to the floor by use of a special adhesive.

Merits :

Linoleum floor covering has the following advantages :

(*i*) It is washable, dustproof and can be easily cleaned with just a swab of damp cloth.

(*ii*) It has cushioning effect which gives comfort to the users.

(*iii*) It reduces noise effectively.

(*iv*) It forms a long-wearing and durable surface if laid and maintained properly.

(*v*) It gives a decorative floor finish which looks very attractive.

(*vi*) It is economical.

Linoleum floor covering is largely used in residential buildings offices, schools, hospitals, libraries, restaurants, railway carriages and buses etc. It is, however, not recommended in basement floors.

12·16. ACID-PROOF FLOORING

In places like chemical laboratories and plants, storage battery buildings etc. where acids are used or manufactured, it is necessary that the floor topping should be acid-proof. Asphalt blocks, made by moulding under high pressure a mixture of inert crushed rock aggregate and an acid proof asphalt successfully meet the requirements of an acid-proof flooring. The asphalt blocks are first laid on the hard base and then acid-proof asphalt is uniformly spread over the surface of the block. Before the liquid asphalt hardens, fine sand in small quantity is uniformly spread over it and surface is finished smooth and levelled.

Questions for Review

1. Describe briefly the type of floor used for different types of buildings and state the reasons for their choice. Give detailed specification for coloured Terrazzo flooring.
2. Describe in brief the types of floors that are normally adopted in modern buildings. Explain the advantages and disadvantages of each type. Suggest the type of floor you would recommended for a hospital ward measuring 12 X 12 m.
3. How are the following materials used in construction of floor ? Give an idea of the relative cost of each type of flooring in U.P.

   (*a*) Linoleum. (*b*) Cork.

   (*c*) Rubber. (*d*) Clay tiles.

   (*e*) Marble. (*f*) Stone.

   (*g*) Slate
4. Give the specifications for flooring of a food grain storage godown where trucks of 5 tonne capacity would come in and where the sub-soil water-level is within 1·5 m. of the existing ground level.
5. Describe briefly with sketches the type of flooring you would adopt for the following. Justify the selection.

   (*i*) Operation theatre in a modern hospital.

   (*ii*) Recreation hall of a high-class hotel.

   (*iii*) Public laterine.

(*iv*) Warehouses.

(*v*) Dancing halls.

6. What type of flooring you would adopt for the following.

(*a*) Grain store. (*b*) Factory hall.

(*c*) Recreation hall of a high-class hotel.

(*d*) Dancing hall. (*e*) Materials testing laboratory.

7. Write short notes on :

(*a*) Granolithic flooring. (*b*) Mosaic flooring.

(*c*) Terrazzo flooring. (*d*) Asphalt flooring.

(*e*) Rubber flooring (*f*) Tiled flooring.

8. Under what circumstances , cement, concrete flooring is recommended ? What treatment would you suggest if the cement concrete flooring is to be provided in the following situations.

(*a*) Basement of a building to be situated in a water-logged ground.

(*b*) Children ward of a modern hospital.

(*c*) Bathroom.

9. What type of flooring would you recommended for the following :

(*i*) Lecture halls of a modern college. (*ii*) Drawing halls.

(*iii*) Laboratories. (*iv*) Hostels.

(*v*) Workshops.

10. A wooden floor is to be provided in the ground floor room. Draw neat sketches showing the various parts.

11. What factors affect the design and specification of a floor finish ?

Describe the types of floor you would recommend for any four of the following, assuming that they are to be built on hard moorum soil ;

(*i*) A large school hall.

(*ii*) A warehouse.

(*iii*) A drawing room of a spacious residential bunglow.

(*iv*) Floor of an R.C.C. reservoir at ground level.

(*v*) A garage.

12. What kind of floor finishing will be required for :

(*a*) Warehouse. (*b*) Walkways.

(*c*) Hospitals. (*d*) Library.

(*e*) Dance hall.

Briefly describe the method of construction in each case.

13. Describe four commonly adopted type of floorings that are recommended in modern buildings giving their merits and demerits.

14. Explain the method of construction of cement concrete floor on ground at a site having low value of bearing capacity.

13

Stairs

13·1. DEFINITION

A stair may be defined as series of steps suitably arranged for the purpose of connecting different floors of a building. It may also be defined as an arrangement of treads, risers, stringers, newel posts, hand rails and baluster, so designed and constructed as to provide an easy and quick access to the different floors, rendering comfort and safety to the users. The enclosure containing the complete stairway is termed as staircase.

Stairs may be made from various materials like timber, stones, bricks, steel, plain concrete or reinforced concrete. The selection of the type of material to be used depends upon the aesthetical importance, funds available, durability desired and fire resisting qualities expected.

13·2. LOCATION OF STAIRS

The location of stairs in a building requires careful consideration. In the advent of fire or any such calamity, stairs provide the only means of communication and as such they are so located as to serve the purpose for which they are provided. In public building it should be located near the main entrance and in residential building it should be placed centrally so as to provide easy access from all the rooms and to maintain privacy at the same time.

13·3. TECHNICAL TERMS

The common technical terms used in stair construction are briefly discussed below :

(1) **Steps.** A portion of a stairway comprising the tread and riser which permits ascent or descent from one floor to another.

(2) **Tread.** The horizontal upper part of a step on which foot is placed in ascending or descending stairway.

(3) **Riser.** The vertical portion of a step providing support to the tread.

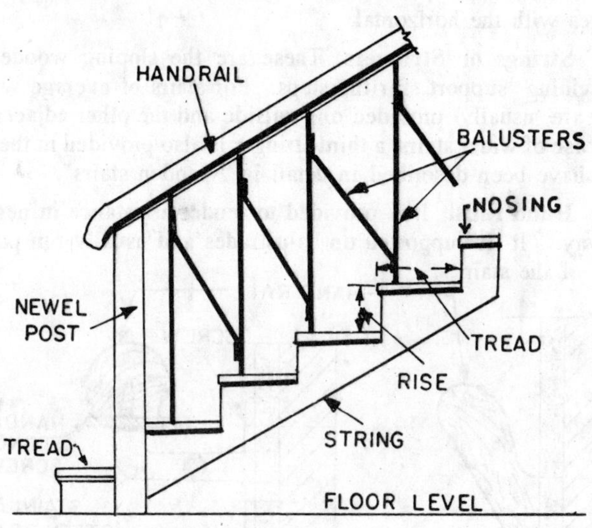

Fig. 13·1

Part section of stair showing its various components

(4) **Flight.** A series of steps without any platform, break or landing in their direction.

(5) **Landing.** A platform or resting place provided between two flights. A landing extending right across a stair case is termed as half space landing and the one extending only half across a stair case is called a quarter-space landing.

(6) **Nosing.** The outer projecting edge of a tread is termed as nosing. Nosing is usually rounded to give good architectural effect to the treads and makes the stair-case easy to negotiate.

(7) **Scotia.** It is a moulding provided under nosing to beautify the elevation of the step.

(8) **Line of Nosing.** It is an imaginary line touching the nosing of each tread and is parallel to the slope of the stair.

(9) **Flier.** It is a straight step having a parallel width of tread.

(10) **Winders.** These are tapering steps which are provided for changing the direction of a stair.

(11) **Going of Step.** The run of a step in a stair or the width of the tread between two successive treads.

(12) **Rise of Steps.** It is the vertical distance between the upper surface of the successive treads.

(13) **Soffit.** It is the under surface of a stair.

(14) **Pitch or Slope.** It is the angle which the line of nosing of the stair makes with the horizontal.

(15) **Strings or Stringers.** These are the sloping wooden members providing support for the steps. For stairs of average width, the stringers are usually provided one outside and the other adjacent to the wall. In case of wider stairs, a third stringer is also provided in the middle. Stringers have been described in detail in "wooden stairs".

(16) **Hand rails.** It is provided to render assistance in negotiating a stair-way. It is supported on balustrades and usually run parallel to the slope of the stair.

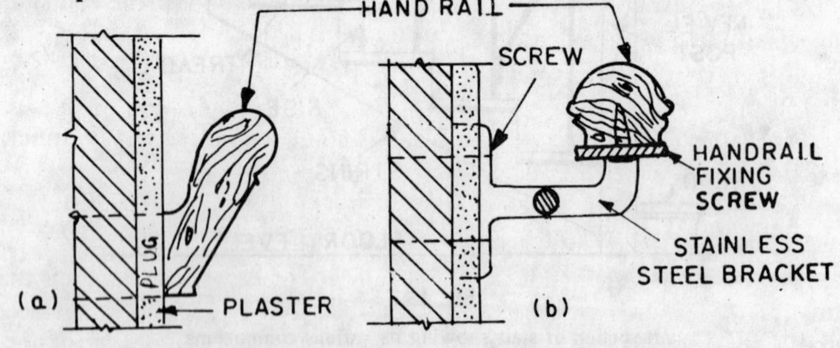

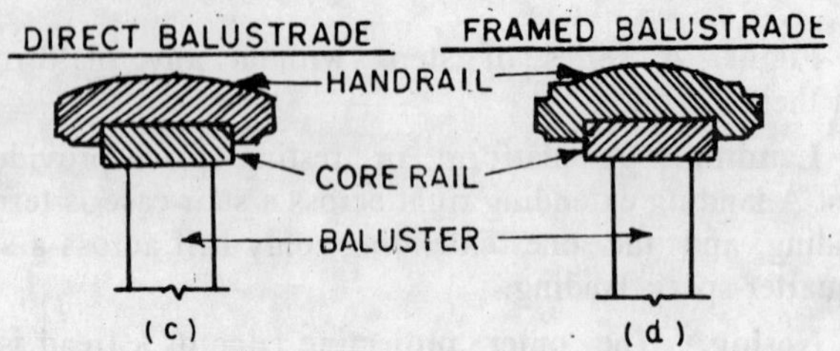

Fig. 13·2 Different types of handrails

(17) **Baluster.** It is a wooden, metal or masonry vertical member supporting a hand rail.

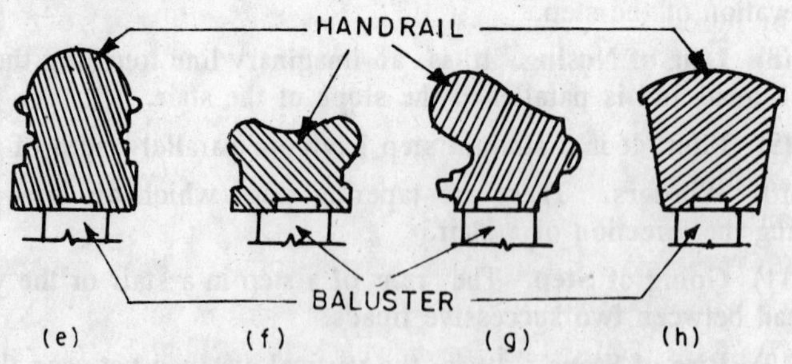

Fig. 13·3 Different types of wooden handrails

(18) **Balustrade.** It usually consists of a row of balusters surmounted by a rail and is provided to perform the function of a fence or guard for the users of the stair-way.

(19) **Newel.** It is a wooden or metallic post supporting the hand rail and is usually provided at the hand, foot or at points where the balustrade changes its direction.

(20) **Head room or Head way.** It is the clear vertical distance between the tread of a step and the soffit of the flight or the ceiling of a landing immediately over it.

(21) **Spandrel.** It is the triangular framing under the outside string of an open string stair.

13·4. REQUIREMENTS OF A GOOD STAIR

A well-planned and designed stair should provide an easy, quick and safe mode of communication between the various floors. The general requirement of a good stair may be divided into the different heads, described below :

(1) **Location.** It should be so located that sufficient light and ventilation is ensured in the stairway. If possible it should be located centrally so as to be easily accessible from the different corners of the building.

(2) **Width of a Stair.** Width of a stair varies with the situation and the purpose for which it is provided. In public building, where there is a regular traffic of people using the stair-way, its width should be sufficient while in a residential building it may be just the minimum. The usually adopted average value of the stair width for public and residential building is 1·8 m and 90 cm respectively.

(3) **Length of Flight.** For the comfortable ascent of stair-way the number of steps in a flight should be restricted to a maximum of 12 and a minimum of 3.

(4) **Pitch of Stair.** The pitch of long stair should be made flatter by introducing landings to make the ascent less tiresome and less dangerous. In general, the slope of stair should never exceed 40° and should not be flatter than 25°.

(5) **Head room.** The head room or the clear distance between the tread and the soffit of the flight immediately above it should not be less than 2·14 m.

(6) **Materials.** The stair should preferably be constructed of materials which possess fire-resisting qualities.

(7) **Balustrade.** The open well stairs should be provided with balustrade so as to minimise the danger of accidents.

(8) **Landing.** The width of landing should not be less than the width of stair.

(9) **Winders.** The introduction of winders in stair should be avoided as far as possible. They are liable to be dangerous and involve extra expense in construction. They are difficult to carpet and are especially unsuitable for public buildings. However, where the winders cannot be dispensed with, they should preferably be provided near the lower end of the flight. Thus instead of quarters space landing, three winders may be used and for a half-space landing, five winders or four radiating risers may be adopted.

(10) **Step proportions.** The rise and tread of each step in a stair should be of uniform dimensions throughout. The ratio of the 'going' and the ''rise'', of a step should be so proportioned as to ensure a comfortable access to the stairway.

13·5. FIXING THE "GOING" And "RISE" of A STEP

The size of a tread and riser varies with the situation of the stair and the purpose for which it has to be provided. In general, the following thumb rules are considered as a guide to obtain a satisfactory result :

(1) (Going in cm) + (2 x Rise in cm) = 60

(2) (Going in cm) x (Rise in cm) = 400 to 410 approx.

(3) Adopt the standard size of going and rise as 30 cm and 14 cm respectively and for each 25 mm deducted from going, add 12 to 13 mm to the rise.

In addition to empirical rules described above, the following rules must be kept in view while deciding the step size. The results given below are the ones which are practically adopted in the stair construction :

(1) For residential buildings the size of step commonly adopted is 25 cm x 16 cm.

(2) For public buildings the adopted size of step varies from 27 cm x 15 cm to 30 cm x 13 cm.

13·6. CLASSIFICATION OF STAIRS

The different forms of stairs may be classified under the following main heads :

(*i*) *Straight stairs*	(*ii*) *Dog-legged stairs*
(*iii*) *Open-newel stairs*	(*iv*) *Geometrical stairs*
(*v*) *Circular stair*	(*vi*) *Bifurcated stairs.*

(*i*) **Straight Stair.** The most obvious use of the straight stair is to form an access to entrance, porch or portico. Straight stairs cannot be avoided in places, where the stair-case hall is long and narrow and the possibility of any other form of stair may not be practically possible. In this form of stair, all the steps rise in the same direction. If the ascending is steep, the straight flight may be broken at an intermediate landing.

(*ii*) **Dog-legged Stair**. It consists of two straight flights of steps with abrupt turn between them. Usually, a level landing is placed across the two flights at the change of direction. This type of stair is useful where the width of the stair-case hall is just sufficient to accommodate two widths of stair.

(*iii*) **Open-newel Stair**. It consists of two or more straight flights arranged in such a manner that a clear space called a 'well' occurs

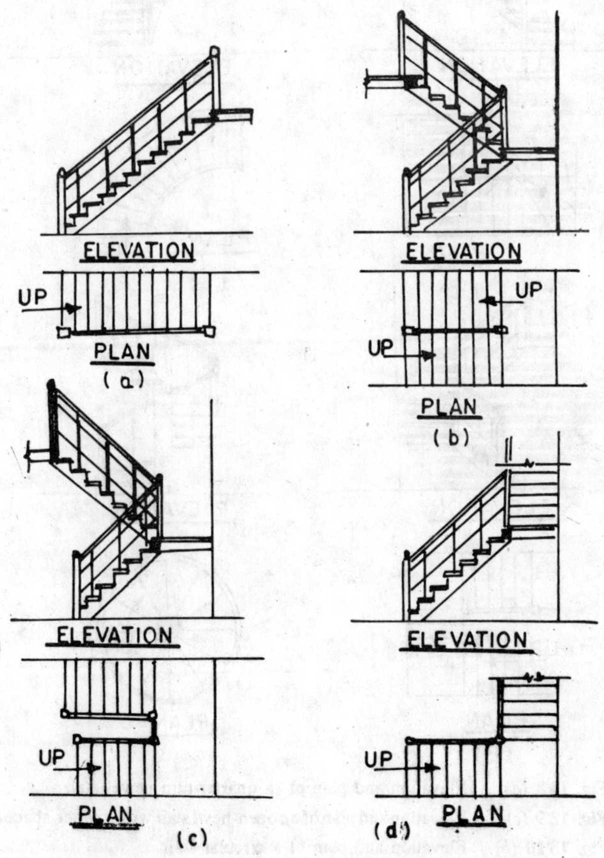

Fig. 13·4 (*a*) Elevation and plan of a straight stair
Fig. 13·6 (*b*) Elevation and plan of a dog-legged stair
Fig. 13·5 (*c*) Elevation and plan of a open-well stair
Fig. 13·7 (*d*) Elevation and plan of a quarter turn newel

between the backward and the forward flights. If the width of the stair case hall is such that it becomes difficult to accomodate the number of steps in the two flights, without exceeding the maximum allowable limit of steps in each flight, a short flight of 3 to 6 steps may be provided along the width of the hall. In this state, the intermediate short flight will have quarter space landing on its either side.

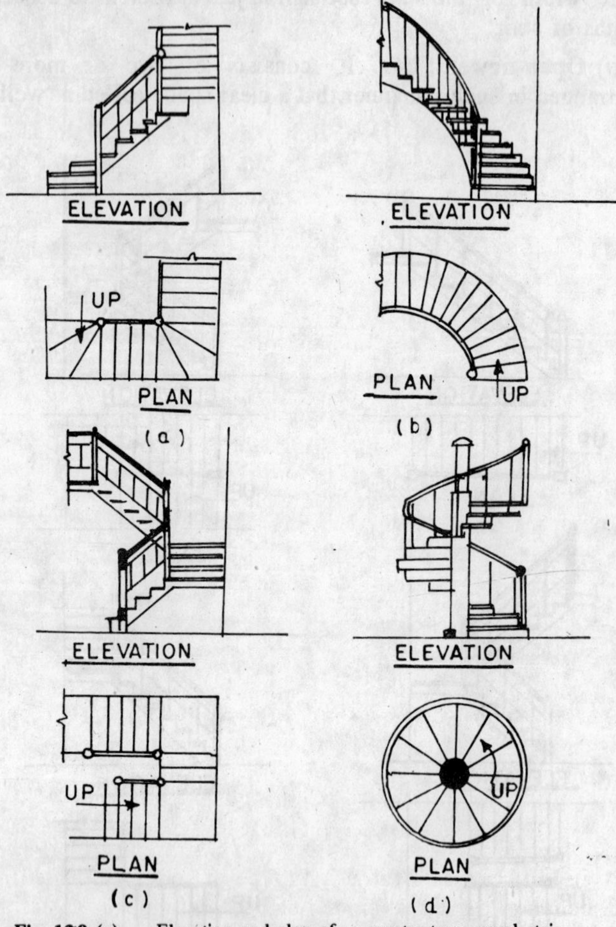

Fig. 13·8 (*a*) Elevation and plan of a quarter-turn newel stair

Fig. 13·9 (*c*) Elevation and plan of an open-newl stair with quarter-space landing

Fig. 13·10 (*b*) Elevation and plan of a circular stair

Fig. 13·11 (*d*) Elevation and plan of a spiral stair

(*iv*) **Geometrical Stair**. This is similar to the open-newel stair with the difference that the open well between the forward and the backward flight is curved. In this form of stair, the change in direction is obtained through winders.

(*v*) **Circular Stair**. Circular stair is commonly provided at the backside of a building for rendering access to its various floors for service purposes. The circular stairs are commonly constructed in R.C.C., cast iron or stone. In this form of stairs all the steps radiate from a newel post or well hole, in the form of winders.

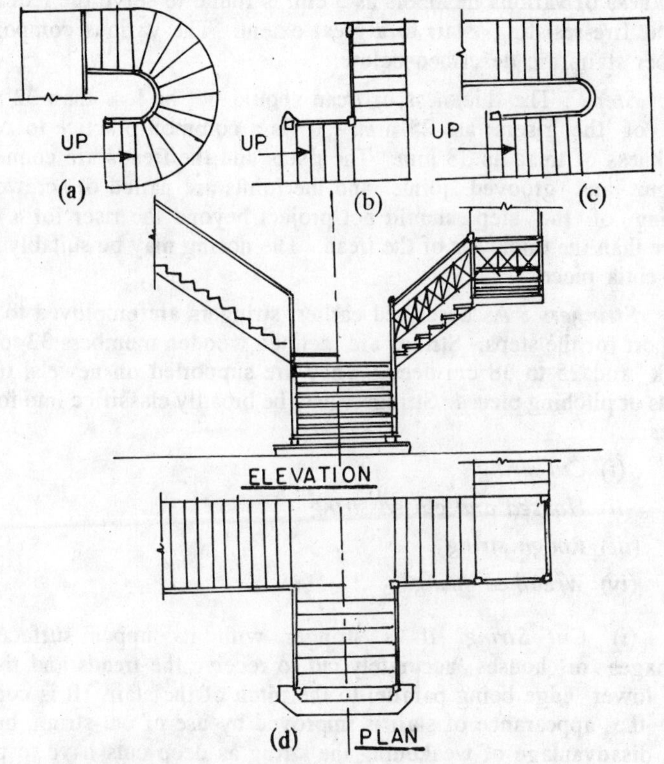

Fig. 13·12 (*a*) Plan showing half-turn geometrical stair

Fig. 13·13 (*b*) Plan showing quarter-turn stair with quarter-space landing

Fig. 13·14 (*c*) Plan showing open-well geometrical stair

Fig. 13·15 (*d*) Elevation and plan of a bifurcated stair.

(*vi*) **Bifurcated Stair**. This type of stair is provided in modern public buildings. In this type of stairs, the flights are so arranged that there is a wide flight at the start which is sub-divided into narrow flights at the mid-landing. The two narrow flights start from either side of the mid-landing.

13·7. STAIR OF DIFFERENT MATERIALS

The commonly adopted materials in the construction of stairs are wood, stone, steel, cast-iron, plain concrete or reinforced concrete, brick work or reinforced brick work. The stair of different materials have been described on the next page.

13·8. WOODEN STAIR

Timber is mainly used for stair construction where fire-resistance is not of much consideration. Timber stairs are cheap, easy in construction and maintenance and light in weight. Timber stair constructed from fire-resisting hard wood, like oak, teak, mahogany etc., using minimum thickness of various members as 5 cm. is found to serve the requirements of a fire-resisting stair to a great extent. The various components of timber stairs are described below :

Steps : The thickness of tread should not be less than 32 mm and that of the risers as 25 mm. It is a common practice to adopt the thickness of tread as 38 mm. The risers and the treads are connected by tongue and grooved joints and the joints are nailed or screwed. The nosing of the step should not project beyond the riser for a distance more than the thickness of the tread. The nosing may be suitably finished by scotia piece.

Stringers : As described earlier, stringers are employed to provide support for the steps. Strings are inclined wooden members 32 to 50 mm thick and 25 to 38 cm deep. They are supported on newels, trimming joists or pitching pieces. Stringers may be broadly classified into four main types:

(*i*) *Cut string*

(*ii*) *Housed and closed string*

(*iii*) *Rough string*

(*iv*) *Wreathed string.*

(*i*) *Cut String.* It is stringer with its upper surface having carriages or houses accurately cut to receive the treads and risers and the lower edge being parallel to the pitch of the stair. It is considered that the appearance of stair is improved by use of cut-string, but it has the disadvantage of weakening the string as deep cuts have to be made in it.

(*ii*) *Housed or Closed String.* It is a stringer having its top and bottom edges parallel to the slop of the stair. It has grooves cut on its inside to receive the treads and risers of the steps. The treads and risers are usually nailed, glued and wedged to the stringer.

(*iii*) *Rough String.* Rough strings are introduced in wider stairs for the purpose of rendering intermediate support to the steps.

(*iv*) *Wreathed string* : It is curved or geometrical stair string. It may be of cut or closed type.

Winders : In narrow stairs, the treads of winder steps are supported on risers which act as cantilever beams fixed in the staircase wall. In such cases, the riser should be of sufficient thickness to perform

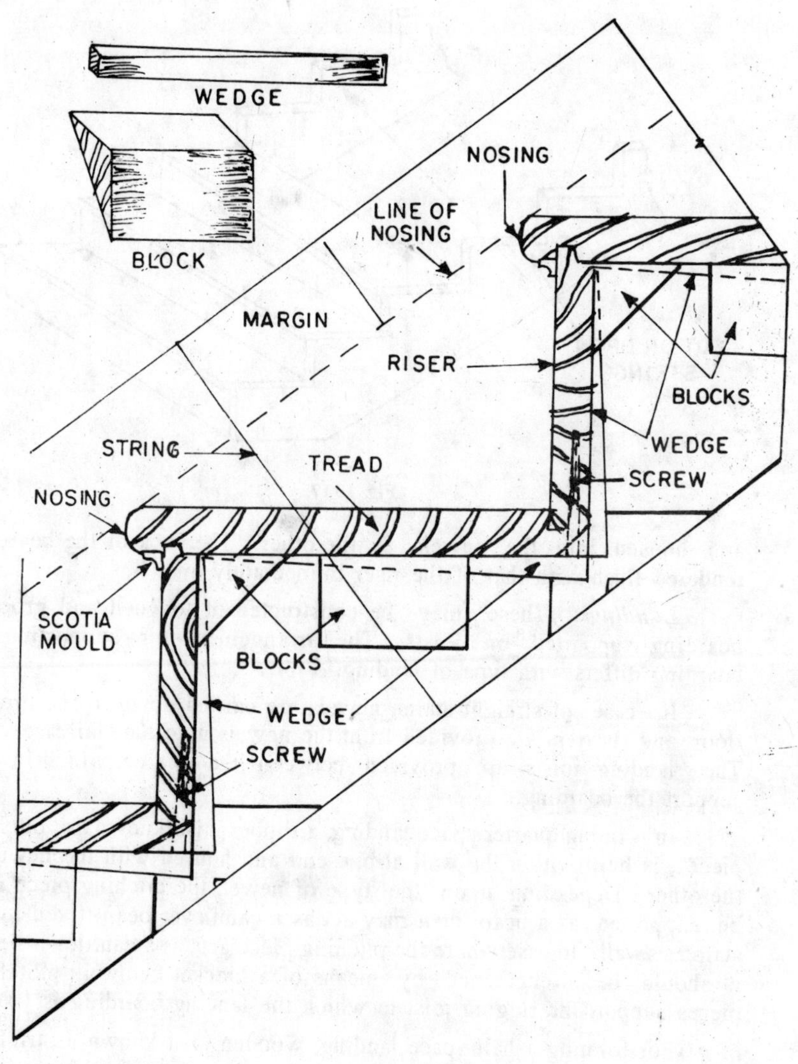

Fig. 13·16

Wooden stair

the function of a cantilever beam. At its free end, the cantilever riser beam is housed into the newel post. For wider stairs the winders are supported by means of bearers built into the wall at one end and framed

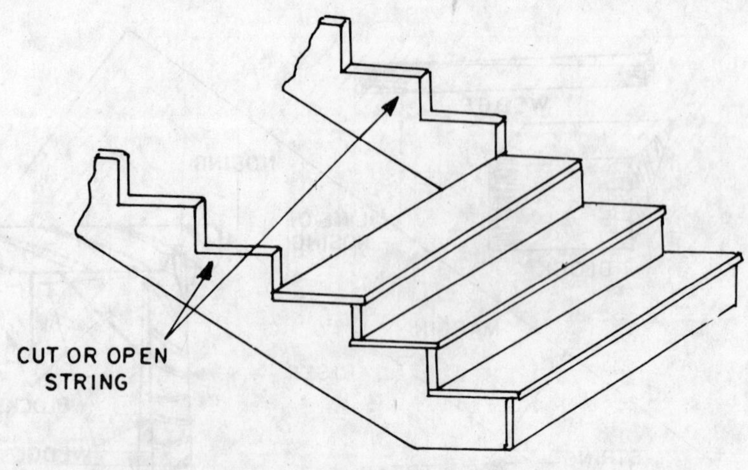

Fig. 13·17

and housed into the newels at the other. The back of the bearer is rendered flush with that of the riser immediately over it.

Landings: These may be constructed of tongued and grooved boarding supported on joists. The arrangement for supporting the boarding differs with type of landing.

In case of straight stairs, newels are carried down to the ground floor and bearers are provided from the newels upto the staircase wall. The landing joists are provided between the bearers which in turn support the boarding.

In forming quarter-space landing, a timber joist, known as "pitching piece", is built within the wall at one end and housed with the newel at the other. Depending upon the type of newel, the pitching piece may be supported as a beam or it may act as a cantilever beam fixed in the staircase wall. In case where the pitching piece acts as a cantilever beam, it should be strengthened by means of a bracket below it. Pitching pieces support the ridging joist to which the landing boarding is fixed.

For forming a half-space landing, wooden joist known as trimmer is fixed across the width of the staircase. Landing joists or bridging joists are tennoned into the trimmer at one end and are supported on the staircase wall at the other end.

13·9. STONE STAIRS

Stair made out of hard, strong, resistant to wear stones are found to be durable and fire resisting. They are commonly constructed in workshops, warehouses and other public buildings. Their usage in residential building is generally restricted to outside stairs. Being heavy in

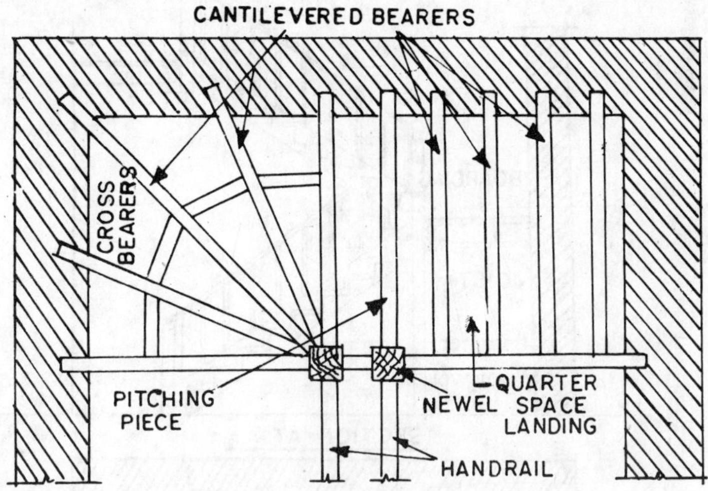

Fig. 13·18

Plan on left side shows arrangement of bearers for supporting winders and the plan on right hand side shows the frame work for quarter space landing

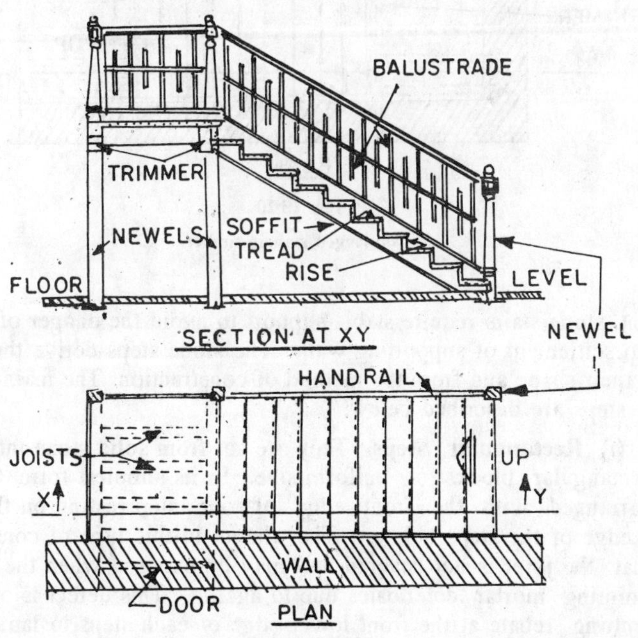

Fig. 13·19

Straight flight stairs

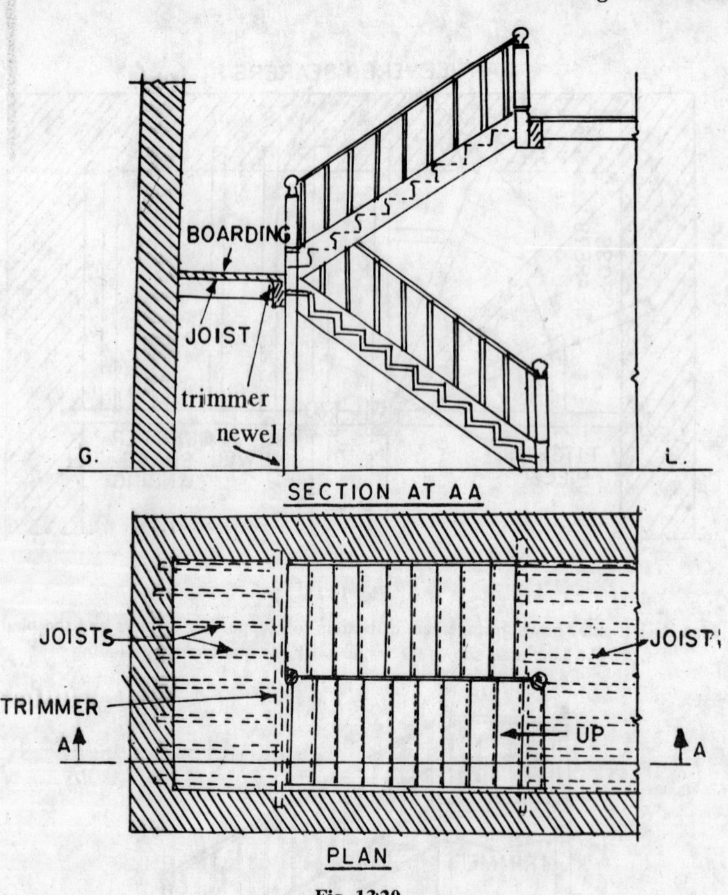

Fig. 13·20

Dog-legged stair in timber

weight, stone stairs require stable support to avoid the danger of damage due to settlement of supporting walls. The stone steps derive their name from their shape and from the method of construction. The main types of stone steps are described below:

(*i*) **Rectangular Steps.** They are cut from solid stone into square or rectangular blocks of uniform size. In its simplest form, the steps are arranged with the front edge of one step resting on the upper back edge of the steps below. The defect in this type of construction is that the joint is not covered and may be exposed when the bedding or pointing mortar deterioates due to ageing. This defect is overcome by cutting rebate at the front lower edge of each steps to lap over the back edge of lower step.

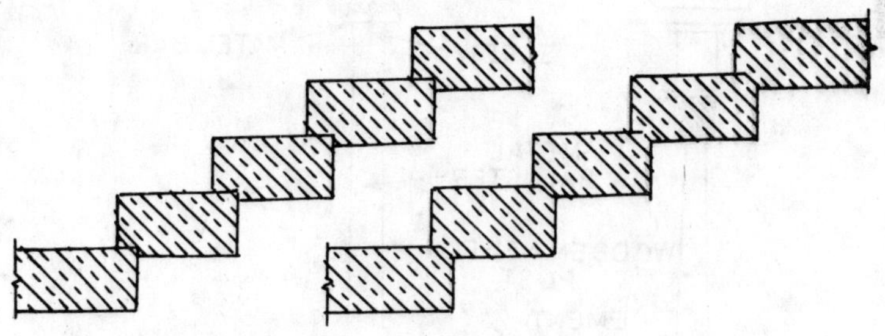

Fig. 13·21–13·22

Rectangular steps

(*ii*) **Built up Steps.** They consist of treads and rises formed of thin sawn stone slabs. They are often used as facing for concrete or brick steps. The minimum thickness of tread, if supported at end is restricted to 5 cm.

(*iii*) **Spandril Step.** A spandril step, excepting the ends, which are built into the wall, approximates to a triangle in cross-section. The under-side of the steps or the soffit usually exhibits a plain surface. This gives a greater head-room. The steps are rebated to fit on the one's below.

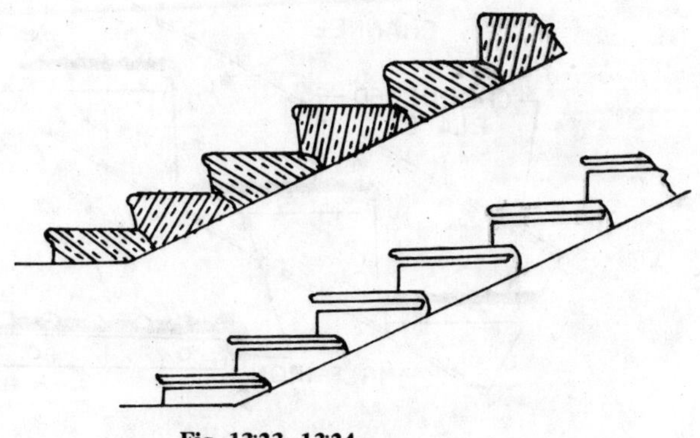

Fig. 13·23–13·24

Spandril steps

13·10. METAL STAIRS

They are usually fabricated from steel and cast iron and their use in generally restricted to factories, godowns etc. In its simplest form, a metal stair consists of steel stringer to which steel angles are riveted or welded and metal treads are provided over them. Sometimes concrete is filled in steel grating to from the treads. The details of metal stair are shown in figure 13·27.

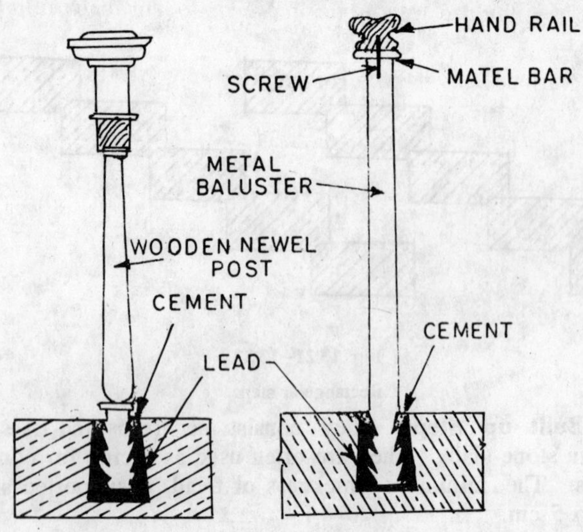

Fig. 13.25—13.26

Methods of fixing balusters in stone steps

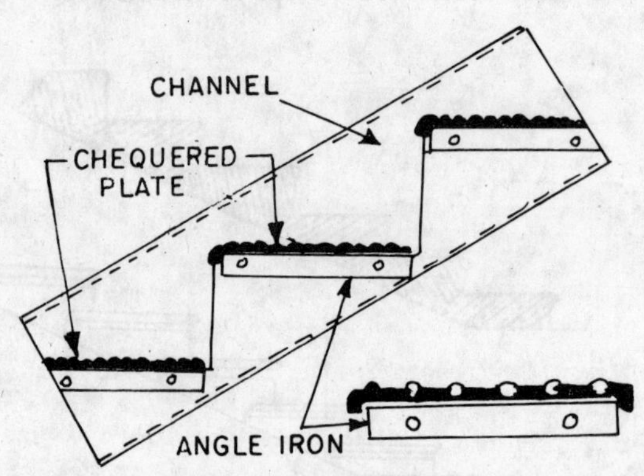

Fig. 13·27

Metal stair

13·11. REINFORCED CONCRETE STAIRS

Reinforced concrete stairs predominate the stairs made from wood, stone or metal. This is because of the various advantages R.C.C. has over other materials. Reinforced concrete is perhaps the most suitable of all the said materials for the construction of stairs. R.C.C.

stairs can be moulded in any desired form to suit the requirements of the architect.

The various advantages of reinforced concrete stairs are given below:

(*i*) They have requisite fire-resisting qualities to a great extent.

(*ii*) They are durable, strong, pleasing in appearance and can be easily rendered non-slippery.

(*iii*) They can be designed for greater widths and longer spans.

(*iv*) They can be easily cleaned.

(*v*) The cost of maintenance is almost nil.

(*vi*) In a framed structure of reinforced concrete, R.C.C. stair in perhaps the only choice.

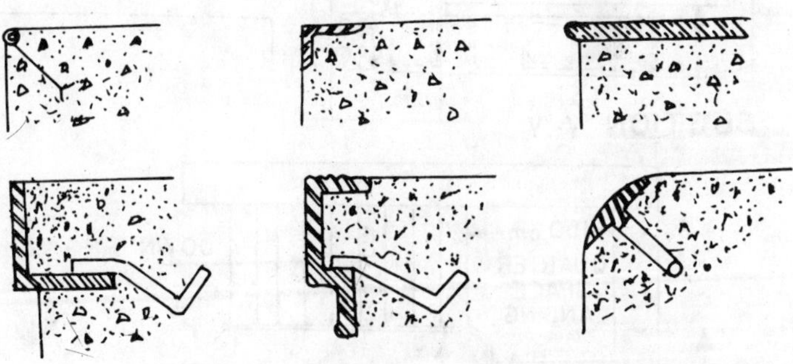

Fig. 13·28—13·33

Different methods of protecting the nosing of concrete steps

Reinforced concrete stairs can be pre-cast or cast-in-situ. To change their appearance, R.C.C. stairs are often covered with thin slabs of stone, marble tiles or with terrazzo finish. In order to prevent the nosing of the step getting cracked or knocked off, it should preferably be protected by metallic strips or angles as shown in Figs. 13·28-13·33.

Design Principle. The choice of the type of stair to be adopted depends very much upon the size of the staircase hall and the loading condition. In short, the design of the reinforced concrete stairs may be broadly divided into following categories :

(*i*) R.C.C. steps cantilevering out from the stair- case wall and acting both as supporting beam and stair

(*ii*) R.C.C. steps spanning between two inclined R.C.C. beams.

(*iii*) An inclined slab of the full width of the stair and supported between beams or landing slabs at either end and the steps cast monolithic with the inclined slab.

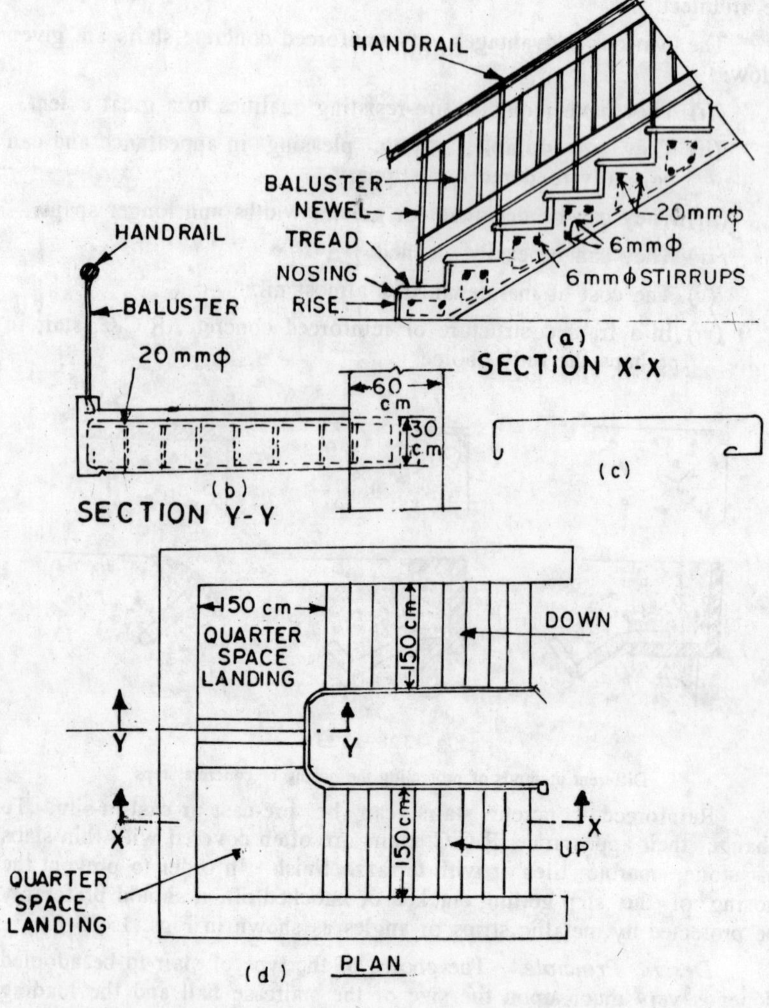

Fig. 13·34. Details of an R.C.C. stair

The various details of R.C.C. stairs are shown in figure 13·34.

13·12. ESCALATORS

A power-driven, inclined, continuous stair way, used for raising or lowering passengers, is called an escalator. The stair way has continuous operation without the need for operators. The main components of an escalator are a steel trussed frame work, hand rails and an endless belt of

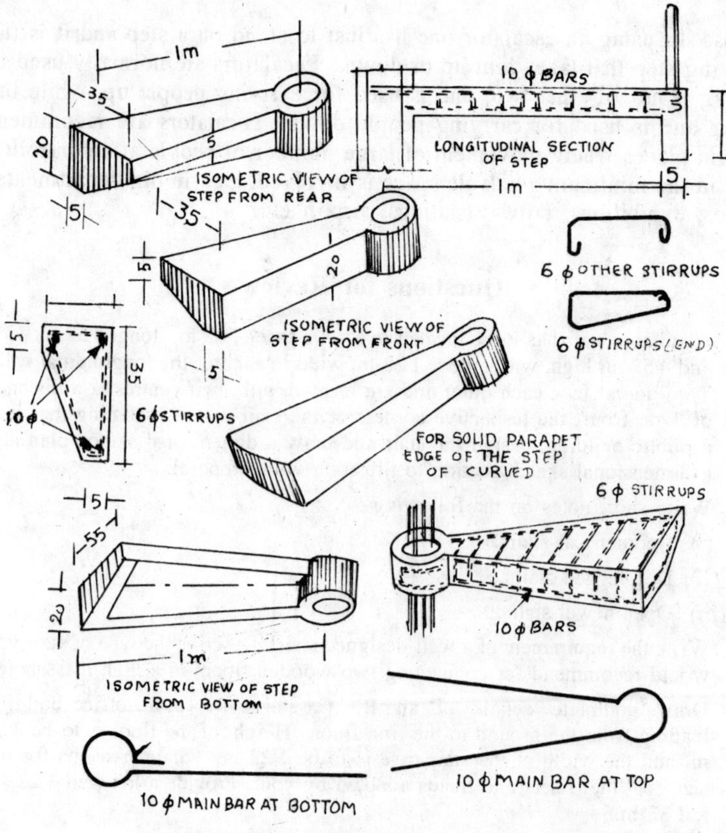

Fig. 13·35 Structural details of a pre-cast R.C.C., spiral stair

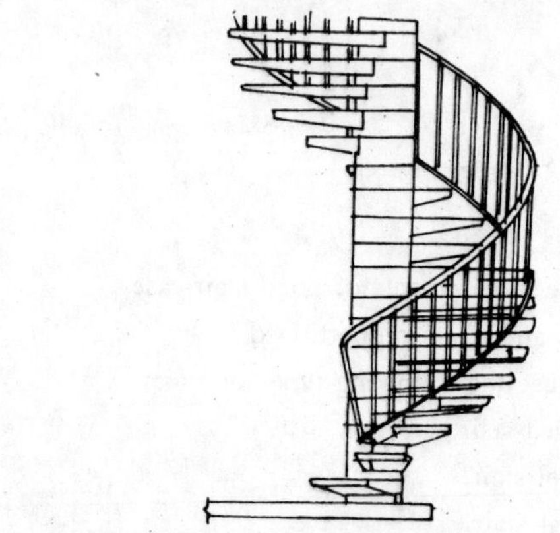

Fig. 13·36 Elevation of a spiral stair

steps. In using an escalator one has just to stand on a step and it is the moving step that takes him up or down. Escalators are normally used in pairs. Out of the two, one is used for carrying people up, while the other one is used for carrying people down. Escalators are recommended in places where movement of large number of people at a controlled rate in the minimum available space is involved, *e.g.*, in big departmental stores, exhibitions, railway stations, airport etc.

Questions for Review

1. A suitable stair has to be located in a stair-case, 7·62 m. long, 4·57 m wide and 4·57 m. high, with a door 1·22 m. wide in each of the longitudinal wall. The doors face each other and are located with their centres at a distance of 1 m. from the respective corners of the stair-case. Assuming that it is a public building, design the stair and draw a dimensional sketch plan and a dimensional sketch section to illustrate your proposal.
2. Write short notes on the following :
   (*i*) Geometrical stairs.
   (*ii*) Dog-legged stairs.
   (*iii*) Open-newel stairs.
3. Write the requirement of a well-designed stair. Describe the type of stair you would recommend for connecting two wooden floors in a high class hotel.
4. Draw complete details of an R.C.C. stair-case for an office building leading from the ground to the first floor. Height of the floor is to be 4·27 m. and the width of the stair case is to be 1·22 m. State reasons for the size of the risers and treads adopted by you. Provide a lift well 1·22 m. x 1·83 m.
5. What do you understand by the following ? Draw rough sketches to support your answers wherever necessary :
   (*a*) Landing
   (*b*) Nosing
   (*c*) Stringer
   (*d*) Winders
   (*e*) Balustrade
   (*f*) Spandril.
6. (*a*) State briefly the requirements of good stair-case.
   (*b*) How are treads and risers proportioned ?
   (*c*) Where do you use the following types of stairs :
      (*i*) Dog-leggged stair.
      (*ii*) Open newel stair.
      (*iii*) Geometrical stair.

7. Sketch clearly large view to show details of the following :
   (*i*) Treads with rounded nosing.
   (*ii*) Tongued risers.
   (*iii*) Wall or outer string.
   (*iv*) Newel, hand rail and balusters.

8. A two-storeyed building is to have an R.C.C. stair from ground to first floor roof. The size of the stair-case is 4·27 m. x 3 m. and there is one door opening in the short wall and a window opening on the opposite wall.

   Design the stair-case, giving the details of foundations, R.C. slab, arrangement of risers and treads along with top finishing. The width of the stair is 1.22 m and the height of each storey is 3·36 m.

9. Write short notes on the following :
   (*i*) Metal Stairs.
   (*ii*) Stone stairs.
   (*iii*) R.C.C. stairs.
   (*iv*) Spiral stairs.

10. A straight flight stairs has to be provided for connecting two floors. The vertical distance between the floors in 3 m and the width of stair has to be restricted to 75 cm. for want of space. Draw sectional plan and elevation of the designed stair, showing clearly the constructional details.

11. What are the different types of stairs used in buildings ? Which of them will be most suitable for a public building ? Describe in detail the construction of a helical stair-case and state where it is commonly used.

12. What are the limitations on different types of stair-cases in regard to their rise and tread. How would you choose them for :
    (*a*) House
    (*b*) School
    (*c*) Hospital
    (*d*) Railway station.

13. Discuss the various considerations that are made in planning of stair-cases. Illustrate the different types of stair-cases generally used, indicating their suitability for specific use.

14. Define the following terms as applicable to a stair-case with the help of an enlarged sketch :
    (*i*) Newel
    (*ii*) Line of nosing
    (*iii*) Baluster
    (*iv*) Nosing
    (*v*) Soffit
    (*vi*) Stringer.

14

Upper Floors

14·1. GENERAL

Depending upon the arrangement of beams and girders of the frame-work for supporting the flooring and the materials used in the construction of the entire system, the upper floors may be broadly classified into the following different types :

(1) Timber floors.

(2) Timber floors supported on rolled steel joists.

(3) Flag stone floors resting on steel joists.

(4) Jack arch floors.

(5) Reinforced cement concrete floors.

(6) Ribbed floors.

(7) Pre-cast concrete floors.

14·2. TIMBER FLOORS

There are three main types of timber floors :

(1) *Single floors*

(2) *Double floors*

(3) *Framed floors*

(1) Single floors. They are the cheapest form of wooden floors. They are easy in construction and are frequently used in domestic buildings. These floors are constructed by fixing 25 to 30 mm. thick planking or boarding to wooden joists known as "bridging joists" spaced at 30 to 40 cm. apart. The bridging joists are placed along the short span of the room and are nailed to timber wall plates fixed in the walls on either end. The span for single floor is limited to 3·66 m., as for greater span, the joists become very deep and consequently this system proves uneconomical. The depth of joists depends upon the loading condition

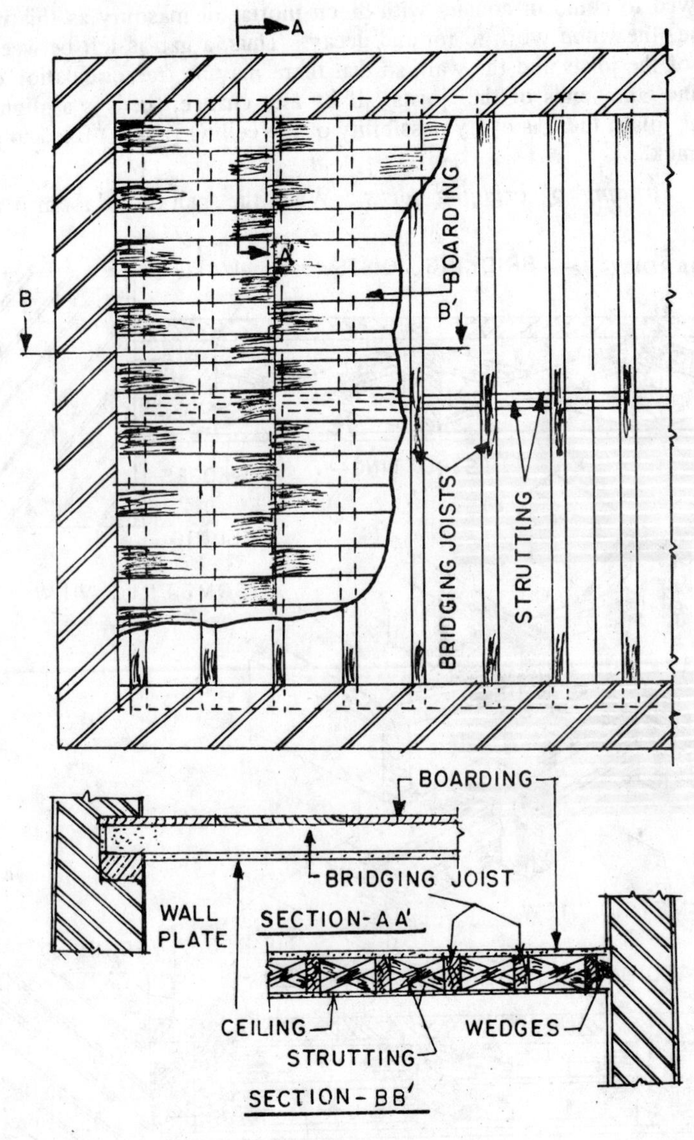

Fig. 14·1

Plan and sections of wooden single floor

of the flooring. The joists are designed both for strength and rigidity. In other words, the joists provided should be such that they may safely carry the floor loads without deflection. The joists should not be allowed to come in contact with green mortar or masonry as this would cause the wood work to rot and decay. Thus, a gap is left between the end of the joists and the wall, so that there may be free circulation of air around the ends of the joists. If by any chance, there is a slight sag in the joists, there is every possibility of the ceiling (if any) fixed to joists to crack.

Strutting of bridging joists : When the span of the room is more

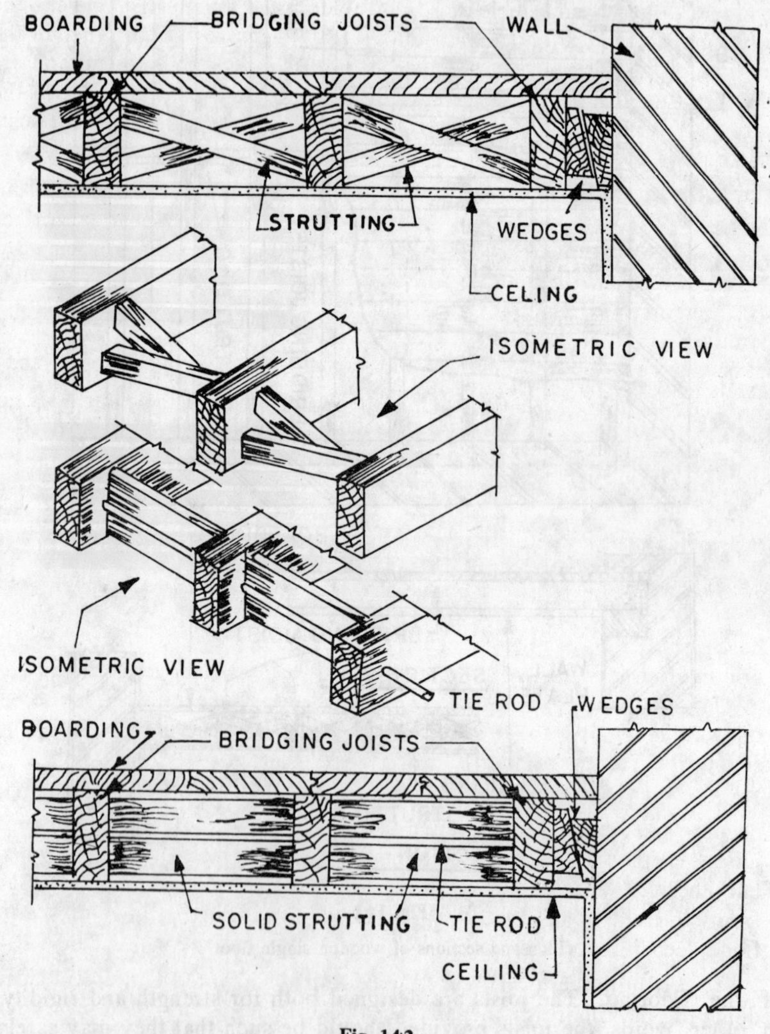

Fig. 14.2
Sectional details and isometric views of Herring bone and solid strutting

than 2·44 m. the bridging joists have a tendency to buckle. In deep joists this tendency of tilting or buckling sideways is prominent. To prevent this, the joists are strutted apart in the midspan all along the length of floor. There are two different types of struttings, namely :

(*i*) Herring bone strutting

(*ii*) Solid strutting

Herring bone strutting consists of a system of inclined pieces of timber 2.5 to 3.8 cm. thick and 5 cm. wide fitted diagonally between the bridging joists in such a manner that they cross each other in the middle of the joist. At their ends, the timber pieces are splayed and are secured to the joist by nailing. To give additional strength to the strutting joist, wedges are inserted between the walls and the end joists of the strutting.

Solid strutting is not so strong as herring bone strutting. This consists of short pieces of timber 2·5 to 3·8 cm. thick, fixed between the joists in continuous row. The length of each piece being equal to the centre to centre spacing of the bridging joists and breadth being about 20 to 25 mm. less than the depth of the bridging joists. To strengthen the strutting, a 13 to 25 mm. ϕ steel rod, placed by the side of strutting, is passed through the mid-depth of all the joists and is tightened at its ends.

Methods of Fixing Ceiling. The underside of the floor does not provide a good appearance for the users of ground floor. In order to have the pleasing effect of plain ceiling, the bridging joists must be concealed. Ceilings of wood, jack board, floor board, etc. are fitted to the bottom of joists for this purpose. Metal lathe and plastered ceiling are also commonly provided. The various methods of fixing ceiling to the joists are described below :

(*i*) The ceiling boards which may be of wood, jack-board, fibre board or other similar material are directly nailed to the underside of the bridging joists. This is the simplest and cheapest form of ceiling. Due to the vibrations arising in the floor joists on account of the floor usage, this ceiling is liable to crack or open at joints. This type of ceiling is not sound-proof unless specially treated.

(*ii*) In the second method, every third or fourth joist is made deeper and the ceiling is nailed to the ceiling joist fixed to the deeper joists. Although this type of ceiling renders the floor more sound-proof, but the increased depth of bridging joists (which is not necessary from structural considerations) required for fixing ceiling, makes the floor deeper and costly.

(*iii*) In the third method, the ceiling is supported entirely independent of the flooring. Separate beams called "ceiling beams" are laid parallel to the bridging joists at suitable distance apart. The ceiling, strong and rigid enough to carry ceiling loads, is fixed in the wall plates

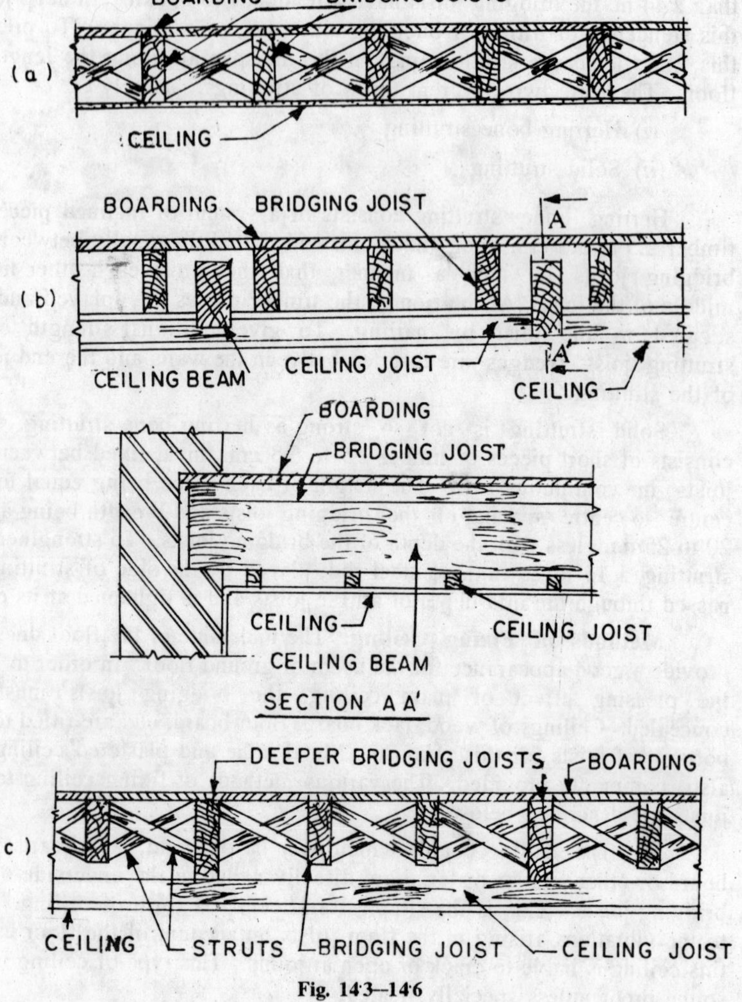

Fig. 14·3—14·6

Methods of fixing ceiling to wooden floors

in the end walls and is kept about 20 to 25 mm. below the bottom face of the boarding. Ceiling joists are fixed at right angles to the ceiling beams. Ceiling of required material is then fixed to the ceiling joists as shown.

Method of Sound-Proofing of Timber Floors. In its simple form, the method of sound-proofing of timber floors consists in dividing the space between the bottom of the boarding and the ceiling by means of insulating material. The insulating material obstructs the sound waves and does not allow them to travel from floor to the ceiling. The base for

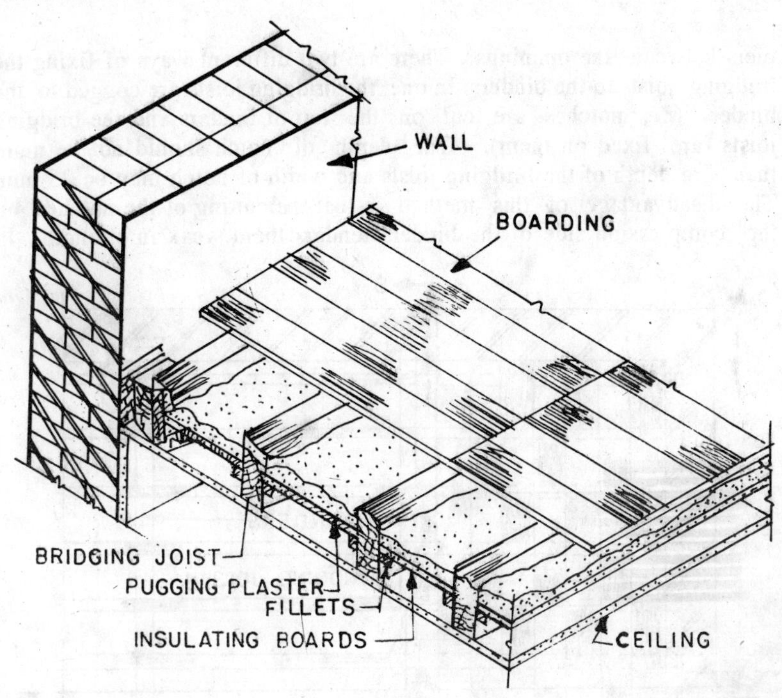

Fig. 14·7

Sect ion of a wooden floor showing method of making it sound proof

storing the insulating material is made by fixing small timber pieces (fillets) on either side of each bridging joist which supports the insulating boards. Upon these boards, insulating material, which may be sand, dust, asbestos, cork, slag wool, etc. is filled as shown in the figure 14·7.

(2) **Double Floors.** In case the shortest span of the room is more than 3·66 m. and it is desired to provide a timber floor, double type of timber floor may be constructed. The span of double floors is restricted to 7·32 m. beyond which framed floors have to be adopted. In double floors the bridging joists do not span the whole room, but rest on other joists placed at right angles to them. The joists are called binding joists or binders and are placed across the short span. Thus, in double floor, the bridging joists support the boarding and the binders support the floor load through bridging joists. The centre to centre spacing of the binders varies from 1·8 to 2·4 m. At their ends, the binders rest or stone on concrete templates embedded in the walls. As far as possible, the binders should not be placed immediately over door or window openings. In case this cannot be avoided, they should rest on lintel or beams (of stone or R.C.C.) which may span over

piers between the openings. There are two different ways of fixing the bridging joists to the binders. In one, the bridging joists are cogged to the binders (*i.e.* notches are cut on the top of binders and the bridging joists are fixed on them). The depth of notch should not be more than $\frac{2}{3}$rd depth of the bridging joists and width of notch may be 25 mm. The disadvantage of this method is that the cutting of the notches on the compression side of the binders renders them weak in strength. In

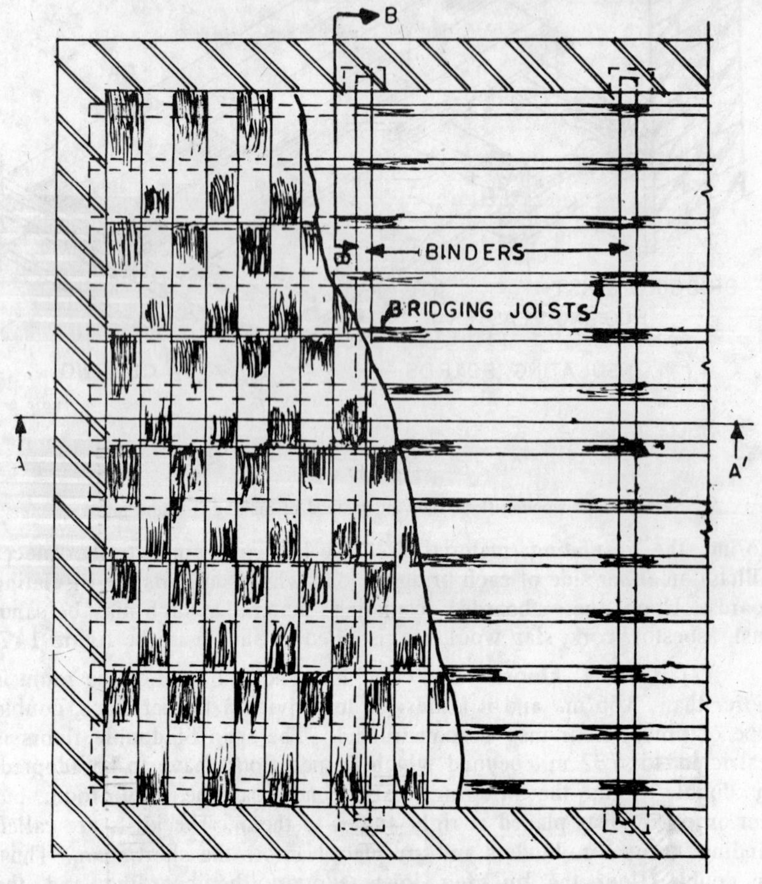

Fig. 148

Plan showing arrangement of supporting boarding in a wooden double floor

the second method, timber pieces or fillets are fixed on either side of each binder and the bridging joists are cut as shown, so as to rest on fillets. This method of fixing is better than single floor construction and has better sound insulating property. The plastered ceiling fixed to the binders through the formwork of fillets and ceiling joists, is not liable to crack in

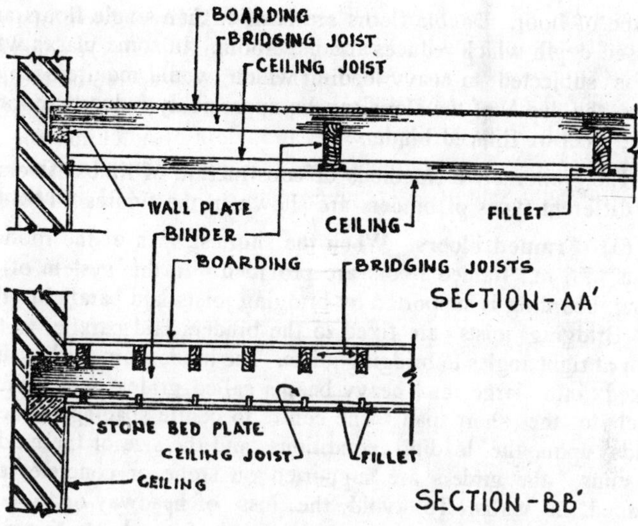

Fig. 14·9

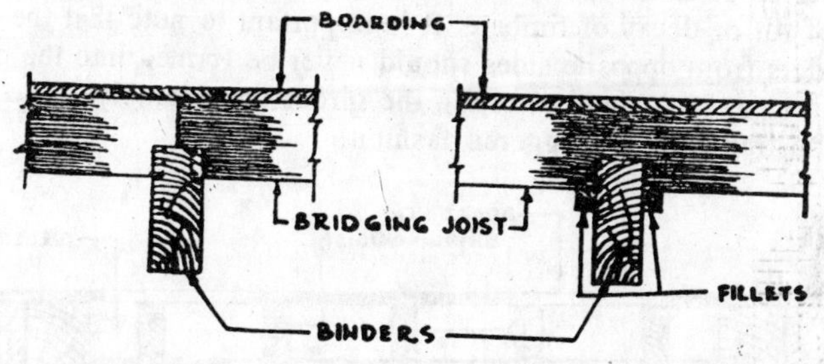

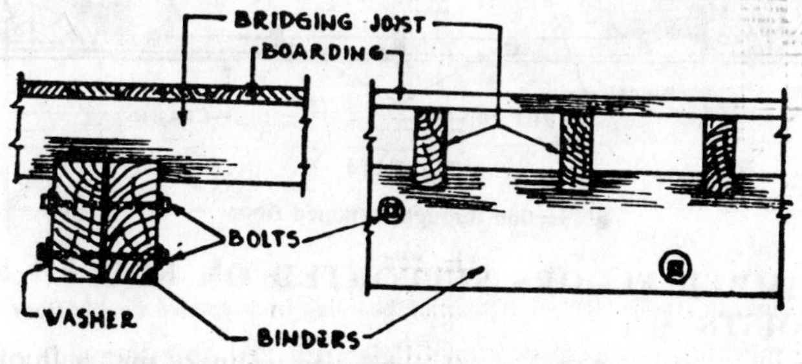

Fig. 14·10–14·13

Details showing different methods of supporting bridging joists in timber double floor construction

this type of floor. Double floors are costlier than single floors and have increased depth which reduces the headroom. In some places where the floor is subjected to heavy loading which would require deep wooden binders, the depth of the floor can be appreciably reduced by providing R.C.C., steel or flithced binders.

The alternative methods of construction of timber floors fixture using different types of binders are shown in the figures 14·10—14·13.

(3) **Framed Floors**. When the shortest span of the room or hall exceeds 7·6 m., framed floors are provided. In this system of wooden flooring, boarding is supported by bridging joists laid parallel to the short span. Bridging joists are fixed to the binders laid parallel to the long span or at right angles to bridging joists. The binders are further supported or fixed into large and heavy beams called griders which are placed parallel to the short span. The centre to centre spacing of the girders depends upon the loading conditions and the size of the binders. At their ends, the girders are supported on stone or concrete templates embedded in walls. To avoid the loss of headway or to reduce the thickness of floor, the binders are normally framed into the girders by means of tusk and tenon joint. A gap is left between the walls and end of grider so as to ensure free circulation of air at the ends to avoid the defect of rot or decay of timber. It is important to note that the ends of two binders from opposite sides should never be framed into the girder at one place, as it would weaken the girder considerably. The binders should be framed in a staggered fashion.

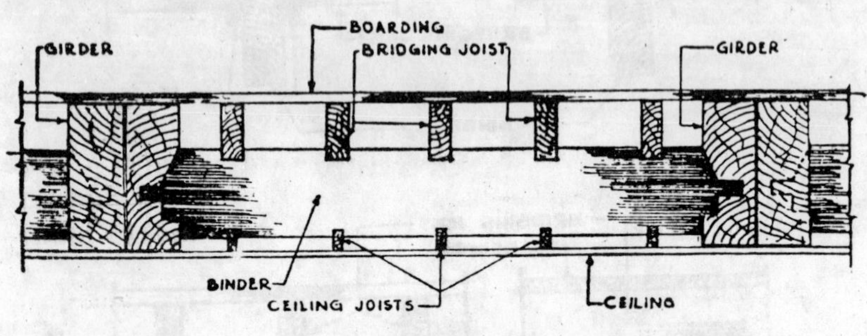

Fig. 14·14

Section through a framed floor

14·3. TIMBER FLOORS SUPPORTED ON ROLLED STEEL JOISTS

This system of flooring is similar to ordinary timber floor except that wooden joists are replaced by rolled steel sections which are available in longer lengths. This enables the use of lighter sections of wooden member which ultimately makes the floor economical. In

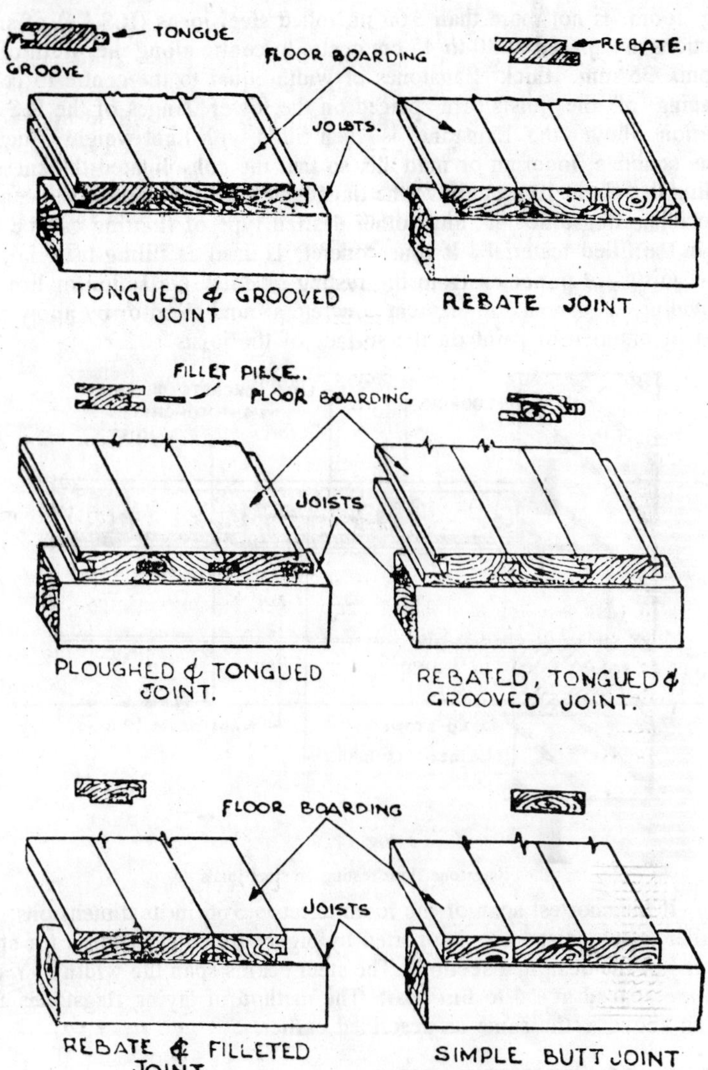

Fig. 14·15

Views showing different types of joints used in floor boarding.

order to reduce the depth of floor, the joists supporting the flooring are notched over the top flanges of the girder and are made to rest upor angle iron cleats bolted on either side of each rolled steel joist.

14·4. FLAGSTONE FLOORS RESTING ON STEEL JOISTS

It is a cheap and easy form of floor construction. If the width of the room is not more than 3·66 m, rolled steel joists (R.S.J's) of small sections are spaced at 30 to 45 cm centre to centre along the width of the room. 38 mm. thick flagstones of width equal to the centre to centre spacing of the joists are placed on the lower flanges of the R.S.J'S. Portion above the flagstone is then filled with light weight concrete, lime concrete, moorum or mud etc. so that the consolidated thickness of filling lies about 13 mm above the flanges of joists. Finally, tiles, cement concrete, flagstone or any other desired type of flooring can be laid upon the filled material. If lime concrete is used as filling material, the steel joists are protected from the rusting or corrosive action of lime by providing a screed of cement concrete around them or by applying a coat of bitumen or paint on the surface of the joists.

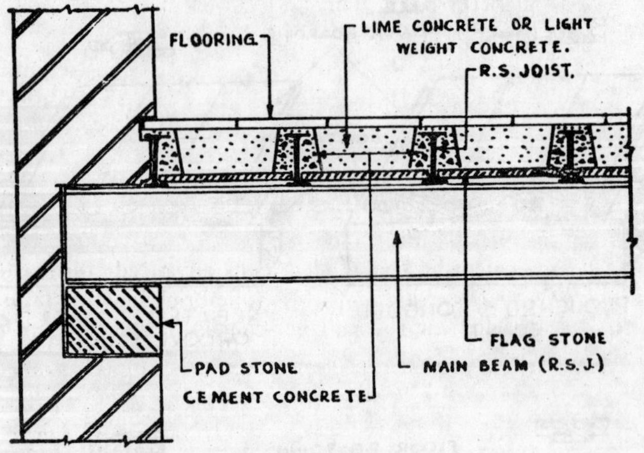

Fig. 14·16

Flagstone floor resting on steel joists

If the shortest span of the room exceeds 3·66 in its dimensions, the smaller steel joist are supported by bigger rolled steel joists (or steel beams) of the designed section. The steel beams span the width of room and are spaced at 2·4 to 3m apart. The method of laying flagstones and filling above is the same as described earlier.

14·5. JACK ARCH FLOORING

This type of flooring may be constructed by using brick or concrete. The arches span between the lower flanges of rolled steel joists spaced at centre to centre distance not exceeding 1·2 to 1·5 m. The arch is given a rise of $\frac{1}{12}$th of the span. As the arch action of floor develops tension in the supporting walls, mild steel ties (not less than three in a bay), are provided in end spans at suitable distance apart. The diameter of the tie rods varies from 18 to 25 mm. and they are usually spaced 1·8 to 2·4 m. apart. The tie rods are anchored into each wall with a stout washer. The floor may be finished with tiles, cement concrete mosaic or any other desired finishing after filling the space above the arch with lime concrete or other filling material. The ceiling in this type of construction is not plain and this is considered a drawback of this flooring.

Method of construction of brick Jack arch flooring : The centring of the arch consists of a 38 mm thick segmental piece of timber with a chord length equal to the span of the arch and the curved portion of the arc forming the soffit of the arch curve. The centering board is cut slightly at its ends and is laid on edge with the curved portion upwards on the lower flanges of the rolled steel joists at a distance of 7·6 cm from the wall. The selected bricks are then laid on edge on the centering board from both the ends of the arch. The end bricks are cut suitably to the required shape to enable them to fit in the joists properly. In order to prevent the joists from the rusting action of lime, the joint next to the joist is filled with cement mortar. In the first arch ring, the brick are arranged in length of 20 and 10 cm alternately. This is necessary to secure a bond between the first and the successive arch rings. The key brick is laid in stiff mortar and is filled in the arch as tight as practicable. The centering board is then advanced further by light hammer blows in a new position which is (one brick length) 20 cm ahead of the previous one and the second arch ring is constructed.

The successive rings are constructed in a similar manner. The entire work is kept wet for 10 days. After this, the space is filled with lime concrete, moorum or light weight concrete or any other desired fill. Finally, the required type of flooring is laid above the consolidated filling. The following precautions are necessary to be observed in the brick arch construction.

(1) Only first class bricks, which are well wetted before hand, should be used in the work.

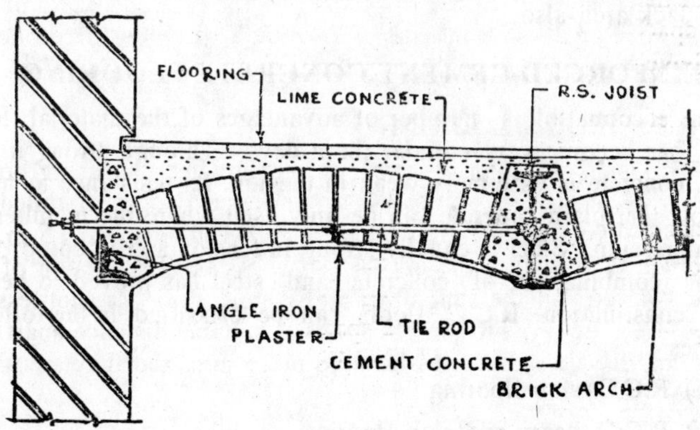

Fig. 14· 17 Brick Jack arch-flooring

(2) It should be ensured that the steel joists are well secured to the supporting wall before starting the arch construction.

(3) If lime concrete is used for filling haunches of the arch, the R.S.J's are protected from the rusting action of lime concrete by providing a screed of cement mortar around them.

(4) The work should be cured properly for at least 10 days.

(5) Filling above the arch and construction of floor should start only after curing is complete.

Construction of concrete Jack arch flooring : The centering for concrete arch construction consists of 3 mm thick mild steel plate bent to the shape of arch soffit. The steel plate has holes at 75 cm centre to centre all along the length of the plate at each end. In order to support the mild steel centering two No's 12 mm. diameter iron rods of suitable length are employed. Each rod is hooked at one end so as to form eye large enough to allow a 12 mm. diameter rod to pass through. Each rod is passed through the eye of the other and the straight ends of the bars are then passed through the holes in the mild steel plate and are supported on the flanges of the rolled steel joists at either end. Thus the centering plate gets supported on the lower flanges of the joist through the iron bars at every 75 cm along the length of joists. A wooden packing piece is tightly driven between the rods and the mild steel plate in order to make the centering more rigid.

Concrete of the specified mix is then laid on the top of the centering plate to the required depth and consolidated properly. Desired type of flooring is then laid over the concrete base and finished. The flooring is kept wet for at least 10 days. After the curing is complete, the centring is removed by removing the wooden packing piece and then hammering the eyes in the rods towards each other. The under-side of the arch is normally plastered to improve the appearance of the surface. The precautions for brick arch construction hold good for concrete Jack arch also.

14·6. REINFORCED CEMENT CONCRETE FLOORING

On account of a number of advantages of the material, R.C.C. flooring has become very popular these days. Through strong in compression, concrete is extremely weak in tension. Its resistance to tension is so low that plain concrete can be only used where the member is in pure compression. Steel is equally strong in tension and in compression. Thus the combination of concrete and steel has proved to be ideal for floor construction. R.C.C. floors can be classified in the following main types :

(*i*) R.C.C. slab flooring.

(*ii*) R.C.C. beam and slab flooring.

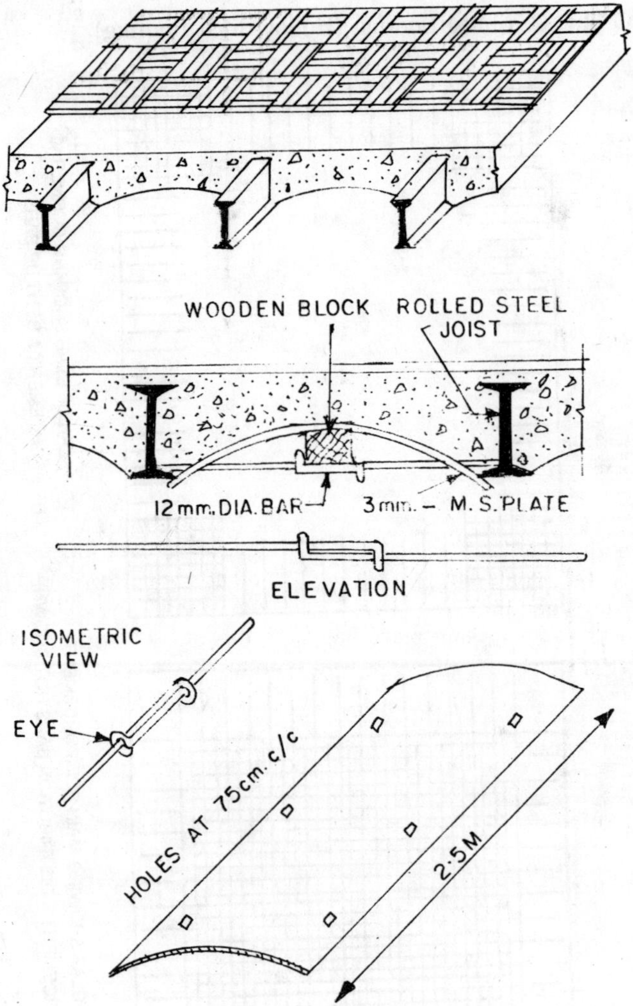

WOODEN BLOCK
ROLLED STEEL JOIST
12mm. DIA. BAR
3mm. – M. S. PLATE
ELEVATION
ISOMETRIC VIEW
EYE
HOLES AT 75cm. c/c
2·5M

Fig. 14·18 Concrete Jack arch flooring

DISTRIBUTION BARS.
WALL
MAIN BARS
WALL

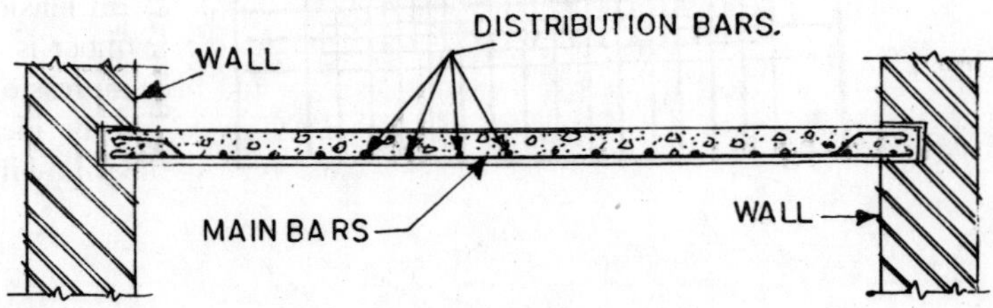

Fig. 14·19

Section through a one-way reinforced concrete floor slab

(*iii*) Flat slab flooring.

(*iv*) R.C.C. ribbed flooring (Hollow tiled flooring).

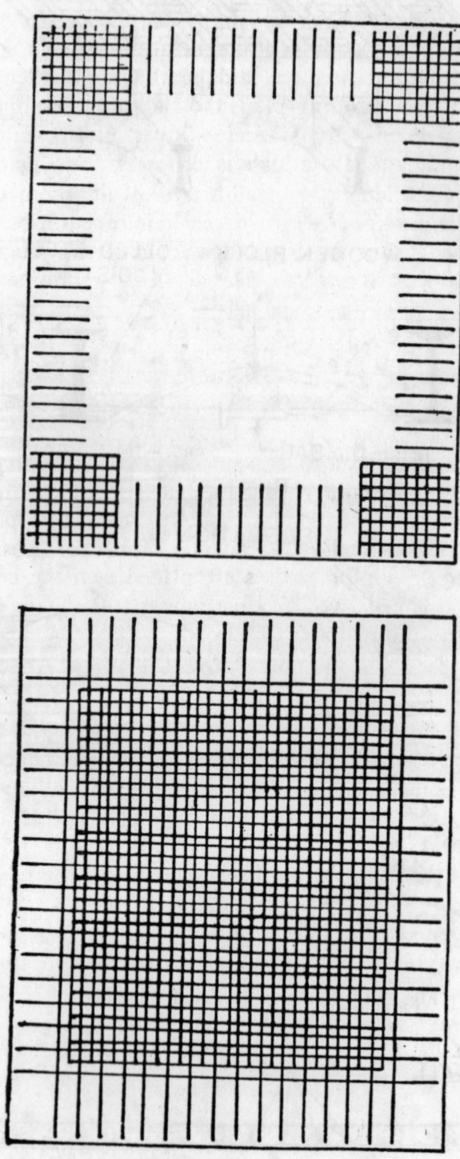

Fig. 14·20
Details of Two-way reinforced concrete slab

(*i*) **R.C.C. Slab Flooring.** For small spans, say upto 3·6 m., which are not subjected to heavy loadings, a simple slab may suffice. When the ratio of length of the room to its breadth is greater than 2, the floor slab is designed to span along the width of the room as a one way slab. In this case, the main reinforcement of the slab span along the width, while distribution bars are laid parallel to the length of room. Depending upon the building plan, the slab may be simply supported on the walls, or it may be continuous over intermediate walls. The design of a continuous floor slab is different from that of a simply supported one. If the length to breadth ratio of the room is less than 2, the floor slab is designed as a two-way reinforced slab. In this case, main reinforcements are provided both along the length as well as along the breadth of the room. The details of reinforcements in such types of slabs are shown in Fig. 14.20.

The maximum bending moment for which the slab should be designed varies with the nature of the slab (whether one way reinforced or two-way reinforced) loading conditions, the number of spans and the end conditions of the slab. The bending moment which causes tension at the bottom of the slab, usually near the centre of the span, is called positive or sagging bending moment and the bending moment which causes tension at the top of the slab, usually over the supports, is called negative or hogging bending moment. Thus for positive bending moment, the main reinforcements are placed near the bottom face of the slab while for negative bending moment, the main reinforcement are placed near the top of the slab.

(*ii*) **R.C.C. Beam and Slab Floor**. For bigger span and heavy loading conditions or in situations where intermediate walls are not provided, to reduce the span of the floor slab, R.C.C. beam and slab construction is adopted. The beams act as intermediate walls and the floor slab is designed as a continuous slab monolithic with the beams. The beams thus used are known as *T*-beams because a part of the floor slab assists in resisting the compression developed in the beam. The projecting part of the beam below the slab is called the rib of the beam. In a framed structure, the floor is usually supported on a system of main and secondary beams. The floor slab is laid continuous over secondary beams which are framed into and supported by main beams.

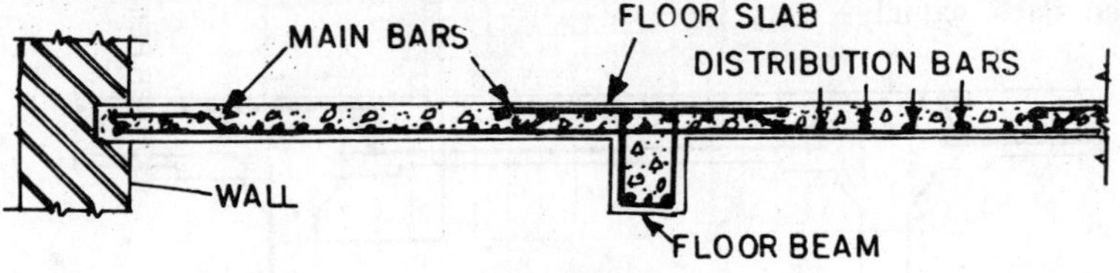

Fig. 14·21

R.C.C. beam and slab floor

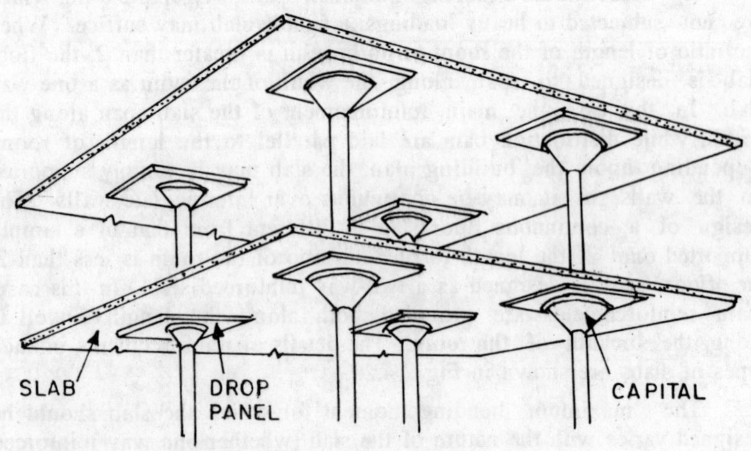

Fig. 14·22

View showing flat slab construction

(*iii*) **Flat Slab Flooring.** Flat slab, also known as beamless slab, is a type of construction in which the flooring slab is directly supported on columns without the agency of beams or girders. They are provided in theatres, factories, mills and other public buildings. The flooring consists of R.C.C. slab, 13 cm. or more in thickness and supported on R.C.C. columns arranged in such a manner that they form square of nearly square panels. The size of panel should not exceed 9 m. x 9 m. There must be at least three rows in two directions at right angles. The ratio of length to breadth of a panel should not be more than 4:3 and the length and breadth of any two adjacent panels should not differ by more than 10% of the greater length. Sometimes, instead of providing uniform thickness of slab, a portion of slab is thickened symmetrically around the column. The thickened portion is called drop. The flat slab may be designed either with or without drops. At their head, 4 cm. below the drop, the columns are enlarged so as to form diameter equal to 0·21 to 0·25 of the span. This flared up portion of column head is called capital. There are two different systems of providing reinforcement in flat slabs, namely—

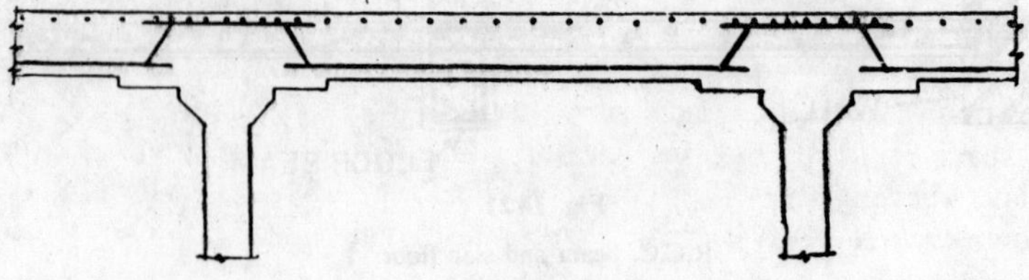

Fig. 14·23

Section through the column strip

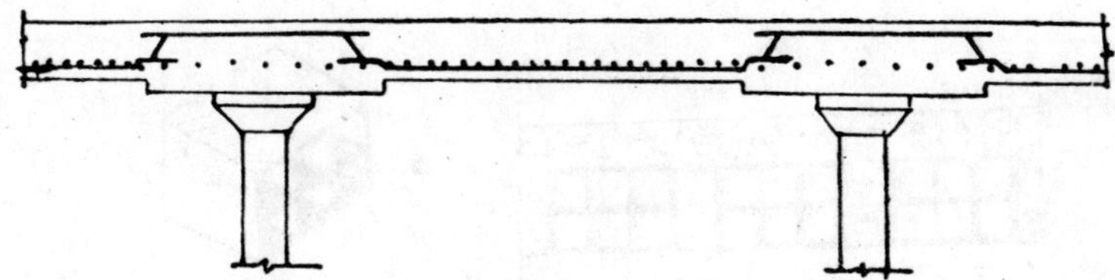

Fig. 14·24

Section through the midstrip

(*i*) Two-way system.

(*ii*) Four-way system.

In two-way system, reinforcements are carried from column to column parallel to the side of panels and a two-way system of reinforcement is placed in the central portion of the slab.

In four-way system, reinforcement is placed in four directions and it consists of two bands of main steel reinforcement carried directly from column to column, and two other bands placed diagonally across the panel from column to column.

Advantages : The advantages of flat slab flooring may be summarised as below :

(1) Since the slab is constructed without any beam or girder, the floor has flat ceiling which gives a good appearance.

(2) Due to elimination of beams and girders there is a saving in storey height.

(3) Due to flat ceiling, better lighting facilities are available.

(4) The formwork is simple and economical.

(5) It is structurally economical, especially when the live loads are more than 500 kg/m^2.

14·7. RIBBED FLOOR

It is a form of reinforced concrete floor slab construction in which, the portion of concrete below the neutral axis of slab is removed so as to form ribs at short distances apart. The ribs act monolithically with the concrete topping and behave like T-beams. Thus, such a slab has an appearance of T-beam and slab construction with series of small closely spaced ribs. In order to obtain plain ceiling effect, the space between the ribs is often filled with hollow block made out of clay, fiber glass, PVC or concrete. This type of slab is also termed as *hollow block slab.*

Hollow block slabs are economically used in buildings like hospitals,

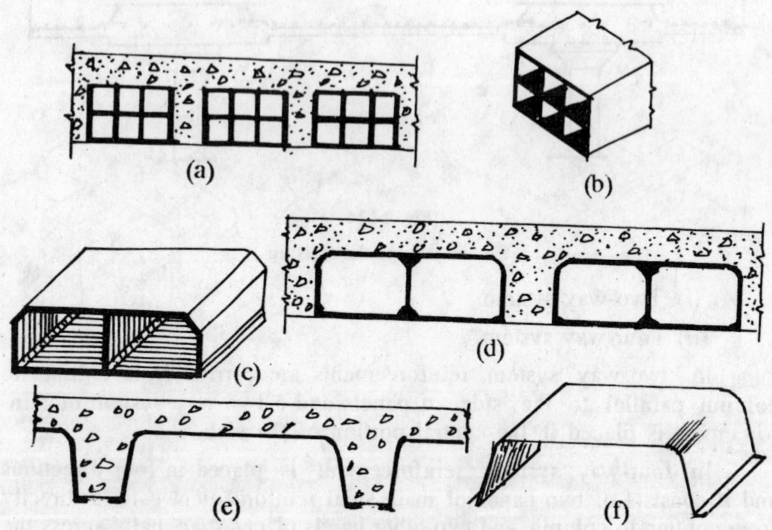

Fig. 14·25—14·30

Fig. 14·25—Section showing ribbed floor

Fig. 14·26—14·27—View of tiles used for filling the space between two ribs

Fig. 14·28—14·29—Ribbed-floor with steel forms

Fig. 14·30—View of a metal pan

hotels, school, offices etc., when live loads are fairly small and the spans are comparatively longer. The advantages of ribbed slab construction can be summarised as under.

(*i*) The dead weight is reduced as hollow block replace a portion of solid concrete thereby effecting economy.

(*ii*) It provides better thermal insulation.

(*iii*) Electrical, plumbing and other similar service installations can be concealed without affecting the plain ceiling and at no extra cost.

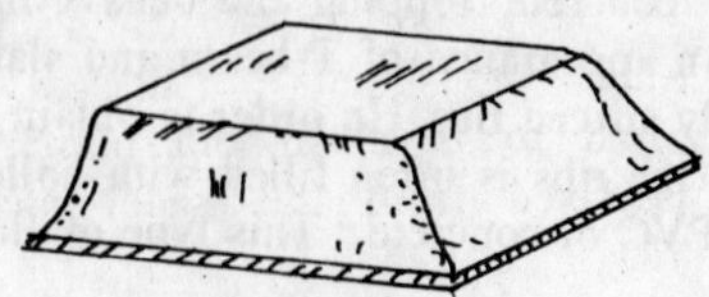

Fig. 14·31

View of mould for two-way ribbed slab construction

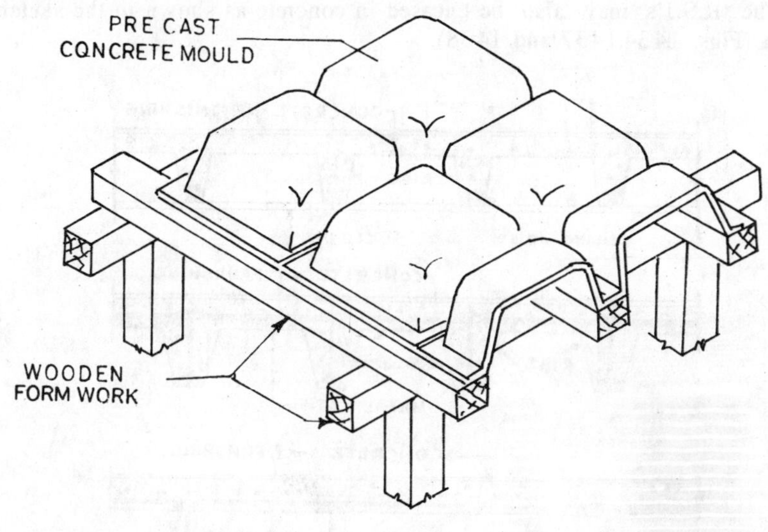

Fig. 14·32

View showing arrangement of wooden form work for supporting the moulds for two-way ribbed slab construction

The thickness of top slab in ribbed slab construction is kept very small and slabs having ribs in one direction are considered suitable for moderately light and uniformly distributed loading conditions. In case of heavy uniformly distributed loads or for concentrated loads, it is necessary to provide ribs in two directions at right angles to each other. Two way ribbed slab is constructed by use of dome-shaped steel or precast concrete moulds. The moulds are supported on timber formwork and they function as complete shuttering for casting ribs (in two directions) as well as top slab. The reinforcement for the ribs and top slab is placed in position and the concrete is poured directly over the mould. After curing, the steel moulds are removed, thereby resulting in a two-way ribbed slab having waffle-like ceiling. In case precast concrete moulds are used, it is not necessary to remove them after curing as they can be left in position as part of slab construction (For design details please refer chapter 15 of author's book on Treasure of R.C.C. Designs).

In case it is desired to have smooth plastered flat ceiling, a continuous metal lath is fixed from the under-side of the ribs and plastered.

In places where the loading conditions are such that the depth or R.C. ribs will work out to such a value that it cannot be adopted on account of certain practical consideration, rolled steel joists are placed at suitable distance apart to support the floor (Refer Figs. 14·35–14·36). The R.S.J's may also be encased in concrete as shown in the sketch in (Figs. 14·34,14·37 and 14·38).

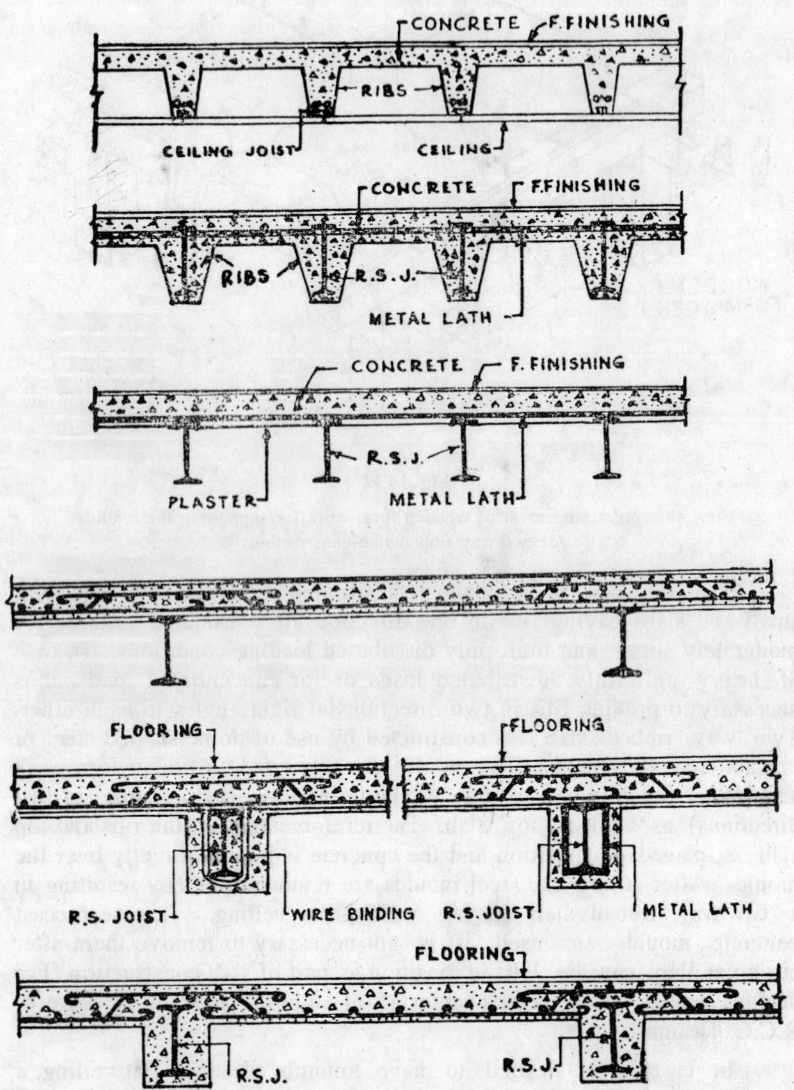

Fig.14·33–14·38
Alternative methods of supporting R.C.C. floor slab

14·8. PRE-CAST CONCRETE FLOOR

With the development of pre-cast concrete constructions technique, the pre-cast units like floor slab beams or girders, lintels etc. are being successfully manufactured by various factories. The pre-cast units of the designed depths are used in the floor construction. The members are made of such sizes as can conveniently be handled and transported. The advantages and disadvantages of this type of construction are given as under:

Advantages :

(1) The construction is quick.

(2) It is fire-proof.

(3) It is sound-proof.

(4) The members are light in weight which render the foundations economical.

(5) It does not require formwork in construction.

(6) No time is wasted for curing of the floor.

(7) It is economical.

(8) It has good thermal insulation properties.

Disadvantages :

(1) It requires fairly uniform spacing of beams throughout the structure which becomes practically difficult.

(2) In order that the pre-cast members may be able to resist handling stresses, great care has to be taken in their construction and design.

(3) There is wastage due to breakages in transportation of the members from factory to the site of work.

(4) Careful supervision and skilled labour is required for manufacturing the units.

(5) The forms used for casting the units are normally costly and as such the construction is economical only when the units are manufactured on a large scale.

Questions for Review

1. Sketch and name different types of joints that are used in the construction of different types of wooden floors :
2. Draw with neat sketches following types of floors :
   (*i*) Filled joists floor.
   (*ii*) Jack arched floor.
   (*iii*) Double framed timber floor.

3. Describe the disadvantages of a beamless floor. Draw a view to illustrate the following :
   (*i*) Capital.
   (*ii*) Drop-panel.
   (*iii*) Flat slab.
4. Draw to a suitable scale the plan and section of a teak wood flooring at first floor level, to be constructed over a 3·66 x 4·54 m. room which has all four walls 1½ brick thick. Give approximate sizes of teak wood supports and boarding.
5. Write short notes on :
   (*a*) Timber floors,
   (*b*) Jack arch floors.
   (*c*) Flag-stones floors supported on steel joists.
   (*d*) Ribbed floors.
   (*e*) Reinforced concrete floors.
6. A hall 10 m x 7·6 m. is to be roofed over with a R.C.C. floor slab. Suggest the type of construction you would adopt. Explain fully the process of carrying out the whole work. Draw rough sketches wherever considered necessary.
7. Describe the method of construction of a two-way ribbed floor giving sketches of mould and arrangement of supporting moulds for casting the slab.

15

Lintels and Arches

15·1. LINTELS

Openings for doors, windows or cupboards etc. must be bridged over by some means to support the masonry above. The support to the load above the opening is usually given by providing an arch or a lintel. Lintel acts like a beam and transfers the load vertically to the supporting walls. The ends of lintel are built into the masonry so as to convey the weight carried by them to the masonry in jambs. The amount of bearing which the supports provide for the ends of lintel is an important factor. As a general rule, the bearing of a lintel at its ends should be 10 cm. (depth of one brick) or 3·8 cm. for every 30 cm. of span, whichever is greater. Lintels are made of various materials such as wood, stone, brick, reinforced brick, steel and reinforced cement concrete.

Classification of lintels by material :

15·2. WOODEN LINTELS

In hilly areas, where timber suitable for the construction of lintel is easily available, wooden lintels are commonly used. In plains, the use of wooden lintel is very rare on account of high cost of the material. Timber if not properly ventilated, is liable to decay. Wood being combustible, wooden lintels cannot be recommended for fire-proof construction. The minimum depth of the lintel should be 7.6 cm. Sometimes a rough arch is constructed above a lintel to relieve the latter of the weight above. When used for large spans, the lintel should be properly designed as a timber beam simply supported at its ends.

15·3. STONE LINTELS

The use of stone lintels is very common in stone districts. Its use is usually confined to stone masonry structures or buildings faced with stones. In general, stone lintel, on account of its high cost and its inability to withstand excessive transverse stresses, is rarely used. The minimum thickness of stone lintel in kept 8 cm. and as a thumb rule, the thickness is taken as 4 cm. per 30 cm. length of span.

15·4. BRICK LINTELS

They are constructed with hard, well burnt, first class bricks laid on end or on edge. Depending upon the span, their depth varies from 10 cm. (depth of one brick) to 20 cm. (length of one brick). They are considered suitable for small openings (generally not more than 90 cm. in span) with light loading conditions. Bricks having frogs are more suitable for the construction of lintels. The frogs, when filled with mortar, form joggles which increase the shear resistance of the end joints and ultimately result in increased strength of lintel.

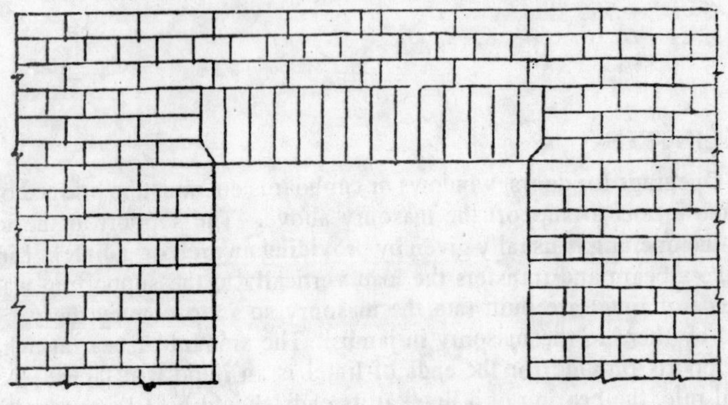

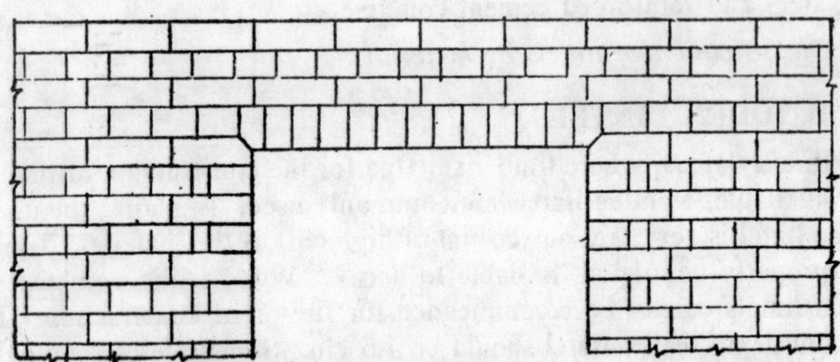

Fig. 15·1--15·2

Different forms of brick lintels

15·5. REINFORCED BRICK LINTELS

When brick lintels are required to be provided over large spans, they are reinforced with steel reinforcements. The depth of the lintel is limited to 10 cm. (depth of a brick) or a multiple of one brick thickness. Bricks are so arranged that 1·9 cm to 3·8 cm wide space is left length-wise for the insertion of reinforcements. The gap is then filled with rich cement mortar or cement concrete.

15·6. REINFORCED CONCRETE LINTELS

On account of their durability, strength and fire resisting properties, reinforced concrete lintels are very common in use. They are economical and easy in construction. They can be used for bigger spans and heavy loading conditions.

R.C.C. lintels are constructed with 1:2:4 cement concrete reinforced with steel bars. They may be pre-cast or cast-in -situ. For smaller spans, say up to 1·2 m, pre-cast lintels are preferred.

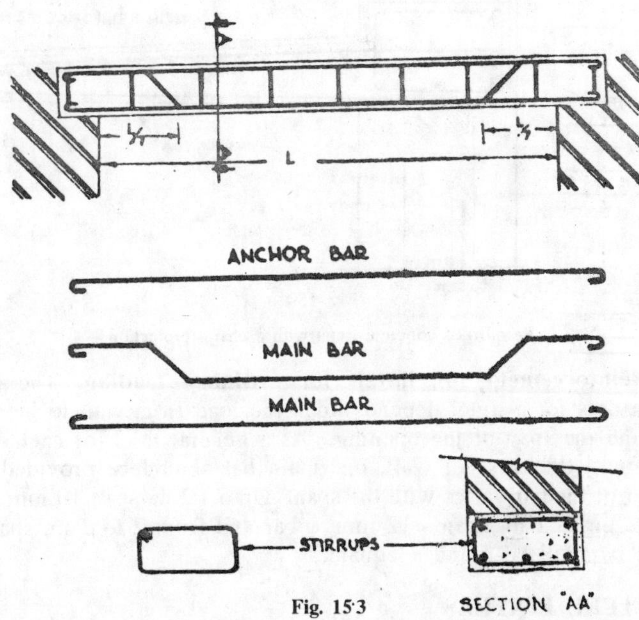

Fig. 15·3

Details of a Reinforced concrete lintel

Pre-cast lintels are manufactured and cured well in advance and as such prove economical since the same mould can be used a number of times. The use of pre-cast. R.C.C. lintels expedites the construction. While laying a pre-cast lintel special care should be taken to see that the marked face of the lintel (soon after pouring concrete in the mould, the top of the concrete surface is marked with some paint or tar so as to distinguish it from bottom face) is kept at top. If the lintel is placed in reverse position, *i.e.,* the marked face pointing downward, disasterous result may take place. For larger spans, lintels are cast-in-situ, as pre-cast lintels will have to be provided with additional reinforcements to guard against the lifting stress. For lintels over large spans subjected to heavy loadings, the depth of the lintel and the area of the reinforcement should be accurately computed.

Depth of lintel for ordinary loading. Up to 1·2 m. length of span, a 15 cm. deep lintel is found to be safe. As a general rule, add 25 mm. for every additional 30 cm. span.

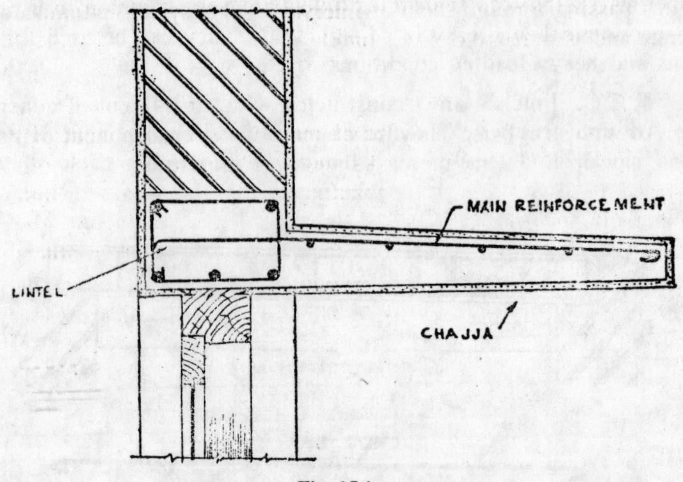

Fig. 15·4

Reinforced concrete lintel with a chajja projection

Reinforcement for lintels for ordinary loading. The number of main bars for a lintel depends upon the load from wall to be carried above and the span of the opening. As a general rule, for each 10 cm. (half brick) thickness of wall, one main bar should be provided. The diameter of the bar varies with the span. Upto 1·2 m. span 10 mm. ϕ bar; from 1·2 m. to 2 m. span - 12 mm. ϕ bar and from 2 to 3 m. span - 16 mm. ϕ bar will be found adequate.

15·7. STEEL LINTELS

When a large opening carrying heavy wall load is to be spanned by a lintel, an R.C. lintel will work out to be very deep. If due to architectural reasons, the depth of the lintel is to be restricted, it is common to use steel lintel. Steel lintel consists of section of rolled steel

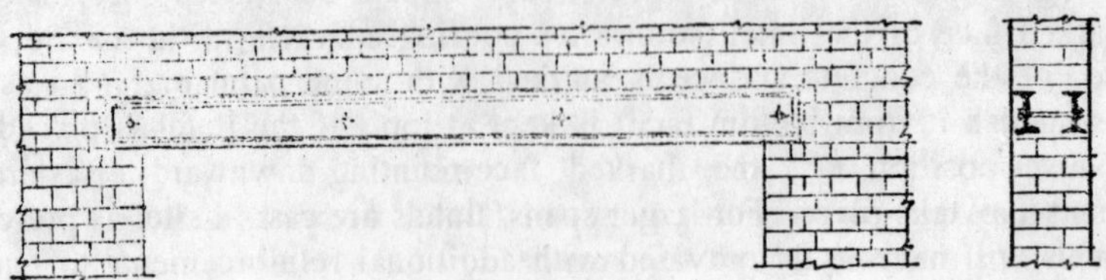

Fig. 15·5–15·6

Elevation and section of a steel lintel

joists, used singly or in a combination of two or three. Built-up lintel consists of three R.S.J's spaced together and connected to each other by bolts passing through them at intervals. This system is embedded in concrete and is cured like R.C. lintel.

15·8. ARCH

An arch may be defined as mechanical arrangement of wedge-shaped blocks of stones or bricks mutually supporting each other and supported at the end by piers or abutments. In common with lintels, the function of an arch is to carry the weight of the structure above the opening. Because of their shape, the blocks support each other by the mutual pressure of their own weight and the structure remains in position by the resistance from the support.

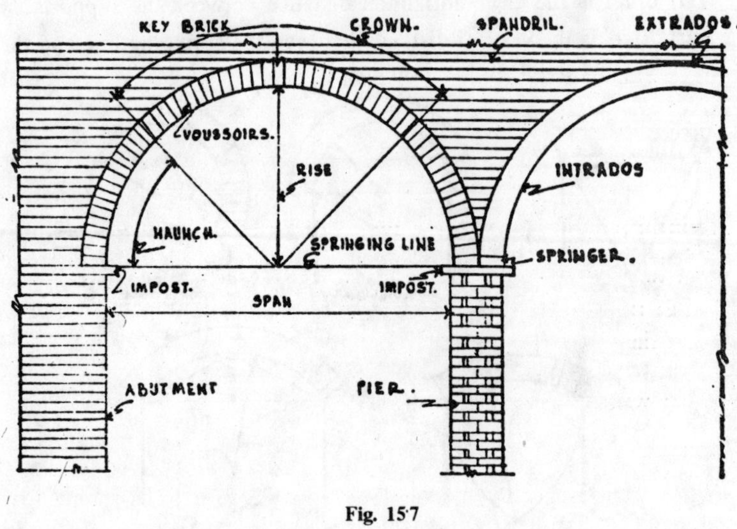

Fig. 15·7
Arch

15·9. TECHNICAL TERMS

The various technical terms used in arch work are described below:

(*i*) *Abutment or pier* is the part of the wall or pier on which the arch rests.

(*ii*) *Arch ring* is a course of stones or bricks having a curve similar to that of the arch.

(*iii*) *Intrados or soffit* is the under-surface of an arch.

(*iv*) *Extrados* is the outer surface of an arch or the outer curve of the arch ring.

(v) *Voussoirs or arch blocks or arch stones* are the wedge shaped units forming the arch.

(vi) *Springing stone or springer* is the first voussoir at springing level on either side of the arch.

(vii) *Springing line* is an imaginary line joining the two springing points.

(viii) *Crown* is the highest point of the extrados or it is the highest part of an arch.

(ix) *Key-stone* is the highest central wedge-shaped block of an arch.

(x) *Skew-back* is the surface of the abutment on which the arch rests.

(xi) *Span* is the clear horizontal distance between the supports.

(xii) *Rise* is the vertical distance between the springing line and the highest point on the intrados.

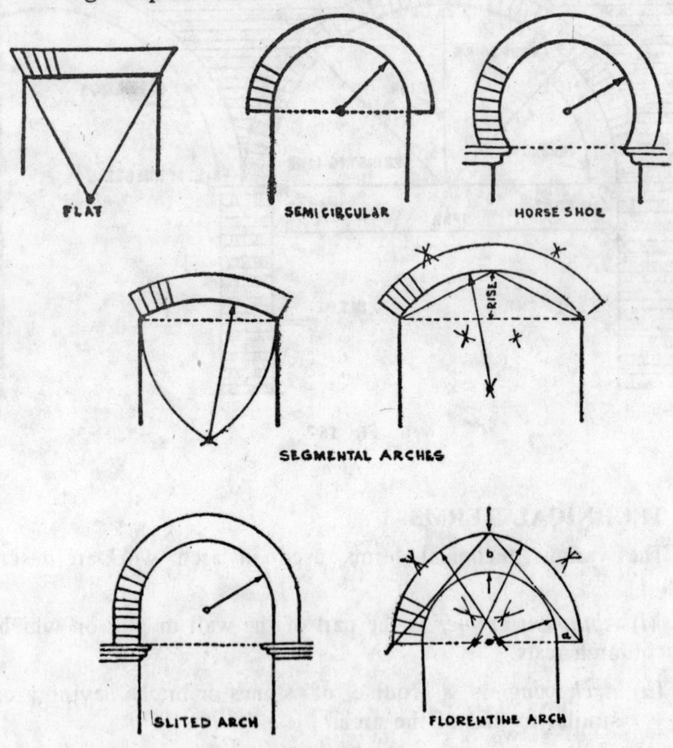

Fig. 15·8

Types of arches

(*xiii*) *Springing points* are the points at the extremities of the intrados.

(*xiv*) *Depth of the arch* is the perpendicular distance between the intrados and the extrados.

(*xv*) *Haunch of the arch* is the portion of the arch situated centrally between the key and skew backs.

(*xvi*) *Spandril* is the triangular walling enclosed by the extrados of the arch, a horizontal line from the crown of the arch and a perpendicular line from the springing of the outer curves.

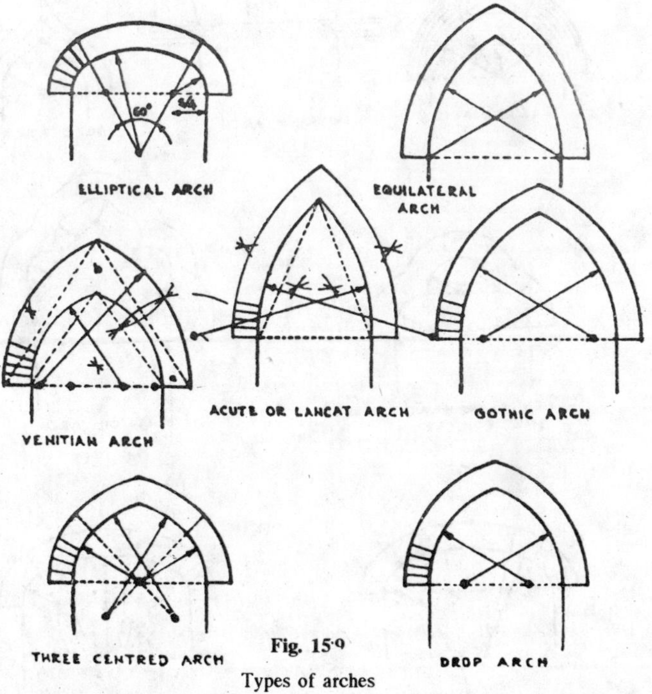

Fig. 15·9
Types of arches

15·10. TYPES OF ARCHES

Arches derive their names from the shape of curve formed by their soffit or intrados. The type of arches commonly adopted in construction work are :

(1) Flat arch

(2) Semi-circular arch

(3) Segmental arch

Fig. 15·10

Types of arches

(4) Relieving arch

(5) Dutch or French arch

In addition, there are various other type like elliptical arch, parabolic arch, four-centred arch or Tudor arch, five centered arch, points arch, etc. The various type of arches have been shown in Figs. 15·8-15·10.

15·11. MATERIALS OF CONSTRUCTION

Arches may be constructed with bricks, stones or concrete. Classification of arches by material is given below :

15·12. BRICK ARCHES

Depending upon the nature of worksmanship and the quality of bricks used, brick arches may be classified as rough arches, axed brick arches, gauged brick arch and so on.

Rough arches . These arches are built with ordinary bricks which are not cut to wedge shape. In order that all the bed joints may be normal to the curve of arch, the joints are made wedge shaped.

Fig. 15·11

Rough brick arch

Thus the joints at the extrados are wider than those at the intrados. The wedge-shaped joints spoil the appearance of the arch and as such rough arches are not recommended for exposed brick-work.

These arches are generally used over wooden lintels in order to relieve the latter of the weight of the masonry above. Hence they are sometimes termed as "relieving arches"

Axed brick arches. In this type of construction, the bricks are cut to wedge-shape. Thus the joints of the arch are of uniform thickness.

Since the wedge-shaped units are not finely dressed, the arch does not present an attractive appearance. It is therefore common to finish the brick work by plastering.

Fig. 15·12
Axed brick arch

Gauged brick arches. As the name suggests, gauged bricks are those which have been accurately prepared to a wedge shape for the arch construction. Hard bricks cannot be used in this type of construction as it is difficult to cut them to true wedge shape accurately. Hence specially made bricks known as "rubber bricks" are used in the arch construction. The bed joints are very thin usually 1·6 mm. thick. Cement mortar is considered too coarse for such joints, hence lime putty is used for binding the blocks. The thin white joints add to the appearance of the finished work.

Bonds in brick arches. For ordinary work, bricks in concentric rings of half bricks are laid with non-continuous radial joints. Each ring is fully completed before constructing the ring-immediately above it. For spans above 1·6 m, the concentric rings of brick work have a tendency to separate and there is a continuous danger of the whole pressure being thrown to one ring, which might get crushed and thus collapse the entire arch structure. This danger is avoided either by laying bricks alternately as headers and stretchers in section with continuous radial joints or by providing key stones at intervals.

15·13. STONE ARCHES

Considering the nature of workmanship and dressing of the material, stone arches may be divided in the following two categories :

1. *Rubble arches*
2. *Ashlar arches.*

Rubble arches. They are constructed with rubble stones which have been hammer-dressed, roughly to required shape and size of the voussoirs. In arch ring up to 38 cm. in thickness, all stones should be of the full depth of the arch. In arches above 38 cm. in thickness, the stones are laid in two rings in alternate course of headers and stretchers. This type of construction is weak in strength and is not attractive in appearance and is as such provided for small openings only.

Fig. 15·13

Semi-circular brick arch showing bond in brick work

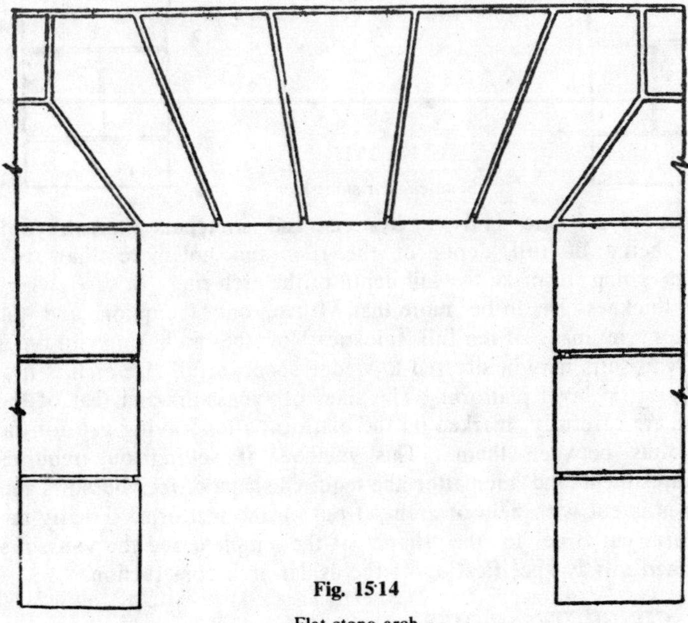

Fig. 15·14

Flat stone arch

Ashlar arches. They consist of arch stones which have been dressed full and true to their proper shapes. In all the cases where the depth of the arch is 60 cm. the voussoirs are made of full thickness of the arch and are set in fine mortar with their radial joints.

Fig. 15·15
Semi-circular stone arch

The stones are laid as headers and stretchers alternately, all headers being of full depth of the ring and not more than two stretchers going to make the full depth of the arch ring. In case where the arch thickness has to be more than 90 cm., only the quions and the key-stones are made of the full thickness of the arch ring. In order that the voussoirs may be dressed to wedge shape, a full size arch is first set out on a level platform. The sizes of voussoirs and that of the keystone are carefully marked on the platform after leaving gap for the mortar joints between them. This method of setting out requires slight adjustment and thereafter the required shape of the voussoirs and key-stones is cut with help of arch set out on the platform. Finally the stones are cut true to the shape of the template and the voussoirs thus formed satisfy specification of the ashlar arch construction.

15·14. CONCRETE ARCHES

They are of two different types namely, concrete block arches and monolithic concrete arches :

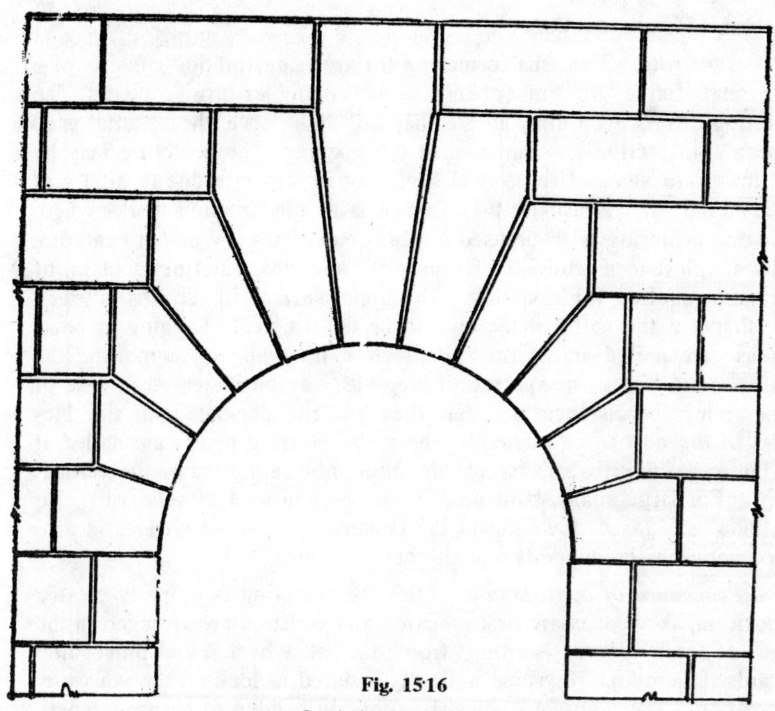

Fig. 15·16
Semi-circular stone arch

Concrete block arches. For small opening in a building, concrete block arches can be provided with advantage. The voussoirs, in the form of cement concrete blocks, are prepared in specially made moulds for the purpose. The moulds of concrete block for the 'key' or skew backs are of slightly different dimensions. The concrete mix usually adopted for block construction consists of one part of cement, two parts of sand and four parts of aggregates. After the concrete in the mould is fully set, the blocks of concrete are removed for curing in a water pond, constructed for the purpose. Generally, the concrete blocks are used without reinforcement. The arch construction with concrete block is similar to that for stone or brick arch construction

Monolithic concrete arches. Monolithic concrete arching is frequently used for roofing of buildings, culverts and bridges. The construction for smaller span with ordinary loading conditions requires no reinforcements. The arching can be done in lime concrete as well as in cement concrete. For roofing arches, a rise of 5 cm. for every 30 cm. of span is usually allowed when lime concrete is used. The minimum thickness of arch should not be less than 15 cm. for spans up to 3 m. As a general rule, add 38 mm. for each additional 30 cm. of span. For larger spans, arches have to be reinforced.

Method of Construction.

Centering. The usual centering for arch construction consists of a horizontal frame of timber known as centre or turning piece. The centering differs according to the shape of arch curve, the material used in arch construction and the span of the opening. The centering may be of timber or steel. Use of steel centering is very rare due to high cost of material and is usually preferred in large construction works where the same centering is to be used a number of times. Wooden centering in its simplest form consists of a pair of solid ribs, cut from a plank of thickness required for centering. The upper surface of the ribs is given the shape of the soffit of the arch to be constructed. Lagging or cross battens are nailed across the ribs at close intervals for supporting the voussoirs at top. The spacing of laggings may be increased in case of stone arch. The distance between the two ribs depends upon the thickness of the wall to be carried by the arch. Bearing pieces are nailed at the bottom of the ribs so as to enable them to be supported on the vertical props. For large spans, built up ribs are used instead of solid ribs. To facilitate easing or lowering of the centering, a pair of wedges is provided between the supports and the bearing piece.

Procedure of construction. After the centering is properly createc in position, skewbacks are first prepared and vousoirs are arranged in the form of arch curves starting from the skewbacks and proceeding towards the crown. Keystone is finally inserted to lock all the voussoirs in position. The voussoirs must be properly bedded to ensure strength

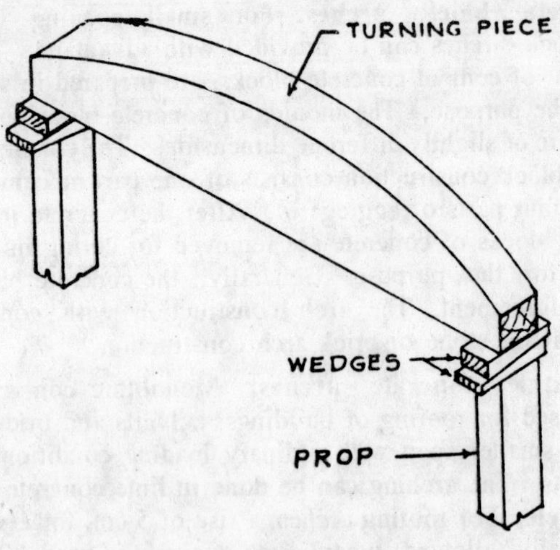

Fig. 15·17

Centering for a segmental arch

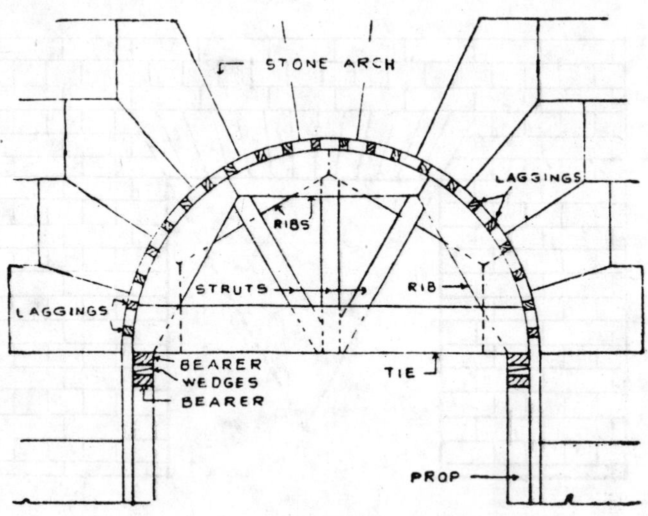

Fig. 15·18

Centering for a semi-circular arch

and stability of the arch. After the key has been placed in position, the centre should be eased by slackening the wedges. During the process of easing, great care should be taken to see that the centre is lowered only by a small height say 3 mm. The lowering of centre is necessary in stone arch construction as it allows the voussoirs to settle upon their beds nicely. No load should be super-imposed upon the arch work before it has fully set. The centre is removed before the construction of masonry above the arch.

15·15. FLAT ARCH

When used over opening of door or window, the flat arch acts similar to a lintel. Flat arch, also known as straight or camber arch, is laid with its bed joints radiating to centre. The skewback is made to rest in an inclined position so as to make an angle of 60° with the springing line. The extrados of the arch is horizontal but the intrados has a straight camber or upward curvature. The camber is usually 2 mm. for every 30 cm. of span and is necessary to avoid optical illusion of sagging. Flat arches are not strong as compared with other forms.

15·16. SEMI CIRCULAR ARCH

The arch derives its name from the shape of the curve given to the arch soffit. The different types of semi-circular arches in stone and brick masonry are shown in Figs. 15·13, 15·15, 15·16, and 15·20.

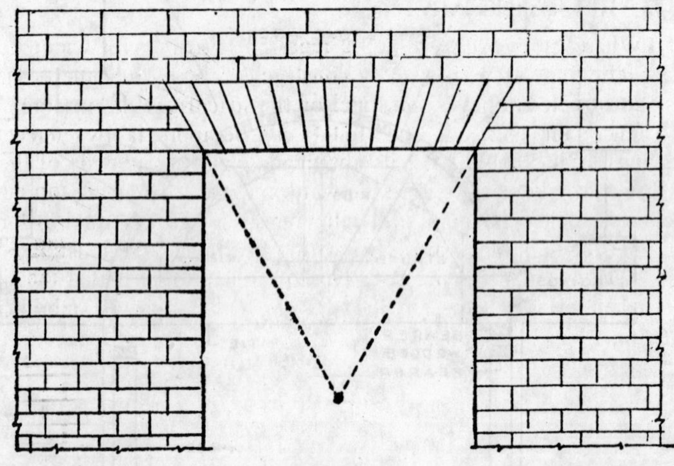

Fig. 15·19

Flat-brick arch

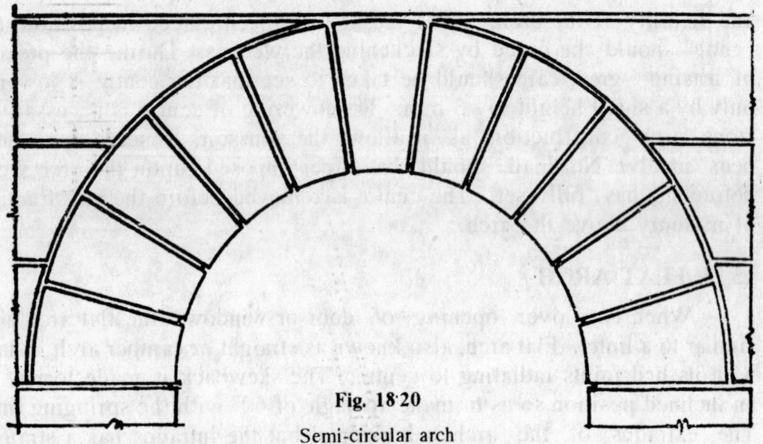

Fig. 18·20

Semi-circular arch

15·17. SEGMENTAL ARCH

This type of arch is segmental in shape and is commonly provided for openings of various sizes. The bed joints of segmental arch radiate from a common point, which lies below the springing line and is equidistant from all points on the arch curve. When used over a lintel, segmental arch is termed as "relieving arch".

15·18. RELIEVING ARCH

Where a wooden lintel is required to span over wider opening, brick relieving arch is usually constructed above the lintel. The arch thus constructed relieves the lintel of the load from the masonry above and hence the size of the lintel can be appreciably reduced. The relieving arch should have its abutments clear of the ends of the lintel so that the skewback slopes outward and away from the top edges of lintel ends. Thus the span of a relieving arch will be equal to the clear span of the opening plus twice the bearing of the existing wooden lintel. The advantage in this type of construction is that the wooden lintel when dacayed can be replaced without affecting the stability of the structure.

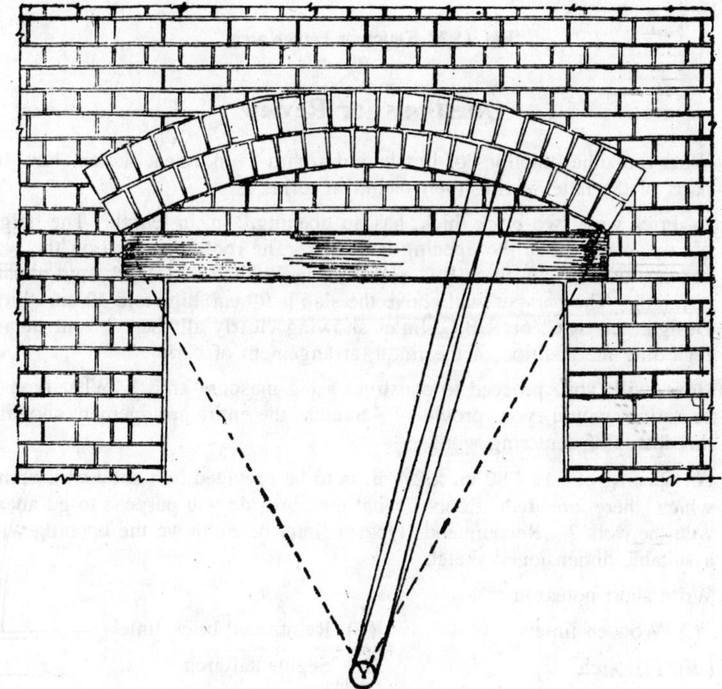

Fig. 15·21 Relieving arch

15·19. DUTCH OR FRENCH ARCH

This type of arch is similar to a flat arch in design but differs in shape and method of construction. This form of arch is not so sound and as such it is only suitable to be provided for small openings.

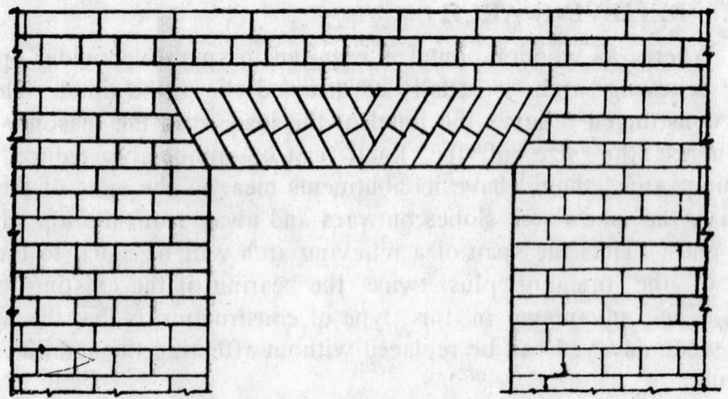

Fig. 15·22 Dutch or French arch

Questions for Review

1. What are the functions of lintels and arches ? In a neat sketch show the parts of the latter giving their technical terms.
2. A brick wall, two-brick thick, has an opening 2 m. in length. The height of the wall above the opening is 1·5 m. The roof supported on the wall consists of a 12 cm. thick R.C. slab finished with 10 cm. thick mud phuska terracing. The parapet wall above the slab is 90 cm. high and 10 cm. thick. Design an R.C. or R.B.C. lintel showing clearly all the relevant details regarding the position of the lintel, arrangement of bars, etc.
3. How will you proceed to construct a big masonry arch ? What type of centering would you provide ? Sketch the entire arrangement showing the details of centering work.
4. An opening of size 1·80 m. x 2·4 m. is to be provided in a solid wall above which there are two floors. What measures do you purpose to go ahead with the work ? Recommend type of lintel beam above the opening with a suitable dimensioned sketch.
5. Write short notes on :

   (*i*) Wooden lintels (*ii*) Reinforced brick lintels

   (*iii*) Flat arch (*iv*) Segmental arch

   (*v*) Dutch or French arch (*vi*) Relieving arch
6. What do you understand by the terms, arches and lintels ? Describe in brief the various types of arches used in modern construction.
7. Define the following terms with regard to an arch.

   (*i*) Abutment (*ii*) Skew-back

   (*iii*) Spandril (*iv*) Voussoir

   (*v*) Key-stone (*vi*) Crown.
8. Make a neat sketch of two adjacent segmental arches of brick-work

consisting of three rings each in two-brick thick wall with a pier 75 cm. thick in between. The span of each arch is 2 m. and the rise is 1/3rd of the span. On your sketch write down the following terms in their proper places :

(*i*) Springing (*ii*) Haunch
(*iii*) Voussoir (*iv*) Skewback
(*v*) Spandril (*vi*) Crown
(*vii*) Intrados (*viii*) Key
(*ix*) Abutment (*x*) Extrados.

9. Draw in plan,elevation and section a stone arch 3 66 m. centre to centre of supporting piers. The piers are 60 cm. square. Show clearly all stones comprising the arch on adrawing and indicate by arrows the following terms:

(*a*) Extrados (*b*) Spandril
(*c*) Haunch (*d*) Springing line
(*e*) Intrados (*f*) Voussoir
(*g*) Key-stone

10. Distinguish between a lintel and an arch. Make a neat sketch of a stone arch and name the various parts. What are the various materials used in the construction of lintels. Sketch an R.C.C. lintel.

11. What are the different types of arches that are used in building construction? Describe any three types in detail with sketches.

16

Carpentry Joints

16·1. GENERAL

Carpentry is the term applied to that form of wood-work in which the wooden members are subjected to loads or thrusts. Depending upon the type of work, the members may be permanently subjected to bear loads as in case of wooden lintels,wooden floor joists etc. or they may be provided to bea. the load in a temporary capacity in the form of scaffolding, centering, shoring etc. Joinery may be defined as the trade in wood work in which skilled labour is required to render the wooden members capable of framing together, it is an art of cutting, preparing and joining the individual pieces of timber, so as to form the frame of the desired shape, size and finish. In India, carpentry and joinery are treated as a single trade. The workman who handles the work of carpentry and joinery is called a *carpenter*. The wooden members may either be jointed by nails, screws, bolts, connections or by cutting the different members in different ways to form groves or projections which when framed together serve the purpose. Sometimes it becomes necessary to use iron castings in the form of straps, gibbs and cotters, etc. to strengthen the joints in wood work.

16·2. TECHNICAL TERMS

The various technical terms used in carpentery or joinery have been briefly described below :

1. **Chamfering.** It consists in taking off the edges of a wooden members. Thus a chamfered member has a sloping or bevelled edge. Chamfering is usually done by means of a *plane.* Chamfering is essentially carried out with the aim of preventing the exposed wood work to have sharp edges, and incidentally it imparts architectural beauty to the member.

2. **Plaining.** The process of taking the shavings of wooden member is called planning. The carpenter's tool used for the purpose is called a "plane". The surface smoothened with a 'plane' is called dressed or rough surface.

3. **Plank.** It is the name given to a heavy, thick and wide wooden piece of timber.

4. **Rebating.** It is the process of cutting a rectangular groove along the edge of a member, so as to enable the edge or tongue of another member to fit in the former.

5. **Housing.** It is the process of cutting groove in a timber piece across its grain, so as to permit the insertion of another member in it. Housing may be plain, shouldered or dove-tailed.

6. **Scribing.** It is the process of framing one moulding into the face of another at an interior angle.

7. **Seasoning.** It is the process of rendering timber free of the sap either by natural or artificial means.

8. **Mitring.** It is the process of joining two wooden members at an angle.

9. **Milled.** It is the term applied to wooden member which has been run through a wood working mill.

10. **Bead.** It is rounded or semi-circular moulding provided on the edges of surface of wood.

11. **Batten.** It is the name given to a narrow strip of wood normally nailed over joints of boards.

12. **Studding.** It is the term applied to the fixing of small timber battens, to timber walls, to which lathes and boards are to be nailed.

13. **Veneering.** It is the process of covering the exposed surface of a wooden member with a thin sheet of a costlier type of good looking wood, to improve the appearance of the member. Veneering is done just for decorating the wood work and it does not contribute to the strength of the veneered member.

14. **Wain-scot.** It is the term applied to wooden panelling on the masonry walls for a height of 60 cm from the floor level.

16·3. CLASSIFICATION OF JOINTS

The various type of joints used in wood framing can be divided into the following categories :

1. *Lengthening, spliced or longitudinal joints*
2. *Bearing joints*
3. *Framing joints*
4. *Angle or corner joints*
5. *Oblique-shouldered joints*
6. *Widening or side joints.*

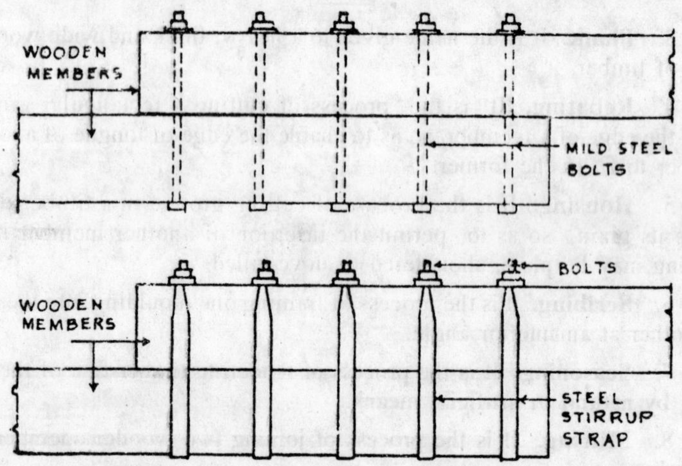

Fig. 16·1 Lapping joint with mild steel bolts

Fig. 16·2 Lapping joint with stirrup straps

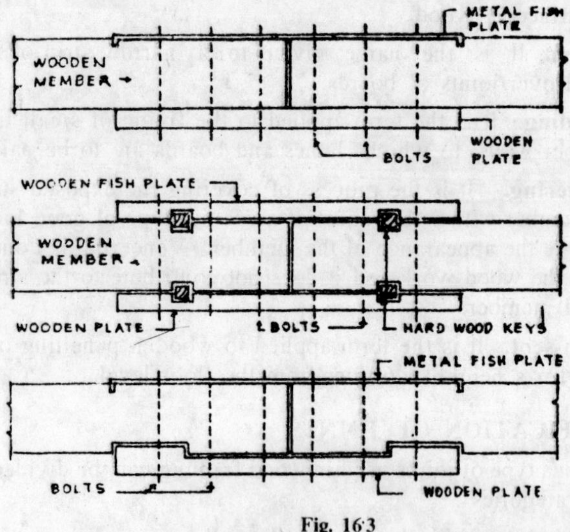

Fig. 16·3

Sectional elevations of various types of fishing joints

16·4. LENGTHENING, SPLICED OR LONGITUDINAL JOINTS

As the name suggests, these joints are used to lengthen a wooden member. The method of lengthening the member varies with its situation in a framed structure. Thus, lengthening joints are further sub-divided into the following types:

(*i*) Lapping joint.

(*ii*) Fishing joint.

(*iii*) Scarfing joints

(iv) Tabling joints.

(*i*) *Lapping joints* : This is the simplest form of lengthening joint and is formed by lapping the end of one member over that of the other and fastening them together by bolting or by using connections. This types of joint should not be recommended for members subjected to large stresses because of the axis of the members at the joint not being in one line. The various types of lapping joints have been shown in Fig. 16·1—16·2.

(*ii*) *Fishing joints* : In this arrangement of jointing wood work, the ends of the two members butt together and the junction is bolted after providing wooden or steel pieces called "Fish Plates" on two opposite faces of the joint. The strength of the joint depends upon the strength of fish plates and bolts used in the joint.

(*iii*) *Scarfing joint* : This system of lengthening wooden member

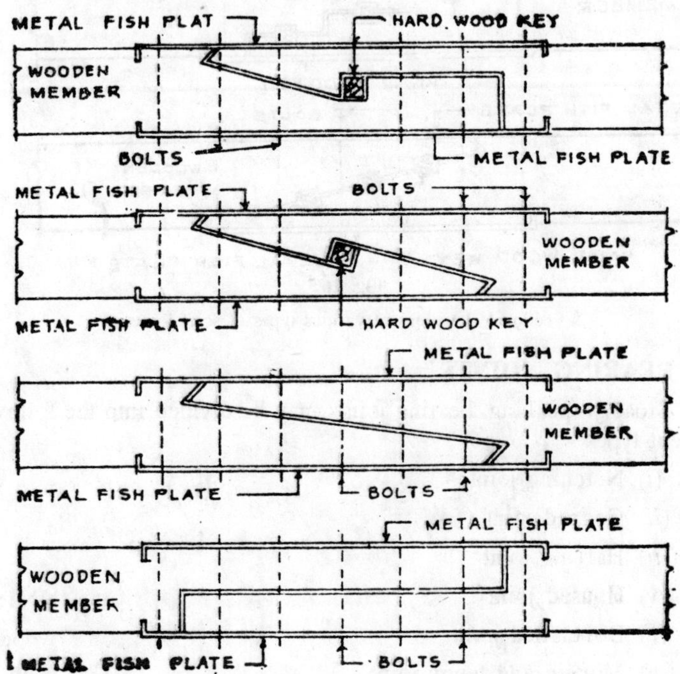

Fig. 16·4

Sectional elevations of various types of scarf joints

is especially suitable in places where it is necessary to maintain the same depth and width of the member throughout its length. To achieve the said condition, a part of the thickness of the timber is cut from each member from opposite sides, so that they lap on each other when jointed. The strength of the joint depends upon the bolts used for jointing. The fish-plates used in the joints are usually bent and let into the timber.

(*iv*) *Tabling joint* : This joint is very much similar to scarf joint in construction and is adopted where the member is subjected to both tension and compression.

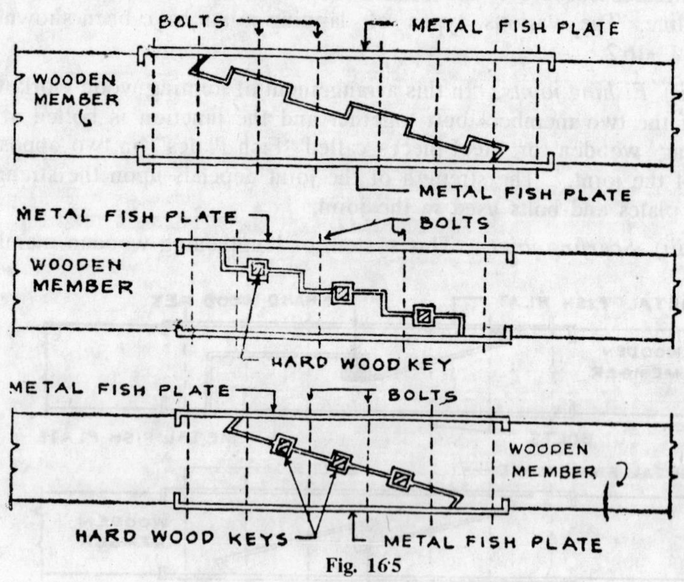

Fig. 16·5

Sectional elevations of various types of scarf joints

16·5. BEARING JOINTS

Broadly speaking, bearing joints may be divided into the following different types :

(*i*) Notching joint

(*ii*) Cogged joint

(*iii*) Halving joint

(*iv*) Housed joint

(*v*) Dovetailed joint

(*vi*) Mortise and tenon joint

(*vii*) Joggle or stump joint

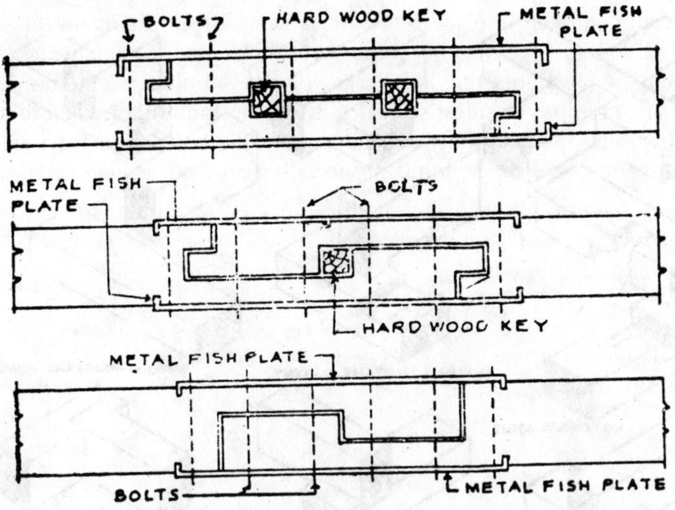

Fig. 16·6

Sectional elevations of various types of tabling joints

(*viii*) Dovetailed tenon joint

(*ix*) Open mortice joint

(*x*) Chase mortise joint

(*xi*) Tusk tenon joint

(*i*) *Notching joint* : This joint is commonly provided at the junction of two members meeting at right angle. Notching consists in cutting a depression to the full width of one member to enable it to be fixed to another member. In some cases both the members are cut to form the joint. In such cases the joint is called double notched joint. It is very much clear from the shape of the joint that the function of notching is to prevent the lateral movement of the two members.

(*ii*) *Cogged joint* : This joint is commonly provided between wall plate and the tie beam of wooden truss where it is essential to utilise the entire depth of the wall plate for bearing the tie-beam. The joint may be single or double cogged. In single cogged joint, a small depression, equal in width to the width of the member to be placed over it, is cut in the lower member. In case of a double cogged joint, a projection is left in the lower member in middle of the recess formed, for receiving the upper member.

(*iii*) *Halving joint* : This joint is used in members crossing or meeting at an angle. In this method of jointing, each member on opposite sides is cut to half of its thickness, at the joint and lapped. Thus

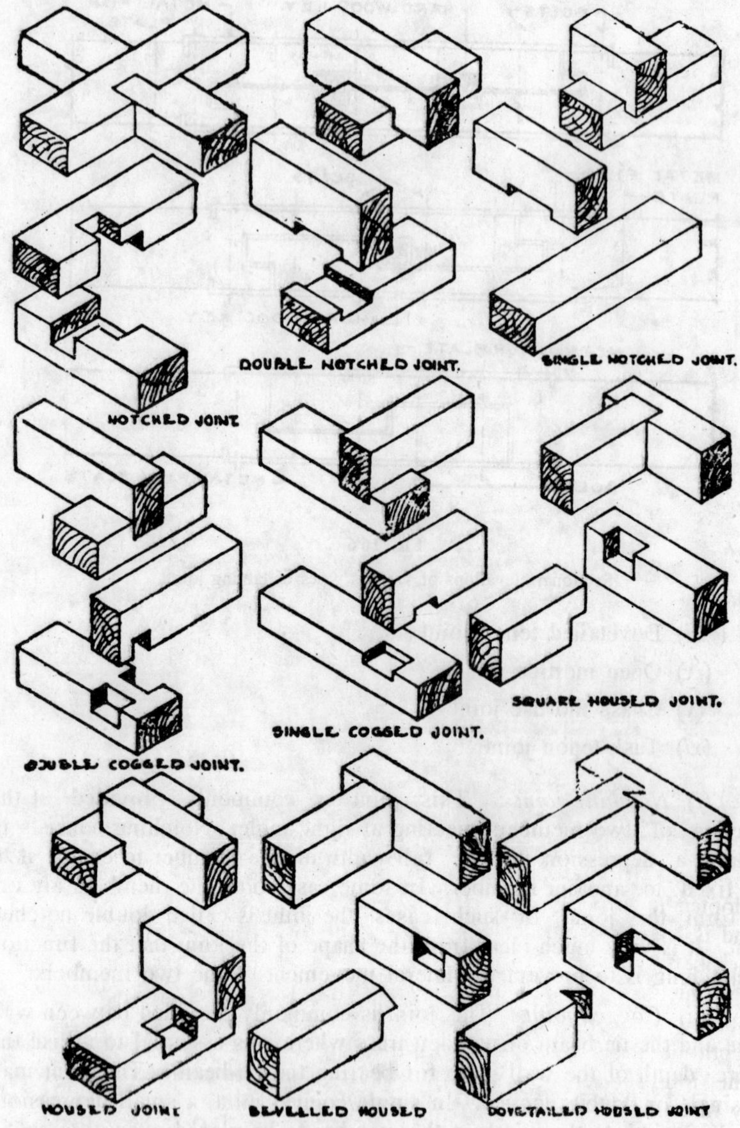

Fig. 167

Views showing various types of bearing joints

if the two members are of different thicknesses, the top face of the member at the joint will be flushed. In case both are of the same thickness, both top and bottom face of members at the joint will be flushed. Sometimes the members are bevelled from consideration of strength. In such cases, the halved joint is termed as bevelled halved joint.

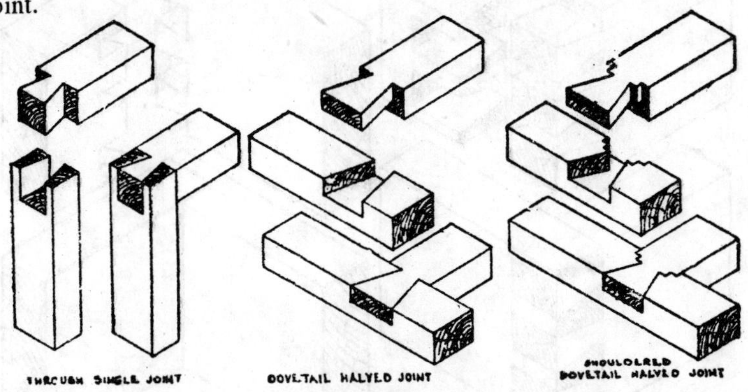

Fig. 16·8

Bearing Joints

(*iv*) *Housed joint* : In the housed joint, the entire end or thickness of one member is let into the notch in another. This joint is commonly adopted in works of heavy nature, where other types of joints may not be practically suitable either from considerations of strength or economy.

(*v*) *Dovetailed joint* : In this method of fastening wood work, wedge-shaped or flaring shaped pieces are cut out of each member and the joint is formed by hooking the projection of one member into the other one. This joint is specially used in cabinet work. In such places, the interlocking tenons which are dovetailed are lapped and the junction is strengthened by means of pins which guard against the possibility of the members getting pulled apart.

(*vi*) *Mortise and tenon joint* : This type of joint is frequently adopted in framing wood work.. It efficiently holds the pieces together and is simple in construction. This joint is formed by cuting one of the members, so as to form a projection termed as tongue or tenon which fits into a slot known as mortise, cut into the other member. The mortise should be of sufficient length, breadth and depth. Generally the depth of mortise and tenon is one-third the thickness of the members. The joint is strengthened by inserting wooden wedges from back or by inserting dowel pins from the face.

(*vii*) *Joggle or stump joint* : This type of joint is commonly used for framing the studs into the sill of a wooden partition wall. Its construction is essentially similar to that of mortise and tenon joint with the only difference that the tenon is cut shorter in length and does not extend to the full thickness of the mortise member.

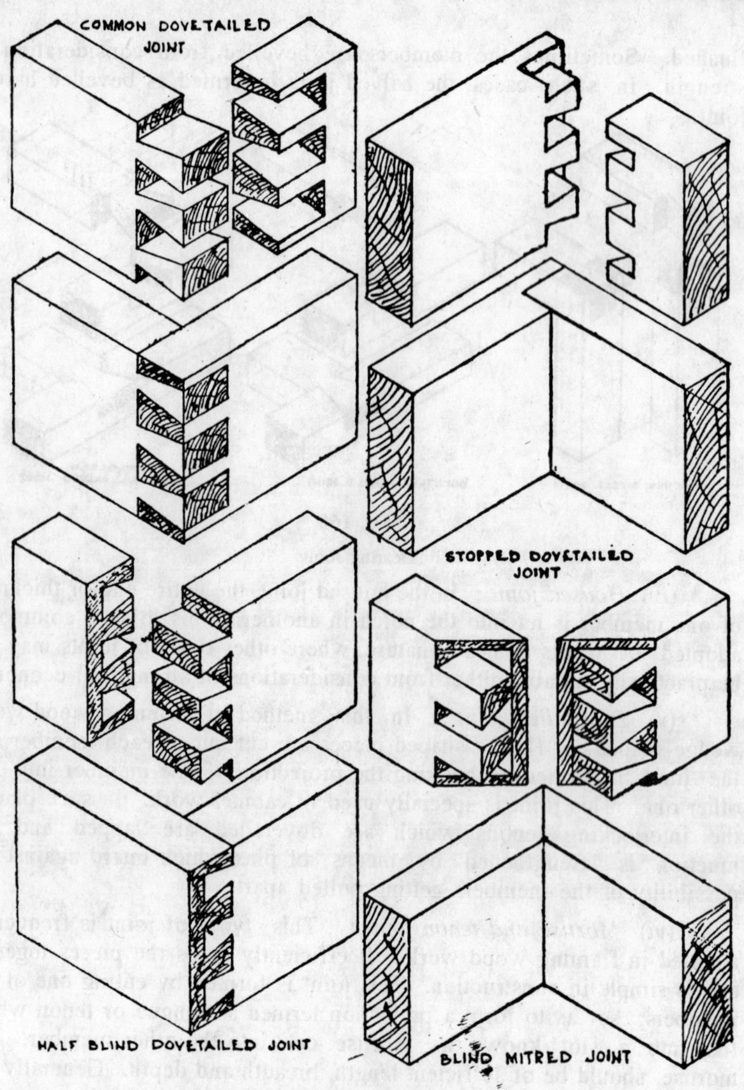

Fig. 169

Bearing Joints

(*viii*) *Dove-tailed tenon joint* : This type of joint is not so common in use on account of its typical shape. Generally this joint is formed by cutting the tenon, dove-tailed in shape and framing it into the mortise of sufficient length, so as to permit the wide part of the member attached to the back of tenon to be fixed inside the mortise. The joint is further secured by a hard-wood wedge.

—

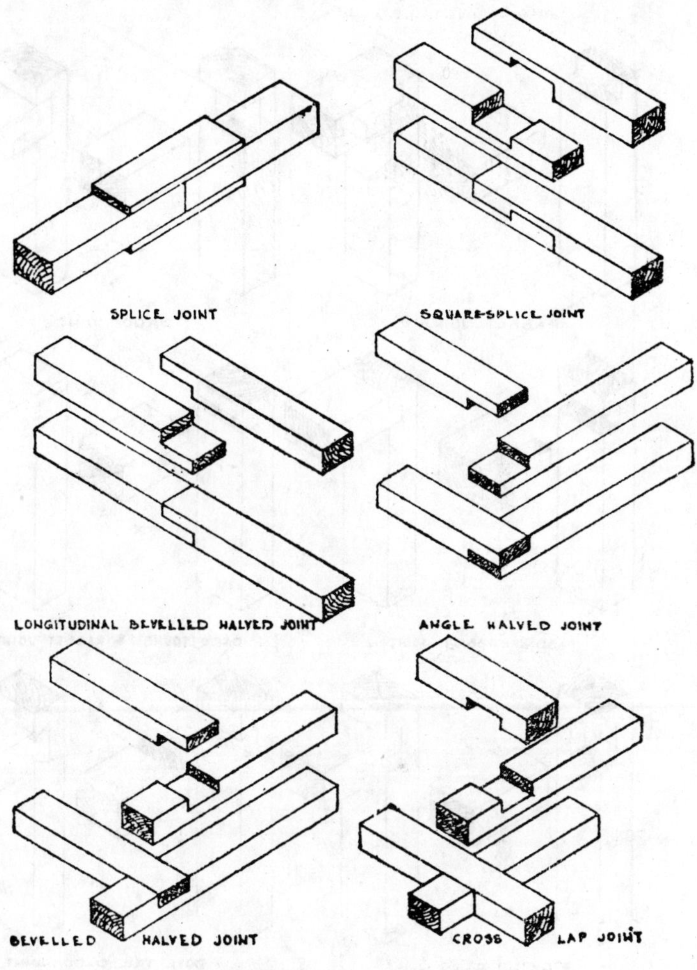

Fig. 16·10
Bearing Joints

(*ix*) *Open mortise joint* : This is a form of mortise and tenon joint in which a metal dowel bar has to be used to secure the mortise and tenon in their respective positions.

(*x*) *Chase mortise joint* : This method of framing woodwork is adopted when it is required to fix wooden members between such timber members as are permanently fixed in position in wooden structure long before.

(*xi*) *Tusk tenon joint* : This type of joint is commonly adopted in the construction of timber floors. In order to obtain maximum rigidity

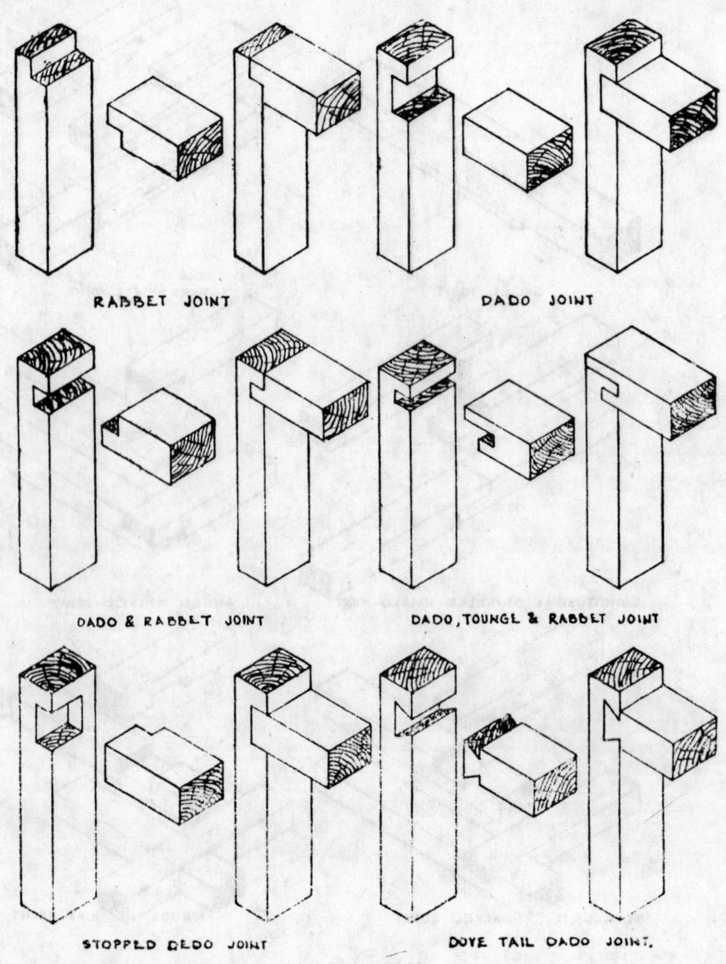

Fig. 16·11
Bearing Joints

and strength, the tenon must be mortised in the centre of the members to be tenoned.

16·6. FRAMING JOINTS

These joints are similar to the various forms of bearing joints described above with the only difference that the method of cutting the grooves and tongues in the members is suitably altered to obtain the desired form of joint. This type of joint is essentially employed in framing doors, windows, partition walls etc. This form of joint varies with

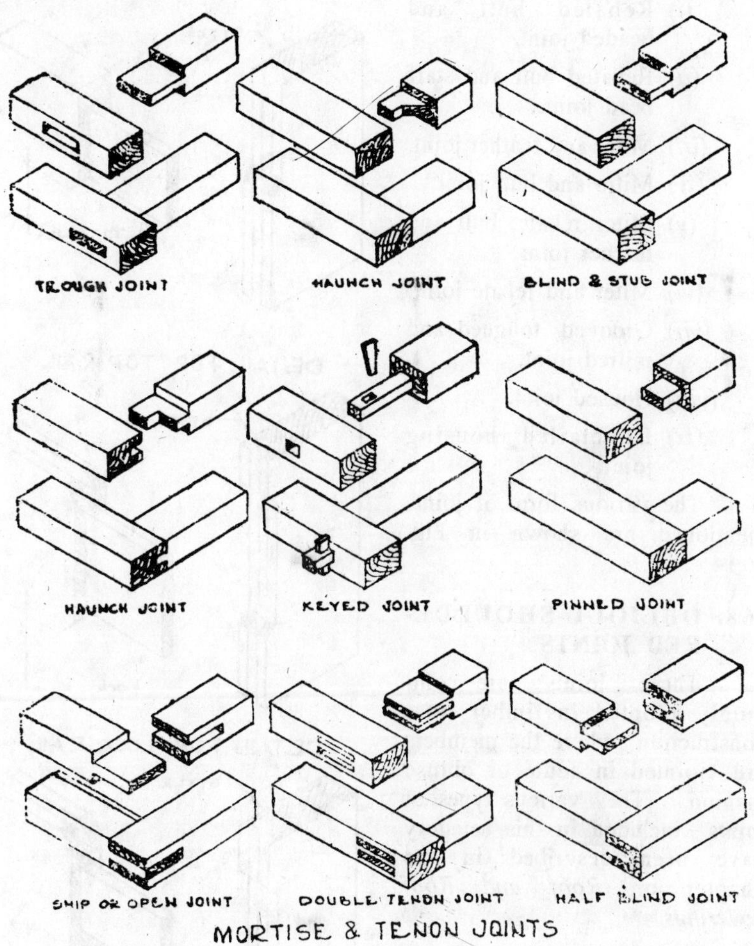

Fig. 16·12

Bearing Joints

the shape of door or window frame adopted and the architectural treatment desired. In general, these joints are just a modified form of mortise and tenon joints. For door frames over 11 cm. in width, double tenon joints are used.

16·7. ANGLE OR CORNER JOINTS

When two members are to be framed so as to form a corner or angular edge, angle or corner joint is used. The joints commonly used are given on the next page :

(i) Rebated butt and beaded joint.

(ii) Rebated butt and staff bead joint.

(iii) Miter and feather joint.

(iv) Mitre and butt joint.

(v) Miter rebate, butt and feather joint.

(vi) Miter and rebate joint.

(vii) Grooved, tongued and mitred joint.

(viii) Housed joint.

(ix) Dovetailed housing joint.

The various form of joints mentioned are shown in Fig. 16·14 :

16·8. OBLIQUE-SHOULDERED JOINTS

These joints are commonly adopted in timber truss construction where the members are jointed in acute or obtuse fashion. The various types of joints included in this category have been described in the chapter on *Roof and Roof coverings.*

16·9. WIDENING OR SIDE JOINTS

The joint between two wooden members placed side by side is termed as widening or side joint. Theses joints are commonly used in flooring, panelling etc. The types of widening or side joints commonly used are given below :

(i) Butt or plain joint.

(ii) Tongued and groove joint.

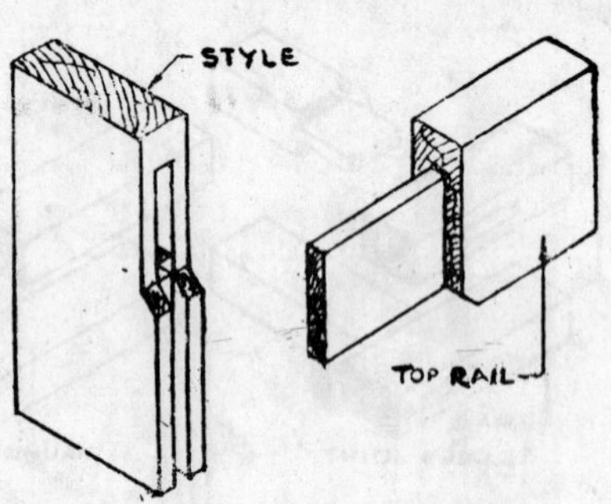

DETAIL FOR TOP RAIL

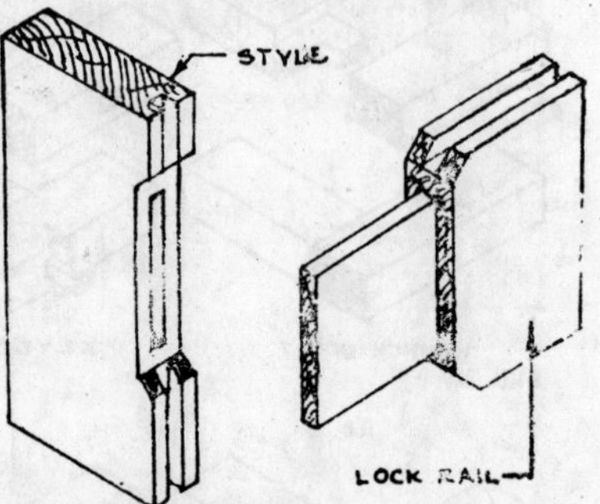

DETAIL FOR LOCK RAIL

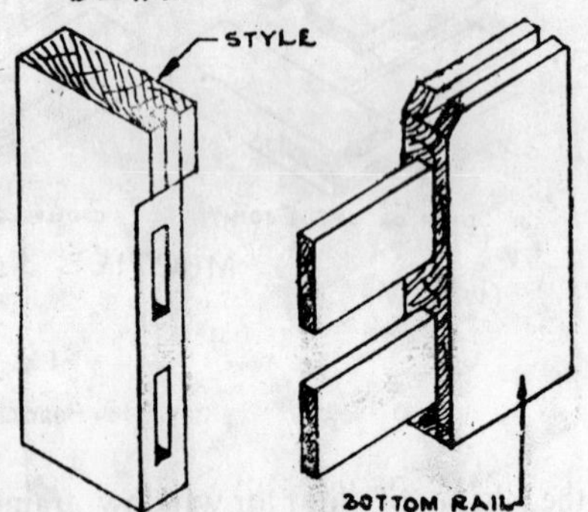

DETAIL FOR BOTTOM RAIL

Fig. 16·13

Framing joints

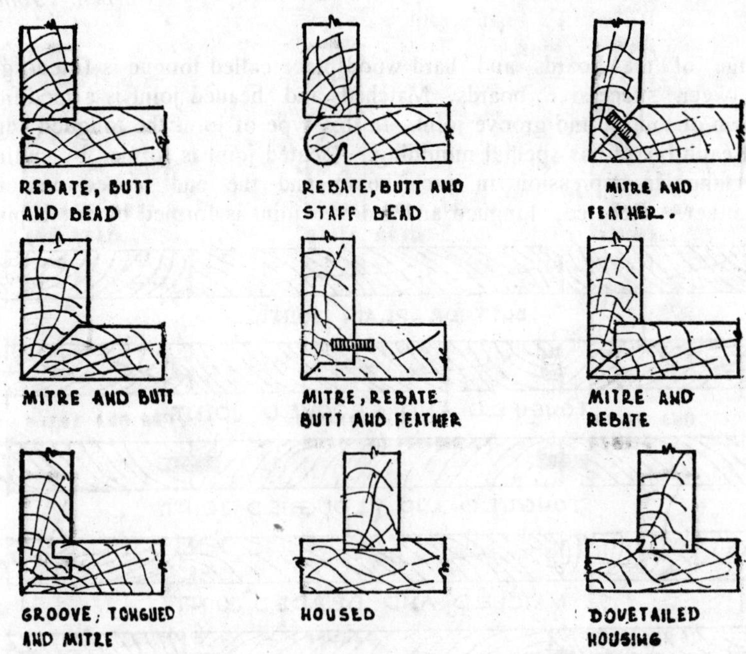

Fig. 16·14

Various types of angle or corner joints

(*iii*) Tongued and ploughed joint.

(*iv*) Matched and beaded joint.

(*v*) Rebated joint.

(*vi*) Rebated, tongued and grooved joint.

(*vii*) Butterfly joint.

(*viii*) Matched and vee joint.

(*ix*) Rebated and filled joint.

(*x*) Splayed, rebated, tongued and grooved joint.

(*xi*) Dowelled joint.

Butt or plain joint is formed by laying the boarding side by side. The boards used in the work must be well, seasoned and sawn in such a manner that the annual rings remain perpendicular to the flooring joints.

In tongued and grooved joint each board is cut so as to have tongue on one edge and groove on the other.

In tongued and ploughed joint similar grooves are cut along the edge of the boards and hard wood piece called tongue is fitted tight between successive boards. Matched and beaded joint is a modified form of tongue and groove joint. In this type of joint the tongued edge of each board has special moulding. Rebated joint is formed by cutting rectangular depression in each board and the end lapped over one another. Rebated, tongued and grooved joint is formed by so shaping

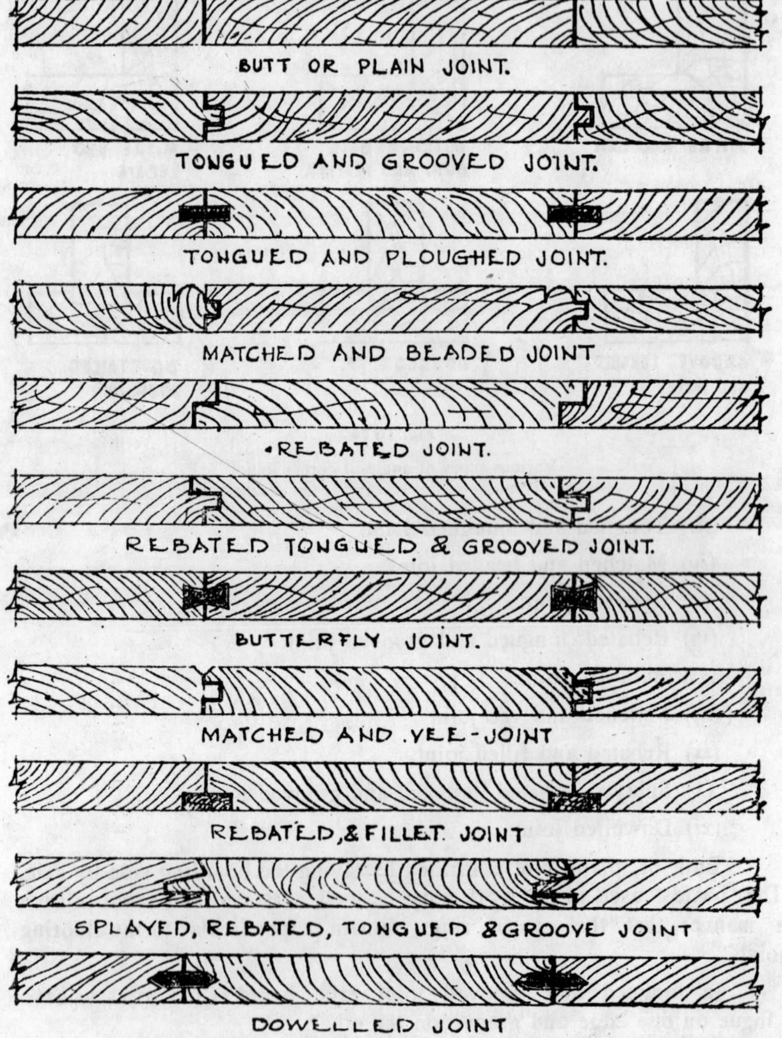

Fig. 16·15

Various types of widening or side joints

the tongued and grooved ends of the members that the rebate is also formed by cutting an additional width on one side of the tongued member. Butterfly joint is similar to tongued and ploughed joint. Matched and vee-joint is similar to matched and beaded joint with the only difference that the moulded tongue of each board is given a vee-shaped moulding. Rebated and filled joint is similar to tongued and ploughed joint with only difference that the hard wood piece is situated at the lower face of the boarding. Splayed, rebated, tongued and grooved joint is similar to rebated tongued and grooved joint except that the one side of the tongue is cut at an angle. Dowelled joint is similar to tongued and ploughed joint.

16·10. FASTENINGS

The fastenings which are commonly used in framing timber work, are briefly given below :

1. **Nails.** A nail may be defined as a slender piece of metal, circular or oval in section. The nails usually have a flat head at one end and a point on the other.

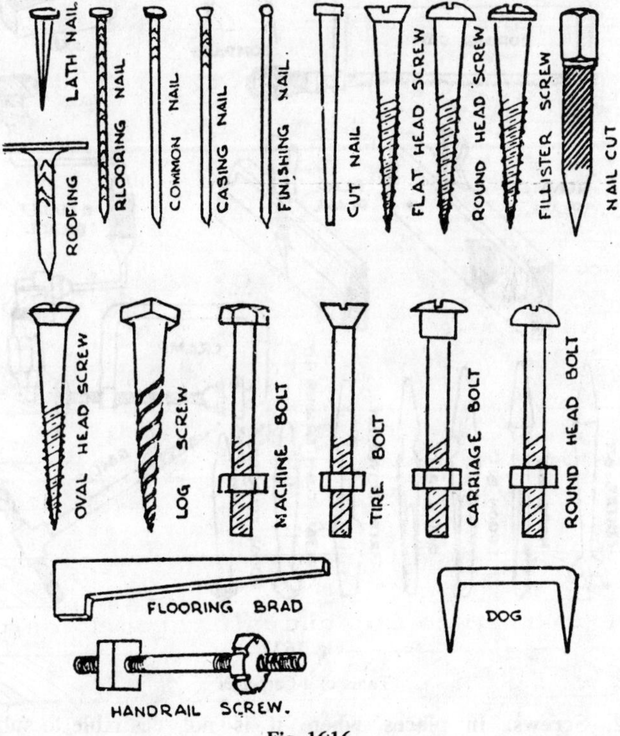

Fig. 16·16

Various types of nails, screws and bolts etc.

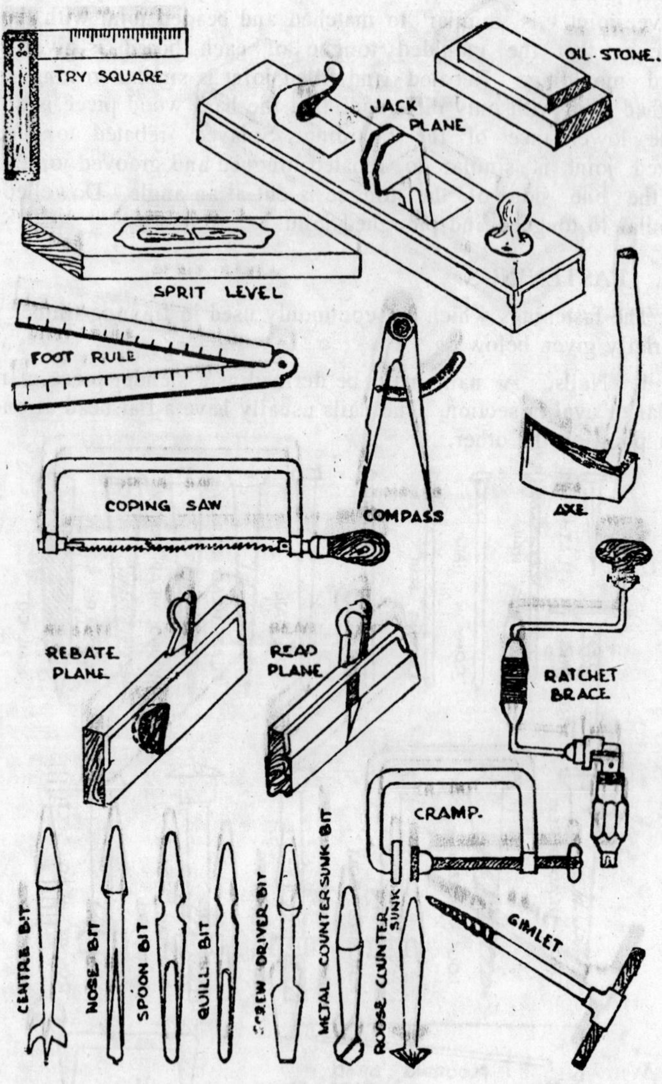

Fig. 16·17
Tools of a Carpenter

2. **Screws.** In places where it is not desirable to subject the joint to vibrations due to nailing, screws are used. The screws thus perform the same function as the nails do. The screws are manufactured in a variety of different shapes and are usually named after the shape of the head of the screw.

3. **Bolts.** Bolts are stronger than nails or screws and are used in places where the fastening is subjected to considerable tensile stresses. Bolts are commonly employed for rigidly holding the members (which may be of wood or steel) at the joint.

4. **Dogs.** These are metallic pieces of round or flat section bent at ends forming skipped hook.

16·11. CARPENTRY TOOLS

The tools commonly used in carpentry work are shown in Fig. 16·17—16·18.

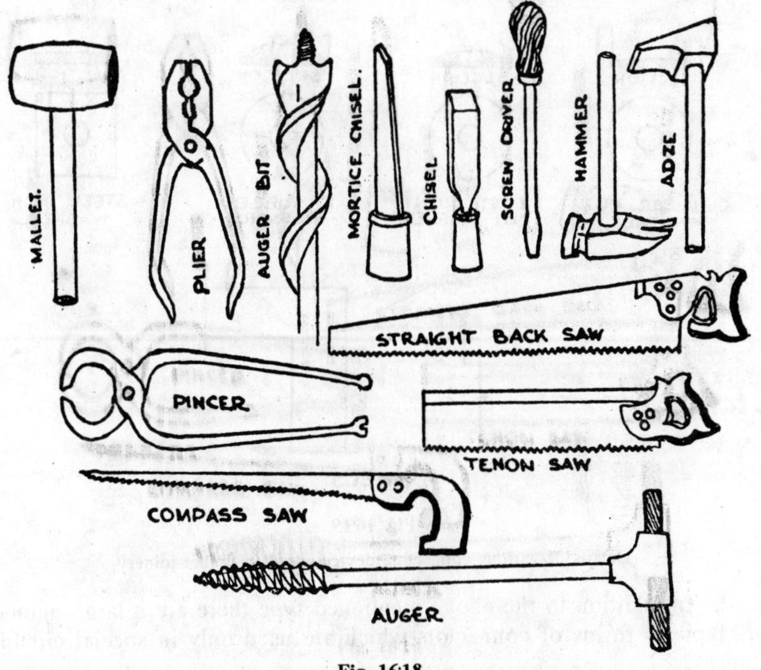

Fig. 16·18

Tools of a Carpenter

16·12. CONNECTORS

With the advancement in timber construction techniques various types of metallic connection are now commonly used for connecting timber pieces in such a way that transmission of stress from one member to another is adquately ensured. Connectors are advantageously used in the construction of timber trusses, arches or other similar structures. They are usually embedded in the individual member in such a way that the danger of joint against sliding is completely guarded. Split ring connector and toothed- ring connector are the two varieties commonly

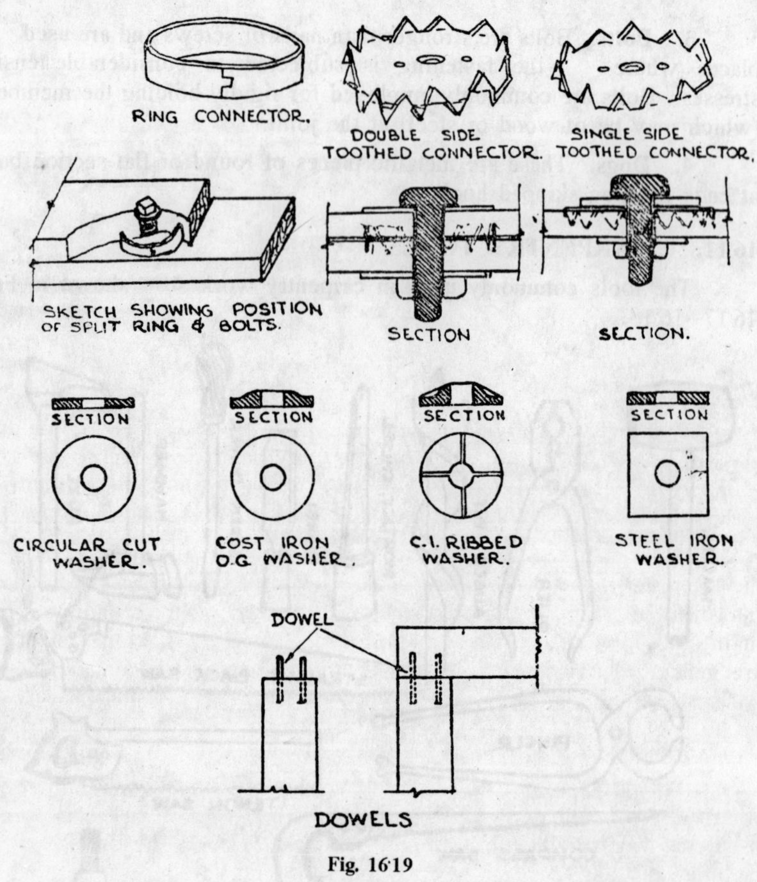

Fig. 16·19

Details of various connecting devices used in timber joinery

used. In addition to the above mentioned type there are a large number of typical forms of connectors which are used only in special circumstances.

Questions for Review

1. Describe briefly, with the help of sketches, lengthening, transverse and framing joint in wood work.
2. What is a bearing joint ? Show with sketches four different types of bearing joint.
3. Write short notes on :
   (*i*) Notching (*ii*) Dove tailing (*iii*) Cogging (*iv*) Halving.
4. Illustrate any five types of carpentry joints used in buildings and state when they are used.

Name various types of fastening used in wood work.

17

Doors and Windows

17·1. GENERAL

A door may be defined as a framework of wood, steel, aluminium, glass or a combination of these materials secured in an opening left in a wall for the purposes of providing access to the users of the structures. It basically consists of two parts *i.e.* (*i*) *A frame and* (*ii*) *shutter* (*leaf*).

The frame is normally made up of timber. This is on account of the fact that timber is easily available everywhere and it can be worked into different shapes without difficulty. However where economy is the main consideration, frames made up of steel or precast R.C.C. members are generally used. Steel or R.C.C. Frames are also recommended in construction on sites infested with termites. Shutters are generally made up of timber framework with panel insets of glass, timber, plywood, block board or a combination of such materials. In case of steel doors, the shutters are made of frame work of rolled steel sections with panel inserts of glass or steel plates. Next to steel, aluminium is the metal largely used in the manufacture of doors. Aluminium doors are considered to be the best as they are rust proof and present an elegant appearance. However since aluminium is costly as compared to timber or steel, use of aluminium doors is normally restricted to public buildings.

For small opening a door is provided with one leaf or shutter and such a door is known a *single leaf door.* In case of wider opening the door should have two leaves and such a door is termed as *double leaf door.* Sometimes a door is provided with shutter on either side of the frame. Shutter on one side of the frame may have wire gauge mesh (in case of kitchen door) and on the other side the shutter may be fully panelled, such a door is known as *double shuttered door.*

17·2. LOCATION OF DOORS

The following guidelines should be kept in view while deciding the location of doors in a building.

(*i*) From consideration of adequate air circulation within the room, the doors should be provided in opposite walls facing each other.

(*ii*) From consideration of proper space utilisation and privacy, doors should as for as possible be located near the corner of a room (say about 20 cm away from the corner).

(*iii*) The location of the door should meet the functional requirements of the room.

(*iv*) The number of doors in a room should be kept minimum to achieve optimum utilisation of space. Large number of doors besides causing obstruction, consume more area in circulation.

17·3. SIZE OF DOORS

The size of door to be adopted for a room depends basically upon the functional requirements of the room. In case of W.C. or bath room where only one person is expected to go at a time a door width of 75 cm is considered adequate. On the other hand, the size of a garrage door should be such that it may permit a car to pass through. In general, the

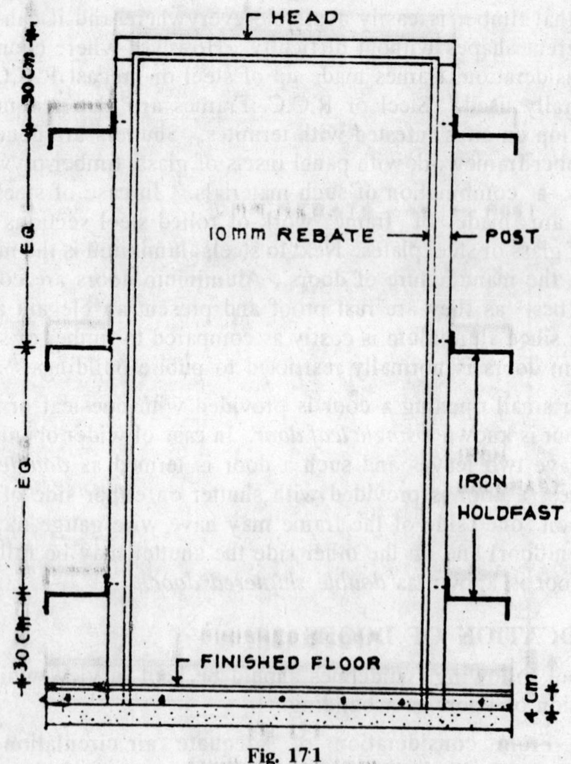

Fig. 17·1

Detail of a door frame

size of the door adopted should be such that it will permit the movement of the largest object likely to pass through the door opening. Commonly adopted size of doors for different types of buildings are given below.

Sl. No.	Type of building	Size of door in mm
1.	Doors of Residential buildings	
	(a) External door	1000 x 2000
	(b) Internal door	900 x 2000
	(c) Door for bath or w.c.	800 x 2000
2	Doors of public buildings like school,	1200 x 2000
	Hospital, Library etc.	1200 x 2100
		or 1200 x 2250
3.	Doors of garrage for cars	2250 x 2250

17·4. DOOR FRAME

Door frame is made up of two vertical members knownas *jambs or posts* and a flat member connecting the jambs at top called *head*. The cross-sectional dimensions of the jamb and the head are normally kept same. The size of door frame is worked out after allowing a clearance of 5 mm. alround for convenience of fixing the frame in the opening. Frames made from timber and steel are described below :

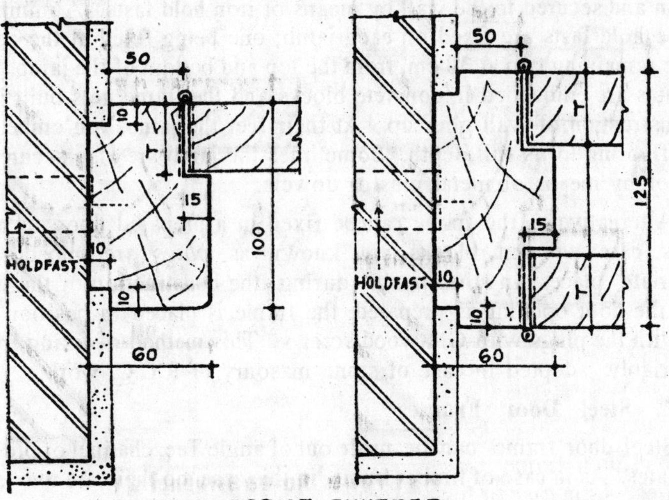

Fig. 17·2 Sectional plan showing details for single and double shuttered door frames

1. Timber Door Frame

In case of doors having one shutter, the size of the jamb or head is taken as 60 mm. thick and 100 mm. wide. In case of doors having two shutters, the width of the frame members is increased from 100 mm. to 120 or 140 mm. Typical sections for the jamb and head of door frame are shown in Fig. (17·2).

The members are rebated for accommodating plaster and the shutter. The rebate for accommodating the plaster should not be less than 10 mm. Similarly the rebate for accommodating the shutter should not be less than 15 mm. in width, the depth being equal to the thickness (T) of the shutter.

Details of frames having (*i*) shutter on one side and (*ii*) shutters on either side (kitchen door having double shutter *i.e.* fly proof shutter on one side and panelled shutter on the other), are shown in Fig. 17·2.

The jambs are usually tenoned into the head and wedged. Sometimes the head projects out by about 7 cm. from the jambs. The projected portion is known as *horn* which helps in securing the frame head to the masonry. However horns create difficulties at the time of repair or replacement of the frame and as such provision of horns is getting obselete these days.

Alternatively the head may be tenoned into the mortices made in the jambs and pinned with 8 mm. ϕ nails or 10 mm. ϕ hard wood bamboo pins.

Method of fixing. Before fixing, all portions of the frame which will come in contact with masonry, are painted with thick coat of coaltar or by any approved type of wood primer. The frame is placed in proper position and secured to the wall by means of iron hold fasts. A minimum of three hold fasts are fixed on each jamb; one being fixed at the centre and the remaining two at 30 cm. from the top and bottom of the jamb. The hold fasts are embedded in concrete blocks and the frame gets built in as the construction of wall goes up. At their feet the jambs are embedded in the flooring to its full depth. Sometimes the jamb are also secured to the floor by means of metal pins or dowels.

Alternatively the frame can be fixed in a prepared door opening. In this case wooden blocks are known as *plugs* are embedded at appropriate places in the jambs during the construction of the wall. After the door opening is prepared, the frame is placed in position and fixed with the plugs with G.I. wood screws. This method of fixing frame is invariably adopted in case of stone masonry or R.C.C. work.

2. Steel Door Frames

Steel door frame can be made out of angle Tee, channel or pressed steel plates. As in case of timber frame, in this case too there are two jamb

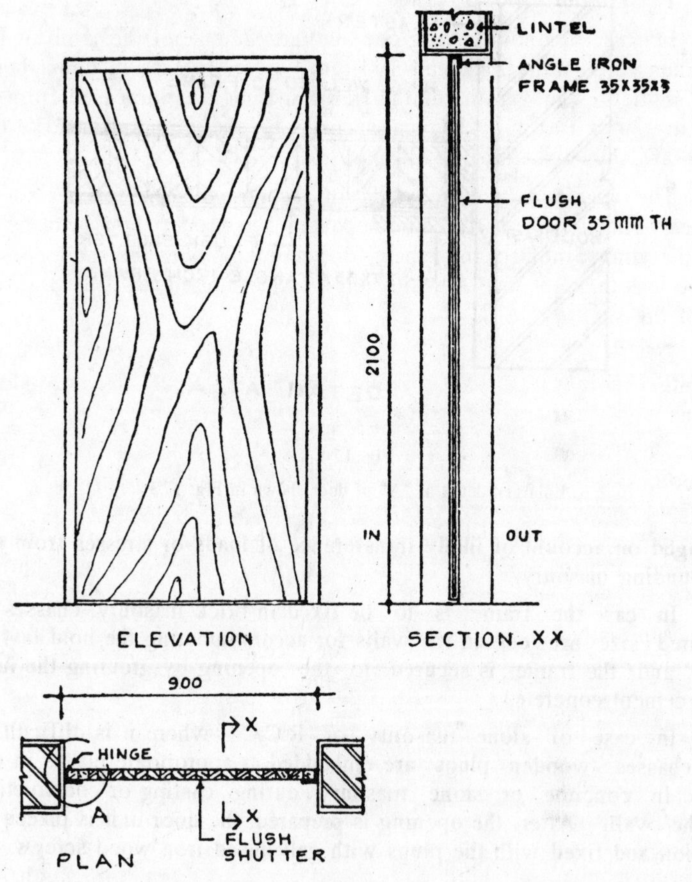

Fig. 17·3

Details of door with Steel frame

posts and a head. At the joints the members are either welded are rigidly fixed together by mechanical means. Sometimes a horizontal member (angle iron) is also provided to fix the jambs at base at floor level. This member is known *threshold*. In case of steel frames the hold fasts or lugs and the hinges are invariably welded to the frame.

Method of fixing. Steel frames are always fixed in prepared door opening *i.e.* after the masonry work for the door opening as well as lintel etc. are complete and fully set. In addition the size of door frame in this case is kept slightly less than the size of the door opening so as to allow adequate clearance between the frame and the opening. This arrangement is necessary to ensure that the steel frame does not get distorted or

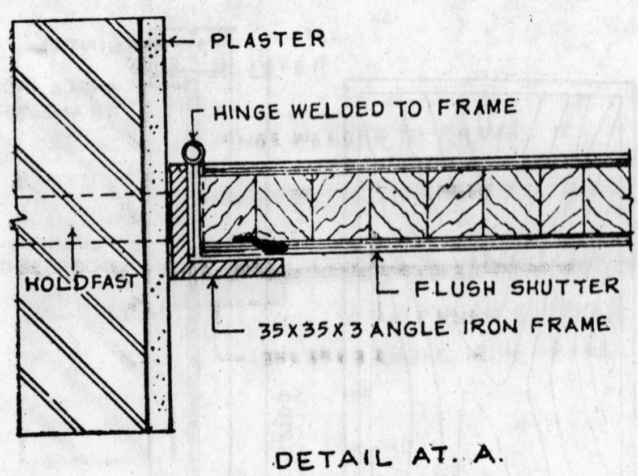

Fig. 17·4

Enlarged detail at "A" of door shown in Fig. 17·3

damaged on account of likely transference of loads or stresses from the surrounding masonry.

In case the frame is to be fixed in brick masonry, chasses of required size are cut in the walls for accommodating the hold fast or lugs, and the frame is secured to the opening by grouting the lugs with cement concrete.

In case of stone masonry or R.C.C., where it is difficult to cut chasses, wooden plugs are embedded at appropriate places in the jamb in concrete or stone masonry during casting or construction of the wall. After the opening is prepared, the door unit is placed in position and fixed with the plugs with galvanised iron wood screws.

17.5. TYPES OF DOORS

Depending upon the type of material used, arrangement of different components, method of construction and nature of working operation, doors can be broadly classified into different types as under :

(1) *Battened and ledged doors.*

(2) *Battened ledged and braced doors.*

(3) *Battened ledged, braced and framed doors.*

(4) *Framed and panelled doors.*

(5) *Glazed or sash doors.*

(6) *Flush doors.*

(7) *Fly proof doors.*

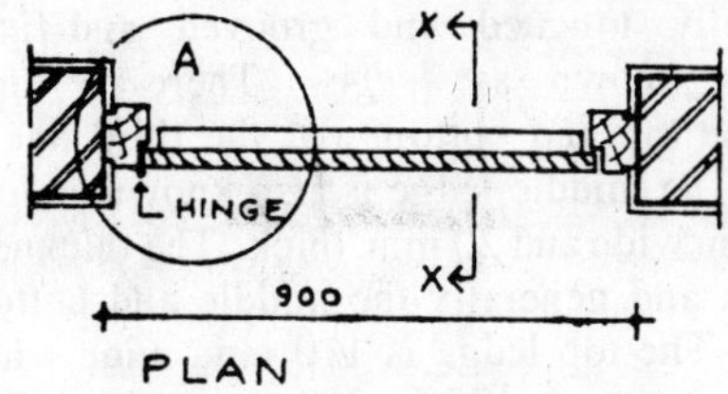

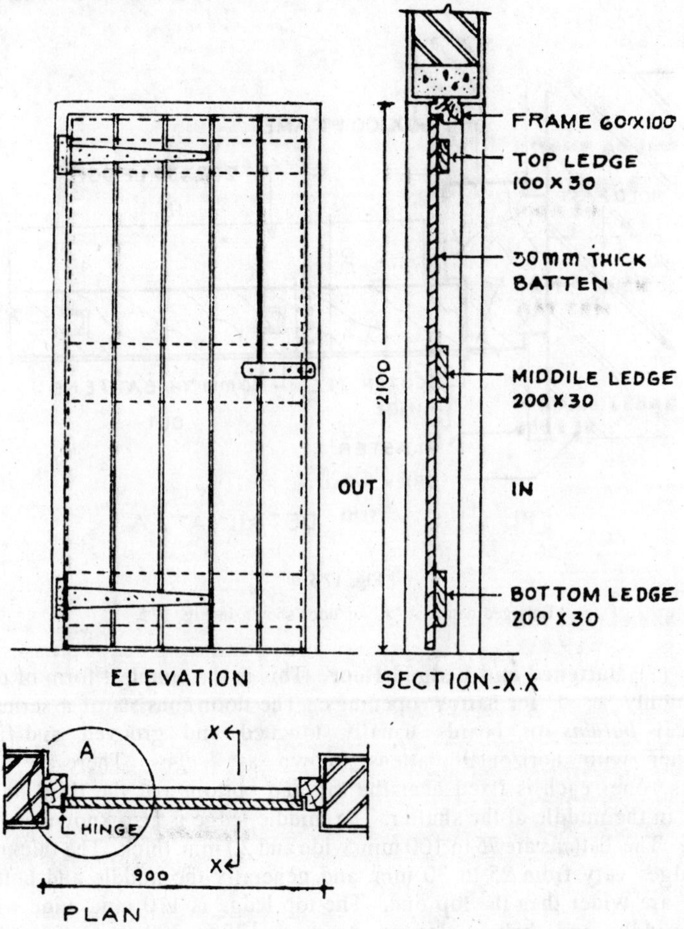

Fig. 17·5

Details of Battened and Ledged door

(8) *Revolving doors.*

(9) *Sliding doors.*

(10) *Swing doors.*

(11) *Collapsible steel doors.*

(12) *Rolling steel shutter doors.*

(13) *Rolling grill doors.*

(14) *M.S. sheet doors.*

(15) *Fire check door.*

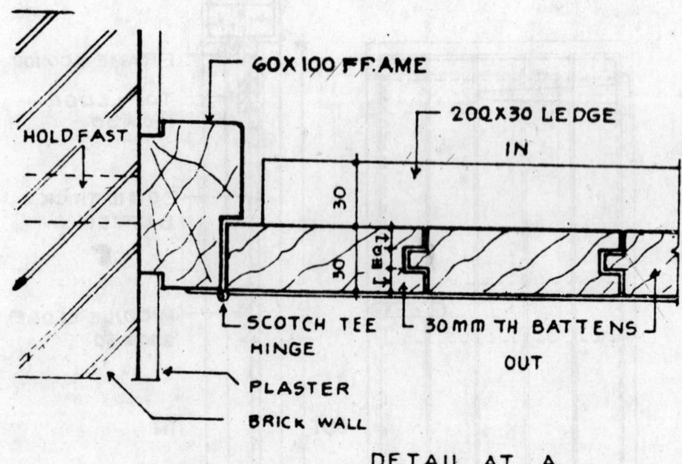

Fig. 17·6

Enlarged detail at "A" of door shown in Fig. 17·5

(1) **Battened and Ledged Door.** This is the simplest form of door commonly used for narrow openings. The door consists of a series of vertical *battens* or boards usually tongued and grooved and fixed together with horizontal battens known as *ledges.* There are three ledges one each is fixed near the top and bottom and the third one is fixed in the middle of the shutter. The middle ledge is also known as *lock ledge.* The battens are 75 to 100 mm. wide and 20 mm. thick. The thickness of ledges vary from 25 to 30 mm. and generally the middle and bottom ledge are wider than the top one. The top ledge is 110 mm. wide while the middle and bottom ledges are each 175 to 200 mm. wide. The door is hung on the frame by iron hinges (called T-hinges or cross garnets) fixed on the ledges.

This type of door, is recommended for W.C's, bathrooms etc. in buildings where economy rather than the appearance is of main consideration.

(2) **Battened ledged and braced door.** This is an improved form of battened and ledged door. In this type of door two additional members are provided in the form of inclined *braces* between the ledges as shown in (Fig. 17·7). The braces are normally housed and not tenoned into the ledges. The braces incline down towards the side on which the door is hung. In this position they act as struts and provide increased rigidity and strength to the door. This type of door can be used for wider opening in situation where appearance is not so important as economy.

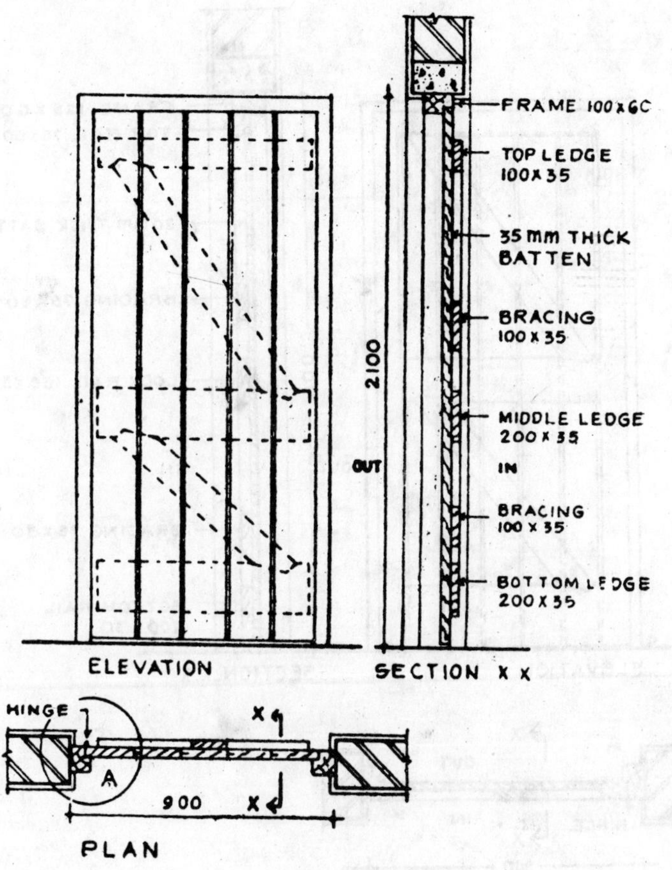

Fig. 17·7

Details of Battened, ledged and braced door

The sizes of different components of the door are as under.

(*i*) Battens : 100 to 175 mm. wide and 20 mm. thick.

(*ii*) ledges : 175 to 225 mm. wide and 25 to 30 mm. thick.

(*iii*) Braces : 125 to 175 mm. wide and having same thickness as that of ledges.

(3) **Battened, ledged, braced and framed door.** This type of door is considered to the superior in strength, durability and appearance to other two type of doors mentioned above. The frame work of this type of door consists of two vertical members called *stiles*, three horizontal members known as ledges or *rails*, vertical battens and two

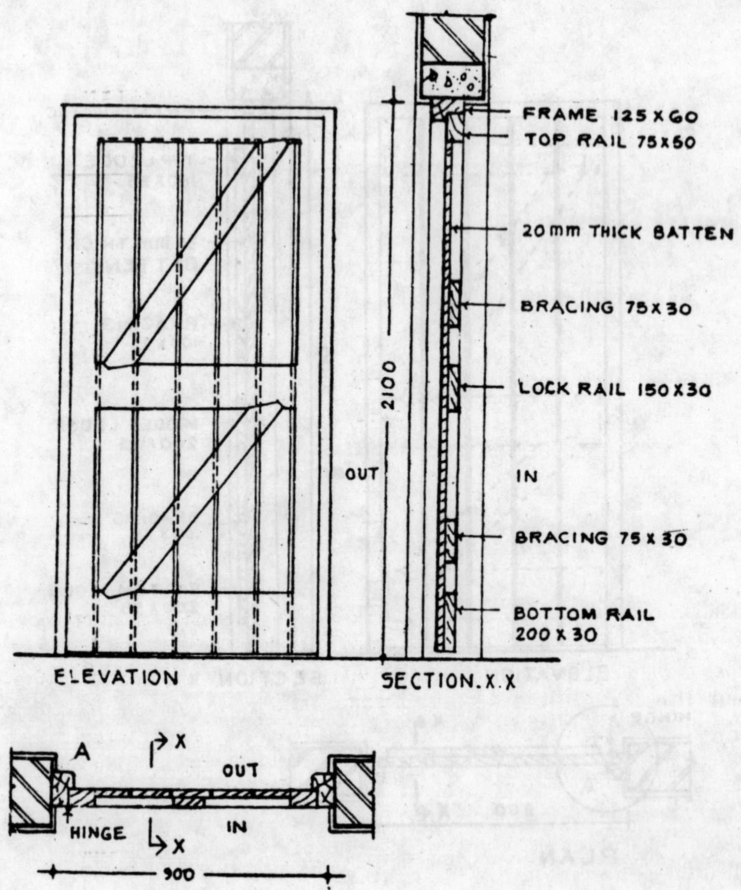

Fig. 17·8

Details of battened, ledged, braced, and framed door

inclined braces. The rails positioned near the top, middle and the bottom of the shutter are morticed and tenoned into the stiles. The battens are generally tongued and grooved and V-jointed. At their upper end, the battens are let into the top rail while at mid height and lower ends they go right over the middle and bottom rails. The side battens are tongued into the stiles. The braces are normally housed into the rails at about 4 cm. from the stiles. In order to have this arrangement of jointing, the thickness of the stiles and top rail is kept same and is equal to the combined thickness of braces and battens.

This type of door is considered suitable for external use as well as

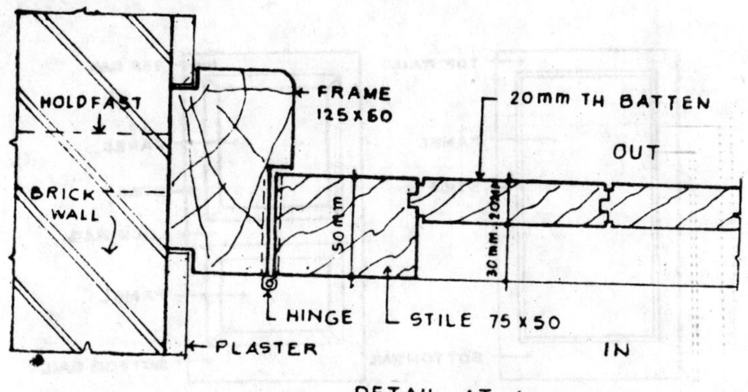

Fig. 17·9

Enlarged detail at "A" of door shown in Fig. 17·8

in situations where the door is likely to be subjected to rough handing.

(4) **Framed and Panelled Door.** This type of door is commonly provided in all types of buildings. This door consists of timber frame work of stiles and rails which are grooved on the inside to receive one or more *panels.* Framed and panelled doors are made in several designs to suit the functional as well as architectural requirements. The door may be *single, double, three, four* or *six panelled* and so on. The panels can be of timber, plywood, block board or hard board. Timber panels are fixed within the grooves in the frame. However, additional timber beading is also provided either on one or both sides to improve the elevation of the door. Commonly adopted designs of framed and panelled doors are shown in Fig. 17·10.

A review of the figures will indicate the following important construction details.

(*i*) The stiles are continuous from bottom to top.

(*ii*) Different rails, *i.e,* top rail, bottom rail, lock rail and frieze rail are jointed to the stiles.

(*iii*) Mullions or muntins, if any, are jointed to the rails.

(*iv*) Bottom and lock rails are of bigger size than the top and frieze rails.

(*v*) The lock rail is so placed that its centre line is at height of 800 mm. from the bottom of the shutter.

(*vi*) The stiles and the rails are mortised and tenoned.

(*vii*) In case, the panel inserts are of timber, the minimum width and thickness of timber panel should not be less than 150 mm.

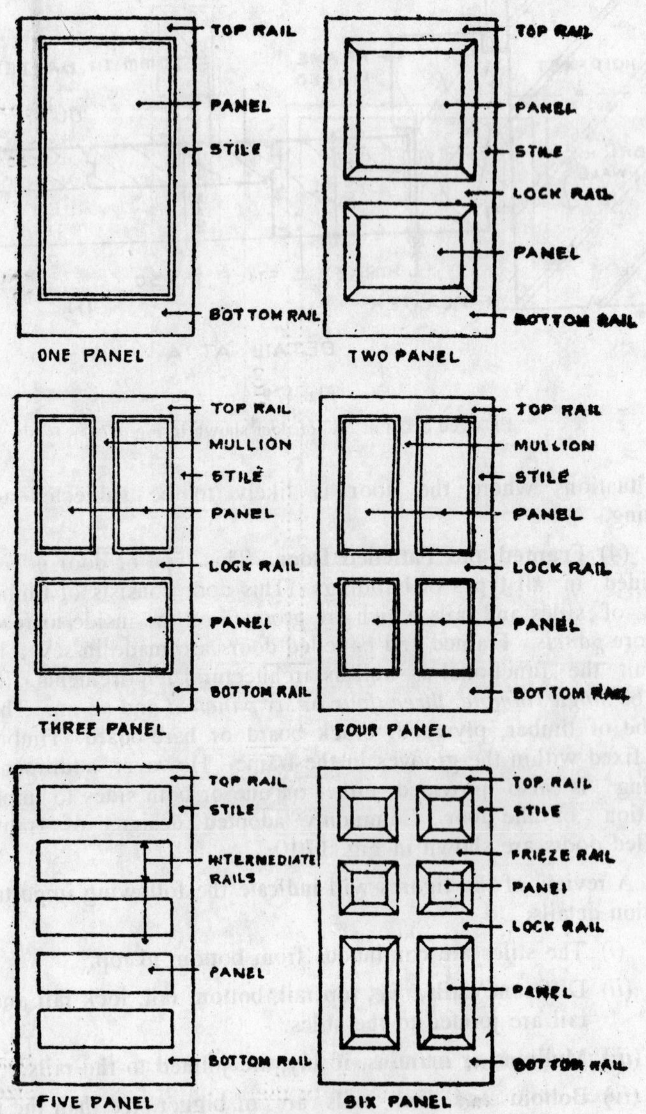

Fig. 17·10
Details of different types of panelled doors

x 15 mm. The maximum area of a single panel in timber should not be more than 0·5 m^2.

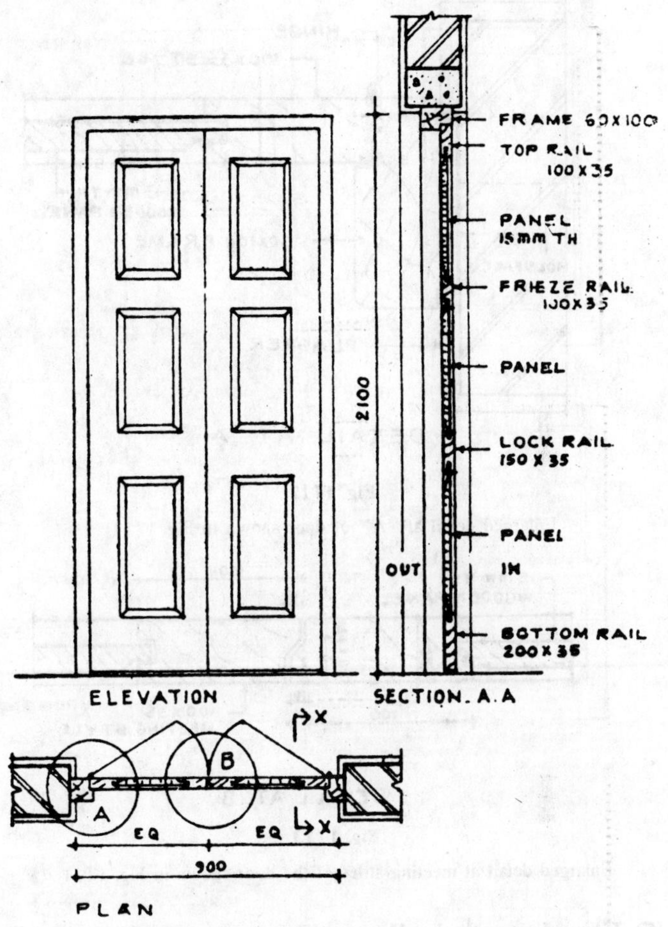

Fig. 17·11
Details of a six panelled door with wooden panels

(*viii*) In case of plywood, particle board or hard board panels, there is no restriction or minimum or maximum size of panel. However, for single panel door, the thickness of plywood, should not be less than 12 mm. and for two or more panelled door it should not be less than 10 mm.

(*ix*) The width of stiles is normally 10 cm. and the bottom and lock rail are 15 cm. wide each. The thickness of shutter can vary from 3 cm. to 5 cm. depending upon the size of door, thickness of panels, the situation of door and type of usage.

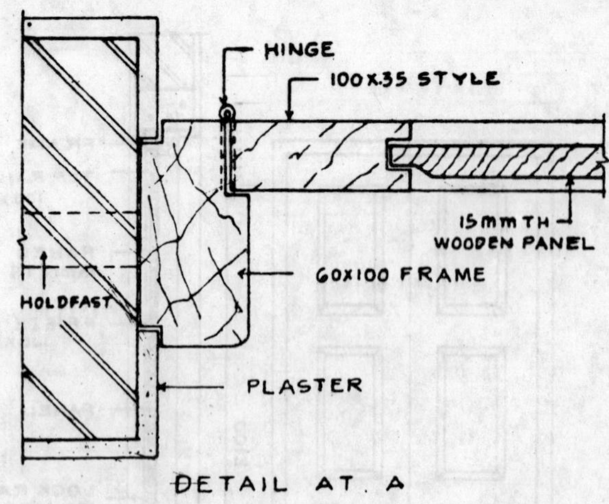

Fig. 17·12

Enlarged detail at "A" of door shown in Fig. 17·11

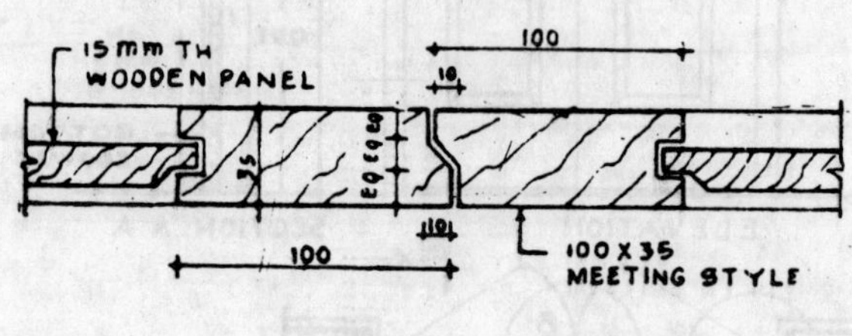

Fig. 17·13

Enlarged detail at meeting stiles of the door shown in Fig. 17·11

(5) Glazed or Sash Door. This type of door is used in residential as well as public buildings like hospitals, schools or colleges etc. with a view to supplement the natural lighting provided by windows or to make the interior of one room visible from the adjacent one. Glazed or sash doors are made in several designs. They can be glazed fully or partly. Fully glazed doors are recommended in situations where sufficient light is required through the door opening as in case of shopping or display centres, lobbies, entrance hall etc. In case of partly glazed doors normally the bottom 1/3rd height of the door is panelled and the remaining 2/3rd height is glazed. In case of partly glazed doors, the stiles are sometimes diminished at the lock rail to improve the elevation and to permit more area of glazing in the process. The diminished stiles are known as gun stock stiles (Fig. 17·18).

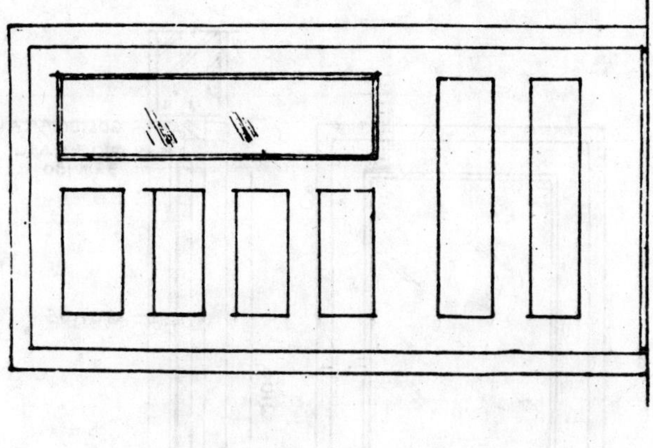

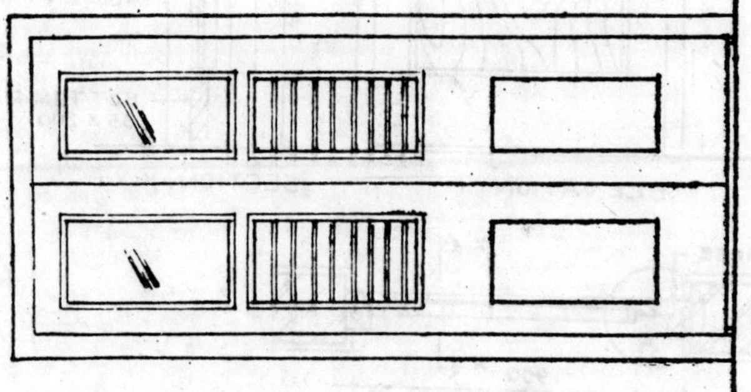

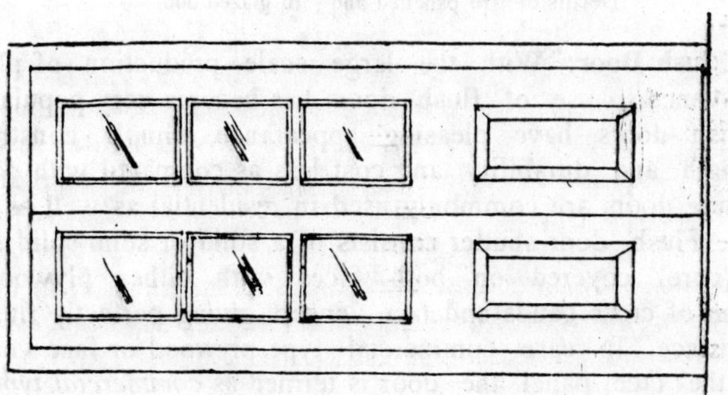

Fig. 17·14

Various forms of panelled and glazed doors

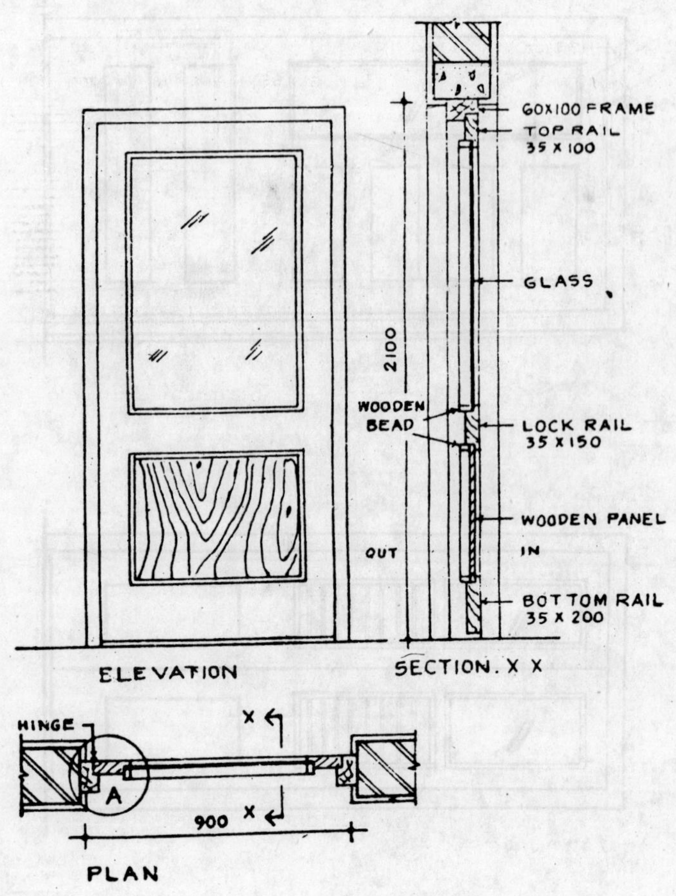

Fig. 17·15

Details of $\frac{1}{3}$rd panelled and $\frac{2}{3}$ rd glazed door

(6) Flush Door. With the large scale production of plywood and face veneers, use of flush doors has become very popular these days. Flush doors have pleasing appearance, simple construction, high strength and durability and cost less as compared with panelled doors. These doors are commonly used in residential as well as public buildings. Flush door shutter consists of a solid or semi-solid framed skeleton (core) covered on both faces with either plywood or a combination of cross bands and face veneers giving perfectly flush and jointless surface. In case commercial type plywood or face veneer is used for the face panel the door is termed as *commercial type flush*

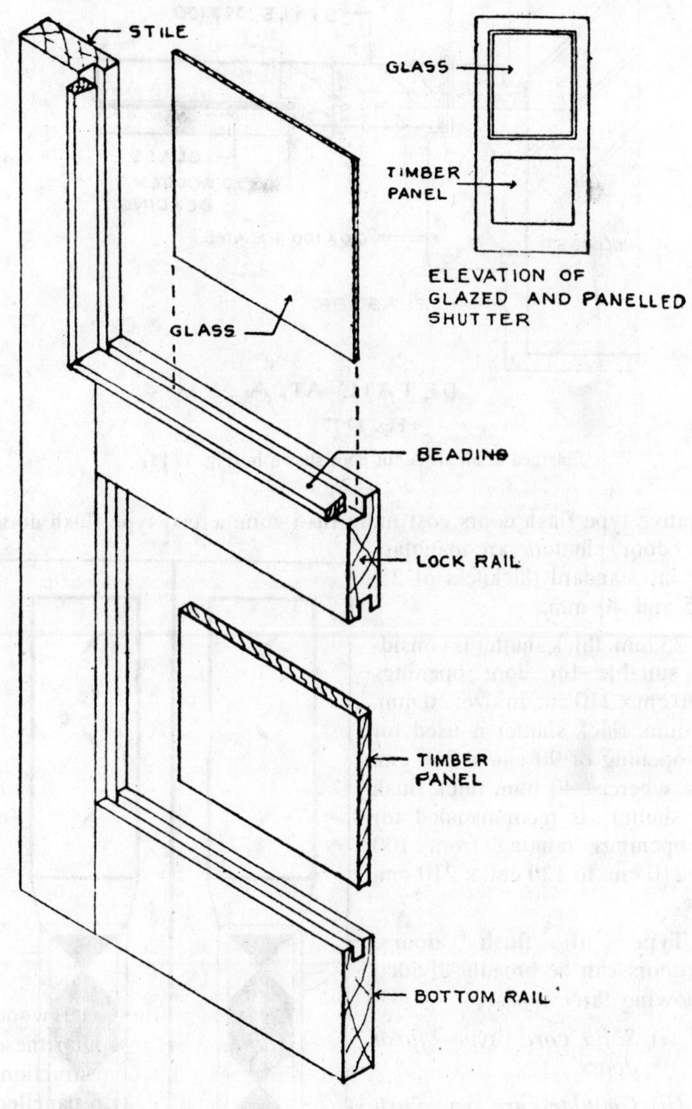

Fig. 17·16

View of the door shown in Fig. 17·15 showing arrangement of fixing glass and timber panel

door. Similarly if decorative type of plywood or face veneer is used for face panel, the door is named as *decorative type of flush door*.

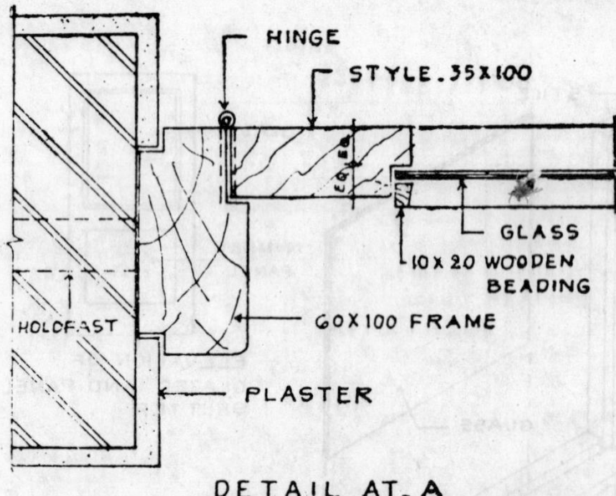

Fig. 17·17

Enlarged detail at "A" of door shown in (Fig. 17·15)

Decorative type flush doors cost more than commercial type flush doors. Flush door shutters are manufactured in standard thickness of 25, 30, 35 and 40 mm.

25 mm. thick shutter is considered suitable for door openings upto 80 cm x 210 cm. in size; 30 mm. or 35 mm. thick shutter is used for door opening of 90 cm. x 210 cm. in size whereas 40 mm. thick flush door shutter is recommended for door openings ranging from 100 cm. x 210 cm. to 120 cm. x 210 cm. in size.

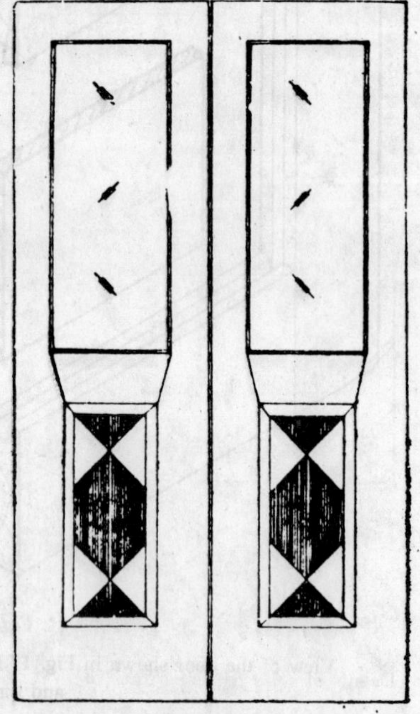

Fig. 17·18

Elevation of a panelled and glazed door with gun stock styles

Type of flush doors. Flush doors can be broadly divided in following three types.

(*i*) *Solid core type flush doors.*

(*ii*) *Cellular core type flush doors.*

(*iii*) *Hollow core type flush doors.*

(*i*) **Solid Core Type Flush Door.** In this type, the wooden frame used for holding the core

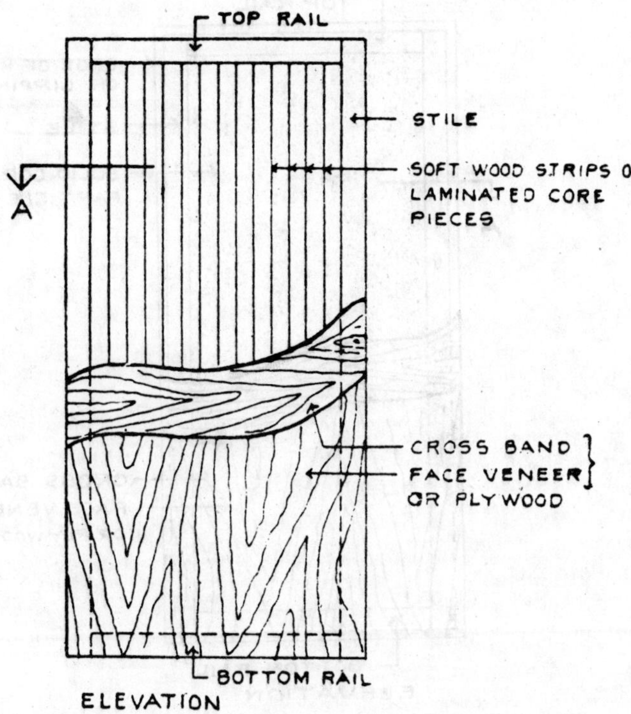

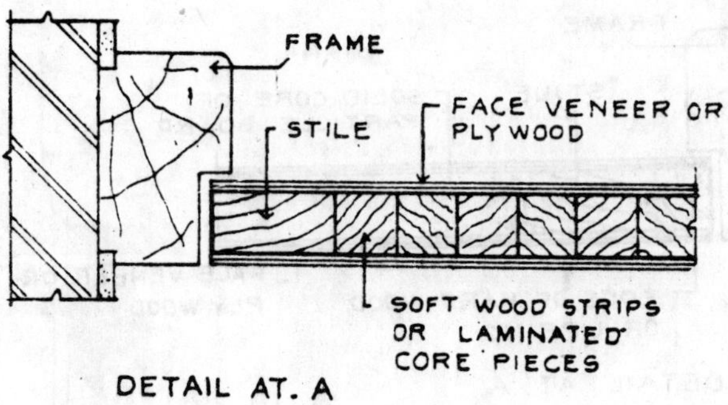

Fig. 17·19

Details of solid core type flash door (solid core consisting of laminated core pieces

consists of stiles, top rail and bottom rail, width of each member being not less than 5 cm. The frame is made out of selected species of timber. Where it is not possible to use same species of timber for the frame,

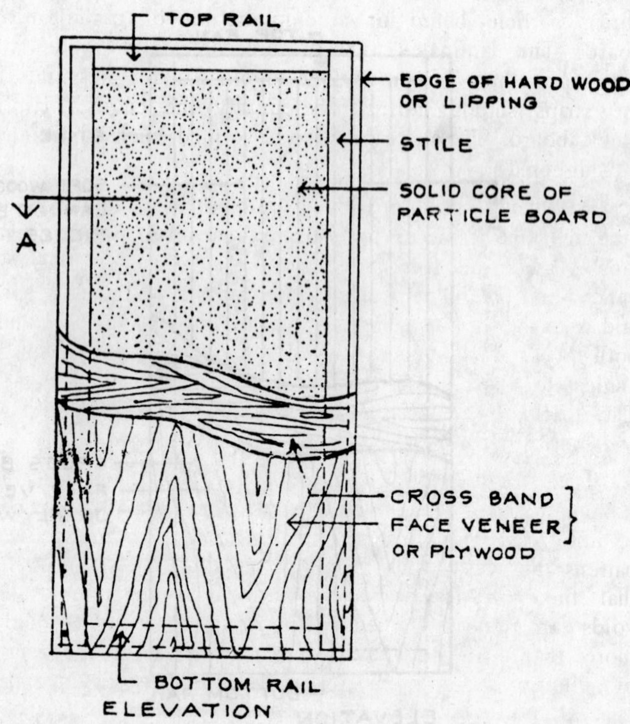

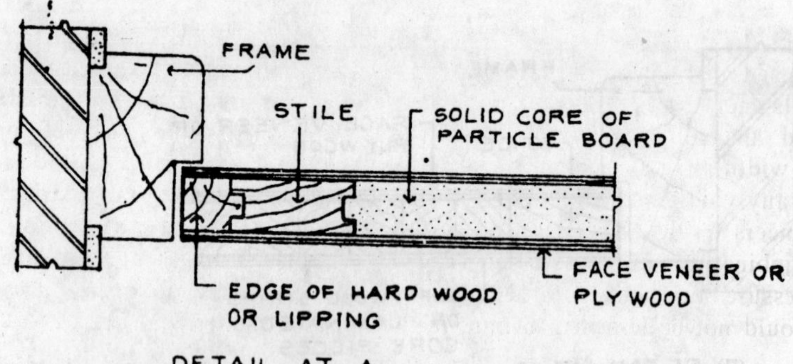

Fig. 17·20

Details of solid core type flush door (solid core consisting of particle board)

it is necessary to provide hard wood *lipping* around the frame to achieve required strength and durability. The width of the lipping should be equal to the thickness of the core and its depth should not be less than 25 mm.

The solid core for the door consists of block board (laminated core), particle board or a combination of particle board and block board. The laminated type of core consists of wooden strips of width not more than 25 mm, glued together under pressure. The length of the strips should normally be equal to the length of laminated core or block board. The strips can also be jointed, such joints should, however, be staggered.

Plywood sheets are glued under pressure to the assembly of core housed in the frame on both faces. Instead of plywood sheets, separate cross bands and face veneers can also be used. However, when cross bands and face veneers are glued separately, the cross bands should be laid with their grains at right angles to those of the core and glued on its both faces. Face veneers are thereafter laid with their grains at right angles to those of the cross bands and are glued to the cross bands. The thickness of the plywood or the combined thickness of cross band and face veneers should not be less than 3 mm.

(*ii*) **Cellular Core Type Flush Door**. In this type, the timber frame is constructed from stiles, top rail and bottom rail each not less than 75 mm. wide. The cellular core is formed by fixing wooden or plywood battens not less than 25 mm. in width to the frame in such a manner that the voids formed do not exeed 25 cm² in area. In addition the voids are uniformly distributed and total contents of the void is not more than 40% of the area of the shutter. The door shutter is formed by gluing, under pressure , plywood sheets or cross bands and face veneers on both faces of core. The thickness of the plywood should not be less than 3 mm.

(*iii*) **Hollow Core Type Flush Door**. In this type, the frame consists of stiles, top rail, bottom rails and minimum two intermediate rails-each not less than 75 mm. in width. The space between the stiles and the rails is divided by fixing wooden battens not less than 25 mm. in width in such a manner that the voids are equally distributed and area of any void remains less than 50 cm². Instead of battens, rolls or strips of veneers can also be used for forming core. Plywood sheets or combination of cross-bands and face veneers are then glued under pressure on both faces of the core. The thickness of plywood used should not be less than 6 mm. in this type of shutter.

(7) Fly Proof Door. This type of door is used to check the entry of flies, mosquitoes, insects etc. into the room and to allow free circulation of air at the same time. The door consists of timber framework of vertical stiles and horizontal rails and the opening of the panels are fitted with fine-mesh galvanised wire gauge. The G.I. wire is fixed by use of nails and timber beading. Generally the door opening in such a case is provided with double shutters. The shutter opening inside the room is fully panelled whereas the fly proof or wire gauge shutter opens out side the room or *vice versa*.

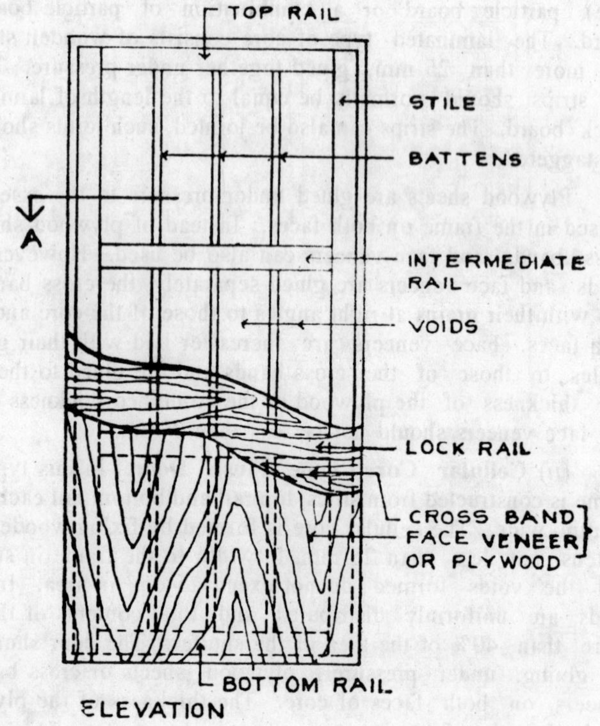

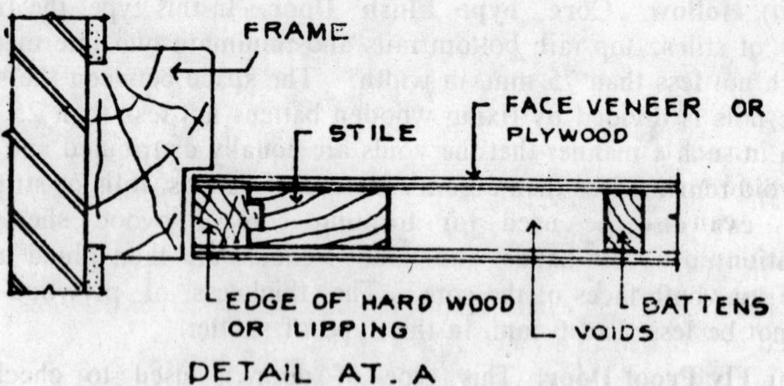

Fig. 17·21

Details of Hollow core type flush door

Fly proof doors are commonly provided for kitchens, canteens, refreshment rooms, cupboards for storing food, sweets, meat etc.

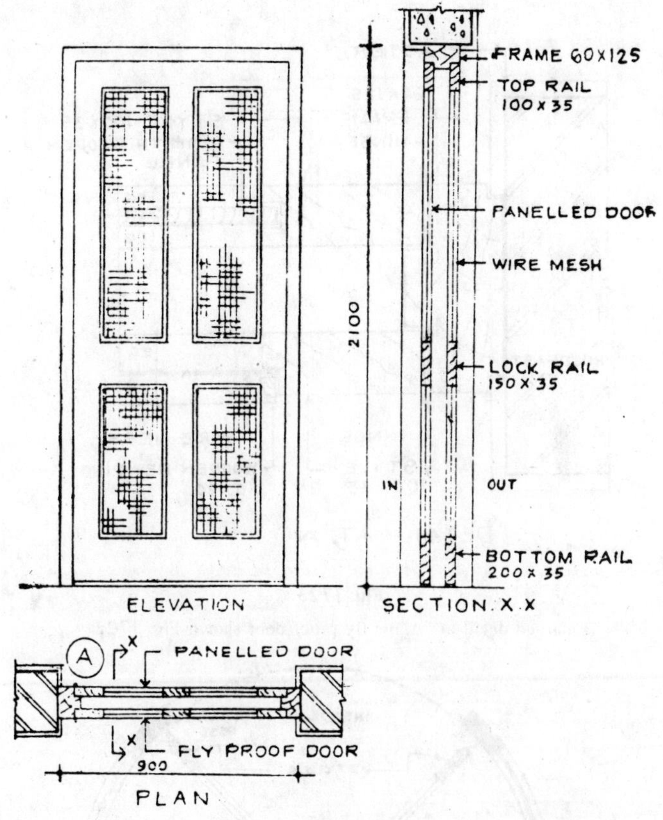

Fig. 17·22 Details fo Fly proof door

(8) Revolving Door. Revolving door provides entrance on one side and exit on the other simultaneously keeping the opening automatically closed when not in use. The door is so assembled as to exclude the wind draught. This type of door is advantageously provided in places where there is a regular foot traffic of people entering in and going out of the building especially when it is air conditioned or situated in a place where strong winds blow for most part of the year. The door essentially consists of four leaves radially attached to a centrally placed mullion in a circular opening. The mullion or the central member is provided with ball-bearing at the bottom and bush bearing at the top so as to enable the door to revolve smoothly without producing jerks. The leaves and the mullion are enclosed in a vestibule. At their edges the leaves are provided with rubber pieces which fit flexibly against the inside face of the vestibule. The attached leaves may

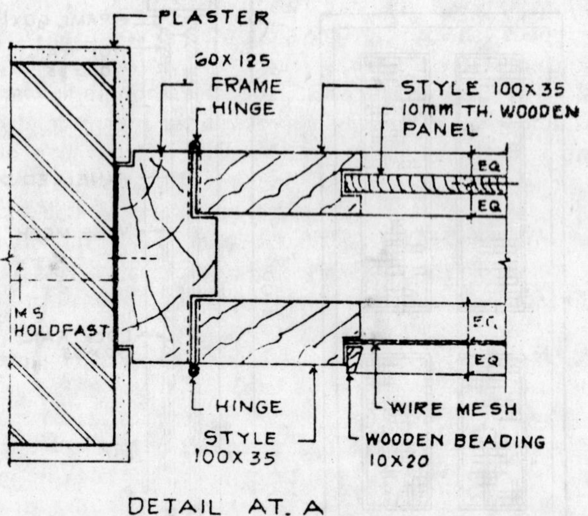

Fig. 17·23

Enlarged detail at "A" of fly proof door shown Fig. 17·22

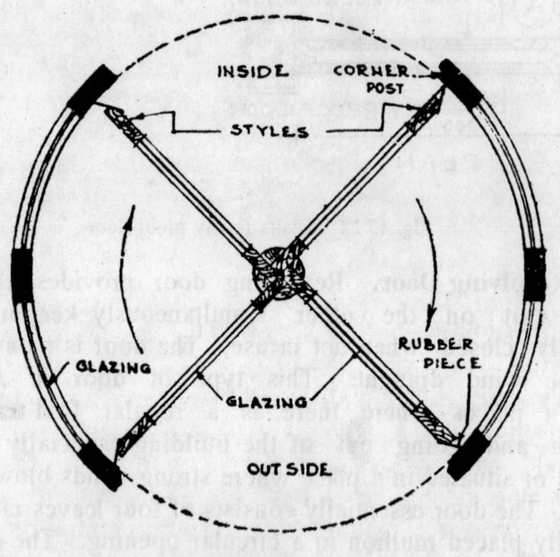

Fig. 17·24

Revolving door

be glazed, panelled or partly glazed and partly panelled. Such doors are commonly provided in hotels, banks, offices and other such important public buildings.

(9) M.S. Sheet Door. This type of door is recommended for railway goods sheds, garrage, godowns etc. where high degree of protection and safety is needed. The door is normally provided with two shutters. The door shutter is fabricated from angle iron or channel section frame which is suitably braced with the angle iron diagonal braces or with m.s. flats placed horizontally. Different members of the shutter frame are rivetted or welded at the junctions. In case of rivetted connection, gusset plates are necessarily used at each junction. Mild steel plates of required thickness are fixed to the shutter frame with rivets or welds. The outer frame of the shutters is provided with flat iron angle cleats for fixing the shutter to the door opening. Each cleat has a 25 mm. ϕ hole in the horizontal leg for supporting the shutter on pin clamps. Two pin clamps are used for each shutter. One end of the pin clamp is embedded in masonry and the other end has a 20 mm. ϕ m.s. pin riveted or welded to the clamp. The pin clamps are so placed that top pin faces downward and the bottom pin faces upwards so that the shutter can not be removed by lifting over the pins. The pins are made to pass through the holes in the cleats of the outer frame of door shutters and that is how the shutters get supported and hence attached to the jambs.

(10) Sliding Door. This type of door is considered suitable for shops, sheds, godowns, garrage etc. and in places where the use of hinges for fixing the shutter is to be avoided. Depending upon the size of the opening, and the space available on either side, the door can have single, double or more number of leaves or shutters. The door is provided with top and bottom guide rails or runners within which the shutters slide. The guide rails run past the opening for a distance equal to the width of the shutter so that when the door is required to be opened, the door shutter occupies a new position parallel to the wall face and clear off the opening. The shutters are also provided with locking arrangements, handles, stopper etc.

(11) Swing Door. This type of door is generally provided in passages of public buildings like offices, banks etc. The door may have single shutter or two shutters. The shutters are fixed with special hinges known as *double action spring hinges* which hold the shutter in closed position when not in use. Since these doors are pushed open, they should have glazed shutters so as to enable the users to see the objects on other side of the door and avoid accidents. As the springs return the door with force, the glazing should preferably be carried out by use of wired glass or it should be protected by other suitable means. Such doors should not be rebated at the meeting stiles. The closing edges of the meeting stiles should be segmental. Alternatively the door shutter should have a peep hole, fixed with glass. The peep hole should be at eye level. Details of a swing door with flush door shutter and peep hole are shown in Fig. 17·28—17·29.

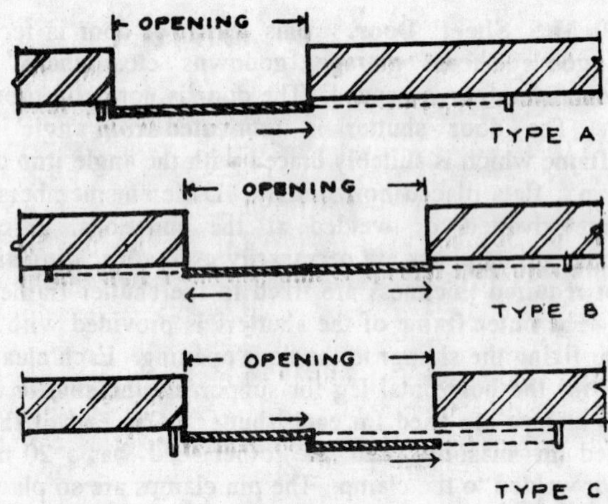

Fig. 17·25

Plan showing alternative arrangements of shutters in a sliding door

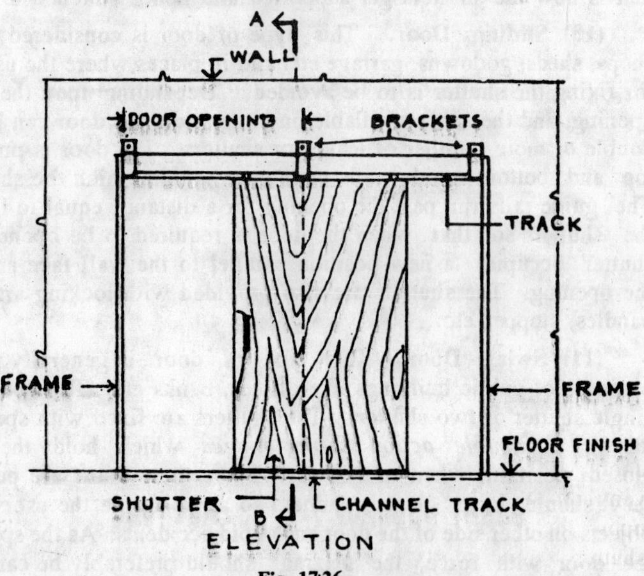

Fig. 17·26

Elevation of a single shutter sliding door

(12) Collapsible Steel Door. These doors are extensively used for shops, garrages, public buildings, godowns etc. and in situations where width of opening is large and provision of hinged shutter becomes difficult for want of space. These doors are also used from consideration

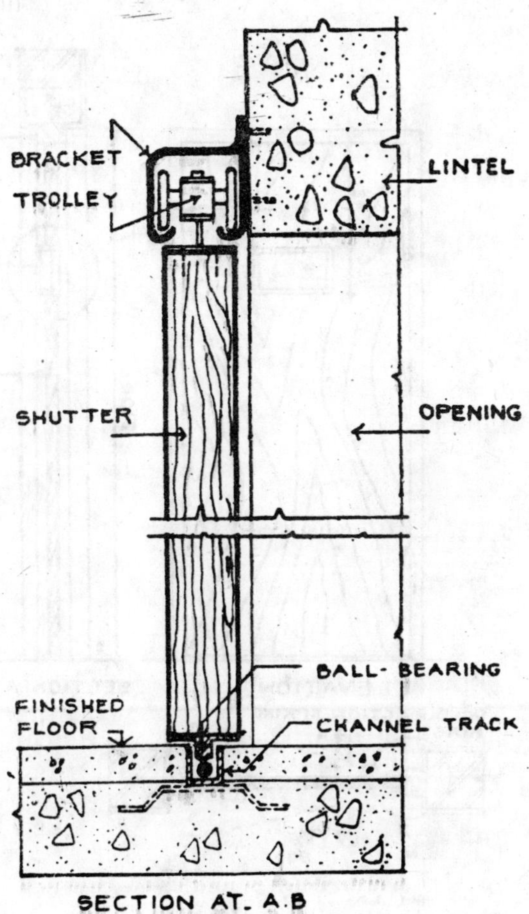

Fig. 17·27

Enlarged details at top and bottom of the door shown in Fig. 17·26

of increased safety and protection of the property. Collapsible steel doors are commonly recommended in situations where light and ventilation are desired even when the opening is closed. Depending upon the size of the opening, the door may have single or double shutters. The door essentially consists of vertical double channels each 20 x 10 x 2 mm. in size and spaced at 10 to 12 cm. apart. The channels are joined together with the hollows of channels on the in side and are braced with flat iron diagonal 20 mm. wide and 5 mm. thick which allow the shutter to open out or get folded. The shutters operate between two T-iron rails, one fixed to the floor and the other to the lintel by means of anchor bolts. The door shutter slides over roller mounted at its bottom and

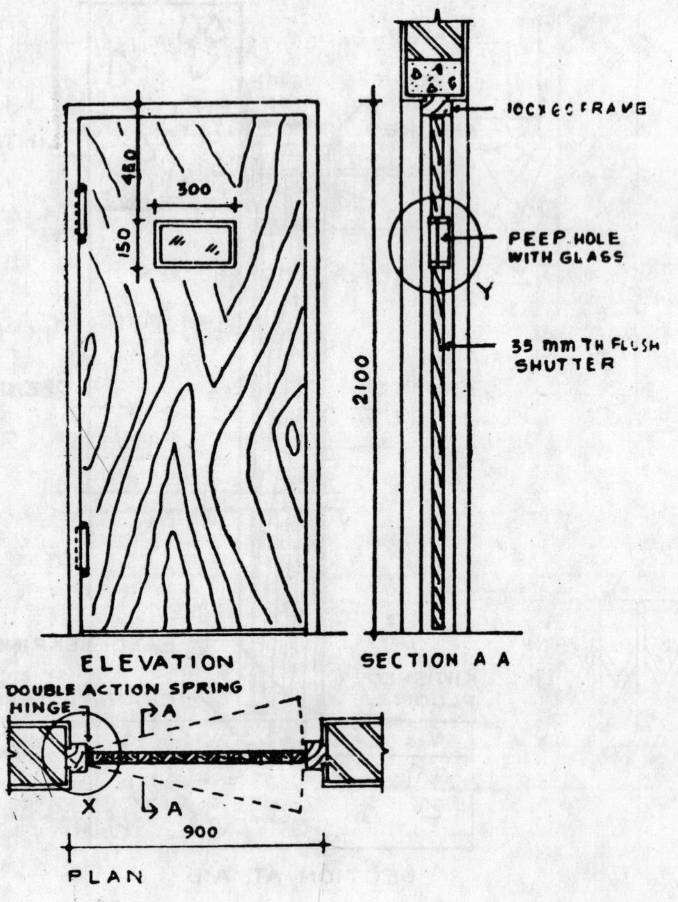

Fig. 17·28
Details of swing door

is held in position by the rails (Fig. 17·30). The door is provided with locking arrangements, handles, stoppers etc. and it can be opened or closed manually by slight push or pull.

(13) Rolling Steel Shutter Doors. These doors are commonly used for show windows, stores, shop fronts and godown etc. The door shutter acts like a steel curtain and provides adequate protection and safety against burglers and fire. The shutter consists of thin steel slabs (also known as *laths or slates*) about 1·25 mm. thick interlocked to each other and coiled upon specially designed pipe shaft (also known as *drum*) mounted at the top of the opening. The door shutter travels in two vertical steel guide channels installed at either end of the opening.

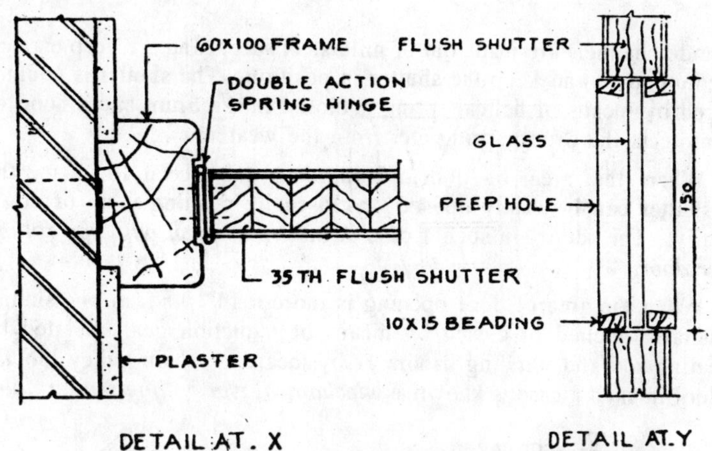

Fig. 17·29

Enlarged details at "X" and "Y" of the swing door shown in Fig. 17·28

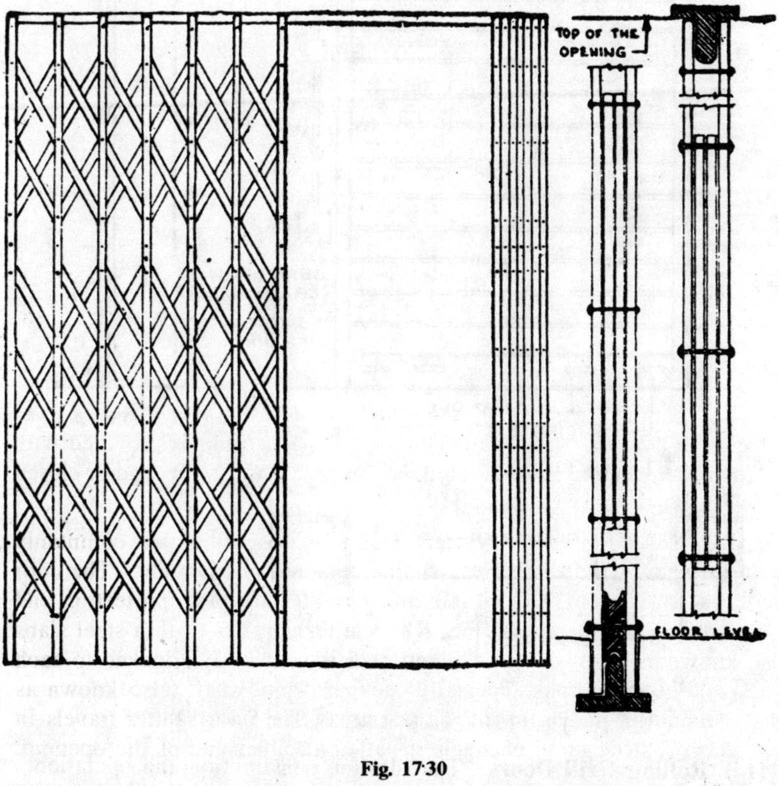

Fig. 17·30

Elevation and details at bottom and top of an opening provided with a collapsible steel door

The guide channels are made out of mild steel sheets and are deep enough to accommodate and keep the shutter in position. The shutter is counter balanced by means of helical spring enclosed in the drum and a hood of steel protects the drum, spring etc. from the weather.

When the area of door opening does not exceed 10 sq. m., the door shutter can be easily opened or closed by pushing it up or down manually. The door in such a case is known as *push pull type rolling shutter door.*

When the area of door opening is more than 10 sq. m. the shutter is generally opened or closed by means of reduction gear operated by connecting rod and winding handle or by means of chain pulley blocks. The door in such a case is known as *mechanical type rolling shutter door.*

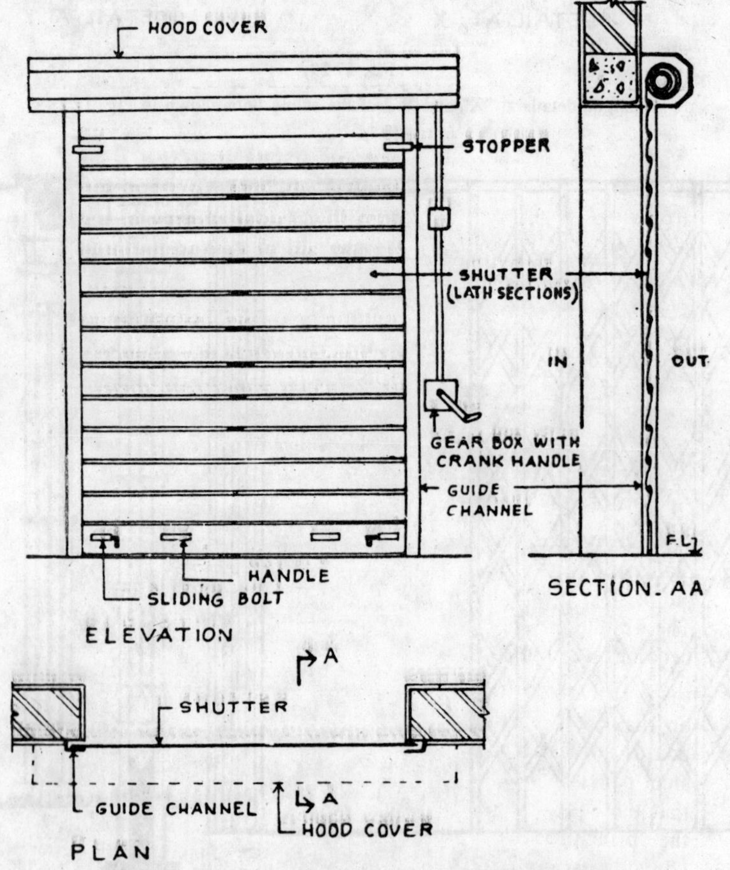

Fig. 17·31 Details of Rolling Steel Shutter Door

(14) Rolling Grill Doors. The design, construction and operation of rolling grill door is similar to the rolling steel door in all respects except the type of shutter. In this case the grill shutter is built of aluminium alloy or steel links of 0.9 mm. thickness assembled on tubes or rods. The grills can also be made out of 8 mm. diameter mild steel or aluminium alloy round bars. The rolling grill shutter can be manufactured in a variety of designs to suit the individual's choice. This type of door is used in situations where besides protection and safety, visibility and ventilation is also desired.

In some cases like transformer rooms, substations etc., certain amount of ventilation is needed through the door along with increased magnitude of proteetion and safety. This is achieved by providing rolling steel door with small portion of the shutter at top or bottom being made up of rolling grills. This type of door is known as *rolling shutter cum grill door.*

(15) Fire Check Doors. Fire check doors are required to control and restrict the spread of fire through the door openings thereby minimising the damage of the adjoining property in the event of fire. By provision of fire check doors it is possible to prevent the spread of fire to communicating rooms and floors and to reduce the chimney effect (which occurs in tall buildings) thereby ensuring desired degree of fire protection to the building. The design of the fire check doors and the materials to be used in their construction should be such that doors are capable of resisting the intense heat of fire without distortion and its thermal conductivity should be sufficiently low to prevent the ignition of the combustiable materials on either side. The fire resistance of the doors should as far as practicable be the same as that for the wall in which it is to be fitted. Normally for openings upto 1·5 meter wide, single leaf door shutter is used and for wider openings two leaves shutters is used. The doors can be of sliding, hinged, manually operated or automatic type.

The fire check doors can be of the following two types :

(1) Steel plate doors.

(2) Metal covered doors.

In case of the steel plate door, the shutter is made out of the steel plates of 6 mm. or more in thickness with stiles and rails on each face made out of steel plate 6 mm. thick and 10 mm. wide riveted or welded together. The arrangement of the plates/rails should be such that the door shutter becomes 18 mm. or more in thickness continuously along each edge.

In case of metal cover doors, the core of the shutter consists of 3 or 4 layers of plained, tounged and groved seasoned teak or yellow pined boards not less than 22 mm. in thickness. The core is incased in the sheet steel panels which have tightly folded joints to exclude air and to ensure that the core does not ignite. This type of door is constructed based on the principle that wood does not burn when air is excluded but it is converted into charcoal. This type of door has added advantage of not only preventing the flames but also the heat passing from one side to another.

Fire check doors are used to safeguard against fire in Warehouses, Hotels, Banks, Departmental Stores and other Public Buildings.

17·6. WINDOWS

A window may be defined as an opening made in a wall for the purpose of providing day light, vision and ventilation. The construction of window is identical to that of door. Window consists of a window *frame* and *shutter*. The frame consists of two vertical members called jambs, one flat member connecting the jambs at top, known as head and another flat member connecting the jambs at their feet. known as *sill.* the shutter can be fully glazed, panelled and glazed or panelled type. Windows are normally provided with two leaves. The selection of size, shape, location and the number of windows to be provided in a room depends upon the following considerations.

(*i*) Size of room to be lighted.

(*ii*) Location of the room and its utility.

(*iii*) Architectural treatment is to be given to the building.

(*iv*) Direction of wind and its speed.

(*v*) Climatic considerations of the site such as humidity, temperature variation etc.

Based on the above, following thumb rules have been framed to determine the area requirements for windows in a room. The size and hence the number of windows can thus be worked out, once total area required for window opening is known.

(*a*) Total area of window opening should normally vary from 10 to 20% of the floor area of the room. In hot and arid regions it could be 10 to 15% of the floor area. When the building is to be located in hot and humid regions, the area of window openings should vary between 15 to 20% of the floor area.

(*b*) In case of public buildings like, schools, offices, factories, hospitals etc. the minimum area of windows should be 20% of the floor area.

(*c*) For adequate natural light, the total area of glass panes in windows should be at least 8% of the floor area.

17·7. DESIGN OF WINDOWS

The points to be kept in view while making provision for windows in a room are :-

(1) The size and number of windows should be sufficient to provide adequate light and ventilation in the room.

(2) Windows should be located opposite to each other where-ever possible. Windows provided on Northern side permit maximum day light without glare.

(3) The window sill should be placed at 75 to 100 cm. above the floor level. This is considered reasonable height for the inmates to have a convenient look outside. In case of bath and W.C., and the other situations where privacy is desired, window sill should be kept at 1·75 m above the floor level. The top of windows and doors should however be at the same height.

(4) Buildings in humid region need special attention for adequate ventilation of the room. The area of window opening in humid region should be maximum and the window should be so located as to permit maximum ventilation of the apartment.

(5) The shutter of windows in external walls should open outside. This checks entry of rain water inside the room.

(6) All external windows, specially the one's on ground floor, should be provided with mild steel round or square bars or steel grills to safe guard against theft.

(7) Windows in external wall should be provided with Chajja projections to prevent the entry of the rainwater in the room. Similarly it is desirable to slope the window sill towards the external face of wall. The sill should preferably be weathered and throated to throw the rain-water off the face of wall.

17·8. CLASSIFICATION OF WINDOWS

Depending upon type of material used, nature of operational movements of the shutters, location and the manner in which they are fixed, windows can be broadly classified as under.

(1) *Fixed windows*

(2) *Pivoted windows*

(3) *Double hung windows*

(4) *Sliding windows*

(5) *Casement windows*

(6) *Sash or glazed windows*

(7) *Louvered or venetianed windows*

(8) *Metal windows*

(9) *Bay windows*

(10) *Clerestory windows*

(11) *Corner windows*

(12) *Dormer windows*

(13) *Gable windows*

(14) *Sky lights*

(15) *Fan lights*

(16) *Ventilators*

(1) Fixed window : In this type of window the glass pane or the glazed shutter is permanently fixed in the opening in the wall. This type of window is used in situations where light or vision alone is needed as no ventilation is possible through fixed windows.

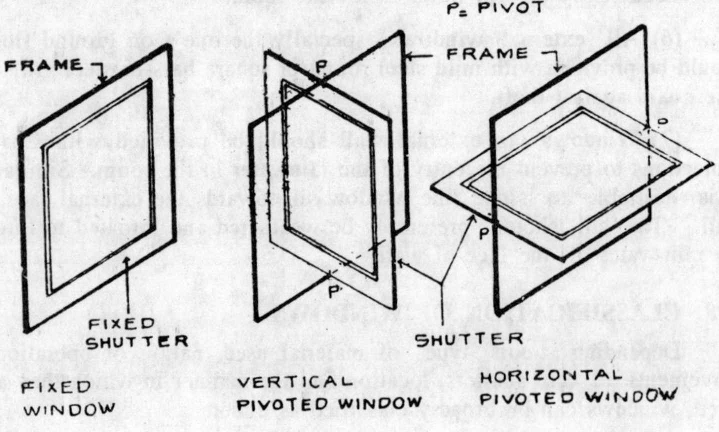

Fig. 17·32

Details of fixed and pivoted windows

(2) Pivoted window : In this type of window the shutter is capable of rotating about a pivot fixed to window frame. The window frame has no rebate. The shutter can rotate horizontally or vertically depending upon the position of pivot.

(3) Double hung window : This type of window consists of a pair of shutters arranged one above the other which can slide vertically within the grooves provided in the frame. A pair of metal weights connected by cord or chain passing over pulleys is provided for each shutter. By this arrangement the window can be opened at top or bottom to desired extent by pulling the metal weights suitably. Thus in this type of window it is possible to have controlled ventilation. In addition, the shutter can also be cleaned easily.

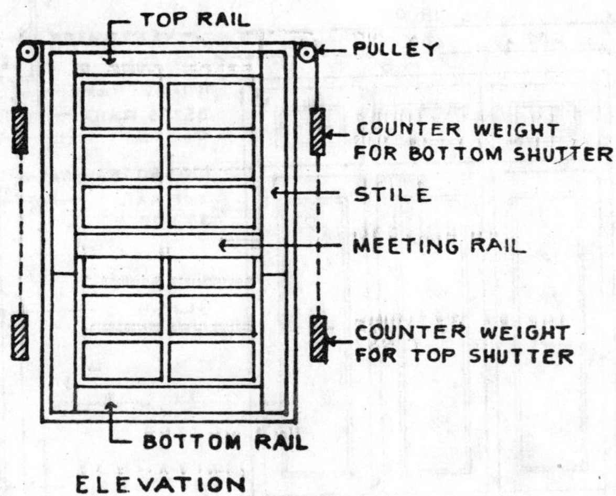

Fig. 17·33

Details of a Double hung window

(4) Sliding window : In this type of window the shutters move on roller bearing and can slide either horizontally or vertically.

(5) Casement window : Windows where shutters open like doors are called casement windows. Similar to doors, this type of window has frame and shutter composed of stiles, top rails, bottom rails, vertical and horizontal sash bars with glass panes. Depending upon the design, the frame of a casement window can also have additional vertical and horizontal members known as mullion and transome respectively.

(6) Sash or glazed window : In this case the window shutter consists of two vertical stiles, top rail and a bottom rail. The panel space between the stiles and the rails is suitably divided by small timber members or bars placed both horizontally and vertically or only horizontally to suit the design requirements. These bars are known as *sash bars* or *glazing bars*. Thus the sash bars divide the total panel space of window shutter into panels of small size. The sash bars have rebates for fixing glass panels. The glass panels are cut 1·5 to 3 mm. smaller in size than the size of the panel to permit slight movement of the sash bars due to temperature variation without damaging the glass panes. The glass panes are secured in position either by putty or by small fillets known as glazing beads.

(7) Louvered or venetianed window : In this type of window the stiles of shutter are grooved to receive a series of louvers which may be of glass or wood. The louvers are set within the grooves in inclined position so that they slope downward to the outside in order to run off the rain water and obstruct the horizontal vision at the

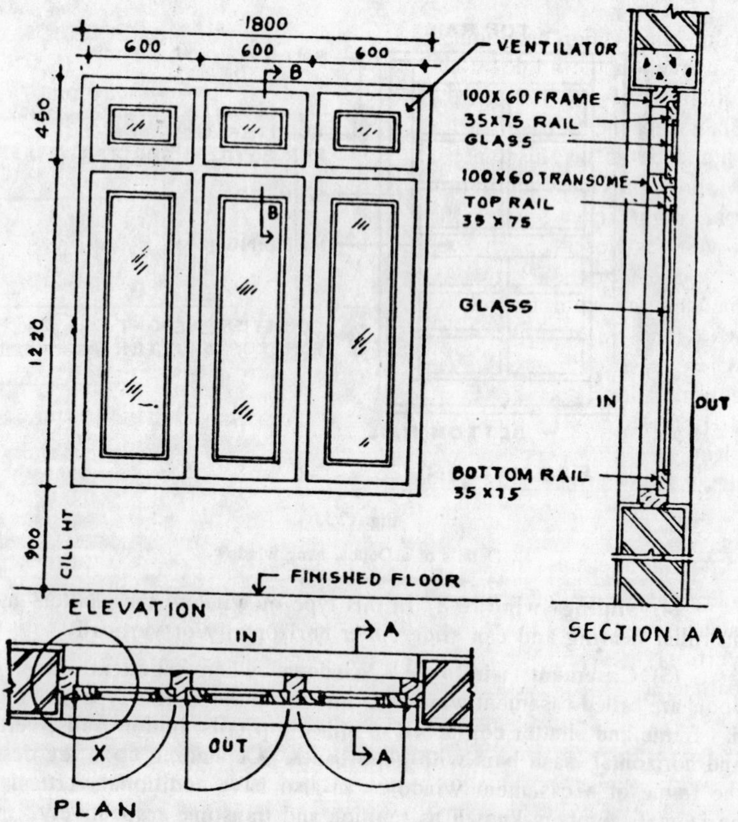

Fig. 17·34 Details of a casement window

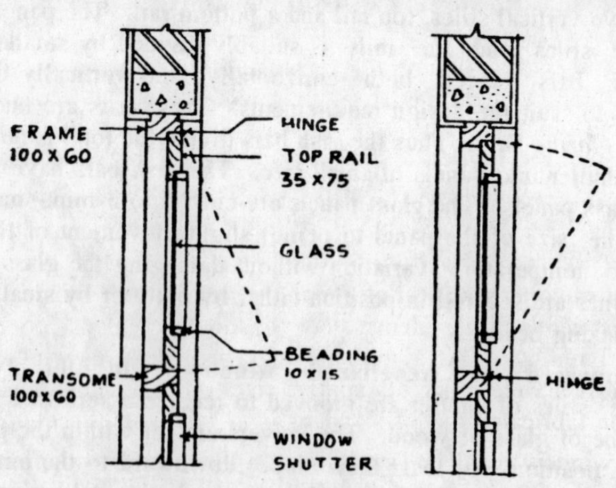

Fig. 17·35

Enlarged section thorugh 'B-B' (Refer Fig. 17.34) showing alternative methods of fixing ventilator shutter

same time. For economical construction the angle of inclination of the louvers or blades to the vertical stiles should be 45°. This type of window provides ventilation and light even when the opening is closed. Such windows are commonly recommended for bathroom, wc's. workshops, and other areas where privacy is the main consideration. It is however, difficult to maintain them clean as the dust gets easily deposited on the louvers. Sometimes the louvers or the blades are pivoted to the frame at both ends and in addition every blade is connected by a small hinge to a vertical batten. When the batten is pulled up, the gaps between the blades are opened and when the batten is pulled down, the gaps between the blades get closed.

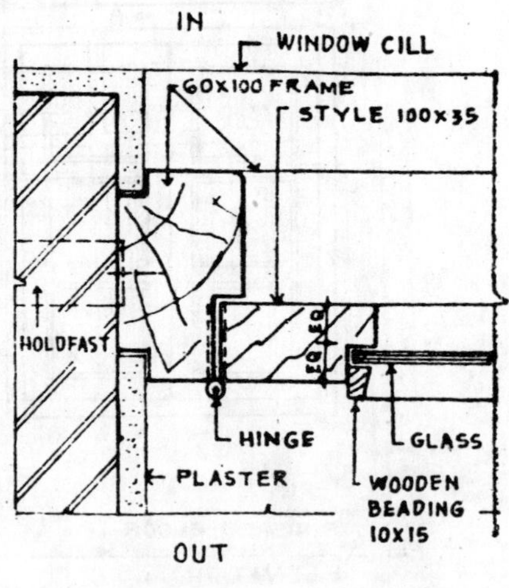

Fig. 17·36

Enlarged details at X (Refer Fi.g 17·34)

(8) Metal window. Use of metal windows is getting increasingly popular these days. Windows made of metals, like mild steel, galvanised mild steel, aluminium, bronze, stainless steel etc. are recommended especially for public buildings like offices, hospitals, schools etc. Bronze, aluminium and stainless steel are considered to be the best as they possess high degree of elegance, finishing, durability and are rust proof as well. However, Bronze and stainless steel being very costly, use of windows made out of these metals, is restricted to posh buildings where cost is not the prime consideration. As compared with bronze or stainless steel, aluminium windows are less expensive and hence they are commonly recommended for buildings where high degree of elegance is required at lesser cost. Aluminium windows are also rust proof, durable and require no maintenance and painting.

Mild steel being cheapest of the above metals, steel windows work out to be most economical. Hence steel windows are extensively used in all types of buildings. As a measure of economy, steel windows are manufactured in a wide range of standard sizes. The window of required design can be fabricated out of standard light rolled steel sections.

Steel window can be fixed direct in the masonry opening in the wall or it may be fitted in a wooden frame fixed in the window opening

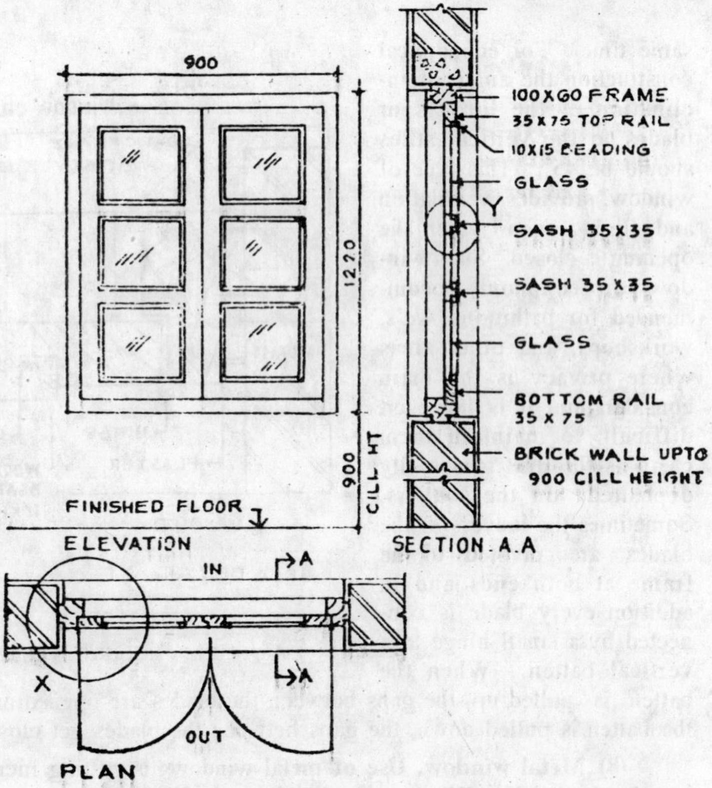

Fig. 17·37 Details of a glazed or sash window

in the wall. In fixing metal windows it has to be ensured that no structural loads, stresses or movements in the structure are transmitted to the window. This is necessary to prevent damage of the window as the thin rolled steel section forming the window unit can not bear such loads or strains. This can be achieved by respecting the following two requirements.

(*i*) The size of window opening is kept slightly more than the size of the window unit so as to allow some clearance between frame and the sides of opening.

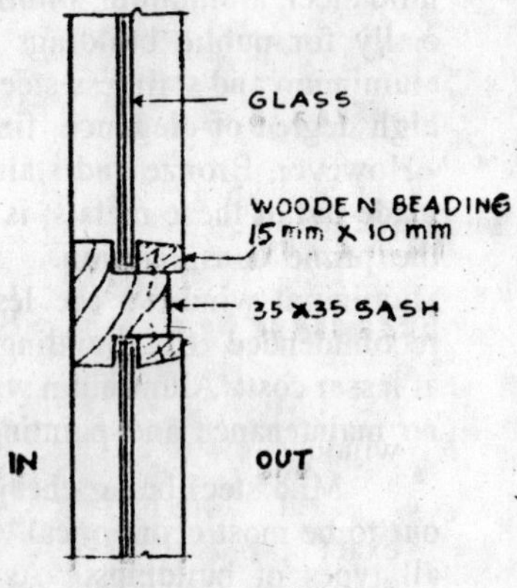

DETAIL AT-Y

Fig. 17·38
Enlarged details at Y (Enlarged detail at X being same as shown in Fig.17.36

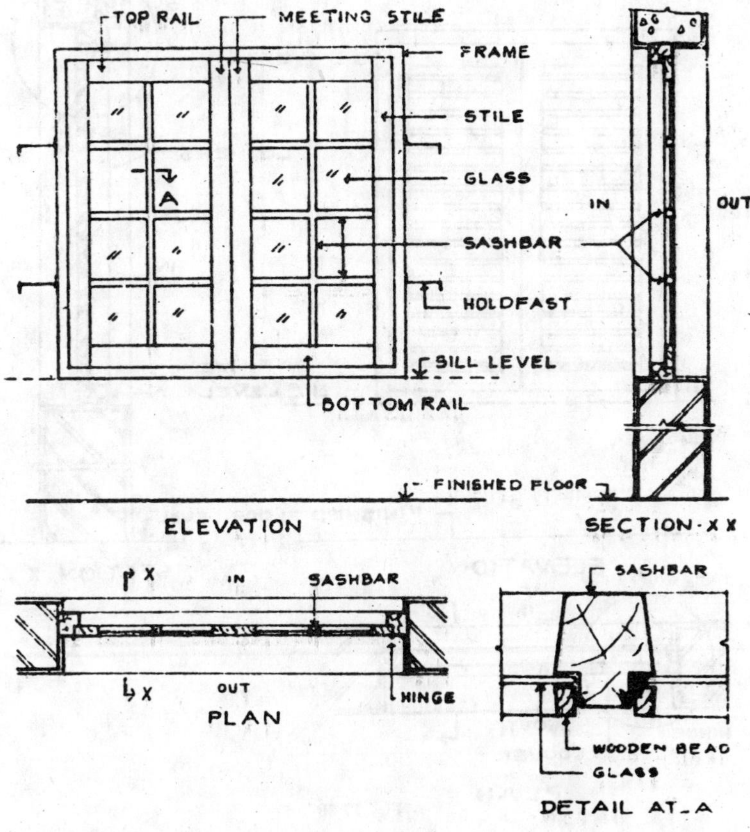

Fig. 17·39

Details of Glazed or sash window for wider openings

(*ii*) The window is fixed into prepared opening *i.e.* only after the items of masonry work for the opening as well as the lintel etc. are complete and fully set.

Method of fixing steel windows. The method of fixing steel windows under different situations is summarised in the following steps.

(1) The prepared opening is cleared of all obstructions and the exact position of the window from the external face of wall (as per drawing) is marked on the jambs with the help of chalk lines. The marking of chalk line is also made along the head and the sill of the opening.

(2) The frame of the steel window has holes at locations where hold fasts or lugs are to be provided. The holes are known as fixing

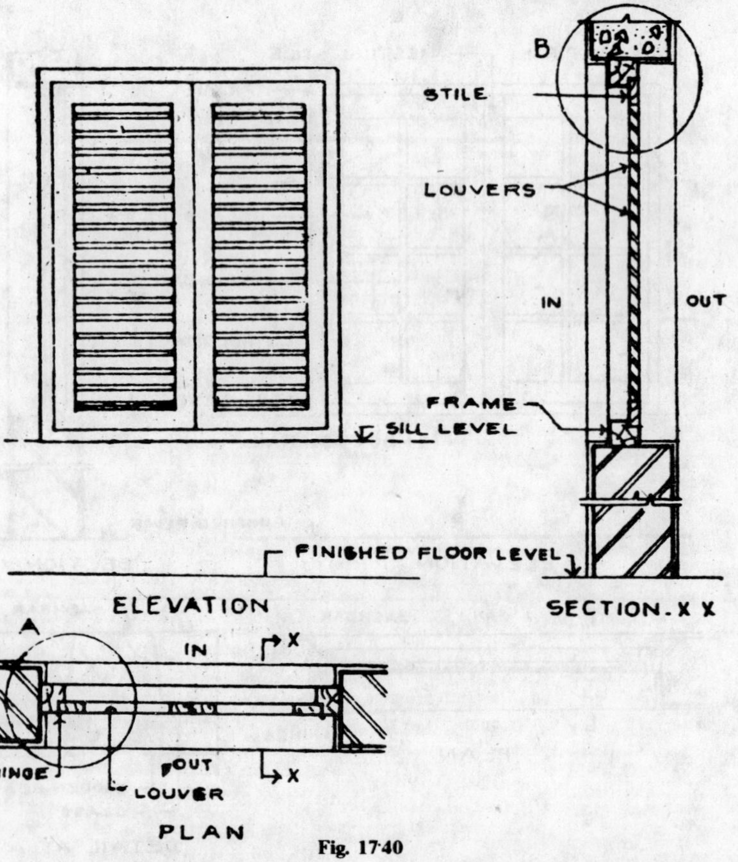

B
STILE
LOUVERS
IN
OUT
FRAME
SILL LEVEL
FINISHED FLOOR LEVEL
ELEVATION
SECTION. X X
A
IN
X
HINGE
OUT
LOUVER
X
PLAN

Fig. 17·40

Details of louvered or venetianed window

FRAME
BEADING
IN
HINGE
OUT
LOUVER
STILE

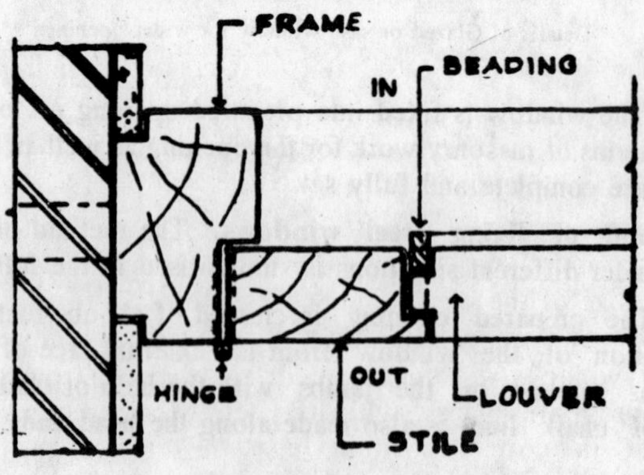

DETAIL AT -A

Fig. 17·41 Enlarged detail at "A" of the window shown in Fig. 17·40

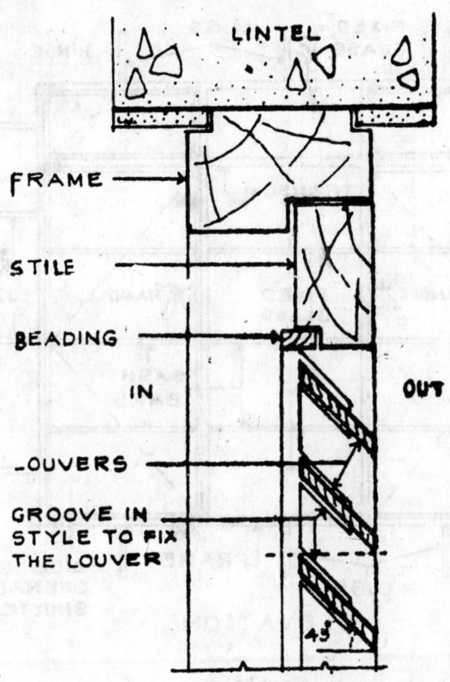

DETAIL AT -B

Fig. 17·42

Enlarged detail at "*B*" of the window shown in Fig. 17·40

holes. The position of the fixing holes is measured from the window unit and marked on the chalk line at corresponding points.

(3) In case of brick masonry, the steel window is invariably fixed to the opening with the help of small sized hold fasts also known as *lugs*. For installing the lugs, holes are cut in the brick work, normal size of the hole being 5 cm. square and 5 to 10 cm. deep.

(4) The steel window unit is thereafter placed in position in the window opening. Since the size of the opening is kept slightly bigger than the size of the window unit, it is necessary to use temporary wooden wedges at jambs, head and still to set the window in correct position. The window is adjusted in correct alignment by striking the wedges at appropriate places with the help of a hammer. After the window is put in correct position, lugs are screwed on tight to the frame at specified positions. Thereafter the lugs are grouted into the holes with cement mortar. After the lugs get firmly set, the wooden wedges are removed and the gap between the window frame and the

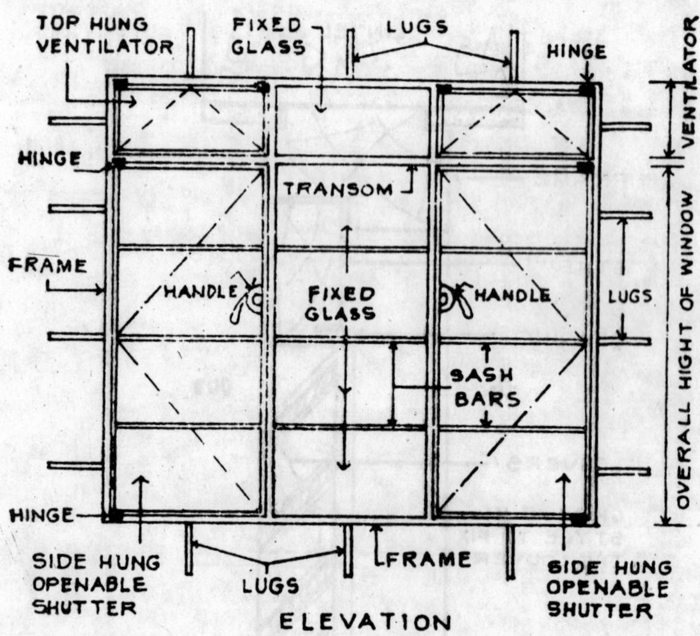

Fig. 17·43

Elevation of a steel window showing location of holdfasts or lugs etc.

jambs, head and still of the opening (also known as *surrounds*) is filled with cement mortar. In case the surrounds are not to be plastered from out side, the gap between the window frame and the surrounds is filled and painted with mastic.

(5) In case of reinforced concrete work or stone masonry it is difficult to cut holes for fixing the lugs in a prepared opening. In such situation, wooden blocks (also known as *fixing plugs*) are embedded in appropriate places in the concrete or masonry during casting or construction of the wall. It is however important that the plugs or set very accurately so as to match the alignment of the fixing holes in the window frame. The window unit is subsequently fixed to the plugs with the help of galvanised iron wood screws.

(6) When the steel window is to be fitted into wooden frame already fixed in the window opening, the wooden frame is invariably rebated to receive the steel window unit. After the window unit is set is position accurately, it is secured to the wooden frame by means of G.I. wood screws.

(7) In case of structural steel work the steel windows are fixed in the opening with the help of special fixing clips or with nuts and bolts.

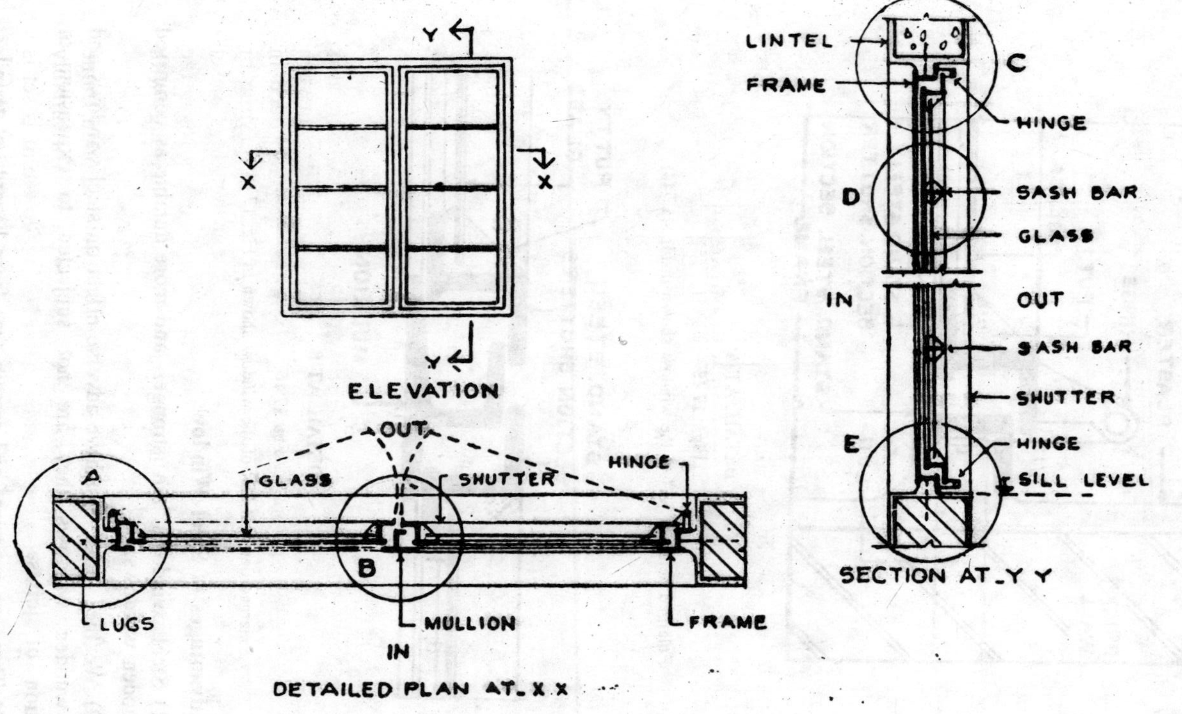

Fig. 17·44 Details of steel windows

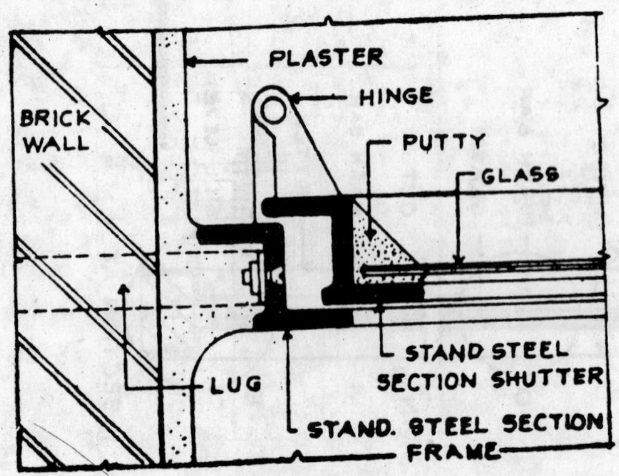

DETAIL AT-A

Fig. 17·45

Enlarged detail at "A" of the window shown in Fig. (17·44)

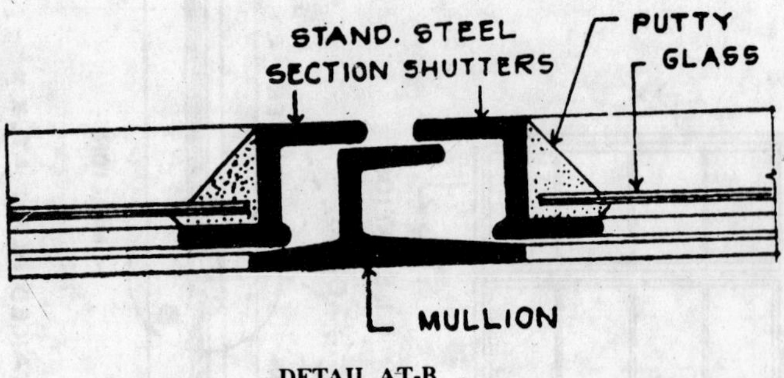

DETAIL AT-B

Fig. 17·46

Enlarged detail at "B" of the window shown in Fig. 17·44

Advantages of Steel Windows

(1) Steel windows are stronger and more durable as compared with wooden windows.

(2) Weather does not have adverse effect on steel windows and unlike wooden windows they are not subjected to expansion or contraction of joints.

(3) They are rot-proof and remain free from hazards of termites, funges or white ants.

(4) They are highly fire resistant.

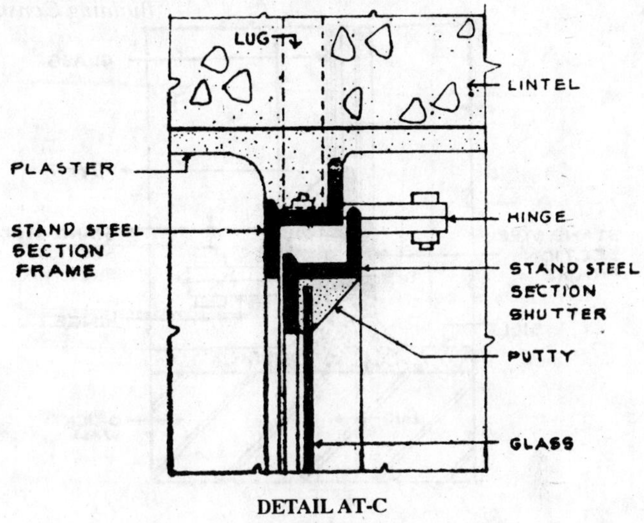

DETAIL AT-C

Fig. 17·47

Enlaged detail at "C" of the window shown in Fig. 17·44

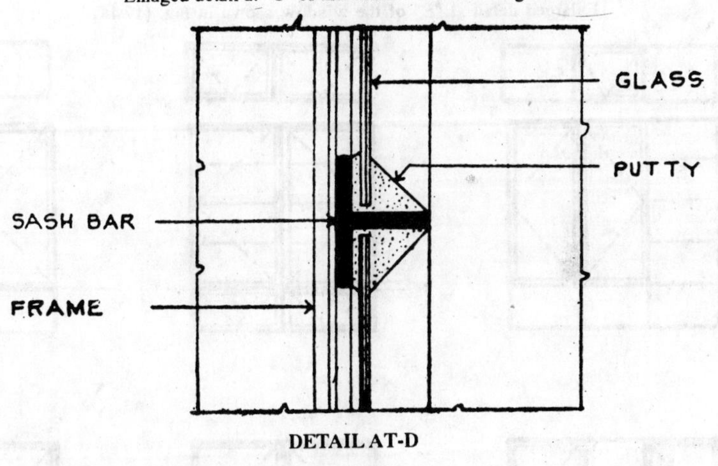

DETAIL AT-D

Fig. 17·48

Enlarged detail at "D" of the window shown in Fig. 17·44

(5) They are fabricated with greater precision and quality control. Hence steel windows present elegant appearance and steam lined fishing which make them superior to wooden windows.

(6) Steel windows (being made up of narrow and thin sections) provide more area for light and ventilation than wooden windows.

(7) Steel windows afford greater facilities for providing different types of openable parts.

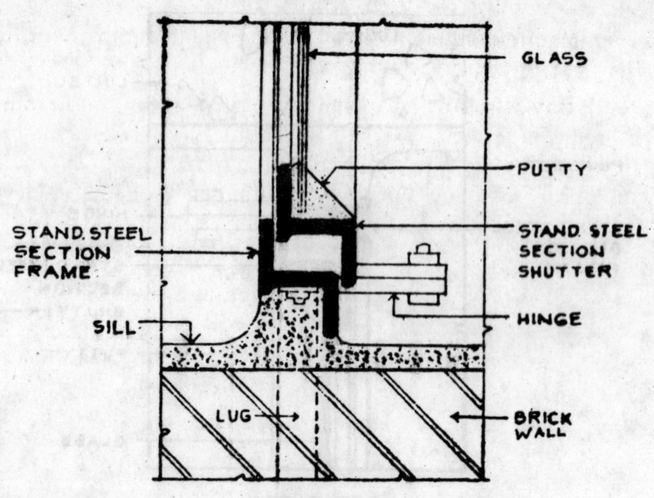

DETAIL AT-E

Fig. 17·49

Enlarged detail at "*E*" of the window shown in Fig. (17·44)

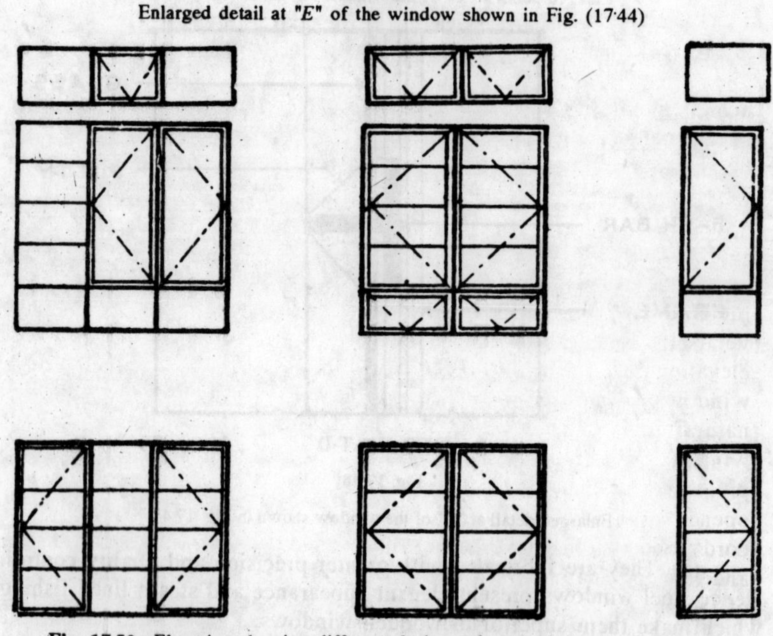

Fig. 17·50 Elevation showing different designs of steel windows

(8) The cost of maintenance of steel windows is almost negligible and they prove economical after considering all the above aspects. They

however, require painting to avoid rusting due to damp or corrosive atmosphere.

(9) **Bay window.** A window projecting outward from the walls of a room is termed as a bay window. Bay window may be square, rectangular or polygon in plan and it is introduced with a view to provide an increased area of opening for admittance of light and ventilation. Bay-window is illustrated in Fig. 17·51.

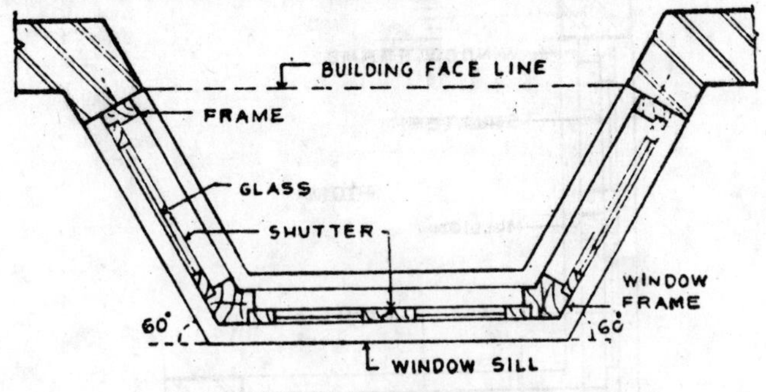

Fig. 17·51

Sectional plan of Bay window

(10) **Clere story window**. This type of window is used to achieve better ventilation and cooling effect in the living or main rooms of a building which have ceiling height greater than the surrounding rooms. Clere storey windows are provided near the top of the roof of the main rooms and they open out above a lean-to-roof or the roof slab of the adjoining verandah. Besides improving the elevation of the building, such windows also provide adequate natural light in the room. The window shutter is horizontally pivoted and the window can be opened or closed by use of two cords, one attached to the top rail and the other to the bottom rail of the shutter.

It is however, important to see that the upper part of the shutter opens towards inside of the room and the lower part opens outside. This is necessary to prevent rain water entering the room.

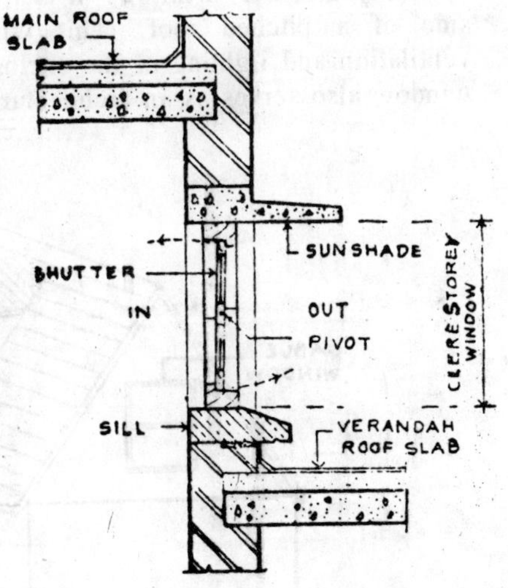

Fig. 17·52

Details of a clere storey window

(11) **Corner window.** As the name suggests this type of window is essentially located in the corner of a room. By use of this window it is possible to ensure light and ventilation from two directions at right angles. In addition it serves as an architectural feature for improving the elevation of the building. Details of corner window are shown in Fig. 17·53.

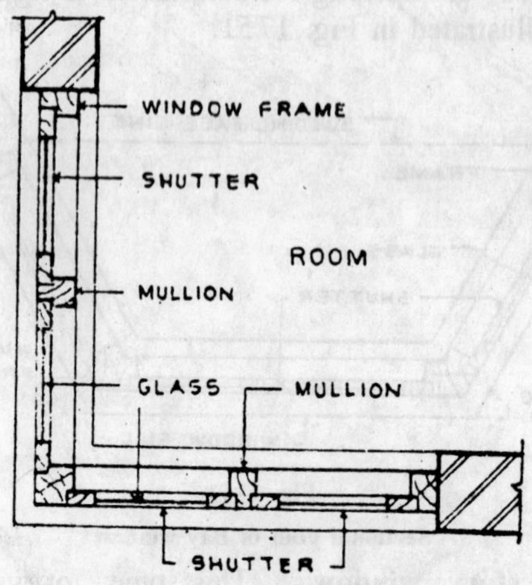

Fig. 17·53

Sectional Plan of a corner window

(12) **Dormer window.** It is a vertical window built in the sloping side of a pitched roof. This window is provided to achieve proper ventilation and lighting of the enclosed space below the roof. Dormer window also serves as an architectural feature to the building.

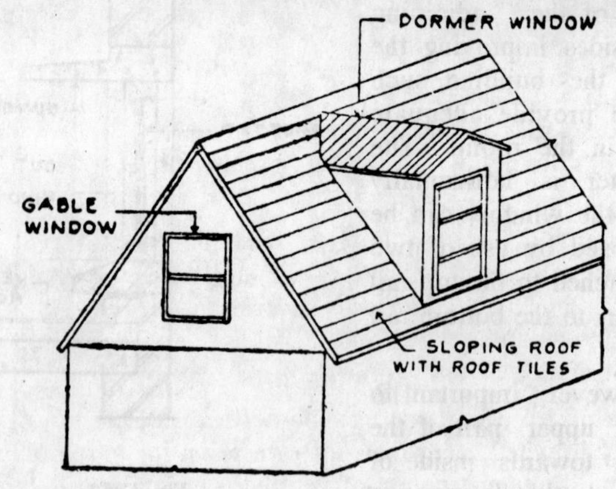

Fig. 17·54

View showing Dormer window and Gable window

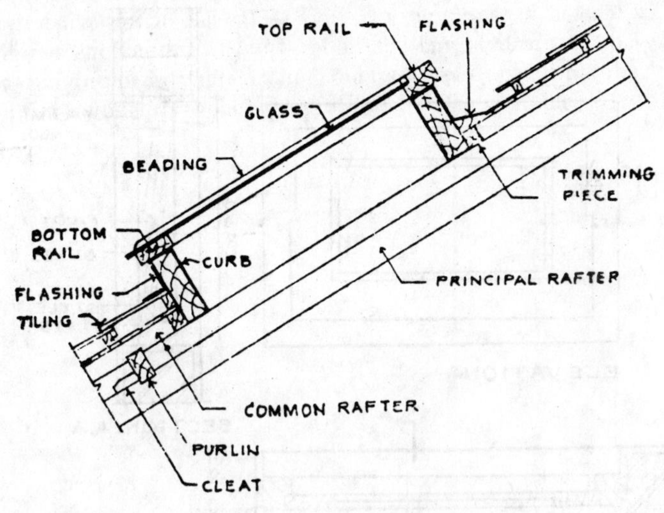

Fig. 17·55

Details of a skylight

(13) **Gable window.** The window provided in the gable end of a pitched roof is known as gable window.

(14) **Sky light.** This is a type of fixed window provided on the sloping surface of a pitched roof, the window being parallel to the sloping surface. The sky light is provided with a view to permit the room below to be fully lighted with natural light. The opening for the sky light is made by cutting the common rafter suitably. The frame work of sky light supporting the glass panels consists of an arrangement of trimming pieces, curb frame, bottom rail and top rail is shown in Fig. 17·55. The opening made for accommodating the sky light is properly treated by lead flashing so as to ensure complete water proofing of the roof surrounding the opening.

(15) **Fan light.** The function of fan light is to ensure cross ventilation in the room even when the door is closed. It also assists in providing natural light in the corridors or passages. Fan light is fitted between the head of the door frame and the transome. The shutter of the fan light may be pivoted at the centre or it may be top hung.

(16) **Ventilator.** A ventilator may be defined as a narrow window of small height provided near the roof of a room for providing ventilation in the room. The construction of ventilator is similar to that of fan light. The shutter of the ventilator is horizontally pivoted and can be

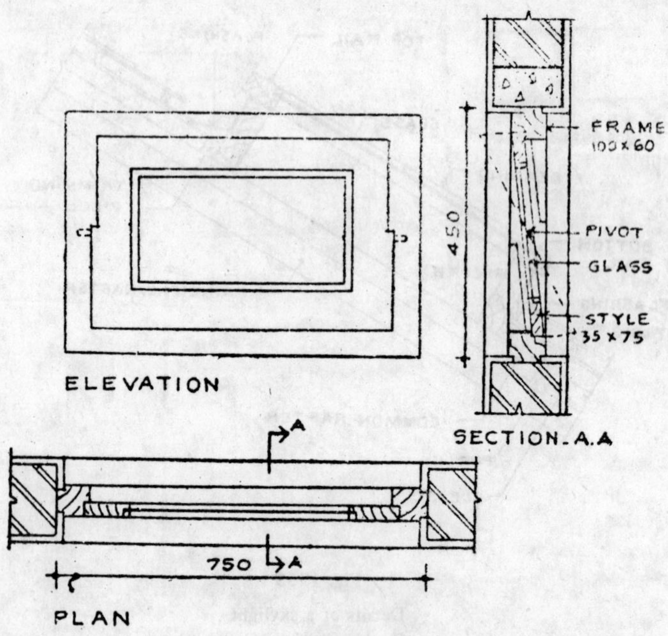

Fig. 17·56

Details of a ventilator

opened or closed by use of two chords, one attached to its top rail and the other to the bottom rail.

17·9. FIXTURES AND FASTENINGS FOR DOORS AND WINDOWS

The fixtures and fastenings for doors and windows can be of iron, brass or aluminium. It is however important to ensure that the item should be reasonably smooth and free from sharp edges and corners. The screw holes in the item should be of counter sunk type. Iron fixtures and fastening are normally black enamelled or copper oxidised. Brass fittings on the other hand are oxidised or finished bright or chromium plated. Aluminium fittings are normally anodised. The fixtures and fastening are manufactured in different shapes and sizes to suit the requirements of the different size and type of doors or windows. The fittings and fastenings can be broadly grouped as under.

(1) *Hinges*

(2) *Bolts*

(3) *Handles.*

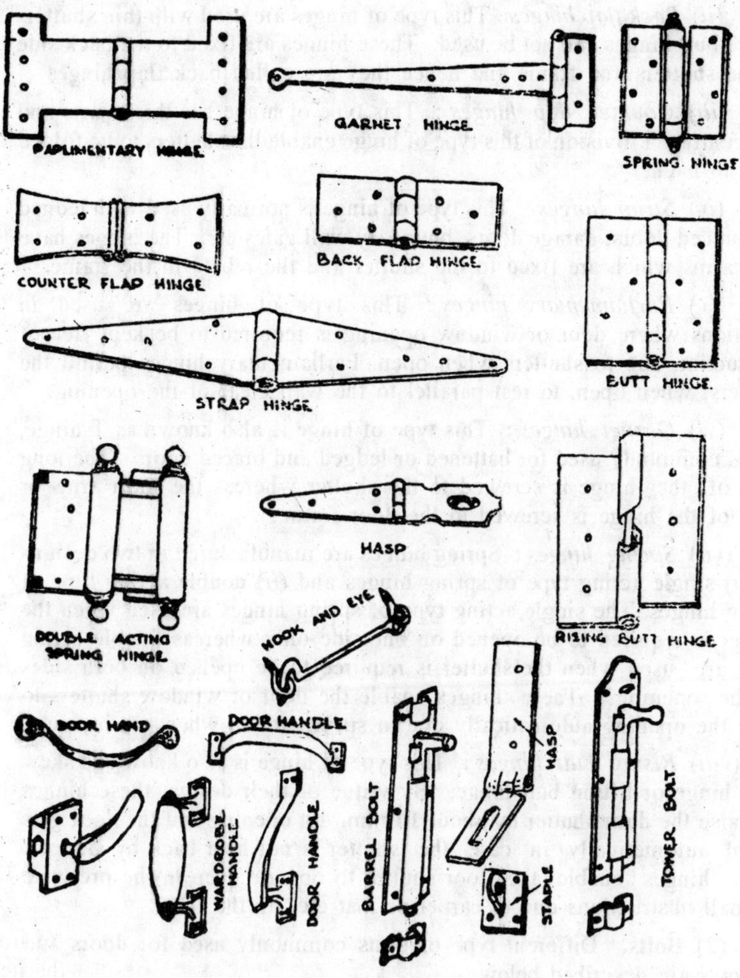

Fig. 17·57

Fixtures and fastening for doors and windows

(1) **Hinges.** Different types of hinges commonly used for doors and windows are described below,

(*i*) *Butt hinges* : This type of hinge is most commonly used for fixing door or window shutters to the frame. The sizes of hinges vary from 50 mm. long x 37 mm wide to 125 mm long x 75 mm wide. One flange of hinge is screwed on to the edge of the shutter and the other flange is screwed to the rebate in the frame. For normal size of door or window three hinges are used for each shutter.

(*ii*) *Back flap hinges*: This type of hinges are used with thin shutters where butt hinges can not be used. These hinges are fixed to the back side of the shutters and frame and hence they are called back flap hinges.

(*iii*) *Counter flap hinges* : This type of hinge has three parts and two centres. Provision of this type of hinge enable the shutters to be folded back to back.

(*iv*) *Strap hinges* : This type of hinge is normally used with ledged and braced doors, garage doors, boundary wall gates etc. The hinges have long arms which are fixed to the shutter and the rebate in the frame.

(*v*) *Parliamentary hinges* : This type of hinges are used in situations where door or window opening is required to be kept free of obstruction due to shutters when open. Parliamentary hinges permit the shutters, when open, to rest parallel to the wall clear of the opening.

(*vi*) *Garnet hinges* : This type of hinge is also known as T-hinge, and is commonly used for battened or ledged and braced doors. The long arm of the hinge is screwed to the shutter whereas the short arms or plate of the hinge is screwed to the door frame.

(*vii*) *Spring hinges* : Spring hinges are manufactured in two designs *i.e.* (*i*) single acting type of spring hinges and (*ii*) double acting type of spring hinges. The single acting type of spring hinges are used when the shutter is required to be opened on one side only whereas double acting type are used when the shutter is required to be opened on both sides of the opening. These hinges enable the door or window shutters to close the opening automatically due to spring action when not in use.

(*viii*) *Rising butt hinges* : This type of hinge is also known as skew butt hinge or lifting butt hinge. By virtue of their design, these hinges can raise the door shutter by about 10 mm. on opening and the door gets closed automatically in case the shutter is not held back by stopper. These hinges enable the door shutter to operate even in the presence of small obstructions due to carpet or mat etc. on the floor.

(2) **Bolts.** Different type of bolts commonly used for doors and windows are described below.

(*i*) *Tower bolts* : In this type, the bolt passes through two or three stapples attached to the base plate. The plate is screwed to inside face of the door shutter and the bolt engages in a metal socket screwed to the frame.

(*ii*) *Barrel bolt* : This type of bolt is similar to tower bolt except that the stapples are replaced by a barrel.

(*iii*) *Hasp and stapple bolt* : In this case the hasp is secured to the shutter whereas the stapple is screwed to the door frame.

(*iv*) *Eye and Hook* : This type of fitting is used for keeping window shutter in position when the window is open. The hook is fixed to the sill of the window frame and the eye is fixed to the bottom rail of window shutter.

(3) **Handles.** Handles are manufactured in a variety of designs. Some of the commonly used types of handles are shown in Fig. 17·57.

Questions for Review

1. Give full details of construction with neat sketches of a corner window of a bed room measuring 6·1 m X 4·9 m X 3·6 m high with 38 cm thick wall on all sides. The window is to be glazed with venetians.

2. (*a*) Draw a neat dimensioned sketch showing plan, elevation and section of a door 2/3 glazed and 1/3 panelled, to fit in an opening 2·29m X 1·22 m in a brick wall. Assume all necessary and reasonable particulars.

   (*b*) Draw sketches of the following types of doors and windows and state where they are generally used.

   (*i*) Ledged and braced doors

   (*ii*) Louvered windows

   (*iii*) Flush doors

   (*iv*) Sky-lights

3. Write short notes on the following with suitable sketches wherever necessary :

   (*i*) Door frame

   (*ii*) Dormer windows

   (*iii*) Rolling shutters

   (*iv*) Swing doors

   (*v*) Clere-storey window

   (*vi*) Panelled doors

4. What are ventilators and where are they placed ? Describe with sketches the different types of ventilators used in buildings.

5. Draw to full scale, plan, elevation and section of a window with fanlight suitable for a bed room. The window is 1 m wide and 1·8 m high in a wall 30 cm thick. Give three half full size details to explain the construction.

6. Describe with neat sketches any five of the following :

   (*i*) Battened and ledged doors

   (*ii*) Battened, ledged and braced doors

   (*iii*) Framed and braced doors

   (*iv*) Framed and panelled doors

(*v*) Glazed and sash doors

(*vi*) Rolling shutters

(*vii*) Revolving shutters

(*viii*) Collapsible steel doors.

7. Design a wooden window fully glazed for a hospital building, 3 m wide and 1·82 m high with shutters side-hung, 1·22 m high and ventilators 60 cm high centrally pivoted. Draw sketches of the cross-section showing dimensions of members and their names. Describe by sketches details of joinery.

8. Write short notes on the following with suitable sketches wherever necessary:

(*i*) Window

(*ii*) Sashes

(*iii*) Fanlights

(*iv*) Venetian shutters

(*v*) Bay windows

(*vi*) Clere-storey windows.

9. Draw to suitable scale, plan, elevation and section of a French window opening out of a bed room on to an open terrace. The size of the opening is 2·4 m x 1·52 m and the wall thickness is 30 cm. The bedroom floor is 8 cm higher than the terrace to prevent water getting in. Give full size detail through head, jambs and bottom rails of the French window.

10. Draw sketches to show the fixing arrangement of metal windows to wooden frame or metallic sub-frames. Sketch the various standard shapes of the windows manufactured in India these days.

11. Draw to 1 :100 scale, the plan, elevation, and section of a door suitable for the front entrance of a cottage. The door is 1·22 m x 2·12 m. Give three-half- full size details to explain the construction.

12. Write short notes and give suitable sketches for any four of the following and mention where they are used in building :

(*a*) Barrel bolts

(*b*) Venetian shutters

(*c*) Glazed windows

(*d*) Collapsible steel gates

(*e*) Rolling shutter.

13. What are the special types of doors and windows used in modern public buildings. Neatly sketch a wooden door one-third panelled and two third glazed the size of door being 2·2 m x 2·1 m.

14. Draw neat sketches to illustrate a 40 mm thick fully panelled door to be installed in an opening 1·20 m x 2·10 m. in size giving method of fixing it in a $1\frac{1}{2}$ thick brick wall. Name different parts of the door. Show details of any one of the carpentry joints.

15. Differentiate between the following with the help of sketches.

    (*i*) Fixed and pivoted windows

    (*ii*) Skylight and ventilator

    (*iii*) Corner window and bay window.

    (*iv*) Gable window and dormer window

    (*v*) Wooden window and steel window.

16. Write short note on door frames and method of their fixing to brick masonry, stone masonry and R.C.C.

17. Describe different types of flush doors giving suitable sketches along with the description.

18. Describe with sketches important types of fixtures and fitting used for doors and windows.

19. Write short notes on the following :

    (*i*) Design of windows

    (*ii*) Advantages of steel windows

    (*iii*) Barrel bolt

    (*iv*) Fly proof doors

    (*v*) Spring hinges

20. Differentiate between the following with the help of sketches.

    (*i*) Panelled door and Glazed door

    (*ii*) Sliding door and Rolling door

    (*iii*) Revolving door and swing door

    (*iv*) Butt hinges and Rising butt hinges.

18

Roofs and Roof Covering

18·1. GENERAL

Roof may be defined as a covering provided over the top of a building with a view to keep out rain, snow, sun and wind and to protect the building from the adverse effects of these elements. A roof basically consists of roof covering materials supported on structural elements installed on the building top. The structural elements may be trusses, portals, flat slab, shell, dome or space frame whereas the roof covering materials may be thatch, wooden single, tiles, slates, A.C. sheets G.1 sheets etc. Roof is considered to be one of the most important item of work in a building and due attention should therefore be paid in the matter of selection of the type of roof and the method of construction.

As it is important to make safe foundations for a building, it is equally important to provide good roof above it. Good foundations protect the building against damage from below the ground and in a similar way good roof prevents deterioration of the building starting from the top.

The choice of the type of roof for a building should be made after paying due consideration to the climatic conditions as well as the availability of selected materials at site. A well planned roof should satisfy the following requirements.

(1) It should be structurally sound and strong enough to carry the anticipated dead and live loads safely.

(2) The roof should be durable against adverse effect of elements like rain, snow, sun, wind etc.

(3) It should have efficient water-proofing and drainage provisions.

(4) The roof should provide desired insulation against heat.

(5) It should provide desired insulation against sound.

18·2. CLASSIFICATION OF ROOFS

Roofs can be broadly classified in the following two main types :

(1) Pitched or sloped roofs.

(2) Flat or terraced roofs.

Pitched or sloped roofs are considered suitable for buildings in coastal regions or in areas where rainfall is very heavy. On the other hand flat roofs are considered suitable for buildings in plains or in hot regions where rainfall is scanty. Sometimes however it becomes necessary to provide roof having shapes other than pitched or flat type to meet the functional as well as aesthetical requirements of the building. For example domes are normally provided as roof for large circular halls used for assembly, gymnasiums etc. Similarly shell roof and folded plate roofs are provided in public buildings like libraries, theatres, recreation centres, factories, workshops etc. where large floor areas are required to be covered without obstruction from columns. Pitched Roof and Flat roof have been delt separately in this chapter.

18·3. PITCHED ROOF

It is the most common form of roof and is generally regarded as the cheapest alternative for covering a structure. Pitched roofs are almost always constructed in wood or steel. *Steel pitched roofs* have been described in the chapter on *"Structural Steel Works."* Wooden pitched roof consists of a system of joists, rafters and purlins arranged in the form of a triangular shaped support known as truss. The lower ends of the rafter rest upon the wall plates and at their upper end they are connected to a common ridge piece.

In pitched roof a slope of less than 1 in 3 is generally not considered satisfactory from drainage point of view. In areas of heavy snow fall steeper slopes of say 1 to $1\frac{1}{2}$ or 1:1 are provided to reduce the incidence of snow load on the roof.

When sloped roofs are to be constructed in hot regions, it is desirable to provide false ceiling from consideration of thermal comfort. The false ceiling should be of insulating material like plywood, fibre board, gypsum board, asbestos cement sheet etc.

The slope of the roof varies according to the span, the climatic conditions of the site, the nature of the covering material and other similar factors.

18·4. TECHNICAL TERMS

The Various technical terms used in timber pitched roof construction are given below :

(1) *Span* : The clear distance between the supports of an arch, beam or roof truss.

(2) *Rise of Roof* : It is the vertical distance between the wall plate and the top of the ridge.

(3) *Pitch of Roof* : The inclination of the sides of a roof to the horizontal is termed as the "Pitch of the roof".

(4) *Eaves* : The lower edge of the inclined roof surface of a pitched roof is termed as eaves.

(5) *Ridge* : It may be defined as the apex of the angle formed by the termination of the inclined surface at the top of a slope.

(6) *Ridge Piece, Ridge Beam or a Ridge Board* : It is a horizontal timber piece provided at the apex of a roof truss. The common rafters are fixed to this piece and are supported by it.

(7) *Valley* : It is the acute angle or a gutter formed by the intersection of two slopes in a pitched roof.

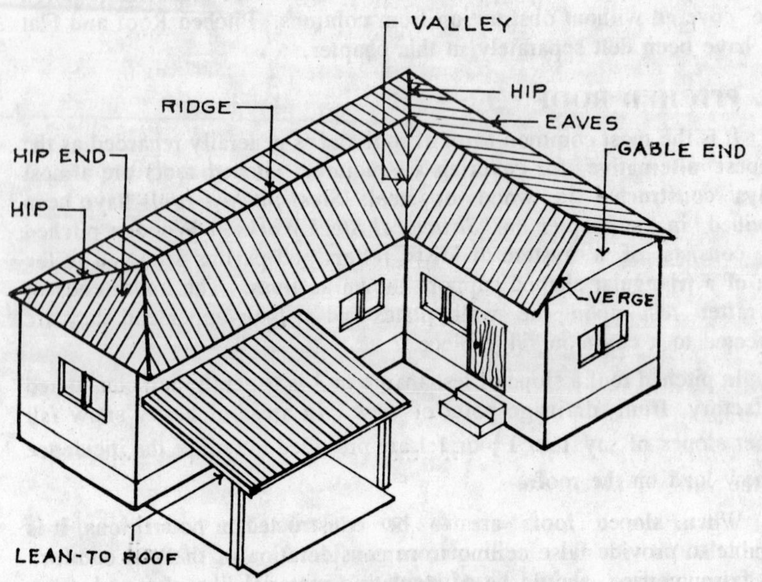

Fig. 18·1

(8) *Hip* : It is a ridge formed by the intersection of two sloped surfaces having an exterior angle greater than 180°.

(9) *Hipped end* : It is sloped triangular surface formed at the end of a roof.

(10) *Gable* : It is the triangular portion of the end wall of a sloped roof formed by continuing the end wall up within the roof. Thus the gable has two slopes with ridge in between.

(11) *Verge* : The edge of the sheets, slates or tiles, projecting beyond the gable end of the sloped roof.

(12) *Barge Board* : It is a timber board used to hold the common rafters forming verge.

(13) *Eaves Board or Facia Board* : It is a wooden plank usually 25 mm. thick and 15 cm. wide fixed along the eaves connecting the feet of the common rafters.

(14) *Common Rafters* : These are inclines wooden members laid from the ridge to the eaves. They are bevelled against the ridge piece at the head and are fixed to the purlins at intermediate points. They are usually spaced 30 to 45 cm. apart.

(15) *Purlins* : These are horizontal members of wood or steel, used to support common rafters.

(16) *Cleats* : These are short sections of wood or steel, nailed, screwed or bolted to the rafters of the truss for supporting purlins.

(17) *Wall Plates* : Timber pieces, which are provided over the top of a stone or brick wall for the purposes of fixing the feet of the common rafters.

(18) *Template* : This is a block of stone or concrete provided under the end of tie beam to spread the load from the roof over a larger area of bearing.

(19) *Hip Rafters* : These are the rafters which form the hip of a sloped roof. These rafters run from the ridge to the corners of the wall to support the covering over the hip. They also receive the end of purlins and jack rafters.

(20) *Jack Rafters* : These are the rafters shorter in length which extend from hip or valley to the eaves.

(21) *Valley Rafters* : These are the rafters provided at slope to support the valley gutters.

(22) *Post Plate* : This is similar to a wall plate with the only difference that it is run continuous, parallel to the face of wall, over posts and supports the rafters at their feet.

(23) *Battens* : Small scantlings which are nailed to the rafters for laying tiles, over the sloped roof.

(24) *Boarding* : They act similar to battens and are nailed to common rafters for laying roofing material above.

18·5. TYPES OF PITCHED ROOFS

Pitched roof can be broadly divided into the following different types.

(1) *Lean-to-roof*

(2) *Coupled roof*

(3) *Couple-close roof*

(4) *Collar roof*

(5) *Scissors roof*

(6) *King-Post roof truss*

(7) *Queen-Post roof truss*

(8) *Mansard roof truss*

(9) *Composite roof truss*

The Various types of pitched roofs are described below :

(1) **Lean to Roof.** This is the simplest type of pitched roof consisting of rafters sloping on one side only. At their upper ends the rafters are nailed to the wooden wall plate , placed on corbel which may be of stone, brick or steel. At the lower end the rafter are notched and nailed to the wooden post plate. The post plate consists of a timber section running parallel to the wall and supported by posts (which may be of wood , brick or stone) at intermediate points. Battens spaced at 15 cm. centre to centre are nailed to the inclined rafter and upon this the covering material is laid. This type of roof is considered suitable for a maximum span of 2·4 metres.

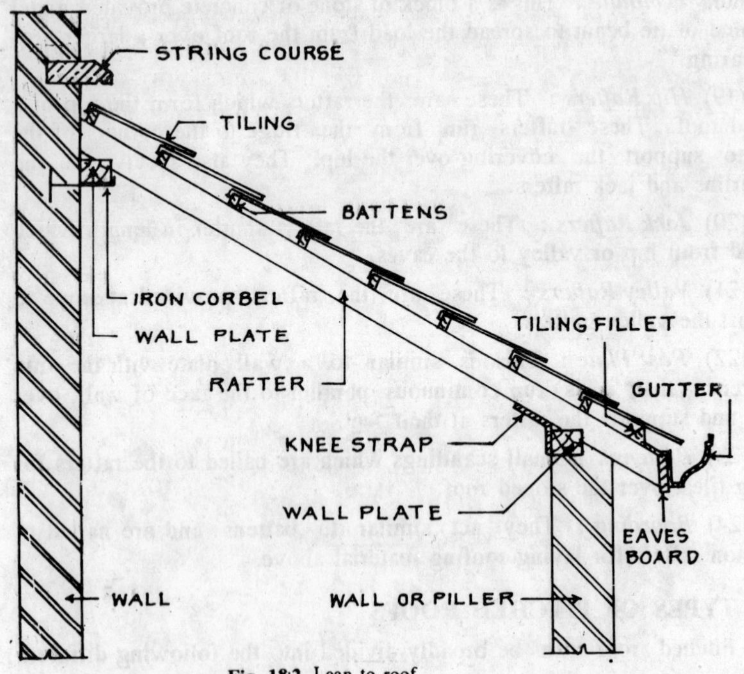

Fig. 18·2 Lean to roof

Lean to roof is commonly used for covering verandahs, out houses, sheds, etc.

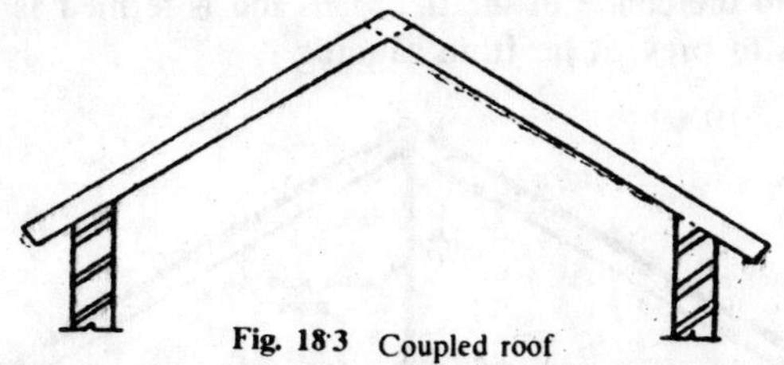

Fig. 18·3 Coupled roof

(2) **Coupled Roof.** Coupled roof is formed by a pair of inclined rafters with their upper ends nailed to a common ridge piece and their lower ends, notched and nailed to the wooden wall plates embedded in masonry on the top of the wall on either end. The rafters are spaced at suitable intervals and the battens are nailed on their top. Upon the frame work thus prepared, roof covering is laid. This type of construction is considered suitable for maximum span of 3·7 meter.

(3) **Couple-close Roof.** In a coupled roof the notches provided at the feet of the rafters prevent them from spreading. If the span of the roof is increased or the loading on the truss is increased gradually, it will be found that at a certain stage the notch will be incapable of resisting the thrust at the end of the rafters and the roof will collapse. Thus, to prevent this, the feet of the rafters are jointed by a tie, which by holding them in position, prevents them from spreading and

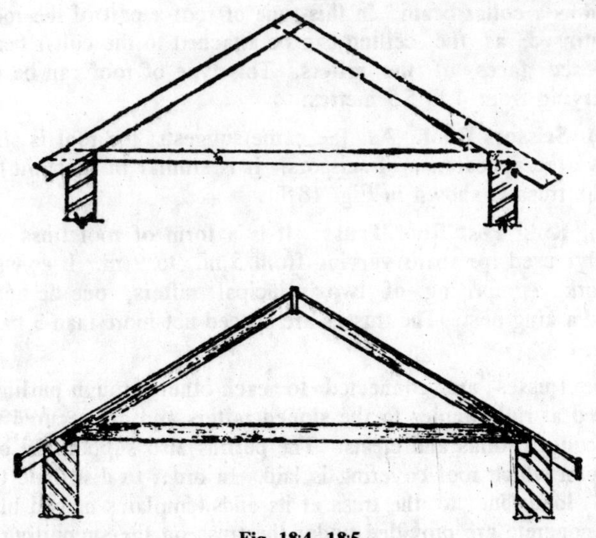

Fig. 18·4—18·5

Different forms of couple-closed roof

thrusting out of the wall. This is called a couple closed roof. With ordinary loading conditions this type of roof can be used for spans up to 5 m. For increased spans a vertical rod is introduced between the ridge piece and the centre of the tie. This rod is termed as *king rod* and its function is to prevent tie from sagging.

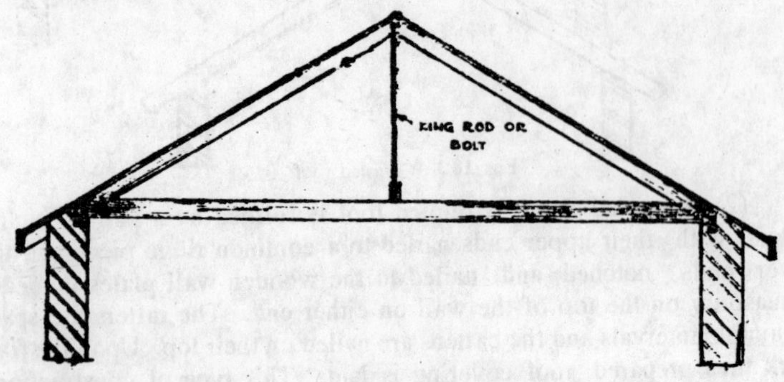

Fig. 18·6

Couple-closed roof with a king rod

(4) **Collar Roof.** With excessive loading conditions or increased spans the rafter of the couple closed roof tends to bend in the middle. This tendency is overcome by fixing the tie near the middle of the rafters. The truss then formed is called collar roof. Thus collar roof is similar to couple, closed roof with the only difference that the tie instead of being placed at feet of the rafters is raised up. The tie in collar roof is termed as a collar beam. In this type of roof a part of the roof space can be utilised, as the ceiling can be attached to the collar beams and the exposed faces of the rafters. This type of roof can be used for spans varying from 4 to 5·5 metres.

(5) **Scissors Roof.** As the name suggests, the roof is shaped so as to have the appearance of scissors. It is similar in function to collar roof. The truss is shown in Fig. 18·8.

(6) **King Post Roof Truss.** It is a form of roof truss which is commonly used for spans varying from 5 m. to 9 m. It consists of a frame-work, comprising of two principal rafters, one tie-beam, two struts and a king post. The trusses are spaced not more than 3 m. centre to centre.

The trusses are connected to each other through purlins which are placed at right angles to the sloped rafters and are secured to them through cogged joints and cleats. The purlins also support the common rafters upon which roof covering is laid. In order to distribute the concentrated load due to the truss at its ends,templates or bed blocks of stone or concrete are provided under the truss on the supporting walls.

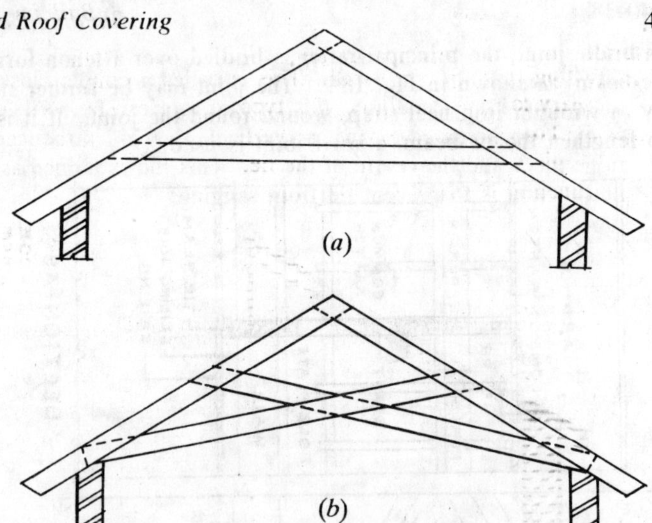

Fig. 18·7–18·8 Collar and scissors roof

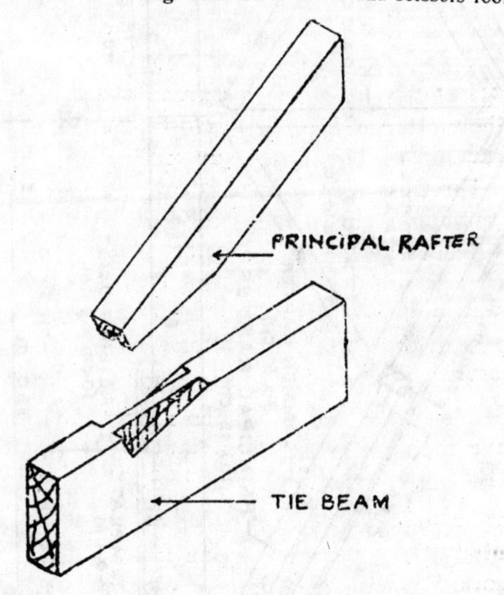

Fig. 18·9

Joint between the principal rafter and tie-beam

The details of joints between various members of king post truss are described below :

(*i*) *Joint between the principal rafter and tie-beam* : At the feet, the principal rafter is jointed to the beam by a "single abutment and tenon joint" or by a bridle joint". In the former, the abutting end is made at right angles to the length of the rafter to ensure direct thrust.

In bridle joint, the principal rafter is bridled over a tenon formed on the tie-beam as shown in Fig. 18·9. The joint may be further strengthened by a wrought iron heel strap, wound round the joint. If it is necessary to lengthen the tie-beam, a scarf joint is used.

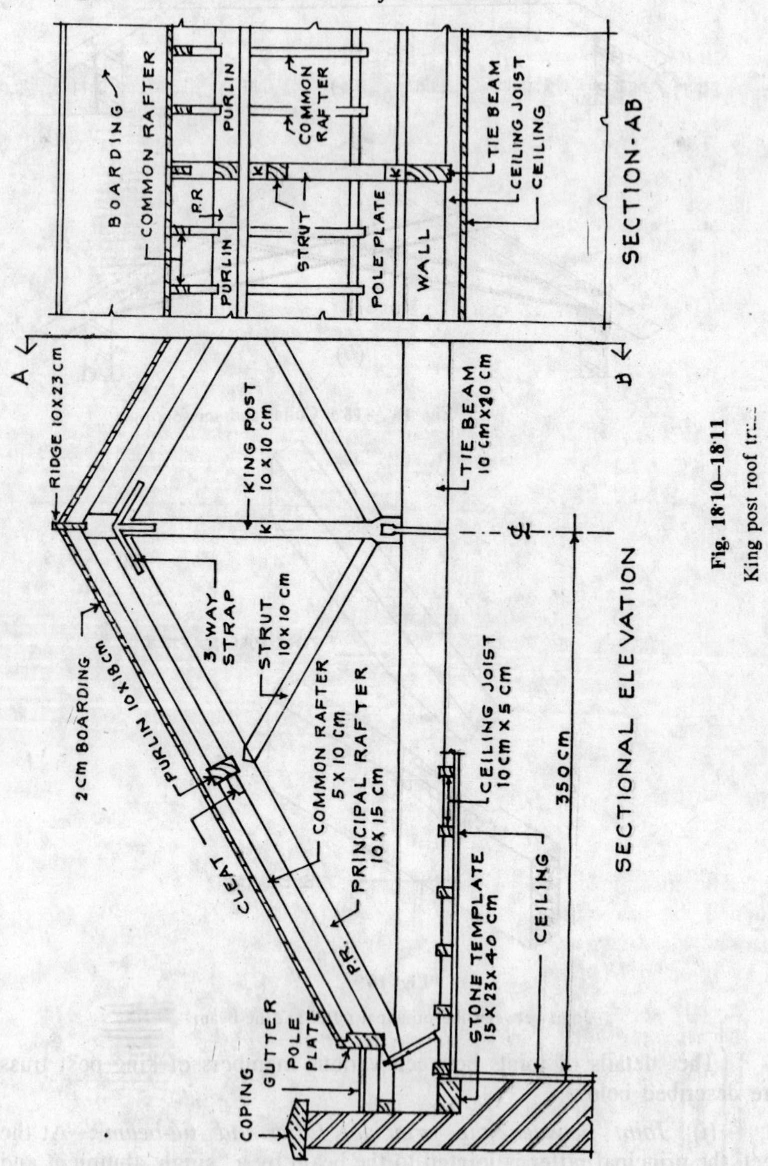

Fig. 18·10—18·11
King post roof tru...

(ii) *Joint between the king post and tie-beam* : The king post is provided with splayed shoulders at feet and is tenoned into the tie

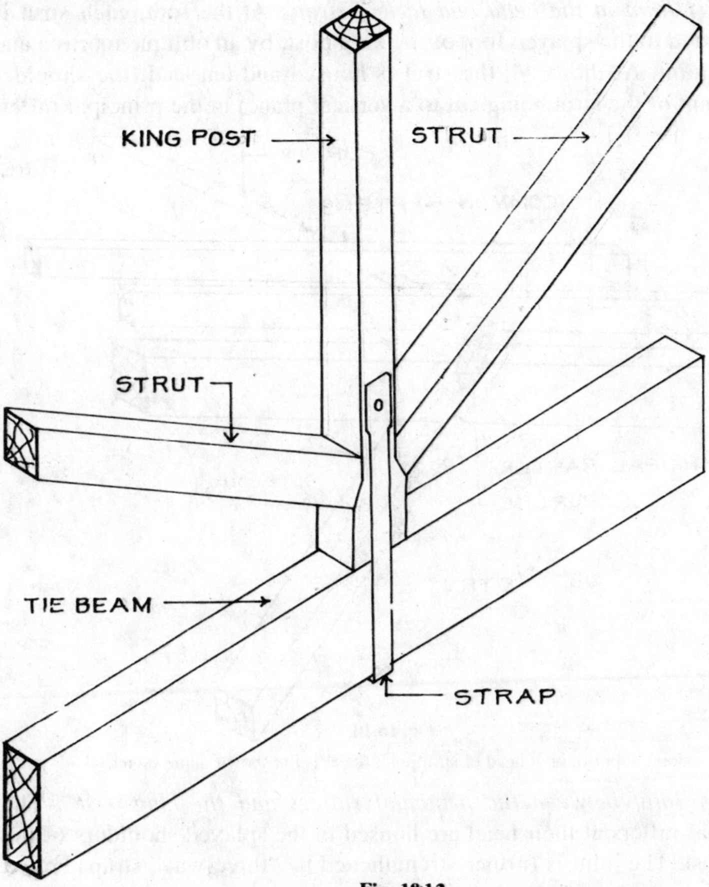

KING POST

STRUT

STRUT

TIE BEAM

STRAP

Fig. 18·12

View showing joint at the feet of king post

beam. The joint between the king post and the tie-beam is further strengthened by wrought iron or mild steel strap. The stirrup strap is taken around the joint and is held up by metal pieces, clips and wedges termed as "*gibs and cotters*".

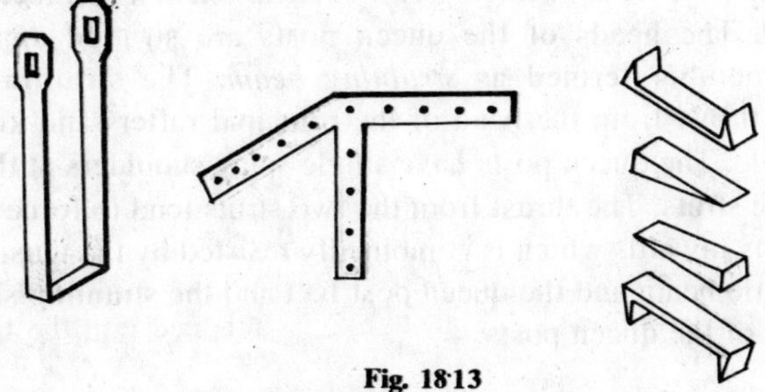

Fig. 18·13

Gibs, Cotters and view of three-way strap

(*iii*) *Joint at the head and feet of strut* : At the foot, each strut is connected to the splayed foot of the king post, by an oblique mortice and tenon joint. At the head, the strut is housed and tenoned (the shoulder abutment of the strut being cut to a vertical plane) in the principal rafter.

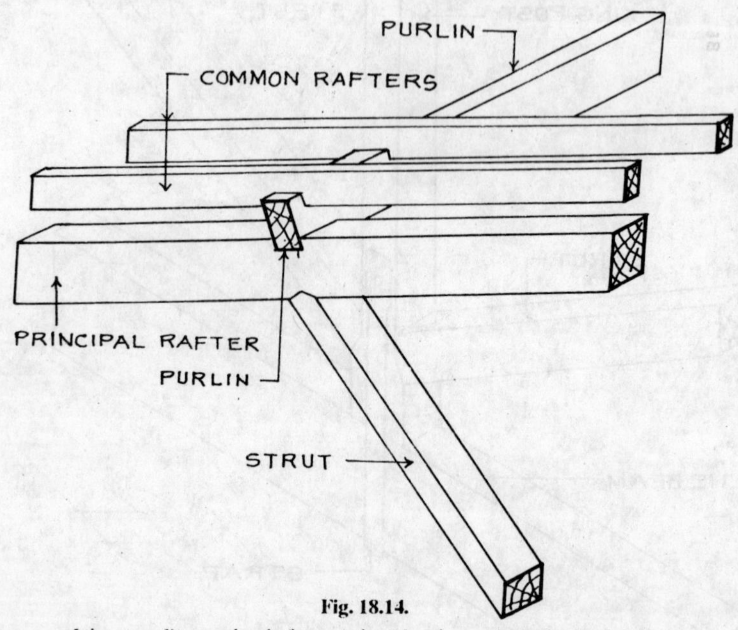

Fig. 18.14.

Joint at purlin or at head of strut point (cleat is not shown in the sketch)

(*iv*) *Joint between the principal rafters and the king-post* : The principal rafters at their head are housed in the splayed shoulders of the king post. The joint is further strengthened by "three-way" straps bolted on each face of the frame.

7. **Queen Post Truss.** For spans greater than 9 m. and less than 14 m., queen post truss is commonly used in timber roof construction. It may be defined as frame work consisting of two principal rafters, two queen posts one straining beam, two inclined struts and a straining sill. It differs from the king-post truss in having two vertical tension members called queen posts. The heads of the queen posts are strutted apart by a horizontal member termed as *straining beam*. The straining beam receives the thrust from the head of the principal rafters and keeps the junction stable. The queen posts have single splay shoulders at their feet to receive the struts. The thrust from the two struts tend to force the feet of queen posts inwards which is combinedly resisted by the tension joint between the tie beam and the queen post feet and the straining sill fixed between feet of the queen posts.

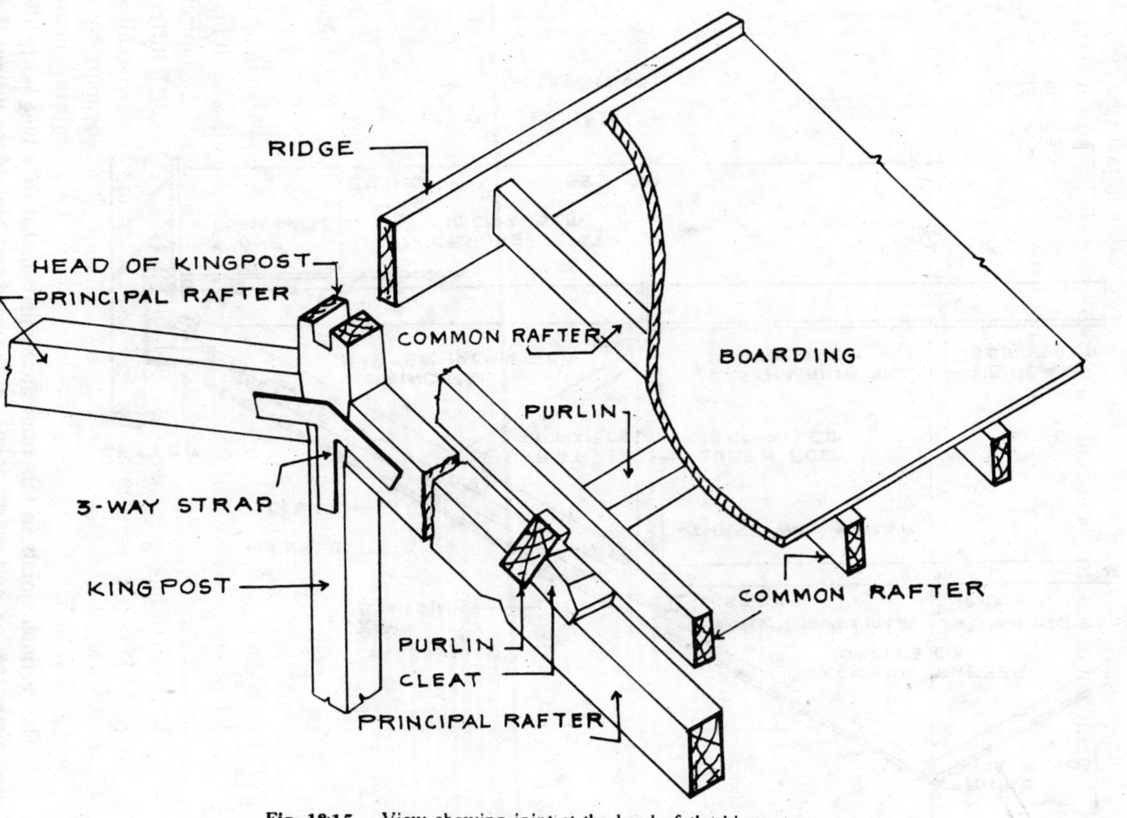

Fig. 18·15 View showing joint at the head of the king post

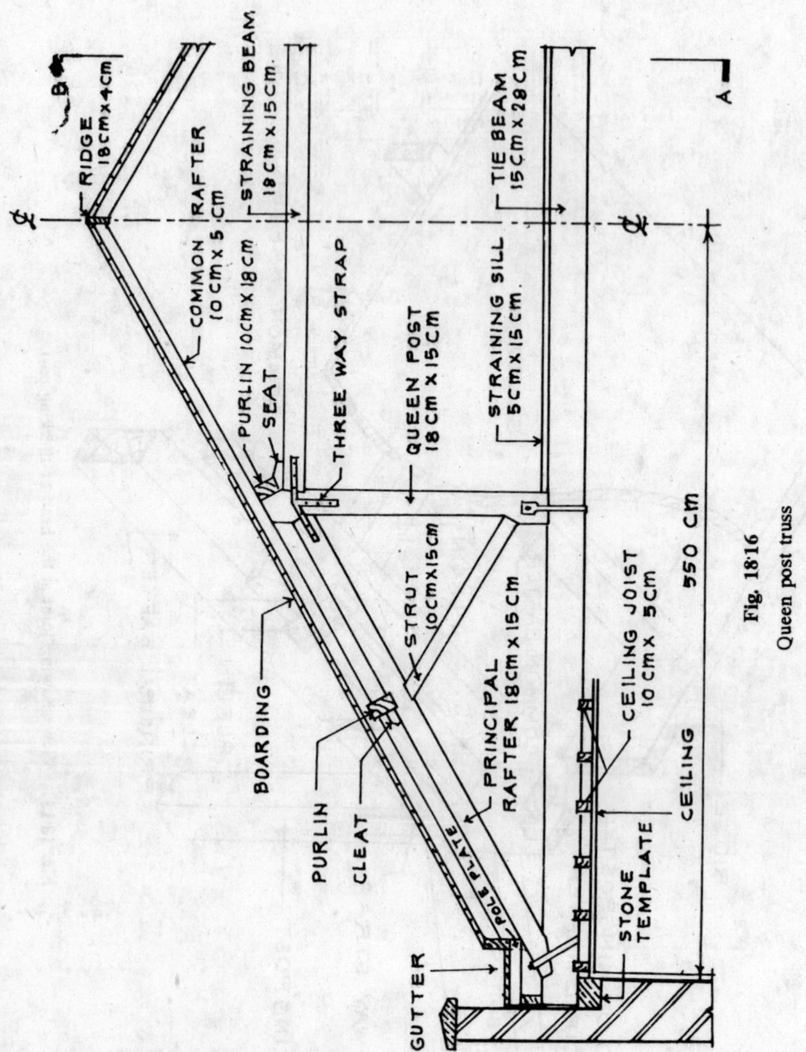

Fig. 18·16
Queen post truss

The various joints in this truss are similar to that of a king post truss with the exception of joints at the head and feet of the queen post which have been described below :-

(*i*) *Joint at the head of queen post* : The junction is formed by meeting of two compression members, *i.e.* principal rafter and the

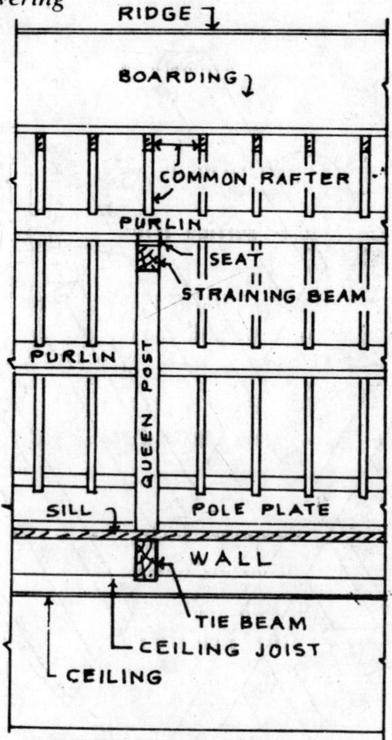

SECTION - A.B

Fig. 18·17

(For elevation of truss refer Fig. 18·16)

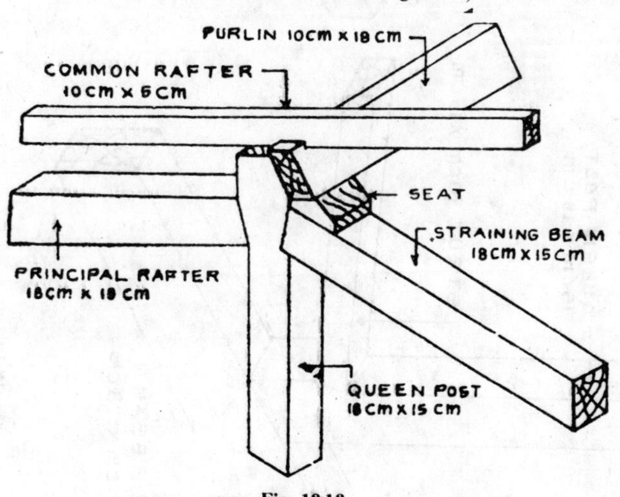

Fig. 18·18

View showing joint at the head of queen post

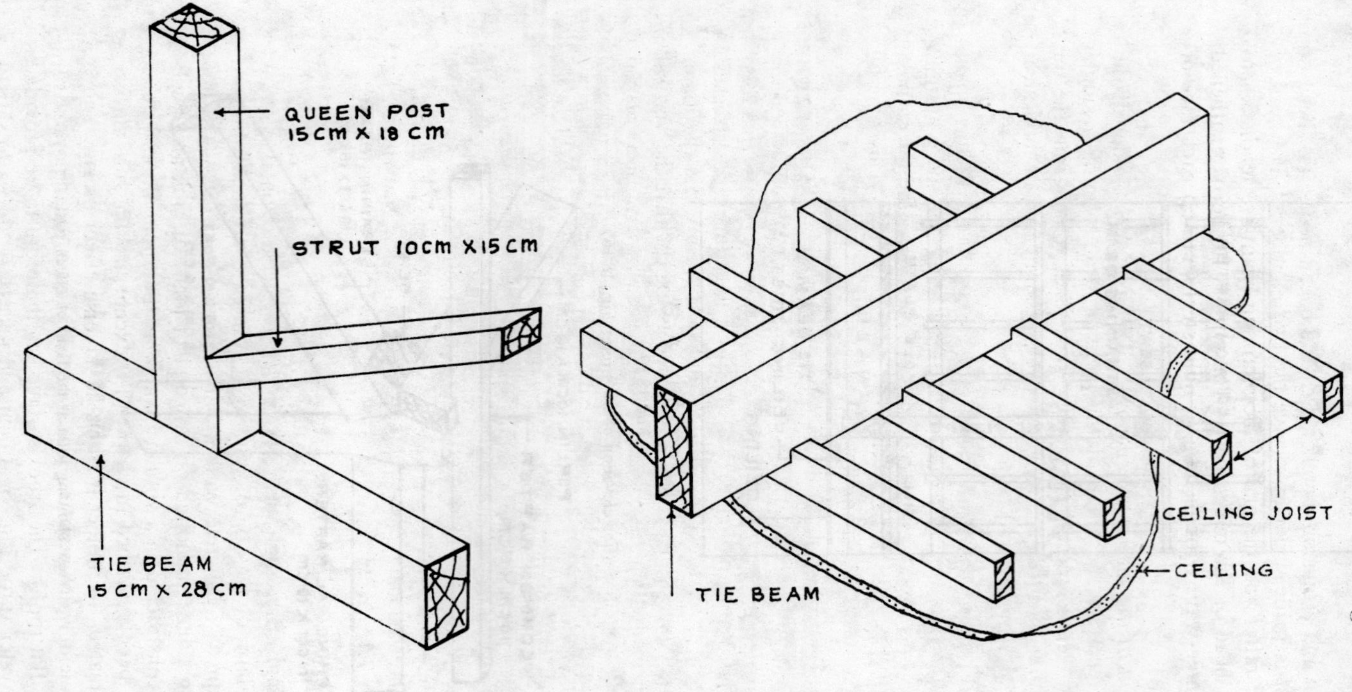

Fig. 18.19
View showing joint at the feet of queen post

Fig. 18.20
View showing method of fixing ceiling to the tie beam of a truss

straining beam and a tension member, *i.e.* the queen post. The head of the queen post is made wider to receive the two members. The head of the principal rafter and the end of the straining beam are tenoned into the widened head of the queen post. The joint is further strengthened by fixing a three-way, wrought iron or mild steel strap on each face of the joint.

(*ii*) *Joint at the feet of queen post.* The splayed shoulder of the queen post receives the tenon of the inclined strut forming a single abutment and tenon joint. The queen post itself tenons into the tie beam. This joint is further strengthened by wrought iron or mild steel stirrup strap held by gibs and cotters.

When it is desired to use wooden trusses beyond 14 m. span, the queen post truss is further modified by introducing one more upright member called *princess.* Diagonal bracing between the queen post and the ends of the tie beams is necessarily done as a rule. This type of frame-work is suitable for spans upto 18 m.

(8) **Mansard Truss.** With the introduction of steel trusses the Mansard roof truss is rarely constructed. This truss has two pitches, the lower varying from 60° to 70° and the upper from 30° to 40°. This truss is more or less a combination of king post and queen post trusses. The upper truss is shaped exactly to a king post truss while the lower truss resembles queen post truss in all respects except that the vertical upright members (which may be called as queen posts) are placed almost near the end of the tie beam. This type of truss provides economy in space and cost, but because of its odd shape, its use has become more or less obsolete these days.

(9) **Composite Roof Trusses.** Trusses built with the combination of wood and steel are termed as composite trusses. In this combination, advantage is taken of the high tensile strength of steel. In the truss frame-work the members in tension are usually replaced by steel rods. The various types of composite trusses are shown in Fig. 18·22–18·24.

18·6. FLAT ROOFS

A roof laid at an angle of less than 10° to the horizontal is know as flat roof. Flat roof may be of reinforced cement concrete, reinforced brick work, precast concrete with (channel units, waffle units, cored units etc), flag stones supported on rolled steel joists or jack arch type. Flat roofs are constructed in a similar manner as the upper floors (which have already been described in chapter 14) except that the roof surface is properly treated to protect it from the adverse effect of rain, snow, heat etc. Efficient water-proofing and roof drainage is an important requirement of flat roof. The roof surface has therefore to be provided with proper water proofing treatment and necessary slopes to satisfy

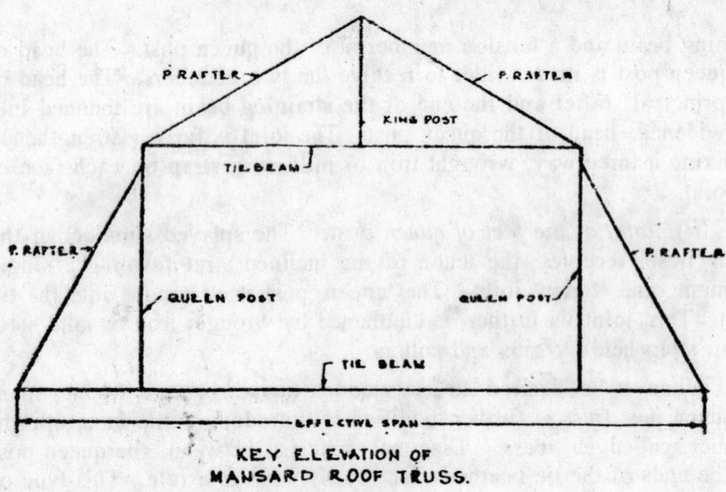

Fig. 18·21

these requirements. In addition a layer of insulating material (mud, lime concrete, mud phuska etc.) is laid over the roof surface to provide adequate thermal insulation to the space below. This layer is known as *terracing or grading*. In case of R.C.C. or R.B.W. roof slab, provided with terracing, a slope of 1 in 40 to 1 in 60 is considered desirable for proper drainage of rain water in areas of moderate rain fall. The slope may be increased, in case of flat roofs in heavy rain fall regions. The slope to the roof can be given either by varying the thickness of the terracing material or by constructing roof slab in slope or by providing part slope in the roof slab and part in the terracing material. In situations where thermal insulation is not essential, the slope for drainage of the roof is provided in the roof slab itself and the roof surface is given adequate water proofing treatment by using bitumen felts etc. In places where the terraced roof is to be used for outdoor living (playing, sleeping etc.) the top surface of the terracing should be made resistant to wear besides making provisions for efficient water proofing and drainage for the roof.

Although flat roofs are comparatively expensive yet keeping in view their numerous advantages, they are commonly constructed these days. The advantages and dis-advantages of flat roof are given below —:

Advantages of flat roof :

(1) The roof can be conveniently used as terrace for playing, gardening or sleeping purposes.

(2) It is easy to tender flat roof fire resistant.

(3) It avoids the enclosure of the triangular space.

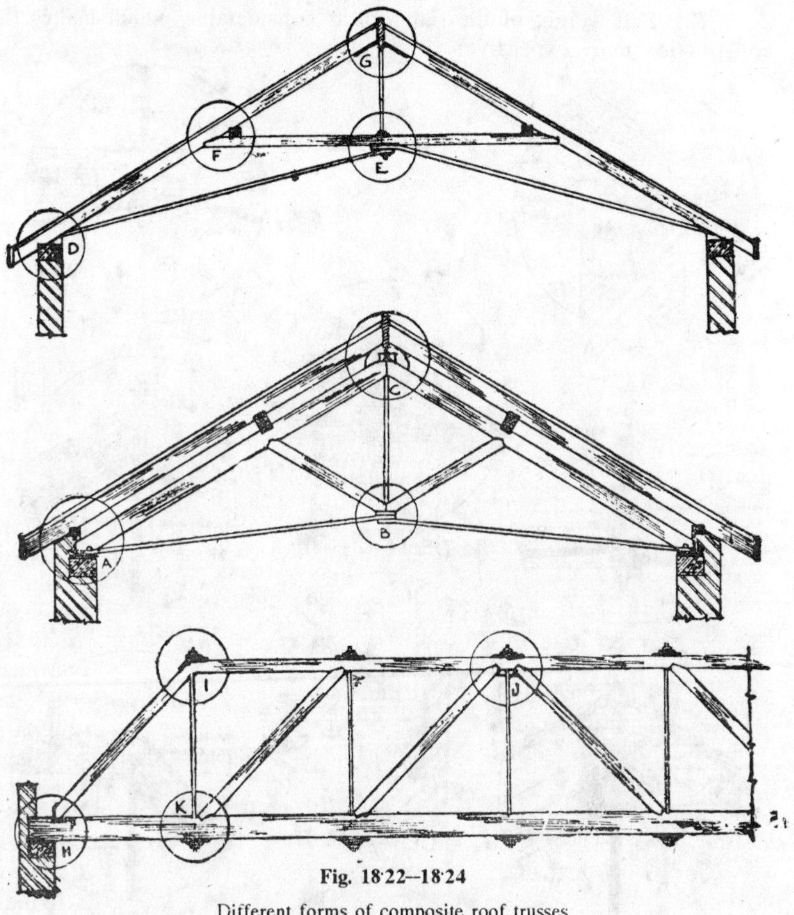

Fig. 18·22–18·24

Different forms of composite roof trusses

(4) It has good insulating properties.

(5) In this age of steel and reinforced concrete structures, the construction of flat roof is considered more simpler and architecturally suitable.

(6) In a multi-storeyed building, flat roof is considered to be the best choice.

(7) Flat roofs require lesser area of roofing material than sloped roofs.

Disadvantages of flat roof :

(1) Their span is restricted and as such they cannot be used for large spans without the introduction of intermediate columns and beams.

(2) Self-weight of the flat roof is considerable which makes the construction more expensive.

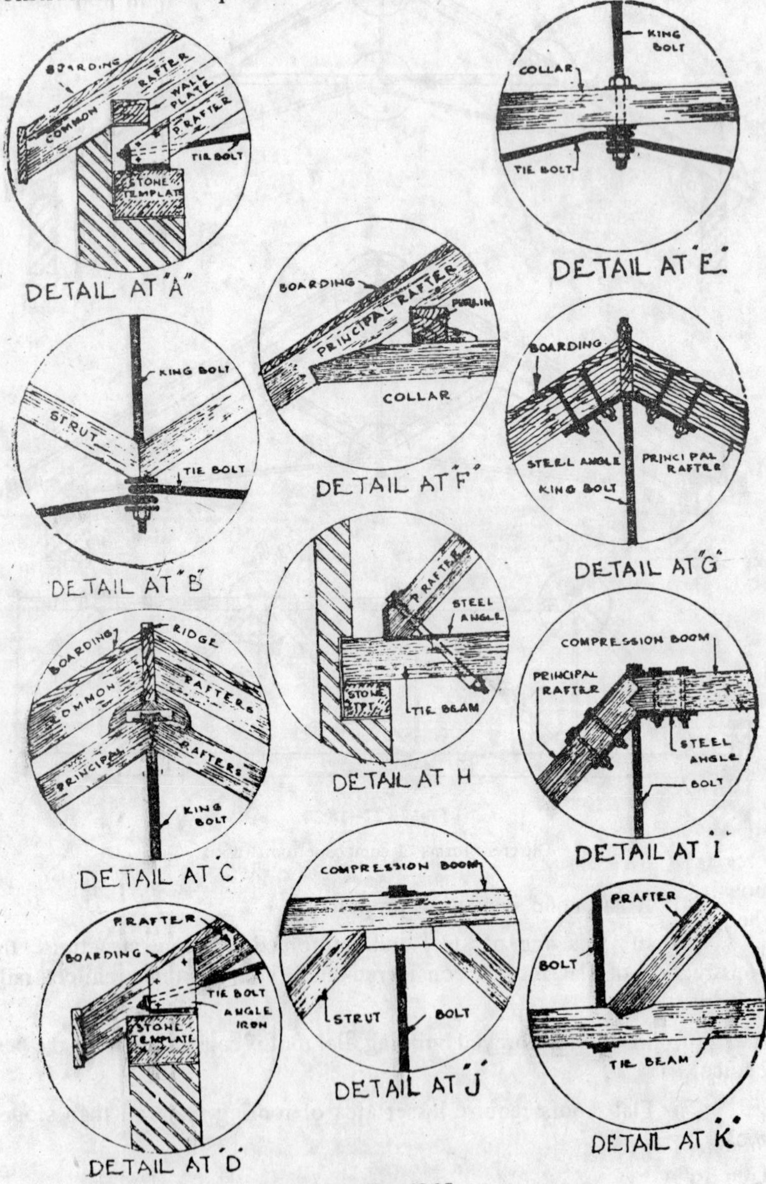

Fig. 18·25

Details of joints at various points of trusses shown in Figs. 18·22–18·24

(3) Flat roofs are exposed to sun and are subjected to voilent temperature changes, which may lead to cracks in the surface of the roof.

(4) The leak in a flat roof is difficult to trace and rectify.

(5) Its initial cost is much more than that of a pitched roof.

(6) Progress of construction of a flat roof is slow as compared with that of a pitched roof.

18·7. TYPES OF FLAT TERRACED ROOFING

The commonly adopted types of terraced roofing are as under.

(1) *Mud Terrace Roofing.*

(2) *Brick Jelly or Madras Terrace Roofing.*

(3) *Mud phuska Terracing with Tile Paving.*

(4) *Lime Concrete Terracing.*

(5) *Lime Concrete Terracing with Tile Paving.*

The different types of terraced roofing are described below :

(1) **Mud Terrace Roof.** This type of roof is extensively used in places where rainfall is less. This is the cheapest form of roof terracing and is fairly water-tight. The terracing is made with white earth mud containing large percentage of sodium salts. The mud terrace roof commonly constructed in Punjab is described below :

The roof in this case consists of T-sections 50 mm. x 50 mm. x 6 mm. spaced at 32 cm. centre to centre and supported on R.S.J. beams of the designed sections. The rolled steel joists, span along the width of the room and are spaced at suitable distance apart. Well burnt tiles measuring 30 cm x 30 cm. x 5 cm. or 30 cm. x 15 cm. x 5 cm. are placed between the flanges of the T-sections, and are set in lime mortar. A 15 cm. thick layer of stiff mud made from good white earth containing large percentage of sodium slats, is then laid over tiles. This layer is continuously beaten till the surface becomes hard and the beater rebounds when struck upon it. The surface is then plastered with mud cow-dung plaster and finally finished with cement cow-dung plaster 1:4 (1 cement : 4 cow-dung). To drain off rain water, the terracing is given a slope of 1 in 30.

In Maharashtra state and Madhya Pradesh, the mud terrace roofing is constructed as described below —:

Teak wood boards 4 to 5 cm. thick are nailed on the top of wooden joists of the designed section and spanning along the width of the room. A 2·5 cm. layer of wood shaving is spread over the boards and covered with layer of well-burnt bricks laid on edge and set in lime or mud mortar. This layer is further-covered with a 8 to 10 cm. thick layer of mud which is beaten hard. The mud layer is finally covered with 2.5 cm. thick layer of white earth containing large percentage of sodium salts. The top layer has to be renewed once a year.

(2) **Brick jelly or Madras Terrace Roofing.** It consists of a course of thoroughly brunt terrace bricks 15 cm. x 8 cm. x 2 cm. laid on edge in lime mortar 1 : $1\frac{1}{2}$ (one lime : $1\frac{1}{2}$ sand) diagonally across the joists of the timber roof. The bricks are immersed in water 24 hours before use and the thickness of joint is kept 10 mm. After the bricks are completely set, a 10 cm. thick layer of brick concrete 2 : $1\frac{1}{2}$ is laid over brick on edge course and beaten down to 7·6 cm. thickness by wooden hand beaters. The beating is continued till the beater makes no impression on the concrete and gets readily rebound by the surface when struck upon it. The whole surface is then cured for about a week by sprinkling lime water. After the concrete has set, three courses of Madras flat tiles, 15 cm. x 10 cm x 12 mm. are laid in lime mortar 1: $1\frac{1}{2}$. The thickness of the joints should not exceed 6 mm. and the vertical joints should be broken. The joints of the top layer are left open to afford key for the plaster. The top surface is then plastered with three coats of lime mortar and finally rubbed to a polished surface. The roof is usually laid to as lope of 1 in 36. The slope of the roof is given to the joists themselves and not by increasing the thickness of terracing.

Types of terracing at S. No.(3) to (5) *i.e.* Mud phuska terracing or grading with tile paving ; lime concrete terracing terracing or grading ; lime concrete terracing with tile paving, have already been described in chapter on Damp Proof Course (Article No. 5·12).

18·8. ROOF COVERING FOR PITCHED ROOFS

Roof covering is a material covering provided over the form-work of roof structure to act as a barrier for the rain, sun, wind or other such elements. The selection of covering material depends upon various factors such as availability of material, its initial cost and cost of maintenance, appearance and durability etc. There are various type of roof coverings and the one's which are commonly adopted in India are given below :

(1) *Thatch covering*
(2) *Shingles*
(3) *Tiles*
(4) *Asbestos-cement sheets*
(5) *Galvanised corrugated iron sheets*
(6) *Slates*

(1) **Thatch Covering.** This form of roof covering is largely used

in low cost houses and villages. It is suitable for rural areas mainly because it is the cheapest form of covering and is simple in construction. The frame-work for supporting the thatch consists of round bamboo rafters spaced at 30 cm. apart and tied with split bamboos laid at right angles to the rafter. The thatch is lightly secured to the frame-work with the help of ropes or twines dipped in tar. For proper drainage of rain water, the thatch covering should be at least 15 cm. thick and laid with a minimum slope of 44°. Thatch is combustible, absorbs moisture and is liable to rot or decay. It harbours rates and the bad smell of rotting materials makes it difficult to stay under the roof. On account of these defects, thatch covering is never allowed in good work.

(2) Shingles. The use of shingles is generally restricted to hilly areas where wood is easily available at low cost. Wooden shingles cut from well seasoned timber are laid in a similar manner as slates or tiles. Shingles are obtained in lengths varying from 30 cm. to 38 cm. and widths varying from 6 cm. to 25 cm. Generally two courses of shingles are laid at the eaves.

(3) Tiles. The tiles are named according to their shape and pattern and are manufactured in somewhat similar manner as bricks. The commonly used tiles are ; Flat pan tiles ; Half round country tiles ; Mangalore tiles, Quilon and other similar tiles. Tiling is one of the oldest methods of roof covering but with the introduction of asbestos cement sheets, its use is gradually declining day by day. The method of laying of a few types of tiles is given below —:

(*a*) *Flat or curved pan tiles and Half-round country tiles* : For laying the tiles, the ground work consists of battens nailed to the common rafters of the pitched roof. The common rafters are usually spaced 22 to 30 cm. apart and battens are fixed to the rafters at 6 cm. centre to centre. The tiles are laid with sufficient overlap. Ridges, hips and valleys are formed with specially moulded tiles specially made for the purpose. The last row of tiles laid near the eave, ridge, valley or hip is set in lime mortar. In order to avoid danger of water leaking through the valleys, lead flashing is normally done.

(*b*) *Mangalore or corrugated Tiles* : In ordinary works, the tiles are laid on battens but for a superior type of construction they are laid on boardings, covered with protective coat of tar or felt. Boarding is directly nailed to the purlin and tiles are laid on battens nailed on the boarding. In this case too, the ridge, valley and hip tiles are moulded into special shape. Ridge and hip tiles are laid dry and finally pointed with cement mortar.

(4) Asbestos cement sheets. Asbestos cement is material obtained by the combination of cement with about 15 percent of asbestos fibres. The sheets made out of this material are light, cheap, impervious, durable and fire resisting. It does not require any protective paint and cannot

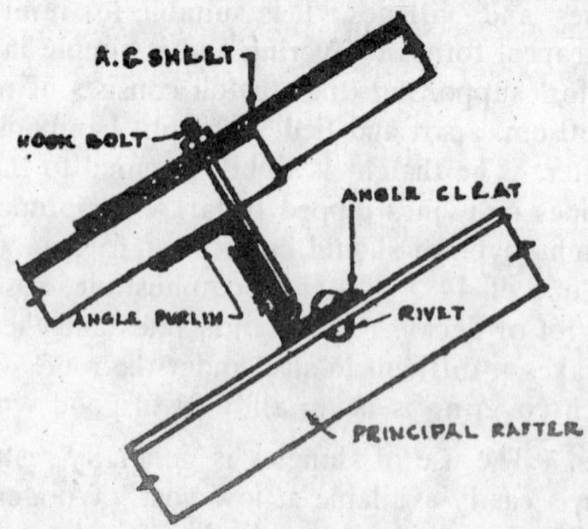

Fig. 18·26

Details of the end lap for A.C. sheets at purlin point

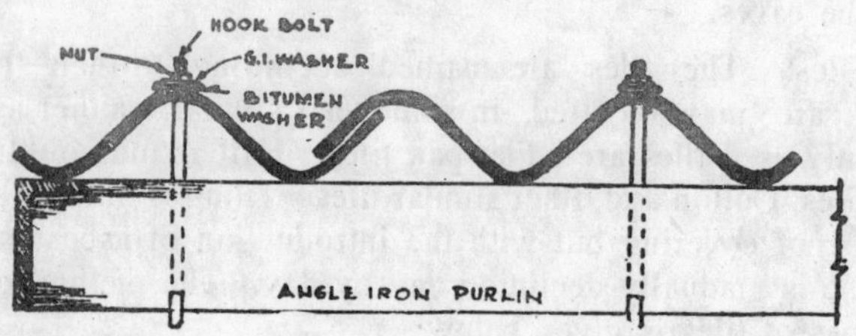

Fig. 18·27

Side lap for "Bix six" sheets

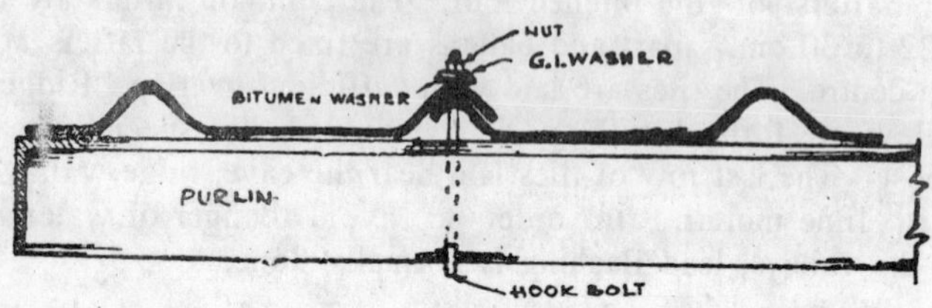

Fig. 18·28

Side lap for "Trafford" sheets

be eaten away by vermins. On account of these properties, asbestos cement covering is commonly adopted as a roofing material for factories, workshops, offices, garrages, generating stations and residential buildings. The two different types of sheets commonly used are "Big six" and "Trafford" sheets. These sheets are available in 2 to 3 m. length.

Fixing of A.C. Sheets. The following points should be carefully noted while fixing A.C sheets.

(1) The sheets should be laid with smooth side upward and the end marked "Top" pointing towards the ridge.

(2) They are usually laid with an end lap of 15 cm: but this lap can be slightly varied to suit the purlin spacing.

(3) The side laps for '*Big six*' and '*Trafford*' sheets consist of half corrugation and one corrugation respectively.

(4) The holes for fixing accessories must be drilled and not punched in the crown of the corrugations.

(5) The diameter of the hole should be 3 mm. greater than the diameter of the screw or bolt to be used.

(6) 8 mm. diameter hook bolts, crank bolts or coach screws are inserted in the 11 mm. diameter drilled holes and screwed lightly. A bitumen washer must be provided under G.I. flat washer with each bolt or screw. When 10 to 12 sheets have been laid, the nuts of the screws or bolts are screwed tight. The screws or bolts on purlin are not screwed very tight.

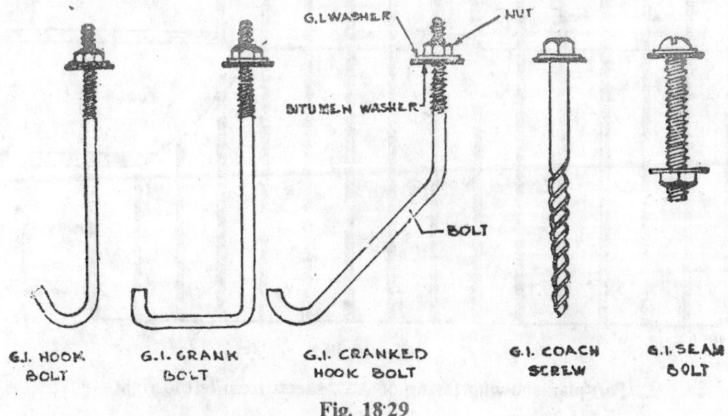

Fig. 18·29

Different forms of bolts and screws used for fixing A.C. sheets

(7) Ridge capping should preferably be secured to the ridge purlin by the same kind of bolts which are used for fixing sheeting. "J" bolts should be used for angle purlin and crank bolts should be used for purlins made up of timber, rolled steel joist, channel etc. Coach screws may be used for timber purlins.

(8) *Mitre* (cut) is necessary to avoid gaps where four sheets meet at a lap.

(*a*) In Big six sheets *mitre* may be defined as a cut formed at the required corner by cutting the corner of the sheet for a length of 15 cm. or the length of end lap whichever is more along the vertical edge and for a width equal to 5 cm or length of the side lap whichever is more along with horizontal edge of the sheet.

(*b*) In Trafford sheets, *mitre* may be defined as a cut formed at the required corner by cutting the corner of the sheet for a length of 15 cm. or the length of the end lap whichever is more along the vertical edge and for a width equal to the width of one corrugation along the horizontal edge of the sheet.

(9) The unsupported overhang of the sheets at the eaves. should not preferably be more than 30 cm.

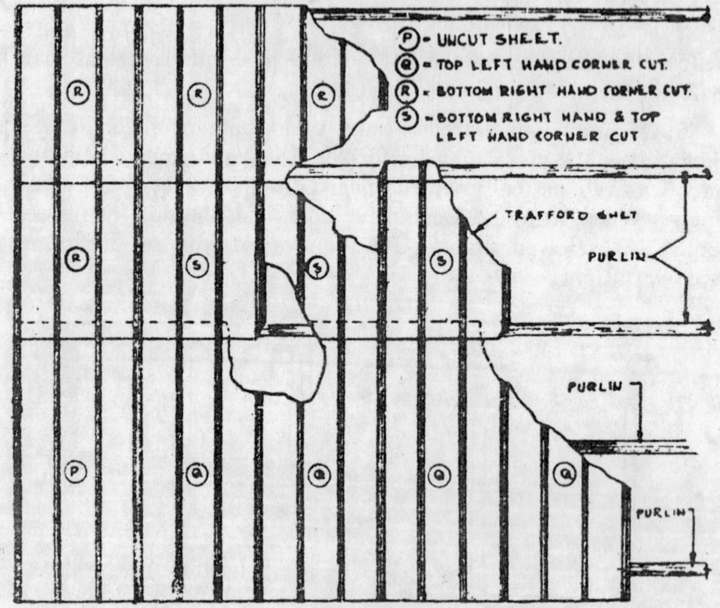

Fig. 18·30
Part-plan showing laying of A.C. sheets from left to right

Laying of 'Big Six' Sheets. Sheets may be laid from left to right or *vice versa*, preferably laying should commence at the end opposite to the direction of the prevailing wind. The method of laying Big six sheets from *left to right* is described below :

The laying of sheets is started at the eaves. In the first row or in the eaves course, first sheet is laid uncut while the remaining sheets have top left-hand corners metred. In the second or intermediate rows; the first sheet in each row has bottom right-hand corner mitered while the other sheets have both top left-hand and bottom right hand corners mitred. In topmost row, all the sheets except the last one have bottom right-hand corners mitred. The last sheet is laid uncut.

If the sheets are to be laid from right to left, the whole procedure will be reversed.

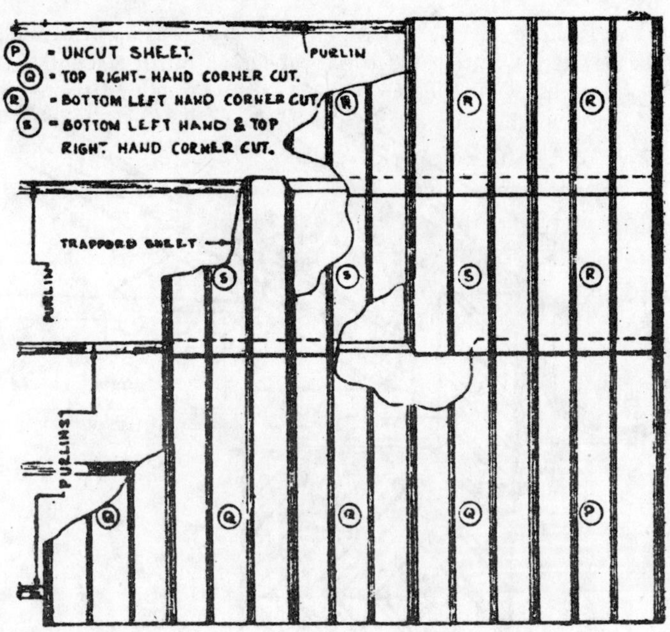

Fig. 18·31

Part-plan showing laying of A.C. sheets from right to left

(5) **Galvanised, Corrugated Iron Sheets.** Ordinarily 22 gauge sheets are extensively used as roof covering material in factories, workshops, generator sheds etc. The sheets are generally not laid at a slope flatter than 1 in 4. These sheets are durable, fire resistant light in weight and require no maintenance. The purpose of corrugation is to give the thin sheet additional strength. Sheets are galvanised to protect them from rusting action of weather. The sheets are laid with end lap of not less than 15 cm. and side lap varying from $1\frac{1}{2}$ to 2 corrugations. The sheets should be fastened above eaves by 25 mm x 6 mm flat iron wind ties. The holes in the sheet crowns may be drilled or punched. However, when four sheets overlap at corners, the holes should be drilled and not punched. The sheets are secured to the purlins by means of galvanised iron hook bolts, screws and nails etc. with curved washers.

(6) **Slates.** The stone used for slate roof covering is obtained from sedimentary rocks of limestone or sandstone. It is seldom confused with the slate obtained from metamorphic rock. Stone slates vary considerably in size, shape and weight. They form the most durable roof covering material, but in India it is rarely adopted. The common size of slates vary from 60 cm x 35 cm to 25 cm x 12 cm. The thickness of the regular size stone slate varies from 1·6 mm to 4·8 mm. The slates may

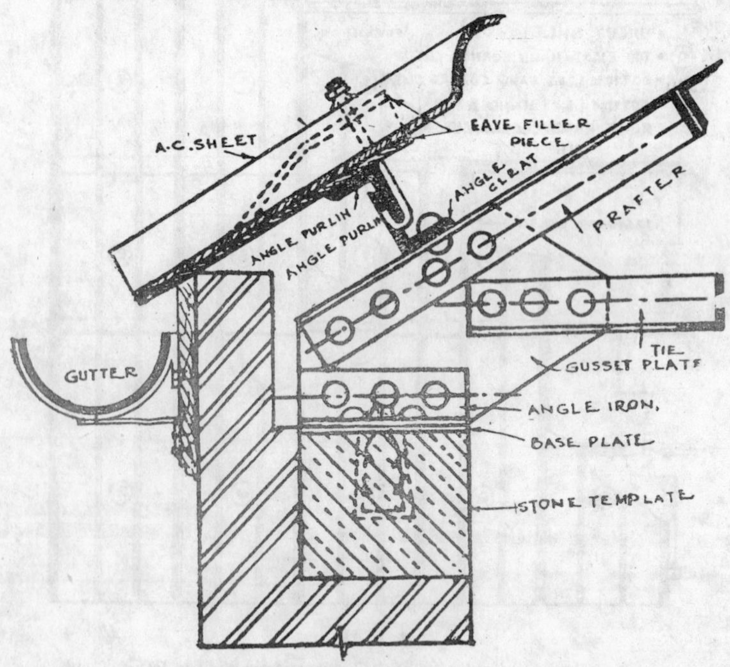

Fig. 18·32

Detail showing the arrangement of fixing A.C. sheets at the eaves level of steel truss

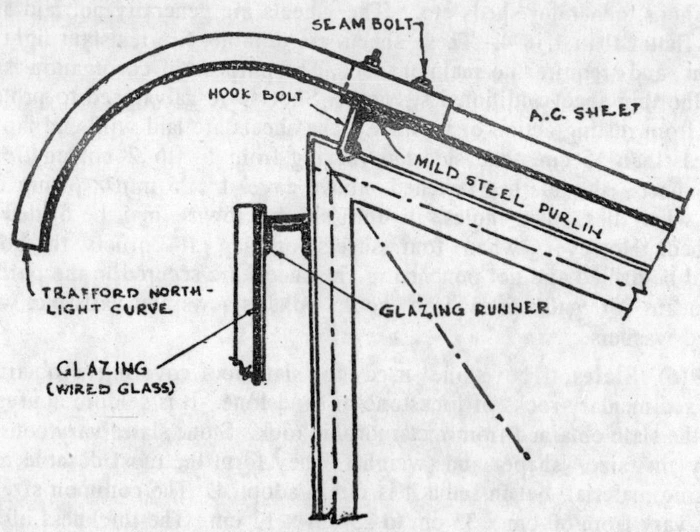

Fig. 18·33

Detail showing A.C. sheet covering for a North light roof truss

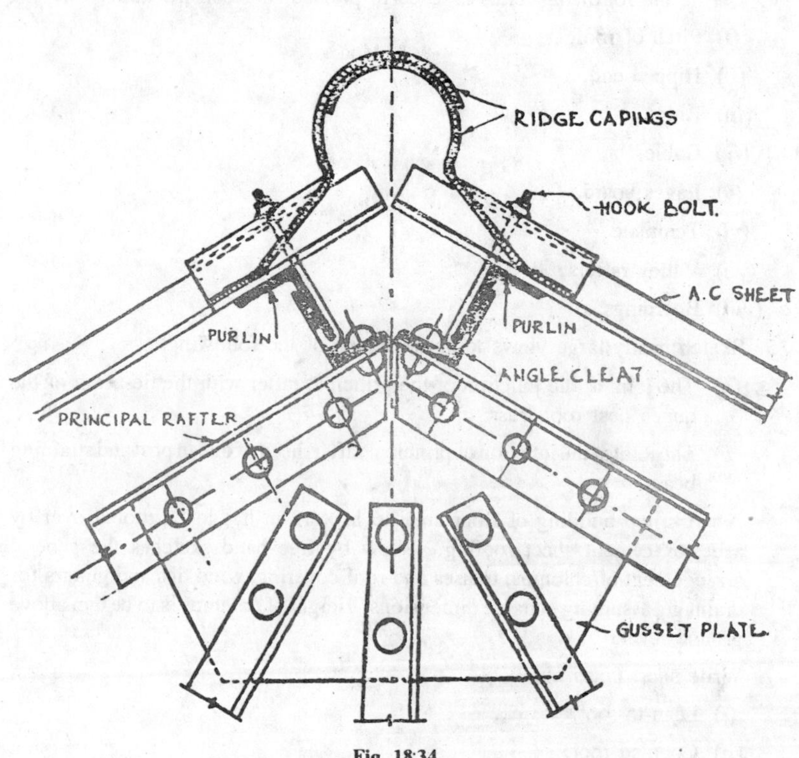

Fig. 18·34

Details showing the arrangement of fixing A.C. sheets at the crown of a steel truss

be fixed on the battens fixed on timber rafters, or they may be laid on wooden boarding fixed on the roof framework. The slates are secured to the base which may be of battens or boarding, through nails of zinc, copper or galvanised iron.

Questions for Review

1. What are the various types of roofs commonly used in India ?

   Describe each stating clearly the distinction between the different types.

2. Draw to scale of 1:25 a lean to roof over a verandah 2·4 m wide and 7·32 m long, with 60 cm high plinth. The timber roof is to be supported by brick $1\frac{1}{2}$ square piers and covered by Manglore tiles. Draw half full size detail at the eaves and at its junction with the main wall, which is $1\frac{1}{2}$ brick thick.

3. Write a detailed note on the different types of the roof coverings for pitched roof which are generally used in India. Describe the method of laying and fixing them with the aid of sketches.

4. Define the following terms as used in pitched roof construction:
   (*i*) Pitch of roof.
   (*ii*) Hipped end.
   (*iii*) Ridge.
   (*iv*) Gable.
   (*v*) Eaves board.
   (*vi*) Template
   (*vii*) Valley rafters.
   (*viii*) Boardings.
5. Sketch neatly large views to show details of the following:
   (*a*) The joint at the junction of the principal rafter with the tie-beam of the queen post roof truss.
   (*b*) The joint at the junction of principal rafter, head of queen post and straining beam.
6. A workshop building of floor area 46 m x 12 m has to be roofed over by asbestos cement sheet roofing. Show by free hand sketches the general arrangement of columns, trusses and roof covering, wind ties and gutters for drainage, assuming suitable dimensions. Height of columns is to be 6 m above ground level.
7. Write short notes on.
   (*i*) Lean-to-roof.
   (*ii*) Coupled roof.
   (*iii*) Coupled-close roof.
   (*iv*) Collar roof.
   (*v*) Mansard roof truss.
   (*vi*) Composite roof trusses.
8. A 2.4 m wide corridor, 9 m long connecting a school building with its physical training hall is to be covered with Mangalore tiled roof. The plinth is 30 cm. high and the roof is supported on brick piers. Draw the plan and cross-section of the corridor to a scale of 1:50 and give quarter full size details of the roof at its junction, with the pier, showing also the eaves.
9. (*a*) Write a short note on the different types of flat roofs and add sketches to illustrate their construction.

   (*b*) It is proposed to provide Madras terrace roofing over an apartment 12 mX 6 m as only teaks wood is available for use as beams and joists in a certain locality.

   Draw longitudinal section and cross-section of the apartment to illustrate your design for its construction.

10. A bungalow has a drawing room measuring 5 m x 7·5 m which is covered with a pitched roof. Draw the plan and cross-section of the roof to a scale of 1:25 and give sizes of the scantlings used.

11. Write short notes on any four of the following :

    (*i*) Laying of asbestos-cement sheets.

    (*ii*) Laying of Mangalore or corrugated tiles.

    (*iii*) Thatch covering.

    (*iv*) King post roof truss.

    (*v*) Queen post roof truss.

    (*vi*) Mansard truss.

12. A hall 8 m x 16 m is required to be roofed over with Mangalore tiles on king post truss. The trusses are spaced at 3 m centre to centre, and are supported on $1\frac{1}{2}$ brick, thick walls on either side. Draw detailed sketches for the following:

    (*i*) Junction of tie beam with principal rafter.

    (*ii*) Arrangement of fixing the truss to the wall supports.

    (*iii*) Joint between the principal rafters and the king-post.

    (*iv*) Joint between the king-post and the tie-beam.

13. Draw to a scale of 1:25 a roof truss over 6 m span and suggest the scantlings which in your openion you consider suitable. The side walls are one brick thick and the roof is to be covered with mangalore tiles.

14. A hall 10 cm x 20 m is required to be roofed over with asbestos cement sheeting on queen post truss. The trusses are spaced at 3 m centre to centre and are supported on two brick thick walls at either end. Sketch the following details:

    (*i*) A lay-out plan indicating clearly the position of trusses.

    (*ii*) A sectional elevation and section of the queen post truss with suitable sizes of the scantlings.

    (*iii*) Joint at the head of queen post.

    (*iv*) Joint at the feet of queen post.

15. An assembly hall 10 m x 30 m is to roofed with Mangalore tiled roof over wooden trusses, rafters etc. Sketch the type of truss you would propose and show in detail (with dimensions) and three joints in the proposed truss.

16. (*a*) What type of roof will you suggest for a room 4·80 m by 7·20 m at a place where rainfall is over 180 cm a year ? Give reasons for your choice.

    (*b*) If you select a pitched roof, draw cross- section of the truss and purlins etc.,showingapproximatesizesofthemembersandjointsbetweenmembers. If you recommend a flat roof, how will you lay it and what precautions will you take to make it leak-proof ?

17. (*a*) A queen-post timber truss is to be made for a span or 12 m. You have

only 20 cm x 13 cm size of timber, 6·5 m long. Draw a neat sketch of the lengthening joint you will provide in making the tie beam.

(*b*) A lean-to-roof with Mangalore tiles is to be provided for a verandah 3 m wide. Draw a neat sketch of the roof showing clearly all the relevant details.

18. Show the construction details of cycle-stand, for keeping 10 cycles. How will you provide protection from sun, wind and rain ? What factors will have to be taken into consideration if the cost is to be minimum ?

19. Write short notes on the following :

(*i*) Madras Terrace roofing

(*ii*) Couple- close-roof

(*iii*) Mud phuska terracing with tile paving

(*iv*) King-post roof truss.

20. Define the following terms as applied to sloped roof construction :

(*i*) Eaves board

(*ii*) Templates

(*iii*) Pitch of roof

(*iv*) Valley rafters

(*v*) Ridge

(*vi*) Common Rafters.

21. What are advantages and disadvantages of flat roof ? Describe the method of providing terracing with lime concrete over R.C.C. flat roof.

19

Structural Steel Work

19·1. SHAPES OF STRUCTURAL STEEL

The cross-sectional properties of a steel member are identified by its shape. The various names given to the members of different shapes are Angle; T-section ; I-section ; Z-section ; Channel ; H-section and so on. Each shape has its particular use in steel structures. The various sections are fabricated form white hot steel by passing it through rolling mills or other machines. An angle section is named by writing the two leg lengths and the thickness of the plate forming the angle. An equal angle-section

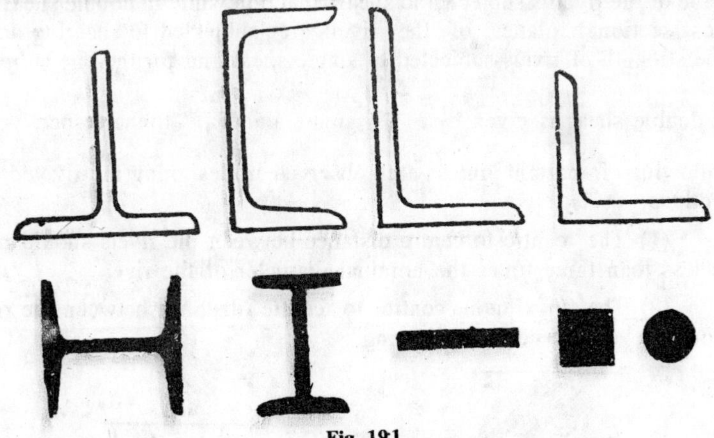

Fig. 19·1

Different shapes of structural steel sections

having length of its each leg as 8 cm and its thickness equal to 12 mm will be written thus : 80 x 80 x 12 mm. In case of I section, the depth of the section is written first. Thus an I-section weighing 34·2 kg/m and having a depth equal to 200 mm and flange width 100 mm will be written as 200 mm x 100 mm x 34·2 kg/m. Similarly, the other shapes of steel section are named by writing their dimensions and weights.

19·2. RIVETS

The sections use in steel structures are connected to each other by riveting or bolting. Welding is another alternative method of connecting the structural sections. The various forms of riveted connections are shown in Fig. 19·2. The rivet is rendered soft by heating and then it is placed in the rivet hole. The shank extends out of the prepared rivet hole and the extended part of the shank is given the shape of a second head by riveting machine. The usual form of rivet head which is cupshaped in appearance is termed as *snap head.* Thus snap head always extends out from the face of the riveted plate. Incase, it is desired to have the second head flush with the plate or member to be riveted, countersunk head is employed. The two different types of heads are shown in Fig. 19·3. Depending upon the nature of connection, the rivets may be subjected to single shear or double shear. In single shear, one cross-sectional plane of the rivet is subjected to shearing action while in double shear, two cross-sectional planes of the rivet are subjected to shearing action. The strength of rivets subjected to single shear and for the one subjected to double shear is given by $\frac{\pi d^2}{4}$ tonne and $\frac{\pi d^2}{2}$ tonne respectively.

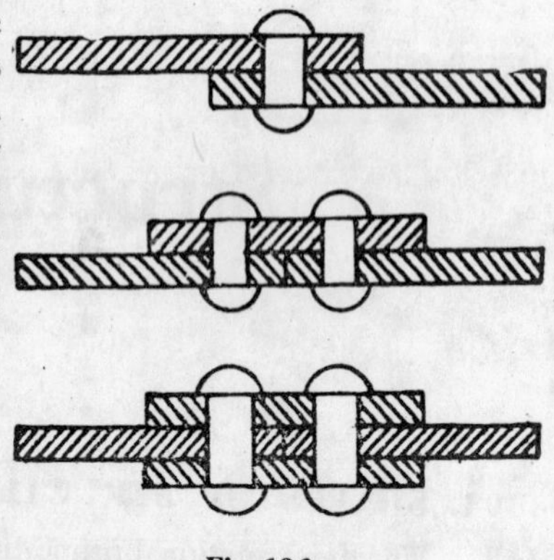

Fig. 19·2

Types of Riveted connections

Following important rules are observed in designing the riveted connections.

(1) The centre to centre distance between the rivets should never be less than three times the nominal diameter of the rivet.

(2) The maximum centre to centre distance between the rivets should never exceed the following :

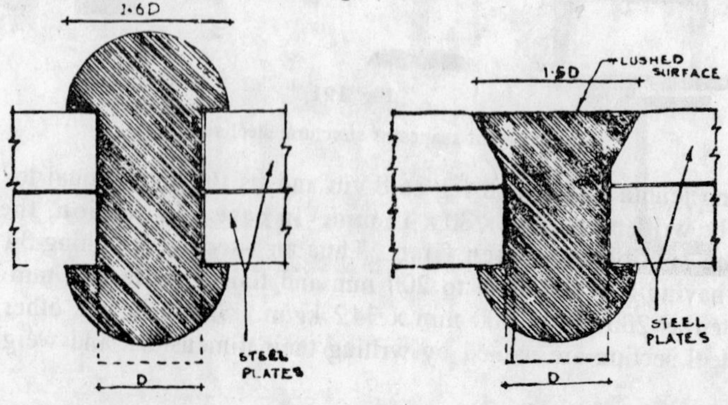

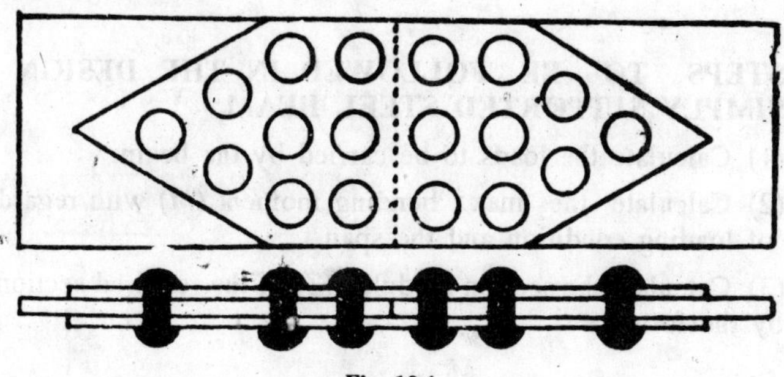

Fig. 19·4

Plan and sectional elevation of a riveted connection

(*a*) In case of members subjected to tension : 10 times the thickness of the thinnest outside plate or angle or 20 cm whichever is less.

(*b*) In case of members subjected to compression : 16 times the thickness of the thinnest outside plate or angle or 15 cm whichever is less.

(3) The distance between the edge of a rivet and edge of the plate etc. should never be less than the diameter of the rivet.

19·3. BEAMS

I-sections or rolled steel joists are commonly used as simple beams. They are suitable for moderate loading conditions and spans. But for heavy loads and longer spans, compound beams are used. They are fabricated from one or more rolled steel joists or riveted with plates and angles. In some cases only plate and angle are riveted together to form a compound beam. Plate girder and box girder are two examples of compound beams. The various forms of steel beams are shown in Fig. 19·5.

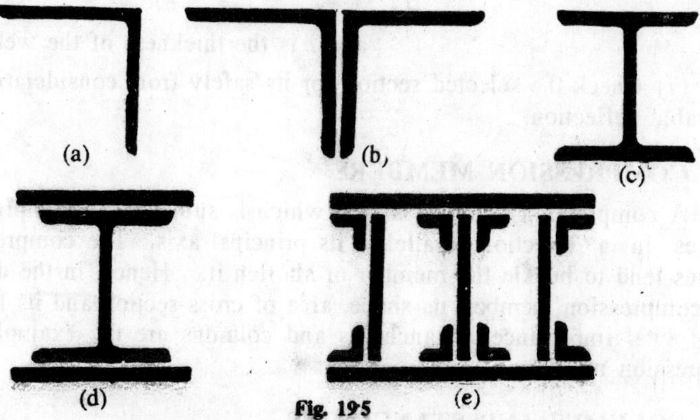

Fig. 19·5

Cross-section of various types of beams

19·4. STEPS TO BE FOLLOWED IN THE DESIGN OF A SIMPLY SUPPORTED STEEL BEAM

(1) Calculate the loads to be carried by the beam.

(2) Calculate the max. bending moment (M) with regard to the nature of loading condition and the span.

(3) Calculate the section modulus (Z) of the required section of the beam by the formula :

$$Z = \frac{M}{f}$$ where f is the safe allowable stress in steel in bending.

(4) From the table on "Properties of section" select a suitable section having its section modulus slightly greater than the calculated one. In making the selection of the section, deeper section should be preferred

(5) Calculate max. shear force in the beam.

(6) Check the section for safe shear stress by the formula :

$$s = \frac{S}{I\,b}\,A\bar{y}$$

Where, s = intensity of shear stress.

S = max, shear force in beam.

$A\bar{y}$ = The moment of the area of the part of the section situated beyond its neutral axis.

Incase of a rolled steel joists, this formula may be reduced to

$$s = \frac{S}{d.t.}$$ where d is the over-all depth of the joist and t is the thickness of the web.

(7) Check the selected section for its safety from consideration of allowable deflection.

19·5. COMPRESSION MEMBERS

A compression member is one, which is subjected to compressive stresses in a direction parallel to its principal axis. The compression stresses tend to buckle the member or shorten it. Hence, in the design of a compression member, its shape, area of cross-section and its length are of vital importance. Stanchions and columns are the examples of compression members.

19·6. COLUMNS AND STANCHIONS

Column is a vertical compression member which is circular in

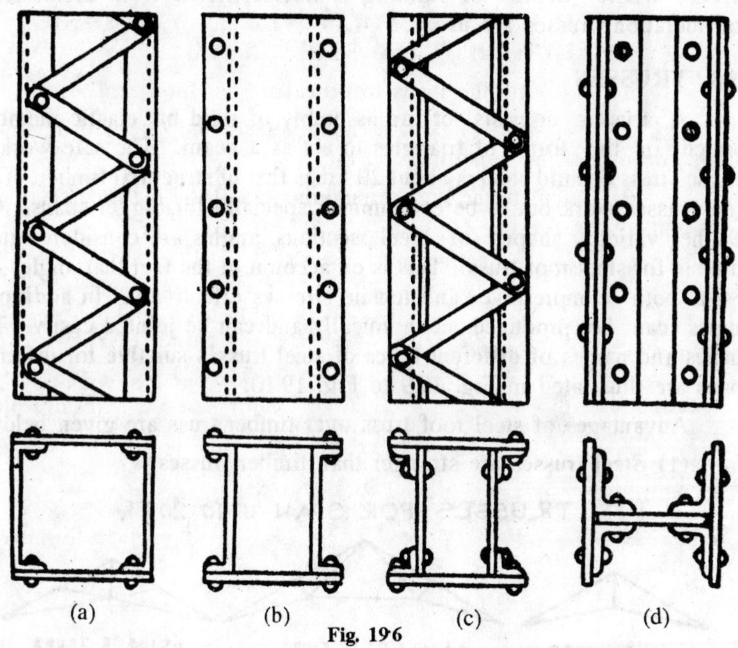

Fig. 19·6
Plan and elevation of various types of stanchions

cross-section. It may be hollow or solid in section. Vertical compression member may be termed as stanchion. Depending upon the height of stanchions, and the loading conditions ; they may be fabricated in various shapes. Long column tends to fail in bending and as such, from consideration of economy, the column should be so shaped that the variation between the maximum and minimum radii of gyration about its two principal axis and the value of its $\frac{\text{area}}{\text{least radius of gyration}}$ should should be as small as possible. Hence, single sections of rolled steel joists, angles or channels etc. are commonly adopted for columns subjected to light loading conditions. For columns required to carry heavy loads, two or more sections of angles channels, rolled steel joists, etc. are joined to form a compound section. The different types of stanchions are shown in Fig. 19·6.

19·7. FRAMING LONG SPANS

For auditoriums, theatres, hangers, large industrial buildings where greater clear distance is necessary, girders, trusses; arches etc. fabricated from steel section are used for roofing. Girders are used in situations where the depth of framing is limited to 1 m. When the designed depth of the beam works out to be more, plate girders are employed. But in places where depth of framing is not restricted from architectural consideration, trusses are used.

19·8. TRUSSES

A trusses consists of an assembly of rigid but elastic members jointed in the form of triangles to act as a beam. The safe working tensile stress of mild steel is about 20 times that of structural timber. Thus steel trusses work out to be economical, specially for bigger spans. Out of the various shapes of steel sections, angles are considered most suitable for steel roof truss. This is on account of the fact that angles can resist both compressive and tensile stresses effectively. In additional angles can be produced economically and can be jointed easily. The shapes and names of different types of steel trusses suitable for different spans are illustrated in Fig. 19·9 to Fig. 19·10.

Advantages of steel roof truss over timber truss are given below :

(1) Steel trusses are stronger than timber trusses.

(a) TRUSSES FOR SPAN UPTO 20 FT.

(b) TRUSSES · FOR SPAN UPTO 30 FT.

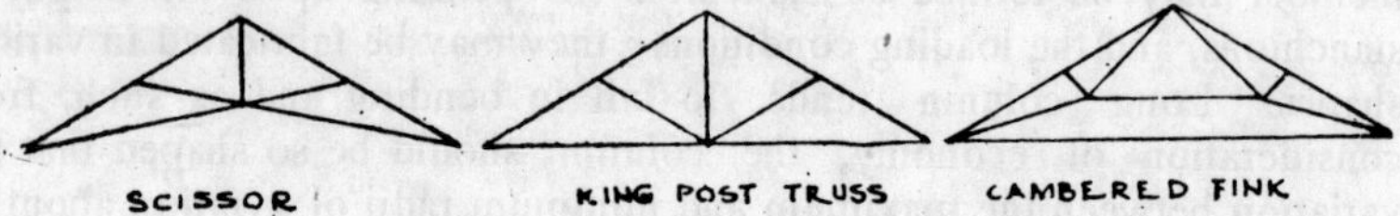

(c) TRUSSES FOR SPAN FROM 30 TO 40 FT.

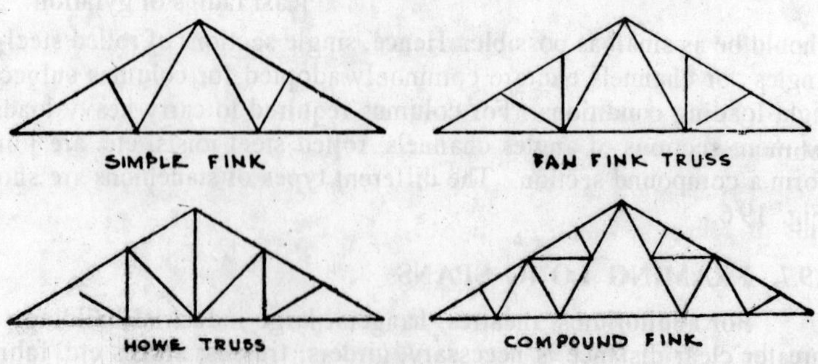

Fig. 19·7

Types of steel trusses suitable for various spans

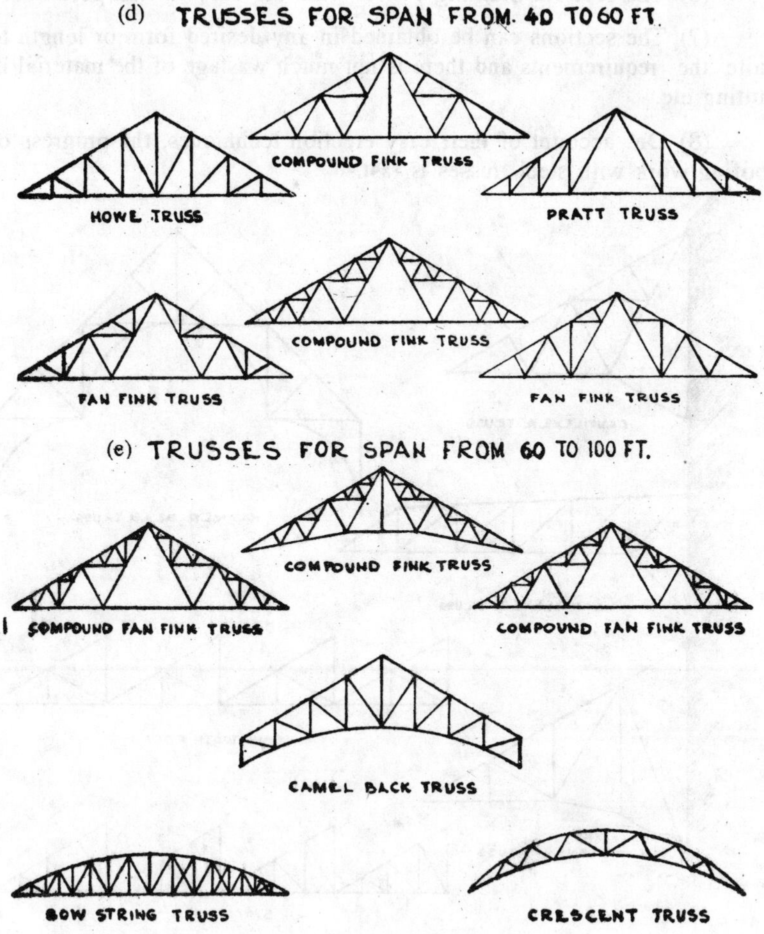

(d) TRUSSES FOR SPAN FROM 40 TO 60 FT.

COMPOUND FINK TRUSS

HOWE TRUSS

PRATT TRUSS

COMPOUND FINK TRUSS

FAN FINK TRUSS

FAN FINK TRUSS

(e) TRUSSES FOR SPAN FROM 60 TO 100 FT.

COMPOUND FINK TRUSS

COMPOUND FAN FINK TRUSS

COMPOUND FAN FINK TRUSS

CAMEL BACK TRUSS

BOW STRING TRUSS

CRESCENT TRUSS

Fig. 19·8

Types of steel trusses suitable for various spans

(2) Steel section forming the truss are light in weight and can be fabricated in any desired pattern to suit the architectural requirement.

(3) There is no danger of the material being eaten away by white ants or other insects.

(4) Steel trusses are more fire-resisting than timber trusses.

(5) They do not have span restrictions and as such steel trusses can be used for industrial buildings and other such structures where large areas are required to be covered without obstructions due to columns etc.

(6) The sections forming a steel truss are easy in transportation.

(7) The sections can be obtained in any desired form or length to suite the requirements and there is not much wastage of the material in cutting etc.

(8) On account of their easy erection techniques, the progress of roofing work with steel trusses is fast.

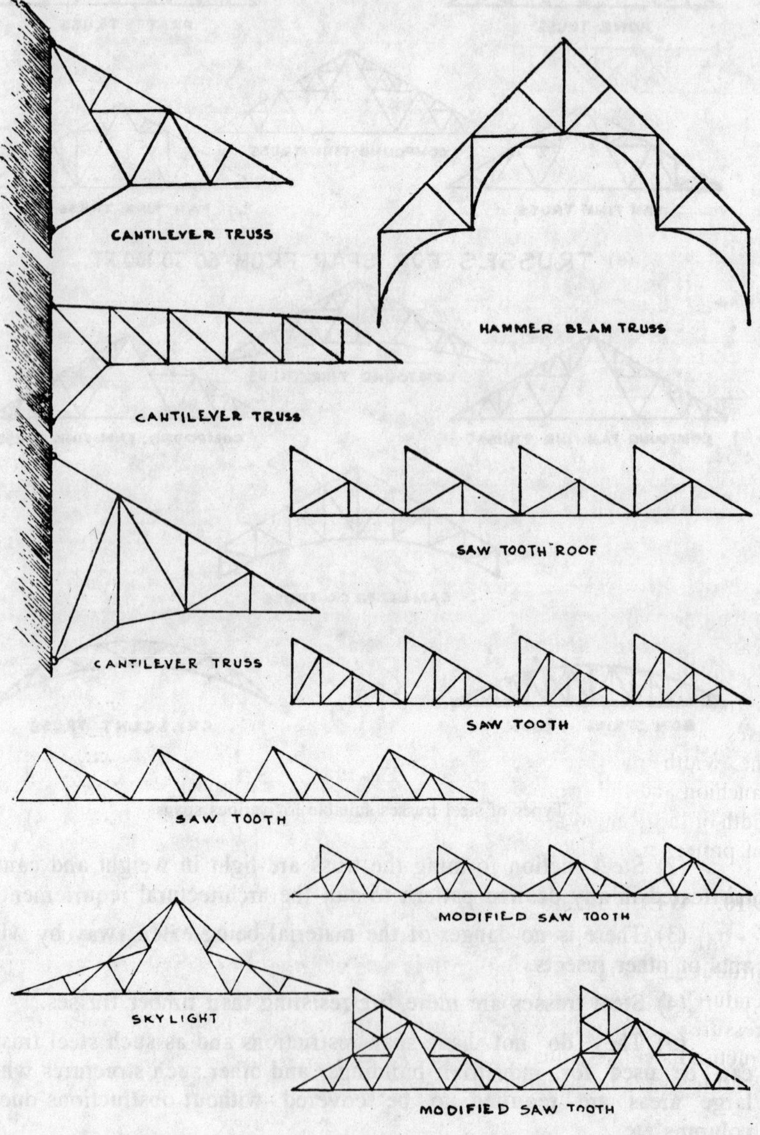

Fig. 199 Various types of steel trusses provided in special places

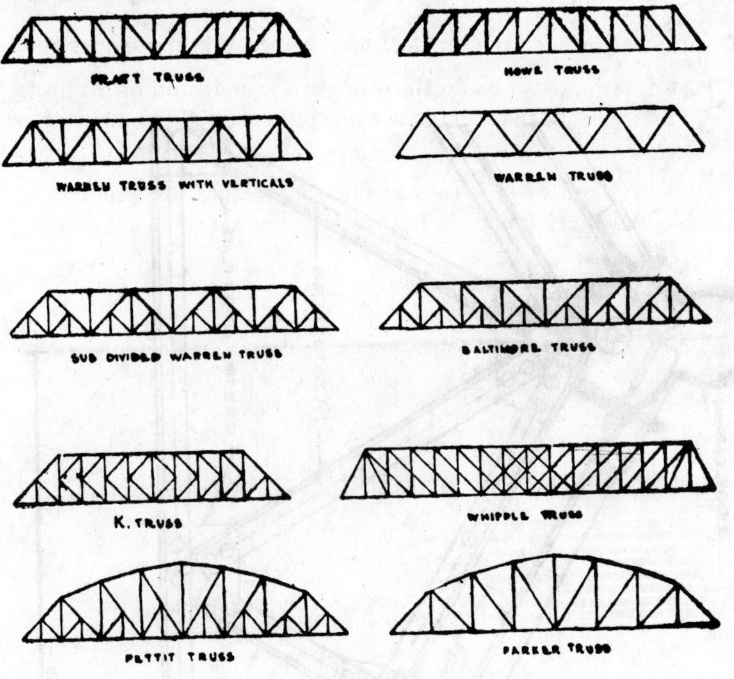

Fig. 19·10

Types of steel framing suitable for longer spans

19·9. STANCHION BASE

Stanchions are riveted to specially built bases. The base of a stanchion consists of an arrangement of base plate, gusset plates, angle cleats, stiffeners, fastenings etc. The stanchion is fully connected to the base plate with the help of the above fixtures and finally the base plate is fixed to the foundation of stanchion (which may be in most of the cases a simple cement concrete block foundation) through rag bolts. The width of the base plate varies from 2 to 4 times the width of the stanchion and the height of the gusset plates vary from 1½ to 2 times the width of the stanchion. To avoid shear and bending, the base plate should not project more than 8 times its thickness beyond the stanchion face.

19·10. WELDING

It is a process of jointing two or more structural members by introducing fused metal into fillets between them or by raising the temperature of their surfaces to the fusion temperature and then applying pressure. It may thus be said to be an alternative method of connecting structural steel sections. With the development of welding it has become possible to render architectural beauty to steel structures also. Structural

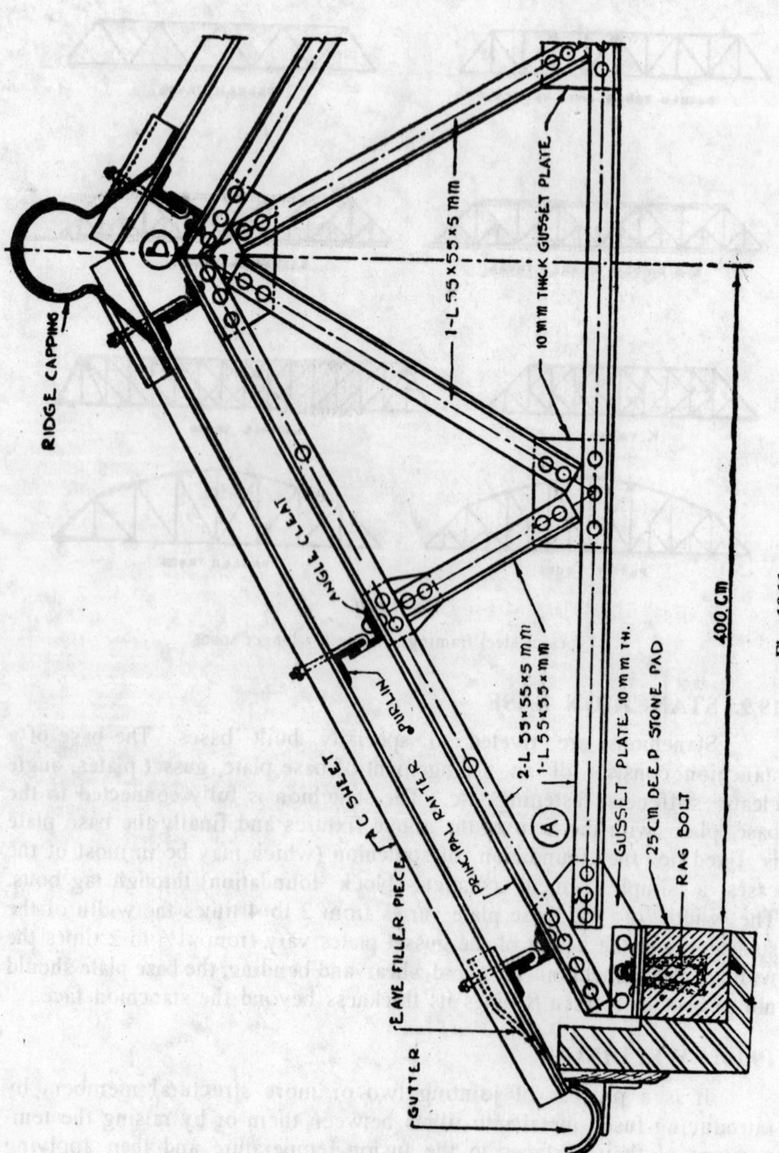

Fig. 19·11 Steel truss suitable for 8 m, spans

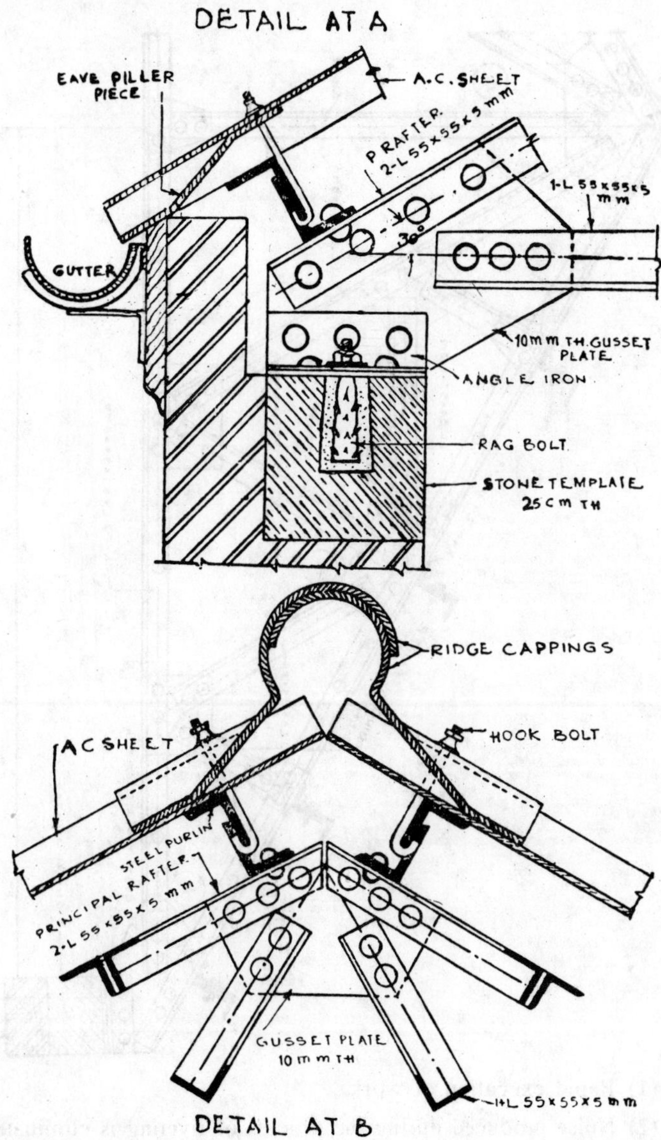

Fig. 19·12–19·13

Details of the truss shown in Fig. 19·11

members of the desired pattern can thus be joined flush to a smooth and pleasing surface. It has a number of advantages over riveted connections some of them are given on 474

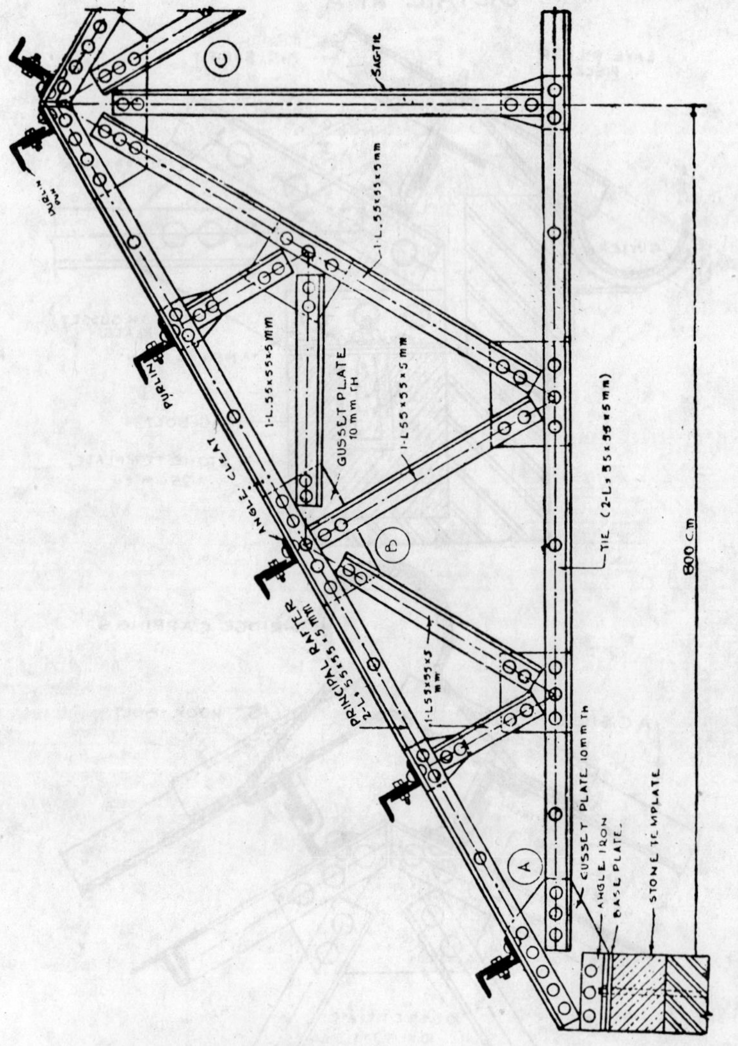

Fig. 1914 Steel truss suitable for 16 m. span

(1) Rapid execution of work.

(2) Noise produced during the process of riveting is eliminated in welding.

(3) It is economical when used on a large scale.

(4) There is a great saving of material as the angle cleats which are necessary in riveted connection, are eliminated in welding.

(5) In welded connection the entire section is considered

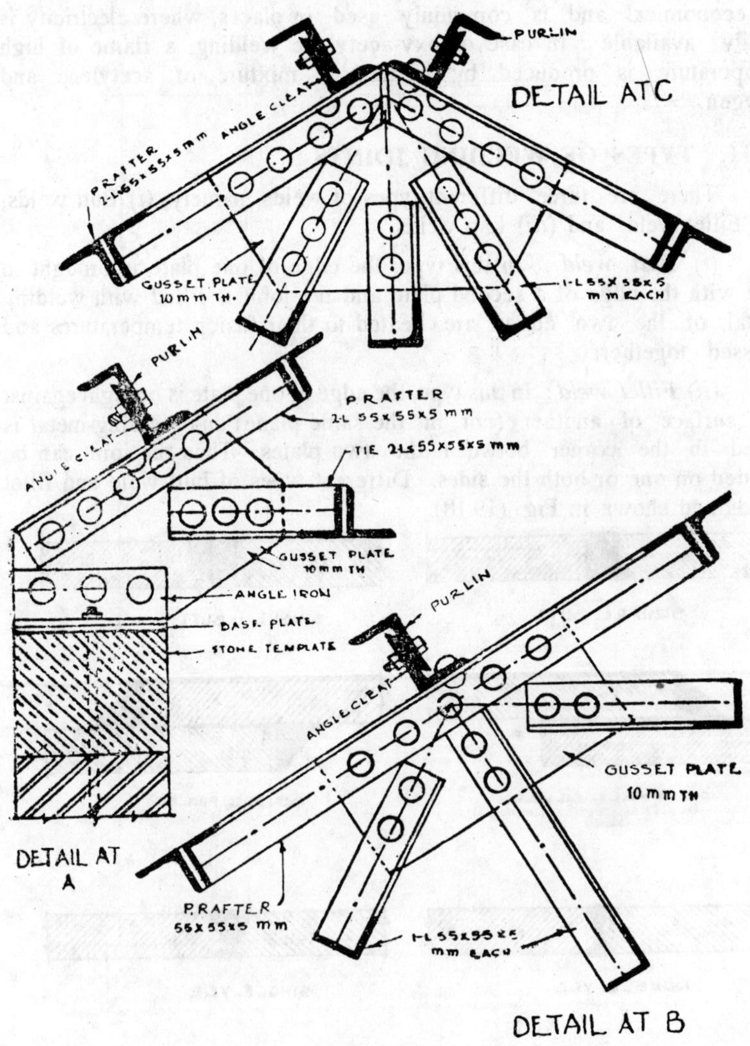

Fig. 19·15–19·17

Details of the truss shown in Fig. 19·14

effective for taking up tensile stresses as there is no reduction in sectional areas on account of holes etc.

There are two different methods of welding commonly adopted. (1) Electric arc welding (2) Oxy-acetylene welding. Electric are welding is economical and is commonly used in places, where electricity is easily available . In case of oxy-acetylene welding, a flame of high temperature is produced by burning a mixture of acetylene and oxygen.

19·11. TYPES OF WELDING JOINTS

There are three different types of welds, namely, (*i*) Butt welds, (*ii*) Fillet welds and (*iii*) Lap welds.

(*i*) *Butt weld* : In this type, the edge of one plate is brought in line with the edge of a second plate and the joint is filled with welding metal or the two edges are heated to their fusion temperatures and pressed together.

(*ii*) *Fillet weld* : In this type, the edge of one plate is brought against the surface of another (not in the same plane) and welding metal is fused in the corner between the two plates. Thus the joint can be welded on one or both the sides. Different types of butt weld and fillet welds and shown in Fig. (19·18).

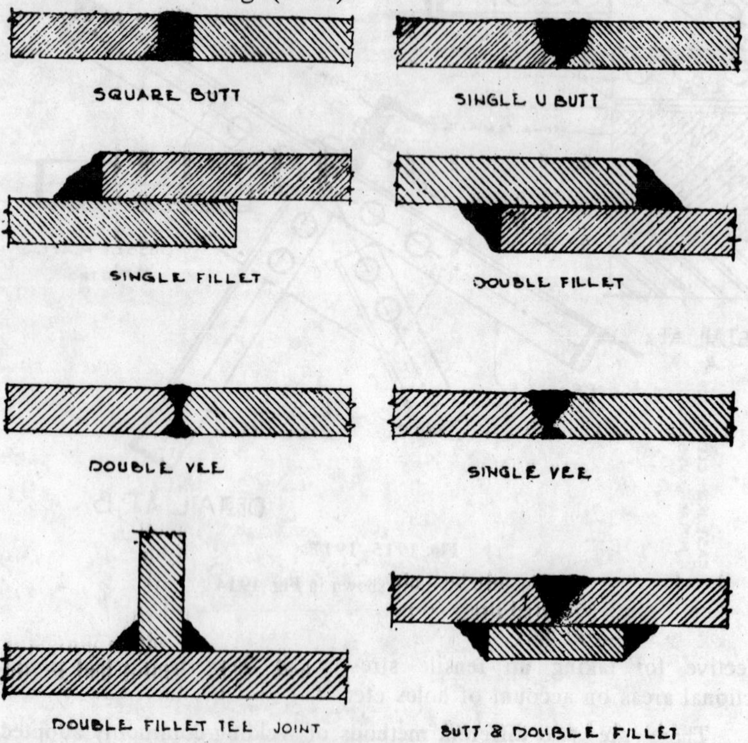

Fig. 19·18 Different types of welded joints

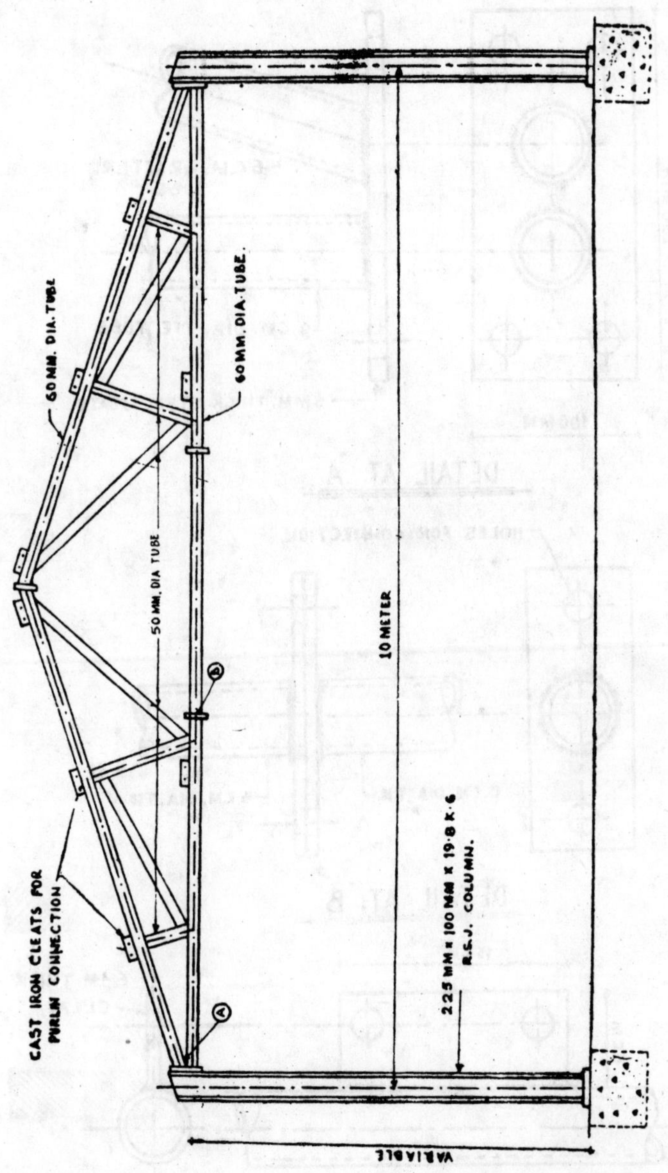

Fig. 19·19

Elevation of a Tubular steel truss suitable for 10 metre span

(*iii*) *Lap weld* : **In this type, the edges of plates are lapped one over the other and the edges of one plate are welded to the surface of the other.**

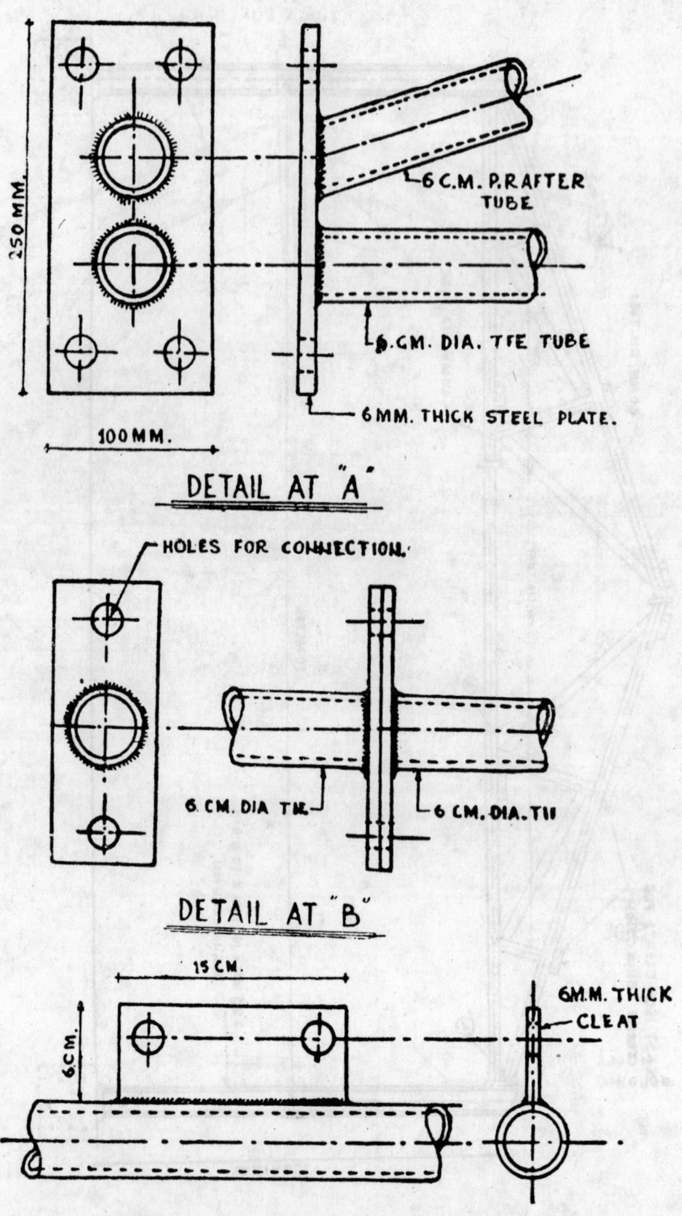

Fig. 19·20
Details of the tubular truss (shown in Fig. 19·19) at various junctions

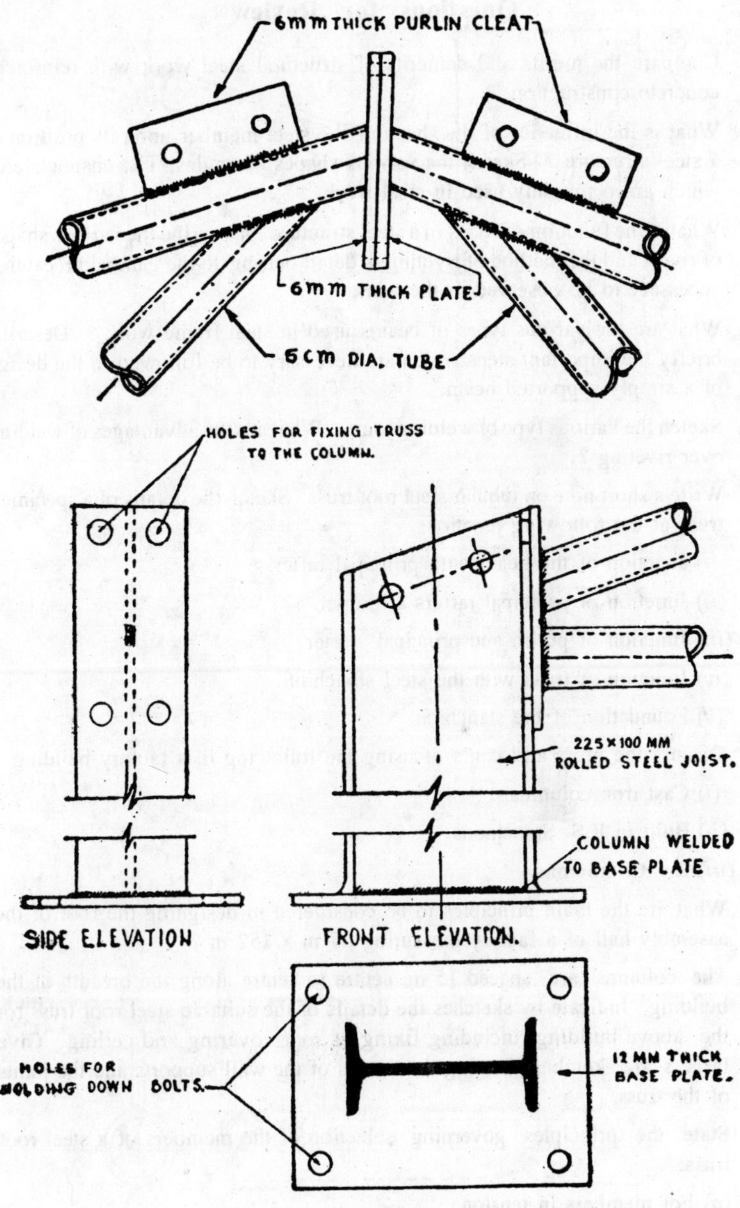

Fig. 19·21

Details of tubular truss shown in Fig. 19·19

Questions for Review

1. Compare the merits and demerits of structural steel work with reinforced concrete construction.
2. What is the influence of the shape of the steel member upon its position in a steel structure ? Sketch the various shapes of angles, T's. channels etc., which are commonly used in steel work.
3. What is the function of rivets in a steel structure ? Describe the various shapes of rivets and the method of riveting in detail outling to the general precaution necessary to be observed in the work.
4. What are the various types of beams used in steel frame work ? Describe briefly the important steps which are necessary to be followed in the design of a simply supported beam.
5. Sketch the various type of welding joints. What are the advantages of welding over riveting ?
6. Write a short note on tubular steel roof truss. Sketch the details of a specimen truss at the following junctions :

   (*i*) Junction of the beam with principal rafter.

   (*ii*) Junction of principal rafters at crown.

   (*iii*) Junction of purlin and principal rafter.

   (*iv*) Junction of truss with the steel stanchion.

   (*v*) Foundation of the stanchion.
7. Discuss the pro's and con's of using the following in a factory building :

   (*i*) Cast iron columns.

   (*ii*) Built in R.S. Stanchions.

   (*iii*) R.C.C. columns.
8. What are the main principles to be considered in designing the roof of the assembly hall of a factory measuring 37 m x 152 m ?

   The columns are spaced 15 m. centre to centre along the breadth of the building. Indicate by sketches the details of the suitable steel roof truss for the above building, including fixing of roof covering and ceiling. Give large scale sketches showing the details of the wall supports and the joints of the truss.
9. State the principles governing selection of the members of a steel roof truss:

   (*a*) For members in tension.

   (*b*) For members in compression.

   Show, by suitable sketches, the fixing arrangements you would recommend for the end bearings of a 12 m. span steel truss supported on brick walls.

10. Describe possible types of failures of riveted joints. Draw sketches to illustrate your answer. On what basis are allowable rivet unit stresses usually determined ?

11. Draw enlarged working details to show the following :

    (*a*) End connections for an angle-iron riveted steel truss of 12 m. span resting on a masonry wall. Rain water gutter and down-take pipes should be also shown.

    (*b*) Purlin connection with principal rafter and asbestos sheet roofing to be shown in section and plan.

12. An asbestos cement sheet roof has to be provided for a workshop building 49 m. x 12 m. Sketch the type of truss you would adopt and detail the junction at the eaves (with gutter) and the crown. Show how the truss is supported on the brick wall. Assume suitable dimensions for all the members.

13. A drawing hall in an office building is to be covered with C.G.I. Sheet roofing over tubular steel truss. The clear span of the truss is 10 m. and the trusses are to be supported on stone masonry columns spaced 3 m. centre to centre. Draw a sectional elevation of the shape of the truss you propose to use and show to a large scale the details of the important junctions of the truss.

14. A factory building 20 m. x 40 m. is to be covered with asbestos cement sheeting fixed on steel truss. The trusses are to be supported on 60 cm. x 60 cm. R.C.C. columns 5.5 m. high above the floor level. The columns are spaced at 4.5 m. centre to centre along the length of the building. Draw an elevation of a truss-suitable for the said span assuming approximate sizes of its various members. Draw details of joint at the following point :

    (*i*) Eaves.

    (*ii*) Ridge (crown).

    (*iii*) Principal rafter and purlin.

    Also show the method of fixing A.C. sheeting to the truss.

15. The roof of a workshop building is to consist of C.G.I. sheets on steel trusses. The span of each truss is 15 m. and the trusses are supported on 60 cm. (3 brick) thick pilasters on either side . Show the details of the truss you would propose for the span of 15 m. Draw to a large scale the details of its fixing arrangement with the pilasters. Adopt suitable sections of the various members of the truss. Draw the details of all the important junctions of the truss.

16. A 6 m. x 6 m. room has to be provided with jack-arch. roofing. Describe, with the help of sketches the formwork and centering arrangements required for the purpose and state how you would proceed to lay the roof.

    You are also required to give a suitable design for the R.S.J.'s and the tie rods, if the roof has to support a live load of 300 kg/m^2. You may assume your own working stresses and any other data required in this connection.

17. An automobile workshop shed of size 40 m. x 10 m. is to be constructed. Suggest a suitable steel roof for this shed. Sketch the details of the roof you purpose, giving the approximate size of the members to be used. Will you require any bracing for this roof ? The height of the columns and the wall is 6 meters.

18. The roof of a theatre consists of steel trusses having clear span of 20 m and the trusses are placed at 4 m. centres. The roof is covered with A.C. sheets on angle iron purlins and the ceiling is of celotex sheets.

    Draw a section showing the details of the truss, roof covering and ceiling. Also give sketches showing the details at wall supports, ridge and other joints of truss.

20

Plastering Pointing and White /Colour Washing

20·1. PLASTERING

This is a process of covering rough surfaces with a plastic material to obtain an even, smooth, regular, clean and durable surface. Plastering conceals defective workmanship and covers up unsound and cheap quality material. Often, plastering is required to provide a satisfactory base for decorating the surface by white-washing, colour washing, distempering or painting. External plastering also termed as 'rendering' is done with the object of improving the resistance of the surface to rain water penetration and other atmospheric influences.

The plastic material or plaster is made by working together a mixture of building materials which may be cement, lime or clay, fine aggregates (usually sand) and water. Certain additives are sometimes added to improve its adhesi veness, durability and lusture. When cement is used as the binding material, the plaster is termed *as cement plaster* and when lime is used as the binding material, it is called *lime plaster*. Sand normally forms the greatest proportion of the constituents of a plaster. Sand controls the shrinkage, porosity, strength and adhesive properties of plaster. Fine sand is often recommended for plastering and it should be so graded that it does not pass by more then 5% through a sieve of 100 mesh (B.S. sieve). Sand used should be clean, sharp and free from deleterious matter. Water used should be clean and fit for drinking purposes. Depending upon the availability of the materials, the choice of plaster for any particular location, is governed by the rainfall, weather conditions and the finish desired.

20·2. PREPARATION OF SURFACE FOR PLASTERING

The durability of plaster depends to a great extent upon its adhesion with the back-ground. The preparation of surface for plastering is therefor of prime importance. All the projections which extend by more than 13 mm. from the general surface of the masonry wall face are knocked off to obtain a uniform surface and this also helps to reduce consumption of plaster. In order to obtain a good key for the plaster with the wall surface, all the joints in the masonry are raked out for a depth of at least 13 mm.

Dust and loose mortar is brushed out of the raked joints. The surface is rendered free from oily1greasy spots. Efflorescence, if any, is removed by brushing and scraping. In case of old walls, the smoothness of the surface is destroyed by hacking it with some tool. The surface is then thoroughly washed with water and kept wet before plastering is commenced.

To ensure required thickness and a true surface, patches of plaster 15 cm. x 15 cm. are first of all applied horizontally and vertically at about 1.8 m. apart over the entire surface. The surface of the patches of plaster (also known as screed) serves as gauges for maintaining even thickness of the plaster being applied. Mortar is then applied on the wall between the screed with trowel.

20·3. LIME PLASTER

Lime used in plastering may be fat lime or hydraulic lime. Fat limes make best plaster as they yield good putty after slaking. Hydraulic lime on the other hand yields harder and stronger plaster, but it may contain some unslaked particles which may slake slowly (may be in 8 to 12 months) on absorbing moisture from atmosphere and damage the plastering by forming blisters. As a precaution the hydraulic lime, if used, should be ground dry with sand ; left for about 2 or 3 weeks and then reground before use.

Mortar for lime plaster is usually prepared by mixing sand and lime in equal proportions. Cement in small quantity is sometimes added to the mixture to improve its strength. Gugal (a kind of fragrant gum) and chopped hemp are sometimes added at the rate of 4.5 kg. and 2.7 kg. respectively to every 2.83 cu. m. of mortar. This treatment prevents the formation of cracks in plaster on drying.

20·4. APPLICATION OF LIME PLASTER

It is usual to provide 13 mm. and 16 mm, thick plaster on plain side and rough side of one brick thick walls respectively. In case of rubble masonry 19 mm. thick plaster is recommended to account for the roughness or uneveness of the surface. Plaster may be applied in one, two or three coats on the prepared wall between the screed with the help of trowel. In the one coat treatment mortar is applied in a uniform surface slightly more than the specified thickness. This is then beaten with thin strips of bamboo to ensure thorough filling of joints and finally brought to a true surface.

During the process of finishing, a solution of lime putty is applied on the surface to make the surface smooth and workable. The finished surface is then cured for at least seven days.

In case it is desired to apply plaster in more than one coat, the first coat, also known as rough coat, is applied on the surface in a layer of 6 to 10 mm. in thickness. In the first coat the mortar is dashed against the surface to serve as a key to hold the next coats.

Second coat is then applied over the first coat after the latter has set and roughened. Mortar for second coat is well pressed and rubbed with wooden straight edge to obtain the desired surface. The thickness of layer of mortar in 2nd coat varies between 6 to 10 mm.

The third or finishing coat is applied 5 days after the 2nd coat. This coat does not exceed 3 mm. in thickness and usually consists of mortar made with cream of white or fat lime (neeru) with fine white sand taken in the ratio of 1:2. This is well rubbed with wooden float and then finished with a trowel to a smooth and even surface.

20·5. CEMENT PLASTER

Cement plaster forms an ideal coating for external renderings. It is specially suited for damp conditions (bathrooms, etc.) where non-absorbent wall surface is essential. Cement plaster is usually applied in one coat. The thickness of coat can be 12 mm, 15 mm or 20 mm depending upon the site conditions and type of building. Sometimes, when the thickness of plaster is more than 15 mm. or when it is desired to have a finer finish, plaster is applied in two coats.

In case of plastering with one coat, cement plaster is applied on the prepared surface between the screed with mason's trowel. The surface is levelled by means of wooden float and straight edges and finally polished with a trowel.

In case plastering is required to be done in two coats, the first coat is applied as described above with the only difference that the plastered surface is not polished. The surface of the first coat when set (not dry) is roughened with a scratching tool to form a key for the 2nd or finishing coat. The finishing coat is then applied over the first coat surface (made rough and damp), within 48 hours. This coat is well troweled and the surface finished smooth. The plaster should be kept wet (cured) for at least 7 days.

20·6. MUD PLASTERING

This type of plastering is commonly seen in kuchha construction in villages and in other structures of temporary character. This is the cheapest form of plastering. Mud to be used for plastering should be made from earth free from grass roots, gravel,stone grit etc. All the clods in the earth should be broken and reduced to a fine powder. The earth is then mixed with plenty of water in a pit, adding, chopped straw, hay or hemp at the rate of 33 kg./cubic metre of earth. This mixture of earth and straw, well flooded with water, is kept for a least 6 days. During this period, the mixture is worked up at intervals with foot or phawras so as to convert it into a homogeneous mass.

The surface to be plastered is prepared exactly in the same manner as that for lime or cement plaster. Mud plaster is generally applied in two coats, the first coat being 18 mm thick while the thickness of the second coat is kept 6 mm. The plaster is dashed against the wall and worked with a straight edge and float. The second coat is applied only when the first coat has set (not dry).

No curing is done in this case. Cracks formed during drying or portions which sound hollow when tapped, are repaired by opening such patches out and then treating the surface with a mixture of clay and cow-dung.

20·7. STUCCO PLASTERING

Stucco is the name given to a decorative type of plaster which gives an excellent finish. The surface plastered with stucco compares very much with marble finish both in smoothness and elegance. Stucco plaster can be used for interior as well as exterior surfaces. It is usually laid in three coats making the total thickness of the plaster to about 25 mm. The first coat is called the *scratch coat* ; the second a finer coat, also known as the *brown coat,* and the third is called the white coat or *finishing coat.*

Each coat should be permitted to dry thoroughly before the next is applied.

Stucco for interior walls : In this case the scratch coat consists of a coat of lime plaster 12 mm. in thickness. The second coat or the brown coat consists of lime plaster which is made richer in lime and is 10 mm. in thickness, and the finishing coat which is 3 mm. thick consists of a mixture of finest lime and well-ground white stone (which may be marble or quartz). The finishing coat is carefully polished with a bag of linen containing moist chalk in the first instance and then oil and chalk. The surface is finally rubbed with oil only and finished smooth and bright.

Stucco for exterior wall : In this case the mortar for the rough coat is prepared by mixing cement and sand in the ration of 1:3 and then adding 10% by weight of hydrated lime to the mixture. The thickness of the first coat in this case is also kept 12 mm. The second coat has the same composition as the rough coat but it is 10 mm. in thickness. The finishing coat which may be 3 to 6 mm. in thickness consists of a mixture of cement and sand taken in proportions of 1:2 to 1:2 ½ White or coloured cement may be used in finishing coat to obtain the desired shade.

20·8. PLASTER ON LATH

Lathing may be either of wood or that of expanded metal. Wooden lath is used for forming a ground for plastering timber partition walls and ceilings of timber floors. Wooden laths are liable to decay or get attacked by ants and hence they are rarely used. Wooden lath consists of thin strips of well seasoned wood, about 25 mm. in width and 90 to 120 cm. in length. Depending upon the strain due to load of plaster, thickness of the strips vary from 3 mm to 13 mm. The thicker strips are used in the ceilings where the strain is maximum while the thinner ones are used for vertical sides of the wall. The strips are nailed in parallel lines to the wooden frames. 10 mm. gap is provided between two successive strips for plaster. The strips are fixed so as to break joints at the butt ends. The plastering is then done in the usual manner.

Metal lathing is available in a wide range of designs. They may be classified as (1) Expanded metal ; (2) B.R.C. fabric ; (3) Ribbed expanded-with ribs forming an integral part of the sheets, (4) perforated ; (5) Dovetailed and so on. Plain or expanded metal lathing is commonly used in residential buildings. The lath is stretched tightly as much as possible and then fixed to the wooden supports by G.I. staples. In case of concrete or masonry surface, wooden plugs have to be embedded for fixing the lath. The lath sheets must be over-lapped not less than 25 mm. at the sides and ends and they should be fastened by galvanised lacing wire running through the meshes. The fixing of the lath is so planned that the overlaps do not occur at corners. The lath is always bent at the corners. After fixing, the lath is plastered in the usual manner.

20·9. SPECIAL TYPES OF FINISHING

The plastered surfaces may be finished in a variety of forms. The types of finishing treatments, usually adopted for external surfaces, are briefly given below :

(*a*) *Smooth cast* : It is a finish which presents levelled and smooth surface. The mortar for the finish is made by mixing cement and fine sand in ratio of 1:3. The mortar may be applied in one or two coats and finished by working with a wooden float. Steel floats are not recommended for external rendering as they give a very smooth finish which is liable to crack and craze due to exposure to atmospheric conditions.

(*b*) *Rough cast* : It is a finish in which the mortar for the final coat contains a proportion of fairly big size coarse aggregates. The wet plastic mix is thrown on the surface by means of a scoop or trowel, and is left in rough condition. The texture desired is regulated by the size of the coarse aggregate which generally varies between 3 to 12 mm. Rough cast finish is normally 10 mm. thick and it is applied over a 13 mm. thick under layer of plaster. The top layer is applied after one or two days after the under layer is laid. The mortar for the finish is made by mixing cement fine sand and coarse aggregates in the ratio of 1:1 ½:3. Rough cast finish is water-proof, durable and resistant to cracking and crazing. This type of finish is considered particularly suitable for buildings subjected to heavy rainfall and strong winds.

(*c*) *Pebble dash* : It is a finish in which small pebbles or crushed stones of suitable size (usually varying from 10 to 20 mm.) are thrown on to a freshly applied final coat of mortar and left exposed. The mortar for the final coat is usually made by mixing cement and coarse sand in the ratio of 1:3 and is laid in about 1 mm thickness. The pebbles used for dashing should be well washed and thrown wet on to the final coat while it is still plastic. To ensure good bond between the dashing and the final coat, the pebbles may be tightly tapped into the mortar with a wooden float or with the flat of trowel. This finish possesses all the advantages of the rough cast finish and is recommended in similar condition.

(*d*) *Scrapped finish* : It is a finish in which the final coat after being levelled and allowed to stiffen for a few hours, is scraped with a steel straight edge, old saw blade or other such tool to remove the surface skin. Various types of scrapped finishes can be obtained by using different types of tools in different manner. In this case, the final coat is usually 6 to 12 mm. thick of which about 3 mm. is removed in the scrapping process. The scrapping is done after the setting has taken place. In the process of scrapping, the surface skin of the mortar is removed to expose the aggregates and the texture obtained depends upon the grading of aggregates used in the final coat. This type of finish is less liable to crack than plain finishes.

(*e*) *Textured finish* : In this finish, ornamental patterns or textured surfaces are produced by working with various tools on the freshly applied final coat. This type of finish possesses all the advantages of the rough case finish.

The various finishes described above (smooth cast finish being an exception) require little maintenance except cleaning of fungus growth and lodgement of dirt.

20·10. SPECIAL MATERIALS USED IN PLASTERING

1. *Plaster of Paris* : It is also known calcium sulphate plaster and is obtained by heating Gypsum. Gypsum occurs in the form of natural rock which is usually composed of one part of calcium sulphate and two parts of chemically combined water of crystallization. When the water of crystallization is driven off from gypsum stone through calcination or heat treatment, the product obtained is called the plaster of Paris. When water is added to this powdery product, it forms a plastic mass, which sets rapidly. Plaster of Paris adheres well to gypsum, wood, metal lath, masonry or other plastering surfaces. The setting time of the plaster of Paris can be controlled by the addition of certain salts. It is mostly used in combination with ordinary lime for making good small defects in plastered

surfaces. Since gypsum is slightly soluble in water, plaster of Paris is not used for external works.

2. *Keene's Cement* : This is hardest and densest form of gypsum plaster. It is obtained by calcining selected form of gypsum. It is pure white in colour and takes a high glass like polish. It is used for highly decorative plastering work and ornamental work. It is used alone for extra-hard finish or it is mixed with lime putty for medium hard finish.

3. *Barium Plaster* : This is essentially made from barium sulphate and is applied in the form of final coat to serve as a protection around X-ray rooms.

4. *Acoustical plaster* : This is gypsum mixture, which is employed as a final coat to serve the purpose of a sound-repellent finish. The minute passages in the form of tiny openings in the plaster coat, absorb sound. The plaster is usually applied in two coats, each 6 mm. in thickness. The surface must be finished uniformly in a manner that promotes porocity. Wooden float is preferred to a steel trowel in finishing.

20·11. DEFECTS IN PLASTER WORK

The following defects may occur in plaster work :

1. *Cracking* : This consists in formation of cracks in plaster work which may be on account of the following :

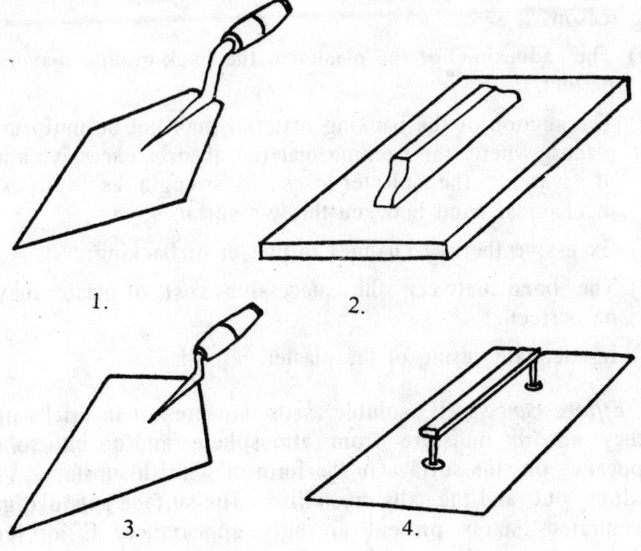

Fig. 20·1

Figs. (1) and (3). Different types of steel trowels

Fig. (2) Wooden float.

Fig. (4) Metal float

(*i*) Old surface not being properly prepared.

(*ii*) Movements in the backing either on account of thermal expansion or due to shrinkage caused by the drying of the backing material. To avoid this the backing surface should be allowed to dry before applying the plaster.

(*iii*) Movement in the plaster itself on account of the expansion (in case of gypsum plaster) or shrinkage of the plaster coat (in case of lime-sand plaster) during drying.

(*iv*) Excessive shrinkage of plaster due to the application of mortar in thick coats.

(*v*) Workmanship and the method of application of the plaster. Rough textured finishes are generally less likely to show cracking than smooth finishes.

2. *Blowing or blistering of plaster* : This consists in formation of small patches of plaster swelling out beyond the plastered surface and are chiefly due to improper slaking of lime particles in the plaster. After the application of plaster, the unslacked particles in the lime start slaking and thus the defect takes place. Properly slaked lime should therefore, be used to prevent blistering.

3. *Falling out of plaster* : This defect may arise due to the following reasons :

(*i*) The adhesion of the plaster to the back-ground may not be perfect.

(*ii*) The suction of the backing material may not be uniform. At places where the backing material absorbs excessive amount of water, the plaster loses its strength as well as the mechanical bond between the two surfaces.

(*iii*) Excessive thermal changes in plaster or backing.

(*iv*) The bond between the successive cost of plaster may not be perfect.

(*v*) In-adequate curing of the plaster.

4. *Efflorescence* : If soluble salts are present in bricks or the mortar they absorb moisture from atmosphere and go into solution which appears on the surface in the form of whitish substance as the moisture dries out and the salts crystallise. The surface gets disfigured and concentrated stains present an ugly appearance. Efflorescence may occur on brick work or on the surface of plaster.

Efflorescence on the surface is unlikely to cause any real damage to the brick unless the quantity of soluble salts present is abnormally high. However, if the formation of crystals (on drying) of the salt solution takes place within the brick, it can result in the disintegration of the material and cause severe flaking of the plastered surface.

Efflorescence on surface can be removed to some extent by brushing and washing the surface repeatedly.

In case of brick work, salts can be removed, by applying a solution or zinc sulphate and water and then burshing off the surface when dry.

In another method a solution consisting of one part of hydrochloric or sulphuric acid diluted with 5 parts of water is applied on the surface with the scrubbing brush. The surface is then thoroughly washed and rendered clean and dry.

POINTING

20·12. GENERAL

In exposed brick or stone masonry, mortar joints are considered to be weakest spots for giving access to rain water or dampness. Pointing consists in raking out joints in brick work or in stone masonry to depth of about 13 mm. and filling the same with mortar of slightly richer mix. This treatment not only protects the joint from the adverse effect of atmosphere but also magnifies the appearance of the surface by exhibiting the pattern of the joints, their thickness, colours and texture prominently.

Preparation of surface. Pointing should preferably be done whilst the mortar in the joints is still green. This has the advantage of giving a good bond between the soft and wet mortar in the masonry and the mortar used for pointing. The racked joints, both in case of old or new work, are cleaned of all loose mortar or dust by using brushes. The joints as well as the wall surface is well washed and kept wet for a few hours before pointing.

20·13. MORTAR

Pointing may be done by using lime or cement mortar. The mortar for lime pointing is made by taking equal part of fat lime and fine sand and then grinding the mixture thoroughly, in a mortar mill. The mortar for cement pointing is made by mixing cement and sand in proportion of 1:2 or 1:3.

20·14. APPLICATION OF MORTAR

The mortar is pressed into the raked joints with a pointing trowel, and rendered flush sunk or raised according to the type of pointing required. The desired type of pointing is then made with the proper tools and the surface of the masonry cleaned of all superfluous mortar. The pointing should be kept wet for about a week. The pointing lines should be regular, truly horizontal and vertical. In case of random rubble masonry, where the joints are slanting, the line of joints from different directions should preferably meet at or near the junctions.

20·15. TYPES OF POINTING

The various types of pointing commonly used are described below. The choice of the particular type depends upon the type of masonry, nature of the structure and the nature of the finish desired.

(1) *Flush pointing* : In this type of pointing, the mortar is pressed into the raked joints and finished off flush with the edges of the bricks or stones, so as to give a smooth appearance. The edges are then neatly trimmed with a trowel and straight edge. This is the simplest type of pointing and is extensively used in brick work and stone masonry face work.

(2) *Cut or weathered or struck pointing* : In this type of pointing, the mortar is first pressed into the raked joints. While the mortar is still-green, the top of the horizontal joints is neatly pressed back by 3 to 6 mm. with the pointing tool. Thus the joint is finished sloping from top of the joint to its bottom.

(3) *V-grooved pointing* : This type of pointing is made similar to keyed or grooved pointing by suitably shaping the end of the steel rod to be used for forming the groove.

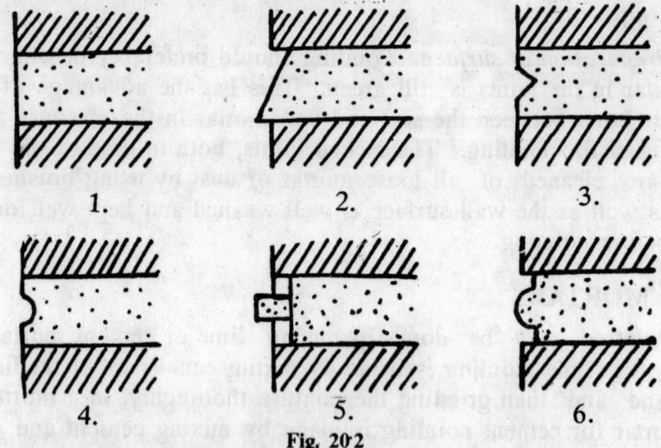

Fig. 20·2

Fig. 1. Flush pointing

Fig. 2. Cut or weathered or struck pointing

Fig. 3. V-grooved pointing

Fig. 4. Keyed-or grooved pointing

Fig. 5. Tuck-pointing

Fig. 6. Beaded pointing

(4) *Keyed or grooved pointing*. In this type of pointing, the mortar is pressed into the raked joints and finished off flush with the face of the wall. While the pressed mortar is still green, a groove is formed by running the bent end of a small steel rod (6 mm. in diameter) straight along the center line of the joints. The vertical joints are also finished in the same manner.

(5) *Tuck pointing* : In this type of pointing, the mortar is first pressed in the raked joints and there after it is finished flush with the face of the wall. While the pressed mortar is still green, the top and bottom edges of the joints are cut parallel so as to have a uniformly raised band about 6 mm. high and 10 mm. in width.

(6) *Beaded pointing* : In this type of pointing, the mortar is pressed in the raked joints and finished off flush with the face of the wall. While the pressed mortar is still green , a steel rod having its end suitably shaped is run straight along the centre line of the joints to forrn the beading.

WHITE/COLOUR WASHING

20·16. PREPARATION OF WHITE-WASH

White-wash is made from pure fat lime (white stone lime) or shell lime. The unslaked lime is dissolved in a tub with sufficient quantity of water, and it is thoroughly stirred with a wooden pole, until it attains the consistency of thin crèam. As a rough guide about 5 litres of water should be added to each kg. of lime for making the cream. This mixture is allowed to stand in the tub for a period of about 24 hours and then strained through a clean coarse cloth. Clean gum dissolved in hot water is then added in the proportion of 4 kg. of gum per cu. m. of thin cream.

To prevent glare and to obtain pleasing effect, Indigo (Neel) in proportion of 3 kg. of Indigo per cu. m. of thin cream is also added. In order that the coat of white-wash may stick well to the surface, alum or common salt taken in the same proportion as gum, is also sometimes added.

20·17. PREPARATION OF SURFACE

Before applying white wash on new walls, the wall surface is thoroughly cleaned, brushed and rendered free from mortar droppings and other foreign-matter. If the surface to be coated is sufficiently smooth, it should be rubbed with sand paper to ensure proper adhesion of white-wash.

In case of re-white washing an old surface, all loose pieces and scales should be scrapped off. When heavy scaling has taken place the entire surface should be scrapped by means of steel wire brushes or steel scrapers. The old loose white-wash is removed by rubbing with sand paper and the holes in wall are filled with lime putty. The plaster repairs to the wall should be done well in advance so that they are dry before white washing. The greasy spots should be given a coat of a mixture of rice water and sand so that the wash may stick to the surface. In kitchens or other places where the walls have been discoloured by smoke or due to same other reasons, the surface should be given a coat of a mixture of wood ash in water before applying white-wash.

20·18. APPLICATION OF WHITE-WASH

White-wash may be applied in specified number of coats until the surface presents a smooth and uniform finish. Three coats are considered necessary for new work and one or two coats for old work. Usually moonj brushes are used for white-washing. Each coat consists of a stroke of the brush applied from the top downwards, another stroke applied from the bottom upwards over the first stroke, and similarly one stroke is applied horizontally from the right and another from the left. Each coat should be allowed to dry before next one is applied. The finished dry surface should not show any signs of cracking or peeling and it should not readily stain the hand when rubbed. As a measure of protection before start of work, all doors, windows, floors, articles or furniture etc., should be protected from being splashed upon by while wash.

20·19. COLOUR WASHING

Colour wash is prepared by adding mineral colours (which are not affected by lime) of desired shade to the prepared white-wash. The colour wash is applied in the same manner as white-wash. The mixture is kept constantly stirred with a bamboo stick while it is being used. Only colour wash, sufficient for the day's work, should be prepared at a time. Prior to the application of colour-wash on new surfaces, a coat of white-wash is applied which acts as a priming coat. In case of old prepared surface, a coat of colour wash is applied over the patched and repair work before start of colour washing. Two or more coats of colour wash are then applied on the entire surface till it presents a uniform tint of shade. The surface should not appear powdery and it should not readily stain the hand when rubbed.

Questions for Review

1. What is the object of plastering. State in brief the different types of plasters adopted for internal and external finishing of wall surfaces.
2. Write short notes on the following :
   (*i*) Stucco plastering
   (*ii*) Mud plastering
   (*iii*) Scrapped finish.

3. Differentiate between cement plaster and lime plaster.
4. Write short notes on the following :

   (*i*) Plaster of Paris

   (*ii*) Keene's cement

   (*iii*) Barium plaster.
5. Write short notes on the following :

   (*i*) Cracking

   (*ii*) Blistering of plaster

   (*iii*) Rough cast plaster

   (*iv*) Pebble-dash.
6. What is the object of pointing brick work ? Describe the operation of pointing giving details regarding the preparation of surface, the mortar and its application completely.
7. Write short notes on the following :

   (*i*) Flush pointing

   (*ii*) Grooved pointing

   (*iii*) Tuck pointing

   (*iv*) Cut or weathered or struck pointing.
8. What is the difference between the terms-colour-washing and white-washing ? Describe in detail the complete operation of white-washing of plastered wall surface.
9. Describe the procedure of carrying out plastering in cement mortar on new and old wall separately.
10. Describe the procedure of white-washing giving details of various stages in proper sequence.

21

Paints, Varnishes and Distempers

21·1. PAINTS

Paint is applied on timber, metal, brick or other materials in the form of a liquid which, on drying, forms a thin film on the painted surface. The essential function of the paint film is to provide protection or decoration or both. Paint is indispensable for the proper maintenance and preservation of wooden structures against warping, checking or decay. Most of the metals corrode if not painted at suitable intervals. Painting enables the masonry and concrete work to resist checking and disintegration, thereby extending the life of structure. Decoration, sanitation or improved illumination are the additional factors which often make painting necessary.

Paint is composed of two main components, *viz.*, pigment and organic binder. Pigment is the solid ingredient of the paint and is used in the form of a fine powder. Organic binder is the liquid part of the paint and it is called vehicle. The vehicle varies with the type of paint. It is usually linseed oil for an oil paint, varnish is kalsomine in case of enamel paints and it is water in case of cold water paints.

21·2. CONSTITUENTS OF AN OIL PAINT

An oil paint is generally made up of six main constituents given below:

(*a*) *A base*

(*b*) *An inert filler or extender*

(*c*) *Colouring pigment*

(*d*) *A vehicle*

(*e*) *A solvent or thinner*

(*f*) *A drier*

(*a*) **Base :** It is generally a metallic oxide and is used in the form of powder. Base is essentially a pigment which forms the chief ingredient of a paint. The most important purpose of adding base or pigment in a paint is to make an opaque coating to hide the surface to be painted. In addition, it makes the film of coating resistant to abrasion and prevents shrinkage cracks likely to be formed in the film, during drying. White lead, red lead, zinc oxide, iron oxide and titanium whites are the bases commonly used.

(*b*) **Inert filler or extender :** It is a cheap pigment added in a paint to reduce its cost. In addition, it modifies the weight of the paint and makes it more durable. With proper formulations, the inert filler pigments may be helpful in contributing useful properties to the paint film. The commonly used inert fillers or extenders are : Barytes (Barium sulphate), lithopone, silica, silicate of magnesia or alumina, whiting, gypsum, charcoal etc.

(*c*) **Colouring pigments :** As the name indicates, it is a white or coloured pigment, mixed in a paint to get the desired colour of the paint.

(*d*) **Vehicle :** It is a liquid which acts as a binder for the various pigments, *viz.*, bases, extender and colouring pigment. The vehicle makes the paint in the state of a fluid and thus helps to spread its ingredients uniformly over the surface to the painted. This forms as elastic, resistant to abrasion and reasonably impermeable film on drying. Refined linseed oil is a commonly used vehicle in oil paints. Tung, perilla, soyabean, fish, sunflower, tobacco seeds are also being used as vehicles in various combinations with linseed oil or without linseed oil.

(*e*) **Solvent or thinner :** It is a liquid which thins the consistency of the paint and evaporates after the paint film has been applied. It imparts to the paint film good properties such as brushability, smoothness and easy flow. Turpentines, pure oils, petroleum spirit and highly solvent naptha are commonly used as solvent or thinner.

(*f*) **Drier :** These are materials containing metallic compounds and are used in small quantities for accelerating the drying of paint film. They act as catalysts for the oxidation and polymerization of the vehicle used in the paint. Litharge, lead acetate, manganese dioxide and cobalt are the driers commonly used. Not more than 10% (by volume) of the drier should be used in an oil point. If used in excess, especially in the final coat, they tend to destroy the elasticity of the paint which finally leads to its flaking.

21·3. CHARACTERISTICS OF A GOOD PAINT

A good paint should have the following characteristics :

1. It should stick well to the surface and should be able to seal the porous substrata.

2. Its consistency should provide easy workability.
3. The thickness of the paint film should be adequate for good protection and decoration of the surface.
4. The paint film should dry rapidly.
5. The dried paint film should be able to withstand the adverse weather effect for a long time without losing gloss.
6. It should offer resistance to failure by checking, cracking and flaking.
7. It should possess good moisture resistance.
8. Its colour should not fade with the passage of time.

21·4. BASES

The pigments commonly used as the principal constituents for form--ing the body of various types of paints are briefly given below :

White lead : This is perhaps the cheapest pigment and is thus extensively used for ordinary painting works. It is sold both in powder form as well as in the form of a stiff paste made by mixing it with linseed oil. It has good binding and spreading power. It is very poisonous and is blackened when subjected to an atmosphere contaminated with sulphur compounds. Paints with white lead base can be easily tinted. It is not recommended for painting of iron works as it does not check rusting of iron. White lead paints are best suited for wood work.

Red lead : It is an oxide of lead, bright red in colour and is usually sold in powder form or in the form of paste made by grinding it with linseed oil. Red lead mixed with linseed oil is considered to be the best primer both for wooden surface as well as structural steel work. It forms a tough and durable film which sticks well to the surface.

Lead pigment paints have excellent moisture and mildew resistance properties. They have good adhesion and the paints film fades only on account of checking and cracking.

Zinc white or Zinc oxide : It is a fine white zinc powder which is available both in dry form as well as in the form of paste made by mixing it with linseed oil. This pigment is used as a base for all the zinc paints. Zinc white mixed with linseed oil forms a white paint, which has a good hiding and spreading power. It is not effected by sulphur compounds and is not poisonous. It forms a costly paint, less workable and less durable than the white lead paints. It amalgamates readily with the binding oil during the process of drying. It finally gets very hard forming a paint film which is brittle and has a tendency of surface cracks.

Iron oxide : It is a pigment produced from haematite ore. The pigment is obtained by grinding and levigating the red or brown haematite. When mixed with vehicle oil, it forms a cheap and durable paint, which is exclusively used for the priming coat on iron or structural steel.

Titanium white or titanium dioxide : This pigment is chemically inert and is not affected by heat, or light. Due to its high refractive index, titanium white possesses excellent hiding power and as such it is generally used as an under-coat in all types of exterior or interior organic coatings. It can be used with all types of vehicle oils.

Lithopone : This pigment is obtained from the precipitate formed by mixing equal quantities of the solutions of barium sulphate and zinc sulphate under carefully controlled conditions. It is a dense white pigment, having a good covering power, and it tends to turn yellow when exposed to sunlight. The paint made with lithopone as its base is generally used as under-coat. It is also used in cheap enamels. It should not be allowed to come in contact with water.

21·5. VEHICLE

The vehicles commonly used in an oil paint are briefly described below:

Linseed oil : This is extensively used as a vehicle in oil paints. It is obtained by pressing crushed flax seeds. The drying of the oil takes place due to its oxidation when exposed to air. It may be used either in its raw state or refined state or in boiled form.

Raw linseed oil : It is a transparent vehicle, yellowish in colour, sweet in taste and has very little smell. It has inferior drying qualities and is generally used for inside painting work. It is normally used for delicate tints. The drying qualities of this vehicle can be improved by adding about one pound of white lead to every gallon of the oil and then allowing, it to settle for a period of one week.

Boiled linseed oil : The drawback of slow drying of raw linseed oil is overcome by boiling the raw oil. During the process of boiling, red lead and litharge are added to the oil in proportion of one pound each to one gallon of raw oil. The boiled linseed oil is thicker in consistency and its colour is darker than that of the raw linseed oil. Boiled linseed oil is inferior to raw linseed oil from consideration of elasticity and power of penetration. Being darker in colour, it is not recommended for white or light-coloured paints.

Double boiled linseed oil : Its colour and transparency is exactly similar to that of raw linseed oil but it smells slightly different. It has quick drying qualities and is generally used for painting plastered surfaces.

Poppy oil : It is prepared by pressing poppy seeds. It is used for very delicate colours. It is expensive and is thus used for internal painting of light shades. It is inferior in drying qualities as compaired with raw linseed oil. The raw oil is not suitable for use in paints and as such it is boiled and suitable materials are added to make it possible to be used as vehicle.

Tury oil : It is extracted by pressing the nuts of tury tree. It is a vehicle better than linseed oil and is largely used for making superior paints and varnishes. It stands well the attack of alkalies.

Nut oil : It is a cheap vehicle and a generally used for painting ordinary works of temporary character. It is colourless, less durable and it dries very quickly. It is used as a vehicle for white or light coloured paints.

21·6. THINNER

Turpentine is the most common thinner used in the preparation of an oil paint. Turpentine is a spirit obtained by distilling balsam, which is the resin or gum obtained from pine tree. Turpentine thus obtained is termed as spirit of turpentine, sap turpentine or oil of turpentine. Turpentine is an inflammable, volatile, colourless liquid, having a pungent smell. It is used in paints and varnishes to dilute them and to increase their flowing properties. It is added in more quantity in the priming coat on wood-work in order to increase ability of the paint to penetrate into the pores of the woodwork, while in the outer or final coat it is used in much less quantity. If used in excess, it effects the durability of the paint and on evaporation it leaves an excess of colour not mixed in the oil. The usage of turpentine as thinner is adopted only when dull or flat finish without gloss is desired. Weather has adverse effect on turpentine, and as such it should normally be recommended for indoor paint work. Benzene and solvent naptha are the thinners which are cheaper and are sometimes used as substitutes for turpentine.

21·7. COLOURING PIGMENTS

Colouring pigments are used to give the desired colour to the paint. In general, for white, black and very dark shades, the base itself is chosen to be a colour pigment. In case of light or bright colour, the desired shade is obtained by mixing pigment to white paint.

The colouring pigments generally used are given below. The desired colour may be obtained by using the pigments, singly or in combination of two or more:

Black pigments : Lamp black ; vegetable black ; ivory black ; graphite.

Brown pigments : Raw umber : burnt umber ; burnt sienna ; vandyke brown.

Blue pigments : Prussian blue ; indigo ; ultramarine blue ; cobalt blue.

Yellow pigments : Chrome yellow ; yellow ochre ; zinc chromate; barium chlorate ; sienna.

Red pigments	;	Indian red ; carmine ; red lead ; vermillion ; venetian red; tuscan red.
Green pigments	:	Verdigris ; green earth ; chrome green ; emerald green.

21·8. PREPARATION OF PAINTS

Mostly white lead oil paint forms the basis of all the paints of light or bright colours. The base which is usually white lead is ground in linseed oil to the consistency of paste. Alternatively the base may be obtained in the form of the paste directly form the market.

The base in the state of stiff paste is then softened or "broken up" by adding linseed oil in a small quantity at a time and then stirring the paste with wooden paddle. If a coloured paint is desired the proper colouring pigment is ground very fine and mixed with linseed oil to the consistency of paste separately. Driers, if used should also be separately ground in linseed oil and the paste mixture of the three *i.e.* the base, the colouring pigment and the drier, is further thinned in to the consistency of cream by adding more linseed oil and stirring it well. The mixture is then strained through a fine canvas or a fine sieve. The paint is thinned to the desired consistency by the addition of oil or turpentine or both, just before it is to be used.

The paint thus prepared, should be used and consumed at the earliest. If the prepared paint has to be kept unused for some time, it should be covered with a film of water to protect it from drying or skinning. Ready-mixed paints of varying composition and colour are now produced by many manufacturers such as Shalimar Paints, Solignum Paints, Holes Paints Ltd, Johnson & Nicholson (India) Ltd. etc. Ready-mixed paints are generally too thick for direct application and as such they are thinned to the desired consistency by adding suitable thinners and by stirring it well.

21·9. TYPES OF PAINTS

The various types of paints commonly used can be broadly divided in the following categories :

(*i*) *Aluminium paints*
(*ii*) *Anti-corrosive paints*
(*iii*) *Asbestos paints*
(*iv*) *Bituminous paints*
(*v*) *Bronze paints*
(*vi*) *Cellulose paints*
(*vii*) *Casein paints.*
(*viii*) *Cement based paints*

(*ix*) *Enamel paints*

(*x*) *Oil paints*

(*xi*) *Rubber base paints.*

(*i*) **Aluminium Paints.** Aluminium paint is used for painting wood work and metal surfaces. Aluminium powder forms the base in this paint. In this paint the base is held in suspension and bound by, either quick-drying spirit varnish, or slow drying oil varnish to suit the requirements of the surface to be painted. This paint is well established for its good weather-resisting and water-proofing qualities. It is highly heat reflective, resistant to corrosive action of sea water, and stands well the atmosphere contaminated with acidic fumes. Because of its brilliant silvery shining texture, the paint has the advantage of being visible in darkness. Aluminium paint is commonly used for painting metal roofs, silos, machinery, electric and telegraphic poles or towers, oil or gas storage tanks, hot water pipes or tanks etc.

(*ii*) **Anti-corrosive Paints.** This paint is generally used as a metal protection paint for preserving structural steel work against the adverse effect of fumes, acids, corrosive chemicals and the ravages of rough weather. Several types of anti-corrosive paints are manufactured these days. The pigments used in the various types of paints are dry red lead, sublimed blue lead, zinc oxide and iron oxide, zinc dust, zinc chromate etc. Linseed oil is generally used as the vehicle. Sometimes, driers and inert fillers are also added to modify the requirements of the coatings.

(*iii*) **Asbestos Paints.** This type of paint is specially suitable for patch work or stopping leakage in metal roofs. It is also used for painting gutters, spouts flashings, etc. in order to prevent them from rusting. Asbestos or fibrous coating is sometimes used as damp-proof cover for the outer face of the basement walls.

(*iv*) **Bituminous Paints.** Bituminous paints are alkali resistant and are chiefly used for painting exterior brick work, and plastered surfaces. They are also used for water-proofing and protection of iron and steel and are commonly applied on iron water mains and structural steel work under water or on fabricated iron and steel products. Such paints usually consist of asphalt, bitumen or pitch dissolved in mineral spirit or naptha. When required, drying oils may also be added to the paint to modify its properties. This type of paint is obtained in black colour but its colour can be modified by incorporating certain pigments like red oxide etc. These paints, however, deteriorate when exposed to the direct rays of the sun.

(*v*) **Bronze Paints.** This type of paint is often used for painting interior or exterior metallic surface. On account of its high reflective property, it is commonly applied on radiators. Aluminium powder (or aluminium bronze) or copper powder (or copper bronze) are pigments commonly used in this paint. The vehicle used in the paint is usually nitro-cellulose lacquer.

(*vi*) **Cellulose Paints.** This type of paint is made form celluloid sheets and amyl-acetate substitutes. For making a superior type of paint, Nitro-cotton is used. It dries very quickly and possesses the additional advantages of hardness, flexibility, and smoothness. It can be cleaned easily and it remains unaffected by hot water, smoky or acidic atmosphere and stands well the wear and tear and ravages of rough weather. It is much superior to ordinary house paint and on account of its high cost, its usage is generally restricted to paints of motor cars, aeroplanes or other such situations.

(*vii*) **Casein Paints.** Casein is a protein substance extracted from milk curd which is mixed-up with a base consisting of white pigment like whiting, titanium, lithopone etc. to from the paint which is usually available in powder or pasty form. On account of its high opacity, the paint can be applied on new plaster work without any danger of scaling or alkali burning. The paint can be tinted in any desired shade of colour. It is usually applied on walls, ceiling, wall boards, cement block construction etc. to enhance the appearance of the surface. When used for exterior surfaces of cement, brick or stone masonry, a little quantity of drying oil, or varnish is added to the paint to make it weather well.

(*viii*) **Cement-based Paints.** This is a type of water paint in which white or coloured pigment and cement forms the base. No oil or other organic matter is used in making this paint. This type of paint is available in packed powdered form under different trade names such as snow-cem etc. This paint can be made by adding paint powder to water to obtain thick paste and thereafter diluting the paste with water to brush-able consistency. Normally equal volume of paint should be mixed only in such quantity which can be used within an hour of its mixing as after that the paint gets spoiled due to setting of cement.

For new works, cement paint is applied in three coats whereas for old works one or at the most two coats may be sufficient. In any case, the second coat should be applied only after the first coat has set for 24 hours. The surface to be painted must be well wetted and the successive coats must be sprayed over with water during the process of drying.

This paint is applied to exposed or plastered brick masonry and concrete work. On drying, it forms a decorative, strong, durable and water resistant film on surface.

(*ix*) **Enamel Paints.** Enamel paint is made by adding pigments like white lead or zinc white to a vehicle which is a varnish. Colouring pigments may also be mixed to obtain the enamel paint in any desired colour. On drying, it forms a smooth glossy, relatively hard and permanent solid thin film. Different types of enamel paints are now manufactured in a variety or colours and are available in readymade form these days. Enamel paints are used both for interior as well as exterior painting. They are not affected by atmosphere and stand well the adverse effects of weather. The enamel paint film presents a glossy surface but, if desired, an eggshell or flat finish may also be obtained by adding turpentine in small or large quantity. It is commonly used for painting porches, decks, stairs, concrete surfaces etc.

(*x*) **Oil Paints.** This type of paint can be used for almost all surfaces such as wooden, masonry, metal etc. Oil paint basically consists of two main components *viz.* a base and a vehicle. Oil paints are manufactured in different shades and grades. They possess all the characteristics of good paint and as such are very common in use.

(*xi*) **Rubber Base Paints.** This type of paint has excellent acid, alkali and moisture resistant properties. It can be readily used for application on new concrete and lime plastered surfaces. The paint is made from rubber treated with chlorine gas (chlorinated rubber) which is dissolved in suitable solvents and mixed with other pigments. It is commonly used as protective coating on cement concrete floors and interior or exterior masonry surfaces.

PROCESS OF PAINTING DIFFERENT SURFACES

21·10. PROCESS OF PAINTING ON NEW WOODWORK

The correct preparation of surface plays an important role in preserving the properties and life of the paint. The process of painting on new wood work can be divided into the following different stages :

1. *Preparation of surface*
2. *Knotting*
3. *Priming*
4. *Stopping*
5. *Second or surface coats or undercoatings*
6. *Finishing coat.*

1. Preparation of surface. Wood work to be painted must be properly seasoned, clean, dry and free from dust. The surface should be smoothened by rubbing it with a fine grade sand paper or glass paper used obliquely across the grains and rendered free from grease.

2. Knotting. Knots exude resin. Knotting is the process of sealing the knots by suitable means so that the resin from the knots may not destroy the paint film by way of cracking, peeling or brown discoloration. The three different methods of knotting commonly adopted are (*i*) Patent knotting, (*ii*) Ordinary or size knotting (*iii*) Lime knotting.

(*i*) *Patent knotting* : This consists in sealing the knots by applying one or more coats of shellac or aluminium varnish on the knots.

(*ii*) *Ordinary or size knotting* : This consists in sealing the knots by applying two coats of varying ingredients. The first coat consists of red lead ground in water to which strong glue size is mixed and the fluid is applied hot on the knots. The second coat which consists of red lead ground in oil and thinned with boiled oil and turpentine in applied soon after the first coat is dry.

(*iii*) *Lime knotting* : This consists in keeping the knots covered with hot lime for about 14 hours. After this period, the lime is scrapped off and the knots are treated with the method of ordinary size knotting.

3. **Priming.** This consists in applying first coat or the priming coat on the surface to fill the pores of wood by penetrating the primer inside the wood. The main function of the priming coat or the primer is to serve as the foundation for the subsequent coats. Primers have definite functions to perform and their choice depends on the material being used and hardness or condition of wooden surface to be coated. In general, the ingredients of the priming coat should preferably be the same as in the subsequent coats but their proportion may vary. The composition of primer for interior and exterior wood-work may consist of the following.

(1) *For interior wood-work*

Red lead	--------------	0·25 kg.
White lead	--------------	3·5 kg.
Boiled linseed oil	--------------	0·50 litre
Raw linseed oil	--------------	0·50 litre
Litharge	--------------	0·05 kg.

(2) *For exterior wood-work*

Red lead	--------------	0·04 kg.
White lead	--------------	4·5 kg.
Raw linseed oil	--------------	2·25 litre
Litharge	--------------	0·09 kg.

For ordinary works primer can be prepared on site by mixing red lead, raw linseed oil and turpentine in the ratio 0·71 kg. : 0·71 kg : 1 litre.

The priming coat is generally applied before fixing wood-work in position.

4. **Stopping.** This consists in filling up all nail-holes, cracks, open joints, dents and other such defects with the putty. Stopping is done soon after the priming coat is dry. Before filling putty, the primed surface is rubbed down with pumice stone or glass paper. After the putty has dried, the surface is once again rubbed with glass paper or pumice stone to level the same. The putty should be made by mixing powdered chalk

with pure linseed oil to the consistency of a paste, well kneaded and worked, till smooth and soft. It should be coloured to match the shade of the finished coat.

5. Second and succeeding coats or under-coatings. After the primary coat is dry, second coat or under coatings are applied on the primed surface. The colour of the under-coat should approximately be of the same shade as that of the desired finish. It should be a thin opaque film, free from brush marks. Two or more under-coatings are sometimes necessary to achieve the object. The final appearance of gloss and durability of the finishing coat depends very much upon the quality of under-coating. Sufficient time must be allowed for each coat to dry before applying the succeeding coat. For good results, the dry coating is rubbed down with pumice stone or glass paper and cleaned before applying the succeeding coats.

6. Finishing coat. This is the last coat applied on the surface after the under-coat is properly dry. This coat is applied very carefully in a workman like manner, so that the finished surface is free from thin patches and brush marks.

21·11. PAINTING OLD WOOD WORK

In case the old paint is in good condition but dirty and its removal is not considered necessary, the surface should be cleaned of all dirt and grease by washing it with lime and water. The cleaned surface is rubbed with pumice stone and the holes and cracks, if any, are filled with glazier's putty or wood putty.

If the old painted surface is in bad condition and the paint has badly blistered and flaked, it should be completely removed by use of a blow lamp or one of the patent paint removers. While removing paint with a lamp, care should be taken to see that neither the paint nor the background gets charred and the flame is played upon the surface in such a manner that the paint only gets soft. The softened paint is then removed with a stripping knife following the flame as it is moved up the surface. Paint remover should be used in places where removing the old paint with blow lamp is not considered desirable. The paint remover is applied on the surface with a brush. Under the action of the paint remover, the paint film will lift up and wrinkle and can be easily removed with the help of a sharp instrument.

Old paint can also be removed by applying caustic soda solution (formed by dissolving caustic soda with 48 times its volume of water), on the surface and then stripping it with a sharp instrument. After removing the old paint, the cleaned surface is given two or three coats of the paint to obtain the desired finish.

21·12. PAINTING IRON AND STEEL WORK

Prevention of rust is one of the important requirements of painting on metal surfaces. Painting on steel, iron, etc, requires utmost care in the preparation of the surface for painting.

The mill scale and rust is removed by scrapping or brushing with steel wire brushes while the oil, grease and dust may be removed by washing the surface with petrol, benzene or lime water. The clean surface is then given the priming coat which usually consists of an oxide-of-iron paint. Priming coat consisting of red lead is considered to be the best. Two or more coats of the desired paint are then applied over the primed surface with a brush or by spraying. Care should be taken to see that each succeeding coat is applied only after the previous coat has dried completely. For repainting iron or steel, the same procedure as given in repainting old wood work is adopted.

21·13. PAINTING OF GALVANISED IRON WORK

Galvanised iron is one of the most troublesome metal surfaces to paint. This is on account of the adhesion difficulty the surface offers, when organic finishes are used. Galvanised iron should not preferably be painted until it has been exposed to weather for a year. This is done because after sufficient weathering the zinc coating on the galvanised surface gets oxidised, forming a thin film of zinc oxide on the surface to which paint easily adheres. If, however, weathering before painting is not possible from consideration of time factor, the new galvanised surface is treated with (*i*) a solution prepared by mixing (40) gm. of copper acetate in a litre of water or (*ii*) with a solution prepared by mixing 13 gm. each of muriatic acid, copper chloride, copper nitrate and ammonium chloride dissolved in one litre of water. The quantity of solution thus formed will be sufficient to cover an area of 250 to 300 sq. m. The galvanised iron surface treated with either of the solutions gets black (due to chemical reaction of copper present in the solution on zinc) and ensures proper adhesion of paint. A priming coat consisting of red lead mixed with linseed oil and turpentine in equal proportions is then applied on the treated surface. When priming coat is dry, the surface can be coated with desired oil paint.

21·14. PAINTING PLASTERED SURFACE

The greatest danger in painting plastered surfaces, particularly if they are relatively fresh, is the free alkalies present in the lime or cement. The free alkalies discolour the paint, destroy its oil and adversely effect the drying properties of the coating. Therefore, preparation of the surface before painting is extremely important. The plastered surface should not ordinarily be painted until it has dried completely. If possible, leave the walls unpainted for at least first 12 months, so that they may dry out completely. The walls may, however, be white-washed or

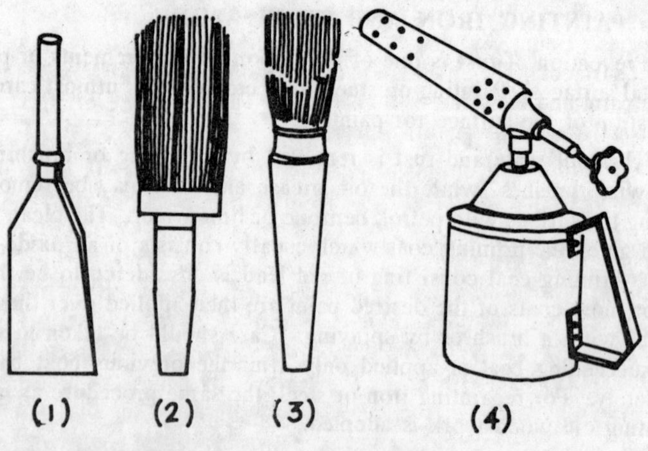

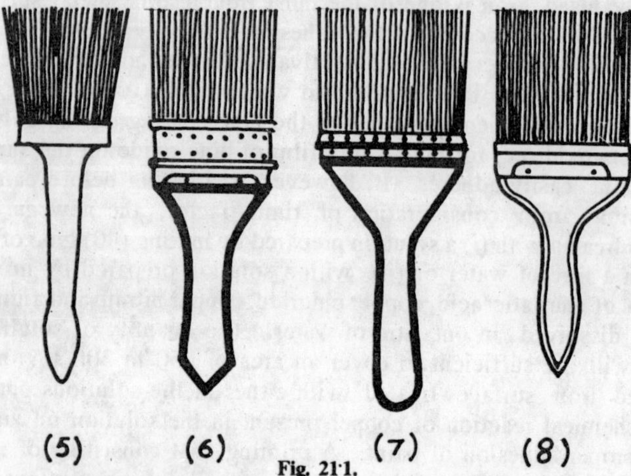

Fig. 21·1.

Fig. 1. Scrapping tool.
Fig. 2. Paint brush.
Fig. 3. Sash brush
Fig. 4. Burning stove.
Fig. 5. Varnish brush.
Fig. 6. Distemper brush.
Fig. 7. White-colour wash brush.
Fig. 8. Wall paint brush.

colour washed for this duration. In new work, however, the surface should be washed with a dilute solution of zinc sulphate to neutralise the free lime on the walls. Before primer is applied the holes and undulations should be filled up with plaster of Paris and surface rubbed smooth. Primer for plaster work may be prepared by mixing equal parts of white and red lead in boiled linseed oil. After the primed surface has dried, two or more coats of the desired paint may be applied in the usual way.

With the advent of emulsion paints, new plastered surfaces can be painted quite easily. These paints have good alkali resistance and sealing properties. Being permeable to moisture, emulsion paints can be applied on wet surfaces. The emulsion paint dries on the evaporation of the water in the paint and the dried film gets hard. The next coats are then applied to get a flat velvety-smooth finish.

In addition, painting of plastered or concrete surface can also be carried out by use of cement paints, already described earlier.

21·15. PAINTING DAMP WALLS

Take a mixture consisting of 10 litres of paraffin, 8 litres of benzoline and 6 kg. of pale resin in a vessel and shake them well, so that they get completely dissolved. Add 10 kg. of whiting (powdered chalk) to the mixture and grind it well. The paint thus formed may be applied in one or more coats on damp walls in the usual manner. On drying, it will form a dry hard film on the surface painted. The mixture is kept air-tight to prevent drying.

21·16. PAINTING WITH BRUSHES

A good brush should have springiness in the bristles. Any new brush may have loose bristles. Prior to painting, the loose bristles should be removed by taping the brush gently against the palm of hand, twirling it rapidly between the palms of hand and finally putting the loose bristles out. The brush should not be dipped too deeply in the paint, and the excess paint should be removed by gently rubbing the brush against the inside surface of the paint pot. After use, the brushes should be cleaned very thoroughly with kerosene oil. When a brush has been cleaned and is not to be used again, it should be immediately wrapped in an ivory paper or kept in raw linseed oil so that only the bristles are covered by the oil film.

21·17. SPRAY PAINTING

Painting with the spray gun is considered to be a highly developed and efficient method of applying all types of protective coatings. It result, in rapid execution of work. There is also economy in labour and the consumption of material as only two coats of spray are considered sufficient to obtain the desired finish. For elegance, perfection and durability, spray painting is recommended on metal or woodwork. The process of painting with spray machine may be summarised as below.

The paint is put into container and forced by air pressure through a hose to a metal gun. The gun is provided with a spray nozzle and a hose conveying compressed air. When the trigger of the gun is pulled, both the air and the fluid valves get opened. The compressed air strikes the paint and converts the liquid fluid into atomized paint, which is forced out in the form of a spray from the nozzle.

In operating the spray gun, care should be taken to see that the spray is at right angles to the surface and at all times, the gun is kept at the same distance from the surface to be painted. This distance may vary from 15 to 20 cm. Spray painting should, however, be done only under dry condition.

VARNISH

21·18. GENERAL

Varnish is a clear, pale solution of a resinous substance (like amber, common resin, copal, lac or shellac) dissolved in either oil, turpentine or alcohol. The solution on drying forms a hard, transparent, glossy film on the varnished surface. Varnish plays an important role in finishing wooden surfaces of doors, window, floors etc. Varnish is applied on unpainted furniture and other woodwork to decorate the surface, without hiding the beautiful grains of wood and to protect the surface against the adverse influence of the atmosphere. Painted surfaces are varnished to enhance the appearance of the paint and to increase the durability of the paint film.

The preparation of varnish is a tedious job and as such as far as possible readymade varnishes should be used for all iobs.

21·19. TYPES OF VARNISH

Depending upon the solvent used, varnishes may be classified as below:

Oil varnish : It is a cooked mixture of a drying oil (usually linseed oil) and resin (usually copal) dissolved in a volatile liquid thinner (turpentine or benzine). Oil varnish is used both for interior as well as exterior works. It takes about 24 hours to dry.

Spar varnish : This derives its name from its utility on spars and other exposed parts of ships. It forms a weather-resistant coating when used on exposed surfaces like rail road coaches, exterior wood work or metal surfaces. Spar varnish should not be used indoors as it gives a stickily effect in warm weather.

Flat varnish : It is sometimes desired that the varnished surface should present dull appearance. To attain this, manufacturers make varnishes which dry with a reduced gloss by adding such material as wax, metallic soap and finely divided silica.

Asphalt varnish : This is made by dissolving melted hard asphalt in linseed oil and thinned with turpentine or petroleum spirit. This varnish is chiefly used to give a black shop coat to fabricated iron and steel products.

Spirit varnish : This is a solution of a resin or other film forming material dissolved in a volatile liquid. It dries quickly (as the volatile liquid evaporates leaving behind the resin used) in the form of a hard glossy but somewhat brittle film. Spirit varnishes are mostly used for wooden furniture. French polish, Lacquer and Shellac varnishes fall in this category.

21·20. FRENCH POLISH

It is a spirit varnish. In its simplest from French polish is made by dissolving 0.15 kg. of pure shellac in 1 litre of methylated spirit and straining the solution through a double thickness of coarse muslin. Suitable pigment may be added to get the required shade. It is mostly used for high class furniture and other wood work.

Method of applying French polish : The wooden surface is cleaned of all dirt and dust and all unevennesses are rubbed down smooth with sand paper. All the visible knots should be covered with a preparation of red lead and glue size laid on while hot. Holes and indentations on the surface should be stopped with glaziers putty. The surface is then given a coat of wood filter made by mixing whiting (ground chalk) in methylated spirit at the rate of 1·5 kg. of whiting to 1 litre of spirit. The surface is again rubbed down perfectly smooth with glass paper and wiped clean. On this prepared surface, French polish is applied with a pad of woollen cloth covered by a fine cloth. The pad is moistened with the polish and then rubbed hard on the surface in a series of over- lapping circles so that the polish is applied sparingly but uniformly over the entire area. The surface is then allowed to dry. Several coats will be necessary (each succeeding coat being applied in the same way after the previous one has dried) before desired finish is achieved.

Furniture polish : The following mixture will make an excellent furniture polish :

Material	*Qty. required*
Linseed oil	16 litre
Methylated spirit	2 litre
Vinegar	1 litre
Turpentine	1 litre
Copal varnish	1 litre
Muriatic (Hydrochloric) acid	0·75 litre

The oil should be heated first and then the remaining materials mixed up well to form the polish.

Lacquer : It is considered to be the fastest drying finish. **On drying** it gives tough and durable finish which is easy to rub, sand or patch. It can be applied on furniture, floors, linoleum etc. It performs the function of a varnish. Lacquers are available in a variety of colours and can be brushed or sprayed. Since they have low solid contents, the surface requires several coats to attain a film thickness equivalent to varnish or enamel coat.

Stains: It is a liquid preparation which is used to change the colour of various types of cheap quality wood to give them the appearance of wood of superior quality. Dyes are the basis of most stains. Aniline, a derivative of coaltar is the base generally used in stains. The stains are named after the vehicle, with which they are applied. Thus when water is used as vehicle it is called water stain and if oil is used, it is called oil stain and if a alcohol is used for dissolving the dye, it is called spirit stain.

Water stain is the cheapest and can be applied easily. If applied on wood work, water stain causes the fibres of wood to rise and makes the coated surface rough which is not desirable. Spirit stains on the other hand dry very quickly and require skilled painter to apply them satisfactorily. Oil stains are generally prepared by mixing pigments like chrome yellow, burnt sienna, umber etc. to the oil together with some proportions of turpentine and a drier. Oil stains are considered most suitable, for coating hard wood surfaces.

21.21 METHOD OF APPLYING VARNISH

After knotting and stopping, the prepared wood work should be treated with glue sizing to seal the pores in wood and to prevent absorption of the oil in the varnish by the surface. The treatment consists in applying a hot coat of thin clean glue. When dry, the surface should be rubbed down smooth with sand paper and cleaned. A second coat of thin clean glue size (which is nearly cold) is then applied on the surface. If the wood work is to be stained, the colouring pigment should be mixed with second coat of glue size. The sized surface is again rubbed down smooth with a fine sand paper and cleaned. Two or more coats of varnish are then applied on the cleaned surface. Usually the under-coat consists of a flat varnish. This coat, when dry, is rubbed down to produce a smooth surface and then the second coat of varnish is applied.

DISTEMPERS

21·22. GENERAL

Distempers may be defined as water paints consisting of whiting (powdered chalk), some colouring pigment (if desired) and glue size mixed in water. Distemper may give either a washable or non-washable surface according to the medium used. They are cheaper than paints and varnishes and are easier to work. They act as sealers over porous surfaces and are generally used over plastered surfaces to which a priming coat of whiting has been applied. Distempers for exterior use on cement concrete, stucco and brick surfaces are provided with weather resistant ingredient during the process of manufacturing. They are more durable and present smoother and distinctive appearance than an ordinary colour/white wash. The cost of distemper compares favourably with that of white—wash in the long run and hence its application should be given preference. It should only be used in dry climates as in wet conditions, it gives poor results. Ready mix, washable distempers are now available in a variety of different shades in the form of a stiff plaster or dry powder in sealed tins.

21·23. DISTEMPERING WITH DRY DISTEMPER

The new plaster is allowed to dry for at least two months before applying distemper. However, incase of old surfaces, all dust, dirt, scales and greasy marks should be removed and all the holes and cracks should be filled with plaster of Paris mixed with dry distemper of the colour to be used. A coat of whiting is then applied over the prepared dry-clean surface which acts as the priming coat for the distemper. The dry distemper is then converted into a liquid by mixing it with water and stirring it well. As a rule 0.6 litre of water should be used per kg. of distemper. Distemper is then uniformly applied over the entire surface with broad stiff distemper brushes in horizontal strokes followed immediately by vertical ones. Second or third coats should only be applied after the previous coat has dried.

On newly-plastered walls, distemper should be applied in at least two coats. When the old wall surface is very smooth and is in good condition and priming coat is not felt desirable, a coating of warm glue should be applied prior to the application of distemper.

21·24. OIL BOUND DISTEMPER

It is form of distemper in which the drying oil is rendered mixable with water. The distemper in pasty form is available in different shades in sealed tins. When required, it is diluted or thinned with water. On drying, the oil content of the distemper hardens and forms a durable coating which can be cleaned (when dirty) by washing. However, for oil bound, washable distemper, the surface is given a coat of cement primer or any other primer recommended by the manufacturers prior to the application of distemper.

21·25. CASEIN PAINTS

Casein is a protein substance derived from milk curds. Casein paint is made by working a mixture of finely ground casein (which acts as a binder) with a base of white pigment which is usually slaked lime, and is sold in powdered or paste form in sealed tins. Casein paints, on account of their high opacity, give pleasing results in one coat. They are applied on new plaster without any danger of blistering, scaling or any adverse effect of free alkali. They are manufactured in various colours and are used on plastered surfaces, stucco, fibre board, cement blocks etc.

WAX POLISHING, OILING AND GLAZING

21·26. WAX POLISHING

Wax-polished surface gives a highly pleasing luster. It is applied on varnished surface to modify its elegance, and to protect the undercoat. Wax polishing is mostly used for polishing cement concrete, mosaic or terrazzo floors. The polish may be prepared by mixing bees wax, linseed oil, turpentine and varnish in the ratio of 2 : 1 : 12 by weight. Bees wax and linseed oil are first heated over a slow fire. When the wax is completely dissolved, the mixture is cooled down slightly and turpentine and varnish are then added and the mixture stirred well.

The polish is applied on the prepared wood-work, with a clean soft pad of cotton cloth and the surface is then rubbed continuously for about half an hour. The subsequent coats are applied when the earlier coat is dry and rubbed in the similar manner till the desired polished surface obtained.

21·27. OILING WOOD WORK

Oiling may be done to wood-work to increase its durability and to make it appear more pleasing. The oiling mixture may be prepared as below:

A mixture comprising of 1·5 kg. of double boiled linseed oil and 5 kg. of bees-wax is heated over a slow fire till the wax is melted. After the mixture has cooled, .5 kg. of turpentine is added to the mixture. The mixture is applied freely with brushes till the surface can absorb no more of the oil. This mixture when applied in two coat will cover an area of 75 sq. m. of the surface.

A mixture made by using country sweet oil to which equal parts of vinegar and turpentine have been added, may also be used for oiling wood-work.

21·28. COAL TARRING

Wood or iron work to be coated is first cleaned of all dust, rust, scales-etc. in the similar fashion as is essentially done before painting. The mixture for coal tarring is prepared as described below.

A mixture prepared by taking 200 gms. of unslaked lime and one litre of tar is heated till it starts boiling. The hot mixture is thinned by addition of litre of country spirit. The mixture is best applied while it is hot. When it is possible to remove iron-work, it should be (if possible), heated to nearly red hot stage and then brushed over with the prepared hot mixture of coat tar. If possible, the article to be tarred should be dipped in the hot mixture of tar for better results. However, if this is not possible, the hot mixture of coat tar is applied with the help of a brush on the prepared wood or iron work.

21·29. GLAZING

Glazing consists in securing glass, in prepared opening in iron, steel, masonry or wood work. Depending upon the size of the opening to be glazed, glass panes, weighing not less than 5·5 kg. per sq. m. should be used for area of opening upto 0·3 sq. m. For area greater than 0·3 sq. m. the thickness of the glass should be such that its weight is not less than 7·2 kg. per sq. m. The glass panes are so cut (with the help of a glaziers diamond) that they fit slightly loose in the rebates of the sashes.

In wood work, the glass panes are generally fixed with the help of glaziers putty (prepared by mixing one part of white lead with three parts of finely powdered chalk and then adding boiled linseed oil to the mixture which is finally kneaded into a stiff paste).

The glazing is done only after the frames have been primed and prepared for painting so that the wood may not absorb the oil of putty.

A thin layer of putty is first applied along the inner edge of the rebate for bedding the back of the glass panes. Glass pane is then put in position and secured in position by brads (small nails without head). Putty it then applied in the rebate, uniformly sloping from the inner edge of the rebate. If desired, panes may also be fixed with wooden beading having mitred joints.

21·30. GLAZING STEEL FRAMES

In glazing steel doors, windows etc., the glass used should not in any case weight less than 6.9 kg./sq. m. Special metal sash putty (prepared by adding gold size varnish to glazier's putty for wood work at the rate of 5.64 litre/100 kg. of putty), should preferably be used for fixing glass panes. Prior to fixing the glass pane, putty is applied between the glass pane and glazing bars and the glass pane is put in position. Putty is then applied over the glass pane which is stopped 1.5 mm. from the sight line of the back rebate. When the size of the opening for glass pane exceeds 600 × 300 mm, the glass should be secured by special spring clips which are inserted in holes already provided in the window steel frame before applying the front putty. Putty should normally be painted within 3 weeks to avoid its cracking.

Questions for Review

1. What are the chief constituents of an oil paint ? Explain the functions of each.

2. Name the constituents and state the uses and methods of application of the following :

   (*a*) French polish (*b*) Wax polish

   (*c*) Distemper (*d*) Painter's brushes

3. What are varnishes ? Write short notes on the following :

   (*a*) Oil varnish (*b*) Spar varnish.

   (*c*) Flat varnish (*d*) Asphalt.

   (*e*) Spirit varnish.

4. Write short notes on any four of the following :

   (*a*) Anti-corrosive paints (*b*) Bituminous paints.

   (*c*) Double-boiled linseed oil (*d*) White lead.

   (*e*) Lithopone (*f*) Poppy oil.

5. Distinguish between the following :

   (*a*) Plastering and painting.

   (*b*) Painting and varnishing.

   (*c*) White-washing and distempering.

6. Write an explanatory note on constituents of paints and characteristics of a good paint.

22

Plain and Reinforced Cement Concrete

22·1. GENERAL

Cement concrete may be defined as building material obtained by mixing cement, aggregates and water in suitable proportions and then curing this plastic mixture to a hard mass. Hardened concrete resembles stone in weight, hardness and strength to a great extent. The properties of concrete depend on the quantity and proportions of the ingredients used in the mix and the control exercised during various operations starting from mixing of aggregates to its placement in formwork and curing. Plain cement concrete has considerable strength in compression but it has very little strength in tension. Hence the use of plain cement concrete is normally restricted to situations where high compressive strength and weight are the primary requirements. Plain cement concrete is therefore used in the construction of bed blocks, massive gravity dams, dockwalls, gravity walls, columns, and arches etc.

It is seen that in most of the cases, structural members are subjected to bending, shear and torsion stresses. Inspite of its low tensile strength, it has been found possible to use cement concrete in construction of members subjected to high tensile stresses. For this purpose, steel reinforcement is placed in the tensile zone of the member before it is concreted. The material thus obtained is termed as *reinforced cement concrete or R.C.C.* The principal materials used in making concrete are described below.

22·2. CEMENT

Cement in its broadest term means any substance which acts as a binding agent for materials. Cement as applied to construction and engineering is produced by calcining at high temperature an intimate mixture of calcareous, siliceous and aluminous substances and crushing the resulting clinkers to a fine powder. Cement is the most expensive ingredient in concrete and it is available in a variety of different forms. The properties of cements depend upon the chemical composition, the process of manufacture and the degree of fineness to which they are ground. When cement is mixed with water, a chemical reaction takes place as a result of which the cement paste first sets and then hardness to a stone like mass. Depending upon their chemical composition, setting and hardening properties, cements can be broadly divided in the following two categories :

(1) *Portland cement*

(2) *Special cement.*

22·3. PORTLAND CEMENT

This is the most widely used type of cement and is so named because of the resemblance of it properties with a well-known natural stone quarried at Portland (U.K). Joseph Aspdin, a Yorkshire brick-layer is regarded as the discoverer of Portland cement.

Composition of Portland Cement

Lime, silica and alumina are considered to be the three principal constituents of cement. In addition, most of the cements contains iron oxide,magnesia, sulphur trioxide and alkalies in small proportions. Cement is manufactured by burning to white heat an intimate mixture of the above ingredients and then grinding the resulting clinkers with gypsum to an extremely fine powder. The magnitude of the different constituents in the composition of cement are :

Lime (CaO)	...	...	60 to 67%
Silica (SiO_2)	...	...	17 to 25%
Alumina (Al_2O_3)...	...	...	3 to 8%
Iron oxide (Fe_2O_3)	...	...	0.5 to 6%
Magnesia (MgO)	...	...	0·1 to 4%
Sulphur trioxide (SO_3)	...	...	1 to 2·75%
Alkalies (Soda and Potash)	...	...	0·5 to 1%

Setting and Hardening of Portland Cement

If a well-burnt Portland cement clinker is ground to a fine powder and examined, it will be found to be composed of four important mineral compounds given below :

			Abbreviated form
1.	Tri-calcium silicate ($3\ CaO.SiO_2$)	...	C_3S
2.	Di-calcium silicate ($2\ CaO.SiO_2$)	...	C_2S
3.	Tri-calcium Aluminate ($3CaO.Al_2O_3$)	...	C_3A
4.	Tetra-calcium Alumino ferrite ($4\ CaO.Al_2O_3.Fe_2O_3$)	...	C_4AF

If the powdered Portland cement clinker is mixed with 2 or 3 percent of gypsum and water is added to the mixture to form a paste, it will be observed that the soft paste will slowly lose its plasticity and will gradually stiffen and harden into a solid mass. During the process of hardening from paste to a hard mass, cement undergoes two types of sets. In the first set, also known as the *initial set*, the cement paste loses its plasticity, and becomes sufficiently coherent to withstand a certain arbitrary pressure. After the initial set, cement hardens still further, becomes much more rigid and in course of time it becomes strong enough to be capable of withstanding higher intensity of pressure. At such a stage, the cement is said to have undergone *final set*.

The setting and hardening of cement paste is mainly due to the hydration and hydrolysis of the four compounds mentioned above. Out of the four, tri-silicate acts as the best cementing material and greater its percentage in the cement, the better cement it will make. The aluminates (Tetra-calcium alumino ferrite and tricalcium aluminate) are the first to set and harden and are responsible for the initial setting of cement. Thus the rapidity of the initial set of cement depends upon the proportion of the aluminates present. Setting time can however be regulated by the addition of 1 to 3% of gypsum.

The strengthening effect of the aluminates is maximum during the first 24 hours and it does not extend beyond 28 days. During reaction with water, the aluminates generate considerable heat which contributes towards many desirable properties of the material. A cement having low percentage of aluminate compound exhibits less tendency towards volume changes and formation of cracks, generates less heat, resists sulphate solution attack better and possesses high ultimate strength. Tri-calcium silicate is the next to set and harden and it imparts first marked strength in 7 or 8 days. Di-calcium silicates reacts with water at a very slow rate. Its influence on strength and hardness is small during the first 28 days but after about one year, it contributes nearly as much strength as tri-calcium silicate does.

Although the process of setting and hardening continues simultaneously, yet there is a difference between the two. Setting is the change of cement paste from a plastic state to a stiff solid state. Immediately after setting, the mass is not strong and has very low compressive strength. With the passage of time, when the hydration proceeds, the product becomes harder and its compressive strength increases. This is termed as hardening of cement.

Types of Portland Cement

Portland cement can be further sub-divided into the following different types. The difference in the properties of the various types of cement are basically on account of variance in the proportions of its ingredients and degree of fineness to which the clinkers are ground.

(*i*) *Ordinary Portland cement*

(*ii*) *Portland Pozzalana cement*

(*iii*) *Rapid hardening Portland cement*

(*iv*) *Low-heat Portland cement*

(*v*) *White or coloured Portland cement*

(*vi*) *Sulphate resisting Portland cement*

(*vii*) *Water-repellent Portland cement*

(*viii*) *Water-proof Portland cement*

(*ix*) *Portland blast furnace cement*

(*i*) **Ordinary Portland cement**. This is the basic type of cement which is used on large scale in all general type of construction work. The details regarding composition and properties of this type of cement are highlighted in IS : 269.

(*ii*) **Portland Pozzolana cement.** This can be defined as a general purpose cement which can be used in all situations where ordinary Portland cement is considered suitable. Portland Pozzolana cement is manufactured either by grinding together, Portland cement clinkers and Pozzolana with addition of gypsum or calcium sulphate or by intimately and uniformly blending Portland cement and fine Pozzolana.

The term Pozzolana means a silicious material which while in itself possesses little or no cementitious properties but which will in finely divided form and in the presence of water react with calcium hydroxide (liberated during the hydration of cement) to form compounds possessing cementitious properties. The Pozzolana commonly used in the manufacture of Portland Pozzolana cement in India is burnt clay, shale or flyash. The manufacture of Portland Pozzolana cement by grinding of Portland clinker with Pozzolana does not present any difficulty. On the other hand, the manufacture of Pozzolana cement by blending of Portland cement with fine Pozzolana requires extra special care to ensure that the blending achieved is intimate and uniform. As such manufacture by blending process is recommended only where grinding process works out to be uneconomical. As per Indian Standard Code (IS : 1489) the proportion of Pozzolana may vary from 10 to 25% by weight of Portland Pozzolana cement. The properties of Portland Pozzolana cement are closely similar to that or ordinary Portland cement. It has same 7 days compressive strength value as ordinary Portland cement. This shows that by use of Portland Pozzolana cement the time of removal of formwork and supports remains same as applicable to ordinary Portland cement. Portland Pozzolana cement produces lower heat of hydration and has greater resistance to attack of chemical agencies than ordinary Portland cement. Concrete made with Portland Pozzolana cement is thus considered particularly suitable for construction in sea water, hydraulic works and for mass concrete works. Keeping in view the shortage of ordinary Portland cement in the country, the Indian standard code recommends the adoption of Portland Pozzolana cement as a suitable substitute for ordinary Portland cement for plain and reinforced concrete works in general building construction.

(*iii*) **Rapid Hardening Portland cement.** The main difference between the ordinary Portland cement and this type of cement is that the latter attains greater strength at an early stage. The property of high early strength is achieved almost entirely due to higher degree of fineness in grinding, burning at higher temperature and increased lime content in the composition of cement. The strength development of this cement at the age of 3 days and 7 days is almost the same as the respective 7 days and 28 days strength of ordinary Portland cement with the same water cement ratio. The main advantage of using rapid hardening cement is that the formwork can be removed earlier and reused in other areas thereby effecting considerable saving in cost of formwork. It also permits faster completion of the work and thereby the structure can be **put to use early.**

(*iv*) **Low heat cement.** Considerable quantity of heat is evolved during the chemical reaction that takes place during the setting and hardening of cement. In case of ordinary structures, it gets lost by dissipation from the surface of the member and as such produces no harmful effect. However, in case of mass concrete work such as dams, massive retaining walls, bridge abutments etc. heat of hydration generated (by use of ordinary Portland cement) proves dangerous. In such cases, the rate of loss of heat from the surface is much lower than the build up of heat inside the concrete mass which results in setting up tensile stresses during curing and finally leads to cracking of concrete mass. It is seen that the heat of hydration can be reduced by making suitable alteration in the chemical composition of cement. In low heat Portland cement tri-calcium aluminate and tri-calcium silicate which have the maximum contribution towards heat generation are kept to the minimum and the amount of di-calcium silicate is increased. Low heat Portland cement has not only lesser and slower heat generation but also greater resistance to cracking, less rapid strength development with equal strength at advanced ages and greater sulphate resistance due to decreased tri-calcium aluminate. Such a cement is, however, not suitable for structures of ordinary nature.

(*v*) **White or coloured Portland cement.** White cement is a type of ordinary Portland cement which is pure white in colour, and hence very useful for cast stone, terrazzo flooring or face plaster (stucco) to walls and other ornamental works. It is also used for traffic curbs, bridge rails and aerodrome markings. It is available in the market under various trade names as "snow crete", "Atlas" "Medusa" and "Silvicrete" etc. The strength of white cement is slightly less than that of ordinary cement and it is 4 to 5 times costlier than the latter.

In this case special care is taken to exclude iron oxides which impart grey colour to the cement during manufacture. Raw materials used are china clay and pure limestone. Only oil is used as fuel for burning and the rotary kiln is lined with special fire clay bricks so that the mixture does not come in contact with iron or steel during its manufacture.

Coloured cements are produced by adding 5 to 10% of a suitable mineral pigment to ordinary Portland cement or white cement before it is ground. The pigments used should be stable under the action of sunlight and weather, be free from soluble salts, must be in finely divided state and should not effect the quality of cement nor get effected by it.

Various metallic oxides are generally used as pigments. Coloured cements are sold in the market under the trade names "colour-crete"' "Rainbow" and "Snowcem" etc. and they are used for decorative' finish in swimming pools,, fountains, floors, tennis courts, etc.

(*vi*) **Sulphate-resisting Portland cement.** This type of cement is used in the construction of foundation in soil where sub-soil contains very high proportion of sulphates. The constituents of sulphate resisting Portland cement are so adjusted that the tri-calcium aluminate (C_3A) and tetra calcium aluminate-ferrite (C_4AF) are kept very small. Cements with such composition have excellent resistance to sulphate attack.

(*vii*) **Water-repellent Portland cement.** This type of cement is chiefly used in water-tight concrete and water-tight renderings to check moisture penetration is basements etc. and for coloured renderings and stucco. It contains a small percentage of waterproofing material uniformly mixed with the cement and is manufactured under the name "Aquacrete". This cement is prepared by using ordinary or rapid hardening cement and white cement. Considerable care is needed to avoid reduction in strength if such a cement is used in concrete.

(*viii*) **Water-proof Portland cement.** Water-proof cements are prepared by mixing with ordinary or rapid hardening cements, a small percentage of some metal stearates (Ca, Al etc.) at the time of grinding. Concrete made with such a cement is more resistant to penetration by water and some oils than that made from ordinary Portland cement and as such it is used for the construction of water-retaining structure like tanks, reservoirs, retaining walls, swimming pools, dams bridges, piers etc. It is also more resistant to the corrosive action of acids and alkalies or other harmful salts usually present in industrial waters.

(*ix*) **Portland blast furnace cement.** In this case, the normal cement clinkers are mixed with upto 65% of blast furnace slag for the final grinding. The resulting cement is quite cheap and has almost the same properties as Portland cement. Its early strength is less than that of normal cement but becomes the same after 28 days. In addition, it is more resistant to dilute acids and sulphate soils. It has a lower heat evolution but the speed of hardening falls off very rapidly when the temperature decreases. Also, it requires an extended curing time owing to its slower rate of strength development. In view of its having low heat of hydration and better resistance to attack by sulphates and alkalies, concrete made with this type of cement can be used with advantage in mass concrete work such as dams, foundations and abutments of bridges, retaining walls etc. and for construction in sea water.

22·4. SPECIAL CEMENTS

Special cements are generally manufactured by adding certain admixtures to the cements during the process of mixing or grinding. Special cements include.

(*i*) *Quick setting cement*

(*ii*) *Calcium chloride cement*

(*iii*) *High alumina cement*

(*iv*) *Slag cements*

(*v*) *Expansive cement*

(*i*) **Quick setting cement.** When concrete is to be laid under still water or in running water, quick setting cement is used advantageously. The setting action of such a cement starts within five minutes as it becomes stone-hard in less than an hour. Quick setting cement is mixed with a small percentage of aluminium sulphate and is ground much finer than ordinary Portland cement. It also contains small percentage of gypsum.

(*ii*) **Calcium Chloride cement.** Calcium chloride accelerates the rate of strength development when added upto 2% in the ordinary cement. It also imparts setting and hardening properties to the cement in cold weather. A higher percentage, however, causes excessive shrinkage of concrete and also adversely affects the steel reinforcements in case of R.C.C. works.

(*iii*) **High alumina cement**. This cement is manufactured by fusing together 40 % each of bauxite (aluminium ore) and lime ; 15% of iron oxide and 5% of silica, magnesia etc. and then grinding the resulting clinkers to a very fine powder. The fineness of grinding is higher than that of Portland cement, 8 to 10% of this cement being held on a sieve of 4900 opening per sq. cm. It has the following special advantages over other types of cements :

(*a*) It has an initial setting time of about 3½ to 4 hours.

(*b*) It hardens so rapidly that at the end of 24 hours, it has a compressive strength as high as can be obtained from other cements after a minimum period of seven days.

(*c*) It has excellent resistance to action of sulphates.

(*d*) It has good resistance to acid water, sea water etc.

(*e*) It offers excellent resistance to fire.

The rapid hardening quality of this cement permits the early removal of formwork and in places where the repeated use of formwork is possible, it greatly reduces the cost. However, on account of high cost of bauxite, this cement is 3 or 4 times costlier than normal setting Portland cement.

(*iv*) **Slag cements**. Slag cements are produced by grinding blast furnace granulated slag with a hydraulic binding agent like Portland cement clinkers, lime or gypsum. When Portland cement is used with slag, it is called *Portland-slag cement* : when lime is used, it is called *lime slag cement* and so on. The setting time of the slag cement is adjusted by adding 1 to 5% of gypsum, during the process of grinding. The various slag cements are briefly given below :

(*a*) *Portland-slag cement.* The percentage of the slag component of this cement varies from 40 to 70%. It offers reasonably high resistance to corrosion and possesses excellent adhesion properties with steel reinforcement. It is used in the manufacture of prefabricated reinforced concrete units. It is recommended in situations where cement is liable to be subjected to the action of mineralised water.

(*b*) *Lime-slag cement.* The percentage of the slag component of this cement varies from 70 to 90%. It does not resist air and frost well and as such its usage is generally restricted to damp situations.

(*c*) *Sulphate-slag cement.* The percentage of slag component in this cement varies from 80 to 90% and about 5% of Portland cement clinkers are also added to the mixture of slag and gypsum. Being highly resistant to the action of sulphates, it is generally recommended for use in marine works or works under the ground.

(*v*) **Expansive cement** : This is a form of hydraulic cement which expands during hardening and setting. Concrete made with ordinary portland cement shrinks while hardening and has tendency to develop shrinkage cracks. This can be overcome by adding expansive cement to the concrete mix made from ordinary cement. The concrete thus made will have no tendency of shrinkage or expansion (shrinkage being compensated by expansion). Expansive cement can be used for the construction of roof slabs, canal lining, tunnel lining, concrete parchment etc. When concrete made with expansive cement is used for the constiuction of roof, it is not necessary to provide water-proofing treatment to the roof slab. When expansive cement concrete is used for the construction of canal lining, tunnel lining, concrete parchment etc., it is possible to carry out concreting in long stretches with reduced number of joints. Grout made with expansive cement also known as "no shrunk grout" is used in grouting holding down bolts or other anchorage arrangements for the steel structures, machines with the concrete foundations etc.

22·5. STORAGE OF CEMENT

Storage of cement requires extra care to preserve its quality and fitness for use. To prevent its deterioration it is necessary to protect it from rain, sun, winds and moisture.

Storage of cement is also a noteworthy item, because proper arrangements for storing the cement have to be made in the factories before sale or on large construction projects before use.

Moisture is the first and the greatest danger to be guarded against. Cement has great affinity for moisture and hence it should be stored well shielded from moisture laden current of air. The exposed cement is attacked by air setting which gradually spreads. It gives rise to the formation of lumps. If the lumps formed are so hard that they cannot be powdered by passing between the fingers, it should be concluded that the cement has been rendered useless for any sound construction.

Cement should be stored in dry, leakproof and moisture proof sheds. The shed for the storage of cement should have minimum number of windows and close fitting doors. The cement bags should be stacked on wooden planks placed about 150 to 200 mm above the floor of the shed and a space of 450 to 600 mm should be kept allround between the exterior walls and the stacks.

To prevent possibilities of lumping under pressure, the maximum height of stack should not exceed 15 bags. The width of stacks should not be more than four bags or 3 m. In stacks more than 8 bags high, the cement bags should be arranged alternately lengthwise and cross-wise so as to ensure stability of the stacks and to prevent danger of the bags in stack toppling over.

In monsoon or in situations when it is necessary to store cement for unusually long periods, the cement stacks should be completely enclosed by 700 gauge polyethene sheet or some other water-proofing membrane materials.

The stack should be arranged in such a manner that cement bags can be removed on principle of first come first served *i.e.* oldest cement should be taken out first.

22·6. TESTING OF PORTLAND CEMENT

The properties of concrete or mortar largely depend upon the quality of cement used. The quality of cement can be tested in the laboratory by conducting following test based on IS : 269.

(1) *Chemical composition*

(2) *Fineness*

(3) *Soundness*

(4) *Setting time*

(5) *Tensile strength*

(6) *Compressive strength.*

(7) *Heat of Hydration*

(1) **Test for chemical composition.** Chemical composition of Portland cement, whether ordinary or rapid-hardening, should be as given below :

(*a*) Ratio of Percentage of lime to percentages of silica, alumina and iron oxide, when calculated by the formula $\frac{CaO - 0{\cdot}7\ SO_3}{2{\cdot}8\ SiO_2 + 1{\cdot}2\ Al_2O_3 + 0{\cdot}65\ Fe_2O_3}$	Not greater than 1.02 and not less than 0.66
(*b*) Ratio of percentage of alumina to that of iron oxide	Not less than 0.66
(*c*) Weight of insoluble residue	Not more than 1·5%
(*d*) Weight of magnesia	Not more than 6%
(*e*) Total sulphur content, calculated as sulphuric anhydride (SO_3)	Not more than 2.75%
(*f*) Total loss on ignition	Not more than 4%

Test for chemical composition of low heat cement :

Low heat Portland cement should have all the requirements given in (*b*), (*c*), (*d*), (*e*) and (*f*), above except that the percentage of lime (after deducting the amount necessary to combine with sulphuric anhydride present) should not be more than $2{\cdot}4\ SiO_2 + 1{\cdot}2\ (Al_2\ O_3) + 0{\cdot}65\ (Fe_2\ O_3)$ nor less than $1{\cdot}9\ (SiO_2) + 1{\cdot}2\ (Al_2O_3) + 0{\cdot}65\ (Fe_2O_3)$.

(2) **Test for fineness.** The rate of hydration and hydrolysis of cement depend upon its fineness. Finer cements have quicker action with water and develop high early strength. Fineness of cement may be tested by the method of sieving or by the method of air permeability. When tested by one of said methods, the cement should comply with the following requirements:

	Ordinary Cement	*Rapid hardening cement*	*Low heat cement*
	(1)	(2)	(3)
Method (1) :			
After sieving, the residue by weight on an I.S. Test Sieve No. 9. not to exceed. percentage.	10	5	—
Method (2) :			
Specific Surface (sq. cm./g) by Air permeability Method (as described in IS : 4031) *not less than*	2250	3250	3200

(3) **Test for soundness.** Due to the presence of free lime and magnesia, cement undergoes volumetric changes after setting as these ingredients go on slaking very slowly. This expansion or volumetric instability is called *unsoundness* and may produce cracks, distortion and disintegration of concrete. Soundness test is designed to accelerate the slaking process of the ingredients of cement and to determine the resulting expansion in a short time. Expansion of all types of portland cements when tested by Le chatelier's method in accordance with IS : 4031 should not exceed 10 mm.

(4) **Test for setting time.** The principal aim of this test is to make a distinction between normal and quick setting cement and also to detect the degree of deterioration in cement due to storage. The setting time for different types of cement by use of Vicat's apparatus in accordance with IS : 269 should be as under :

	Ordinary cement	*Rapid hardening cement*	*Low heat cement*
(*a*) Initial setting time *not less than*	30 min	30 min.	60 min.
(*b*) Final setting time *not more than*	10 hrs.	10 hrs.	10 hrs.

(5) **Test for tensile strength.** Tensile strength of cement is tested by fracturing six standard briquettes (made from the test cement mortar) separately and taking the average of the six test results. The average strength of the briquettes after 1 day, 3 days and 7 days immersion respectively should be as under :

	Ordinary cement	*Rapid hardening cement*
(*a*) After 1 day (24 hrs.) *not less than*	–	20 kg./sq cm
(*b*) After 3 days (72 hrs.), *not less than*	20 kg./sq cm	30 kg./sq cm
(*c*) After 7 days (168 hrs.) *not less than*	25 kg/sq cm	–

(6) **Test for compressive strength.** Similar to the tensile strength test, the compressive strength of cement is tested by fracturing 3 cubes (made from test cement mortar) by compression separately and taking average of the three results. The average compressive strength of the cubes tested after 1 day, 3 days, 7 days, and 28 days immersion respectively should be as under :

	Ordinary cement	*Rapid hardening cement*	*Low heat cement*
(*a*) After one day (±30 min.) *not less than*	–	160 kg./cm^2	–
(*b*) After 3 days (±60 min.) *not less than*	160 kg/cm^2	275 kg/cm^2	100 kg/cm^2
(*c*) After 7 days (±2 hrs.) *not less than*	220 kg/cm^2	–	160 kg/cm^2
(*d*) After 28 days (±4 hrs.) *not less than*	–	–	350 kg/cm^2

(7) **Heat of hydration.** This test is applicable only in case of low heat cements. The heat of hydration of low heat portland cement should be as under :

(*i*) 7 days *not more than* 65 calories per gram

(*ii*) 28 days *not more than* 75 calories per gram.

22·7. AGGREGATES

Aggregates are those chemically inert materials which when bonded by cement paste form concrete. Aggregates constitute the bulk of the total volume of concrete and hence they influence the strength of concrete to a great extent. Aggregate is derived from igneous, sedimentary and metamorphic rocks or is manufactured from clays, slags etc. The properties of concrete are directly related to those of its constituents and as such aggregate used in a concrete mix should be hard, strong, dense, durable. free from injurious amounts of clay, loam, vegetable and other such foreign matter. The presence of deleterious substances such as coal, lignite clay lump, soft fragments of foreign materials and other materials prevent proper adhesion of cement on the surface of aggregates and thus affect the properties of concrete adversely.

Depending upon their size the aggregates are classified as :

(*i*) fine aggregate, and (*ii*) coarse aggregate.

(*i*) *Fine aggregate.* The material below 4·75 mm. size is termed as fine aggregate. The sum of percentage of all types of deleterious materials in fine aggregate should not exceed 5%. Natural sand or crushed stone dust is the fine aggregate chiefly used in concrete mix. Sand may be obtained from sea, river lake or pit, but when used in a concrete mix, it should be properly washed and tested to ascertain that total percentage of clay, silt, salts and other such organic matter does not exceed specified limit.

Sea sand should not be used in its natural state for R.C.C. work, as the salts present in the sea water are liable to attack the steel reinforcement. River sand is usually suitable but mostly it is contaminated with mud and thus it is always advisable to wash such sand before use. Pit sand is generally suitable, but it is liable to contain silt or other organic impurities. Hence before using the pit sand in important works, all its impurities must be removed.

Sand as obtained from the above sources may be round or angular in grains. Angular grained sand has good interlocking property which results in a strong mortar while rounded grained sand does not afford sufficient interlock in the matrix.

(*ii*) *Coarse aggregates.* The material whose particles are of such size as are retained on I.S. seive No 480 (4.75 mm) is termed as coarse aggregate. The size of the coarse aggregate used, depends upon the nature of work. The maximum size may be 20 cm. for mass concrete, such as in dams etc. and 63 mm. for plain concrete work. For R.C.C. construction aggregate having a nomional size of 20 mm are generally considered satisfactary. The types of aggregate used in cement concrete work are briefly described below.

Crushed hard stone and gravel are the common materials used as coarse aggregates for structural concrete. Coarse aggregates are usually obtained by crushing granite, gneiss, crystalline lime stone and good variety of sand stone etc. As far as possible, flaky and elongated pieces of stone should be avoided. Broken bricks is cheap aggregate for plain concrete, but it renders the mix, weak in strength. It is not used in R.C.C. work on account of the possibility of the reinforcement getting rusted due to the high porosity of the aggregates. Clinker slag, coal ashes, coke-breeze are also used as aggregates for light weight and insulating concrete where great strength is not desired. Gravel (as obtained from pit or river) or crushed stone contain high percentage of fine material and in this state it is only used for unreinforced work. The sum of percentage of all type of deleterious substances in coarse aggregate should not exceed 5%.

1. Grading of Aggregates

To make a strong dense, workable and impervious concrete, the mixing of its ingredients has to be such that cement used in sufficient to fill the voids in the fine aggregates (sand) and mortar thus produced is sufficient to fill the voids in the coarse aggregates. To achieve this, the fine aggregates as well as the coarse aggregates should be graded suitably so that the smaller particles can occupy the voids between the larger particles and hence the percentage voids is reduced considerably. The reduction in voids results in corresponding reduction in the quantity of mortar required to fill the voids in the coarse aggregates and hence there is reduction in the quantity of cement required to make the concrete. Thus, grading of aggregates is also necessary to have economy in the production of concrete.

The size of different I.S. sieves used for grading of aggregates are as under :

(*a*) *Fine mesh wire cloth type sieves* :

3·35 mm, 2·36 mm, 1·18 mm, 600-micron, 300-micron, 150-micron, 75-micron.

(*b*) *Square hole perforated plate type sieves* :

80 mm, 6·3 mm, 50 mm, 40 mm, 31·5 mm, 25 mm, 20 mm, 16 mm, 12·5 mm, 10 mm, 6·3 mm, 4·75 mm.

2. Grading Limits for Fine Aggregates

The grading limits for fine aggreges as given in IS : 383 are repdroduced in Table 22·1. From the review of the table, it can be seen that gradings for fine aggregates (sand) are divided into four zones.

TABLE 22·1

GRADING LIMIT FOR FINE AGGREGATES

IS Sieve designation	*Percentage passing*			
	Grading Zone I	*Grading Zone II*	*Grading Zone III*	*Grading Zone IV*
(1)	*(2)*	*(3)*	*(4)*	*(5)*
10 mm	100	100	100	100

(1)	*(2)*	*(3)*	*(4)*	*(5)*
4.75 mm	90 – 100	90 – 100	90 – 100	95–100
2.36 mm	60 – 95	75 – 100	85 – 100	95–100
1.18 mm	30 – 70	55 – 90	75 – 100	90–100
600 micron	15 – 34	35 – 59	60 – 79	80–100
300 micron	5 – 20	8 – 30	12 – 40	15–50
150 micron	0 – 10	0 – 10	0 – 10	0–15

The grading becomes progressively finer from grading zone I (coarse sand) to grading zone IV (fine sands). Sands of grading zones I to IV are mostly used for concrete and R.C.C works.

The use of fine aggregates falling in grading limits of zone IV is normally restricted to special mix of concrete.

3. Grading limits for Coarse Aggregates

In case of coarse aggregates the material brought at site may be single sized aggregates (*i.e.* ungraded) of nominal sizes 63 mm, 40 mm, 20 mm, 16 mm, 12·5 mm and 10 mm or it may be in the form of graded aggregates of nominal size 40 mm, 20 mm, 16 mm and 12·5 mm. The grading limits of the two different types of coarse aggregates as per IS 383 are given in Table 22·2.

4. All-in-aggregates. It is observed that in some places available aggregate comprise of a mixture of fine and coarse aggregate. Such aggregates are termed as *all-in-aggregates.* In case of all-in-aggregates it is not necessary to separate the fine and coarse aggregates but it may be necessary to carry out suitable adjustment to meet the requirement of grading.The grading limits for all-in-aggregates as specified in IS : 383 is given in Table 22·3.

TABLE 22·3

GRADING LIMITS FOR ALL-IN-AGGREGATE

IS Sieve designation	*Percentage passing for all-in-aggregates of*	
	40 mm nominal size	*20 mm nominal size*
80 mm	100	—
40 mm	95 to 100	100
20 mm	45 to 75	95 to 100
4.75 mm	25 to 45	30 to 50
600 micron	8 to 30	10 to 35
150 micron	0 to 6	0 to 6

TABLE 22·2

GRADING LIMITS FOR COARSE AGGREGATES

IS Sieve designa-tion	*Percentage passing for single sized aggregates of nominal size of*						*Percentage passing for graded aggregate of nominal size of*			
	63 mm	***40 mm***	***20 mm***	***16 mm***	***12.5***	***10 mm***	***40 mm***	***20 mm***	***16 mm***	***12.5 mm***
80 mm	100		—	—	—	—	100	—	—	—
63 mm	85 to 100	100	—	—	—	—	—	—	—	—
40 mm	0 to 30	85 to 100	100	—	—	—	95 to 100	100	—	—
20 mm	0 to 5	0 to 20	85 to 100	100	—	—	30 to 70	95 to 100	100	100
16 mm	—	—	—	85 to 100	100	—	—	—	90 to 100	—
12.5 mm	—	—	—	—	85 to 100	100	—	—	—	90 to 100
10 mm	0 to 5	0 to 5	0 to 20	0 to 30	0 to 45	85 to 100	10 to 35	25 to 55	30 to 70	40 to 85
4.75	—	—	0 to 5	0 to 5	0 to 10	0 to 20	0 to 5	0 to 10	0 to 10	0 to 10
2.36 mm	—	—	—	—	—	0 to 5	—	—	—	—

22·8. WATER

Water to be used in the concrete work should have the following properties :

(*i*) It should be free from injurious amounts of oils.

(*ii*) It should be free from injurious amounts of acids or alkalies or other such organic or inorganic impurities.

(*iii*) It should be free from iron, vegetable matter, or any other substance which is likely to have adverse effect on concrete or reinforcement.

(*iv*) It should be fit for drinking purposes.

Functions of water. The functions of water in the concrete mix may be summarised as on the next page:

(*i*) It acts as lubricant for the fine and coarse aggregates.

(*ii*) It acts chemically with cement to form the binding paste for the coarse aggregate and reinforcement.

(*iii*) It is necessary to flux the cementing material over the surface of the aggregate.

(*iv*) It is employed to damp the aggregate in order to prevent them from absorbing water vitally necessary for chemical action.

(*v*) It enables the concrete mix to flow into moulds.

22·9. WATER CEMENT RATIO

The ratio of the volume of water to volume of cement used in a concrete mix is termed as water cement ratio. As a result of experiments it is observed that for a given proportion of ingredients in a concrete mix, there is almost a fixed amount of water (*optimum*) which gives maximum strength. A small variation in the quantity of water causes much wider variation in the strength of concrete. In case the water used is less, the resultant concrete will be comparatively dry, difficult to place in position and may pose problems in compaction. Moreover, with less water complete setting of cement cannot be ensured and hence the strength of concrete get reduced appreciably. On the other hand, in case the water is more it would result in formation or excessive voids and honey-combing in the set concrete, thereby reducing its density, strength and durability. Thus, water cement ratio serves as a yard stick for obtaining concrete of desired strength. *The lower the ratio, the greater is the strength of the mix.* A rich mix of concrete gives higher strength than a lean mix, not because of more cement but it is on account of the fact that concrete can be used with a lower water cement ratio.

22·10. MEASUREMENT OF MATERIALS

Accurate measurement of cement, fine aggregate and coarse aggregate is the primary requirement of producing good concrete. The method of measurement adopted for different ingredients of concrete is detailed below.

1. Measurement of Cement

Cement should always be measured by weight and not by volume. It is on account of the fact that the quantity of cement when loosely filled in a container weighs less than the cement filled tightly in the same container. It has been observed that one cubic metre of loosely filled cement weighs 1120 kg. and the same volume weighs 1602 kg. when cement is filled by tamping. As per IS : 1911 one litre of ordinary Portland cement may be taken to weigh 1·44 kg. Cement is normally supplied in bags, each bag of cement weighing 50 kg. Under normal conditions the volume of cement in one bag is considered to be equivalent to 34·72 litres, say 35 litres.

2. Measurement of Fine Aggregates

Fine aggregate *i.e.* sand is normally measured by volume in case of ordinary works whereas in case of all important projects, it is measured by weight.

When dry sand comes in contact with moisture, thin film is formed around the particles, which causes them to get apart from each other. This results in increasing the volume of sand. This phenomenon is known as *bulking of sand*. Bulking of sand increases with increase in its moisture content to some extent. It is seen that when the moisture content in sand is about 4% by weight of dry sand, the increase in its volume may be of the order of 25 to 30% depending upon the grading. Further addition of water causes the film around the particles to break and thus the volume of sand gets reduced. When the quantity of water in sand becomes more than 20% or the sand is fully immersed in water its volume becomes the same as that of dry sand.

The percentage of bulking of the sand due to moisture can be calculated by simple field method.

The sample of damp sand is poured in a graduated glass cylinder of 250 ml. capacity. The quantity of sample should be such that after consolidation by shaking it should reach the 200 ml. mark. The cylinder is then filled with water and the sand is stirred well. The quantity of water should be sufficient to submerge the sand. It will be seen that the level of sand in the cylinder drops below the 200 ml. mark. Suppose the surface of the submerged sand is at the mark "*Y*" ml. The percentage of bulking of the sand due to moisture can thus be calculated by the formula,

$$\% \text{ bulking} = \frac{200 - Y}{Y} \times 100$$

After determining the percentage bulking the actual volume of corresponding dry sand can be worked out by subtracting from the measured volume of sand, the increase in volume due to bulkiness.

Thus when moist sand is measured by volume, proper allowance should be made on account of *bulking.* This is necessary to ensure that the amount of sand actually used in concrete is the same as it would have been if dry sand was used.

Similarly, when sand is measured by weight it is necesary to make allowance for the hygroscopic moisture in sand.

When fine aggregate is measured by weight, the unit of weight used is kilogram or tonne. For volumetric measurement, the unit applicable to measure capacity, liquid and dry material is litres and for solid it is cubic metres.

Thus when fine aggregate is measured by volume, the unit used is litres, (*one cubic metre = 1000 litres*).

3. Measurement of Coarse Aggregate

Coarse aggregate may be measured either by volume (litres) or by weight (kg). In case of wet aggregate, allowance for surface water should be made while determining the quantity of water to be used in the mix.

4. Measurement of Water

Water may be measured by weight or by volume. The normal practice, however, is to measure it by volume. *i.e.* in litres. The quantity of water to be used for a particular grade of concrete should be calculated after taking into account the moisture already present in the fine and coarse aggregates. The amount of water present in the aggregate should be subtracted from the specified quantity of water to the used in the concrete.

Similarly depending upon the nature of material (Brick aggregate etc.) and state of its dryness, sometimes, the aggregates may absorb moisture from concrete. In such cases, it would be necessary to find out the magnitude of absorption of the aggregates and allowance made by increasing the specified quantity of water accordingly. The methods for determining the absorption and moisture content of aggregate are given in detail in IS : 2386 (Part III).

The surface water carried by average aggregate as given by IS : 456-1978 is produced in Table 22·4.

TABLE 22·4

S.No.	Aggregate	Approx. quantity of surface water in litre/m³
1	Very wet sand	120
2	Moderately wet sand	80
3	Moist sand	40
4	Moist gravel or crushed rock	20 to 40

22·11. GRADES OF CONCRETE

In IS : 456—1964, seven grades of concrete were specified as M 100, M 150, M 200 M 250, M 300 M 350 and M 400. In the designation of the mix letter M represented the mix and the figures 100, 150 etc. represented the *works cube compressive strength of* 15 cm cubes at 28 days expressed in kg/cm^2.

In IS : 456—1978, the various grades of concrete have now been specified as M 10, M 15, M 20, M 25, M 30 M 35 and M 40. In the designation of the mix letter M represents the mix and the figure 10, 15 etc. represent the specified *characteristic strength* of 15 cm cube at 28 days expressed in N/mm^2. The characteristic strength is defined as the strength of the material (crushing strength of concrete cubes at 28 days) below which not more than 5% of the test results are expected to fall. The details of specified grades of concrete are given in table below:

TABLE 22·5

GRADES OF CONCRETE

Designation of grade of concrete	Specificed characteristic compressive strength at 28 days in N/mm²
M 10	10
M 15	15
M 20	20
M 25	25
M 30	30
M 35	35
M 40	40

Note 1. M 5 and 7.5 grades of concrete may be used for lean concrete bases./mud mat and for simple foundation for masonry walls.

Note 2. Grade of concrete lower than M 15 shall not be used in reinforced concrete.

It may be noted for guidance that M 10 ; M 15 ; M 20 ... correspond approximately equal to 1 : 3 : 6, 1 : 2 : 4, 1 : 1½ : 3 of nominal mix concrete.

22.12. PROPERTIES OF CONCRETE

Strength, durability, and workability may be considered as the main properties of concrete. In addition, good concrete should be able to resist wear and corrosion and it should be water-tight, compact and economical.

1. Strength

The concrete must be strong enough to withstand without injury all the imposed stresses with the required factor of safety. When the concrete mix has been designed on the basis of maximum permissible water-cement ratio, keeping in view the requirements of durability, it will develop the required strength if properly placed in position and cured. After placing, concrete should not be allowed to dry rapidly because moisture is very much essential for the development for its high strength. Curing temperature also has a great effect on the strength development of concrete. To develop a given strength, longer time of moist curing is required at lower temperature than is necessary while curing is done at higher temperature. The strength of various concrete mixes should be in accordance with detail given in Art. 22.11.

Increase in Strength with age : The structural design of R.C.C. members is generally based on characteristic compressive strength of 15 cm cube of concrete at 28 days as specified in Table 22.5. It is, however, important to note that the process of hydration of cement in concrete continues even after 28 days and consequently concrete continues to harden with age although at a decreasing rate. Thus, permissible strength in concrete can be suitably increased with age. According to provision in IS : 456—1978 in situation where it can be ensured that a member will not receive its full design load/stress within a period of 28 days after casting (e.g. the foundations and lower columns in multistorey building) the characteristic compressive strength given in Table 22.5 may be increased by the factors given on page 538.

2. Durability

It is the property of concrete by virtue of which it is capable of resisting its disintegration and decay which may be caused due to :

(*a*) Use of unsound cement

(*b*) Use of less durable aggregate

(*c*) Entry of harmful gases and salts through the pores and voids present in the concrete.

(*d*) Freezing and thawing of water sucked through the cracks or crevices by capillary action.

Minimum age of member when full design load/stress is expected (months)	*Age factor*
1	1.0
3	1.10
6	1..15
12	1.20

Note 1. No increase in respect of age at loading should be allowed where high alumina cement concrete is used.

Note 2. Where members are subject to lower direct load during construction, they should be checked for stresses resulting from combination of direct load and bending during construction.

(*e*) Expansion and contraction resulting from temperature changes and alternate drying and wetting.

The resistance of concrete to chemical attack, weathering, abrasion and fire etc. depends mostly upon conditions of exposure, grade of concrete used, quality of its materials and the extent of voids and pores present in the concrete mass. The amount of cover provided over reinforcement and the degree of imperviousness of concrete mix also influences the durability of the concrete.

Hence concrete can be made durable by using good quality of materials (cement, aggregate and water), by reducing the extent of voids by suitable grading and proportionating the materials, by using adequate quantity of cement and low water cement ratio thereby ensuring concrete of increased permeability. In addition, thorough mixing, desired placing, adequate compaction and curing of the concrete is equally important to have durable concrete.

3. Workability

Despite all its importance, workability is the most elusive property of concrete and is quite difficult to define and measure. It is simplest form a concrete is said to be workable if it can be easily mixed, handled, transported, placed in position and compacted.

Lubricating effect of the cement-paste which in turn is solely governed by the degree of dilution, affects the workability of a concrete mix. A workable concrete mix must be fluid enough so that it can be compacted with minimum labour. A workable concrete does not result in bleeding or segregation. Bleeding of concrete takes place when the excess of water in the mix comes up at the surface and segregation is caused when coarse aggregates have a tendency to separate from the fine aggregates. Workability of a concrete mix is appreciably affected by the following.

(*i*) By reducing the proportion of coarse aggregate : It should be well understood that the finer is the grading the greater will be the workability.

(*ii*) Shape of the aggregates has a great effect on the workability of concrete. Rounded aggregate renders a concrete mix more workable than the angular one.

(*iii*) By increasing the quantity of water to such limits that the water cement ratio is maintained.

(*iv*) By increasing the quantity of cement.

(*v*) By the process of mixing.

Evidently the requirement of workability varies according to the nature of job and the obstruction to the full flow of concrete caused by the spacing and nature of reinforcement. The workability of concrete can be measured by one of the following three tests.

1. Slump test
2. Compaction factor test.
3. Vee-Bee test.

1. **Slump Test.** Slump test is commonly adopted for ordinary; concrete works. This test is performed with the help of a vessel of the shape of the frustum of a cone and open at both ends. The top and bottom

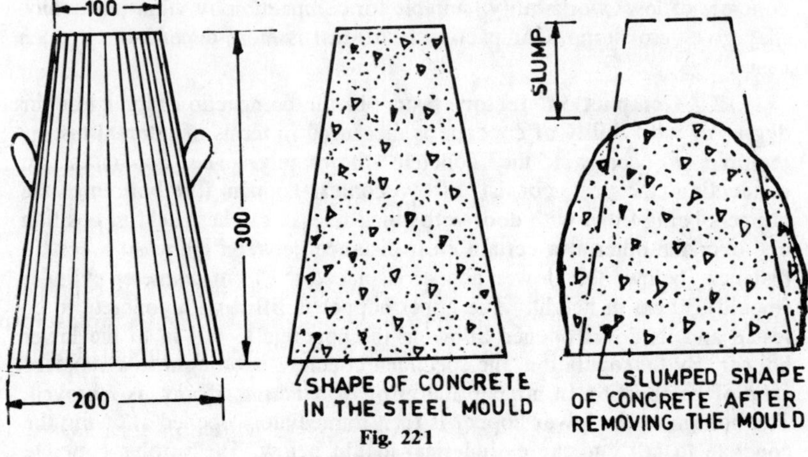

Fig. 22·1
Slump Test

diameters of the vessel should be 10 cm. and 20 cm. respectively and it should be 30 cm. in height. The vessel is placed on a flat non-absorbent surface and then filled with specimen concrete mix in four different layers of equal-thickness. Each layer is tamped 25 times by the point of a 16 mm. ϕ rod 60 cm. in length. The strokes are applied uniformly over the entire area with a force that the rod just penetrates the full depth of the layer being compacted. Immediately after the vessel is completely filled, it is raised vertically, care being taken not to disturb the filling. The concrete filling is allowed to subside or settle. The vertical settlement recorded in terms of mm. is known as slump. The table below gives the values of slump usually adopted for various types of concrete.

TABLE 22·6

SLUMPS COMMONLY ADOPTED FOR VARIOUS CONCRETE MIXES

S.No.	*Types of concrete*	*Slump adopted in mm.*
1	Concrete for road works	20 to 30
2	Beams and slabs	50 to 100
3	Columns, retaining walls and thin vertical members, etc.	75 to 150
4	Impermeable work	75 to 120
5	Vibrated concrete	12 to 25
6	Mass concrete	25 to 50

The slump test gives satisfactory results for concrete mix of medium to high workability. It however, does not give correct indication for concrete of low workability (suitable for compaction by vibration) which may give zero slump. Compaction factor test is more dependable in such cases.

2. **Compaction factor test.** In the compaction factor test the degree of workability of concrete is measured in terms of internal energy required to compact the concrete thoroughly. The apparatus for conducting the test consist of two conical hoppers (the bottom of the hoppers being fitted with door with quick release catches) held in position one over the other at a certain clear distance between them. At a certain distance below the lower hopper is placed a 15 cm diameter cylinder (mould) 30 cm in height. The upper hopper is filled with concrete to be tested and its doors opened allowing the concrete to fall in to the lower hopper. By this operation, the specimen concrete is brought to a standard state of compaction in hopper and error in handling, if any, is removed. The bottom of the lower hopper is then immediately opened allowing the concrete to fall into the cylinderical mould below. The surplus concrete

appearing above the top of the mould is then removed with a trowel. The mould with concrete is weighted and the weight of the concrete found out.

To calculate the theoretical weight of the same volume of specimen concrete without any air voids the mould is filled with dry concrete (in which the cement and aggregate are mixed in the same proportions as specimen concrete) and reweighed. The value thus obtained is considered to be the weight of fully compacted concrete. The compaction factor is calculated by dividing the observed weight of concrete by the theoretical weight.

A compaction factor of 0·95 represents flowing concrete having high workability ; 0·92 plastic concrete having medium workability ; 0·85 stiff plastic concrete having low workability and a compaction factor of 0·75 represents stiff concrete having very low workability.

3. **Vee-Bee test** This test is carried out is such a manner that the specimen concrete in the test receives more or less same treatment in respect of the method of placing as it would in actual execution of the work. This test is preferred for finding workability of stiff concrete mix having very low workability.

In this test a Vee-Bee time of 5 to 3 seconds represent stiff plastic concrete having medium workability, 10 to 5 seconds represent stiff concrete of low workability and Vee-Bee time of 18 to 10 seconds represent very stiff concrete having very low workability.

22·13. METHODS OF PROPORTIONATING CONCRETE MIXES

The aim of evolving different methods of proportionating concrete is to get strongest and the densest mix with least amount of cement. The different methods of proportionating concrete are :

(1) By arbitrary standards

(2) By minimum void method

(3) By trial or by maximum density method

(4) By fineness modulus method

(5) By water cement ratio law

1. By Arbitrary Standards

In this method the aggregate are so proportioned as to have minimum voids when mixed. This condition is assumed to be satisfied when the quantity of fine aggregate is just sufficient to fill the voids of coarse aggregates and the quantity of cement is just sufficient to fill the voids in the fine aggregates.

As a result of experiments and experience it has been possible to fix certain arbitrary ratios in which cement, fine aggregate and coarse aggregate should be mixed to produce concrete suitable for a particular type of work. In general buildings works (beams, slabs, columns etc.) subjected to ordinary stresses, cement, sand and coarse aggregate are taken in the ratio of 1 : 2 : 4. For heavily stressed members or for water tight construction the ratios may be 1 : 1 : 2, 1 : 1 : ½ : 3 or 1 : 2 : 3 etc. Recommended concrete mix for various types of works are given in table 22.7.

TABLE 22·7

RECOMMENDED CONCRETE MIXES FOR VARIOUS TYPES OF WORK

S.No.	*Type of work*	*Mix Recommended*
1.	For ricl mix concrete work in long span arches, heavily loaded columns, beams etc.	M 25 (1 : 1 : 2)
2.	For water retaining structures, long piles columns and other members subjected to medium loads, roof slab without terracing, etc.	M 20 (1 : 1½ : 3)
3.	Reinforced concrete beams, slabs, columns, wall etc. in general building work subjected to ordinary loads, foundations, retaining walls, bunkers, silos, roads, pavements, bridges, etc.	M 15 (1 : 2 : 4)
4.	Mass concrete foundation for walls, piers etc. Mass concrete work in culverts, retaining walls, compound walls etc.	M 10 (1 : 3 : 6)
5.	Mass concrete work under heavy duty floors, for massive walls etc.	1 : 4 : 8
6.	Lean concrete work (mud mat) under reinforced concrete foundation of all types	1 : 5 : 10

The concrete produced by the above method is also termed as nominal mix concrete. In proportionating concrete by this method, quantity of cement is measured by weight. The quantity of fine and coarse aggre-gates may either be measured by volume or by weight.

Although in this method there is not much control over the strength, yet due to simplicity in the design, this method is widely used for all small works of moderate importance. In large and important projects controlled concrete should be used (refer article 22·14)

2. By Minimum Void Method

This method is based on the principle that to obtain dense concrete the quantity of fine aggregate should be slightly in excess of the voids in the coarse aggregate and the quantity of cement should also be slightly more than the voids in the fine aggregates. In its simplest form the voids in the fine and coarse aggregates are separately found out with the help of graduated cylinder and water. Let V_1 be the volume of water in the cylinder. To this a known volume of the specimen aggregate say V_2 is added and the volume of the mixture recorded. Let the volume of mixture be V. Then percentage voids (x) in the aggregate is given the eqution.

$$x = \frac{V - V_1}{V_2} \times 100$$

The concrete mixture is finally decided by using approximately 10% more sand and about 15% more cement than the percentage voids in coarse and fine aggregates respectively.

3. By Trial or by Maximum Density Method

This method is based on the principle, that a concrete mix which is formed by so proportionating its aggregates as to give heaviest weight for same volume, yield the densest concrete. Hence in this method a box is filled with varying proportions of fine and coarse aggregates. The proportion which gives heaviest weight is adopted for the work.

4. By Fineness Modulus Method

The fineness modulus of an aggregate is the term used to indicate the average size of particles in the aggregates. It is represented in the form of an index number. Thus the coarser the aggregate, the higher will be the value of fineness modulus.

In this method the samples of both coarse and fine aggregates are passed through a set of nine standard sieves and the percentage of the sample retained on each of the said sieves is determined. The total of these percentages divided by 100 gives the fineness modulus of the sample. Certain values of fineness modulus for the mixed aggregates have been accepted as giving the best results. Thus if p is the desired fineness modulus for a mix of fine and coarse aggregate, and p_1 and p_2 be their respective fineness modulus, then the amount of the aggregate (w) to be mixed with one part of coarse aggregate is given by the equation :

$$w = \frac{p_2 - p}{p - p_1}$$

Example. *To illustrate the above explanation, we may consider two samples, one of coarse aggregate and the other of the fine aggregate. It is desired that the aggregates should be so proportioned as to have a fineness modulus of 5·34. The samples are passed through the set of nine standard sieves and the percentage retention on each sieve is recorded in a tabular form given on the next page.*

S. Sieve No.	Coarse aggregate (10 kg) Weight retained in kg.	Total weight retained (comula-tive) in kg.	% weight retained	Fine aggregate (1 kg) Weight retained in kg.	Total weight retained (cumula-tive) in kg.	% weight retained
80 mm	—	—	—	—	—	—
63 mm	—	—	—	—	—	—
40 mm	0.6	0.6	6	—	—	—
20 mm	0.4	1.0	10	—	—	—
10 mm	5.2	6.2	62	—	—	—
4.75 mm	3.0	9.2	92	0.07	0.07	7
2.36 mm	0.4	9.6	96	0.07	0.14	14
1.18 mm	0.2	9.8	98	0.15	0.29	29
600 micron	0.2	10	100	0.22	0.51	51
300 micron	0	10	100	0.36	0.87	87
150 micron	0	10	100	0.13	1.00	100
			$\Sigma = 664$			$\Sigma = 228$

Sol. Thus the fineness modulus of fine aggregate $= \dfrac{288}{100}$

or $p_1 = 2{\cdot}88$

and the fineness modulus of coarse aggregate $= \dfrac{664}{100}$

or $p_2 = 6{\cdot}64$

As per formula, the amount of fine aggregate(w) to be mixed with one part of coarse aggregate is given by

$$w = \frac{p_2 - p}{p - p_1}$$

$$= \frac{6{\cdot}64 - 5{\cdot}34}{5{\cdot}34 - 2{\cdot}88}$$

$$= \frac{1{\cdot}3}{2{\cdot}46}$$

$= 0{\cdot}528 = 52{\cdot}8$ per cent

$= 53$ per cent, say

∴ The ratio of the fine aggregate to coarse aggregate

= 53 : 100

5. By Water Cement Ratio Law

Abram, as a result of experiments, established that the proportion of water to cement in any mixture has the greatest influence on the strength of the concrete. According to water cement ratio law, "*for any given condition of test, the strength of well compacted concrete with good workability is dependent only on the water cement ratio.*"

The workability of concrete in addition to other factors like grading and proportionating of aggregates, proportion of cement, the efficiency of mixture etc. also depends upon the quantity of water used in the mixture. According to Abram Law the strength of the mixture increases not merely by increasing the quantity of cement in the mix but it is on account of reducing the water cement ratio. The strength of concrete in accordance with the law is given by modified formula

$$P_{28} = \frac{984}{4^x}$$

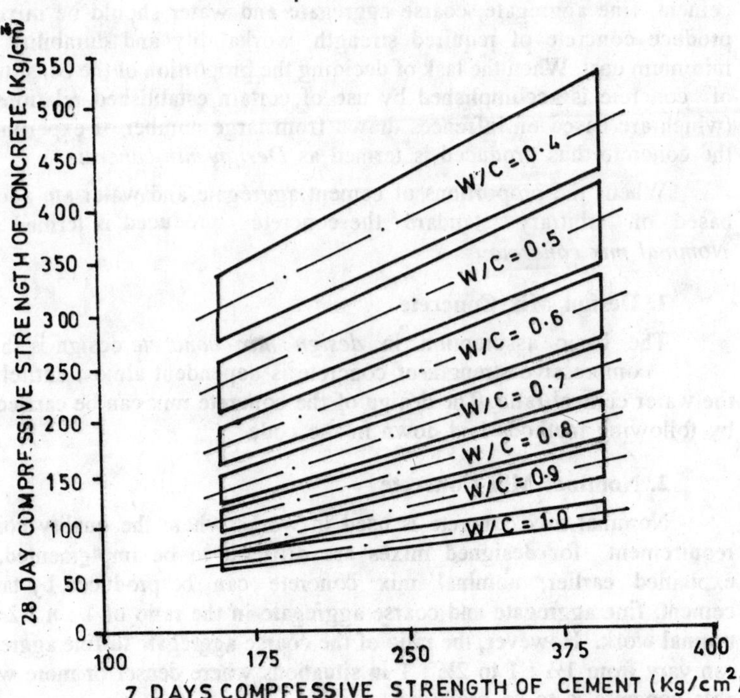

Fig. 22.2

Where P_{28} = cylinder crushing strength of concrete in kg/cm^2 after 28 days curing.

x = water cement ratio by volume.

Tests have been carried out to establish relation between water cement ratio and the compressive strength of concrete using ordinary Portland cement.

Fig. 22·2 shows the graph which has been evolved on the basis of results. Concrete mixture can be designed by use of the graph. Depending upon the required compressive strength of concrete, choose suitable water cement ratio from the graph. Choose suitable fineness modulus of combined aggregate and determine the ratio of the fine aggregate to coarse aggregate. Mix, thus designed should be tested to know the exact value of compressive strength and if necessary, adjust the ratio of the cement to aggregate to achieve the desired strength and workability.

22·14. MIX DESIGN

The aim of mix design is to determine the proportions in which cement, fine aggregate, coarse aggregate and water should be mixed to produce concrete of required strength, workability and durability with minimum cost. When the task of deciding the proportion of the constituents of concrete is accomplished by use of certain established relationships (which are based on infrences drawn from large number of experiments) the concrete thus produced is termed as *Design mix concrete.*

When the proportions of cement,aggregate and water are adopted based on arbitrary standard the concrete produced is termed as *Nominal mix concrete.*

1. Design Mix Concrete

The basic assumption in *design mix concrete* design is that compressive strength of concrete is dependent almost entirely on the water cement ratio. The design of the concrete mix can be carried out by following methods laid down in the code.

2. Nominal Mix Concrete

Nominal mix concrete is used in works where the quality control requirement for designed mixes are difficult to be implemented. As explained earlier, nominal mix concrete can be produced by taking cement, fine aggregate and coarse aggregate in the ratio of 1 : n : $2n$ for normal work. However, the ratio of the coarse aggregate to fine aggregate can vary from 1½ : 1 to 2½ : 1 in situations where denser or more workable concrete is to be produced.

TABLE 22·8

CONCRETE MIX PROPORTIONS FOR NOMINAL MIX CONCRETE

Grade of concrete	*Total quantity of dry aggregates by mass per 50 kg of cement, to be taken as the sum of the individual masses of fine and coarse aggregates (Maximum) in kg*	*Quantityof water per 50 kg of cement (Maximum) in litres*	*Proportion of fine aggregate to coarse aggregate (By mass)*
M 5	800	60	Generally 1 : 2 but subject to an upper limit of 1 : 1½ and a lower limit of 1 : 2½
M 7.5	625	45	
M 10	480	34	
M 15	350	32	
M 20	250	30	

Note. The proportions of the fine to coarse aggregate should be adjusted from upper limit to lower limit progressively as the grading of the fine aggregate becomes finer and the maximum size of coarse aggregate becomes larger. Graded coarse aggregate should be used.

TABLE 22·9

PERMISSIBLE STRESSES IN CONCRETE

Grade of concrete	*Permissible stress in compression (in N/mm²)*		*Permissible stress in bond (average) for plain bars in tension τ_{bd} (in N/mm²)*
	Bending σ_{cbc}	*Direct* σ_{cc}	
M 10	3.0	2.5	—
M 15	5.0	4.0	0.6
M 20	7.0	5.0	0.8
M 25	8.5	6.0	0.9
M 30	10.0	8.0	1.0
M 35	11.5	9.0	1.1
M 40	13.0	10.0	1.2

As a general guidance, it may be noted that for an average grading of fine aggregate the proportions shall be 1 : 1½, 1 : 2 and 1 : 2½ for maximum size of coarse aggregates 10 mm, 20 mm and 40 mm respectively.

22·15. PRODUCTION, PLACING, COMPACTING AND CURING OF CONCRETE

The operations involved in the production and execution of the item of concrete can be broadly divided in following stages :

(i) *Batching of materials*

(ii) *Mixing of concrete*

(iii) *Transportation of mixed concrete*

(iv) *Placing of concrete*

(v) *Compaction of concrete*

(vi) *Curing of concrete*

1. Batching of Materials

In order to ensure uniformity in the quality of concrete it is necessary that all materials that go into its production are measured accurately within permissible tolerances. Batching of material can be done manually or by using batching plant. The choice of system to be adopted for batching depends upon the magnitude of concreting work involved and the rate at which concrete is required to be produced.

While batching cement is always measured by weight, the weight of one bag of cement is taken as 50 kg. Fine and coarse aggregate are measured by volume in case of manual batching and by weight in case of mechanical batching *i.e.* by batching plant.

In case of manual batching the quantity of coarse and fine aggregates are so adjusted as to require one bag of cement for one batch. In this case wooden boxes (farmas) are normally used for gauging aggregate by volume. The material is filled loosely into the wooden boxes and the extra material in the box is struck flush with box top. In case sand used in the work is moist necessary allowance on account of bulking should be made while measuring sand in boxes. If the coarse aggregates used in the work are moist the quantity of water carried by them should be deducted out of the quantity of water required for the mix.

Batching of materials by use of batching plant is carried out for all important works involving large amount of concreting work and requiring controlled conditions for production. Depending upon the size of work, the batching equipment can vary from a small wheel barrow scale, feeding a portable mixer to a complicated automatic or semi-automatic plant.

2. Mixing of Concrete

During the process of mixing great care is taken to ensure that the materials are maintained in a uniform condition. There are two different types of concrete mixing

(*a*) Hand mixing.

(*b*) Machine mixing.

(a) Hand mixing. This method of mixing concrete is resorted to when the quantity of concrete to be used in a work is insufficient to warrant the necessity of machines. This method is used with advantage in places, where machinery cannot be used on account of their non-availability or in situations where noise is not desirable. Hand mixing is done on a clean, hard and impermeablae surface. Cement and sand are first mixed with the help of shovels until the mixture attains a uniform colour. Aggregates are then added to this mixture and the whole mixture is then turned by shovels until the stone pieces are uniformly spread throughout. After this, computed quantity of water is poured into the heap from can fitted with a nose. The mass is then turned until a workable mixture is obtained.

It is usual practice to use 10 percent more cement in case of hand mixing than that specified in the design mix. This is done to take care of lesser efficiency and inferior results of hand mixing than that achieved of machine mixing.

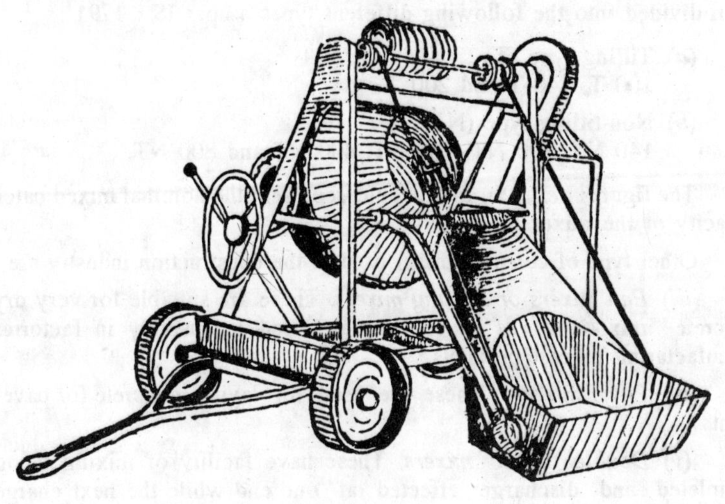

Fig. 22.3
Batch type concrete mixer.

(b) Machine mixing. The machine used for mixing concrete is termed as concrete mixer. Two types of concrete mixers are in common use.

(*i*) Contonuous mixers
(*ii*) Batch mixers

Continuous mixers are employed in massive construction, where a large and continuous flow of concrete is desired. The process of feeding the mixing plant is more or less automatic. The machine requires careful supervision so as to obtain the concrete mix of desired consistency.

In Batch type of concrete mixer, the desired proportions of materials are fed into the hopper of a drum in which the materials get mixed by the series of blades of baffles inside the mixer. Batch mixers are further divided into two main types-*viz*, tilting drum type and closed drum type. In the first type the components of concrete mix are fed in the revolving drum in a tilting position and after some time the concrete mix is discharged by tilting the drum in the opposite direction. In the latter type the drum remains rotating in one direction and is emptied by means of the hopper which tilts to receive the discharge.

While using the mixure, coarse aggregates should be fed first, sand and then cement should be put afterwards. In the revolving state, the components get mixed while water is poured with the help of a can. The concrete should be mixed for at least 1 ½ to 2 minutes.

Depending upon their capacity, the batch type concrete mixers have been divided into the following different types as per IS : 1791 :

(*a*) Tilting type (T) :
100 T, 140 T and 200 T.

(*b*) Non-titling type (NT) :
140 NT, 200 NT, 280 NT, 400 NT and 800 NT.

The figures *i.e.,* 100, 140 ...... etc. represent the nominal mixed batch capacity of the mixer in liters.

Other type of concrete mixer used in the construction industry are :

(*iii*) *Pan mixers or stirring mixers.* These are suitable for very dry concrete mix and as such they are used extensively in factories manufactaring precast products.

(*iv*) *Paver mixers.* These are used for laying concrete for pavements.

(*v*) *Double drum mixers.* These have facility of mixing being completed and discharge effected at one end while the next charge is loaded and half mixed at the other end of the drum.

3. Transportation of Concrete

Concrete should be transported from the place of mixing to the place of final deposit in the shortest possible time so that there is no segregation or loss of ingredients and the concrete remains workable while being placed in position. Different methods adopted for transportation of concrete are : (1) By steel pans (2) By hand carts/wheel barrows (3) By dumpers (4) By trippers (5) By truck mixer and agitator lorries (6) By hoists/skips (7) By cable-ways (8) By cranes (9) By pumping.

The selection of the mode of transportation of concrete is made keeping in view the size of the work as well as the rate at which concrete is required to be produced/transported. For small jobs, the concrete from the mixer is normally transported manually as headload in steel pans. Hand carts or wheel barrow are used where ground is not dry and distance of travel is not much. In case distance of travel in large (say upto 5 km) tripper can be used for haulage of concrete. Truck mixer and agitator lorries are normally used for transporting ready mix concrete (concrete which is mixed at a central plant). Hoists/skips are used for haulage of concrete in construction of tall structures. In this system the skip is fed by wet concrete from the mixer, it is raised upto the required height or level and discharged into a receiving hopper. From there concrete is handled manually in steel pan or in wheel barrows. Cranes are used for transportation of concrete for construction of tall structures like chimneys, silos, multi--storeyed buildings etc.

Cableways are commonly used for transporting concrete for the construction of large and massive works like dams etc. The cableway is supported on two towers on either end of the dam site and concrete is carried in a skip or bucket which can travel along the cableway covering the entire length of the dam. The skip which is bottom-discharging type is lowered just above the working zone and wet concrete is poured in forms with minimum impact.

Transportation of concrete by mode of pumping is resorted to when concreting work is to be carried out on large areas (on ground or at higher elevations) or in congested sites. In this system, concrete is conveyed from a central mixing plant to the place of pouring, by an arrangement of pumps and pipe line. The pipe line is made up of steel tubes each 3 m in length and 100 to 125 mm in diameter. With the help of special type of pumps available for the purpose, concrete can be pumped for a maximum distance of about 400 m horizontally and a about 80 m vertically. This is considered to be a very fast and efficient method of transportation of concrete.

4. Placing of Concrete

Concrete should be placed and compacted immediately after mixing. The arrangement for the conveyance of the concrete mix should be so planned that the mixed mass is used within 30 minutes. This is necessary to prevent the danger of concrete getting its initial set. Thus the quantity of concrete prepared in one batch should be such that it can be used within the said period. Before laying concrete, the shuttering should be cleaned of all dust and debris. In order to prevent the shuttering absorbing the water from the concrete or getting struck to it, a coat of crude oil or grease etc. is usually applied to the shuttering before concreting. In placing concrete, care should be taken to see that it should not be thrown from heights. Concrete should be laid in layers

15 to 30 cm. in thickness, and each layer should be properly compacted before laying the next one.

5. Compaction of Concrete

Consolidation of concrete should proceed immediately after placing. The function of consolidation of concrete is to expel the air bubbles in the mass and make it impermeable enough to attain desired strength. The concrete mass should be consolidated or compacted till the cream of the cement starts appearing on the surface. Over-compaction may lead to segregation of concrete while under-compaction may leave air-voids in concrete. Consolidation may be done by hand or by mechanical device.

Hand compaction may be done by rodding, tamping or hammering. Tamping is usually adopted for compacting concrete for slabs or other such surfaces ; rodding is done for thin vertical members while hammering is done for massive plain concrete works.

Mechanical compaction is done by the use of vibrators. Compaction of concrete by vibration is considered essential for all important works specially in situations where reinforcements are congested or the member is required to have exposed concrete surface finish. When vibrators are used, leaner but stiff, concrete mix should be used to obtain greater durability and higher strength. Vibrators are of four types :

(*i*) Internal vibrators (*Immersion Type*)

(*ii*) External vibrators

(*iii*) Surface vibrators

(*iv*) Vibrating tables

(*i*) *Internal vibrators* are commonly used on large works for flat surfaces (floor slab etc.). The internal vibrator consists of a metal rod like vibrating head which is immersed in the full depth of concrete layer. Two types of immersion vibrators are available in the market, *viz.* (1) flexible shaft driven type (2) motor-in-head type. The flexible shaft type may be operated by compressed air or motor or gasoline engine whereas motor-in-head type can be operated either electrically or pneumatically. The vibrator is kept immersed in one location till the surrounding concrete is consolidated completely. This can be judged visually by the formation of a circular shaped laitance around the vibrator. For stiff mix the time of vibration at one location should not be less than 90 seconds.

(*ii*) *External vibrators* are placed against the concrete formwork and vibrating force for compaction is conveyed to the concrete through the formwork. External vibrators are used mostly for thin sections of members to be concreted or in situations where internal vibrators can not be used with ease.

(*iii*) *Surface vibrators*-are mounted on platforms or concrete screeds and are generally used to compact and finish bridge floors, road slabs etc.

(*iv*) *Vibrating tables*-are used for consolidation of precast units. When concrete is placed on such tables, mechanical compaction takes place which has the following advantages as compared with hand compaction.

(*a*) Leaner mix of concrete can be used to achieve the same strength. Hence it results in saving cement to the extent of 10 to 15%.

(*b*) It permits compaction of narrow or thin sections where hand compaction may be difficult.

(*c*) The resultant concrete is more impermeable, dense and free from honey combing.

(*d*) Compaction is faster which results in economy in time and labour.

6. Curing of Concrete

Curing of concrete is one of the essential requirements of the process of concreting. Curing is the process of keeping the set concrete continuously, damp for some days in order to enable the concrete gain more strength. It has been established that the strength of concrete increases with the age provided it is kept damp. During the process of curing the concrete absorbs the water necessary for the complete chemical action to reach its required strength. The strength of concrete increases more rapidly in the first few days after setting and afterwards the rate of increase in strength goes on retarding. The period for which curing should be continued depends upon atmospheric conditions such as temperature, humidity and wind velocity. In general, the process of curing should be continued for 7 to 10 days. In coid weather, however, the concrete should be cured for at the least 14 days, as the rate of hardening of cement, is low in such conditions. Efficient curing increases both impermeability and the durability of concrete and as such it needs special attention when the concrete is exposed to water pressure or severe weather conditions. Correct curing also increases resistance of concrete to abrasion and reduces shrinkage. Curing may be done by adopting the following methods :

(*i*) By covering the exposed surface with moist sand, earth, gunny bags etc. and sprinkling water at intervals to keep the covering wet.

(*ii*) Floors or other horizontal surfaces may be cured by impounding water in earthen or sandy bunds in squares over the entire area. The depth of water in the enclosed squares may be maintained as 5 to 8 cm.

(*iii*) Columns, walls or other such vertical surfaces are usually cured by wrapping gunny bags or canvas over the surface and keeping the wrapped material continuously wet by sprinkling water or by some other means.

It has now been discovered that during the process of curing, increase in temperature increases the rate of increase in strength of concrete. This property is made use of in curing small precast members by steam. This enables the concrete to gain required strength in less time.

22·16. CONCRETING UNDER WATER

The deposition of concrete under water is a complicated problem. Concreting under water may be necessary in marine works or in case of deep foundations. As there is a continuous danger of loss of cement and segregation of concrete, placing of concrete under water is usually confined to unreinforced (plain concrete) construction. In case the site is such that the provision of cofferdams, caissons or other such dewatering measures is difficult anyone of the methods mentioned below may be selected for concreting under water :

(*i*) By tremie

(*ii*) By skip or bottom opening bucket

(*i*) *Tremie.* Tremie is the name given to a steel pipe 15 to 30 cm. in diameter and sufficiently long enough to reach the bed of water shed keeping its one end above the water level. The pipe is fitted with a hopper at its upper end for pouring concrete inside the pipe. The tremie is equipped with a check valve at its lower end. When concrete is poured in the hopper of the pipe, it displaces the air and water from the pipe and reaches the bed. During the process of concreting the tremie is gradually raised keeping the lower end of the tube submerged in the concrete mass already poured below. In order to compensate the possible loss of cement under water, rich mix of concrete should always be used. Concrete thus laid under water should never be compacted or consolidated.

(*ii*) *Skip or bottom opening bucket.* Cubical hinged bottom bucket or cylindrical bucket equipped with a rolling gate or a skip, is also sometimes used for concreting under water. In either case, it is necessary that the bucket or skip is lowered to the bed of working place before the concrete is discharged. This is necessary to minimise the concrete loss during the process of placing.

22·17. PLACING CONCRETE IN COLD WEATHER

The effect of cold weather on wet concrete is to lower the rate at which the chemical reaction occur while the cement is setting, and also to produce destructive forces due to the expansion caused by freezing of the water in the concrete. Thus cold weather adversely affects the setting and hardening of concrete and hence delays the development of ultimate strength in concrete. The strength after 28 days may be only 50% of the ultimate strength if concrete is cured at 0°C and it may be only 30% of its ultimate strength if it is cured at –10°C. In concreting during cold weather unless proper steps are taken it would be necessary to keep the forms in positions for longer periods which will result in extra cost due to labour, cost of material and due to delay in construction of the project.

Hence, while concreting in cold weather, precautions are taken to keep the temprature of concrete above a minimum value both while placing the concrete in position in the formwork as well as while curing it.

Besides this it is necessary to insulate the formwork by fixing boards or tarpaulins etc. on the outside of the formwork battens.

In moderate or less severe conditions the problem can be tackled by increasing the quantity of cement in the mix or by use of rapid hardening cement or by use of high alumina cement. Use of accelerators like calcium chloride is also sometimes advocated to produce increased internal temperature during setting and hence to achieve early setting of concrete.

However, the method commonly adopted for concreting in cold weather is to heat up the aggregates or water or both with a view to keep the temperature of the resultant concrete less than 38°C. Pre-heating of mixing water is considered to be most effective in this regard. The water is heated so as to have temperature of 60°C at the time of mixing. Aggregate should be heated only upto 15°C in case the temperature of the mixing water is 60°C.

Similarly, adequate temperature for concrete in cold weather can be obtained by increasing the temperature of only sand to about 40°C, if mixing water is taken at 60°C. Aggregates may be heated by embedding steam pipes. For small jobs, aggregates can be heated with the help of steel drums embedded in heaped aggregates and filled with fire.

In cold weather all concrete surfaces should be covered as soon as the concrete is placed in position with a view to keep the heat in and to provide necessary protection form outside frost. This is achieved by covering the concrete surface with dry blankets of straw, cement bags, tarpaulins etc. In case of extreme frost conditions, or when the air temperature is near or below freezing point, it is essential to provide heated, enclosures to keep the temperature of the air surrounding the concrete surface up. The enclosures can be heated by means of steam, hot water pipes or stoves etc.

22·18. PLACING CONCRETE IN HOT WEATHER

For normal concreting operations the temperature of concrete should vary between 15°C to 25°C to enable it to attain its full ultimate strength within 28 days. In case concreting is carried out in extreme hot weather conditions, the rate of hydration of cement gets accelerated and the water in the concrete mix dries out rapidly. This in turn results in formation of stresses which cause shrinkage cracks and damage to the concrete.

The adverse effects of hot weather can be minimised by taking following precautions while concreting in extreme hot weather conditions.

(*i*) By adding retarding agents in concrete to lower the rate of hydration of cement.

(*ii*) By keeping the temperature of the ingredients of concrete down. Temperature of aggregate may be lowered by stockpiling the aggregates in shade and also by sprinkling cold water on the stock piles.

(*iii*) By using cold water for making concrete. In certain cases, crushed ice may be used to bring down the temperature of water and to obtain concrete mix having temperature within desirable limits.

(*iv*) The mixed concrete should be transported and placed in position within the shortest possible time, so as to minimise the losses due to evaporation and rise of temperature.

(*v*) The formwork and the reinforcement should be sprinkled over with cold water just before the placement of concrete. This prevents the loss of water from concrete due to absorption by these materials.

(*vi*) The work should be executed with utmost speed, since delays will effect the concrete adversely due to hot weather conditions.

(*vii*) After placing in positions, the concrete surface should be adequately protected from exposure to direct sun rays. The concrete should be continuously cured with sufficient water during the first few hours and thereafter the surface should be kept moist for the full duration of curing.

22·19. WATER-PROOFING OF CONCRETE

A well designed concrete mix properly laid, compacted and cured is in itself impermeable. As far as possible the addition of water-proofing compounds should be avoided for normal working conditions. A dense concrete having minimum possible air voids shall, therefore, be the essential consideration in the making of damp-proof concrete. The following conditions should be satisfied to achieve the object :

(*i*) Use the best available material.

(*ii*) Proportionate the aggregates by fineness modulus method.

(*iii*) Use only as much quantity of water as is necessary to obtain the desired workability.

(*iv*) Ensure complete mixing of the concrete.

(*v*) Complete supervision during laying and compaction.

(*vi*) Complete curing.

However, in certain cases like water retaining structures, structures in water-logged ground or in damp atmosphere, it may be necessary to take further steps to ensure water-tightness. This includes the addition of certain water-proofing compounds like Permo, Pudlo, Sika etc. in the concrete mix during the process of mixing. This is termed as integral water-proofing. The various other measures adopted to make the concrete impermeable are described in the chapter on 'Damp-proofing and water-proofing' in this book.

REINFORCEMENT

22·20. REINFORCEMENT

The material which can be used as reinforcement in R.C.C. work should have the following characteristics:

(1) It should be able to develop perfect bond with concrete.

(2) Its co-efficient of thermal expansion should be nearly same as that of concrete.

(3) It should have high tensile strength.

(4) Concrete should not produce any harmful effect on the embedded material.

(5) It should be easy to cut, bend, bind or weld.

(6) It should be easily available.

It is seen that steel meets all the above requirements and as such it is the only material which is used on large scale in R.C.C. works.

The various types of steel that are commonly used for making different forms of reinforcement are Mild steel, Medium tensile steel, High tensile steel and Hard drawn steel. These differ from each other in their chemical composition and other properties like, ultimate tensile strength, yield points, etc.

Fig. 22·4
Deformed bar

The various forms in which steel is used as reinforcement in R.C.C. work are : round bars, deformed bars, twisted bars, sqaure bars and flats. Sometimes expanded metal fabric or fabric made by welding or weaving steel wire in the form of oblong or square mesh are also used as reinforcement in slabs, shells and concrete roads. For works of large dimensions like massive foundations, etc., sections like rolled steel beams, channels or angle iron are also used as reinforcement.

Mild and medium tensile steel bars or round section are most commonly used in R.C.C. work. The diameter of round bars (in mm) used in normal building work are : 6, 8, 10, 12, 16, 18, 20, 22, 25, 28, 32, 36 and 40.

Bars of greater diameter, *i.e.*, 45 mm and 50 mm, are only used in exceptionally heavy foundations, large girders, or counterforts etc.

With the introduction of deformed bars and twisted bars, the use of plain round bars is gradually declining. Deformed bars or high yield strength bars (HYSD) are furnished with lugs, ribs or other form of surface deformations for the purpose of increasing their bond strength with concrete. It is seen that the process of twisting a plain or a deformed bar results in the following improvements in its properties.

(*i*) Considerable increase in yield stress.

(*ii*) Increase in tensile strength.

(*iii*) Increase in bond strength.

Twisted plain or deformed bars not only have high yield stress but also have bond strength which is 40% more than that of plain round bars. On account of increased bond strength such bars do not need end hooks, and require reduced length for overlaps etc. thereby effecting reduction in the cost of reinforcement and labour.

22.21. PERMISSIBLE STRESSES IN REINFORCEMENT

The permissible stresses in steel reinforcement as per IS : 456—1978 are given in Table 22.10.

HYSD bars having yield stress (0.2% proof stress) equal to 415 N/mm^2 and permissible tensile stress equal to 230 N/mm^2 are exceedingly being used in R.C.C. work. These bars are also commonly known as Fe 415 HYSD bars.

In case of structures requiring use of still higher tensile strength of reinforcing bars, another form of HYSD bars known as Fe 500 HYSD are used. These bars have a yield stress equal to 500 N/mm^2 and have permissible tensile stress equal to 275 N/mm^2.

It is observed that HYSD bars when used in combination with high strength concrete, *i.e.*, M 20 above prove to be more economical.

TABLE 22·10

PERMISSIBLE STRESSES IN STEEL REINFORCEMENT

Type of stresses in steel reinforcement	Permissible stresses		
	Mild steel bars conforming to Grade 1 of IS : 432(Part-I)—1966 or deformed mild steel bars conforming to IS : 1139—1966	*Medium tensile steel conforming to IS : 432 (Part 1) —1966 or deformed medium tensile steel bars conforming to IS : 1139—1966*	*High yield strength deformed bars conforming to IS : 1139—1966 or IS : 1786—1979 (Grade Fe 415)*
(1)	*(2)*	*(3)*	*(4)*
1. Tension (σ_{st} or σ_{sa})			
(a) Upto and including 20 mm	140 N/mm²	Half the guaranteed yield stress subject to a maximum of 190 N/mm²	230 N/mm²
(b) Over 20 mm	130 N/mm²		230 N/mm²
2. Compression in column bars (σ_{sc})	130 N/mm²	130 N/mm²	190 N/mm²
3. Compression in bars in a beam or slab when compressive resistance of concrete is taken into account.	The calculated compressive stresses in the surrounding concrete multiplied by 1.5 times the modular ratio or σ_{sc} whichever is lower.		
4. Compression in bars in a beam or slab where the compressive resistance of concrete is not taken into account			
(a) Upto and including 20 mm	140 N/mm²	Half the guaranteed yield stress subject to a maximum of 190 N/mm²	190 N/mm²
(b) Over 20 mm	130 N/mm²		190 N/mm²

Note 1. For high yield strength deformed bars of grade Fe 500, the permissible stress in direct tension and flexural tension shall be 0.55 f_y. The permissible stress for shear and compression reinforcement shall be as for Grade Fe 415.

Note 2. For welded wire fabric conforming to IS : 1566-1967 the permissible value in tension σ_{st} is 230 N/mm².

Note 3. The yield stress of steels for which there is no clearly defined yield point, should be taken to be 0·2 per cent proof stress.

Note 4. When mild steel conforming to Grade II of IS : 432 (Part-I)-1966 is used, the permissible stresses shall be 90 per cent of the permissible stresses in column 2 of Table 22.10 or if the design details have already been worked out on the basis of mild steel conforming to Grade I of IS : 432 (Part-I)-1966 ; the area of reinforcement shall be increased by 10 per cent of that required for Grade I steel.

22·22. CHARACTERISTIC STRENGTH OF STEEL REINFORCEMENT

The term characteristic strength of steel reinforcement means that value of the strength of steel below which not more than 5% of the test results are expected to fall. The characteristic strength of different type of steel reinforcement or (f_y) is taken as the value of the minimum yield stress (or 0.2% proof stress) for the type or grade of steel used in the manufacture of the reinforcement. The value of (f_y) or the characteristic strength for three commonly used type of reinforcing bars are given in Table 22·11.

TABLE 22·11

CHARACTERISTIC STRENGTH OF STEEL REINFORCEMENT

S.No.	*Type of reinforcement*	*Yield stress or 0.2% proof stress*	*Characteristic Strength* (f_y)	*Permissible tensile strength* σ_{st}
1.	Mild steel bars conforming to Grade I of IS : 432 (Part I) or deformed m.s. bars conforming to IS : 1139.	250 N/mm² (average)	250 N/mm²	140 N/mm² (for bars upto 20 mm ϕ) 130 N/mm² (for bars over 20 mm ϕ)
2.	High yield strength deformed bars (HYSD bars) conforming to IS : 1109 or Grade Fe 415 of IS : 1786.	415 N/mm²	415 N/mm²	230 N/mm²
3.	High yield strength deformed bars conforming to Grade Fe 500 of IS : 1786.	500 N/mm²	500 N/mm²	275 N/mm²

22·23. HANDLING OF REINFORCEMENT AT SITE

Whatever be the magnitude of R.C.C. work, it is necessary to prepare a bar bending schedule based on the structural drawing prior to start of handling reinforcement at site. Bar bending schedule is a descriptive list containing details regarding the exact shape, dimension and diameter of each and every bar together with the number of bars of each shape.

Bars are cut to the desired lengths and then bent cold in accordance with the details given in bar bending schedule. Before placing the bars in position in the formwork it is necessary to ensure that the reinforcement is clean and free from loose mill scale, loose rust, oil or other coating. This precaution is necessary to meet the requirement of good bond between concrete and steel for monolithic behaviour.

The reinforcement should be placed accurately in position and maintained in position by tying bars at junction with binding wire or by welding. To ensure proper cover to reinforcement, small precast cover blocks made out of cement mortar are used. The cover blocks are inserted below the reinforcement mesh and tied to it with the help of binding wire, prior to concreting. In addition, precaution should be taken to prevent the displacement or distortion of reinforcement during concreting.

REINFORCED CEMENT CONCRETE

22·24. DEFINITION

Though strong in compression, concrete is extremely weak in tension. Its resistance to tension is so low that plain concrete can only be used where the member is in pure compression. Steel on the other hand is equally strong in compression and tension, but it is apparent that while a long steel bar can develop its full strength in tension, it cannot resist equal amount of compressive force, owing to its buckling due to the slenderness. Thus, the combination of concrete and steel has proved to be ideal, as the two material are used to take up the stresses they are most suitable for. Such a combination of steel and concrete is called *'Reinforced cement concrete'* or simply reinforced concrete.

22·25. PROPERTIES OF REINFORCED CONCRETE

The combination of steel and concrete becomes practicable or workable or account the following three reasons :

(1) The concrete while setting grips very fast the surface of the steel bars. Therefore the concrete is able to transmit to the steel bars those stresses which it cannot resist itself.

(2) The co-efficient of linear expansion of concrete and steel are almost the same. Therefore, no internal stresses are set up within reinforced concrete due to variations in temperature.

(3) The coating of cement grout on the surface of steel bars protects them form corrosion and does not produce any adverse chemical effect.

22·26. ADVANTAGES OF REINFORCED CONCRETE

(1) It is economical in ultimate cost.

(2) Its monolithic character gives much rigidity to the structure.

(3) It is durable and fire-resisting. It does not rot or decay and is not attacked by termites.

(4) It is almost impermeable to moisture.

(5) The cost of maintenance of a reinforced concrete structure is almost nil.

(6) The materials used in the reinforced concrete constructions are easily obtainable.

(7) The flexibility of the reinforcement and the fluidity of the concrete make possible the reinforced concrete members to be moulded in any desired shape. Thus reinforced concrete lends itself to a wide range of architectural expressions.

22·27. BASIC ASSUMPTIONS

The following assumptions are made for design purposes :-

(1) A transverse section plain before bending remains plain after bending.

(2) The moduli of elasticity of steel and concrete are constant.

(3) All tensile stresses are taken up by steel reinforcement only and the resistance of concrete to tension is nil.

(4) The bond between concrete and steel is supposed to be sufficient to make them act jointly to resist tensile or compressive forces.

22·28. WHY STEEL RODS ARE USED AS REINFORCEMENT

The fact which makes it possible to combine steel and concrete is that concrete contacts on setting in air and if a steel rod is embedded in a mass of wet concrete , it will be found that considerable force is necessary to pull it out when the concrete is set. If the steel section be in the form of a plate, although it will resist removal when the concrete is set, yet it can be knocked off by sharp blows. In the first case, the concrete grips the steel, while in the second it only adheres. The grip depends upon the strength of concrete used in the work as well as the perfection with which the concrete has been mixed, placed in position, compacted and cured. Besides this, grip also depends upon the condition of the surface of the rod (whether it is smooth or rough). Specially shaped bars (Ribbed bars etc.) have therefore been introduced from time to time with the object of increasing the grip. It is on account of similar co-efficient of expansion of the two materials, superior bond value high tensile strength and less cost of steel that steel rods are used as reinforcement in R.C.C. work.

22·29. CAUSES OF FAILURE OF REINFORCED CONCRETE STRUCTURES

A reinforced concrete member can fail mostly in the following cases:

(1) When the member is subjected to excessive tension, so as to exceed the permissible stress in steel.

(2) When the loading is such that the compressive stress in concrete exceeds its safe permissible value.

(3) When the concrete is subjected to excessive shear.

(4) On account of the slipping of the steel bars from concrete.

(5) Due to bad quality of materials used, shrinkage, creep or thermal effects.

(6) When the member is subjected to extremes of temperature, aggressive liquids or gases.

22·30. GENERAL DESIGN REQUIREMENTS

An efficient design of a reinforced concrete structure requires that the materials be economically selected, proportioned and arranged to carry the required loads without developing stresses which are in excess of the allowable working stresses. Due consideration should be given to the effect of continuity on the distribution of bending moments and shears due to monolithic construction. Correct estimation of all types of loads should be made before taking up the design proper. The design requirements may be summarised into the following steps :

(1) To decide suitable arrangement of beams, columns, foundations etc. for the structure on the layout plan.

(2) To determine all types of loads and forces to which the structure is likely to be subjected to.

(3) To analyse the structure and to calculate, stresses, moments and shears etc. in the members.

(4) To work out safe section and area of steel for different members.

(5) To provide the calculated area for tensile, compressive, shear or temperature reinforcement in correct positions so as to ensure that the stresses remain within permissible limits.

(6) To provide sufficient anchorage for all reinforcements, so that the anchorage and bond requirements are fully met.

Questions for Review

1. What are the chief ingredients of plain cement concrete ? Describe in detail the properties of each.

2. Write short notes on :

(*i*) Rate of setting of cement (*ii*) Rate of hardening of cement.

(*iii*) Heat evolution of cement (*iv*) Resistance of cement to chemical action.

3. Write short notes on :

(*i*) Normal setting cement. (*ii*) Rapid hardening cement.

(*iii*) Quick-setting cement. (*iv*) High-alumina cement.

(*v*) White and coloured cement.

4. What influence do the following have on the strength of cement concrete ?

(*i*) Size of aggregates. (*ii*) Shape of aggregates.

(*iii*) Proportion of aggregates.

5. What are the functions of water in a concrete mix ? Describe the properties of water which are necessary to be used to get good concrete.

6. What do you understand by the term 'bulking of sand' ? Why does it occur ? How do you make allowance for it ?

7. What do you understand by the term, "Water cement ratio" ? Explain how the quantity of water affects the strength of a concrete mix.

8. Write short notes on the following properties of concrete :

(*i*) Strength. (*ii*) Workability.

9. What do you understand by slump test ? Describe the complete test in detail.

10. How do you measure the following in making a concrete mix ? Justify your mode of measurement :

(*i*) Cement.

(*ii*) Sand.

(*iii*) Water.

11. What effect do the grading of aggregates have upon a concrete mix. ? Describe the grading of aggregates by the following methods :

(*i*) By trial.

(*ii*) By fineness modulus method or sieve analysis method.

(*iii*) By minimum void method.

(*iv*) By arbitrary standards.

12. Describe various methods of mixing concrete. Give sketches to illustrate the method.

13. Write short notes on :

(*i*) Workability of concrete. (*ii*) Water-cement ratio.

(*iii*) Slump test. (*iv*) Continuous mixers.

(*v*) Batch mixers.

14. What do you understand by the term 'curing' ? Why is it essential to cure concrete? What are the various methods commonly adopted in curing ?

15. What conditions should be satisfied to get a water-proof concrete? What are the different methods used in the water-proofing of concrete ?

16. Write short notes on :

(*i*) Concreting underwater (*ii*) Proportionating of concrete by weight.

(*iii*) Rapid hardening cement. (*iv*) Water-cement ratio.

(*v*) Workability of concrete. (*vi*) Measurement of coarse aggregates.

17. Write an essay on causes of failure of concrete structures.

18. What is the commonly adopted concrete mix for the following :

(*i*) Concrete for water retaining structures.

(*ii*) Concrete for beams, slabs and columns.

(*iii*) Concrete for foundations of heavy machinery.

(*iv*) Concrete for foundations of ordinary building.

(*v*) Mass concrete work.

19. Write the general specifications of following for R.C.C. work :

(*i*) Cement. (*ii*) Sand. (*iii*) Water.

20. What do you understand by quality control of concrete ? An R.C.C. 'T' beam roof over a hall 9 m. x 9 m. with beams 3 m. centre to centre has to be laid. Explain how you will plan to execute the work giving particular reference to various types of joints necessary. Give list of labour and T & P you would arrange for the purpose. What will you do if your mixer suddenly fails after doing 25% of the work ?

21. Assuming that you are the architect of a large R.C.C. building in which the quality of materials, method of mixing and standard of workmanship are of vital importance, describe the precautions and investigations you would suggest when the work is in progress.

22. What is understood by "designing of a concrete mix" ? How is it done ?

(*a*) Explain the design of such a mix. for an R.C.C. underground water tank.

(*b*) How is the consistency of concrete determined ? Explain the method.

23. Attempt any two of the following :

(*i*) How would you proceed to work out the proportions of materials for a high grade concrete of particular strength ?

(*ii*) What is "vibrated concrete" and in what respect is it superior to ordinary concrete? Describe different types of vibrators commonly used for concrete works of different types.

(*iii*) Describe at least two methods you consider best for rendering concrete of a water-tank leaf roof.

24. Describe the importance of the following in R.C.C. work :

(*a*) Water-cement ratio.

(*b*) Bulking of sand.

(*c*) Slump test

(*d*) Curing.

25. (*a*) What is the function of water in the making of cement concrete ?

(*b*) How does water-cement ratio affect the strength of concrete ?

(*c*) State different methods by which concrete placed in position is compacted with their relative merits and where each method is used.

26. (*a*) Write a brief note on setting of cement.

(*b*) What are the various characteristics of the aggregate that control the quality of cement concrete ?

23

Formwork

23·1. GENERAL

Centering, shuttering or formwork is a sort of temporary construction provided for laying cast-in-situ concrete to required shape. A good formwork should satisfy the following requirements.

(*i*) It should be strong enough to withstand all types of dead and live loads such as selfweight, weight of reinforcement, weight of wet concrete, loads due to workmen, construction equipment, other incidental loads and forces caused by placement and consolidation of concrete, imposed upon it during and after casting of concrete.

(*ii*) It should be rigidly constructed and efficiently propped and braced (both horizontally and vertically) so as to retain its shape without undue deflection.

(*iii*) The joints in the formwork should be tight against leakage of cement grout.

(*iv*) The formwork should be constructed in such a manner that it may permit the removal of various parts in desired sequence without jarring or damaging the concrete.

(*v*) The material of the formwork should be cheap, easily available and should be suitable for re-use several times.

(*vi*) The formwork should be set accurately to the desired line and levels and should have plain surfaces.

(*vii*) The formwork should be as light as possible.

(*viii*) The material of formwork should not warp or get distorted when exposed to sun, rain or water during concreting.

(*ix*) The formwork should rest on firm base.

23·2. ECONOMY IN FORMWORK

It may be noted that the total cost of concrete construction includes the cost of formwork as well. Construction of formwork involves considerable expenditure in terms of cost of material, cost of labour for fabrication, erection and removal of formwork and time element. In case of buildings, the cost of formwork may range between 30 to 40% of the cost of concrete work. In case of special structures like bridges, tall chimneys, dams etc., the cost of formwork may range between 50 to 100% of the cost of concrete work, or even more. Since formwork contributes nothing towards the stability of the finished structure, its cost should be kept barest minimum consistent with safety. Following steps should be followed to effect economy in the cost of formwork.

(1) The building should be planned in such a way that there are minimum number of variations in the size of rooms, floor area etc. so as to permit re-use of the formwork moulds repeatedly.

(2) The scheme of the formwork should be efficiently planned and suitably designed to determine the most economical but safe sizes of different components including the supports or props.

(3) The formwork should be constructed in such a way that timber (where used) is cut to the minimum and it can be struck off with ease and re-used with least damage.

The quality of finished concrete surface depends to a great extent upon the quality of formwork used in the work. In case the formwork is made out of rough timber and is not constructed properly, the resultant concrete surface will be irregular and full of defects. The expenditure involved in rectification of defects in such works is invariably more than the saving made in the cost of formwork.

Thus to effect economy, the formwork should be made with good quality material and should be constructed properly, besides following the steps referred above.

23·3. MATERIALS FOR FORMWORK

Formwork can be made out of timber, plywood, steel, precast concrete or fibre glass, used separately or in combination. The type of material to be used for formwork depends upon the nature of construction as well as the availability and cost of material. In early stages, formwork was made up of timber alone. With the introduction of steel forms, use of timber formwork is getting reduced specially in case of major construction projects and in situations where large number of re-uses of the same forms are possible. However, for small works involving less number of re-uses, timber formwork proves economical. Fibre glass is used mainly for making moulds for repetitive castings of pre-cast concrete products. Moulds made up of pre-cast concrete, fibre glass and aluminium are used in cast-in-situ construction such as waffle slabs or members involving curved surfaces.

23·4. TIMBER FORMWORK

Timber used for the formwork should satisfy the following requirements.

(*i*) It should be well seasoned.

(*ii*) It should be light in weight.

(*iii*) It should be easily workable with nails without splitting.

(*iv*) It should be free from knots.

The sizes of timber sections for different components of form-work depend upon the span of the slab or beam, floor to floor height and the centre to centre spacing of the centering supports.

The size of timber planks or joists that can be adopted for different components of formwork for shuttering of 4·5 m. span and 3·5 m. height are given in table No. 23·1 (The minimum sizes of sections are applicable for 1 m. centre to centre spacing of the centring props whereas the maximum sizes should be adopted when centring props are spaced at 1·2 m. c/c.) In case of situations which do not get covered within these limits, the formwork should be designed in detail and the sizes of various components worked out after considering all type of dead and live loads and other relevant factors.

For normal construction work where repetitive use of shuttering is possible, the quantity of timber shuttering can be worked out on the assumption that one set of shuttering can be used 10 to 12 times.

TABLE 23·1

SIZES OF MEMBERS FOR TIMBER FORMWORK

S.No.	*Component of formwork*	*Size* (*Varying according to the spacing of centering props form I metre to 1·2 metres*)
(1)	(2)	(3)
1.	Sheeting for Slabs, Beam and column side and Beam bottom	25 mm to 40 mm. ... Thick
2.	Joists, ledges	50 x 75 mm to ... cross-section 50 x150 mm.
3.	Posts	75 x 100 mm. to ..." 100 x 100 m. ...
4.	Ballies	Not less than 100 mm at mid-length and 80 mm at thin end

The timber used for shuttering for exposed concrete work should have smooth and even surface on all faces which are to come in contact with concrete. In situations where concrete surfaces are not exposed, as in case of foundation etc., undressed timber can be used to effect economy. The timber planks for the shuttering should be jointed by providing tongued and grooved joints so as to ensure adequate tightness against leakage of cement grout.

23·5. PLYWOOD FORMWORK

Use of plywood instead of timber planks is getting popular these days. In this case resin bonded plywood sheets are attached to timber frames to make up panels of required sizes. The panels thus formed can be easily assembled by bolting in the form of shuttering. This type of shuttering ensures quality surface finish and is specially recommended in works where large exposed areas of the concrete are to be constructed such as floor slabs, faces of retaining walls etc. The cost of plywood shuttering compares favourably with that of timber shuttering and it may prove to be cheaper in certain cases in view of the following considerations.

(*i*) It is possible to have perfectly plain and smooth surface (without joint marks) by use of plywood shuttering. Thus the expenditure on surface finishing can be saved.

(*ii*) By use of large size panels it is possible to effect saving in the labour cost of fixing and dismantling.

(*iii*) Number of re-uses are more as compared with timber shuttering. For estimation purposes, number of re-uses can be assumed as 20 to 25.

23·6. STEEL FORMWORK

This consists of panels fabricated out of thin steel plates stiffened along the edges by small steel angles. The panel units can be held together by two or more clamps or bolts provided along each edge and the shuttering can be assembled and kept in alignment by use of horizontal or vertical centring of timber or steel. The panels can be fabricated in large number in any desired modular shape or size. The usual size for wall or slab panel varies from 60 cm x 60 cm to 60 cm x 120 cm.

Steel forms are mostly used in large construction projects or in situations where large number of re-uses of the same shuttering are possible. Steel forms are extensively used in repetitive castings of pre-cast concrete products. This type of shuttering is considered most suitable for circular or curved shaped structures such as tanks, columns, chimneys etc. and for structures like large sewer, tunnels and retaining walls.

Although steel shuttering costs more initially but in view of its various advantages it may work out to be economical for a medium sized work in the long run.

Advantages of steel formwork over timber forms.

(1) Steel forms are stronger, more durable and have longer life as compared with timber forms.

(2) They can be put to sufficiently larger number of re-uses. For estimation purposes the number of re-uses can be assumed to vary from 100 to 120.

(3) Steel forms can be installed and dismantled with greater ease and speed which results in saving in labour cost for this item of work.

(4) The quality of exposed concrete surface obtained by use of steel forms is excellent and it needs no further treatment. On the other hand construction carried out by use of timber formwork invariably requires plastering to obtain desired finish of the concrete surface. Thus there is saving in the cost of finishing the surface by use of steel forms.

(5) There is no danger of the formwork absorbing water from the concrete and hence the chances of honeycombing are minimised.

(6) They are not liable to shrink or distort and hence it is possible to achieve better workmanship and higher degree of accuracy by use of steel forms.

23·7. CONSTRUCTION OF FORMWORK

The construction of formwork normally involves the following operations.

(*i*) Propping and Centering.

(*ii*) Shuttering.

(*iii*) Provision of camber.

(*iv*) Cleaning and surface treatment.

(*i*) *Propping and Centering.* The props used for centering may be of steel, timber posts or ballies. Pillars made up of brick masonry in mud mortar are also sometimes used as props. In case wooden ballies are used as props, they should rest squarely on wooden sole plates laid either on ground or on brick masonry pillars in mud mortar. The wooden plate should have area of at least 0·1 sq. m. and it should be 40 mm. in thickness. Double wedges are essentially provided between the sole plates and the timber props with a view to permit accurate adjustment of the shuttering prior to concreting operation and to allow easy removal of shuttering afterwords. In case brick masonry pillars are used at props, the wooden sole plates are provided at the top of the pillars, and the double wedges are inserted between the sole plates and the bottom of shuttering.

(*ii*) *Shuttering.* As decribed earlier the shuttering can be made up of timber planks, or it may be in the form of panel units made either by fixing plywood to timber frames or by welding steel plates to angle

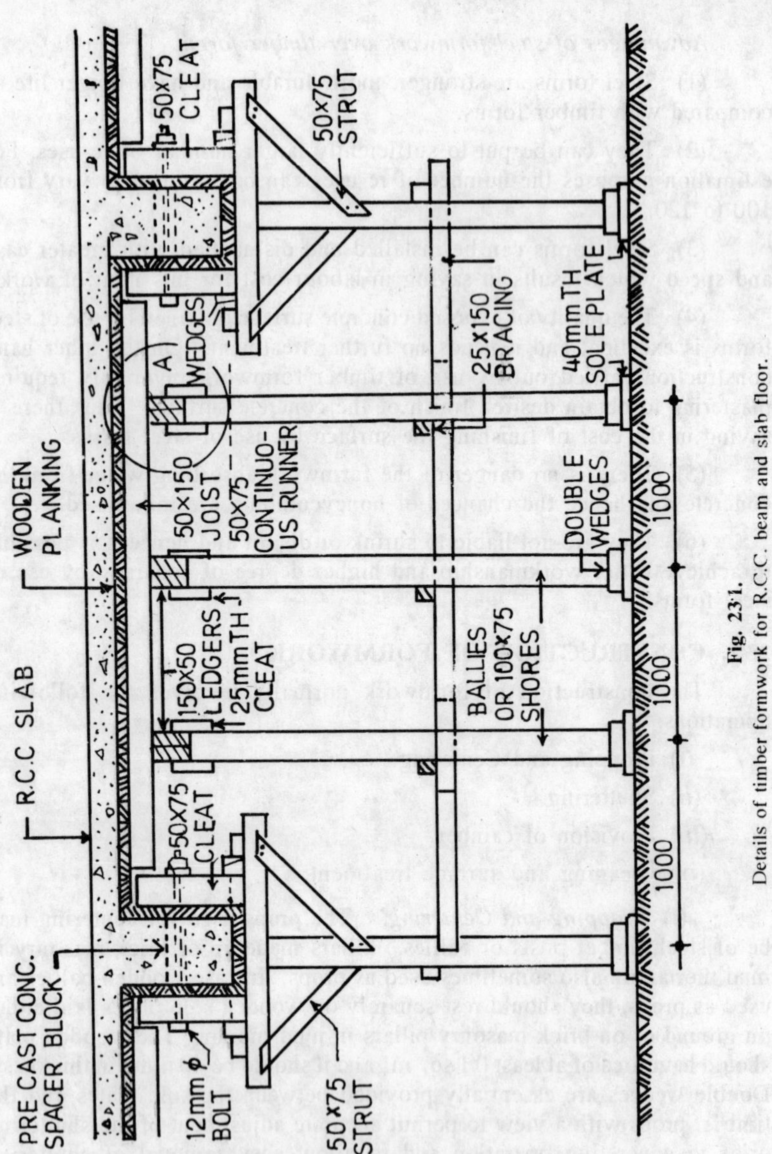

Fig. 231. Details of timber formwork for R.C.C. beam and slab floor.

framing. In any case, the shuttering should be constructed in such a manner that the joints should be tight against leakage of cement grout.

Sometimes the architect specifies provision of moulding in slab or columns or provision of chamfers at corners of beams and columns. All such mouldings, bevelled edges, chamfers etc. required to be formed in

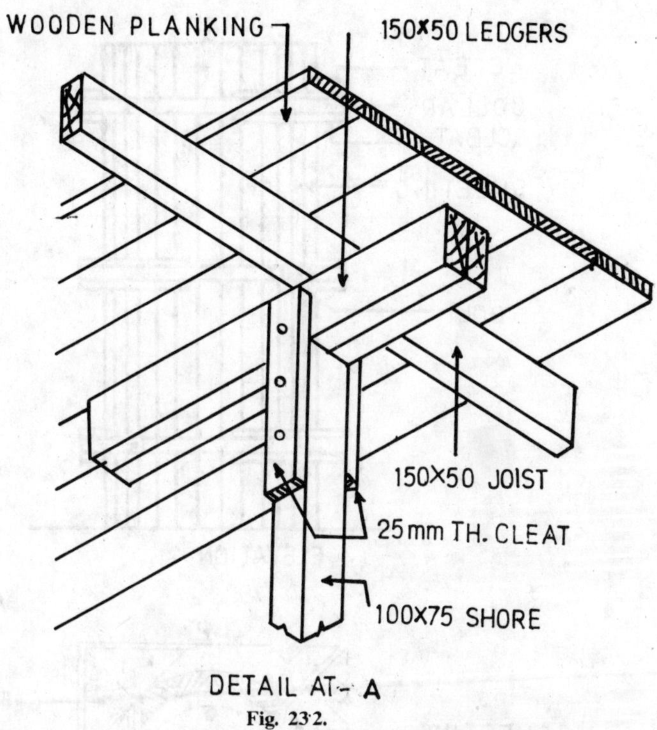

Fig. 23·2.

Enlarged detail at A. (for sectional elevation refer Fig. 23.1.)

the concrete work should be made in the shuttering itself. Similarly the opening for fan clamps or other fittings should be provided in the shuttering as per details in the drawing.

(*iii*) *Provision of camber.* Certain amount of deflection in structure in unavoidable. It is therefore desirable to give an upward camber in the horizontal members of the concrete structure (specially in members having long spans) to counteract the effect of deflection. The provision of desired camber should be made in the formwork itself during its erection. Camber for slabs and beams may be adopted @ 4 mm. per metre of span. In case of cantilevers, the camber at free end should normally be taken @ 1/50th of the projected length of the cantilever.

(*iv*) *Surface treatment.* Before laying concrete the formwork should be cleaned of all rubbish particularly the sawdust shavings and chippings etc. All surfaces of timber shuttering that are to come in contact with concrete should be well wetted with water. This is necessary to prevent the chances of dry shuttering timber absorbing water from the concrete which may cause warping, swelling and distortion of timber besides resulting in defect of honeycombing in concrete.

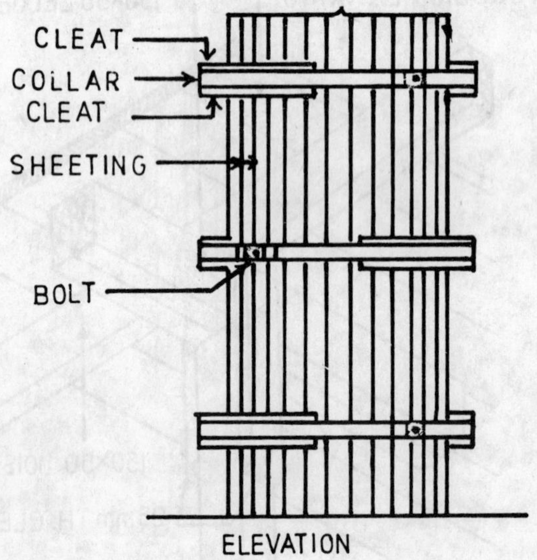

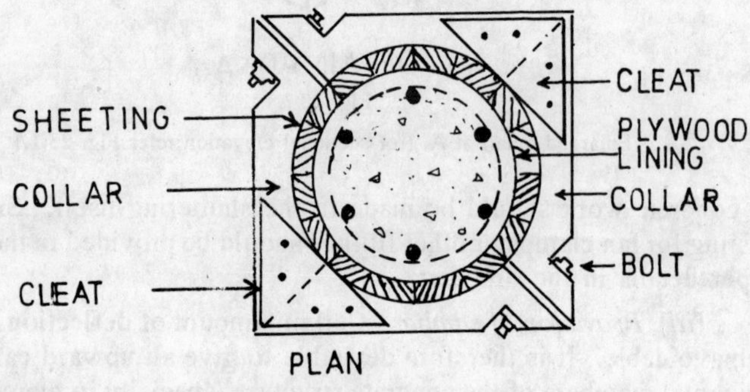

Fig. 23·3.

Details of timber formwork for circular R.C.C. column.

Similarly steel forms that have been exposed to hot weather should be cooled by watering before laying concrete.

In addition, all surfaces of shuttering which are to come in contact with concrete should be given a good coating of raw linseed oil or soft soap solution or any other material so as to prevent the concrete getting struck to the formwork and thus facilitate easy removal. It is, however, important to note that the coating should not come in contact with reinforcement. The selected coating should be impermeable and should neither get flaky on exposure to weather nor stain the surface of concrete.

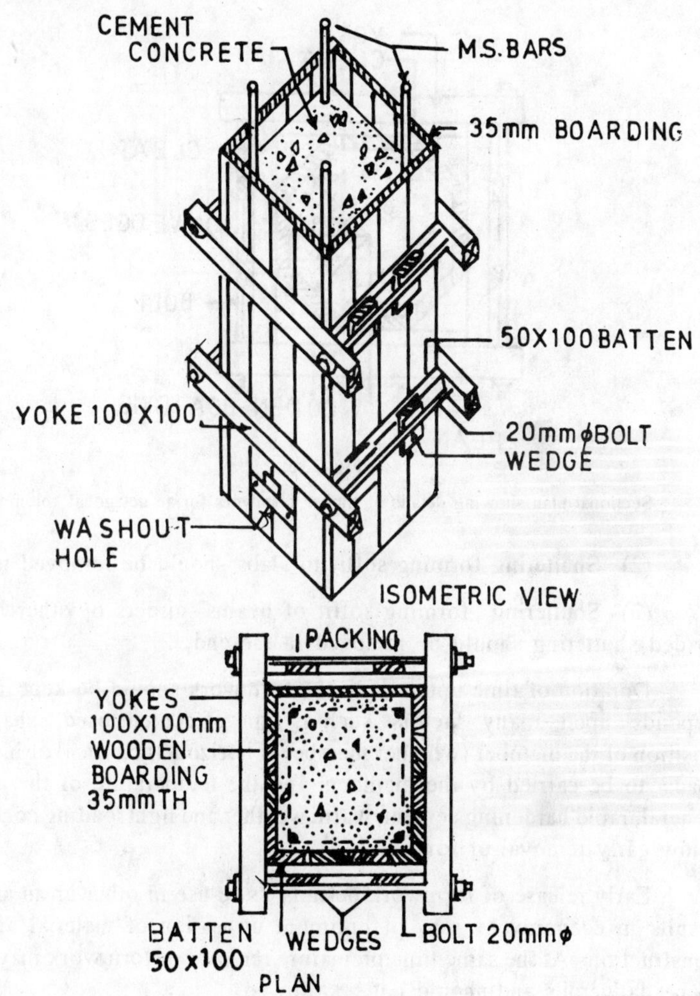

Fig. 23·4.

Details of timber formwork for a square or rectangular R.C.C. Column

23·8. ORDER AND METHOD OF REMOVING FORMWORK

The formwork should be planned and constructed in such a manner that it is possible to remove the different components in the following order of sequence.

(1) Shuttering forming vertical faces of walls, beams and columns sides (which bear no load but are used only to retain the concrete) should be removed first.

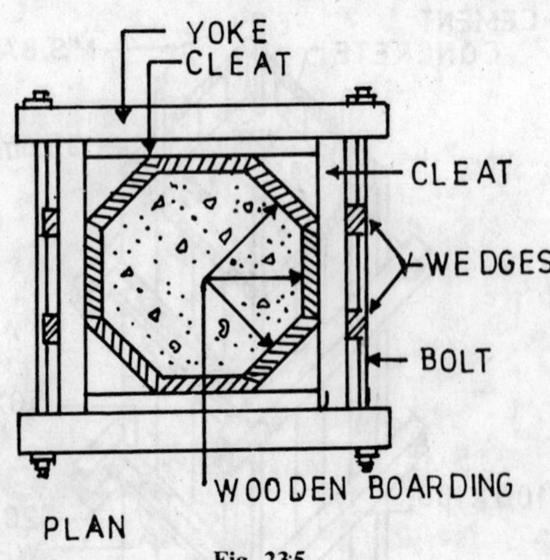

Fig. 23·5.

Sectional plan showing details of timber formwork for an octagonal column.

(2) Shuttering forming soffit to slabs should be removed next.

(3) Shuttering forming soffit of beams, girders or other heavily loaded shuttering should be removed in the end.

Duration of time upto which the formwork should be kept in place depends upon many factors such as type of cement used, shape and position of the member (whether member is horizontal, vertical or inclined) loads to be carried by the members and the temperature of the air. In general, rapid hardening cement, warm weather and light loading conditions allow early removal or formwork.

Early release of formwork permits its re-use in other areas and this results in economy by way of optimum utilisation of material and fast construction. At the same time premature removal of formwork may prove to be dangerous and should not be allowed.

The formwork should, under no circumstances, be allowed to be removed until the concrete reaches a strength of at least twice the stress to which the concrete may be subjected to at the time of removal of formwork.

In normal conditions (where air temperature is above 20°C and where ordinary portland cement is used) the formwork may be struck or removed after the expiry of the period indicated against respective member in Table No. 23·2.

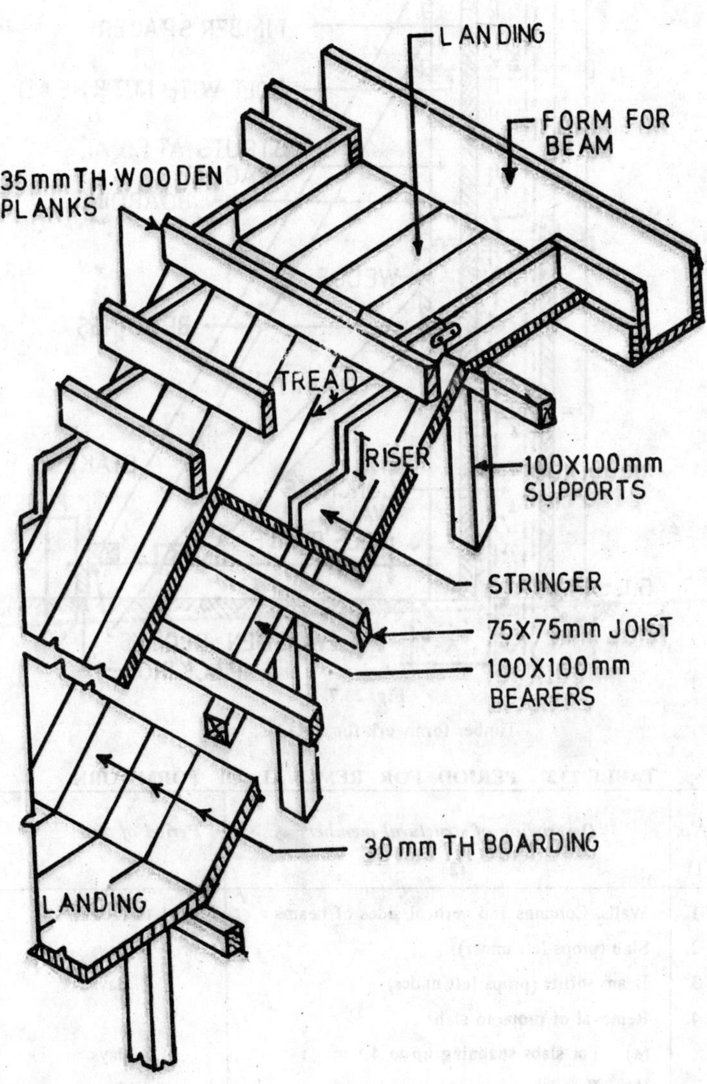

Fig. 23·6.

Details of timber formwork for a stair

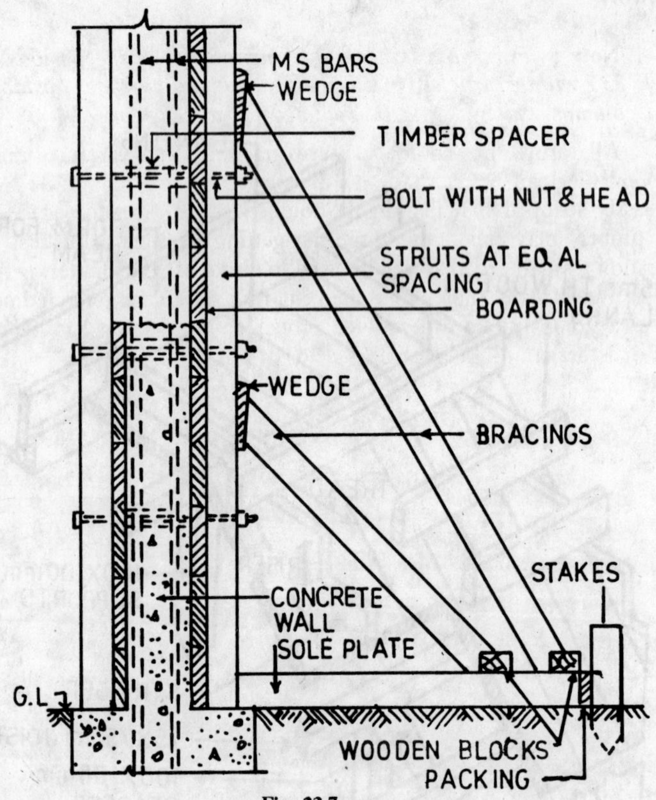

Fig. 23.7.

Timber formwork for a R.C.C. wall

TABLE 23·2 PERIOD FOR REMOVAL OF FORMWORK

S.No. (1)	*Description of structural member* (2)	*Period of time* (3)
1.	Walls, Columns and vertical sides of beams	1 to 2 days
2.	Slab (props left under)	3 days
3.	Beam soffits (props left under)	7 days.
4.	Removal of props to slabs :	
	(*a*) For slabs spanning up to 4.5 m.	7 days.
	(*b*) For slabs spanning over 4.5 m.	14 days.
5.	Removal of props to beams and arches	
	(*a*) Spanning upto 6 m.	14 days
	(*b*) Spanning over 6 m.	21 days

Note : *For rapid hardening cement 3/7th of the periods given in table 23·2 will be sufficient in all cases except vertical sides of slabs, beams and columns which should be retained for at least one day.*

All formwork should be eased gradually and carefully in order to prevent the load being suddenly transferred to concrete. The forms can be easily stripped by loosening the double wedges. Sometimes the swelling of timber may result in shuttering getting jammed at places. In such situation a small crowbar can be used to strike off the shuttering. In case of cantilever slabs and beams, the centering should be removed only after necessary construction to provide proper counterweight has been carried out and the same has attained sufficient strength.

Questions for Review

1. What do you understand by term "Formwork" ? Describe the following with sketches.
   (*i*) Formwork for square and round columns.
   (*ii*) Formwork for beam and slab construction.
   (*iii*) Formwork for R.C.C. stairs.
2. A hall 18 m × 6 m. with masonry walls is to be roofed over by R.C.C. beams and slab. The beams are 3 m. apart. Show by suitable sketches the arrangement of shuttering and centering required for the construction of the roof. Assume suitable dimensions of beam and slab and the other components of the formwork.
3. Sketch the plan and sectional elevation showing details of centering and shuttering suitable for casting the roof of an auditorium 12 m. × 20 m. in size and 6 m. high. The R.C.C. beams 30 cm wide 75 cm. deep are spaced at 3.5 m. c/c and span parallel to the width of the auditorium. The roof slab is 15 cm thick Assume suitable sections of the various components of the formwork.
4. Write short notes on the following :
   (*i*) Plywood formwork.
   (*ii*) Steel formwork.
   (*iii*) Order and method of removing formwork.
5. Describe the construction of formwork with special reference to the following :
   (*i*) Centering.
   (*ii*) Shuttering.
   (*iii*) Camber.
   (*iv*) Cleaning and surface treatment of shuttering.
6. What are requirements of a good formwork ? Describe the steps that should be taken to effect economy in the expenditure on formwork.

24

Provision of Joints in Structures

24·1. INTRODUCTION

It is well known that almost all building materials expand or contract due to change in temperature and variation in moisture content. The magnitude of such expansion or contraction depends upon the type of material used in construction and the extent of variation in the temperature and moisture content. In a continuous structure, the cumulative effect of the dimensional changes in the building elements on account of the above may be considerable. In case components of such a structure are not allowed free movements, internal stresses will be set up which may result in formation of cracks. This may in turn endanger the stability of the structure. Such a problem gains special importance in tropical countries like India, where large variations occur in the atmospheric temperature and humidity. Thus special provisions should be made to control or isolate thermal and other movements to avoid danger to a structure. This is achieved by breaking the continuity of a structure, by introducing joints at regular intervals. These joints sub-divide the building into smaller units and also permit free movement of each unit.

24·2. TYPE OF JOINTS

Joints can be divided into the following different categories.

(*i*) *Expansion joints*

(*ii*) *Isolation joints*

(*iii*) *Contraction joints*

(*iv*) *Sliding joints*

(*v*) *Construction joints.*

24·3. EXPANSION JOINTS

These joints are provided to accommodate the expansion of adjacent building parts and to relieve compressive stresses that may otherwise develop. Total expansion (I) of a structure can be obtained by the formula :

(I) = co-efficient of linear expansion x length of the structure in centimetre x change of temperature in degree F...cm.

Based on this formula, it is seen that a structure 30 m. long expands linearly by about 10 mm. through a temperature change of 50°F. In general, buildings of ordinary size and regular shape in plan can resist the stresses caused by dimensional changes and as such do not require provision of expansion joints. However provision of one or more expansion joints is considered necessary when the continuous length of a structure exceeds 45 m. There is wide divergence of opinion regarding proper spacing of expansion joints. In India, the recommended maximum centre to centre spacing of expansion joints is 30 m. In addition, expansion joints should be provided where a structure changes direction for instance in L,T, H and U shaped structures. Where two or more expansion joints are considered necessary it is desirable to locate the joints in such as way that the distance of joints from the nearest continuous corner does not exceed 15 m.

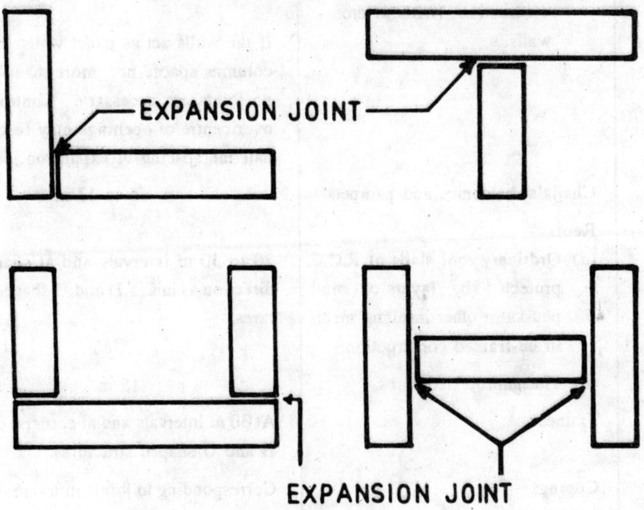

Fig. 24·1
Plans showing location of expansion joint in buildings of different shapes

The joint is formed by providing an initial gap between the adjacent building parts which gets widened or shortened while accommodating the contraction or expansion of the building. The width of the gap should be sufficient to accommodate the maximum thermal-induced movements likely to be encountered. Normally it varies between 10 mm to 40 mm.

There should be complete discontinuity of masonry, reinforcement or concrete at the joint and the gap should permit unobstructed movement of adjacent parts.

The recommended spacing of expansion joints for different types of works as per I.S.3414-1168 is given in table 24·1.

TABLE 24·1

RECOMMENDED SPACING OF EXPANSION JOINTS

S. No.	Description of item	c/c spacing of joint
1.	Walls	30 m
	(*a*) For load bearing walls with cross-walls at intervals. Traditional type of one-brick thick or more.	
	(*b*) Walls of warehouse type construction (without cross-walls.	30 m If the walls act as panel walls between columns spaced not more than 9 m c/c. no joints are necessary. Control joints over centre of openings may be given at half the spacing of expansion joints.
2.	Chajja's, balconies and parapets.	6 to 12 m
3.	Roofs	
	(*a*) Ordrinary roof slabs of R.C.C. protected by layers of mud phuska or other insulting media in un-framed construction.	20 to 30 m intervals and at changes in direction as in L,T,H and U-shaped structures.
	(*b*) The unprotected slabs.	15 m
4.	Frames	At 30 m intervals and at corners of L, T, H and U shaped structures.
5.	Copings	Corresponding to joints in the roof slabs.

1. Materials used for Expansion Joints

Attempt should be made to treat the expansion joint in such as manner that it provides neat appearance and serves as architectural feature in the building. The detailing of the joint should also be made as simple as possible. All that is needed is to fill the gap of the joint with suitable compressible material, so as to ensure wind and water tightness. The materials used for treating an expansion joint are briefly discussed below:

(*i*) *Joint filler* : It is a readily compressible material which can accommodate the expansion of adjacent parts of the joint and has ability to regain 75 % of its original thickness when pressure is released. In addition it should be durable, resistant to decay (due to weathering, termites etc.) and rigid enough to permit formation of straight joint.

Joint filler can be made out of a variety of materials such as bitumen, bitumen containing cellular material, cork strips, natural or cellular rubber, expanded plastic, mineral fibre, coconut, pith etc.

(*ii*) *Sealing Compound* : The function of the sealing compound in a joint are to seal the joint against passage of moisture and to prevent the ingress of dust, grit or other foreign matter into the joint. It should be insoluble, non-toxic and tintless and should be readily workable for placement in position.

Mastic or hot-applied bituminous sealing compound is commonly used for this purpose.

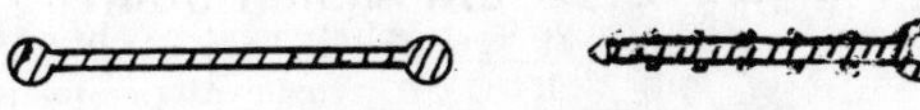

Fig. 24·2

Types of rubber or P.V.C. water bars

(*iii*) *Water bar* : The function of waterbar is to seal the joint against passage of water. The water bar may be of natural and synthetic rubber, polyvinyl chloride (PVC) or metal. The rubber or PVC water bars are manufactured in a variety of different forms such as dumbell type, centre bulb type etc. The metal water bar consists of thin strips of copper, aluminium or other corrossion resistant metal. Galvanished iron sheet water bars are also recommended in view of the high cost of copper. However, G.I. water bar should not be used under corrosive conditions. The width of the metal water bar normally varies for 15 to 20 cm. and thickness should not be less than 0·56 mm. gauge. The metal water-bar should have a U or V fold in the middle to allow for expansion and contraction at the joint.

METALLIC WATER BAR

Fig. 24·3

Types of metallic water bars

2. Provision of Expansion Joints in Walls

As indicated earlier the maximum centre to centre spacing of the expansion joints in all continuous external and internal masonry walls

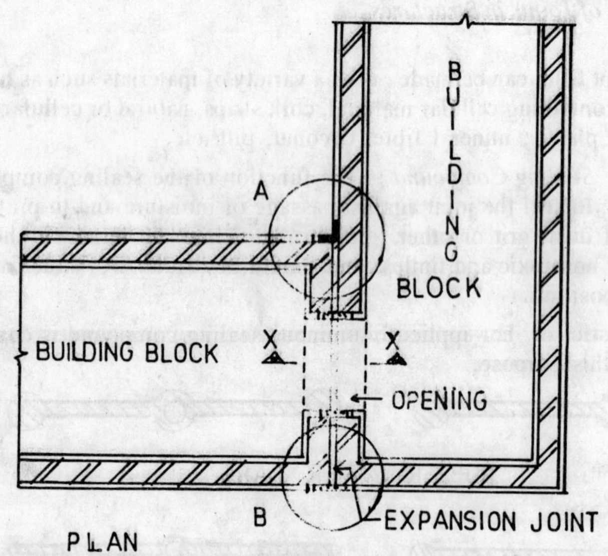

Fig. 24·4

Plan showing location of expansion joint between two building blocks

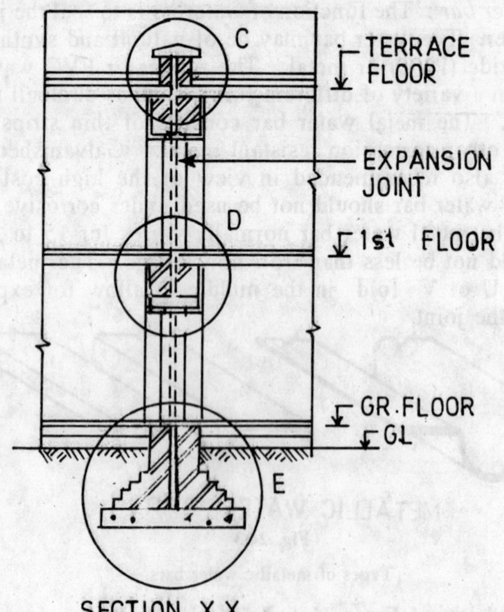

Fig. 24·5

should not exceed 30 m. The width of the joint should preferably be not less than 15 mm. The joint should be provided from top of the wall to the top the concrete foundation. (Ref. figure 24·5 and 24·11). In case the masonry walls rest on pile foundation, the expansion joint should be taken upto the top of grade beam of the pile foundation system.

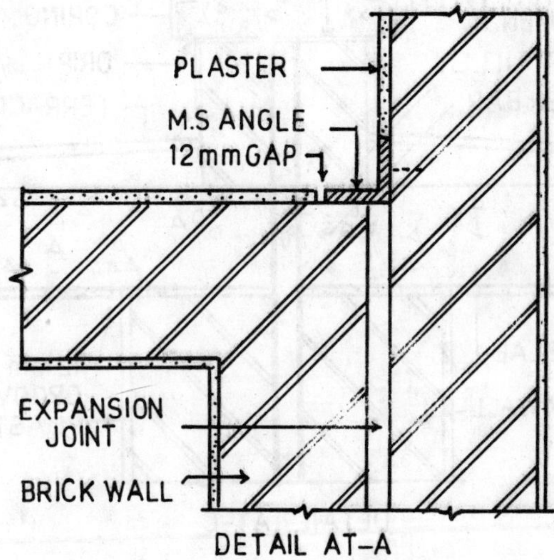

Fig. 24·6

Details of expansion joint between walls (For plan refer Fig. 24·4)

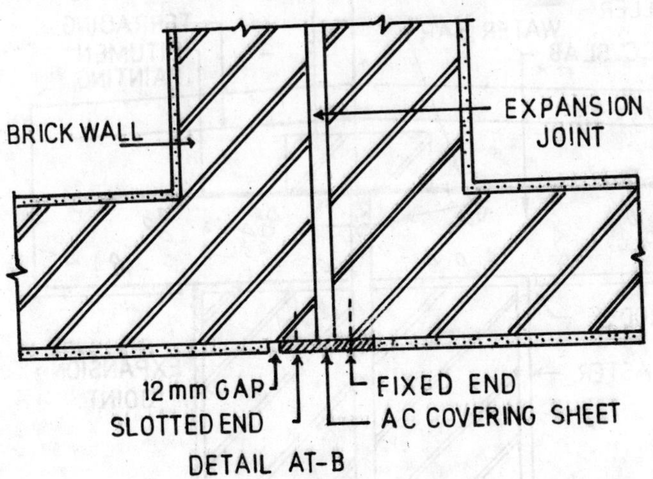

Fig. 24·7

Details of expansion joint between walls (For plan refer Fig. 24·4)

Alternative methods of treating an expansion joint in walls are shown in Fig. 24·6 & 24·7. From aesthtic consideration, it is necessary to cover the joint with covering sheets which may be of aluminium, hard

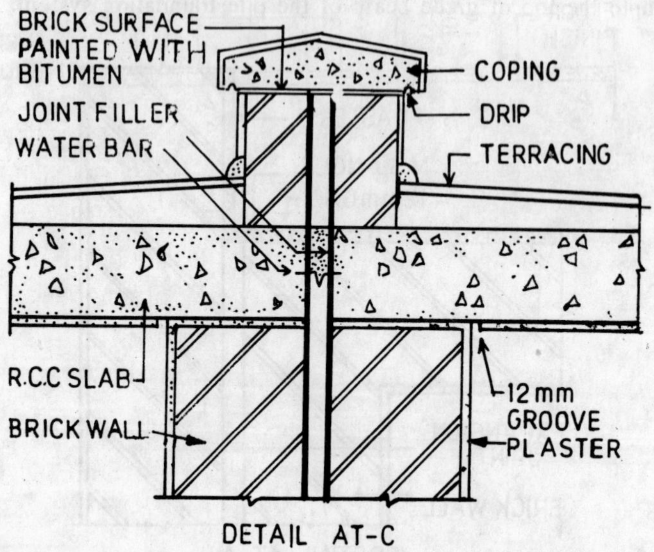

Fig. 24·8 Details of expansion joint at roof level (For section refer Fig. 24·5)

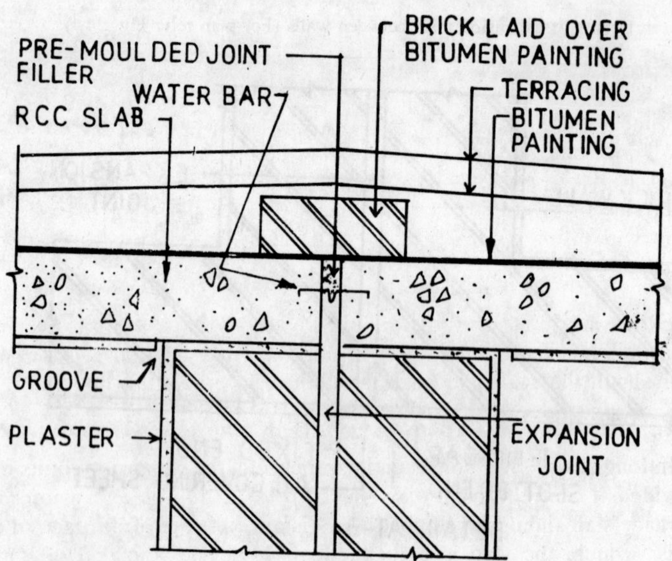

Fig. 24·9 Alternative method of treating expansion joint at roof level (Alternative detail at C)

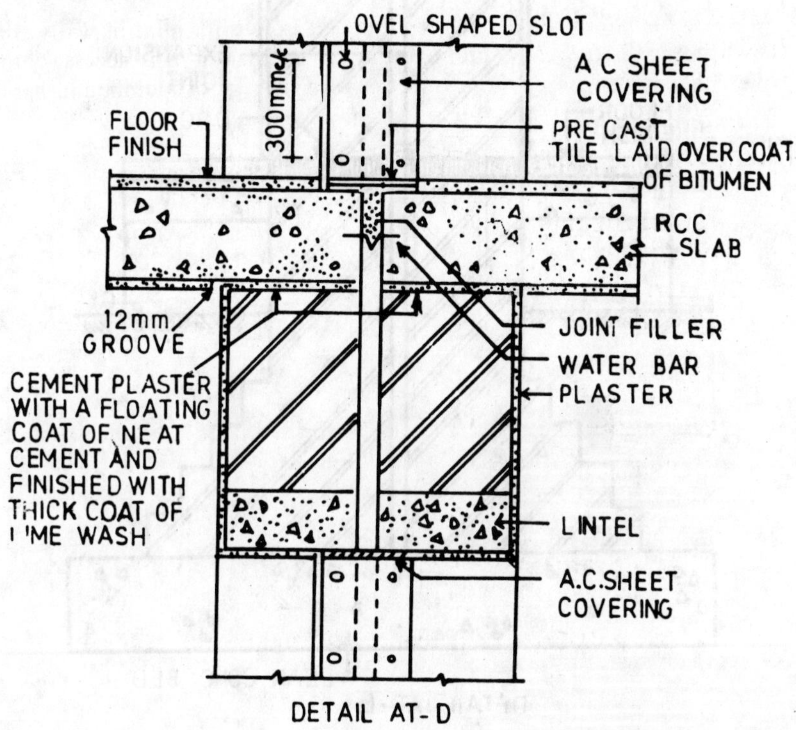

Fig. 24·10

Details of expansion joint at first floor level (For section refer Fig. 24·5)

board, timber plank or AC. Sheet. The sheet covering is firmly fixed to the wall on one side of joint with the help of rawl plugs and screws provided at 30 cm c/c and fixing through oval shaped slots to the wall on other side of joint. The oval shaped hole permits the movements at the joint without damaging the covering sheet.

3. Provision of Expansion Joints in Roof and Floors

Expansion joints in roofs should invariably have a joint filler and waterbar to avoid leakage of water through the joint. Similarly, the joint in floor should be sealed to prevent rubbish accumulating therein.

4. Provision of Expansion Joints in Chajja's, Balconies etc.

In long chajja's, balconies and parapets etc. the expansion joints may be provided at centre to center spacing of 6 to 12 m. The expansion joint in sunshade slab should be stopped at a distance of 5 cm from face of the wall is which the slab is embedded. However, the percentage of

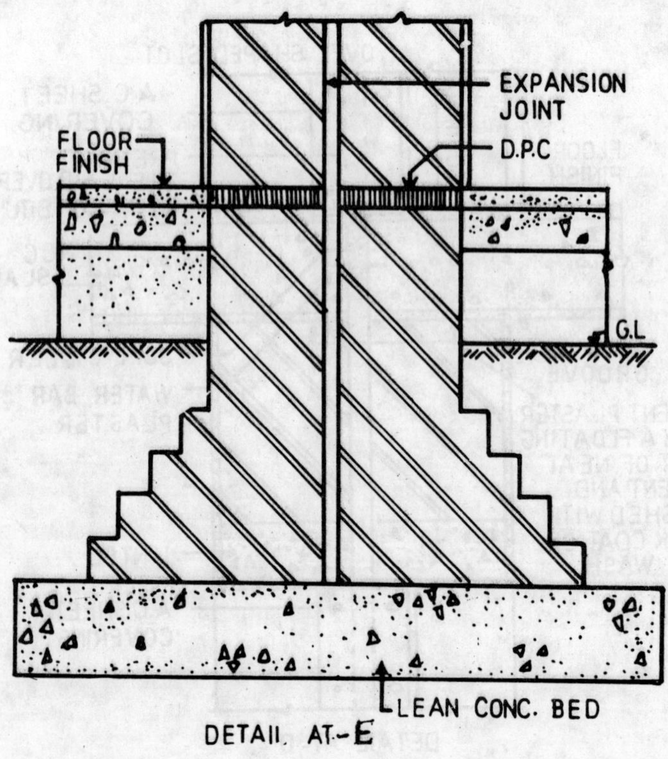

Fig. 24·11

Details of expansion joint at foundation level (For main section refer Fig. 24.5)

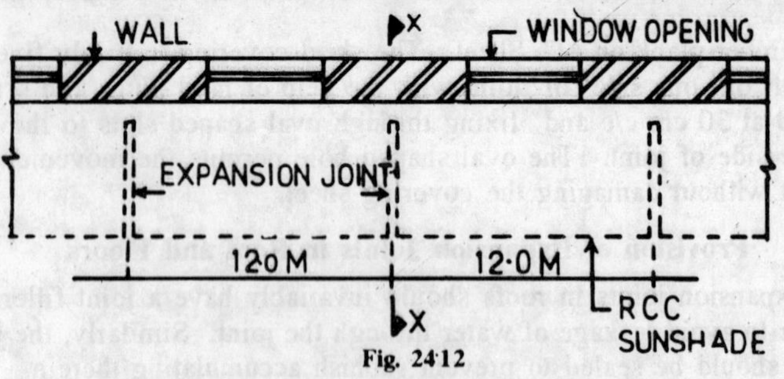

Fig. 24·12

Plan showing location of expansion joint in sunshade.

distribution reinforcement in embedded portion alongwith the 5 cm wide strip of slab (from the face wall) should be increased to 0.3% of the cross sectional area. The gap of the expansion joint in the chajja slab should not be filled with any material.

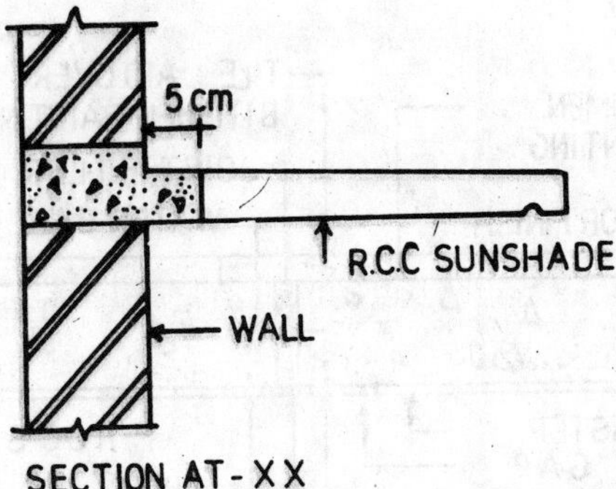

Fig. 24·13

5. Provision of Expansion Joints in Verandah Slab

In case of covered verandah slabs, the expansion joint may be spaced from 12 to 14 m c/c. The joint should not extend beyond the wall and its gap should be sealed with water bar and sealing compound.

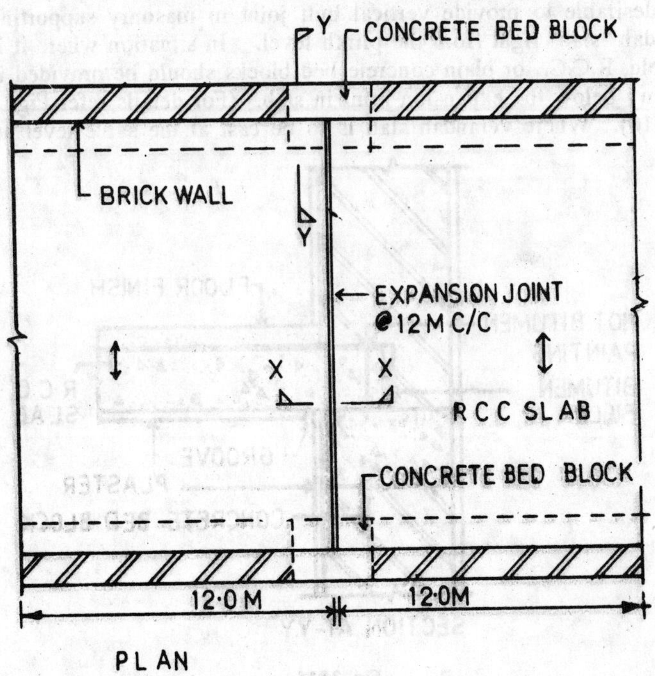

Fig. 24·14

Plan showing expansion joint in a varandah slab

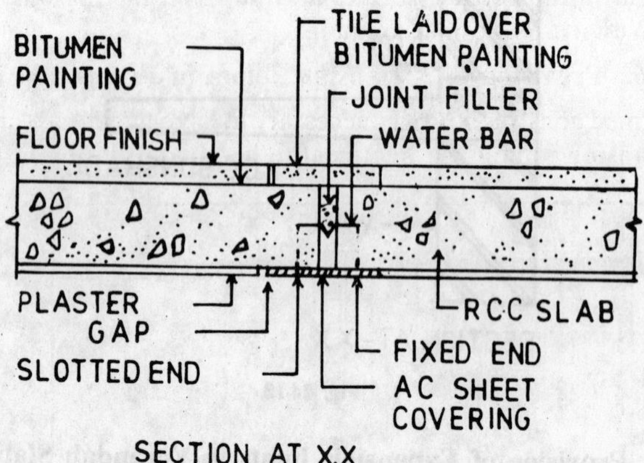

Fig. 24·15

(For Plan refer Fig. 24.14)

To prevent cracks in masonry above or below the expansion joint, it is desirable to provide vertical butt joint in masonry supporting the verandah slab right from the plinth level. In situation where it is not possible, R.C.C. or plain concrete bed blocks should be provided in the masonry below the expansion joint in slab. (For details refer Fig. 24·14 to 24·16). Where verandah slab is to be cast at the same level and in

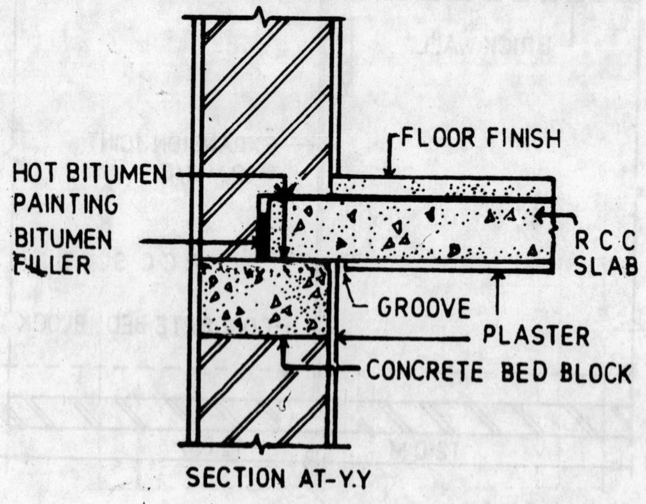

Fig. 24·16

(For plan refer Fig. 24.14)

continuation with the floor slab of the room, the percentage of distribution reinforcement in the portion of the slab resting on the masonry should be increased to twice its normal amount.

6. Provision of Expansion Joints in Framed Structures

In framed construction, it is necessary to have two fames one on either side of expansion joint. The treatment to be given to the joint is same as

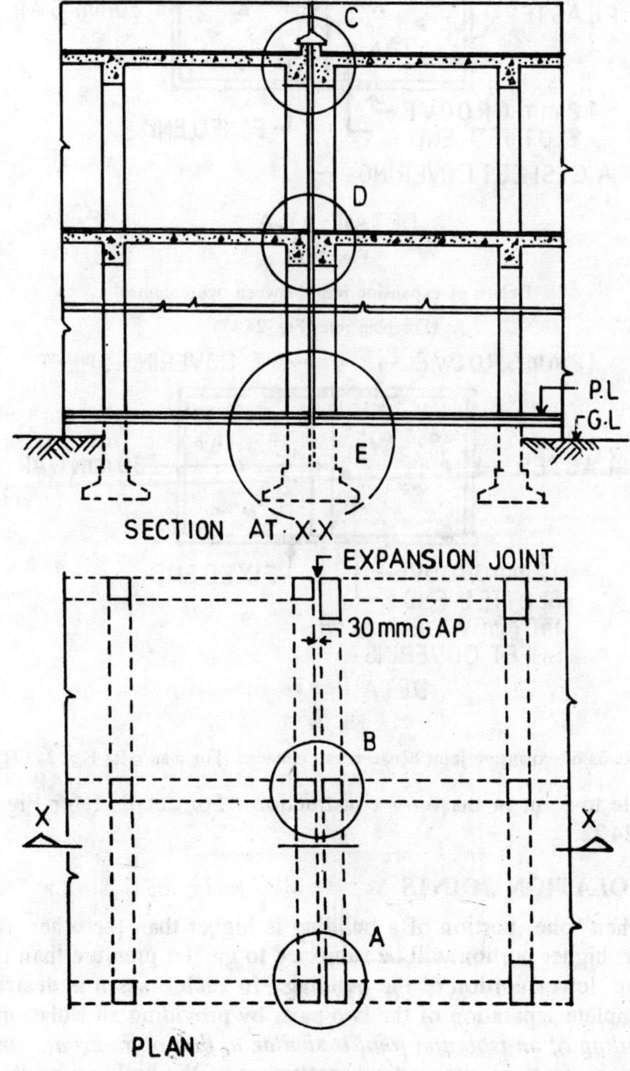

Fig. 24·17

Plan and section of a R.C.C. framed structure showing location at expansion joint

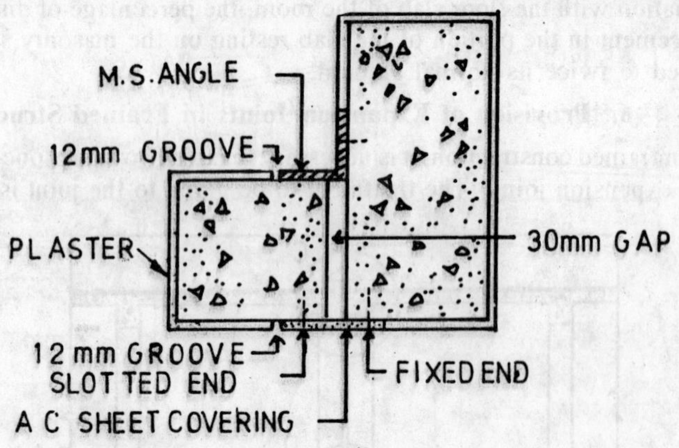

Fig. 24·18.

Details of expansion joint between two columns

(For plan refer Fig. 24·17)

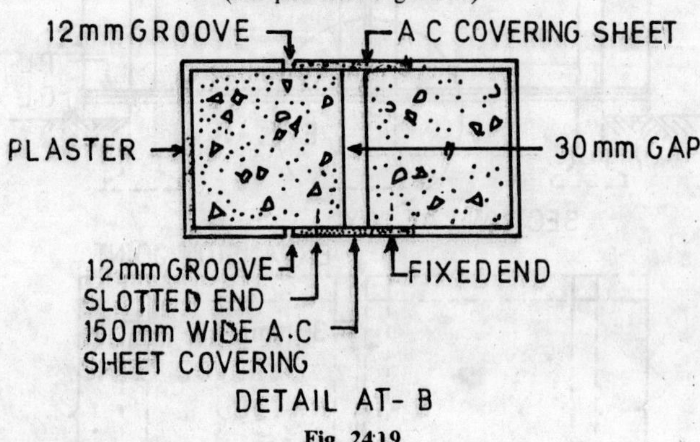

Fig. 24·19

Details of expansion joint between two columns (For plan refer Fig. 24·17)

applicable to joint in masonry construction. For details refer Fig. 24·17 to Fig. 24·22.

24·4. ISOLATION JOINTS

When one portion of a building is higher than the other, the soil below the higher portion will be subjected to greater pressure than the soil below the lower portion of the building. In such cases it is desirable to have complete separation of the two parts by providing an isolation joint. *The detailing of an isolation joint is similar to that of an expansion joint.* This arrangement permits vertical movement of the building parts due to difference in the unit soil pressure without causing any damage to the structure.

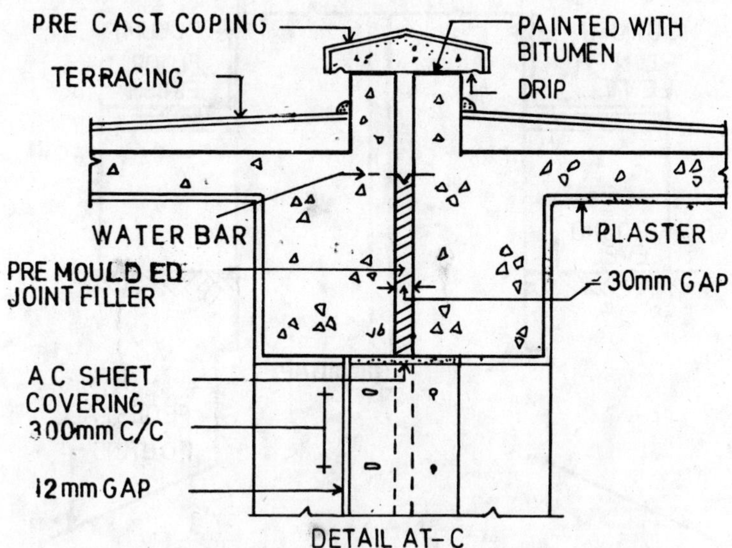

Fig. 24·20

Details of expansion joint at roof level (For section refer Fig. 24·17)

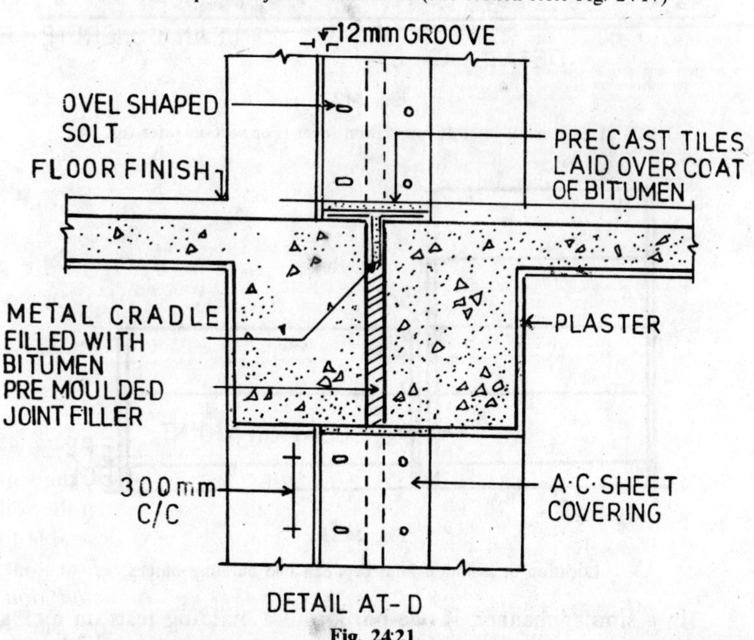

Fig. 24·21

Details of expansion joint at first floor level (For section refer Fig. 24·17)

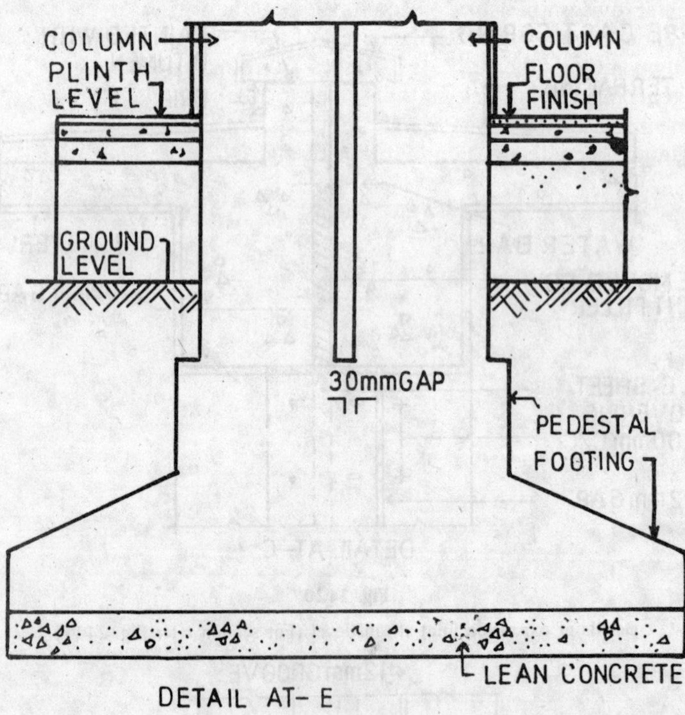

Fig. 24·22

Details of expansion joint at foundation level (For section refer fig. 24·17)

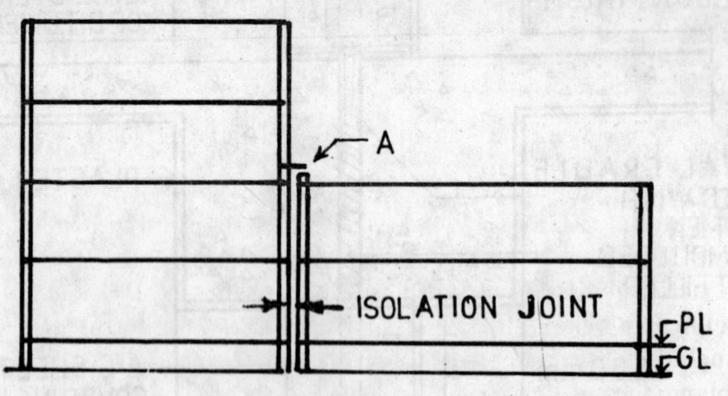

Fig. 24·23

Location of isolation joint between two building blocks

In a similar manner, if one portion of a building rests on rock and the adjacent part on compressible clayey soil, the two parts should be separated by providing isolation joint.

Provision of isolation joint also becomes necessary to separate out the foundation of machines from rest of the structure to avoid transference of dynamic or other types of stresses from the machine to the structure. In general, an isolation joint is provided to take care of horizontal or vertical movements of the adjacent building parts without causing damage to the building. For details refer Fig. 24·23 and 24·24.

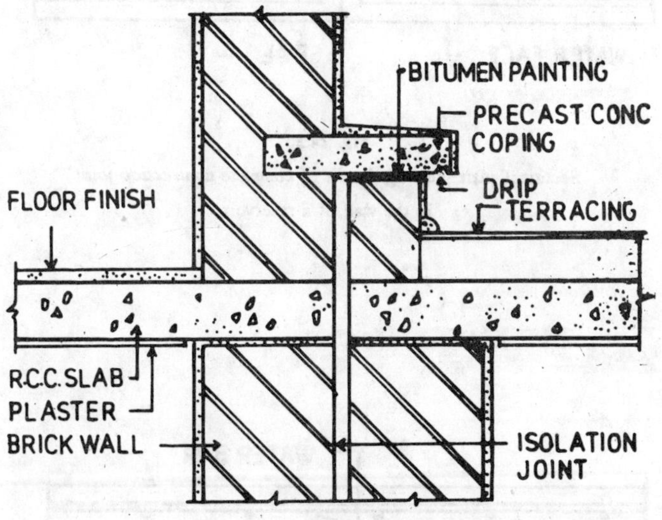

Fig. 24·24

Details of isolation joint (For elevation refer Fig. 24·23)

24·5. CONTRACTION JOINTS

We know that when concrete sets and hardens in air it shrinks in volume. The magnitude of shrinkage or contraction is almost directly proportional to the quantity of water in the mix. Hence concrete mix having more water shrinks more than a concrete mix having low slump. The contraction caused due to drying shrinkage results in development of tensile stresses in concrete. When such tensile stresses exceed the tensile strength of concrete, it results in formation of random cracks. Contraction joints are installed to allow for the shrinkage movement in the structure. In general, building continues to dry out during its life time and as such dimensional changes due to shrinkage should be taken care of in its planning and detailing.

This joint may be either a *complete contraction joint* in which there is complete discontinuity of both concrete and steel (refer Fig. 24·25) or it may be *partial contraction joint* in which there is discontinuity of concrete but the reinforcements continue across the joint. (Refer Fig. 24·26). It may be noted that in both the cases no gap is provided for the

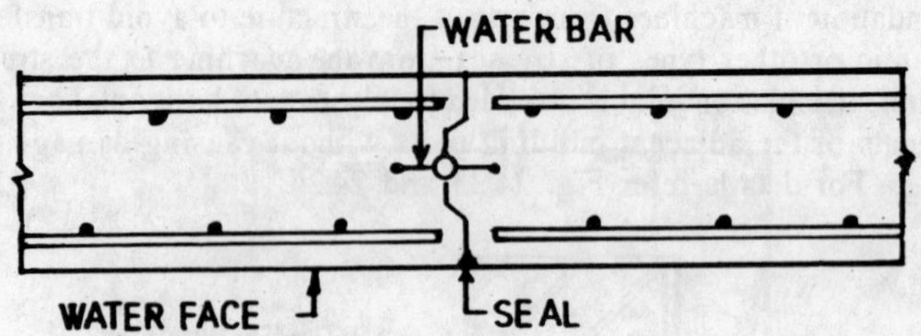

Fig. 24·25

Sectional plan showing *details of complete contraction joint* in the wall of a reservoir

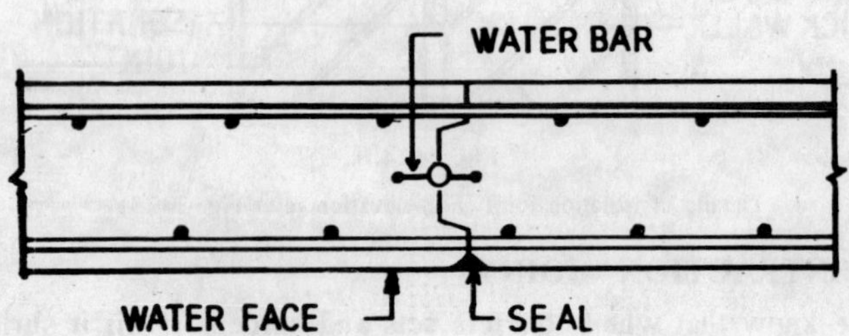

Fig. 24·26

Sectional plan showing *details of partial contraction joint* in the wall of a reservoir

joint but only complete or partial separation of the adjacent sections is created. The third form of contraction joint is *dummy joint.* In this form of joint a groove is created in the concrete member to act as a joint, the depth of the groove being 1/3 to 1/5th of the total thickness of the member.

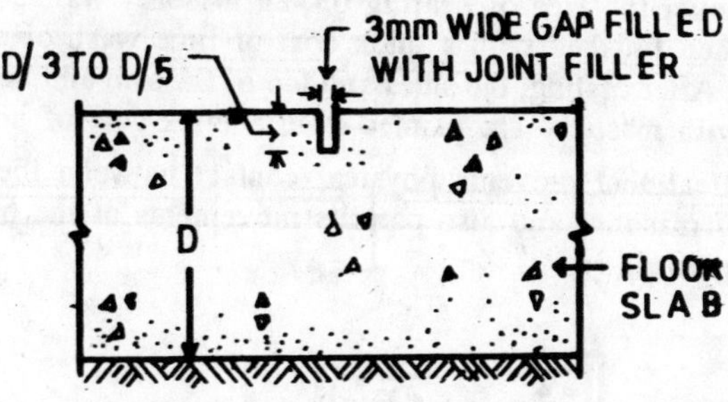

Fig. 24·27

Dummy joint (contraction joint) in a factory floor slab

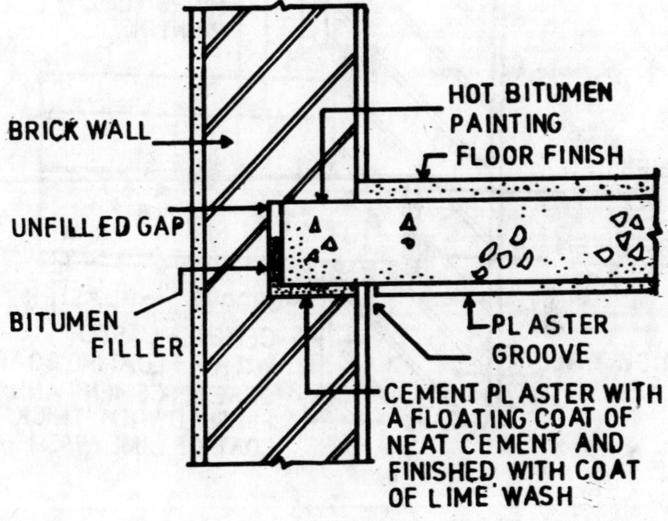

Fig. 24·28

Details of sliding joint at first floor level

24·6. SLIDING JOINTS

This joint is provided between two parts of a structure when one part has a tendency to slide over the other due to variations in temperature and moisture content. A typical example of this joint is the treatment given to the R.C.C. slabs and beams on the masonry walls. In order to facilitate free movement of the components at the joint, prior to the casting of slab, a 6 mm thick layer of rich cement plaster (1:3) finished with neat cement is provided over the area of bearing on the masonry walls. The bearing plaster is then finished with a thick coat of lime wash or covered with kraft paper. After casting, the sides and top of the slab and beams to come in contact with masonry are painted with a thick coat of hot bitumen.

This treatment prevents physical contact between the surfaces of concrete and masonry and also permits movements of the building parts by sliding action.

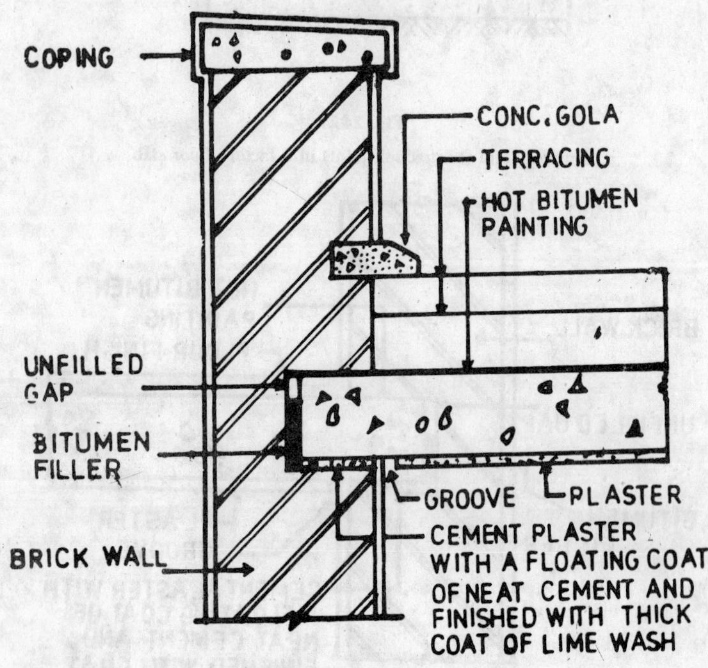

Fig. 24·29
Details of sliding joint at roof level

24·7. CONSTRUCTION JOINTS

Joint provided at locations where the construction is stopped either after the days work or due to any other reason is known as construction joint. Installation of construction joint become necessary to ensure proper bond between the old work and the new one. If it is possible to plan construction in such a way that the day's work end at an expansion or constraction joint, the need for constraction joint will not arise. However, often it is difficult to complete the concreting for large work in one operation and as such special measures are necessary to be taken to achieve prefect continuity between the old concrete that has already set and hardened and the new concrete to be laid subsequently.

From consideration of structural slability, it is necessary to ensure that construction joints are provided at pre-determined locations. Following factors should be kept in view while deciding the installation of construction joint.

(1) The construction joint should be located along or near the plane of minimum shear.

(2) The joint should be vertical or horizontal as the case may be except that, in an inclined or curved member, the joint should be at right angle to the axis of the member.

(3) The concrete at the vertical joint should be finished against a stop-board especially fixed for the purpose. As far as possible a key of tongue and groove type should be provided at all construction joints. The key is normally formed by fixing bevelled edged strip of wood to a stop-board. In case of large works, key made up of sheet metal is used to permit repetitive use. There is a tendency to provide more depth for the key. It is however, seen that key formed by use of a 25 x 50 mm. bevelled edged strip is sufficient for slabs ranging from 12 cm to 18 cm. in thickness.

The stop-board has to be slotted to allow reinforcemet to extend from the old concrete to the new one. The details of formation of construction joint in a slab are shown in Fig. (24·30).

Prior to laying new concrete the stop-board is removed and the face of the concrete that has already hardend is treated and prepared as under.

(*a*) If the new concrete is to be laid within 48 hours, the face of the old concrete is cleaned with wire brush and water and a coat of cement mortar (of the same composition as that of concrete) is applied before placing new concrete.

(*b*) If the new concrete, is to be laid after 48 hours, the face of the old concrete is first roughened by chiselling. It is then cleaned of scum, loose aggregates and other foreign matter, wetted and a coat of cement mortar (of the same composition as that of the concrete) is applied before placing new concrete.

Location of Construction Joints for Different Members

The recommended position for construction joints in different structural components is as under.

(*i*) *For slabs* : In case of slabs supported on two sides, the construction joints should be vertical and parallel to the main reinforcement. Alternatively, it can also be provided at the middle of span at right angles to main reinforcement.

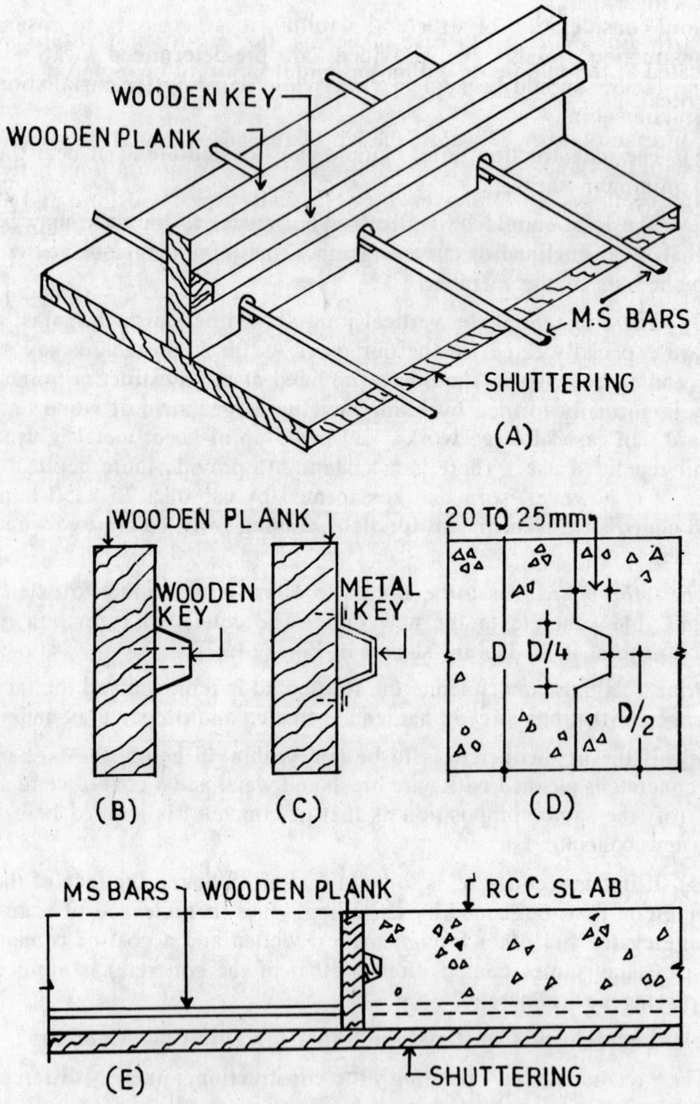

Fig. 24·30

Details relating to formation of construction joint

A :— View showing stop board with wooden key fixed to slab shuttering

B, C :— Alternative detail of keys

D :— Section showing dimensions of key

E :— Section showing stop-board in position for forming construction joint

In case of two way slabs, the joint can be provided near the middle of the either span.

(*ii*) *For Beams* : In case of beams, the construction joint should be located at the middle or within the middle third of span and it should be vertical.

In case of L or T beams, the rib of the beam should be concreted along with the flange slab. In case, however, a construction joint between slab and beam becomes unavoidable (especially in case of long and deep beams) the rib of the beam can be concreted upto 25 mm below the level of soffit of the slab and the construction joint located there.

(*iii*) *For columns* : In case of columns, the construction joint should be formed horizontally by stopping the concrete in column about 75 mm below its junction with lowest soffit of the beam or bottom of haunching (if any). The operation of concreting above the construction joint should be taken up after a gap of at least 4 hours.

(*iv*) *For walls* : The horizontal construction joint in walls should be located at the top of plinth, top or bottom of window opening or at any other convenient height.

In case of concrete or R.C.C. walls, the location of horizontal construction joints is governed by the following factors.

(*i*) Convenience in setting the formwork.

(*ii*) Ease of access for compaction and supervision of concreting.

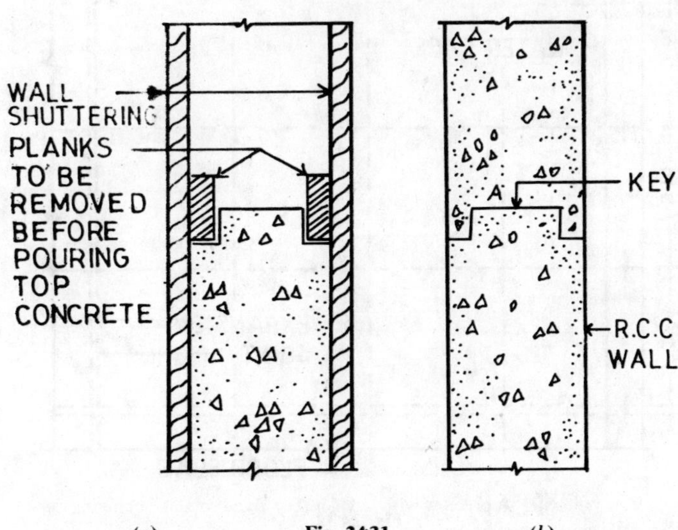

(*a*) **Fig. 24·31** (*b*)

(*a*) Details of planking for construction joint in wall

(*b*) Shape of joint after casting

The continuity at the joint is achieved by the formation of key. The key is formed by use of bevelled edged planks on the inner face of wall shuttering (Refer. Fig. 24·31). Prior to pouring concrete above the joint, the bevelled edged planks are removed and the surface of old concrete is prepared in manner as decribed earlier in the article.

24·8. PROVISION OF JOINTS IN WATER OR LIQUID RETAINING STRUCTURES

The phenomena of movements due to temperature, shrinkage etc. deserves extra special attention in the case of water, or liquid retaining structures. Such movements, if restained, can produce stresses sufficient to cause cracking of the structure even before it is filled with water or liquid. This is on account of the fact that concrete shrinks maximum when it is allowed to dry out completely. Keeping this aspect in view it is recommended that water reservoir should be filled at slow rate and at an early stage. To ensure water tightness, mix of concrete used in the construction should not be weaker than M 20. In addition it is necessary to make adequate provision of contraction and expansion joints to

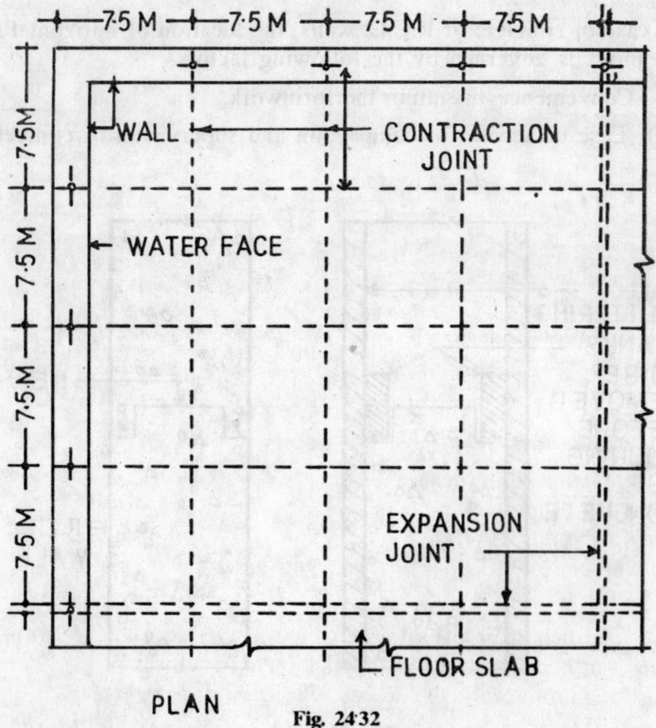

Fig. 24·32

Plan showing location of expansion and contraction joints in a reservoir

control shrinkage and thermal movements. The provision of joints in the different components of reservoirs should be as under.

(*a*) *Joints in wall* :

(*i*) It is necessary to provide vertical contraction joints at a maximum centre spacing of 7·5 m. The joint may be in the form of partial contraction joint as shown in Fig. 24·33 or in Fig. 24·26. This arrangement

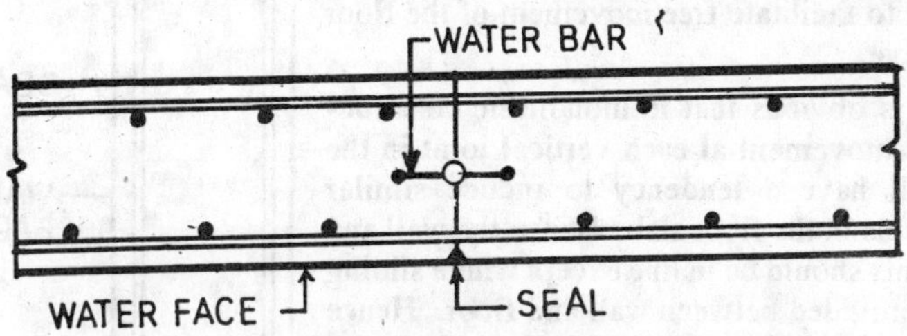

Fig. 24·33

Sectional plan showing details of contraction joint in the wall of reservoir

permits accumulation of shrinkage movements at pre-determined location (*i.e.* at contraction joints) and thus avoids random shrinkage cracks. It is therefore necessary to provide water stops at the joints to take care of small movements and to ensure watertightness of the joint.

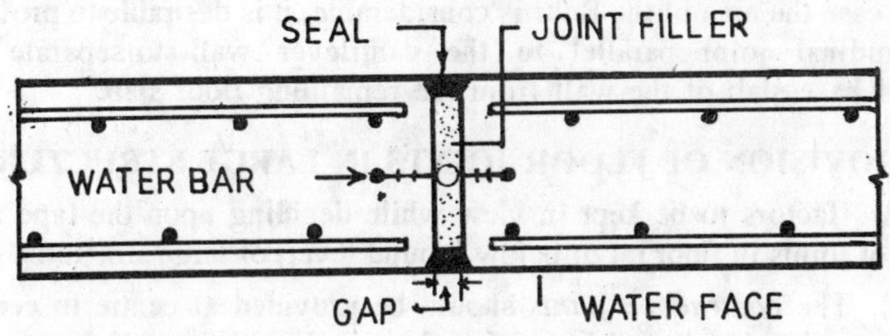

Fig. 24·34

Sectional plan showing details of vertical expansion joint in the wall of reservoir

(*ii*) Provision of vertical *expansion joints* should be made in walls at a spacing of not more than 30 m. Since expansion joints also function as contraction joint, there should be 3 contraction joints (at c/c spacing 7·5 m) between any two expansion joints to meet the requirements of (*i*) above.

(*iii*) The height of lift while concreting the wall should normally not exceed 2 m and hence a horizontal joint is the form of *construction joint* has to be introduced between any two lifts. This is explained in art. 24·7 (Fig. 24·31).

(*b*) *Joints in floors* : To permit free movement, the floor slab of reservoir resting at or below ground level is cast over a bottom layer of concrete normally of M 10 mix. The thickness of this layer of lean concrete should not be less than 75 mm. A layer of building paper or other suitable material is provided over the lean concrete to facilitate free movement of the floor slab above.

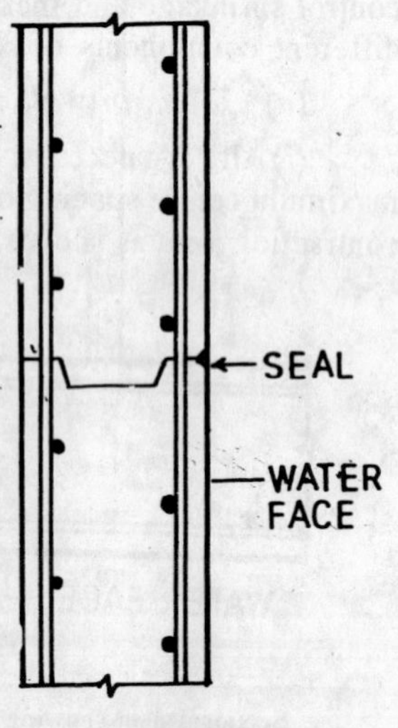

Fig. 24·35
Horizontal construction joint in the wall of reservoir

It is obvious that in monolithic construction, the movement at each vertical joint in the wall will have a tendency to induce similar movements in the floor slab. Hence the wall and floor joints should be in line except where sliding joint is provided between wall and floor. Hence the centre to centre spacing of *contraction* and *expansion* joints in floor should be 7.5 m and 30m respectively. The joints should be provided in two direction at right angles to each other. The contraction joint in floor should normally be of complete contraction type (reinforcement discontinued) and as far as possible, the floor should be cast alternatively in chess board fashion.

In case the area of the floor is considerable, it is desirable to provide a longitudinal joint parallel to the cantilever wall to separate the structural base slab of the wall from the remaining floor slab.

24·9. PROVISION OF FLOOR JOINTS IN LARGE STRUCTURES

The factors to be kept in view while deciding upon the type and location of joints in floor (at or below ground level) of large structures are:

(1) The *contraction joints* should be provided at centre to centre spacing varying between 4·5 to 7·5 m in two directions at right angles. Thus the maximum size of floor panel should normally be not more than 7·5 x 7·5 m.

(2) In case the floor slab is reinforced, it is desirable to discontinue the reinforcement at the contraction joint.

(3) The floor panels should be cast alternatively in chess board fashion. The adjacent slab panels should preferably be laid after a gap of seven days. The joints should thereafter be sawn to a depth of about 20 mm and filled with molten lead or joint sealing compound.

(4) *Isolation joints* Should be provided at junction of the floor with columns, walls etc. This takes care of the horizontal movements due to contraction as well as vertical movements due to difference in unit soil pressure under floor and columns, walls etc.

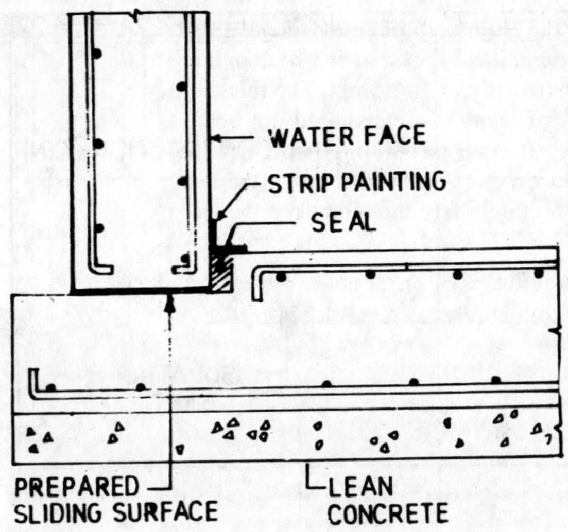

Fig. 24·36

Details of sliding joint between the wall and floor of reservoir

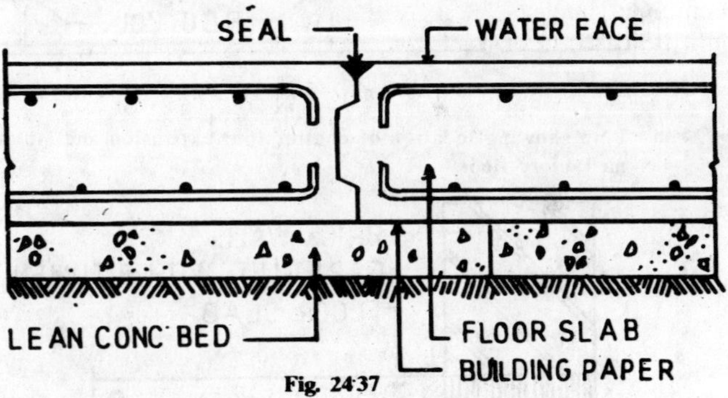

Fig. 24·37

Details of construction joint in the floor of reservoir

(5) To accomodate thermal movements, *expansion joints* should be provided at intervals of about 30 m. Typical arrangement of floors joints is detailed in Figs. 24·38 to 24·41.

In case of floor slab of factories where there is possibility of wheeled trolleys moving across the joints, the edges of concrete face should be protected by steel angles as shown in Fig. 24·40.

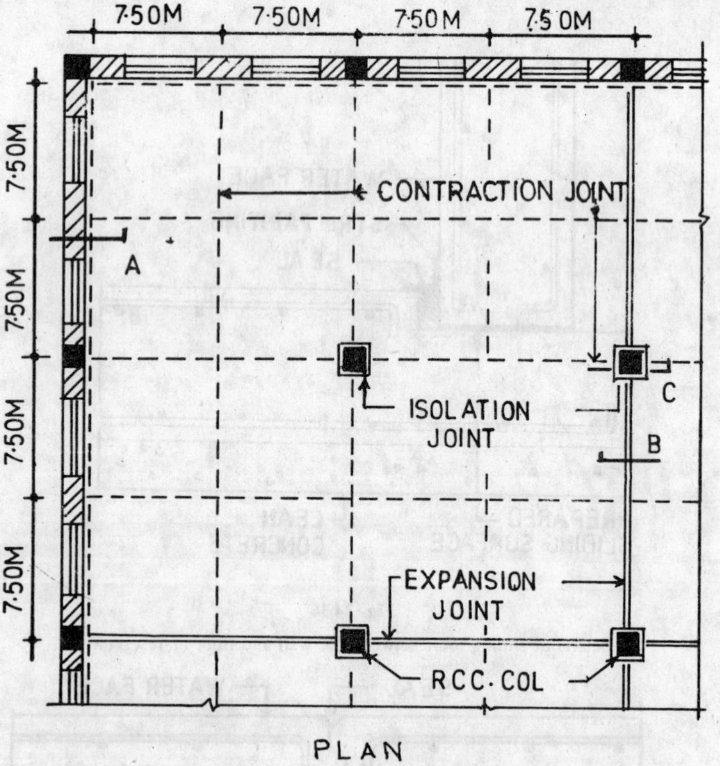

7·50M 7·50M 7·50M 7·50M
7·50M 7·50M 7·50M 7·50M
CONTRACTION JOINT
A
C
ISOLATION JOINT
B
EXPANSION JOINT
RCC·COL
PLAN

Fig. 24·38 Plan showing location of contraction, expansion and isolation joints in na factory floor

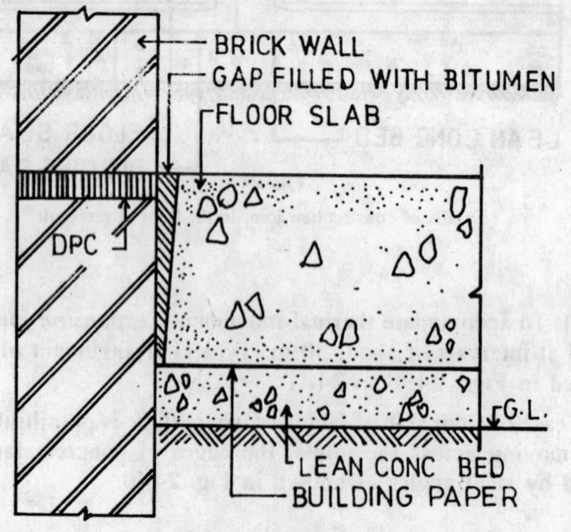

Fig. 24·39 Details of isolation joint between floor slab and wall

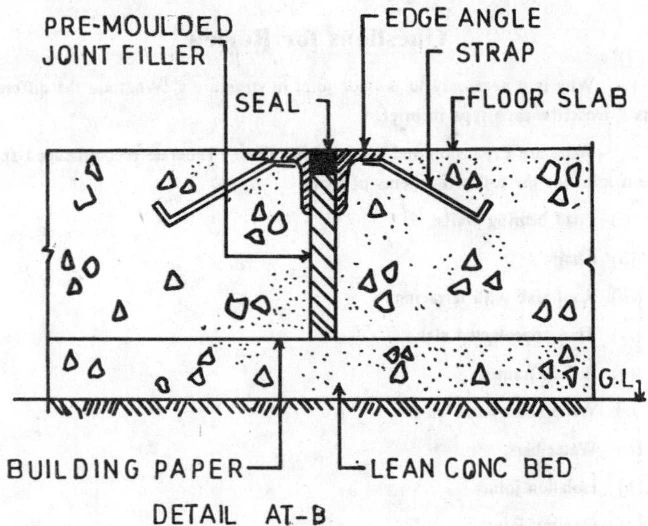

Fig. 24·40

Details of expansion joint in the floor slab (for plan refer Fig. 24·38)

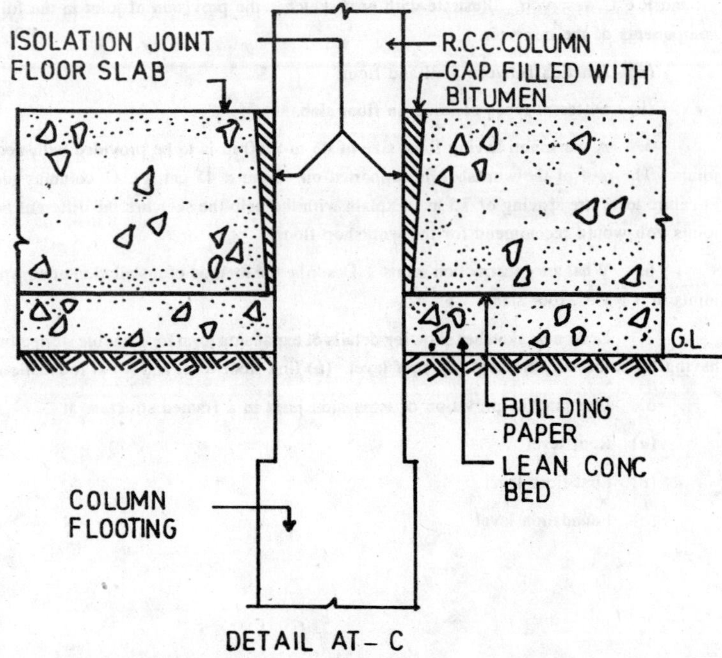

Fig. 24·41

Details of isolation joint between R.C.C. column and the floor slab (for plan refer Fig. 24·38)

Questions for Review

1. Why is it necessary to provide joint in structure ? What are the different types of joints ? Describe each type in brief.

2. What are expansion and contraction joints. Tabulate recommended spacing of expansion joint for the following items of work.

(*i*) Load bearing walls.

(*ii*) Chajja's.

(*iii*) Roof slab with terracing.

(*iv*) Thin unprotected slab.

(*v*) R.C.C. frames.

3. Write short notes on :

(*a*) Water bars.

(*b*) Isolation joints.

(*c*) Dummy joints.

(*d*) Sliding joints.

4. Describe the different types of joints which are normally provided in large under ground R.C.C. reservoir. Illustrate with neat sketches the provision of joint in the following components of the reservoir.

(*a*) Junction between wall and floor.

(*b*) Different types of joints in floor slab.

5. A workshop having floor size of 45 m x 60 m is to be provided with necessary joints. The roof of the workshop is supported on 45 cm x 45 cm R.C.C. columns provided at centre to centre spacing of 7·5 m. Explain with the help the sketches the different types of joints you would recommend for the workshop floor.

6. What are construction joints ? Describe the method of installation of construction joints in a R.C.C. roof slab.

7. Draw neat sketches showing details of expansion joint for a double storied building having brick masonry walls at (*a*) roof level. (*b*) first floor level and (*c*) at foundation level.

8. Illustrate the provision of expansion joint in a framed structure at

(*a*) Roof level

(*b*) First floor level

(*c*) Foundation level

25

Fire Protection

25·1. GENERAL

There is now a fast growing awareness among the modern architects and engineers to plan, design and construct the buildings in such a manner that the safety of the occupants is assured to the maximum possible extent in the event of outbreak of fire in the building. Once the fire starts, it tends to ignite all the combustible materials in the surrounding areas and in case it is not checked it may spread to other parts of the building. If the fire is not controlled within the reasonable period, it may lead to the collapse of the entire structure. It is not possible to attain absolute safety from fire but by adopting certain measures it is definitely possible to provide reasonable degree of safety from fire. From the study of the fire accidents in the past it is analysed that fire causes damage to the building and its contents, whereas the byproduct of the fire *i.e.* the smoke and the hot toxic gases cause maximum damage to the human life. Smoke reduces visibility and the hot toxic gases produce suffocating effect and the combined effect of the two is to bring about mass panic which in turn causes maximum danger to life.

Thus, besides making a building with fire resistant materials, it is equally important to plan it in such a way that the risk of panic due to smoke and hot gases in minimised. The nature of damages or dangers caused by fire can be broadly divided into the following types.

(*i*) *Danger to life of occupants.*

(*ii*) *Danger of spread of fire to other areas.*

(*iii*) *Danger of damage to property (Building and its contents)*

1. Danger to Life of Occupants

This danger can be taken care of by providing adequate means of escape for the occupants so that they can reach the place of safety in the shortest possible time in the event of a fire. Thus the building should be so planned that it has sufficient provisions for safe exit from within the building by way of definite escapeways, passages, corridors and stairs etc.

leading to a street or to an open space. Lifts and escalators are not considered as exits. This is on account of the fact that from the exprience in the past it is seen that even well designed lifts failed to operate successfully in the event of fire. This issue needs special attention in case of multistoreyed buildings. In case there is a fire in a particular storey, the exit provisions in the building should be such that the occupants of the floor immediately above and below that floor should be able to have safe exit in reasonable short time. Thus the size of the corridor, width of staircase and the location of the staircase halls should be suitably decided keeping the above requirement in view. National Building Code has covered the topic of fire protection in detail. Some of the important requirements relating to this issue as given in the code are highlighted below :

(*i*) Every tall building should have a minimum of two stair-cases.

(*ii*) The travel distance (*i.e.* the distance an occupant has to travel to reach an exit) for different types of buildings should be as under

(*a*) Residential, educational, institutional and hazardous ...22·5 m

(*b*) Assembly, mercantile, industrial, and storage ...30 m

(*c*) Business ...45 m

(*iii*) No revolving door should be treated as exits except in residential and mercantile occupancies.

(*iv*) Lifts and escalators should not be considered as exits.

(*v*) A separate fire lift should be provided for use of fire brigade in the event of fire.

2. Danger of Spread of Fire to Other Areas

This danger can be taken care of by palnning and designing the building in such a manner that the fire can be contained in a limited area for sufficient time to enable the occupants in the danger zone to escape to place of safety. To meet this requirement, the building should be designed in the form of series of compartments enclosed by floors and walls. Each compartment should be so created that it is capable of restraining fire within the compartment for appropriate duration of time. Besides this, each stairwell, elevator and service shaft (in case of multi-storeyed building) should also be suitably treated so as to enable it to function in the form of vertical compartment or cell extending from top to bottom of the building.

3. Danger of Damage to Property (Structure and Its Contents)

This danger can be minimised by use of fire resisting materials and by carrying out the construction of the structural elements like columns, beams, floors, load bearing walls, lintels, arches etc. in such a way that at the time of fire they should continue to function as structural elements till such time as would enable the occupants to escape to safe places. The degree of fire resistance necessary to prevent penetration by fire of any structural element is related to the fire load *i.e.* amount of combustible materials (like furniture, upholstery, carpets, curtains, clothes, paper etc.) present in the building per sq. m. of its floor area. Keeping in view the high cost of construction of a fire resistant building, it is not desirable to make all buildings fully protected against fire. The degree of fire resistance to be provided depends to a great extent upon the use of the building as well as the extent of the number of people it accommodates. In general, buildings like schools, hospitals, cinema, theatres, multi-storeyed office buildings etc. should be designed to offer a greater degree of fire resistance as compared to a warehouse or a godown.

National Building code has classified construction into four classes namely type 1, type 2, type 3 and type 4 on the basis of fire resistance offered by building components for 4 hours, 3 hours, 2 hours, and 1 hour respectively. Thus, in construction of type 1 all the structural components of the building are constructed in such a manner and are of such materials that, in case of out break of fire, they can withstand fire as integral members of the structure for a duration of 4 hours.

Hence proper selection of material and the method of construction play an important role in minimising danger from fire to the property.

25·2. FIRE-RESISTING PROPERTIES OF MATERIALS

The fire resisting properties of the different materials commonly used for construction purposes are described below.

1. **Timber.** Timber has unique property of self-insulation and slow burning and offers considerable resistance to fire. When subjected to fire, timber first gets charred to certain depth and there after, this charred layer serves as insulation to check the spread of fire to the inner portion. Additional fire resistance is achieved through impregnation of timber with large quantities of fire-retarding chemicals. The commonly used chemicals for this purpose are ammonium phosphate and sulphate, borax and boric acid, zinc chloride etc. This treatment retards increase in temperature during fire ; decreases rate of flame spread and enables easy fire control. Usually 32 to 48 kg of chemical per cubic metre of wood are used for moderate protection and 80 to 96 kg. per cubic metre for high protection. During the recent years, a number of fire resistant paints have also been introduced for rendering the timber fire-resistant. Paints of asbestos, magnesium sulphate, ferrous-oxide etc. have been found to be very much effective.

2. **Brick.** First-class bricks moulded from good clay can stand exposure to fire for a considerable length of time. The properties of bricks which render them fire resistant are the size of bricks, the method of construction and the component of fire-resistive material in bricks *i.e.* clay etc. It has been well established that brick masonry construction is most suitable for safeguarding the structure against fire hazards.

3. **Stone.** The usage of stone in a fire resistive construction should be strictly restricted to the minimum. Hot stone when subjected to sudden cooling develops cracks and can lead to failure of structure. Granite, when exposed to severe fire, explodes and disintegrates. Lime stone is least recommended as it crumbles and gets ruined by an ordinary fire. Only compact sand-stone having fine grains may be used as it can stand the exposure to moderate fire without serious cracks.

4. **Steel.** Steel although incombustible has a very low fire resistance value. With the increase in temperature, the co-efficient of elasticity of the metal falls appreciably rendering the structural members soft and free to expand. When the members in this state come in contact with water used for extinguishing the fire, they tend to contract, twist or distort and thus the stability of the entire structure is endangered. It has been noticed that unprotected steel when subjected to fire, causes the collapse of the structure. Hence in a fire resistant construction, structural steel members must be suitably protected by covering them with materials like brick, terra-cotta, concrete etc.

5. **Concrete.** The influence of fire on concrete varies with the nature of its coarse aggregate and its density. It has been found that aggregates obtained from igneous rocks containing higher calcerous content, tend to crack when subjected to fire. Coarse aggregates like foamed slag, cinder and brick are best suited for a concrete which has to be fire resistive. It has been noticed that in an average fire, the concrete surface gets disintegrated for a depth of about 25 mm. This is due to the dehydration of mortar in concrete by the fire. Hence in a reinforced concrete fire resistant construction, the thickness of clear cover should be more.

6. **Glass.** Glass conducts heat faster than metal. Because of its low thermal conductivity, the change in volume on account of expansion or contraction is very small and as such it may be considered to be a good fire resisting material. However when subjected to sudden and extreme variation of temperatures, it fractures or cracks. By the use of glass reinforced with steel wire netting, such cracks can be minimised. Even when the cracks are formed, the fractured glass remains in its original position, as the embedded wire holds the cracked portion and does not allow it to fall. The reinforced glass has a higher melting point than the ordinary glass, and as such it is commonly used for fire resisting doors, windows, sky-lights etc.

7. **Asbestos.** Asbestos is a fibrous mineral which is combined with portland cement to form a material having great fire resistive value. Asbestos cement products are largely used for the construction of fire-resistive partitions, roof etc. On account of its low co-efficient of expansion and property of incombustibility, the structural members blended with asbestos cement possess great resistance to cracking, swelling or disintegration when exposed to fire.

8. **Plaster or Mortar.** Plaster or mortar is incombustible and as such, by suitable choice of the type of mortar, the walls or ceiling of a building can be made more fire resistant. Cement mortar is preferred to lime mortar as the latter is liable to calcine. The resistance of the plaster to fire hazards can be increased by using it in thicker layers or reinforcing the plaster with metal laths. Gypsum plaster when applied over structural members like steel columns etc, makes the latter have good fire-resistive qualities.

25·3. FIRE RESISTANT CONSTRUCTION

In a fire resistant construction, the maximum use of non-combustible materials should be encouraged. All the structural elements such as beams, columns, lintels, arches, floors and roofs, load-bearing walls or partition walls etc. should be constructed in such a way that they should continue to function as structural members at least for the period which may be sufficient for the occupants to escape. The following additional points should be kept in view while designing a fire-resistant structure :

(1) The load-bearing walls or columns of masonry should be thicker in section so that they may successfully act as fire barrier for a considerable time.

(2) As far as possible fire resisting material should be used in the construction of flooring. If the usage of material which is likely to be damaged cannot be avoided either due to financial or other considerations, the following precautions should be observed :

(*i*) In case of wooden floors, thicker joists spaced at greater distance apart should be used.

(*ii*) Fire stops should be provided in wooden floors at suitable intervals.

(*iii*) Flooring made from materials like concrete, brick, ceramic tiles etc., is considered to be most suitable for fire-resistant construction. In case, cast iron, wrought iron or combustible materials like rubber, cork,carpet etc. have to be used in flooring, such materials should be protected by covering of insulating material like ceramic tiles, terracotta, bricks etc.

(3) Reinforced concrete framed structures should be preferred to steel structures. As steel is liable to twist, sag or distort under heavy fire, the metal should be protected by using concrete, hollow clay tiles, bricks, metal lath and plaster, etc. The treatment given to steel columns is shown in Fig. 25·1. The cover of the protected material like tiles or bricks etc. all round the steel members should be at least 10 cm. from all sides. In any case, the metal flanges should be protected with not less than 5 cm. layer of concrete. The cover of concrete for reinforced concrete members like beams, or columns should be sufficient to enable the members function

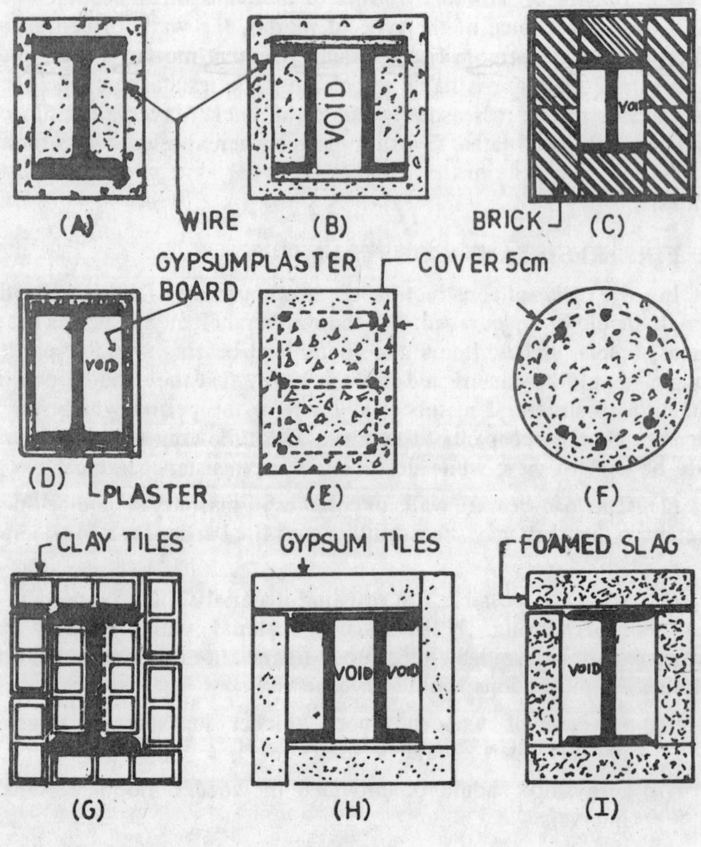

Fig. 25·1

Various types of fire-proofing treatment given to steel and concrete columns

satisfactorily, under fire for a maximum time. The concrete cover outside the main reinforcement should be atleast 5 cm. for very important structural members, like columns, griders, trusses etc. 38 mm for ordinary beam, or long span slabs, arches etc. and 25 mm. for partition walls and short span slabs.

(4) The partition walls should similarly be of fire-resistant material. In general, R.C.C., R.B.C, metal lath and plaster, asbestos cement board, hollow concrete or clay tiles should be used in the construction of partition walls. In case, wooden partition has to be made fire resistant, it should be covered with metal lath and plaster.

(5) The load-bearing walls, as well as non-load bearing walls, should be plastered with fire-resistive mortar.

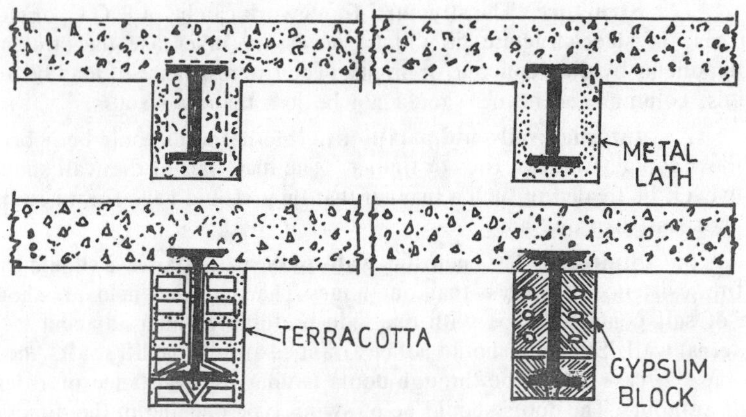

Fig. 25·2

Various types of fire proofing treatment given to beams

(6) The ceiling should be directly attached or suspended from the floor joists and it should be made fire-resistant by fixing asbestos cement boards, fibre boards, metal lath and plaster etc. to its framework.

(7) The doors, windows or exposed sides should be glazed and fitted with reinforced glass panels.

(8) As far as possible, use metal frames for doors with steel shutters.

(9) Due consideration should be made for providing suitable means of escape for the occupants of the structure. In single storey building, the provision of an accessible window, large enough to permit the occupants to get out through the opening should be ensured. In case of multi-storeyed structures, the location of staircase should be decided carefully. For stairs of important buildings, a higher standard of the fire resistance is required. R.C.C. staires are considered best from all considerations. The staircase hall which acts like a vertical shaft connecting the various floors, should be surrounded with the enclosure walls made up of fire-resistive materials. The lift halls should also be similarly treated. All the doors or windows of the staircase hall or lift well should préferably be of metal frame with steel shutters.

(10) As far as possible, flat roof should be encouraged. In case, the sloping roof is considered necessary, either due to site conditons or due to architectural reasons, the ceilings of the sloped roof should be made up of fire-resistive material.

25·4. FIRE PROTECTION REQUIREMENTS FOR MULTI-STOREYED BUILDINGS

Main fire protection requirements for multistoreyed building can be summarised as under.

1. **Structure**. The structural framework can be in R.C.C. or steel. In case of steel structure, it will be necessary to encase the structural components by tiles, concrete or bricks. The fire resistance of load bearing walls, columns and beams should not be less than two hours.

2. **Internal walls and partitions.** Internal walls could be in brick, hollow concrete blocks, tiles or timber. The materials of the wall should, however, be treated in such a manner that they should have fire resistance of not less than one hour.

3. **Staircase**. The enclosing walls of the main staircase should have a fire resistance of not less than one hour. The staircase enclosure should be of self contained type with one side preferably being adjacent to an external wall. Staircase should not be arranged around the lift shaft. Access to the staircase should be through doors having fire resistance of at least half-an-hour. The doors should be of swing type opening in the direction of the escape. The door of the stair case enclosure at ground floor should open out into an open space. Adequate number of staircases should be provided to meet the travel distance requirements of the type of building.

4. **Fire escape or external stairs.** In addition to the main staircase it is necessary to make suitable provision of fire escape in the form of external stairs. All fire escapes should be directly connected to ground floor and their entrance should be away from the internal staircase of the building.

5. **Basements**. The basement should be property ventilated. The main staircase of the building should preferably terminate at ground floor and as far as possible the access to basement should be through a separate staircase. The staircase of basement should be of enclosed type and should have fire resistance of not less than two hours. In case the basement is to be used for car parking, installation of electric sub-station, storage of combustible material, etc it is necessary to install sprinkler system for extinguishing fire.

6. **Water-storage tank.** There should be a provision of an underground water storage tank of one to two lakhs litres capacity exclusively meant for fire fighting purposes. The tank should have arrangement of replenishment of the water either by the municipal mains or by tube well.

7. **Fire detection and Extinguishing System.** All multistoreyed buildings should have adequate provisions for fire detection and fire extinguishing.

25·5. FIRE DETECTION AND EXTINGUISHING SYSTEMS

In important structures particularly in multistoreyed buildings, besides fire-resisting materials and method of construction, it is obligatory to make provision of fire detection and/or fire extinguishing systems. These systems are described below :

1. Fire Detection System

Also known as fire alarm system, it is an installation provided with the aim of warning the occupants of the building of the out break of fire and to call for assistance at the same time. Fire alarm system can be either manual or automatic.

(*i*) *Manual fire alarm system.* This consists of a hand bell or similar sounding device emitting distinctive sound when struck. Such devices are installed near all the main exits and passages. In the event of fire, the device is sounded by watch man and the occupants are thereby warned to have safe exit in shortest possible time.

(*ii*) *Automatic alarm system.* This type of system, on detection of fire, starts sounding alarms or information to the nearest control point. Besides this, the system can also perform the function of sounding of evacuation gongs, hooters or siren and to transmit information of fire to a remote fire station.

2. Fire Extinguishing System

The commonly adopted fire extinguishing systems are decribed below :

(*i*) *Manual fire-extinguishing equipment.* Portable fire extinguisher is the equipment most commonly adopted in this category. Portable fire extinguisher can be of carbon dioxide type, large foam generation type and so on. Depending upon the capacity, the discharge from a fire extinguisher may last from 20 to 120 seconds. Sometimes buckets full of water and dry sand are also installed at convenient places for taking care of minor fires. These equipment are useful in extinguishing fire as soon as it starts.

(*ii*) *Fire hydrants.* This consists in providing a 150 mm diameter ring main outside in the ground around the periphery of the building. Fire hydrants are provided on the ring main. The ring main is fed from an underground water tank and it is ensured that the water pressure available at each fire hydrant is of the order of 3·5 kg/cm^2.

(*iii*) *Wet riser system.* This consists in providing 100 to 150 mm dia vertical G.I. pipes at suitable locations within the building. The pipes are also known as risers and are fed from an underground water storage tank through a fire pump which ensures supply of water at 3 kg/cm^2 at the topmost outlet. The wet risers have suitable connections at each floors from where constant supply of water can be drawn in the event of fire.

(*iv*) *Sprinkler system.* In this system a network of water supply pipes(normally 20 mm. dia.) are fixed to the ceiling of the floor or roof slab. The centre to centre spacing of the pipes is normally 3 m. The pipes receive supply of water through header pipes normally 40 mm in dia. which in turn are connected and fed from water storage tank. An installation known as sprinkler head is fitted to the pipes in ceiling at regular intervals normally at 3 m c/c. Each sprinkler head is provided with as fusible plug. In the event of fire the fusible plug in the sprinkler nearest to fire, melts by virtue of rise in temperature and water gushes out of the sprinkler head. The water spray falls on the fire and helps in extinguishing the same.

Questions for Review

1. What do you understand by the term 'fire-proof construction' ? Describe the fire-proofing properties of the following materials ?

(*i*) Timber

(*ii*) Stone

(*iii*) Steel

(*iv*) Brick

(*v*) Concrete

(*vi*) Glass

(*vii*) Asbestos

(*viii*) Plaster or mortar.

2. What are the guiding factors that should be kept in view while diciding the structural and architectural requirements of fire-resistive structure ? Describe the fire-resistive properties of the following :

(*i*) Asbestos.

(*ii*) Concrete.

(*iii*) Steel.

3. How will you render a steel column fire-resistant ? Draw different sketches adopting various types of materials you can use to achieve the object.

4. Write an essay on a fire-proof building.

26

Anti-Termite Treatment

26·1. INTRODUCTION

Insects could be defined as ancient race of animals which have been in existence for millions of years. They are capable of survival under most adverse conditions and environments. Man is affected in a multipilicity of ways by insects. Insects can attack his body, his food, crops, fruits, other agricultural products including his stored possessions and even the house in which he lives.

The type of insects known as termites cause maximum damage to the buildings. Termites are divided mainly into following two types :

(*a*) *Dry wood termites*

(*b*) *Subterranean termites*

Dry wood termites live in wood and do not maintain contact with the ground. They normally build nests within the dry timber members like door window frames, wooden furniture etc. and destroy them gradually. They are, however, not as common as subterranean termites and they cause lesser damage to the buildings.

Subterranean termites on the other hand are mainly responsible for causing damage to the buildings and its contents. Unlike dry-wood termites they live in soil and require moisture for their existence. They build underground nests or colonies and form mud-wall tunnels or runways (tubes) which serve as protected shelter for their movements. Sometimes they build nests near ground in stumps of dead trees or create colonies in the form of dome shaped mounds on the ground. It is through these mud wall tubes that they maintain direct contact with the soil for meeting their moisture requirements and conditions of darkness essentially needed for their survival. The termites enter the building through foundations or from ground adjacent to buildings and advance upward through floors destroying everything that comes within their reach. They may also enter the building through cracks and cervices in masonry and joints or cracks in floors in contact with ground. Termites eat cellulosic materials like wood, grass, etc. and also attack materials like leather, plastics, rubber, furniture, furnishings; clothings, stationery etc.

In the absence of adequate data, it is rather difficult to estimate the exact loss due to termite attack. The total figure could, however be assumed to run into several millions of rupees every year.

Once the termites find a foot-hold in or near a building they start spreading from their central nest through underground or over ground colonies or tunnels and their attack may spread in the entire building with the passage of time. The termites breed and spread so fast that they can damage a building in 4 to 5 years. Keeping in view the enormous damage caused by the termites it is desirable to take suitable measures for preventing ingress of termites into the buildings.

In situations where the termites have already attacked a building, suitable treatment should be given to ensure their extermination and to protect the building and its contents from further damage.

The commonly adopted method of preventing attack of termites to building consists in treating the soil beneath the building and around the foundation with soil insecticides. This treatment serves to create a chemical barrier between the ground and the parts of the building in contact with the ground thereby preventing the termites from approaching or attacking the building and its contents.

26·2. TYPES OF ANTI-TERMITE TREATMENT

The anti-termite treatment in buildings may be broadly divided into two categories :

(*i*) *Pre-construction treatment*

(*ii*) *Post-construction treatment*

26·3 PRE-CONSTRUCTION TREATMENT

This treatment is carried out right from the stage of initiating the construction activities for the building. The various operations involved in this treatment could be summarised as under :

(1) *Site preparation.*

(2) *Soil treatment.*

1. Site Preparation

This consists in removing stumps, roots, logs, waste-wood etc. from the site where the building is to be constructed. In case the termites mound are discovered within the plinth area of the building they should be destroyed by use of insecticide solution. For this treatment, holes should be made into the mound at several places by use of crow-bar and the insecticides taken in the form of water suspension or emulsion should be poured into the holes. The quantity of insecticide solution to be used will depend on the size of mound. For a mound having volume of about 1 cu. m, 4 litres of an emulsion in water of one of the following chemicals may be used:

Chemical	*Concentration by weight*
(*i*) *DDT*	*5%*
(*ii*) *BHC*	*0·5%*
(*iii*) *Aldrin*	*0·25%*
(*iv*) *Heptachlor*	*0·25%*
(*v*) *Chlordane*	*0·5%*

2. Soil Treatment

This operation consists in poisoning the soil underneath the building and around the foundations of the buildings with insecticide solution consisting of any one of the following chemicals in water emulsion.

Chemical	*Concentration by weight*
(*i*) *Aldrin*	*0·5%*
(*ii*) *Heptachlor*	*0·5%*
(*iii*) *Chlordane*	*1%*

In order that the soil treatment may be fully effective, the chemical water emulsion should be applied in required dosage on entire area of ground covered by the building. To ensure uniform distribution of the chemical emulsion, a watering can or hand operated compressed air sprayer can be used. The soil treatment should be applied in the following stages :

(1) *Treatment of wall trenches, column pits and excavation for basements :*

(*i*) The bottom surface and the sides (Upto a height of about 300 mm) of the foundation trenches, column pits and basements should be treated by applying chemical water emulsion at the rate of 5 litres per sq. m. of the surface area.

(*ii*) After the foundation for the walls, columns, piers and retaining walls of the basement come up, treat the back fill earth in immediate, contact with each side of the foundation with the chemical emulsion @ 7·5 litres per sq. m. of the vertical surface of the foundation masonry.

The treatment described in (*i*) and (*ii*) above is essentially required to be given to masonry foundation with a view to ensure that the termites do not gain entry into the building through the voids in the joints in the masonry.

In case of R.C.C. columns or R.C.C. basement walls, there is no possibility of voids which can permit entry of termites and hence it is not necessary to start anti-termite treatment right from the bottom of excavation. In such cases, the treatment should start at depth of 500 mm below ground level. Thus the back fill around R.C.C. columns, beams and entire basement walls from a depth of 500 mm upto ground level only should be treated with chemical emulsion @ 7·5 litres per sq. m. of vertical surface.

(2) *Treatment of top surface of plinth filling.* Prior to laying the sand bed or subgrade for the ground floor of the building, top surface of the consolidated earth filling within the plinth walls should be levelled and treated with chemical emulsion to 5 litres sq. m. of the surface. If the consolidated earth fill does not allow the chemical emulsion to seep through, 50 to 75 mm deep holes are made at 150 mm centres both ways with crow bars on the earth surface to permit easy seepage and to achieve uniform absorption of the chemical emulsion by the earth fill.

(3) *Treatment of junctions of the wall and floor.* The junctions of walls and the floors require special attention to ensure effective soil treatment. For this it is important to establish vertical continuity of the poisoned soil barrier on inner wall surface upto top of consolidated earth filling in plinth. This is achieved by making 30 mm wide and 30 mm deep channels at the junction of walls and columns with the earth filling in plinth. Holes are thereafter made in the channel at 150 mm apart upto ground level with crow bar and then chemical emulsion is poured in the channel @ 7·5 litres per sq. m. of the vertical wall or column surface. After the treatment the earth should be immediately tamped back into the holes and channel.

(4) *Treatment of soil along external periphery of building.* Despite the treatments given above termites are liable to gain excess in the building from ground surface around the external periphery of the building. In order to check this, 300 mm deep holes at 150 mm centre are dug all along the external perimeter of the building (parallel to the external wall) and filled with chemical emulsion @ 2·25 litres per linear metre. After the treatment the earth should be tamped back in the holes. In case the building is to have apron on the external walls, the area of consolidated earth filling over which the apron is to be constructed should be treated with chemical emulsion @ 5 litres per sq. m. prior to the construction of the apron.

(5) *Treatment of soil surrounding pipes and conduits.* In case some pipes or conduits enter the soil inside the area of foundation or come in contact with the soil on the external periphery of the building, the soil surrounding pipes/conduits should be treated with chemical emulsion for a distance of 150 mm and 300 mm respectively from the point of entry in the inside/outside area.

(6) *Treatment of expansion joints.* As a special precaution, the expansion joints are given additional treatment by applying chemical

emulsion @ 2 litres per linear metre after the subgrade for floor on either side of expansion joints has been laid. This treatment is in addition to the treatment which is necessarily provided to the structure upto top of consolidated earth fill within plinth in accordance with stage (2) above.

26·4. POST CONSTRUCTION TREATMENT

This treatment is applied to existing buildings which have already been attacked by termites. It is observed that even after their entry in the building, the termites maintain regular contact with their nest in the ground. This important symptom is gainfully utilised in eradicating termites from the buildings. Regular inspection and suitable control measures are necessary to prevent damage to buildings from termites. At times when the termites attack is of minor nature it may only be necessary to break off the shelter tubes to check the damage from termites. In situations where the attack is of mild nature the effected materials may also be removed along with shelter tubes. In case of severe attack it is necessary to poison the soil around and beneath the building besides resorting to the above steps. Thus the type of treatment to be given to eradicate termites from the existing building will largely depend upon the extent of attack and the magnitude of cellulosic and other materials available in the building The various operations involved in eradicating termites from an existing building can be summarised as under :

(1) *Inspection.* Inspection is essentially carried out to estimate the magnitude of spread of the termites infestation in the building and also to detect the root of the entry of termites and the zones in the building which are attacked. The portion of the building in contact with or adjacent to the earth should be inspected first. This includes basements, ground floor, steps leading from ground, walls, columns, areas having damp or humid conditions like bathrooms, lavatories, leaking pipes or drains etc. and the places where wood work is embedded in the floor or wall. The ceilings, wooden panelling, battens for wiring conduits, switch boards are other locations which serve as hide-out for the termites and need careful inspection.

In case of multi-storied buildings, lift wells, casings covering electrical wiring, telephone cable, water supply and soil pipes which serve as convenient and well protected zones for termites infestation should also be inspected carefully.

Wherever the mud walled shelter tubes or the termite runways are detected, they should be removed. Wherever possible, oil or kerosine based chemical emulsion should be injected over the attacked areas of wood work and masonry. At times, structural additions may become necessary to ensure elimination of all direct contacts between the soil and the effected portion of the structures.

(2) *Soil treatment for foundations.* This treatment consists in treating the soil under the building and around the foundations with some

chemical emulsion which can kill or repell termites. In this treatment about 500 mm deep trenches are made along the external peripherial wall of the building with the help of shovel (width of the trench being equal to the width of the shovel) and 12 mm dia to 18 mm dia holes at 150 mm centres are then made in the trenches close to the wall face with the help of iron rod. The holes should preferably extend upto the top of footing of foundations or to a depth of atleast 500 mm whichever is lesser. The holes are then filled with chemical emulsion in water and the back fill earth is also sprayed with the chemical emulsion as it is returned to the trench thereby creating a barrier of poisoned soil along the external periphery of the building. The total quantity of the chemical to be used in this treatment should be @ 7·5 litres per sq. m. of the vertical surface of the masonry in foundation.

In case of RCC frame structure, the chemical treatment shall be applied to the soil in contact with column sides and plinth beams along external periphery of the building for a depth of 500 mm below ground level.

In case the building has masonry or concrete apron, about 12 m dia holes at 300 mm centres should be drilled close to the plinth wall along the apron. The holes should be deep enough to reach the soil below. Chemical emulsion should thereafter be pumped into these holes @ 2·5 litres per linear metre of the length of the apron.

(3) *Soil treatment under floor.* Cracks in floors are the weak spots which permit entry of termites from soil below the floor. The cracks usually occur at the junction of the floor and walls, expansion joints in floor and at construction joints in a concrete floor. Cracks in floors may also develop due to use of unsound materials or on account of defective workmanship. In such cases eradication of termites is achieved by poisoning the soil underneath the floors wherever such cracks are noticed. This operation is generally carried out by drilling 12 mm dia holes at 300 mm centre to centre all along the cracks in the floors in different areas and then injecting chemical emulsion into the holes till the soil below gets fully saturated. The maximum quantity of chemical emulsion may, however, not exceed one litre per hole. The holes in floors are sealed after treatment.

(4) *Treatment of voids in masonry.* It has been seen that termites enter into masonry foundations from soil adjacent/beneath the building and work their way up through voids in masonry joints and gain entry into the interior of the buildings. To prevent the entry of the termites through voids in masonry, 12 mm dia holes at 300 mm centre to centre are drilled at downward angle of about 45° from both sides of walls at plinth level and then chemical emulsion is pumped into the holes until masonry gets fully saturated with the chemical emlusion. The holes are then sealed. This treatment is carried out for all walls (both internal as well as external) having foundation in soil. Treatment of drilling hole and pumping chemical

emulsion should also be carried out at critical locations like wall corners and at places where door and window frames are embedded in masonry in ground floor.

(5) *Treatment of wood work.* Wood work which is badly damaged by termites should be replaced by new timber which is adequately brushed or dipped in oil or kerosene based chemical emulsion. The infested wood work for door and window frames, etc. should be given protective treatment by drilling 6 mm dia holes at 150 mm centre to centre at a downward angle of 45° to cover the entire frame work and thereafter pumping oil based chemical emulsion into the holes. The wood work which is not attacked by termites should be sprayed over with chemical emulsion to prevent possible attack.

Questions for Review

1. Explain the terms :

(*i*) Pre-construction treatment and

(*ii*) Post-construction treatment as applied to anti-termit treatment to building.

2. Write a short note on termites.

27

New Concretes

27·1. INTRODUCTION

Conventional concrete made from portland cement, coarse aggregate, fine aggregate and water is extensively used as an important construction material all over the world and will continue to be in wide use in times to come. However, as a result of continued research and the recent development in material and construction technologies, it has been possible to introduce new concepts in the formulations of concrete. The concrete developed based on *new concept* has been termed as new concrete. New concrete can be broadly divided in the following catagories.

(*i*) Polymer concrete

(*ii*) Superplasticized concrete

(*iii*) Fibre reinforced concrete

(*iv*) Sulphur impregnated concrete

(*v*) Ultra High strength concrete

27·2. POLYMER CONCRETE

We know that cement concrete is a porous material. The presence of pores reduces the strength and durability of concrete and it exhibits a tendency to deteriorate when exposed to action of severe chemicals. As a result of continued research it has been established that addition of polymers in concrete brings about marked improvement in its compressive strength, fatigue resistance, impact resistance, toughness and durability. Polymer concrete is observed to be highly impermeable and resistant to attack by acids, alkalies and other chemicals.

Polymer concrete can be classified in following three catagories :

(*i*) Polymer impregnated concrete (PIC)

(*ii*) Polymer cement concrete or (PCC)

(*iii*) Polymer concrete (PC)

1. Polymer Impregnated Concrete (PIC)

It is a hardened portland cement concrete that has been impregnated with a monomer (low viscosity liquid organic material) and subsequently ploymerised in situ. In this case, the cement concrete is cast and cured in the coventional manner. After the concrete product gets hardened and dried, air from its voids is removed under partial vaccum and a low viscosity monomer (viny chloride, styrene, chlorostyrene etc.) is diffused through the pores of the concrete. The concrete product is then finally subjected to polymerisation by radiation or by heat treatment thereby converting the monomer filled in the voids into solid plastic. Polymer impregnated concrete products on account of their high resistance to wear and tear, impact etc. are used as kerb stones, pre-cast slabs for bridge decks, roads, in marine structures, food processing buildings etc.

2. Polymer Cement Concrete (PCC)

This type of concrete is produced by incorporating an emulsion of a polymer or a monomer in ordinary portland cement concrete. The ingredient comprising of cement, aggregates and monomer are mixed with water and the monomer in the concrete mix is polymerised after placing concrete in position. The resultant concrete has improved workability, strength, adhesion, chemical resistance, impact resistance, abrasion resistance besides having increased impermeability and reduced absorption. Polymer cement concrete can be used with advantage in marine works.

3. Polymer Concrete (PC)

In this type of concrete polymer/monomer is used to act as binder in place of cement. The monomer and the aggregate are mixed together and the monomer is polymerised after placement of concrete in position. It is however, necessary to per-heat the coarse and the fine aggregate while mixing monomer. Polymer concrete serves as a very good dielectric material. It can be used with advantage for the construction of irrigation works.

27·3. SUPERPLASTICIZED CONCRETE

This is a type of concrete in which highly effective plasticisers known as superplasticisers are incorported as admixtures to produce flowing concrete which is more of less self-compacting type. Addition of superplasticisers increases the workability of concrete, permits use of less quantity of water, thereby producing high-strength concrete without increasing cement. Superplasticized concrete can be adopted in the following three possible situation.

(*i*) *When it is required to produce concrete of required strength with reduced quantity of cement.* We know that to achieve concrete of desired strength, it is necessary to maintain, certain fixed value of the water cement ratio. In case the cement content in the concrete is reduced, it would be necessary to have corresponding reduction in the quantity of mixing water to maintain the water cement ratio. Reduction of water can render the concrete mix comparatively dry, difficult to be placed in position and hence unworkable. Addition of superplasticisers in suitable doze, compensates for the reduction of mixing water and makes it possible to maintain the water cement ratio as well as workability of the concrete with reduced quantity of cement.

(*ii*) *When it is required to produce high strength concrete having a very low water cement ratio.* The increase in strength of concrete is generally propotional to the reduction in water cement ratio. It has been established that by adding superplasticiser it is possible to achieve water reduction upto 30% and water cement ratio as low as 0.2. Thus to achieve high strength concrete, the water content of the mix is reduced, maintaining the cement content same and the admixtures of superplasticiser compensates the reduced workability due to use of less water without producing any determental effect either in the plastic or in the hardened concrete.

(*iii*) *When it is required to produce concrete having good flowability without any danger of bleeding, segregation or reduction in strength.* By admixture of superplasticiser it is possible to produce flowing concrete which is self-compacting and self-levelling type. In situations where concrete is to be placed in heavily reinforced structural members, superplasticisers can be added to the concrete (without effecting any reduction in the cement or water content) with the sole aim of increasing the slump. In this process it is possible to achieve increase in slump from 75 to 200 mm without producing adverse effect on the strength of the concrete.

The quantity of super-plasticisers to be added in the concrete mix depends upon the purpose for which it is being used *i.e.*, to achieve good flowability or to produce high strength concrete by using less water. In general, superplasticiser, can be added in concrete in doses upto 0.7% by weight of cement.

27·4. FIBRE REINFORCED CONCRETE (FRC)

Because of its low tensile strength and impact resistance, plain cement concrete is considered to be a brittle material. It has now been established that by addition of small diameter, short length, randomly distributed fibres it is possible to bring about marked improvements in the tensile strength and impact resistance properties of concrete. The fibre could be of steel, glass of asbestos. Concrete with fibre is termed as *fibre reinforced concrete.* The extent of improvement in the properties of fibre reinforced concrete depends upon various factors like material of fibre, their shape, size, pattern of distribution and magnitude (volume percentage) in the concrete mix.

Out of various possible types of fibre reinforced concrete the following two types are mostly being recommended.

(*i*) Steel fibre reinforced concrete

(*ii*) Glass fibre reinforced concrete

1. Steel Fibre Reinforced Concrete

This type of concrete is formed by adding steel fibres in the ingredients of concrete. Steel fibres are normally produced by cutting 10 to 60 mm length of low carbon steel wires 0.25 mm to 0.75 mm in diameter. Besides round fibres use of flat sheet steel fibres is also common. Flat sheet steel fibre are produced by shearing 0.15 mm to 0.40 mm thick steel plates in widths ranging from 0.25 mm to 0.90 mm and length 10 mm to 60 mm. Steel fibres have a tendency to cluster together which creates difficulties in ensuring their uniform random distribution in the concrete. This difficulty is overcome by using fibre-bundles (Fibre loosely assembled in the form of bundles with the help of water soluble glue). The steel fibre in the fibre-bundles separate out during the process of mixing concrete and get distributed in a random fashion in the concrete mix. By addition of 2 to 3% of fibres (by volume) it is possible to achieve two to three times increase in the flexural strength of concrete and substantial increase in explosion resistance, crack resistance and other properties of the concrete. Steel fibre reinforced concrete is considered suitable for the construction of pavements, bridge decks pressure vessels, tunnels lining etc.

2. Glass Fibre Reinforced Concrete

It is observed that strength of glass fibre increases as its diameter is reduced. It is also seen that although small diameter glass fibres are reasonably strong in tension, they are very brittle and cannot be used in long lengths. Moreover glass fibres get corroded due to the effect of alkalie present in portland cement. Thus to utilise glass fibres as micro-reinforcement they are suitably treated (coating with epoxy resign compounds etc.) to protect them from alkali attack. It is seen that addition of 10 % of glass fibre (by volume) brings almost two folds increase in tensile strength and substantial increase in impact resistance of concrete. Investigations are on to use glass fibre reinforced concrete in the manufacture of precast products like spun pipes, wall cladding etc.

27·5. SULPHUR IMPREGNATED CONCRETE

This type of concrete consists of a mixture of sulphur, coarse aggregate and fine aggregate (There is no cement or water). The concrete is formed by adding molten sulphur to the pre-heated aggregate in a mixer. The hot concrete mix is a poured in moulds to from pre-cast units. Sulphur impregnated concrete develops high compressive strength at an early age. It needs no curing and the mould can be stripped soon after the concrete mix get solidified. The main advantage of this type of concrete is that it can be re-used and re-moulded without any wastage. Sulphur impregnated concrete has good chemical durability and high strength. This concrete is more durable in acidic environments but is unstable in alkaline conditions. Due to its brittleness and corrosive action of sulphur on reinforcement, this type of concrete is not suitable for structural work. This concrete is used in industrial plant and for lining canals and tunnels.

27·6. ROLLER-COMPACTED CONCRETE

This type of concrete can be defined as a dry concrete material which has been consolidated by external vibrations using vibratory rollers. It differs from normal concrete basically in its consistancy. The roller compacted concrete should be dry enough to support the load of vibratory rollers and at the same time it should be wet enough to permit proper consolidation by vibrations. This type of concrete is used with advantage in mass concrete work. It can be laid in thin layers of about 25 cm depth in continuous operation. This results in substantial reduction in the heat of hydration and hence the need to perform special concrete cooling operations in mass concrete work can be avoided. The use of roller compacted concrete permits faster construction and saving in labour cost.

27·7 ULTRA HIGH STRENGTH CONCRETE

We are conversent with use of concrete having compressive strength varying from 100 to 300 kg/cm^2. With the introduction of pre-stressed concrete in the construction industry, constant research has been going on to develop concrete having very high or ultra-high strength. As a result thereof it has now been established that by suitable selection of material and by adopting special method of production it is possible to make concrete having compressive strength exceeding 1000 kg/cm^2. Ultra high strength concrete can be produced by improved compaction (by pressure and vibration etc.) and adhesion of cement matrix to aggregates or by adoption of cementatious aggregates. Ultra high strength concrete can be easily produced by use of high quality coarse aggregates, synthetic aluminious fine aggregates and cement. Use of high temperature to increase the lime silica bond and use of spiral wrapping of concrete to induce triaxial stress are the other techniques being tried to develop ultra high strength concrete. Polymer concrete and superplasticised conerete also get covered in the category of ultra high strength concrete to some extent.

By use of ultra high strength concrete, load bearing capacity of columns can be increased considerably. Hence by its use it is possible to adopt slender columns in multi storied buildings which besides looking aesthetically good also permits greater utilisation of useful floor space. On account of its high compressive as well as tensile strength, this type of concrete finds wide application in prestressed concrete.

Questions for Review

1. Write short notes on :

   (*i*) Polymer concrete
   (*ii*) Fibre reinforced concrete
   (*iii*) Sulphur impregnated concrete.

2. What do you understand by superplasticized concrete ? Explain the situations under which this type of concrete can be recommended.

3. Describe (*i*) ultra high strength concrete and (ii) steel fibre reinforced concrete.

28

Acoustics and Sound Insulation

28·1. GENERAL

The sound which causes annoyance, interference with speech, damage to hearing and results in reduction in efficiency of work performance is called noise. Noise may also be defined as a disturbance in an elastic medium which includes solid, liquid and gases. It is a form of wave motion created by a vibrating body and is transmitted in all directions in the form of spherical waves consisting of alternate compressions and rearefactions. The procedure of obtaining an acceptable noise environment at a particular point is called noise control.

28·2 CLASSIFICATION OF SOUND

Depending upon the position of source, sound can be broadly divided in two classes :

(*i*) Airborne sound

(*ii*) Impact sound.

(*i*) *Airborne sound* : An airborne sound is one which is transmitted through air and travels direct to the ear of the person. This type of sound travels from one part of the building to another or from outside of the building to the inside through open doors, windows or other openings or through small gaps around doors and windows.

(*ii*) *Impact sound* : The sound which is transmitted first through the structure is called impact sound. The noise of the footsteps, furniture, movement, dropping of utensils on the floor fall under this category. Impact sounds are troublesome and are often very sharp.

28·3. MEASUREMENT OF SOUND

Human ear is incredibly sensitive. It can react to the rustle of the wind in a leaf in a silent night as well as to the noise of a rock drill, latter being 10,000,000,000 times stronger than the former. Sound is usually measured and expressed in terms of sound pressure levels or *decibles* (d_B).

The formula for writing down the two sound pressures

$$= 20 \log \frac{p_1}{p_2} \text{ decibles}$$

where p_1 = the pressure of the sound wave at source

and p_2 = the pressure of the sound wave at recipient

Decible is therefore used as a convenient unit to measure the magnitude of sound, sound levels or sound insulation (noise reduction achieved by sound absorbing materials). 0·002 Dynes/sq. cm. is the lowest value of sound pressure that an average human ear can perceive and hence this value is taken as zero on the d_B scale. The noise at the threshold of pain in the ear is considered to have a sound pressure of 130 d_B.

Table given below gives an idea about the magnitude of noise levels from some common noisy sources.

S.No.	*Type of source of noise*	*Noise levels in* d_B
1	Light road traffic (side streets)	60–70
2	Medium road traffic (main streets)	70–80
3	Heavy road traffic (highways)	80–90
4	Rail traffic	90–110
5.	Air traffic (at 300 m)	90–110

28·4. TRANSMISSION OF SOUND

When sound is produced in a room it proceeds outward in spherical waves until it strikes the boundaries of the room. Thereafter, the sound waves are reflected, transmitted and absorbed in varying amounts depending upon the characteristics of the walls of the room (thickness, weight, material of wall and nature of its surface) and the frequency of the sound.

28·5. TRANSMISSION LOSS

The reduction in the intensity of airborne sound that takes place during its transmission from the source to the recipient is called transmission loss. Transmission loss or T.L is numerically equivalent to the loss in the intensity of sound expressed is decibles. Hence if 100 d_B

and 60 d_B are the sound levels measured on either side of a partition wall, the transmission loss or sound insulation of the partition wall is equal to 100 d_B – 60 d_B = 40 d_B.

28·6. DEFECTS DUE TO REFLECTED SOUND

The behaviour of the reflected sound plays an important role in sound insulation or acoustical design of a room/hall. The main defects which reflected sound waves may cause in a room are :

(*i*) Echoes

(*ii*) Reverberation

(*i*) *Echoes* : When a reflecting surface is so far away from the source that the sound is reflected back as a distinct repetition of the direct sound, the reflected sound is called an echo. Echoes are produced, when the time interval between the direct and the reflected sound waves is about 1/15th of a second. This defect is particularly common when the reflecting surface is curved in shape. To minimise this defect in curved walls, the walls are covered with highly absorbent material or grills on the face work.

(*ii*) *Reverberation* : When the sound waves get reflected (on striking the walls), a part of the sound energy is converted into heat energy by friction and is absorbed by the walls. Subsequently the reflected waves get inter-reflected from one surface to another until they gradually fade and become inaudible. This phenomenon of undue prolongation of sound by successive reflections from surrounding surfaces, after the source sound has ceased is termed as reverberation. A certain amount of reverberation is necessary to enhance the sound. However, excessive reverberation is damaging to clarity.

Reverberation Time : The reverberation period is the time taken for the sound energy to decay by 60 d_B, after the sound source has stopped. The reverberation time depends principally on the volume and absorption of the room.

Prof. Sabine of Harvard University developed a formula for calculating the time of reverberation which is given by

$$t = \frac{0{\cdot}16\ V}{a_1 s_1 + a_2 s_2 + a_3 s_3 \ldots}$$

$$= \frac{0{\cdot}16\ V}{\text{Sum of the absorption of various absorption components}}$$

where t = time of reverberation

V = volume of room in m^3

$a_1, a_2, a_3 \ldots$ = coefficient of absorption

$s_1, s_2, s_3 \ldots$ = area of absorbing surface

The time of reverberation plays a significant role in achieving desired acoustical condition. If the time of reverberation is too long, it results in overlapping of speech and loss of intelligibility, and if it is too short, it produces the effect of deadness and loss of brilliance of sound. If the time of reverberation works out to a value greater than 3 seconds, it is considered bad, between 3 to 2 seconds as fairly good and between 2 to $\frac{1}{2}$ second as very good. The time of reverberation to be allowed varies with the purpose for which the enclosure is to be used. If the enclosure or auditorium is to be used as a sound film theatre or for public address system, the time of reverberation selected should be short, whereas for concert halls and churches, it should be longer. For enclosures to be used for both speech and music, a value midway between the two should be adopted.

The selection of the correct time of reverberation is called *optimum time of reverberation.*

It is noted that presence of audience in a room reduces the time of reverberation. This is on account of the absorption provided by the audience (due to the clothing worn by the persons). This shows that a theatre will have a greater time of reverberation, when it is empty, than when it is full with audience. To make allowance for this aspect in deciding the time of reverberation, the strength of audience contributing towards the absorption of sound is taken as $\frac{1}{4}$, $\frac{2}{3}$ or full depending upon the type of hall. The desirable optimum time of reverberation worked out after considering the indicated audience strength, is shown in the Table No. 28·1.

TABLE 28·1

S.No.	*Description of building*	*Optimum reverberation by Sabine formula*	*Audience factor to be considered in design*
1.	Cinema theatres	1·3 seconds	Two thirds
2.	Churches	1·8 to 3 seconds	Two thirds
3.	Law courts. committee rooms, conference hall etc.	1 to 1·5 seconds	One third
4.	Large halls	2 to 3 seconds	Full
5.	Music concert hall... ...	1·5 to 2 seconds	Full
6.	Parliament house, Assembly hall, council chamber ...	1 to 1.5 seconds	Quorum
7.	Public lecture hall... ...	1·5 to 2 seconds	One third

28·7. SOUND ABSORBENT

Broadly speaking, the material having hard, rigid and non-porous surface, provide the least absorption, whereas those which are flexible, soft porous and can vibrate, absorb more sound. The efficiency of the sound absorption, however, depends more upon the porosity of the material used as sound absorbent. The term used to express the percentage of the incident sound that can be absorbed by a material is known as *absorption co-efficient* of the material. Thus if the absorption co-efficient of a material is 0.75, this would mean that the material is capable of absorbing 75 % of the incident sound. The absorption co-efficient differs with the frequency of the incident sound. In general, low density materials have higher absorption co-efficient at the higher frequencies than at low frequencies. Table No. 28·2 lists the sound absorption coefficients of commonly used acoustical materials.

28·8 CLASSIFICATION OF ABSORBENTS

Sound absorbents can be broadly classified into the following four categories :

(*i*) *Porous absorbents*

(*ii*) *Resonant absorbents*

(*iii*) *Cavity resonators*

(*iv*) *Composite type of absorbents.*

(*i*) *Porous absorbents* : When sound waves strike the surface of porous material, a part of the waves get reflected while a part enters the pores of the material and is thought to be dissipated into heat energy (produced on account of the friction developed between the sound waves in motion in the restricted pores of the material). The efficiency of this type of absorbent increases with the increase in the resistance offered by the material to air flow, its thickness and the porosity. Slagwool, glass-wool, woodwool, asbestos fibre spray, foamed plastic, and perforated fibreboards are some of the categories of porous absorbents. In general, porous materials are selected mainly to absorb sound having high frequency.

(*ii*) *Resonant absorbents* : In this system, the absorbent material is fixed on sound framing (usually timber) with an air space left out between the framing and the wall at the back. Such an arrangement works most efficient for absorbing sound waves at low frequency. The principle of sound absorption in this method is that sound waves of the appropriate frequency cause sympathetic vibrations in the panel which acts as a diaphragm. The absorption of sound takes place by virtue of the dampening of the sympathetic vibration in the panel by means of the air space behind the panel. Dampening effect of this system can be improved appreciably by placing a porous material in the air space.

(*iii*) *Cavity resonators* : Cavity resonators essentially consists of a chamber with a narrow opening (entrance) in which absorption takes place by resonance of the air in the chamber which gives loss of sound energy. Such an arrangement can act effectively over a single selected frequency. Application of cavity resonator is normally restricted to absorption from individual machine or in similar cases.

TABLE 28.2

SOUND ABSORPTION CO-EFFICIENTS OF ACOUSTICAL MATERIALS

S.No.	*Material*	***Application Co-efficients***		
		Frequency of sound in cycle/sec		
		125	***500***	***2000***
1.	Brick wall painted	0.012	0.017	0.023
2.	Brick wall unpainted	0.024	0.03	0.049
3.	6mm. thick carpet or felt on solid concrete floor	0.07	0.5	0.65
4.	Carpet unlined	0.09	0.20	0.37
5.	Carpet felt lined	0.11	0.37	0.37
6.	Curtains-light (3 kg/sq. m.)	0.04	0.11	0.30
7.	Curtains-medium	0.06	0.13	0.40
8.	Curtains-heavy (5.5 kg/sq.m.)	0.10	0.50	0.82
9.	Chairs-metal or wooden.	0.014	0.016	0.019
10.	Chair with wood veneer seat and back	...	0.023	...
11.	Chair with leather cushion	...	0.18	0.07
12.	Chair-heavy upholstered	... ...	0.24 to 0.28	... .
13.	Compressed paper board	...	0.11	0.11
14.	Floor (concrete)	0.01	0.015	0.02
15.	Floor (timber)	0.05	0.03	0.03
16.	Linoleum, or asphalt laid on concrete floor	0.05	0.03 to 0.08	0.03
17.	Fibre board (12 mm. thick) on battens keeping about 25 mm. air space between the boards and the solid wall	0.3	0.3	0.3
18.	Foamed concrete 19 mm thick	...	0.12	0.20
19.	Glass (upto 32 oz.)	0.035	0.027	0.04
20.	Glass wool, or slag wool (50 mm, thick) fixed on battens secured to solid wall	0.35	0.91	0.95

Contd...

S.No.	Material	Application Co-efficients		
		Frequency of sound in cycle/sec		
		125	500	2000
21.	Plaster, gypsum or lime smooth finish on brick wall	0.013	0.025	0.04
22.	Plaster, gypsum or lime on lath over solid backing	0.02	0.03	0.04
23.	Perforated fibre board tiles (18 mm. thick) bedded solidly	0.2	0.65	0.8
24.	Plywood panelling fixed on battens keeping about 2 mm. air gap between the panelling and the solid wall	0.3	0.15	0.1

(*iv*) *Composite Type of absorbents* : They consist of perforated panels mounted on battens so as to leave a cavity between panels and wall at the back. The panels may be of metal, wood (plywood) hard broad, plaster board etc. The area of holes in the panel should vary between 10 to 20% of the total area of the panel. When the sound strikes the panel, the sound waves pass through the holes and get damped by the reasonance of the air in the cavity. The effectiveness of this system can be increased by placing a porous material like mineral wood etc. in the cavity. This type of absorbent is commonly used, as it is easy to install, economical and it can accommodate wide range of frequencies.

28·9. SOUND INSULATION OF BUILDINGS

Control of noise transmission is essential to minimise the disturbing effect of sound passing from one room to another, through walls, partitions, and floors or ceilings. Good planning in respect of the location of building as well as the placement of quiet and noisy areas in the building itself plays an important roles in controlling noise transmission. The constructional measures to be adopted for noise control and sound insulation are briefly discussed below.

28·10. WALL CONSTRUCTION

The sound insulation rating of a wall is generally governed by the net sound transmission loss it provides and also the efficiency with which it serves as a barrier for speed sound. Weight of the wall is the governing factor in wall insulation. It is seen that a solid one brick thick wall plastered on both sides, proves quite effective as a sound insulation partition wall. It has an average reduction of 50 d_B. It is now, however, possible to have wall made from a suitable combination of materials which are light in weight and yet have high insulation value.

A cavity wall type of construction can be made to have increased insulation value by filling the cavity with some resilient material. In this

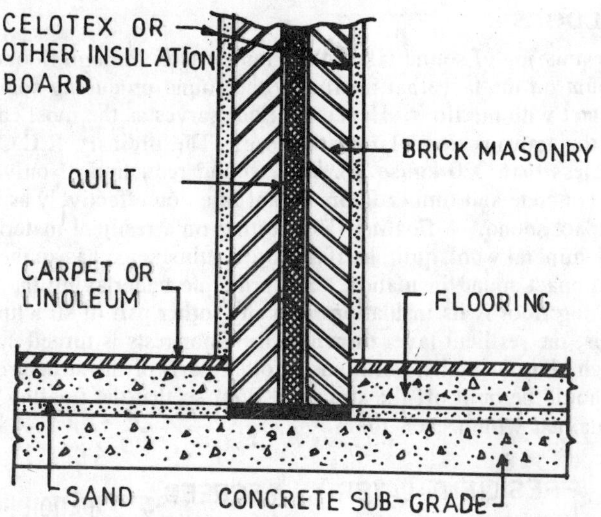

Fig. 28·1.

Details of the junction of cavity type partiton wall and floor for a sound proof construction

type of construction, the cavity should be at least 5 cm. in width and the two wall leaves should be tied by use of only light butterfly wall ties. The insulation value of different types of partition wall is given in Table 28·3. Partition walls having the value of transmission loss of 45 d_B or less are considered adequate for separating critical areas of adjacent dwellings.

TABLE 28·3

TYPICAL INSULATION VALUES OF WALLS

S.No.	*Type of construction*	*Approx. weight in kg. /sq. m.*	*Average T.L. in d_B*
1.	One brick thick wall	488	50
2.	One and a half brick thick wall	708	53
3.	Cavity wall having two leaves, each of half brick thickness, with 5 cm. cavity	488	50 to 53
4.	Cavity wall having two 10 cm. thick leaves of clinker block with 5 cm. cavity.	312	50
5.	Half brick wall with 13mm. thick plaster on both sides	268	45
6.	20 cm. thick hollow dense concrete block wall with 13 mm. thick plaster on both sides	185	45
7.	Partition wall made with gypsum wall board fixed on timber frame work	67	45
8.	75 mm. thick hollow clay block wall with 13 mm. thick plaster on both sides	108	36

28·11. FLOORS

Transmission of sound takes place more easily through floors. This is on account on the fact, that invariably the sound producing source has actual contact with the floor. Hence the floor serves as the most common path for the transmission of impact noise. The ordinary R.C.C. floor weighing less than 220 kg./sq. m. has a sound reduction of only 45dB. Thus bare concrete and timber floors do not function effectively as barrier against impact sound. A floating floor resting on a resilient material like glass wool, mineral wool, quilt, hairfelt, corck, rubber etc., has an increased rating for impact sound insulation. The principle underlaying the design of the floating floor is its insulation from any other part of structure. To achieve this, the resilient layer on which the floor rests is turned up at all edges which abut the walls, partitions or other parts of the structure. The partition should be built off the structural floor so that the floating screed is self-contained within each room.

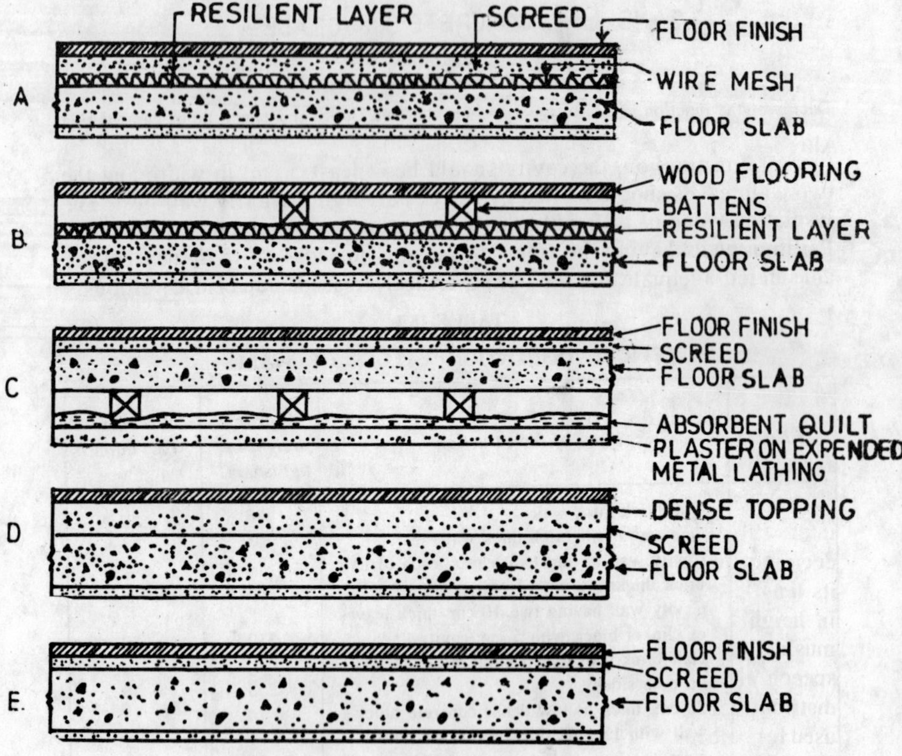

Fig. 28·2

Figures showing different methods of sound insulation of floors

The various types of floating floor construction are shown in fig. 28·2.

The wood raft floating floor as shown in [Fig. 28·2(*B*)] consists of 50 mm deep x 50 mm. wide wooden battens, on which 20 mm thick resilient quilt is laid over the structural floor slab.

The concrete screed floating floor as shown in [fig. 28·2 (*A*)] consists of a 70 mm thick layer of 1 : 1½ : 3 concrete screed laid on a 25 mm. thick resilient layer of mineral wool quilt. The quilt is covered with waterproof paper to prevent the moisture from concrete screed travelling below on the structural floor slab.

Floors with suspended ceilings have an added advantage of insulation against air borne sound, provided a soft floor finish is provided on top to give necessary insulation against impact sound refer [Fig. 28·2(*C*)]

Wooden joist floors : The performance of wooden joist floors is greatly influenced by the amount of indirect or flanking sound, transmitted through the walls. This factor is not important in concrete floors, since the concrete floors are considered heavy, rigid and stiff enough to restrain the vibrations from the walls. In timber floor, this factor can be taken care of by increasing the stiffness and thickness of the wall below the floor. Alternatively, the floor should be made heavy and stiff enough to reduce the vibrations of the walls. The two different types of wooden joist floors which give adequate sound insulation are shown in Fig. 28·3. A noteworthy feature in the insulation of these floors is the floating floor effect achieved by using a 25 mm thick mineral wool or glass wool quilt which is dropped over the joist and turned up at the edges of the boards. The efficiency of the flooring system is improved by increasing its weight with pugging between the joints.

28·12. ACOUSTICAL DESIGN OF AUDITORIUM

The important factors which influence the acoustical design of an auditorium are the volume, the shape and the sound absorption.

The volume of the auditorium should be in proportion to the intensity of the sound that is expected to be generated in the hall. In deciding the volume of the hall, its height plays a significant role than its length or breadth. This is on account of the fact that a small increase in height increases the volume considerably. The volume required for musical concert halls is larger than that required for halls to be used for speech alone. This is necessary from the point of view of proper distribution of musical sound. In case, however, the auditorium is to be used for both musical concerts as well as speech, the volume of the auditorium should be so chosen as to have a value intermediate between the two. The following data may be used as a rough guide for deciding the volume of an auditorium.

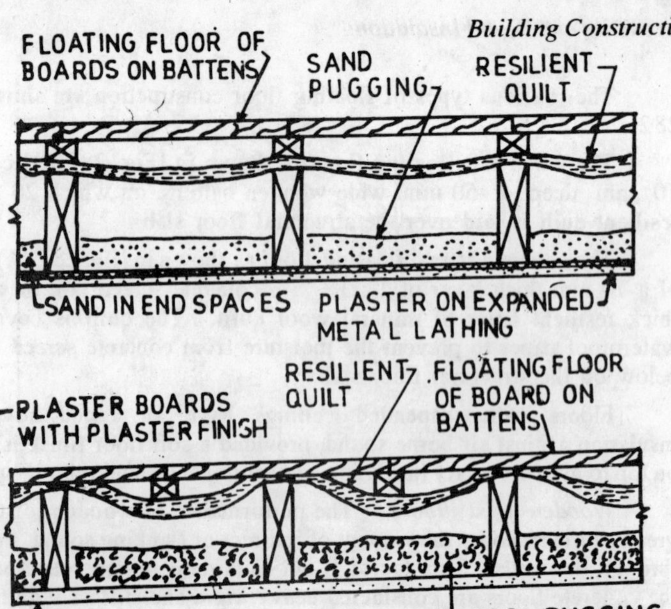

Fig. 28·3

Figures showing different methods of sound insulation of timber floors

(*i*) For cinema theatres = 3·7 to 4·2 cu. m. per person.

(*ii*) For public lecture halls = 2·8 to 3·7 cu. m per person.

(*iii*) For concert halls = 4·2 to 5·6 cu. m. per person.

The shape of the auditorium is the governing factor in avoiding the defects like echoes or other types of reflections of sound waves. On account of the introduction of sound amplifiers, this aspect of planning has become still more important. Since the behavior of sound in a hall is different from that in the open, it is rather easier to create desirable acoustical conditions in an auditorium rather than in an open air theatre. Rectangular, fan, horse shoe, circular or oval are the typical possible shapes of the floor plan of an auditorium.

The side walls and ceiling are advantageously used to provide favourable reflections. The walls of the hall are so shaped and placed as to minimise the possibilities of echoes. Plain walls are normally found suitable. The convex shaped walls are, however, considered best to reduce echoes to a great extent. In general, the ceiling height of the hall should be about $\frac{1}{2}$ to $\frac{2}{3}$ rd of the width. For reducing the focussing effect of a curved ceiling, the radius of curvature of the ceiling should be made at least twice the height or less than half the ceiling height. Ceiling plays a significant role in reflecting the sound to the rear areas of the auditorium. A noteworthy point in the selection and installation of the ceiling is that it should be ensured that the sound waves get reflected either directly or via the walls

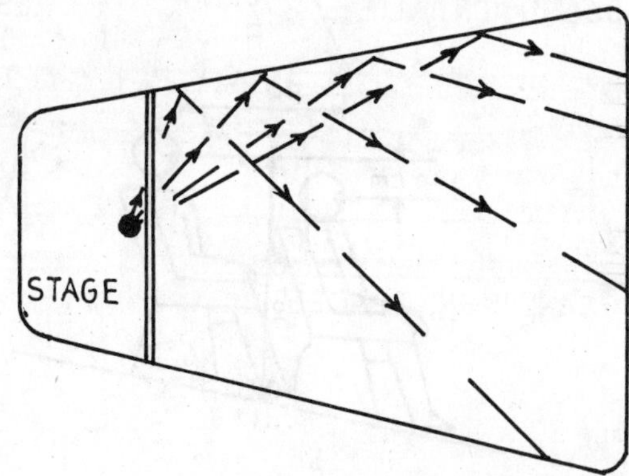

Fig. 28·4

Fan shaped floor plan of an auditorium showing favourable reflections from side walls

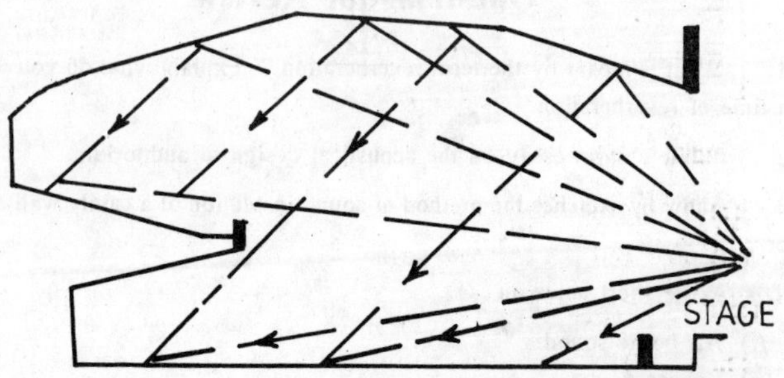

Fig. 28·5

Typical section showing favourable reflections from the ceiling of an auditorium

to the audience in such a manner that the waves do not concentrate at certain spots.

Sound absorbing materials are also used to minimise objectionable reflection of sound. However, to ensure effectiveness of the sound absorbing material, the zones of installations have to be decided very carefully. A variety of acoustical materials are manufactured these days. While making a selection, due consideration should be given to their appearance, light reflection, flame resistance, workability, durability and cost.

In addition, the furnishings and the audience contribute to a great extent to the absorption present in the room. In fact the audience may be largest contributors to the absorption in any auditorium. With a view to ensure optimum absorption from the audience, the seats in the hall are raked so that the heads in one row do not intercept the passage of direct sound to the persons in the row immediately behind. A typical raking arrangement of the seats in auditorium is shown in Fig. 28·6.

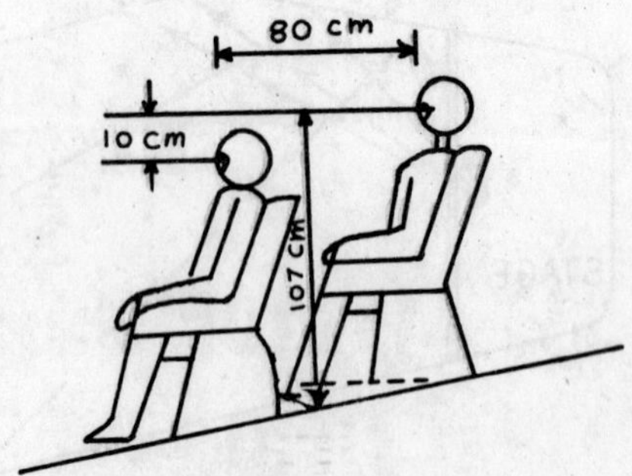

Fig. 28.6

Questions for Review

1. What is meant by the term reverberation ? Explain what do you understand by optimum time of reverberation.

2. White a short essay on the acoustical design of auditorium.

3. Show by sketches the method of sound insulation of a cavity wall and a wooden joist floor.

4. Write short notes on :

(*i*) Air borne sound.

(*ii*) Echoes

(*iii*) Transmission loss

(*iv*) Resonant absorbents.

29

Water Supply and Drainage Systems

29·1. INTRODUCTION

The origin of all sources of water is rainfall. As it rains water can be collected from roof before it falls on ground or as it flows on surface it gets collected in the form of ponds, lakes, streams, river or sea. Water that percolates or seeps into the ground gets stored as ground water which can be tapped in the form of springs shallow wells, deep wells, artesian wells, etc. Adequate supply of potable water is essential for the occupants of buildings.

The municipal corporations or the municipalities are responsible for providing public water supply system which includes collection of water from source of supply, giving necessary treatment to water to make it hygienically safe and potable and finally distribution of water through a network of piping work (trunk mains, street mains etc) . Water from the street main is supplied to the individual building through a service connection. Within the building, water is distributed to different fixture through pipes which may run on surface or concealed in walls or below flooring. Water thus supplied may be used for bathing, cooking, flushing of W.C., washing clothes/utensils/floors etc. In this manner the potable water gets converted into waste water which is drained out into a sewer or other suitable disposal system like septic tank etc.

Plumbing is a general term which broadly includes the system, materials, fittings and fixtures used in a building in connection with supply of water, removal of used water with other liquid and water borne wastes including connected ventilation system as well as drainage of storm water. The various types of fittings and fixtures used in plumbing are termed as *plumbing fixtures.*

29.2. TECHNICAL TERMS

Some of the technical terms used in connection with drainage system are as under ;

1. **Soil appliance.** This includes water closets, urinals, bed-pan washers etc..

2. **Soil pipe.** A pipe which carries discharge from W.C., urinal, or any other soil appliances.

3. **Waste appliance.** This includes wash basins, sinks, bath tubs, washing trough, drinking water fountain etc.

4. **Waste pipe.** A pipe which carries waste water from kitchen, bathroom, floor traps, nahani trap or any other waste appliance.

5. **Rain water pipe**. This is a pipe provided to carry rain water.

6. **Ventilation pipe.** Also known as vent pipe is a pipe which ventilates drainage system. This pipe is open at top and it is connected to a soil pipe or waste pipe at its bottom. This pipe is extended above the roof of the building to permit exit of foul gases into the atmosphere. This pipe only ventilates the system and does not carry any discharge from soil, waste or rainwater pipe.

7. **Stack.** It is a term used for any vertical pipeline of a drainage system.

8. **Drain.** Also known as house drain, is a system of underground horizontal pipes used for drainage of discharge from soil pipes, waste pipes etc. of a single property. Since the drain is laid within the private boundary, the responsibility of its maintenance rests with the owner of the property.

9. **Sewage.** It is combination of discharge from soil pipe, waste pipe with or without rain water.

10. **Sewer.** It is a system of underground pipe belonging to local municipal authority which collects discharge from drains or house drains from outside the private boundary of more than one property.

11. **Trunk sewer.** It is a main sewer which receives discharges from all the smaller sewers and conveys it to a sewage treatment plant or to the point of final disposal.

12. **Inspection chamber and Manhole.** An inspection chamber is a small masonry chamber (usually 60 × 75 × 75 mm in size) provided for inspection, cleaning and repairs of a house drain. Various pipes provided for drainage of the inspection chamber is provided with open U shaped channels (constructed in cement concrete) which receive the discharge from the terminated pipes and conveys it to the sewer through intercepting trap built into the wall of the chamber on the sewer side.

Manholes are also masonry chambers provided at suitable location and on the line of the sewer for the purposes of inspection and cleaning the sewers. Manholes should be of sufficient size to allow a man to work with the cleaning rod or other cleaning apparatus in the pipe sewer. Manholes should be constructed at all changes of directions or gradient and at intervals not exceeding 30 m.

29·3. PLUMBING SYSTEMS

The plumbing systems broadly includes the following.

(*i*) Entire system of water supply and distribution pipes, including fittings and fixtures like taps, valves, tanks etc. used in connection with water supply.

(*ii*) Entire system of sanitary drainage including fittings and fixtures like wash basins, sinks, water closets, urinals, traps, soil types, waste pipes, vent pipes, sewers, septic tanks etc.

(*iii*) Entire system of storm water including collection and carrying of rain water (from roofs, paved areas and ground surface) to a public storm water drain or to a pond or river etc.

29·4. WATER DISTRIBUTION SYSTEM

There are two distinct systems of supply of water to a building from the mains :

(1) *Direct system and*

(2) *Indirect system.*

1. **Direct system.** In direct system also known as upward distribution system, the supply of water is given to various floors in a building directly from the mains which has sufficient pressure to feed all the floor and water fittings at the highest part of the building.

2. **Indirect system.** In indirect system also known as down take supply or downfall distribution system, the water supply from the mains may be drawn either by

(*a*) Feeding water directly into the overhead storage tank provided at roof of the building from where the water is supplied to different floors by gravity or

(*b*) Feeding the water into a underground water storage tank. The water from the underground tank is pumped to overhead storage tank from where the water is supplied by gravity.

29·5. ASSESSMENT OF WATER REQUIREMENTS

The actual quantity of water required by an individual or a family may work out to be very small. However, when families live together forming communities, towns and cities, the requirement of water increases to a great extent. Besides domestic requirements (*i.e.* for drinking, cooking, bathing, washing, flushing etc.) water may be needed for air conditioning, fire fighting, civic use, institutional, commercial and industrial needs etc. In assessing the water requirement due consideration has also to be given to the local needs, people, their habits, standard of living, climatic condition, and other similar factors.

The National Building Code of India recommends that water requirement of 135 litres per head per day may be adopted in design of water supply system for all residential buildings provided with full flushing system for excreta disposal.

The requirements of water for buildings other than residences may be worked out from Table 29.1

TABLE 29.1
WATER REQUIREMENTS FOR BUILDINGS OTHER THAN RESIDENCES

Sl.No.	Type of building	Consumption per day in litres.
1.	Factories where bathrooms are required to be provided	45 per head
2.	Factories where no bathrooms are required to be provided	30 per head
3.	Hospitals (including laundry)	
	(*i*) Number of beds not exceeding 100	340 per bed
	(*ii*) Number of beds exceeding 100	450 per bed
4.	Nurses homes and medical quarters	135 per head
5.	Hostels	135 per head
6.	Hotels	180 per bed
7.	Offices	45 per head
8.	Restaurants	70 per seat
9.	Cinemas, concert halls and theatres	15 per seat
10.	Schools	
	(*i*) Day School	45 per head
	(*ii*) Boarding school	135 per head
11.	Railway stations	
	(*a*) Intermediate stations	
	(*i*) where bathing facilities are provided	45 per head
	(*ii*) where no bathing facilities are provided	23 per head
	(*b*) Junction stations	
	(*i*) where bathing facilities are provided	70 per head
	(*ii*) where no bathing facilities are provided	45 per head
12.	Terminal railway station	45 per head
13.	Terminal bus station	45 per head
14.	International and domestic airports	70 per head

29·6. TYPES OF PIPES

Following types of pipes are commonly used in water supply system.

(1) *Cast iron pipes*

(2) *Steel pipes*

(3) *Galvanised iron pipes*

(4) *Copper pipes*

(5) *Polythene pipes.*

(1) **Cast iron pipes.** Cast iron (C.I.) pipes are extensively used in water distribution mains because they are comparatively cheaper in cost, highly resistant to corrosion and have very long life. C.I. pipes are manufactured by following four methods.

(*i*) Horizontally cast (MC ware pipes)

(*ii*) Vertical cast (Pit cast in sand moulds)

(*iii*) Centrifugally cast in sand lines moulds spun type.

(*iv*) Centrifugally cast in water cooled metal moulds.

Use of Horizontally cast C.I. pipes is now more or less obsolete and as such they are very rarely manufactured these days. Centrifugally cast pipes by virtue of the method of casting have fine grained dense structure and uniform thickness and as such they are widely used. C.I. pipes are cast in length varying from 3m to 6m. C.I. pipes are very heavy and need special care to prevent damage during transportation and making connections.

(2) **Steel pipes.** Steel pipes are recommended for use in water mains in situations where the pipe is subjected to very high pressure (*i.e.* above 7 kg/cm^2) and the diameter of pipe required in large. Steel pipes are stronger and lighter in weight as compared with C.I. pipes. They, however, require adequate preventive treatment to sustain adverse atmospheric conditions.

Steel pipes encased in cement mortar or cement concrete are called Hume Steel Pipes.

(3) **Galvanised Iron pipes.** Galvanised Iron (G.I.) pipes are wrought steel pipes provided with zinc coating. G.I. pipes are most commonly used for water supply work inside the buildings. They are also invariably used in service connections. Mostly screw and socket joints are used for G.I. pipe connections.

(4) **Copper pipes.** Copper pipes are used in hot water supply installations. They have high tensile strength and can therefore have thin walls and they can be bent easily. To enhance their appearance copper pipes are sometimes chromium plated to match with the chromium plated water supply fittings.

(5) **Polythene pipes.** Polythene and P.V.C. pipes are being used increasingly these days for supply of cold water in external and internal plumbing work. They are light in weight, noncorrosive, lower in cost and do not require any threading for connections.

Asbestos cement (A.C.) pipes, Reinforced Concrete (R.C.C.) Pipes. Prestressed reinforced concrete (P.S.C.) pipes, are the among the other commonly used pipes for water supply work. The choice of the type of pipe to be used for any work is made keeping in view the requirement of design, availability of material, cost and other similar factors.

29·7. TAPS, VALVES AND COCKS

The term taps, valves and cocks are used to name different types of fittings required to contrrol the flow of water either along or at the end of a pipeline.

Valves : Valve is a fitting commonly used to control the flow of water along a pipeline. With the introduction of valves it is a possible to isolate any sections of a pipeline for the purposes of inspection, repair of a leak or addition/alteration to the already functioning water supply system. Following types of valves are commonly used in domestic water supply system.

(*a*) *Globe valve*

(*b*) *Gate valve*

(*c*) *Float valve*

(*d*) *Taps and stop cock.*

(*a*) **Globe valve.** This is the most commonly used type of valve for manually controlling or completely closing the flow of water in domestic water installation. This type of valve is normally made up of brass. Globe valve as such or in some modified form is used to control flow of water to wash basins, shower, kitchen sink etc. Fig. 29·1 shows a section through a globe valve. In its simplest form it consists of a disc which is forced down by a screw against a circular seat. The disc and the screw form a single moving part which is operated by a wheel head. This type of valve is normally used for high pressure system where it may be necessary to shut off the water supply completely.

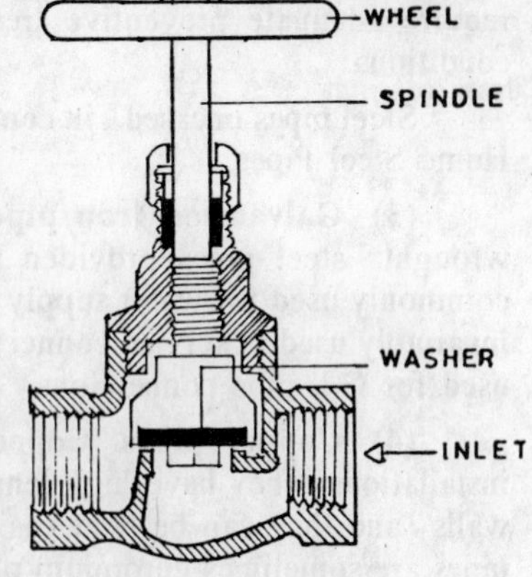

Fig. 29·1

(*b*) **Gate valve.** This type of valve is used to control flow of water or for completely stopping the flow of water in a pipe line. This type is normally used in low-pressure system and it offers much less resistance to flow of water as compared with Glob valve. Gate valve is used for controlling the discharge to the outlet from a storage tank.

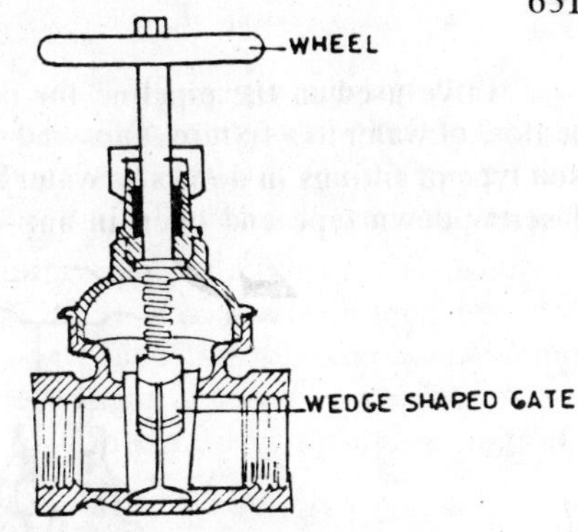

Fig. 29·2.

(*c*) **Float valve or a ball valve.** Float valve is used to supply water to a storage tank or flushing cistern and to automatically shut off the supply when the pre-determined level is reached. The valve is operated by a float which allows the valve to be fully open when it is in lower position. As the water level rises, the float also rises which gradually closes the valve and shuts off the supply of water as soon as the water reaches the full supply level mark.

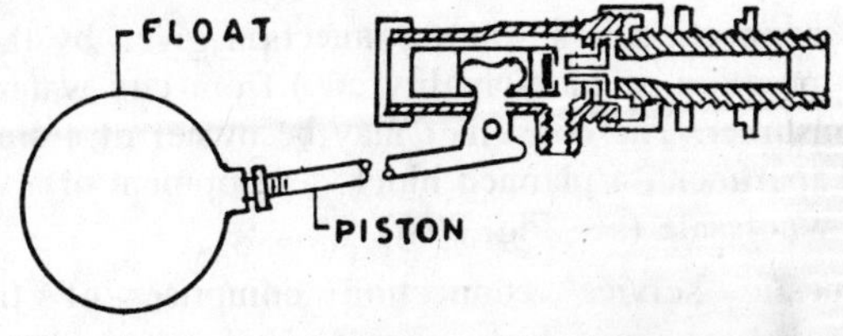

Fig. 29·3. Float valve or a ball valve

(*d*) **Taps and stop cock.** Taps are used at the end of a pipeline for draw off purposes. Taps are also called Bib cock or Bib tap. A stop cock

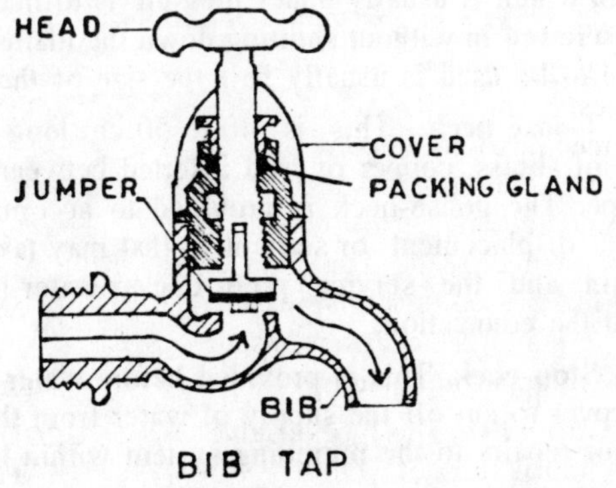

Fig. 29·4

is a valve used in the pipeline for controlling or completely stopping the flow of water to a fixture. Taps and stop cocks are two most extensively used type of fittings in domestic water supply system. They are normally of screw down type and open in anti-clockwise direction.

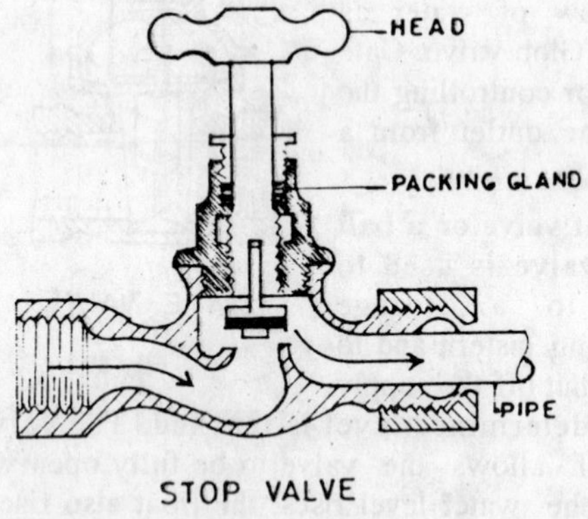

Fig. 29·5.

29·8. SERVICE CONNECTION

Service connection is a water connection given by the local body (municipal corporation or municipality etc.) from city water distribution mains to a consumer. The consumer may be owner of a single house, a multistoreyed apartment, a planned block development or a water district buying water wholesale (see Fig. 29·6).

A domestic service connection comprise of the following components.

(1) **Brass or bronze ferrule.** Ferrule is a special type of appliance made up of brass or bronze. It has a vertical inlet for screwing on to the water main and a horizontal outlet to be connected to service pipe. The water main which is usually under pressure is drilled and tapped and the ferrule is screwed in without shutting down the mains. The normal size of the ferrule to be used is usually half the size of the service pipe.

(2) **Goose neck.** This is 40 to 50 cm long flexible curved pipe made up of brass, copper or lead inserted between the ferrule and the service pipe. The goose-neck is provided to accommodate the possible movement, displacement or settlement that may take place between the water main and the service pipe due to water pressure and prevent damage to the connection.

(3) **Stop cock.** This is provided before water meter in a chamber with a cover to cut off the supply of water from the street main to the building for repairs to the plumbing system within the building.

(4) **Water meter.** Water meter is installed in a chamber provided with a cover for the purpose of measuring the quantity of water used by

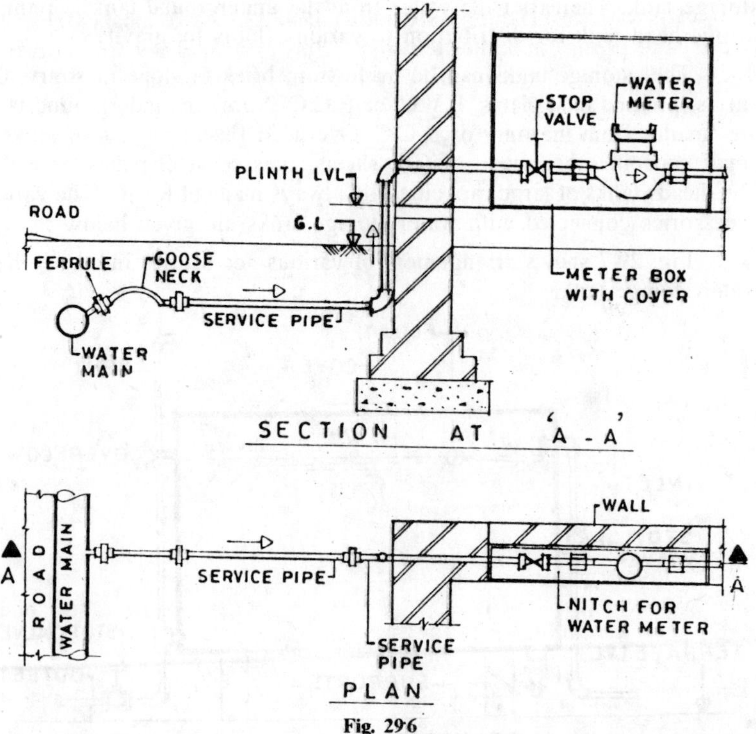

Fig. 29·6

the consumer. The local body raises water bill to the consumer based on consumption recorded by the water meter.

29·9. STORAGE TANKS IN BUILDINGS

Water supply to a building from city mains could be either continuous or intermittent. Normally due to continuously increasing demand and shortage of water, the local authority plan distribution of water in different city zones in two or three shifts (*i.e.* morning, evening and sometimes in afternoon). Even in areas where continuous supply of water is available the pressure of water in the mains may not be adequate to raise the water to upper floors. Thus provision of storage tank is made in a building to ensure availability of water during non supply hours or when the municipal supply is stopped.

In case of multi-storeyed buildings besides meeting demand of water for domestic consumption, it is mandatory to make provision of adequate overhead storage of water for fire fighting requirements. As explained earlier in case the pressure in the mains is not sufficient to feed all floors directly, it becomes necessary to feed water from mains to an underground storage tank. Thereafter the water from the underground tank is pumped to overhead tank for distribution to various floors by gravity.

The storage tanks can be made from brick or stone masonry, G.I. sheets, pressed m.s. plates, P.V.C. or R.C.C. Normally underground tanks are made from masonry or R.C.C. Overhead flushing tanks or tanks of small capacities are made of G.I. sheets, pressed steel plates or P.V.C. Overheads tanks of large capacities are always made of R.C.C. The various accessories connected with water storage tanks are given below :

Fig. 29·7 shows arrangement of various accessories in an overhead water storage tank.

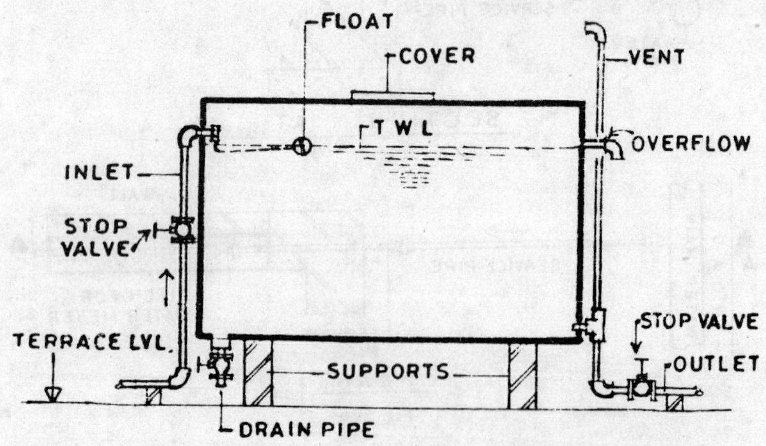

Fig. 29·7

(1) **Ball valve with float.** This is provided at the inlet to the tank to control the flow of water in the tank and to automatically shut off the supply when correct level has been reached.

(2) **Inlet pipe.** The pipe supplying water to the tank is termed as inlet pipe.

(3) **Outlet pipe.** This pipe is installed at 3 to 5 cm above the floor of the tank. The pipe is always provided with a stop valve to stop supply of water to downtake pipe.

(4) **Over flow pipe.** This pipe is provided a little above the inlet pipe to allow the incoming water to overflow in case the ball valve assembly does not function properly and it is not able to shut off the incoming supply of water. This pipe is provided with mosquito proof netting to prevent entry of mosquitoes, flies etc. into the tank.

(5) **Scour pipe.** Also known as drain pipe is provided at the floor of the tank for cleaning the tank.

(6) **Cover.** The manhole cover on the roof of the tank should be tight fitting type to prevent entry of dust, mosquitoes etc. in the tank.

Details of various pipe connections to an R.C.C. overhead (O.H.) water storage tank are shown in Fig. 29·8.

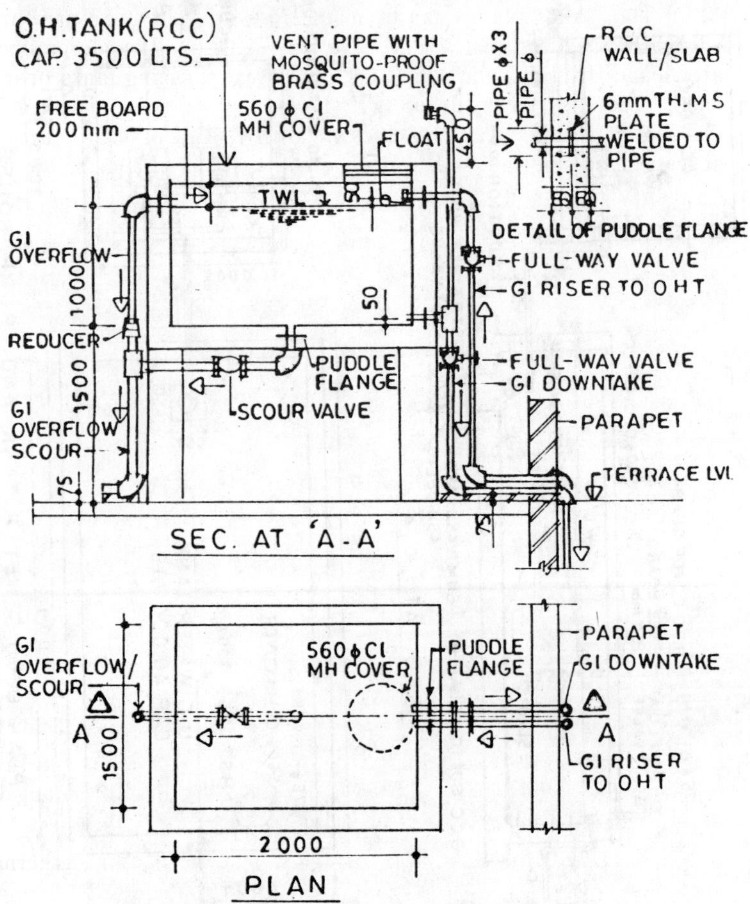

Fig. 29·8

Details of underground (U.G.) water tank and pumphouse are shown in Fig. 29·9.

Water supply scheme for a house : Typical plan for ground & first floor of a house showing details of water supply scheme are given in Fig. 29·10 and terrace plan of the house showing arrangement of water distribution line from overhead (O.H.) tanks to floors below is given in Fig. 29·11.

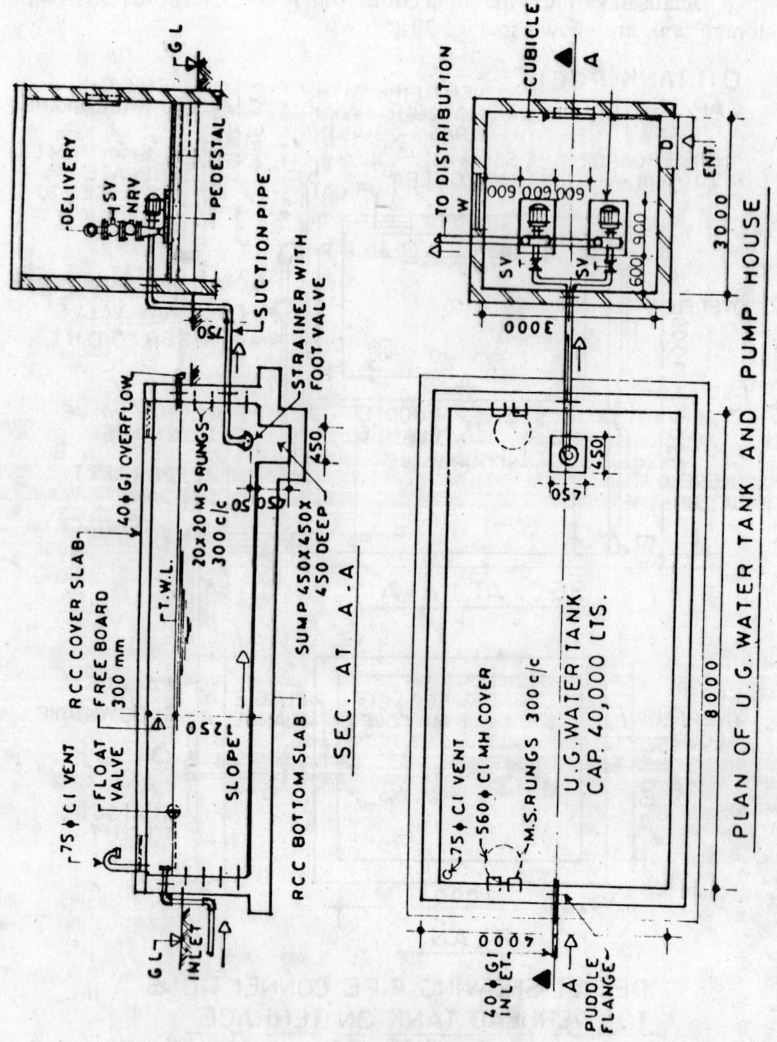

Fig. 29·9

Schematic view (Riser diagram) explains the water supply distribution scheme from O.H. tank to different fittings in the house.

Capacity of storage tank. The quantity of water to be stored depends upon the following

(*i*) Hours of water supply from mains with pressure sufficient to fill up the overhead tank.

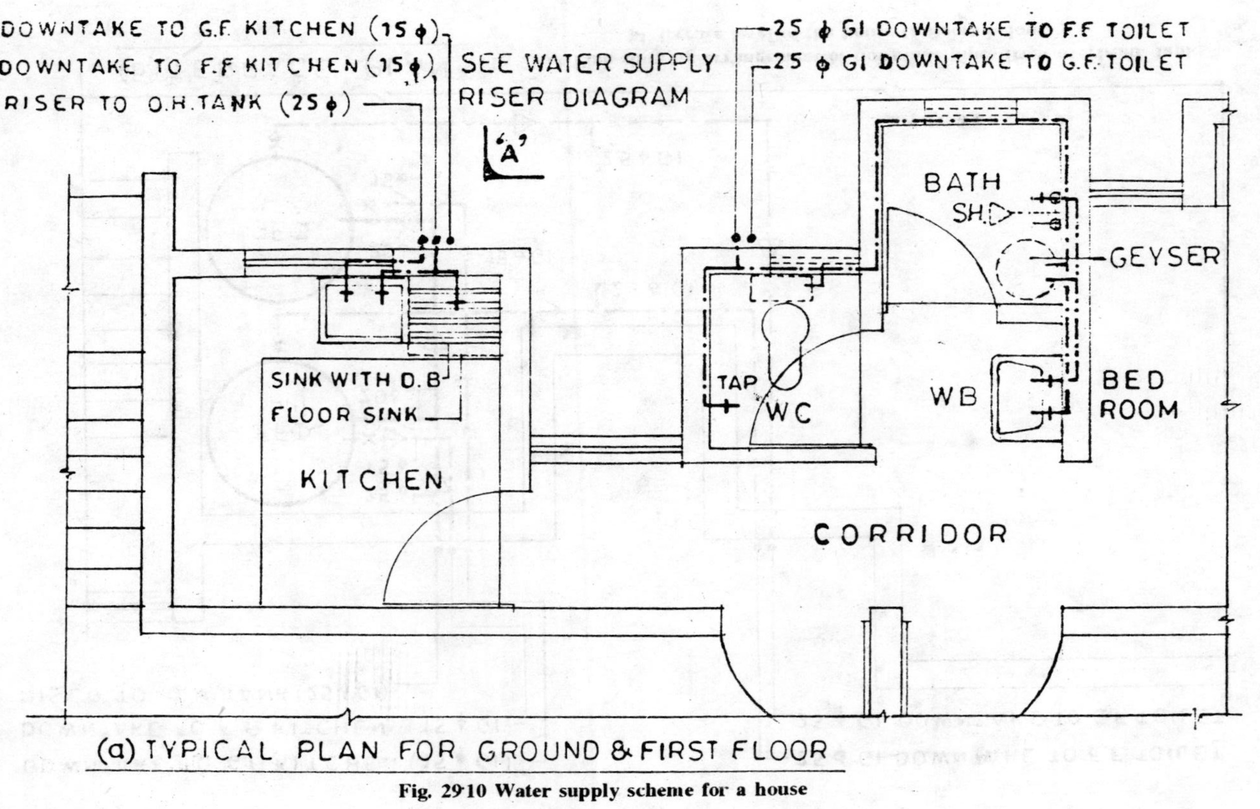

(a) TYPICAL PLAN FOR GROUND & FIRST FLOOR

Fig. 29·10 Water supply scheme for a house

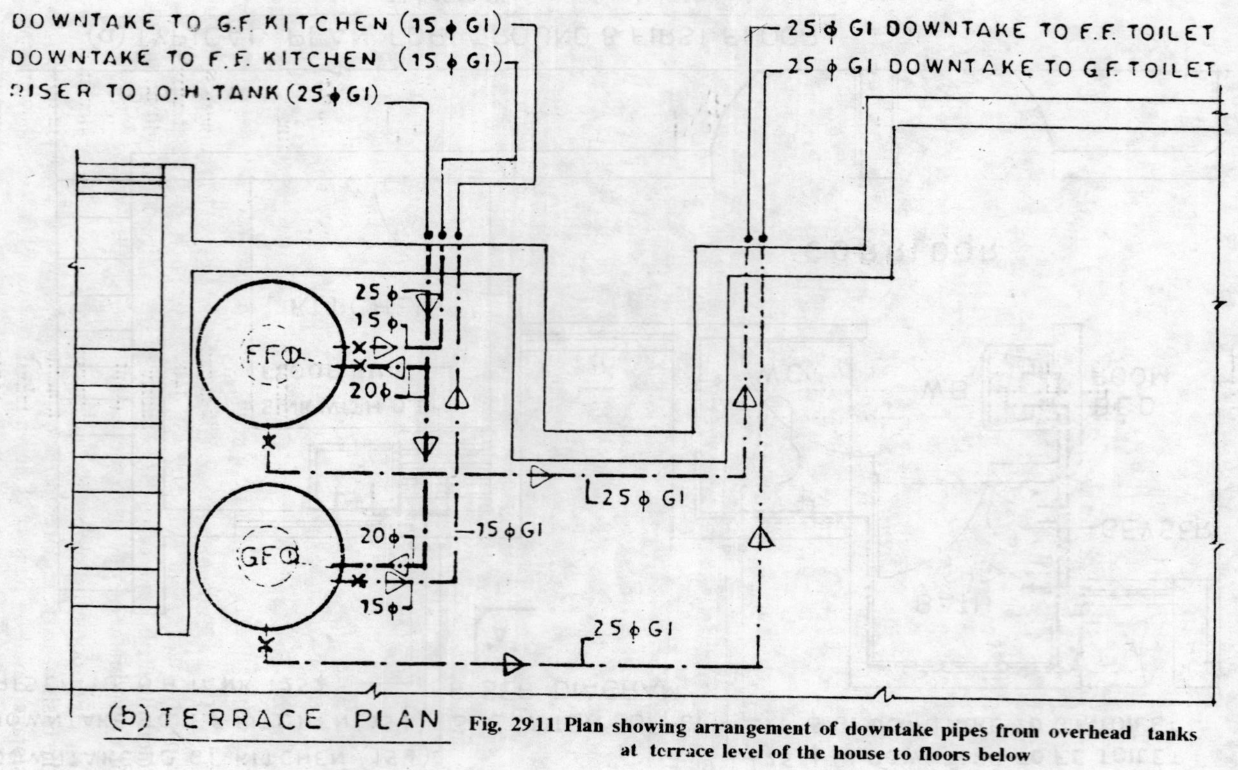

Fig. 29.11 Plan showing arrangement of downtake pipes from overhead tanks at terrace level of the house to floors below

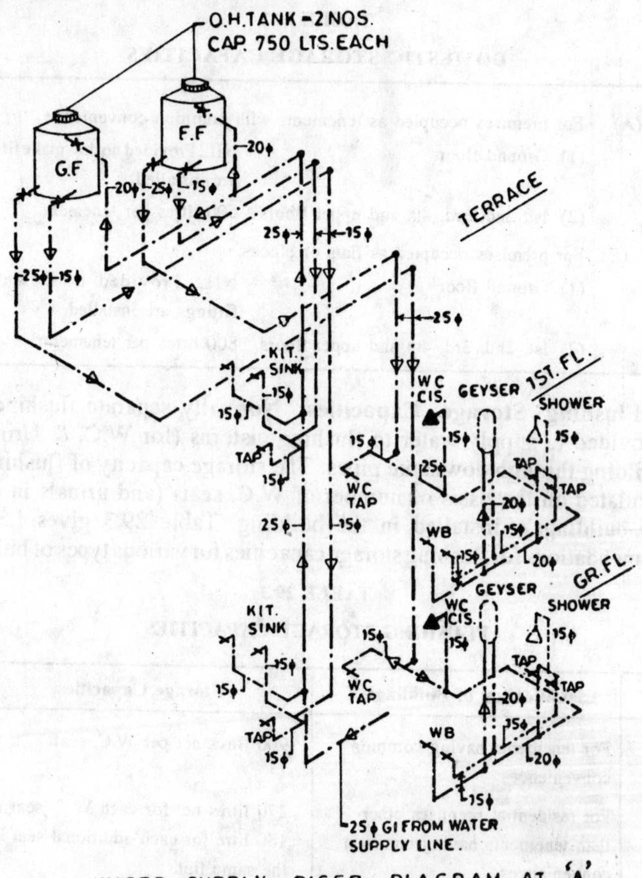

Fig. 29.12

(*ii*) Frequency with which the O.H. tank can be filled during 24 hours.

(*iii*) Rate and regularity of water supply from the mains.

(*iv*) Fire fighting requirement.

The underground storage tank is designed to have a minimum storage capacity equivalent to 50% of daily demand if the water supply is intermittent. In such cases the capacity of the O.H. tank should be 100% or a maximum of one day requirement.

For domestic purposes, the storage requirements are calculated on the basis of basic requirement per tenement, subject to a certain minimum storage based on number of down take fittings such as taps, showers bath tubs, etc. fed from the storage tank. Table 29.2 gives I.S. recommendations for domestic storage capacities.

TABLE 29.2

DOMESTIC STORAGE CAPACITIES

(A) For premises occupied as tenements with common convenience	
(1) Ground floor	NIL. Provided no downtake fittings are installed.
(2) Ist, 2nd, 3rd, 4th and upper floors	500 litres per tenement.
(B) For premises occupied as flats or blocks.	
(1) Ground floor	NIL. Provided no downtake fittings are installed.
(2) Ist, 2nd, 3rd, 4th and upper floors	800 litres per tenement.

Flushing Storage Capacities. Normally separate flushing tanks are provided to supply water to flushing cisterns (for W.C. & Urinals) in the building through downtake pipes. The storage capacity of flushing tank is calculated on the basis of number of W.C. seats (and urinals in case of public buildings) installed in the building. Table 29.3 gives I.S. code recommendations for flushing storage capacities for various types of buildings.

TABLE 29.3

FLUSHING STORAGE CAPACITIES

S.No.	Classification of Buildings	Storage Capacities
1.	For tenements having common conveniences	900 litres net per W.C. seat.
2.	For residential premises other than tenements having common conveniences.	270 litres net for each W.C. seat and 180 litre for each additional seat in the same flat.
3.	For factories and workshops.	900 litres per W.C. seat and 180 litres per urinal seat.
4.	For cinemas, public assembly halls etc.	900 litres per W.C. seat and 360 litres per urinal seat.

DRAINAGE

29.10. GENERAL

As explained earlier, potable water supplied to a buildings is distributed to various areas like kitchen, bath, W.C. etc. through a network of pipes provided with plumbing or sanitary fittings at their terminal ends. Adequate arrangements are required to be made for quick collection, conveyance and disposal of used water from the fittings without any risk to the health of the occupants. It is also essential that the rain or storm water from the roof and paved areas of building and the ground surface is suitably collected and discharged without flooding the area. The term drainage or sewerage includes the system of removal of sullage or waste water (from floor traps, kitchen, bath, and wash basin), soil water (from W.C. and urinals) and storm water from a building and conveying the same upto its ultimate point of treatment and disposal. The system of drainage can be broadly divided in two parts.

(1) *Drainage below the ground.*

(2) *Drainage above the ground.*

(1) **Drainage below the ground.** This comprises of a system of underground house drain, inspection chamber, main drain or sewer, manholes, ventilation shafts etc. provided for conveying the sanitary sewage (soil water and waste water) and storm water for final treatment or disposal.

Underground drainage can be divided into the following three systems.

(*i*) Combined system.

(*ii*) Separate system.

(*iii*) Partially combined system.

(*i*) *Combined system.* In this system the storm water is completely mixed with the sanitary sewage and conveyed through a single drain or sewer.

(*ii*) *Separate system.* In this system the storm water is not allowed to get mixed with sanitary sewage. Two separate drains are provided, one for sanitary sewage and other for drainage of storm water.

(*iii*) *Partially combined system.* In this system a part of storm water (usually run off from roofs, paved yards and streets etc.) is mixed with sanitary sewage and conveyed through sewer and the remaining storm water is conveyed through separate surface drains.

(2) **Drainage above the ground.** This consists of a system of vertical stacks, horizontal branches, floor traps etc. provided for conveying sanitary sewage (soil water and waste water) storm water (rain water) etc. to the underground drainage system for final disposal. This system is also known as house or building drainage system.

29·11. TYPES OF PIPES

The pipes used in drainage system may be of cast iron, reinforced concrete, pre-stressed concrete, mild steel, wrought iron, asbestos, P.V.C.

or stone ware. For drainage work above ground, cast iron, asbestos cement or P.V.C. pipes are generally used. The pipes may be jointed either with cement mortar or lead. Stoneware pipes are commonly used for house drain. For sewers, stoneware pipes, cement concrete pipe or C.I. pipe are normally used.

Minimum size of pipes commonly used in house drainage are as under.

(*i*) Underground horizontal pipes laid to slope ....100 mm dia.

(*ii*) Vertical soil pipe (above ground) ....100 mm dia.

(*iii*) Vertical waste pipe (above ground).........50 to 75 mm dia.

(*iv*) Branch soil pipe ....60 to 75 mm. dia.

(*v*) Vent pipe...50 mm dia.

(*vi*) Rain water pipe ....75 mm dia.

(*vii*) Waste pipe from individual kitchen, sink, washbasins etc. ....40 mm dia. G.I. pipe.

29·12. TRAPS

A trap is a fitting provided in a drainage system to prevent entry of foul air or gases from the sewer or drain into the building. The barrier to the passage of foul air is provided by the water seal in the trap. In its simplest form a trap is merely a double bend or loop in the sanitary fitting, the depth of water seal being the distance between the top

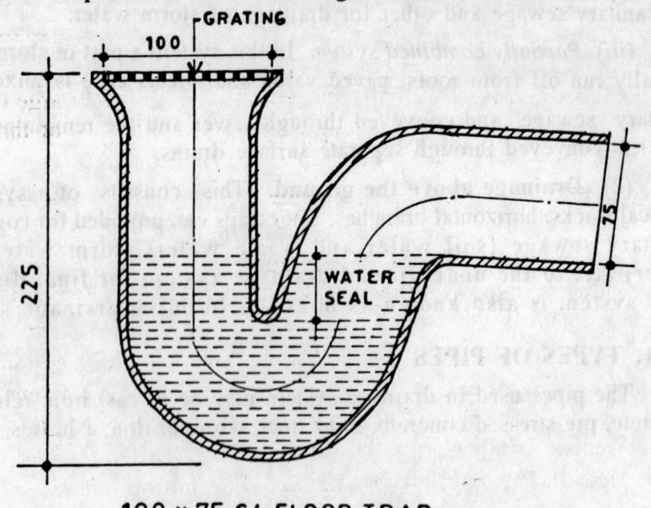

Fig. 29·13

of the first bend and the bottom of the second. The deeper the seal the more efficient is the trap. Depending upon the design of the trap, the depth of water seal vary from 40 mm to 75 mm. The trap should always be fitted close to the waste or soil fitting unless the trap form an integrated part of the fitting as in case of European W.C. (siphonic type)

Essentials of a good trap. A good trap should have the following characteristics :

(1) It should maintain an efficient waterseal under all conditions of flow, both during the water flow as well as in the absence of water flow.

(2) It should be self cleansing.

(3) It should not have any internal projections, angles or contractions so as to permit unobstructed flow through it.

(4) It should have a smooth inner surface so that each part is automatically scoured by flow of water and there is no possibility of dust, dirt etc. getting struck to it.

(5) It should be provided with suitable means of access for cleaning purposes.

Causes of loss or breaking of water seal. The primary object of providing a trap is lost, in case it is not possible to retain its water seal. The water seal in a trap may break due to the following causes.

(1) Evaporation of water in the trap caused on account of not using the appliance for long time.

(2) Use of defective trap, defective installation of trap or development of crack in trap after installation.

(3) Creation of partial vacuum caused due to discharge of another fitting connected to the same stack leading to emptying of the water of the seal by induced siphonage.

(4) Build up of back pressure of sewer gas in the drain forcing up the water of the trap seal.

(5) Pressure on seal of trap due to sudden discharge of water in large quantity into the fitting (bucket full of water into a W.C.) forcing the seal to break due to self siphonage.

(6) Due to capillary action caused by piece of some porous material getting struck at the outlet of the trap in such a manner that one end of the piece remains in water of the seal and the other end of the piece remain hanging over the outlet.

Type of traps. Traps can be made in different shapes and they are normally named after the shape of the letter they resemble. Out of the different shapes, the trap resembling the letter, P (or P-Trap), Q (Q-Trap) and S (S-Trap) are more common. The traps are normally

WATER SEAL
P-TRAP
WATER
SEAL
S-TRAP
WATER
SEAL
Q-TRAP

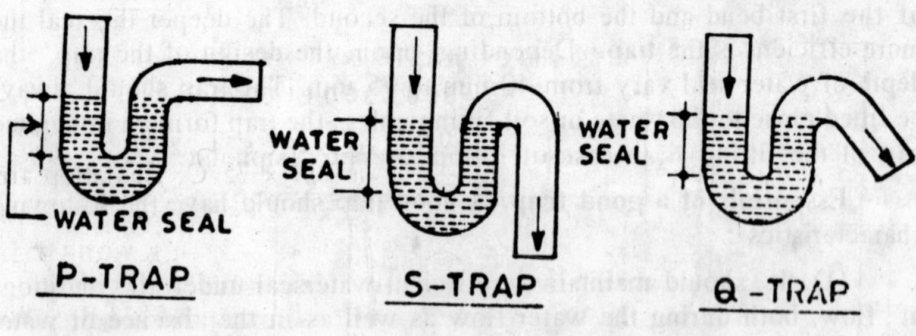

Fig. 29·14

made of cast iron and glazed stone ware. Depending upon the use and location, the various types of traps can be broadly summarised as under

(*i*) *Floor trap or Nahani trap*

(*ii*) *Gully trap*

(*iii*) *Intercepting trap*

(*iv*) *Grease trap*

(*v*) *Silt trap*

(*i*) *Floor trap or Nahani trap* : Trap provided in floors to collect used water from floors of bath room, kitchen or washing floor etc. are

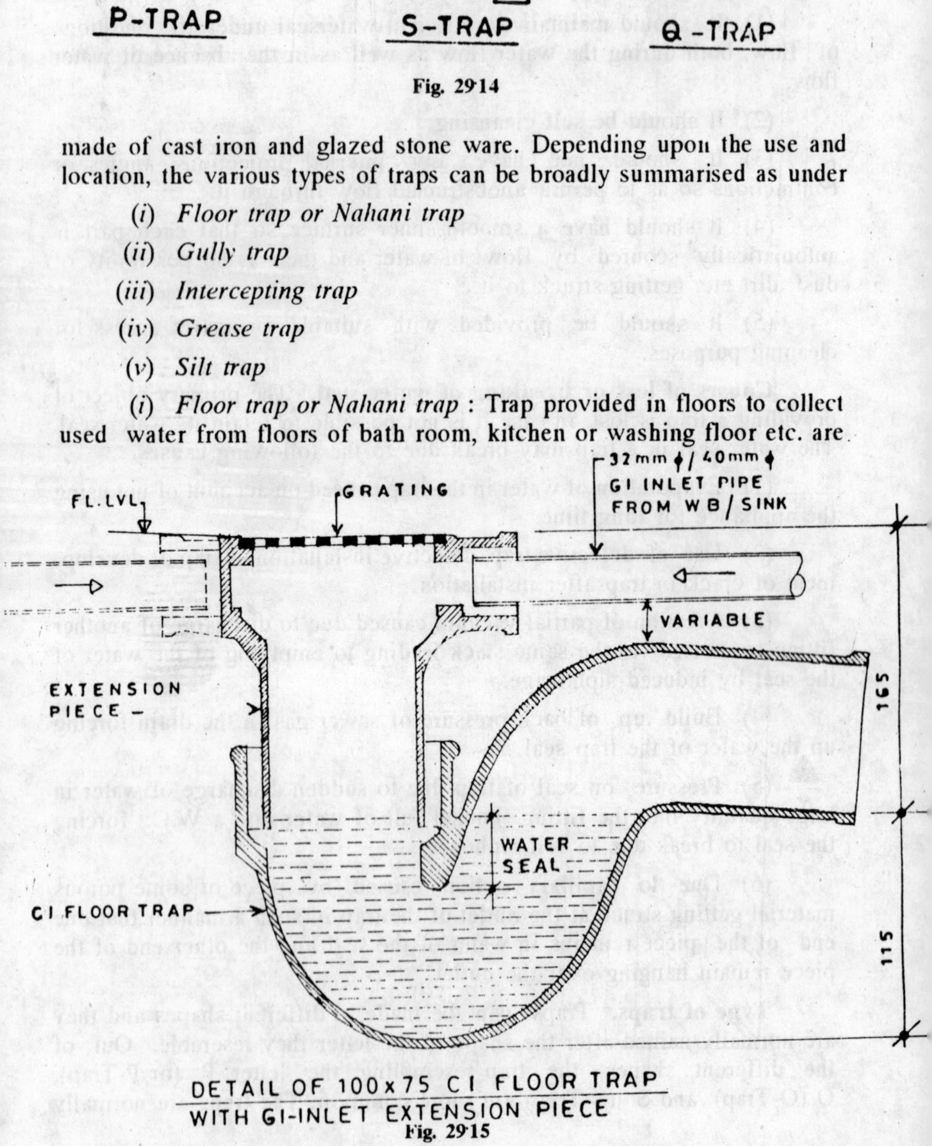

32mm ϕ / 40mm ϕ
GI INLET PIPE
FROM WB / SINK
FL.LVL
GRATING
VARIABLE
EXTENSION
PIECE
165
WATER
SEAL
CI FLOOR TRAP
115

DETAIL OF 100×75 CI FLOOR TRAP
WITH GI-INLET EXTENSION PIECE

Fig. 29·15

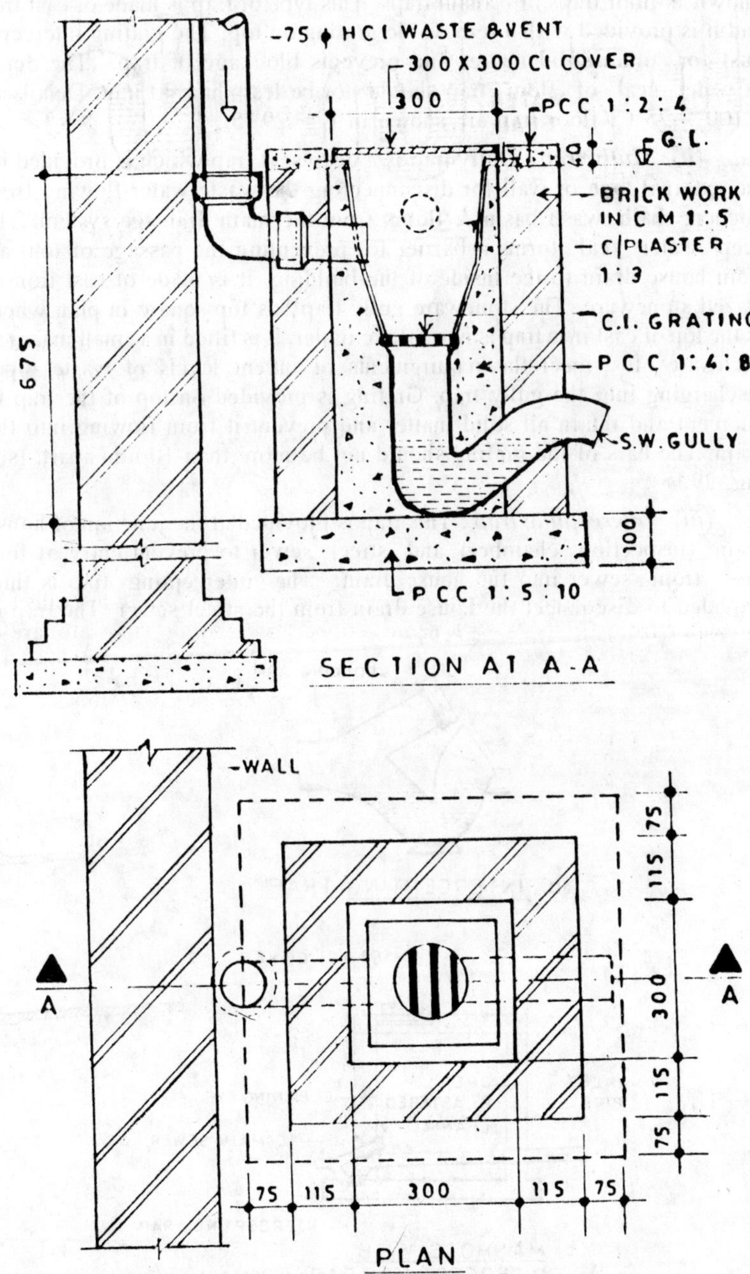
-75 ϕ HCI WASTE & VENT
300 X 300 CI COVER
300
PCC 1:2:4
G.L.
BRICK WORK IN CM 1:5
C/PLASTER 1:3
C.I. GRATING
PCC 1:4:8
S.W. GULLY
675
100
PCC 1:5:10
SECTION AT A-A
WALL
75
115
300
115
75
A
A
75 115 300 115 75
PLAN
DETAIL OF GULLY TRAP

Fig. 29·16

known as floor traps or Nahani traps. This type of trap is made of cast iron and it is provided with a removable grating on top. The grating intercepts dust or other solid matter and prevents blockage of trap. The depth of water seal of floor trap should not be less than 40 mm. Details of a 100 × 75 CI floor trap are shown in Fig. 29·15.

(*ii*) *Gully trap* : Gully trap is a deep seal trap which is provided on the external face of wall for disconnecting the waste water flowing from kitchen, bath, wash basin & floors from the main drainage system. The deep water seal forms a barrier for preventing the passage of foul air from house drain to the inside of the building. It is made of cast iron or glazed stoneware. The stoneware gully trap has top square in plan where as the top of cast iron trap is normally circular. It is fitted in a small masonry enclosure to meet the requirements of invent levels of waste pipes discharging into the gully trap. Grating is provided on top of the trap to intercept and retain all solid matter and prevent it from flowing into the drain. The bars of the grating should not be more than 10 mm apart. (see Fig. 29·16).

(*iii*) *Intercepting trap* : This trap is provided at the junction of house drain (inspection chamber) and street sewer to prevent entry of foul gases from sewer into the house drain. The intercepting trap is thus provided to disconnect the house drain from the street sewer. The trap is

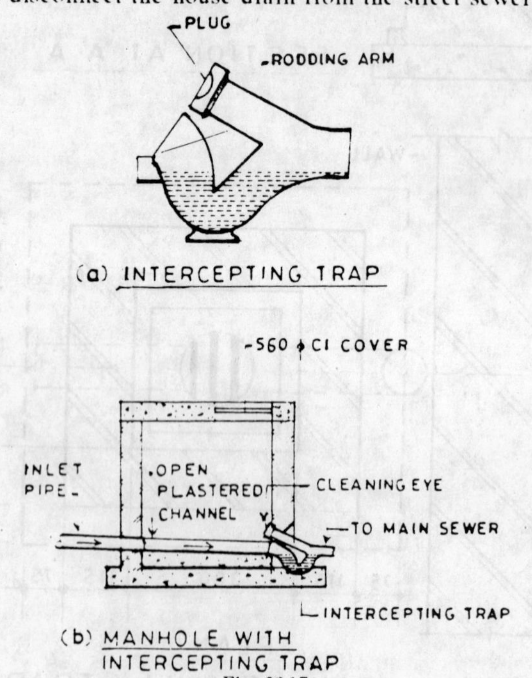

Fig. 29·17

made of glazed stoneware and has a opening at top (known as cleaning eye). The opening is kept closed with a tight fitting plug which is taken out only during cleaning of the trap. It has a deeper seal than normal traps (not less than 100 mm) (see Fig 29·17).

(*iv*) *Grease trap* : Grease traps are provided in large hotels, restaurants or other industries producing large quantity of greasy waste with the primary aim of removing the grease content of waste water before discharging the same into drain. If the greasy or oily matter is not removed, being sticky in nature, it will induce deposition of solids in the drain which can cause obstruction to the flow of water in the drain and may finally result in blockage of drain. Grease trap is a small masonry or cast iron chamber with a T or bent pipe to serve as the outlet. The velocity of waste water flow gets reduced on entering the grease trap (because of sudden increase in area of holding of waste water) and this results in separation of oily or greasy matter from the waste water. The greasy matter appears as floating (in the trap) which is removed periodically with the help of a mild steel tray. The details of the Grease tray are shown in Fig. 29·19.

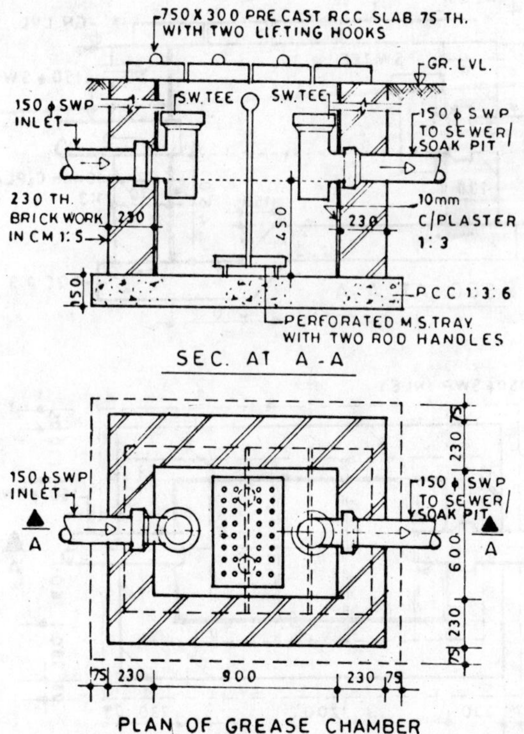

Fig. 29·18

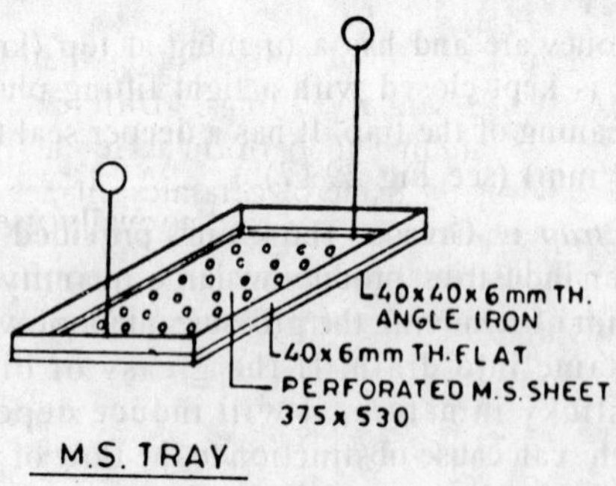

Fig. 29·19

(v) *Silt trap* : Silt traps are provided only in situations where the waste water carries large amount of silt, sand, coarse particles etc. It is a masonry chamber which functions like grit chamber where the silt, sand etc. settle down before the waste water is discharged into the drainage system. The details of the silt trap or silt chamber are shown in Fig. 29·20.

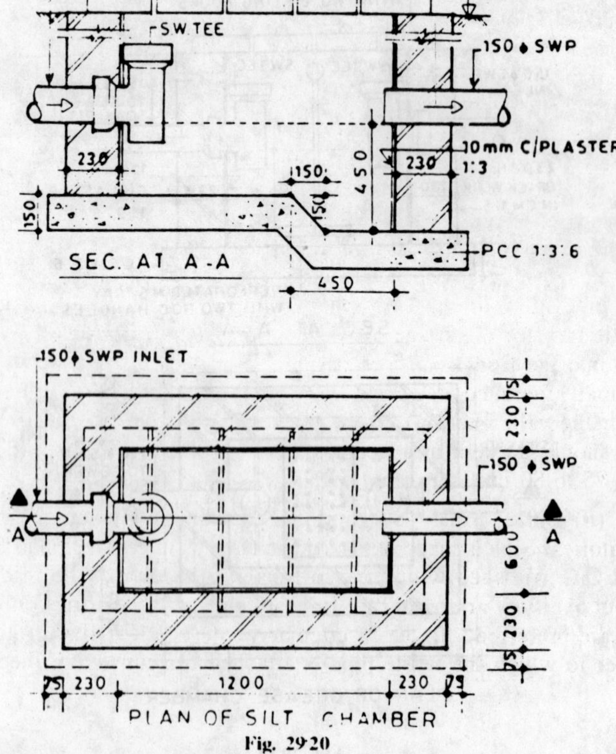

Fig. 29·20

29·13. SANITARY FITTINGS

The fittings or appliance used for collection and discharge of soil or waster matter is termed as sanitary fittings. Different types of sanitary fittings are required in building to perform different type of functions. Sanitary fittings are normally made of ceramics, glazed fire clay, glazed earthen ware or glazed chinaware. The fittings are so designed and shaped that they have non-absorbent surface which can be cleaned easily. The different type of sanitary fittings normally used in buildings are as under :

(*i*) *Wash basin*

(*ii*) *Sink*

(*iii*) *Bath tub*

(*iv*) *Urinal*

(*v*) *W.C.*

(*vi*) *Flushing cistern*

(*i*) **Wash basin.** A wash basin is used for washing hands, face etc. It is made of porcelain vitreoús enamelled steel, or plastic and is available in various pattern and sizes. The type of wash basin normally used in a house has an oval shaped bowl with an overflow slot at top. The wash basin has a flat back and has provision for making holes for installing one, two or even three taps. Normally two pillar taps are provided one for cold water and the other for hot water. It has a circular waste hole for draining out waste water from the basin. A metallic stainer is provided in the waste hole which is connected to a waste pipe either directly or through a bottle trap for discharge of waste water into the floor trap. The basin is normally mounted on 2 angle irons fixed in the wall. The top of the wash basin should be kept at a height of about 75 to 80 cm from floor level.

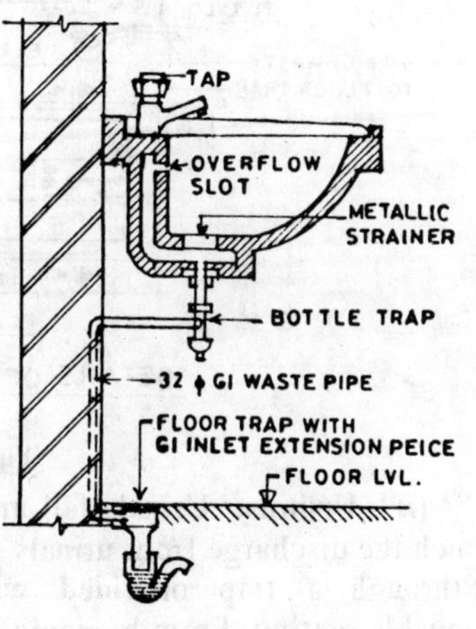

Fig. 29·21

(*ii*) **Sink.** Sink is commonly used in kitchen, hospitals and laboratories. Sink is made of glazed fire clay, stainless steel, plastic, marble and R.C.C. (finished with terrazzo finishing). Sink may be made with or without overflow arrangement. In hospitals and laboratories only vitreous sinks are preferred. It nas a circular waste hole for fixing the metallic strainer to which the waste pipe is attached for conveying the discharge from sink to the floor trap. The kitchen sink is invariably provided with a drain board.

(*iii*) **Bath tub.** A shower is considered to be more efficient and hygienic means of taking bath and as such is widely used in bathrooms by most of the people. However, some people prefer to use bath tub as they find it to be more comfortable and relaxing. Use of bath tub is restricted to certain class of hotels and private residences. Bath tubs are made of enamelled steel, gel-coated fibre glass, reinforced resign, enamelled porcelain, reinforced concrete finished with terra-cotta or marble finishes etc. Bath tubs are provided with holes for fixing hot and cold water connections and have provision for over flow and waste water pipes. The length of bath tubs varies from 1·75 m to 1·85 m, width varies between 0·7 m to 0·75 m and its depth near the waste pipe varies between 0·43 to 0·45 m.

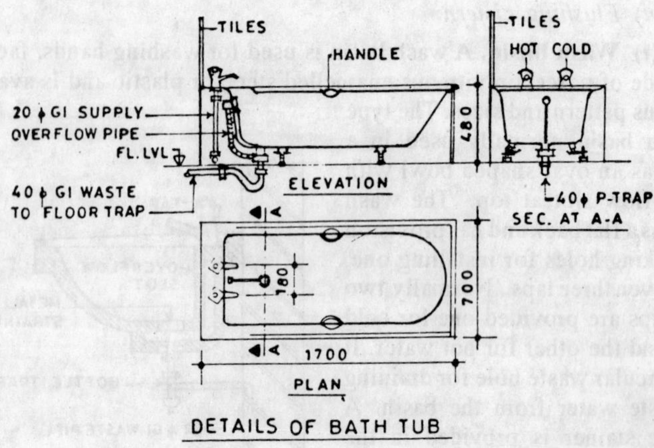

Fig. 29·22

(*iv*) **Urinals.** Urinals fall under the category of soil appliance and as such the discharge from urinals is connected to soil pipe either directly or through a trap provided with gun-metal or bass domed shaped removable grating. From hygienic consideration it is desirable to provide glazed tiles on walls of urinal preferably upto door height. Following types of urinals are commonly used

(*i*) *Bowl Type*

(*ii*) *Slab or Stall type.*

(*i*) *Bowl type* : This is one piece construction with integral flushing box rim having 12 hole suitably distributed for proper flushing. The urinal has an outlet horn at bottom for connecting to the trap and an outlet pipe. In another pattern of bowl type urinal, the appliance has a porcelain trap inbuilt with the pan as a single piece. In case number of urinals are required to be installed in a row, it is necessary to provide vertical partition between two urinals from consideration of privacy.

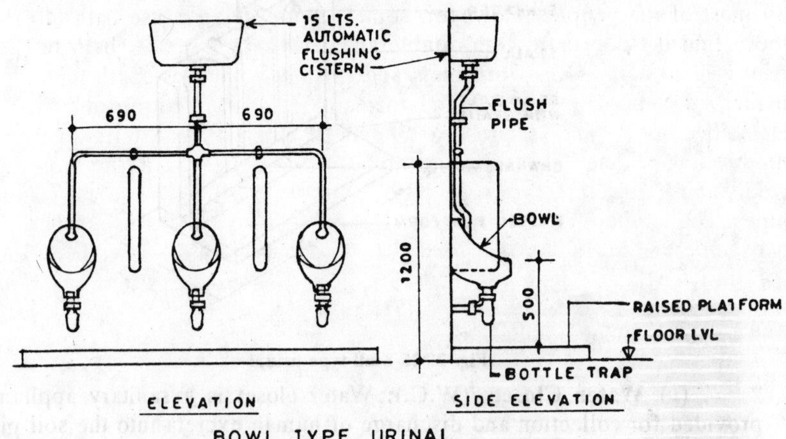

Fig. 29·23 Bowl type urinal

(*ii*) *Slab or stall type :* This type of urinal is manufactured either as a single unit or as a range of two or more units. In case of single unit the width of stall should not be less than 75 cm. The flushing of urine is normally carried out through automatic flushing cistern which operates at regular intervals of 10 to 15 minutes. The discharge from the series of stalls in a row is usually carried through a glazed semi-circular drain which has a sharp fall towards the trap from where it is discharged into the soil pipe.

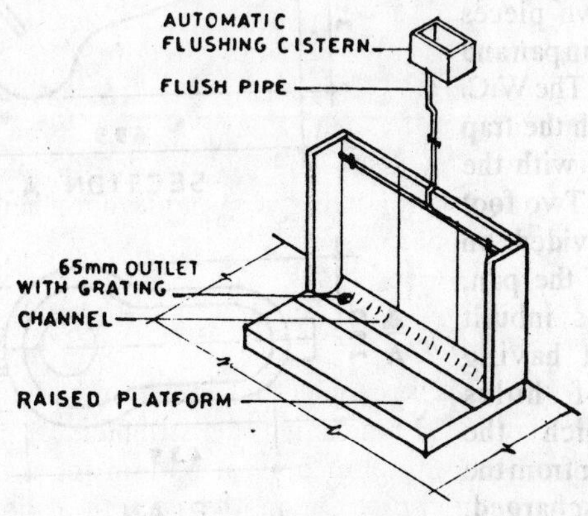

Fig. 29·24 Slab-type urinal

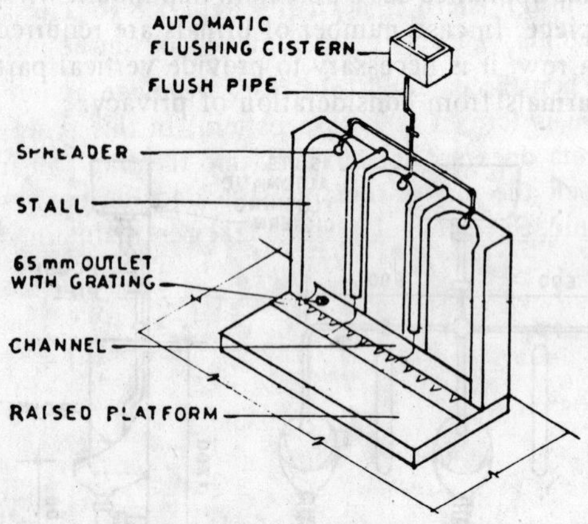

Fig. 29·25 Stall-type urinal

(*v*) **Water Closets (W.C.):** Water closet is a sanitary appliance provided for collection and discharge of human excreta into the soil pipe through a trap. The W.C. is connected to a flushing cistern to flush the excreta from the pan. W.C. is made of glazed earthen ware, fire clay or white vitreous chinaware. The types of W.Cs commonly used are as under :

(*a*) *Indian or squatting Type*

(*b*)) *European type*

(*c*) *Anglo- Indian type.*

(*a*) *Indian or squatting type W.C.*: This type of W.C. is used in squatting position. The W.C. consists of two pieces *i.e.* (*i*) Porcelain pan and (*ii*) a Pors trap. The W.C. pan alongwith the trap is fixed flush with the floor of W.C. Two foot rests are provided on either side of the pan. The pan has inbuilt flushing rim having number of holes through which the flushing water from the cistern is discharged.

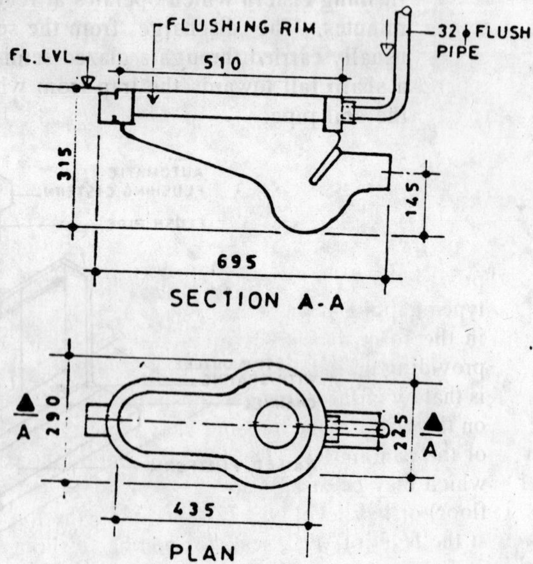

Fig. 29·26

The flushing cistern is normally kept 1.8 to 2 m above the floor level. The contents of the pan are removed by the gravity flush of water.

When the pan and foot rests are made as integral single piece, the appliance is termed as Orissa pattern. In this type while using the W.C. the excreta does not fall directly into the trap and in case the flushing is not proper, the matter remains struck to the pan which is considered unhygienic. Special care is to be taken to maintain cleanliness of the pan.

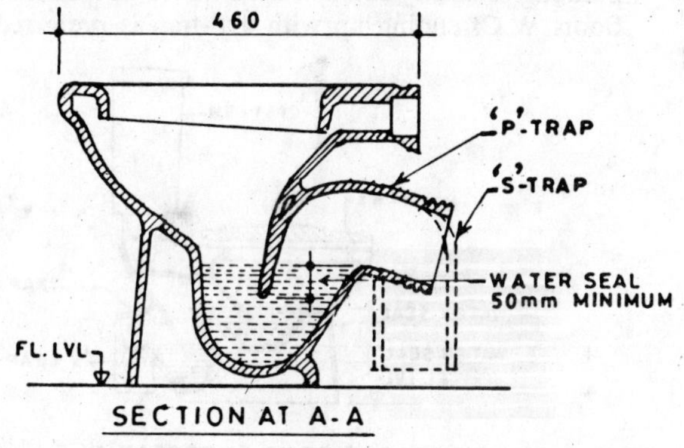

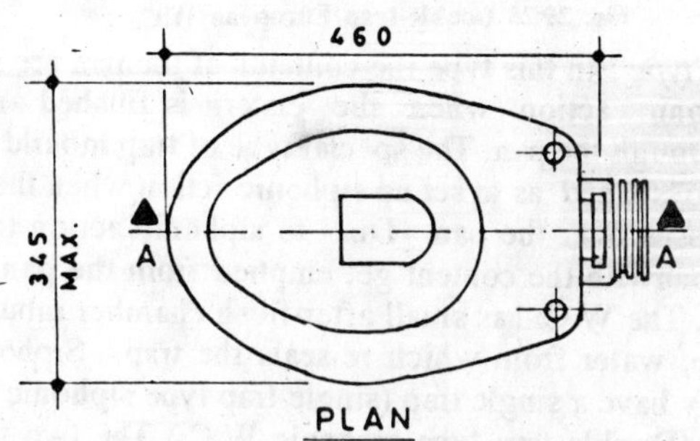

Fig. 29·27 Wash-down type European W.C.

(*b*) *European type W.C.* : This type of W.C. is used in sitting position over a plastic seat hinged to the appliance. This is a pedestal type of appliance with the pan and trap in a single piece. The pan is shaped in the form of a short inverted cone with an almost vertical back and providing minimum fouling area. The main advantage of this type of W.C. is that by virtue of the design/shape of pan, the excreta falls almost directly on the water in the trap and chances of the same getting struck to the sides of the pan are less. The flushing rim of the pan is attached to the cistern which may be of high level type (installed at a height of 1.8 to 2 m from floor) or low level type (30 cm above the top of W.C. seat or resting just at the level of W.C. seat) for getting flushing water. Following two types of European type W.C. are used :

(*i*) Wash-down type and

(*ii*) Siphonic type.

(*i*) *Wash-darn type of European W.C.* : This is the most commonly used pattern in which the content of the pan are removed by the gravity flush of water discharged into the pan through the flushing rim of the pan. For ground floor, normally the W.C. having ending up with S-trap (with outlet pointing vertically down) is used whereas for installation on upper floors W.C. ending up with a P-trap is preferred.

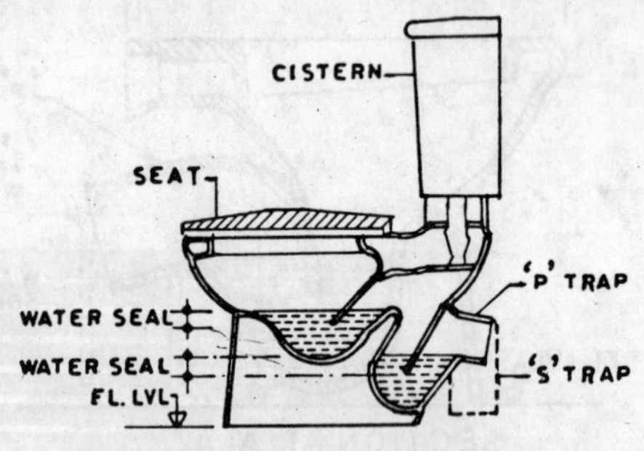

Fig. 29·28 Double-trap European W.C.

(*ii*) *Siphonic type* : In this type the contents of the pan are removed by siphonic action when the cistern is flushed and water passes through the pan. The special type of trap inbuild with the pan is so shaped as to set up siphonic action when the flushed water passes over the pan. Due to siphonic action the entire water alongwith the content get emptied from the pan into the soil pipe. The W.C. has small after-flush chamber inbuilt in the appliance, water from which re-seals the trap. Siphonic type W.C. may have a single trap (single-trap type siphonic W.C) or two trap (Double trap type siphonic W.C.) The two trap type W.C. has larger area of water seal and is more efficient and silent as compared with Single trap siphonic or washdown type of European W.C.

(*c*) *Anglo-Indian type.* This type of W.C. can be used both in squatting position as well as in sitting position and hence it is named as Anglo-India Type W.C. It is a pedestal type appliance with inbuilt trap. The top of the W.C. pan is flared out to provide for foot rest to permit it to be used in squatting position. When it is desired to use the appliance as European type W.C., the plastic seat hinged to the closet is turned on the top of the pan to use it in sitting position. In this the design is such that the fouling area of pan is less and the excreta falls in the water of trap and it is flushed out in manner similar to European typc W.C.

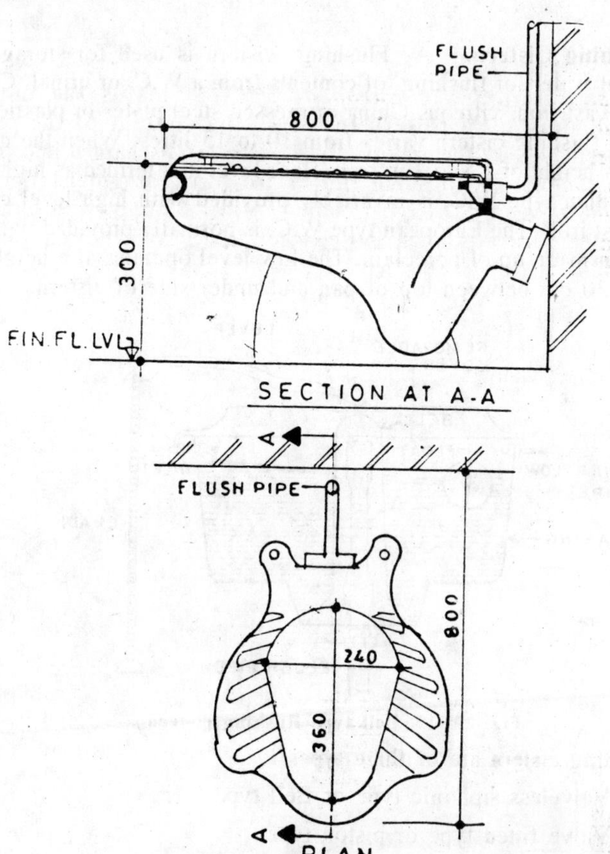

Fig. 29·29 Anglo-Indian type W.C.

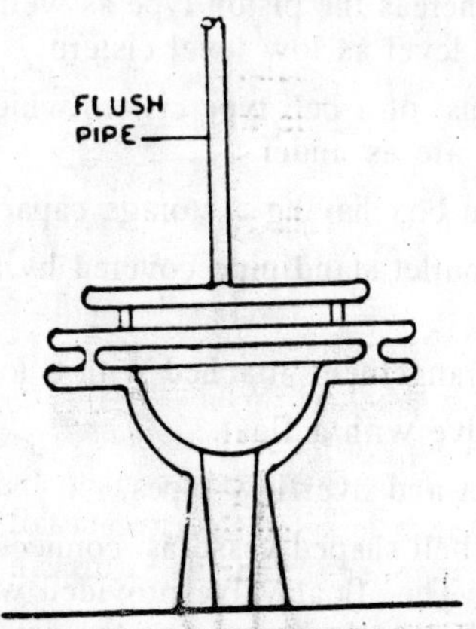

Fig. 29·30 Front elevation of W.C.

Flushing Cistern. A Flushing cistern is used for storage and discharge of water for flushing of contents from a W.C. or urinal. Cistern is made of cast iron, vitreous China or pressed steel plates or plastic. The capacity of flushing cistern varies from 10 to 15 litres. When the cistern is fixed at a height of 1.8 to 2 m from floor level it is termed as high level cistern. Indian type W.C. is invariably provided with high level cistern made of cast iron. The European type W.C. is normally provided with low level cistern made up of porcelain. The low level operates at a height not more than 30 cm between top of pan and under side of cistern.

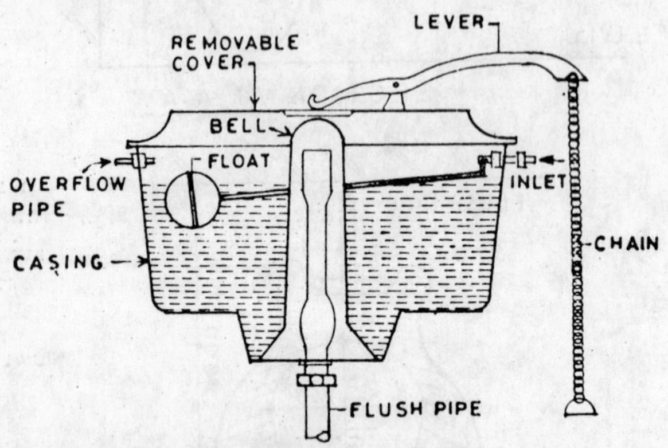

Fig. 29·31 Bell type flushing cistern

Flushing cistern are of three types :

(*a*) Valveless siphonic type or Bell type

(*b*) Valve fitted type or piston type

(*c*) Automatic flushing type

The valveless type or bell type of flushing cistern is used only as high level cistern whereas the piston type as well as automatic type can be used both as high level as low level cistern.

The components of a bell type cistern which is most widely used for Indian type W.C. are as under :

(*a*) A cast iron box having a storage capacity of 10 to 15 litres.

(*b*) A central outlet stand pipe covered by a bell or dome shaped cast iron vessel.

(*c*) A lever arrangement attached with a long chain.

(*d*) A float valve with a float.

(*e*) Inlet, outlet and overflow pipes.

In this case the bell shaped vessel is connected to a chain through a lever arrangement. The float valve provided with a ball float, closes the water supply to the cistern as soon as the pre-fixed level is reached.

For flushing the cistern, when the chain is pulled due to lever arrangement, the heavy bell is lefted up and suddenly released, the water enclosed within the bell is thrown over the top of the inner tube (Central outlet stand pipe) resulting in setting up the siphonic action which causes the entire stored water to get discharge into the flush pipe attached below the cistern.

29·14. SYSTEMS OF PLUMBING

There are four different systems of plumbing for house or building drainage as follows :

(*a*) *Two-pipe system*

(*b*) *One-pipe system*

(*c*) *Partially ventilated single stack system, and*

(*d*) *Single stack system*

(*a*) **Two-pipe system.** This is the traditional system of removing the discharge from sanitary fitting wherein the foul matter from W.C and urinals is discharged into one pipe (known as soil pipe or stack) and the waste water from kitchen, bath, wash basin, floor traps etc. (other than W.C.

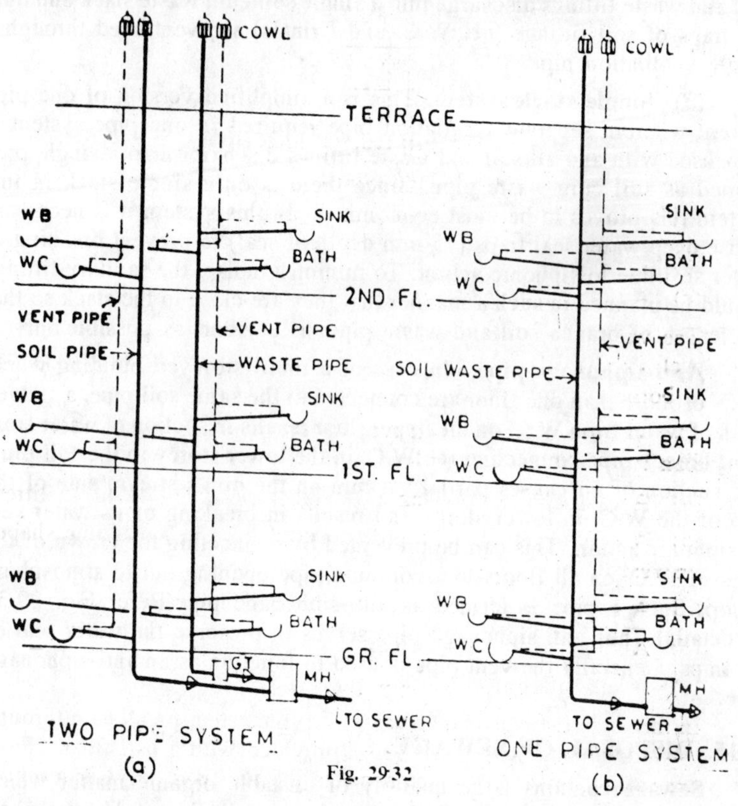

Fig. 29·32

& Urinals) is discharged into another pipe (known as waste pipe or stack). The soil pipe is directly connected to the building drain whereas the waste pipe is connected to the building drain through a gulley trap provided at the foot of the waste pipe. Separate ventilation pipe is connected with each stack. Thus four pipes are necessary in this system.

(*b*) **One-pipe system.** In this system all soil and waste fitting discharge into a single pipe termed as soil cum waste pipe and a separate vent pipe is provided to which all floor traps are connected for ventiliation of the system. There is no need of a gulley trap in this system and the soil cum waste pipe is directly connected to the building drain. This system works out to be economical in case of multi storeyed buildings where the baths, toilets, kitchens etc. are placed near one another on each floor and they are placed above one another on different floors around proper shafts. In this system it is essential that all traps are of deep water seal type (preferably 75 mm depth of seal) and the diameter of ventilation pipe is not less than 50 mm.

(*c*) **Partially ventilated single stack system.** In this system all soil and waste fittings discharge into a single soil cum waste stack and only the traps of soil fittings (*i.e.* W.C. and Urinals) are ventilated through a single ventilation pipe.

(*d*) **Single stack system.** This is a simplified version of one pipe system wherein separate ventilation pipe required in one pipe system is dispensed with and all soil and waste fittings discharge into a single pipe termed as soil cum waste pipe. Since there is only single stack in this system this proves to be most economical. In this system it is necessary to use deep water seal traps (75 mm depth of seal) to prevent breaking of water seal due to siphonic action. To minimise noise, the sanitary fittings should be planned in such a manner that they are close to the stack so that the length of branch soil and waste pipes is as short as possible.

Anti-siphonage pipe. In case of a multi storeyed building where W.Cs of more than one floor are connected to the same soil pipe, a sudden flush of water from W.C. on the upper floor results in suction of water from short branch pipe connecting the W.C. in the lower storey to the soil pipe. The suction of air causes partial vaccum on the down stream side of the trap of the W.C. in lower storey and results in breaking of its water seal by siphonic action. This can be prevented by connecting the crown of the traps of W.C. on all floors to a common pipe opening out to atmosphere at top. Such a pipe is termed as anti-siphonage pipe(Refer Fig. 29.34 for details). Thus anti-siphonage pipe serves to preserve the water seal of the traps. Generally the vent pipe is used to function as an anti-siphonage pipe.

29·15. DISPOSAL OF SEWAGE

Sewage contains large quantity of unstable organic matter which decays and emit offensive smell. Sewage also contains harmful bacteria and

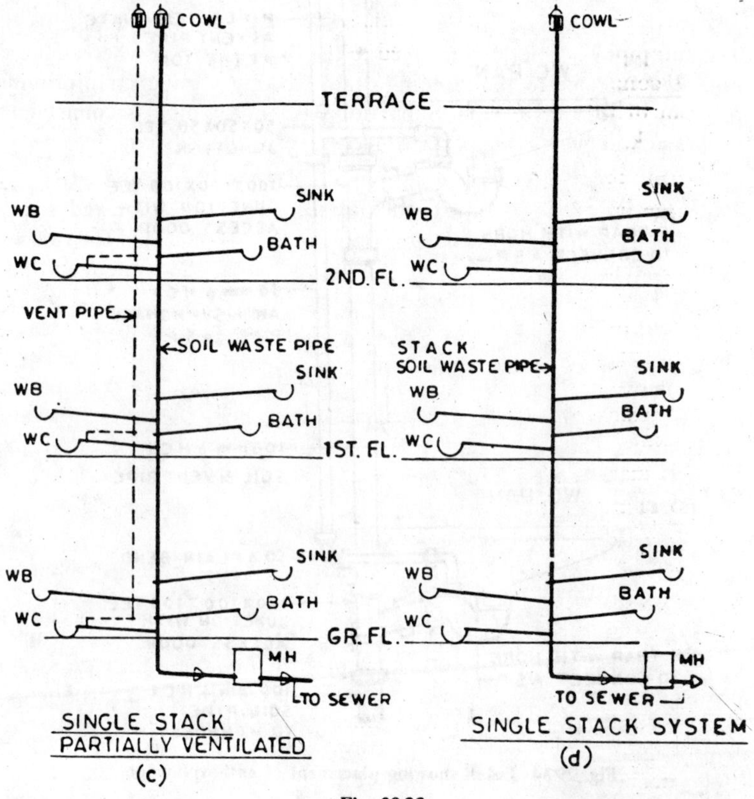

Fig. 29·33

in case it is left in raw state, it can cause real danger to human life. It is therefore, necessary to dispose off the sewage by suitable method to avoid danger of pollution. In case of cities or towns etc. the municipal authorities make necessary provision of sewage treatment plants. However in cases where municipal sewage system does not exist or where the buildings are located in isolated places, sewage from individual house or cluster of buildings can be disposed off by providing *septic tanks*. The effluent from the septic tank could be discharged either (*i*) *into a soak pit* or (*ii*) *by mode of dispersion trenches*. In case of a single house sewage could also be disposed off by providing a *sludge soak pit*.

29·16. SEPTIC TANK

In areas, where municipal underground sewage system is not available, the sewage from residential apartments, small residential colonies and isolated buildings like offices, schools, hospitals, hotels, factories etc. is treated in a septic tank.

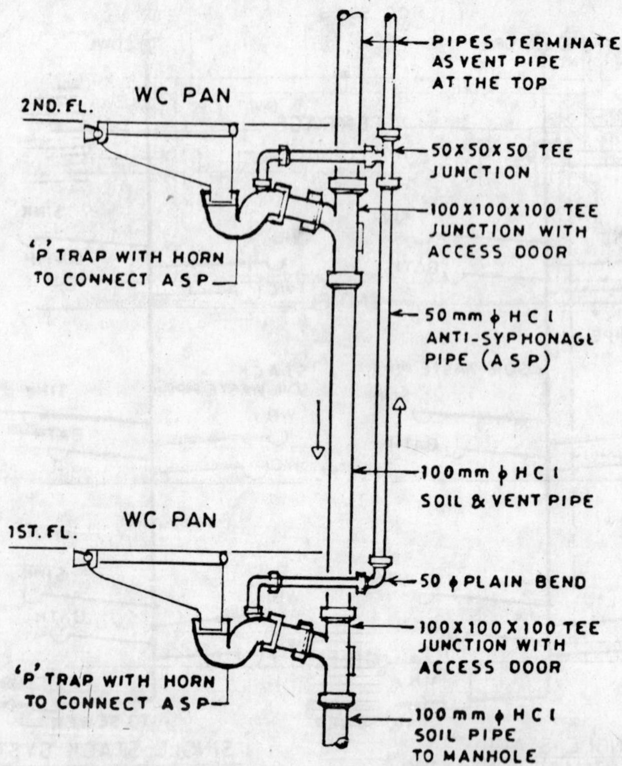

Fig. 29·34 Detail showing placement of anti-syphonage

A septic tank is a combined sedimentation cum digestion tank in which flow of raw sewage is slowed down so that the solids settle down at bottom of the tank by sedimentation. Simultaneously the sewage, in the tank is acted upon by anaerobic bacteria which during the process of digestion, convert the sewage into liquid and gaseous form and there is appreciable reduction in the volume of sludge. The sludge in the effluent thus falls on the tank floor and the scum (lighter matter *i.e.* grease, fats etc.) floats on top. Since the anaerobic action cannot take place in presence of oxygen, it is necessary that the septic tank is covered with a water light top (roof slab).

The effluent that flows out of the tank still contains considerable amount of dissolved and suspended organic solids and other harmful matter and as such it has to be properly disposed off into either (*i*) Soak pit or (*ii*) dispersion trenches for absorption in the soil.

A septic tank may be constructed in brick masonry, stone masonry or concrete. The tank is designed to provide for a detention period of 24

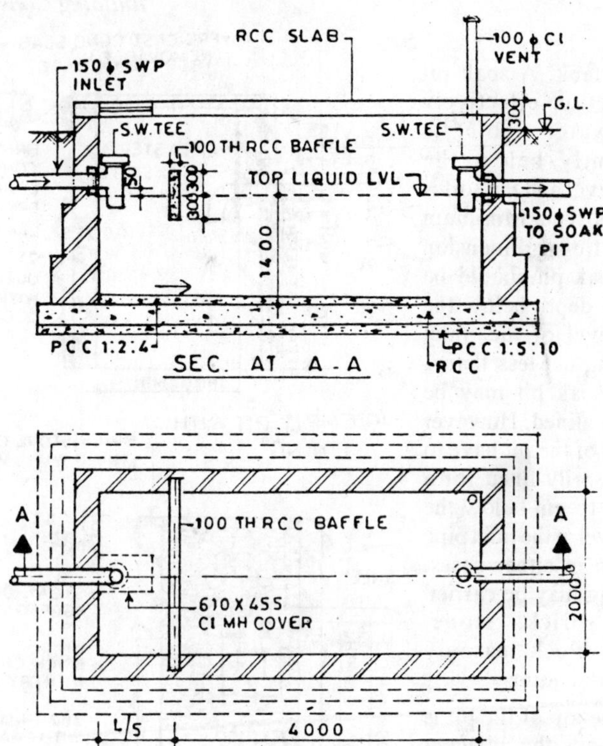

Fig. 29·35

to 48 hours based on an average daily flow of anticipated sewage. It should be water tight and as such all the inner surfaces of its walls, floor and roof slab are plastered with rich cement mortar. The floor of the tank should be in cement concrete and sloped towards sludge outlet. The invert level of the outlet pipe is kept 7 to 10 cm below the invert level of the inlet pipe. To prevent direct currents between inlet and outlet pipes pipe tees are provided in the two pipes. Baffle wall is provided near inlet pipe at a distance of 1/5th the length of tank. A vent pipe is provided on the roof of the tank for discharge of foul gases. For efficient working it is necessary to remove the digested sludge and scum from the tank periodically (say once in a year or so). Access manholes of adequate size are provided in the roof slab for inspection and cleaning.

(*i*) *Soak Pit*

Soak pit is a simple pit of any suitable shape (square or circular shapes are commonly used) dug in the ground for disposal of effluent from a septic tank. A soak pit can function effectively when water table is sufficiently below the ground level and the soil is of porous type. Minimum cross-sectional dimension of the soak pit should be 1 m, and depth below the invent level of the inlet pipe being not less than 1 m. The soak pit may be lined or unlined. However the walls of the pit have to be necessarily lined for a depth of 15 cm below the invent level of the inlet pipe up to top in either cases. The lining may be carried out by brick, stone, concrete block masonry with mortar joints.

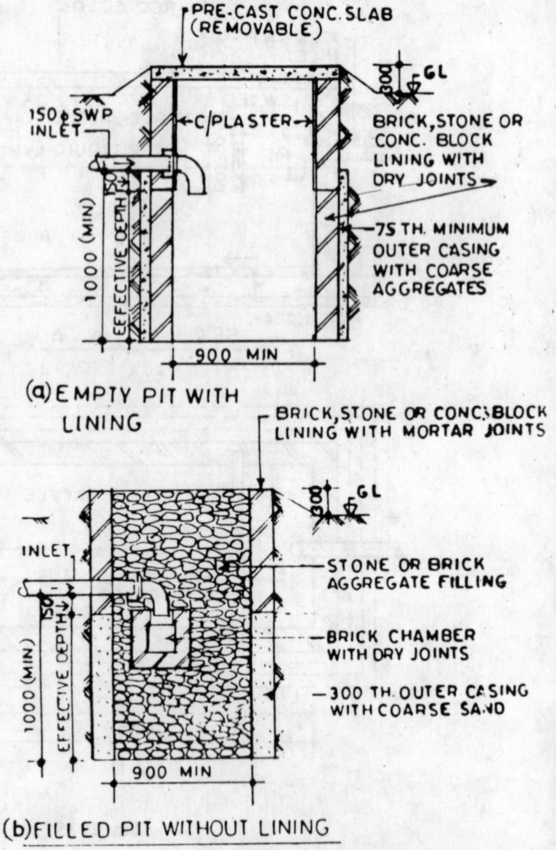

The top of the pit is raised above the adjacent ground to protect it from damage by rain.

(a) EMPTY PIT WITH LINING

(b) FILLED PIT WITHOUT LINING

Fig. 29·36

The lined pits need no filling media and the percolation of effluent into the soil takes place partly by seepage through dry joints of the lining and partly by direct soakage. The lined soak pit is covered with removable pre-cast concrete slabs.

In case of unlined soak pits, the pit is filled with rubble or brick bats.

The construction details of the pit are shown in Fig. 29·36.

(*ii*) *Dispersion Trenches :*

This is an alternative method of disposal of effluent from septic tank by soil absorption system. In this method the effluent from the outlet pipe of septic tank is conveyed into a small masonry distribution chamber through a pipe. From the distribution chamber the efluent is discharged into a number of dispersion trenches through tight jointed 75 to 100 mm dia. stoneware pipes. Each dispersion trench is a narrow and shallow trench about 30 cm to 100 cm wide and from 50 cm to 100 cm deep and

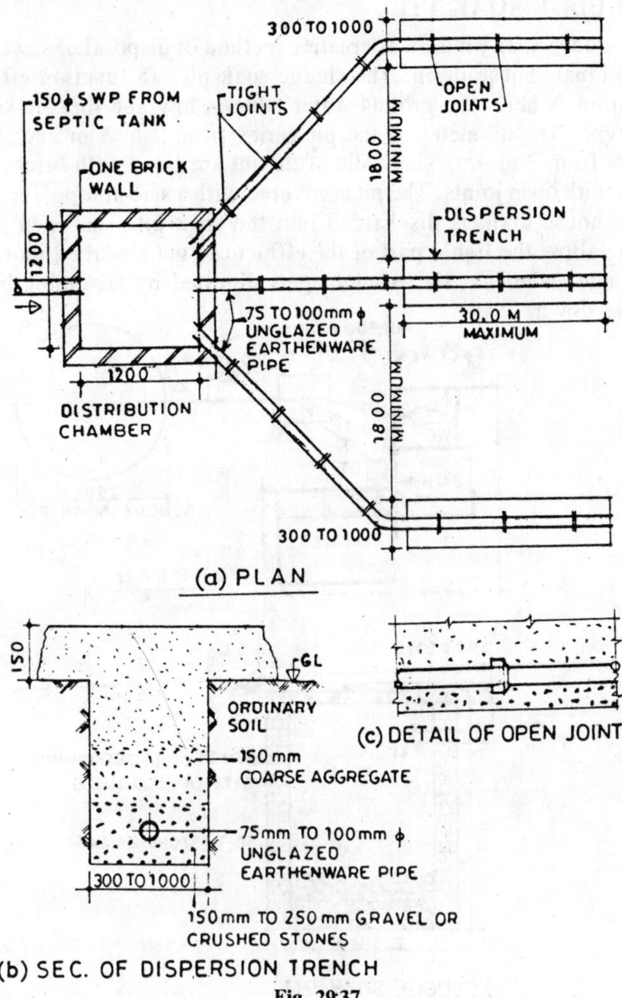

Fig. 29·37

it is excavated to slight gradient. Open jointed unglazed earthenware or concrete pipes, 75 to 100 mm in dia are laid in the trench over a bed of 15 to 25 cm of washed gravel or crushed stones. The remaining depth of the trench is filled with ordinary soil in the form of a mound to prevent flooding due to rain from adjacent ground. The centre to centre spacing of the dispersion trenches should not be less than 2 m and their length should not exceed 30 m.

The effluent flowing through the pipes in the trenches gets dispersed into the soil through the open joints and is finally disposed off by soil absorption.

29·17. SLUDGE SOAK PIT

A sludge soak pit is an alternative method of disposal of sewage from an individual house drain. The sludge soak pit can function effectively in situation where the ground water table is low and the sub-soil is of porous type. The diameter of the pit varies from 2 to 3 m and its depth may vary from 3 to 4m. The walls of the pit are lined with brick or stone masonry with open joints. The pit is covered with a slab at top. The effluent from the house drain is discharged into the soak pit. The open joints in the lining allow the liquid part of the effluent to get absorbed into the soil by seepage whereas the sludge gets digested by anaerobic bacterias and settle down.

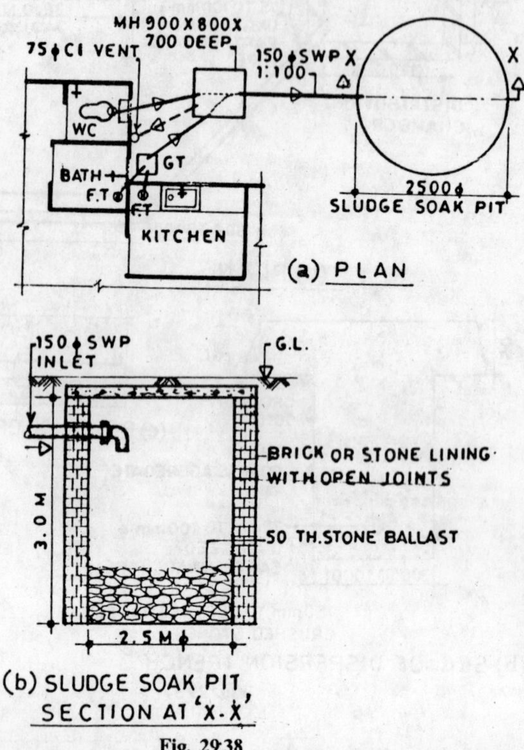

Fig. 29·38

One pit can serve for many years. When the pit gets filled up it is emptied by removing the top slab and it is put back into use again after covering its top. The contents of the emptied pit can be used as fertilizer.

29·18. DRAINAGE SCHEME FOR RESIDENTIAL QUARTERS

The drainage scheme showing disposal of soil and waste water for two different types of residential quarter are shown in Fig 29·39 and 29·40.

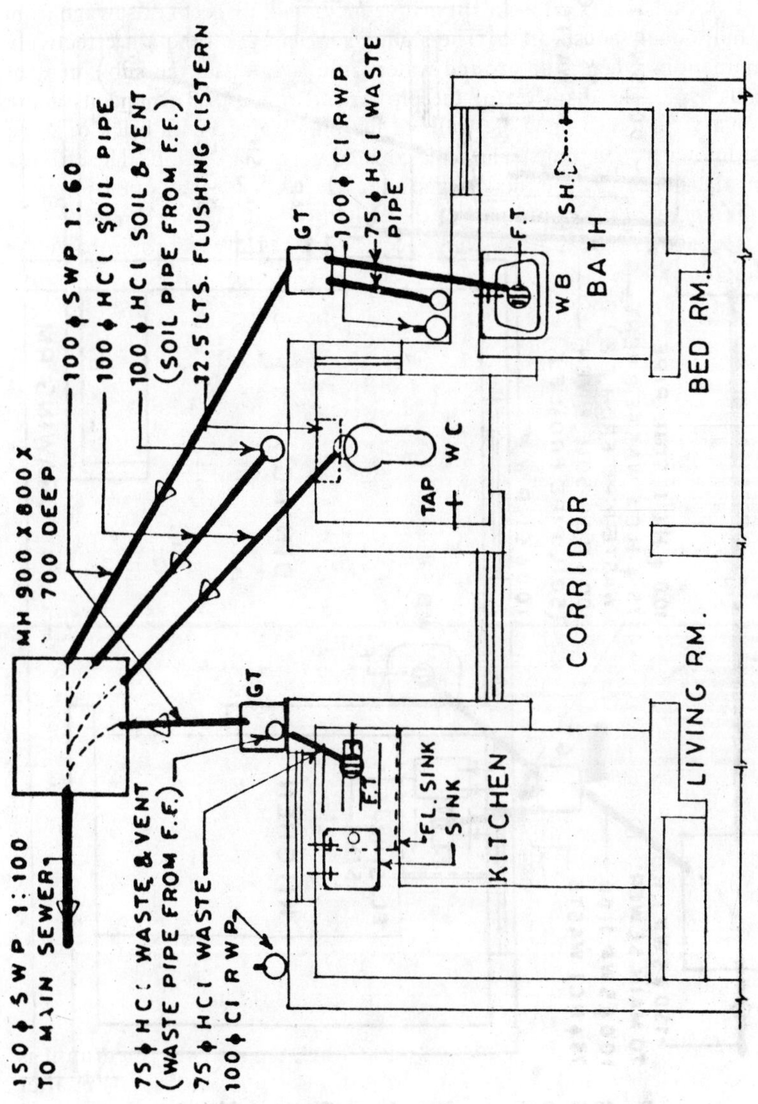

Fig. 29·39 Drainage scheme for a house

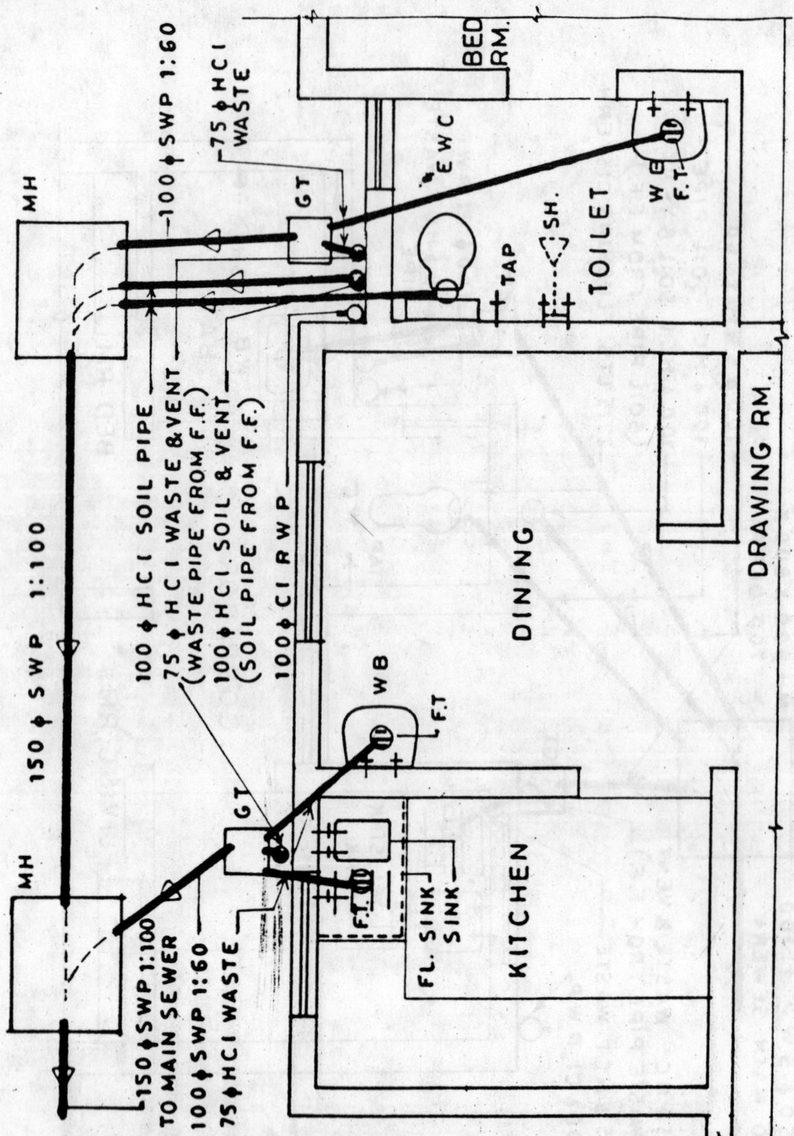

Fig. 29.40 Drainage scheme for another type of house

Questions for Review

1. Write short notes on the following

   (*a*) Anti-siphonage pipe

   (*b*) Flushing cistern

   (*c*) Inspection chamber

   (*d*) Globe valve

2. Explain with the help of sketches different type of traps. What are essential requirements of a good trap ? Enumerate different causes of breaking of water seal of a trap.

3. Explain with the help of sketches following sanitary fittings

   (*i*) Wash basins

   (*ii*) Bath tubs

   (*iii*) Urinals

   (*iv*) Water closets.

4. Explain with the help of diagrams different systems of plumbing commonly adopted for removing the discharge from sanitary fittings in a building.

5. Write short notes on type of pipes used for water supply system.

6. Draw a typical plan showing details of water supply scheme in a public building.

7. Draw typical arrangement of disposal of soil and waste water from a residential building.

30

Specifications for Low Cost Buildings

30·1. GENERAL

The selection of suitable specifications play an important role in controlling the cost of construction of a building. However, the choice of specifications should be such that the building remains structurally sound, functionally efficient and durable.

Following guidelines should be kept in view while deciding specifications for different items of works in a building :-

(*i*) The cost of building materials constitutes about 30 to 35% of the total cost of the building and as such every possible attempt should be made to economise on the cost of materials to be used in the structure. This can be achieved by making maximum possible use of the locally available materials such as bricks, tiles, stone, sand, coarse aggregates, lime, timber etc. In stone districts, where stone in cheaper, stone masonry construction should be adopted whereas in areas where bricks are easily available brick masonry should be specified. Use of locally availble spices of timber, aggregates, sand etc. should be encouraged as it helps in economising in cost of transportation of materials to a great extent.

(*ii*) Due consideration should be paid to the climatic conditions prevalent in the region during different seasons of the year in planning, designing and detailing of building including evolving specification for different items of work. The choice of materials for walls and roofing should be such that adequate thermal insulation is ensured under adverse climatic conditions.

(*iii*) Depending upon the intensity of rainfall in the region, the choice of the specifications for foundations, damp proof course (DPC), walls, external rending on walls, roof, terracing etc.

should be such as to provide adequate resistance against dampness in different components of the structure. In areas subjected to heavy rainfall or snow, sloped roofs prove to be economical as compared with flat roofs.

(*iv*) In the region where the building is liable to be subjected to earthquake, cyclones etc. special provision should be made in the design of the structure and in formulation of the specification for different components so that the structure remains safe under the effect of these forces.

The main items of work in a load bearing wall type of construction can be summarised as under :-

(1) Foundation
(2) Masonry in foundation and plinth.
(3) Damp proof course (DPC)
(4) Masonry in superstructure (load bearing wall)
(5) Partition walls.
(6) Intermediate floors or structural floors.
(7) Roof.
(8) Doors and windows.
(9) Filling in plinth
(10) Flooring.
(11) Terracing.
(12) External rendering on walls.
(13) Internal rendering on walls.
(14) Surface treatment on walls.
(15) Finishes on wood work and iron work.

Economical specifications for different items of works refered above under different site conditions are given below.

30·2. FOUNDATIONS

The factors which influence the choice of specifications for this item of work are-

(*a*) Nature of soil and its bearing capacity.
(*b*) Depth of sub-soil water table.
(*c*) The intensity of load on the main walls.

It is necessary to provide a bed of lean concrete (which may be lime concrete or cement concrete) under the main walls. The function of lean concrete bed is to distribute load from the walls on the supporting ground uniformly and to provide a levelled surface to start the brick work for the main walls.

Under normal site conditions the specifications for the lean concrete under walls should be as under :-

(*i*) 15 to 20 cm thick lime concrete with 40 mm nominal size stone/ over burnt brick aggregates and 40% mortar comprising of — 1 lime : 2 sand/surkhi or 1 lime : 1 surkhi : 1 sand.

OR

(*ii*) 15 cm to 20 cm thick cement concrete 1:5:10 (1 cement : 5 sand:10 stones/over burnt brick aggregates 40 mm nominal size.)

Following aspects pertaining to foundation need special attention :-

(*i*) In case the region is infested with termites, it would be necessary to provide termite proofing treatment in accordance with the IS:6313.

(*ii*) In case the sub-soil at the site is moist with high sub-soil water table (1·5 m or less below foundation level), it is necessary to provided bed of cement concrete of richer proportion (to function as a lean concrete bed).

(*iii*) In case the building is to be cited on soils having low bearing capacity or on soils of filled up nature or on black cotton soil, it would be necessary to give special treatment to the foundation to ensure safe transfer of load from building to the supporting ground. However, in case of black cotton soil, under-reamed pile foundation normally proves to be economical.

(*iv*) For single storeyed building the depth of foundation under normal conditions of soil should not be less than 60 cm below ground level.

30·3. MASONRY IN FOUNDATION AND PLINTH

Load bearing wall type of construction can have walls made out of bricks or stones. Hence the specifications for this item of work comprise of brick masonry or random rubble masonry in cement mortar 1:8 (1 cement : 8 sand) or in lime mortar 1:2 (1 lime : 2 sand) or in cement lime mortar 1 : 1 : 8 (1 cement : 1 lime : 8 sand).

Following aspects pertaining to this item need special attention :-

(*i*) In case of multistoreyed load bearing wall type of construction richer mix of mortar should be used to suit the loading conditions on the masonry.

(*ii*) In cae the sub-soil at the site is moist with high sub-soil water table (1.5 m or less below foundation level), the masonry should be in cement mortar 1:6 or cement lime mortar 1 : 2 : 9.

(*iii*) In regions where building stone is available at reasonable cost it is desirable to adopt stone masonry in foundation and plinth since masonry is less effected by sub-soil water.

(*iv*) For single storeyed construction the width of the foundation under average soil conditions should not be less than 2 bricks plus 20 cm.

30·4. DAMP PROOF COURSE (DPC)

Damp proof course is provided at plinth level of the building to check the rise of moisture from ground to superstructure. The specifications for is item of work depend upon the intensity of rainfall in the area, the depth of sub-soil water table, and the quality and nature of masonry used in the superstructure.

Economical specificatioins for this item of work consist of providing 35 mm thick cement concrete M 15 (1:2:4) using 10 mm nominal size stone aggregates and painted on top with one coat of bitumen of penetration 80/100 using bitumen @ 1·7 kg/sqm.

30·5. MASONRY IN SUPERSTRUCTURE (LOAD BEARING WALLS)

Load bearing walls perform the function of carrying the load from roof and upper floors and transmit the same to the soil through foundations. The walls are also required to withstand the wind and seismic forces besides providing thermal comfort against outside temperature, resistance against rain penetration and insulation against noise. The thickness of the wall to be adopted in a particular situation depends upon the strength of the masonry units (bricks, stone, concrete block etc.) to be adopted, type and mix of mortar used, unsupported length and height of the wall (slenderness ratio), location and dimensions of the openings, location of logitudinal and cross wall, intensity of loading on the wall etc. The method of designing the thickness of walls based on these parameters is given in the National Building Code.

The strength of masonry depends on the crushing strength of the building blocks and the type of mortar used in the work. It is observed that under normal conditions and moderate spans (as in case of residences etc.) it is possible to provide one brick thick load bearing wall upto 4 storeyed construction. Upto 3 storeyed residential buildings 1 brick thick load bearing wall made out of good quality bricks having crushing strength of 75 kg/sqcm can normally be provided without special treatment. However, in case of 4 to 5 storeyed construction with single brick wall there is possibility that in some walls in ground and 1st floor the stress developed in masonry may be of the order 9 to 10 kg/sqcm. Such walls should be constructed with 1st class bricks having crushing strength of 105 kg/sqcm. and in cement mortar 1 : 3 (1 cement : 3 sand).

In case, however, good quality bricks are not available in the area, the thickness of the wall should be designed as per provisions in National Building Code. Under normal conditions, bricks masonry in superstructure can be constructed in lime mortar 1 : 2 (1 lime : 2 sand) or 1 lime : 1 surkhi : 1 sand or in cement lime mortar 1 : 1 : 8 (1 cement : 1 lime : 8 sand) or in cement mortar 1 : 8 (1 cement : 8 sand).

In case the building is to be located in an earthquake zone special provision are required are to be made in the brick masonry in the form of lintel band, vertical steel bars at the jambs of openings, corners and junctions of walls in accordance with the details given in IS : 4326.

In case of stone masonry wall it is observed that 30cm thick random rubble masonry wall can safely be provided upto 2 storeyed residential quarters.

30·6. PARTITION WALLS

Partition walls are non-load bearing walls which are provided to perform the function of space dividers and to cut off noise. In areas where bricks are easily available, half brick partition walls should be provided in cement mortar 1 : 4 (1 cement : 4 sand) with reinforcement comprising of 2 Nos. of 6 mm dia bars at every 4th course and the reinforcement should be bonded and secured at each end into adjoining walls.

30·7. INTERMEDIATE FLOORS OR STRUCTURAL FLOORS

The choice of specifications for the structural floors is made paying due consideration to the requirements of strength, durability and noise reduction. The design of intermediate floor of structural floor is made based on provisions in IS Codes. In normal practice cast-in-situ RCC slab is adopted for this item. However, in residential quarters where the spans are small and the loading conditions are comparatively light, structural floor can be provided by the use of precast RCC cored units, cellular units, channel units, waffle units, etc. By use of precast elements considerable saving can be achieved in the expenditure on this item of work. Besides this by use of precast elements it is possible to reduce consumption of scarce materials of constructed like cement and steel and fasten the speed of work at the same time. In some areas in Punjab economical structural floor for residences are constructed by use of brick tiles with RCC battons. Similarly in some parts of Rajasthan locally available stone stabs (10 cm thick and 25 to 30 cm wide) are economically used as floor slabs for spans upto to 3 metres in residences.

30·8. ROOF

Depending upon the climatic conditions, building can have a sloping roof or a flat roof. The roof should be structurally safe, durable and should provide adequate protection against adverse effects of elements like rain, snow, sun and wind etc. It should provide desired insulation against heat and sound and should have efficient water proofing and drainage arrangements. In areas having heavy rainfall or snowfall sloping roof is preferred to flat roof. In other areas flat roof is provided and it has the added advantage that the roof can be used as terrace for sleeping purposes. Sloping roof can be made by use of timber, structural steel or tubular trusses provided with covering of AC sheets, GI sheets or brick tiles etc. Flat roof can be in the form of cast-in situ RCC slabs or pre-cast units (cored units, channel units, cellular units, waffle units, funicular etc.)

In case of sloped roof it is necessary to provide flase ceiling made out of insulating materials like asbestos sheets, timber, fibreboards etc. to meet the requirements of thermal comfort. In case of flat roof it is necessary to provide terracing for achieving required thermal insulation and for drainage of rain water.

30·9. DOORS AND WINDOWS

Doors are provided to perform the function of access to a building and for inter-communication between different rooms. The normal height of door is taken as 2 m and its width varies from 80 cm (for bath room & WC,) to 90 cm for other rooms in residences. In case of public buildings the normal width of door can be upto 1·2 m. Windows serve the purpose of providing light, ventilation and vision. The area of windows varies from 10 to 15% of the floor area for buildings in hot and arid regions. In case of humid climate it is necessary to provide excessive ventilation to meet requirements of comfort and as such the area of windows in such regions should vary from 15 to 20% of the floor area.

The commonly adopted specifications for door and windows frames and shutter are as under :-

(*a*) Door and window frames :

Door and window frames are normally made in timber. However, from consideration of economy, frames made out of RCC and structural steel are also being increasingly used these days. Commonly adopted size of door and window frames in different materials is as under :-

(*i*) 40 mm x 40 x mm x 6 mm mild steel Tee or 45 mm x 30 mm x 5 mm angle iron frame.

(*ii*) RCC frame 100 x 75 in section.

(*iii*) Wooden frame 100 x 60 mm in section are made from locally available timber such as Deodar, Sal, Teak, Bijasal, Hollock etc. Where 2 set of shutters are provided (one normal door shutter and one fly proof shutter) the section for timber frame should increased to 120 x 60 mm.

(*b*) Door shutters :

Normal practice is to go in for shutters made out of timber. The thickness of the shutter is decided paying due consideration tc width of door, durability and resistance of shutter against wrapping. The internal doors which are not subjected to adverse effects of wheather could have 30 mm thick commerical type flush shutters or 35 mm thick panelled shutters or 35 mm thick particle board shutters. In case of external doors subjected to weather effects, flush type shutters should not be specified since the plywood gets effected by moisture. Panelled door shutter should be recommended under exposed conditions.

(*c*) Window shutters :

Window shutters normally comprise of 30 mm thick fully or partly glazed shutters made from locally available seasoned timber such as Kail, Deodar, Sal, Bijasal, Pillamarudu, Karimarudu, Hollock etc.

30·10. FILLING IN PLINTH

The function of filling in plinth is to raise the height of the ground floor above the level of the outside ground with a view to prevent ingress of dampness in the ground floor. Under normal conditions the plinth in filled with excavated earth and finished with 15 cm thick layer or sand on top. This layer serves the base on which the layer of base concrete under ground floor is laid. A plinth height of 45 cm is considered adequate for normal construction.

30·11. FLOORING

Flooring is provided to have a hard, durable smooth, easy to clean, impervious and dust free surface and the materials used in flooring should be strong enough to resist its wear and tear that take place during use. The treatment to flooring for the ground floor is somewhat different to that for intermediate floor or structural floor. In case of ground floor the water from adjacent ground is liable to rise up by capillery action and make the floor damp. The specification for the floor should therefore be such that the flooring remains free from dampness. It is normal to provide a layer of base concrete under ground floor over 15 cm layer of sand over earth filling in plinth. The specification for the base concrete is as under:-

Base Concrete :

(*i*) 75 mm thick lime concrete with 40 mm nominal size stone/over burnt brick aggregate and 40% mortar comprising of 1 lime : 2 sand or 1 lime : 1 surkhi : 1 or sand.

(*ii*) 75 mm thick cement concrete 1 : 5 : 10 (1 cement : 5 sand : 10 stone/over burnt brick aggreggate 40 mm nominal size).

Floor finishes :

The economical specifications for the floor finish over base concrete in ground floor or over intermediate/structural floors, comprise of 35 mm thick cement concrete M15 grade (1: 2: 4) using stone or overburnt brick aggregates 20 mm nominal size and finished with a floating coat of neat cement @ 2·2 kg/sqm. In case of intermediate floors, it is at times necessary to provide a 20 mm thick cushioning layer of lime mortar (1 lime : 2 sand) or 1 lime : 1 surkhi : 1 sand or cement lime mortar 1:2:9 (1 cement : 2 lime : 9 sand) between the structural floor and the cement concrete flooring to achieve the required noise reduction.

In case of public building or areas requiring somewhat richer specifications it is normal to provide 40 mm terrazzo flooring in 2 layers under layer being 34 mm thick cement concrete M 15 grade (1 : 2 : 4) using stone/over burnt brick aggregates 12 mm nominal size and the top layer 6 mm thick with white or coloured marble chips laid in cement marble powder mix 3 : 1 (3 cement : 1 marble powder) in proportion of 4 : 7(4 cement marble powder mix ; 7 marble chips by volume).

30·12. TERRACING

Terracing treatment is given over flat roofs to perform the combined function of providing thermal insulation and. water-proofing simultaneously. It is necessary to provide the layer of terracing material in slope varying from 1 : 48 to 1 : 60 so as to ensure quick disposal of the rain water. Prior to laying terracing, the top of the roof slab is painted with layer of bitumen of penetration 80/100 using 1·7 kg bitumen for every 10 sq.m. area and impregnated with a coat of coarse sand @ 60 dm^3/10 sqm.

The choice of specifications for the terracing under different climatic conditions should be as under :-

(*i*) *For hot and aride regions :*

Providing 10 cm thick (average) layer of mud phuska laid to a slope of about 1 : 50. This layer is consolidated and plastered with 25 mm thick mud mortar mixed with bhusa @ 35 kg/cum of earth and goberi leaping with mix 1 : 1 (1 clay : 1 cow dung). This layer is covered with flat tile bricks grouted with cement mortar 1 : 3 (1 cement : 3 sand). The joints of the bricks are pointed on top with cement mortar 1 : 3 (1 cement : 3 sand).

In places where good earth is not available, lime concrete terracing may be specified in place of mud phuska and in places where clay tiles are not available 35 mm thick stone slabs or 35 mm thick M 15 grade (1 : 2 : 4) cement concrete flooring may be laid as the topping.

(*ii*) *Hot and humid regions :*

75 mm thick (average) lime concrete terracing laid to a slope of 1 : 50 may be specified.

(*iii*) *In regions of heavy rainfall :*

In case it is necessary to provide flat roof in regions of heavy rainfall, lime concrete terracing layer is covered with a layer of pressed clay tiles or 2 layers of country flat tiles in cement mortar 1 : 3 mixed with 5% crude oil by weight of cement. The joints should be pointed with same mortar.

From consideration of economy it may be desirable not to take the staircase in multistoreyed residences upto the roof level and thus avoid the construction of mumty. In such cases the access to the terrace is provided through *cat ladder* and further economy can be achieved by eliminating parapet walls over the roof slab and instead the RCC roof slab may be projected by about 80 cm from the face of the external walls and the roof slab may be given a gentle slope to reduce thickness of terracing.

30·13. EXTERNAL RENDERING OF WALLS

The external plastering or rendering on walls is carried out with the idea of improving the ability of walls to resist the adverse effects of the atmosphere due to rain, wind, snow etc. and to provide a smooth regular and even base for surface treatment like white washing, colour washing etc.

The economical specifications for external rendering consists of 12 mm thick plaster with lime mortar 1 ; 1 : 1 (1 lime : 1 surkhi : 1 sand) or cement lime mortar 1 : 2 : 9 (1 cement : 2 lime : 9 sand) or 1 : 1 : 8 (1 cement : 1 lime : 8 sand) or with cement mortar 1 : 6 (1 cement : 6 sand)

The thickness of plaster should be increased to 16 mm for the rough surface of single brick wall or stone masonry wall.

Alternatively the external rendering can be in the form of flushed or ruled pointing with lime mortar 1 : 2 (1 lime : 2 surkhi) or cement mortar 1 : 3 (1 cement : 3 sand).

30·14. INTERNAL RENDERING ON WALLS

Internal plastering or rendering is provided to obtain a smooth base for surface treatment like white washing etc. The specifications for internal rendering is identical to that of external rendering with the only different that pointing work is not carried out in case of internal rendering.

30·15. SURFACE TREATMENT ON WALLS

In case of living and dining rooms - 2 coats of dry distemper over a primming coat of whiting are normally specified. Ceiling is always white washed. In case of external walls economical specifications for the item consists in applying 2 coats of colour wash over a base coat of white washing.

30·16. FINISH ON WOOD WORK AND IRON WORK

All wood work and iron work in a building is required to be provided with protected coating to prevent damage/deterioration of the material due to weather effects or usage. The economical specifications for this item consists of one coat of primer and 2 coats of good quality paint.

Questions for Review

1. **What type of specifications do you propose for following items of work in case of low cost buildings.**
   - (*a*) **Masonry in foundation and plinth.**
   - (*b*) **D.P.C.**
   - (*c*) **Intermediate floors and Roof.**
   - (*d*) **Doors and windows.**
2. **Write specifications for low cost buildings in brief ?**

31

Causes and Prevention of Cracks in Buildings

31·1. INTRODUCTION

Majority of cracks occur when the building or its components or the material of which the building is made up of its subjected to forces which are greater than those which it can withstand. Such a situation may arise on account of forces applied externally to the building, internally within the building or forces developed in the materials of the building as a result of chemical changes. Cracks may also occur if the material used in the building is of poor quality and the construction is not carried out in accordance with relevant drawings, prescribed job specifications and in a work-man-like manner. Cracks can be divided in two categories *viz.* (*i*) *Structural cracks* and (*ii*) *Non-structural cracks.*

Structural cracks may arise due to various reasons such as incorrect design, overloading of the structural components, overloading of the soil on which the building is constructed or other similar factors. Structural cracks endanger the stability of the building and may be difficult to be rectified. Extensive cracking of foundations walls. beams, columns or slabs etc. are examples of structural cracks."

Non-structural cracks are generally due to internal forces developed in the building on account of changes in the size of building components due to moisture variation, temperature variation, the effect of gases, liquid and solids on the building components. The non-structural cracks can be repaired provided the reasons for cracks are identified and suitable remedial measures are taken to prevent their re-occurance.

Causes and prevention of non-structural cracks have been dealt in this chapter.

To a layman, even the non-structural crack gives an impression that the structure is getting unsafe. However in view of the reasonably high value of factor of safety adopted in the design of the building, it is very rare that a structure collapses soon after crack is noticed.

In some cases, if remedial measures are not taken, the non-structural cracks may, on account of penetration of moisture in the cracks, result in reducing the durability of the materials (by erosion, chemical attacks, corrosion of reinforcement etc.) and ultimately affect the stability of the structure.

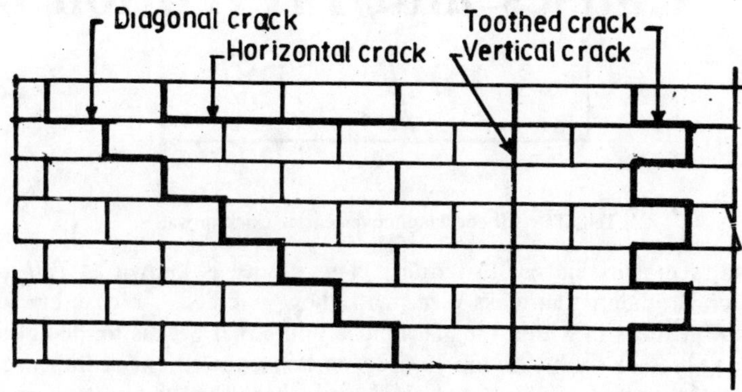

Fig. 31·1.

Types of cracks in masonry

The cracks can be broadly classified as vertical, horizontal, diagonal, toothed cracks, irregulai cracks or random cracks. Closely spaced fine cracks on surface of the matertal is known as crazing. Some cracks may appear to be similar yet cause of the cracking may be different.

From the examination of the portion in which the cracks appear, it is possible to have an idea of the possible factors responsible for the formation of the cracks.

31·2. INVESTIGATIONS RELATING TO CRACKS

A careful study of the location of cracks (starting and finishing points) their width and depth, helps in dealing with the diagnosis of different types of cracks

Immediately after a crack is noticed, it is desirable to observe the same for sometime with a view to know whether it is stable or it is changing in size. Some cracks change in width within a period of few days of formation. Such changes are normally on account of variation in atmospheric temperature. The changes in width of the cracks (the crack may increase in width or expand and reduce in width or contract) can be easily detected with the help of *telltales*.

In this method, a piece of flat glass (about 2 to 3 cm in width and 10 to 12 cm in length) is spanned across the crack and it is stuck to the

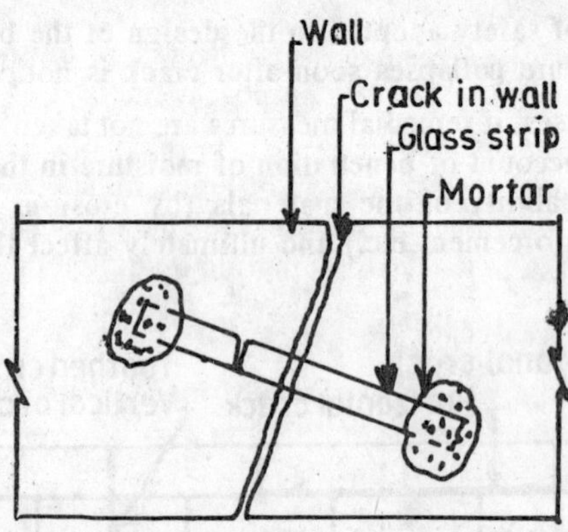

Fig. 31·2.

Tell--Tale for checking movement of crack in wall

wall surface at its ends with mortar. This device is known as *Tell-tale.* If the crack widens the glass piece will show a crack. In case crack in structure reduces in width, the glass piece will either get dis jointed at one end or will crack by buckling. In order to know the extent of increase of decrease in size of width of the crack, it is necessary to use two pieces of flat glass as shown in Fig. 31·3. Both the pieces are spanned across the crack, one above the other and each piece is fixed to the wall with the mortar applied at one end (the other end being free to move). The fixing is done in such a way that the fixed end of the top piece remains opposite to the fixed end of the bottom piece.

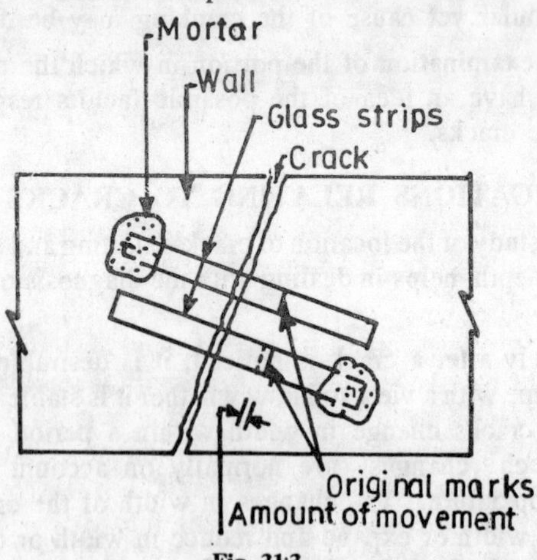

Fig. 31·3.

Tell-Tale for measuring the magnitude of movement of a crack in wall.

After fixing, a line is drawn across the two pieces. Any increase or decrease in width of the crack can be easily observed and measured by this arrangement.

For proper diagnosis, it is also necessary to know whether cracks are superficial or deep. For this purpose a fine steel wire may be used to probe the depth of the cracks.

Many-a-times, the formation of crack may also result in difference of levels on the two sides of the crack. This can be felt by moving a tip of a finger across the crack or by putting a straight edge across the crack.

31·3. CAUSES OF OCCURANCE OF CRACKS

The important causes responsible for occurance of cracks in buildings, are as under :

(1) *Cracks due to moisture changes*

(2) *Cracks due to temperature variation*

(3) *Cracks due to elastic deformation and creep*

(4) *Cracks due to effect of chemical reaction*

(5) *Cracks due to movements of ground*

(6) *Cracks due to vegetation*

The various causes of formation of cracks and measures adopted to prevent them are described below :

(1) **Cracks due to moisture changes.** We know that most of the building materials like bricks, concrete, mortar, stone, timber etc. have pores. Hence these materials increase in size or expand on absorbing moisture and decrease in size or shrink on drying.

Shrinkage on account of drying out of moisture contents in building materials is one of the main factors responsible for cracks in the buildings.

Shrinkage cracks normally appear in weak-sections of the masonry such as door, window openings, staircase walls etc. Such cracks are uniformly and regularly placed. The formation of shrinkage cracks in masonry depends upon the type of material used in the masonry and the strength of the mortar. Compared to clay bricks, concrete blocks and sand lime bricks are more prone to shrinkage cracks. If the masonry work is carried out by use of weak mortar such as cement mortar 1 : 6 or cement lime mortar 1 : 1 : 6, the shrinkage of individual units gets accommodated in the form of fine cracks in the mortar joints around the unit and it does not cause the cracks to appear on the plaster on the wall. On the other hand in case the masonry work is carried out in rich mortar, the individual units cannot shrink independently and the cumulative effect of the shrinkage of the units and the mortar in joints appear in the form of well defined cracks on wall surface.

Similarly in case of concrete, the drying shrinkage results in decreasing the volume and if the component is restrained against free movement (due to volume changes) stresses are developed which may cause cracks.

Cracks due to moisture changes can be prevented by adopting the following measures.

(1) Minimise use of rich cement mortar.

(2) Use lean cement mortar or composite cement-lime-mortar in masonry work.

(3) Use stiffer and leaner concrete mix and allow adequate time for curing and drying.

(2) **Cracks due to temperature variation.** We know that almost all materials expand on heating and contract on cooling. In some materials the changes in temperature can cause appreciable change in their size. The extent of movement due to temperature variation in the building components depends upon a number of factors such as dimensions of the materials, its coefficient of expansion, the magnitude of variation in the temperature etc. The effect of temperature variation in the internal walls and intermediate floors is very small and may not result in cracks. On the other hand the effect is appreciable on the external walls and roof which is directly exposed to sun.

A typical case of occurrence of cracks due to temperature variation is that of roof slab supported on load bearing walls. The roof slab, being exposed to the heat of sun, is subjected to alternate expansion (on getting heated from sun during day time) and contraction (on getting cooled during night). This movement of the slab may result in pushing out top course of masonry and develop horizontal cracks in the supporting walls.

This type of cracks can be prevented by adopting following measures:

(1) Adequate insulation or terracing treatment over the roof slab and by introducing slip joint between the slab and the supporting wall as shown in Fig. 31·4.

(2) Painting top of roof with reflective finish (white wash etc.) can also serve to minimise cracks.

(3) Changes of cracking due to temperature variation can be minimised by introducing expansion, contraction, and slip joints at appropriate locations. In case of structures like, shell roof where it is difficult to provide joints, it is necessary to take the effect of thermal stresses in the structural design of roof to prevent cracks. Details regarding provision of joints in structures are given in chapter 24.

(3) **Cracks due to elastic deformation and creep.** The different components of the building *viz.* wall, column, beam, slab etc. undergo elastic deformation when loaded. The magnitude of such deformation depends upon the type of building materials used in the construction *viz.* brick, cement concrete blocks etc. as well as the spans of the structural

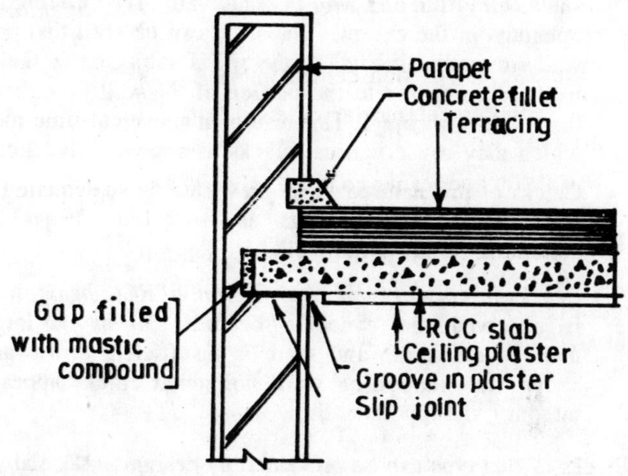

Fig. 31·4.

Details of slip joint for roof slab

members and the loading conditions. It is therefore obvious that in case the brick panel wall in a frame structure is constructed too tightly between the framework columns and beams, it is likely to be subjected to cracks due to shortening and deformation of the RCC frame on account of shrinkage creep. The situations where cracking due to elastic deformation and creep arise, and the *preventive* measure to avoid such cracks are summarised below :

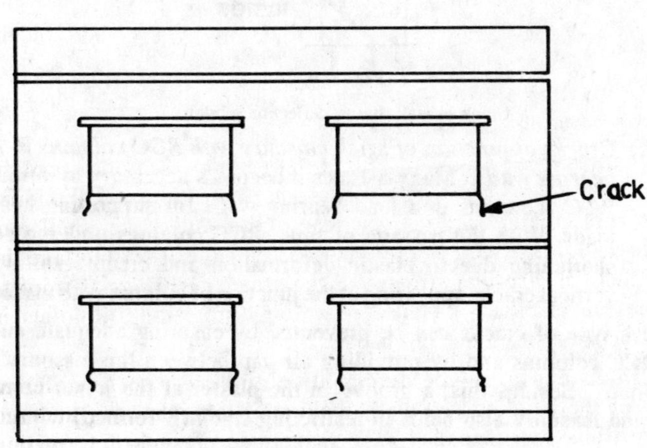

Fig. 31·5.

Cracks in masonry wall loaded unevenly.

(*i*) *Cracks in masonry when a wall is unevenly loaded.* Fig. 31·5. shows elevation of a load bearing wall. The building has large openings in the external walls. It can be seen that portion of wall immediately below the sill is subjected to much lesser loads as compared to the portion of the wall on either side of the window opening. This results in uneven loading of the wall which may cause vertical cracks as shown in the figure.

Cracks of this nature can be prevented by so selecting the size and location of openings as to avoid unequal loading conditions in the same wall.

(*ii*) *Cracks in masonry due to deflection of RCC beam or slab.* It is observed that when a beam or slab having large spans undergo deflection and there is insufficient downward loads above the supporting wall, horizontal cracks appear in the masonry at supports.

Cracks of this type can be prevented by designing the slab or beam in such a way that the depth provided is adequate to restrict the deflection within permissible limits.

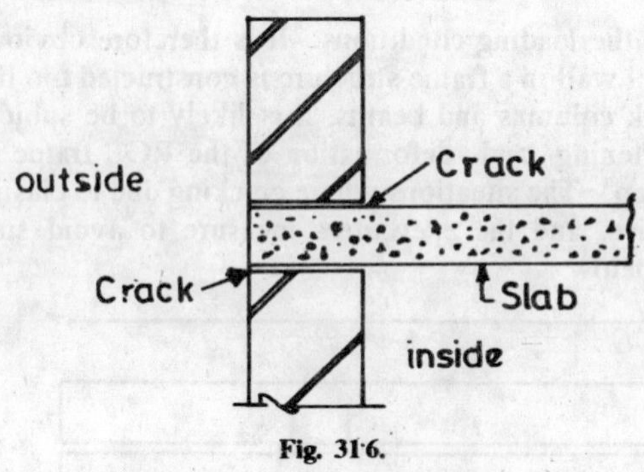

Fig. 31·6.

Crack in wall due to deflection of slab.

(*iii*) *Cracks at junction of brick masonry with RCC columns in load bearing walls.* Many-a-times it becomes necessary to introduce RCC columns in a load bearing walls for supporting heavier loads. With the passage of time, RCC columns undergo some shortening due to elastic deformation and creep resulting in vertical cracks appearing at the junction of column with masonry.

This type of cracks can be prevented by ensuring adequate curing of the RCC columns and by providing air gap between the masonry and the column. Besides this, a groove in the plaster at the joints between column and masonry also helps in restricting crack (if formed) within the groove.

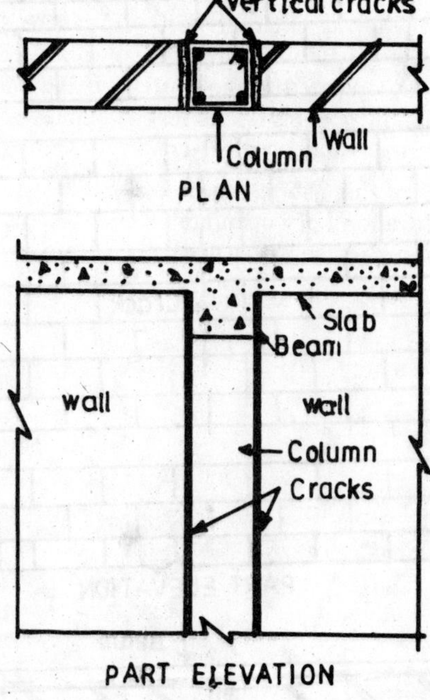

Fig. 31·7.
Vertical crack at junction of wall with column.

(*iv*) *Cracking in brick panel wall in frame structure.* The direction of crack in the tightly built panel wall in frame structure, depends mainly upon the length to height ratio of the panel wall. The cracks could be horizontal or vertical.

Vertical cracks are formed when the length of panel is more and the tightly constructed wall between columns gets compressed due to movement due to moisture, elastic deformation, etc. Many-a-times the brick work gets buckled out.

The horizontal cracks are formed due to shortening of columns on account of elastic deformation, creep and drying shrinkage or due to excessive deflection of upper beams.

The cracks in panel wall in the frame structure can be prevented by adopting following measures :

(*a*) The construction of panel wall should preferably be taken up as late as possible so that most of deformation due to

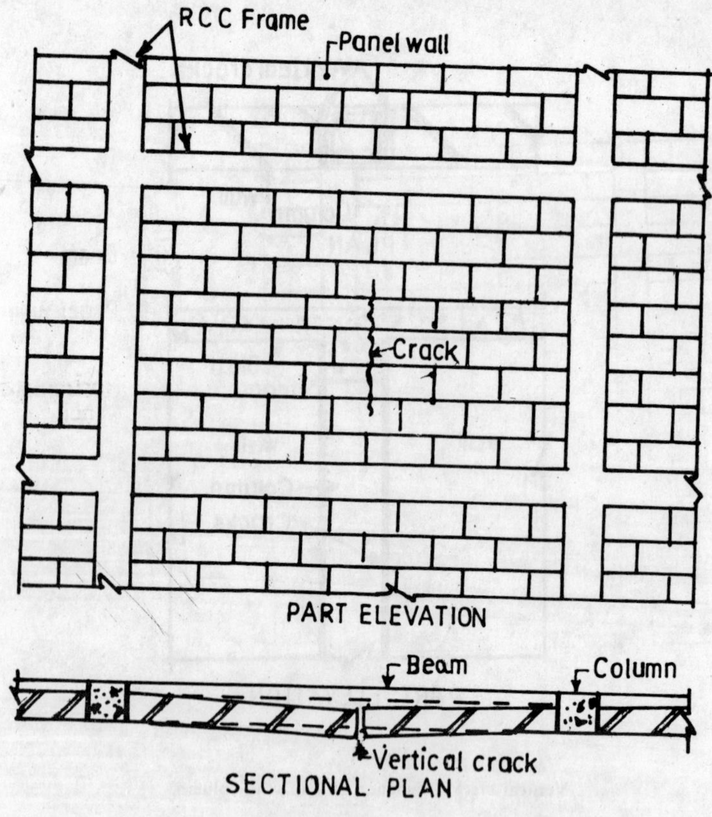

Fig. 31·8.
Vertical crack in Panel wall of a framed structure.

shrinkage, creep etc. in R.C.C, beams and columns is allowed to take place prior to construction of panel wall.

(*b*) The panel walls should not to constructed tightly between the RCC frame. Movement joints should be provided between the top of panel wall and soffit of beam. These joints may be filed up with some compressable jointing material.

(4) **Cracks due to effect of chemical reaction.** The carbon-dioxide present in the air reacts chemically with cement based products (cement concrete, mortar or cement blocks etc.) resulting in appreciable increase in volume of these materials which ultimately leads to cracking. In case of RCC work, the chemical reaction due to *carbonation* can lead to spalling of concrete which may expose the reinforcements. The reinforcement may in course of time be subjected to corrision due to

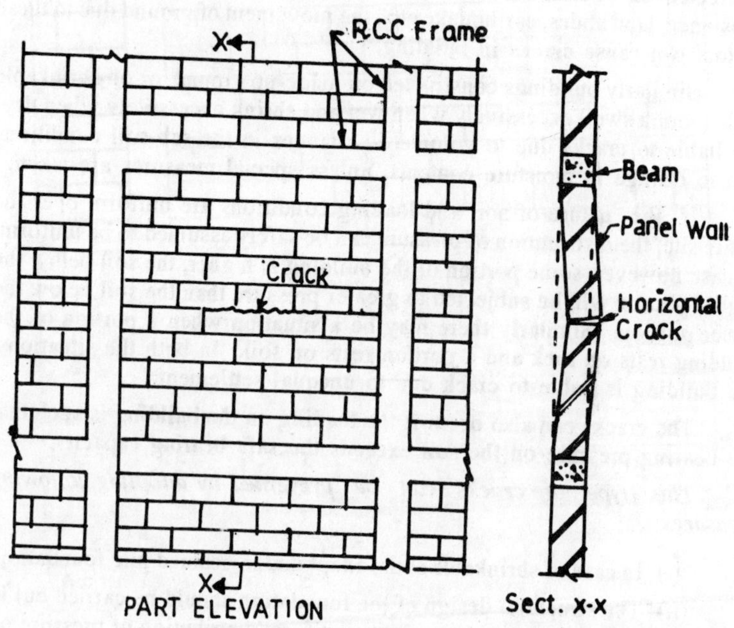

Fig. 31·9.
Horizontal crack in Panel wall of a framed structure.

rusting which may finally endanger to stability of the structure. Similarly soluble sulphates which may be present in soil, ground water or clay bricks, react chemically with portland cement in the presence of water and cause the concrete or mortar joint or brick itself to expand considerably leading to formation of cracks. The sulphate attack besides cracking can also result in weakening of the building components which may affect the stability of the structure.

The cracks due to chemical reaction can be prevented by adopting the following measures :

(*i*) Use dense and good quality concrete and provide increased cover for R.C.C. members to prevent carbonation.

(*ii*) For preventing sulphate attack, avoid use of bricks containing high percentage of soluble salts. In case the subsoil contains high proportion of sulphate salts, use rich cement mortar or concrete or sulphate resistant cement in executing the work.

(*iii*) Since sulphate attack can take place only in the presence of moisture, provide adequate damp proofing treatment to ensure dry conditions.

(5) **Cracks due to movement of ground.** In case where a building is erected on or near an area which is likely to be subjected to mining substance, land slides, earthquake etc., the movement of ground due to these factors can cause cracks in building.

Similarly buildings constructed on made-up ground, or on shrinkable soils (which swell excessively when wet and shrink excessively when dry) are liable to cracks due to volumetric changes in the sub-soil conditions due to change in moisture contents, unless special measures are taken.

If the nature of soil and loading conditions are uniform over the entire site, the distribution of pressure can be safely assumed to be uniform. In case however, some portion of the building is higher, the soil below the higher portion will be subjected to greater pressure than the soil below the lower portion. Similarly there may be a situation when a portion of the building rests on rock and a portion rests on soil. In both the situations, the building is liable to crack due to unequal settlement.

The cracks can also occur if the loading on the building is such that the bearing pressure on the soil exceeds the safe bearing capacity.

This type of cracks can be prevented by adopting following measures:

(*i*) In case of shrinkable soils, adopt under- reamed pile foundation.

(*ii*) The structural design of the foundation should be carried out in such a manner as to achieve uniform distribution of pressure on the ground to avoid differential settlement.

(*iii*) The foundation should be so proportioned that the safe bearing capacity or soil is not exceeded.

(6) **Cracks due to vegetation.** Cracks many-a-times occur due to the existence of fast growing trees in the vicinity of the building. The roots of a tree generally expand horizontally on the sides to the extent to the height of the tree above ground. The effect of the existence of vegetation near the building becomes more damaging when the soil at site is of shrinkable type. The roots of the growing trees causes drying and shrinkage of the sub-soil and this can result in unequal settlement of the foundation leading to cracks.

Such cracks can be prevented by not planting fast growing trees near the buildings.

Questions for Review

1. **Write in brief important causes responsible for occurence of cracks in buildings.**
2. **Explain the investigations you would carry out in dealing with the diagonosis of different types of cracks in buildings.**

32

Preliminary Estimates for Buildings/Projects

32·1. INTRODUCTION

For any project, two types estimates are required to be prepared (*i*) *preliminary estimates and* (*ii*) *Detailed estimates.*

Immediately after it is decided to take up a project, the first information one needs to know is the likely cost of the project. This information is necessary to obtain financial sanction from government/competent authority prior to start of the project. The estimate prepared at the conceptual stage of a project for the purposes of obtaining financial approval is termed as *preliminary estimate.* The preliminary estimate can be prepared without availability of detailed drawings by following a simple method adopted by Central Public Works Department (CPWD).

More than a decade back, CPWD undertook a detailed study of compiling the completion cost of different types of buildings/projects with a view to fix average cost of construction of different types of buildings in the form of plinth area rates per sq. m.

In this study the various types of buildings covered comprised of *non-residential buildings* like offices, colleges, hospitals, dispensaries, schools, community centres, shopping centres and residential buildings. The residential buildings were divided in the following categories :

1. Type 'A' quarter–having a plinth area of 36 sq.m.
2. Type 'B' quarter–having a plinth area of 56 sq.m.
3. Type 'C' quarter–having a plinth area of 84 sq.m.
4. Type 'D' quarter–having a plinth area of 139 sq.m.
5. Servants quarter-having a plinth area of 19 sq.m.

The specifications for C & D type quarters are slightly richer as compared to the specification for A & B type quarters (Refer Table 32·7)

The information thus compiled has been condensed in a tabular form under the following sub-heads :-

(*a*) *Buildings (Civil work)*

(*b*) *Internal services (Plumbing &electrical)*

(*c*) *Water tanks.*

(*d*) *lifts*

(*e*) *Site development including external services*

(*f*) *Proforma for calculations of cost index.*

The details given by CPWD have been slightly rearranged and reproduced in the form of different tables in this chapter.

The provisions under various sub-heads are explained below :-

32·2 BUILDINGS (CIVIL WORKS)

The detail for working out preliminary estimates for buildings (civil works) are given in table 32·1. In this table various types of buildings have been divided in the following categories :-

1. RCC framed construction
2. Load bearing construction

The standard plinth area rates for different types of buildings given in this table have been worked out based on broad specifications for non-residential building and residential buildings as given in table 32·7. In addition following assumptions have been made while fixing the plinth area rates given in the table :-

(*i*) Floor to floor height of a public building. = 3·4 m

(*ii*) Floor to floor height of residential building = 2·9 m

(*iii*) Plinth height of the buildings both public as well as residential = 0·6 m

(*iv*) The depth of foundations of buildings = 1·2 m

(*v*) Bearing capacity of soil = 10 tonne/sq. m.

(*vi*) Maximum size of column grid/module in a framed construction = 35 sq.m.

The plinth area rates given in Table 32·1 cover cost of all items of civil works including cost of foundation based on above assumptions and specifications given in Table 32·7.

In case the floor to floor height or the depth of foundation, or the plinth height of a building for which the preliminary estimate is to be worked out is different from the above referred parameters, provision of extra amount has to be made for such variations as per details given vide item No. 4 in Table 32·1. Again in case a building is to be located in earthquake zone or the building is required to have stronger foundations (strip foundations, pile foundations etc.), provision of additional amount has to be made in the form of extra items based on details given in item No. 4 in the Table. "Similarly in case the specifications to be adopted for some items in a building are different from the once's given in Table 32·7, it will be necessary to make provision of additional amount to take care of the difference in cost between the specifications adopted and the broad specifications prescribed for rates in the table.

32·3. INTERNAL SERVICES (PLUMBING & ELECTRICAL)

The cost of internal plumbing, internal electrical installations and external service connections for different type of non-residential buildings is taken as certain percentage of the *building cost.* The building cost for this purposes is taken as cost of building-civil work vide item No. 1 to 3 of Table 32·1 *plus* cost of one or all of the following extra items (Given in item No. 4 of the table) Whichever is applicable :-

(*i*) Extra due to additional floors (item 4·1·1, 4·1·2)

(*ii*) Extra due to every 0·3 m additional height of floor (item 4·1·3, 4·2·1.)

(*iii*) Extra due to every 0·3 m height plinth (item 4·1·4, 4·2·2)

The percentage for internal services (electric installations, internal water supply and sanitary installations), external; service connections etc. to be considered in the preliminary estimates of non-residential buildings are given in Table 32·2.

In case of residential buildings it is seen that since the number of plumbing fixtures as well as the electrical points in quarters of different specified plinth areas get more or less fixed, the provision of internal water supply & sanitary installations as well as internal electric installations are made in the form of a lumpsum costs for the respective services as per details given in Table 32·2.

In case a building is to be provided with telephone conduits, central call bell system, power wiring etc. extra provision for the same have to be made as per details given in Table. 32·2.

32·4. WATER TANKS

Water tanks are invariably provided in buildings. The water tank may be in form of an under ground sump or an overhead tank or both. Table 32·3 gives rates in rupees per litre for different types of tanks based on which necessary provisions towards cost of tank of required capacity can be made in the estimates.

32·5. LIFTS

Multistoreyed buildings have provision of lifts. The passenger lifts are of different speeds and capacities. Besides passenger lifts some buildings have goods lift as well. Table 32·4 shows the lumpsum amount to be adopted towards cost of lifts depending upon its capacity and speed.

32·6. SITE DEVELOPMENT INCLUDING EXTERNAL SERVICES

The rates of different items of works which can be covered under heading of site development are contained in Table 32·5. In working out the preliminary estimated cost of site development for a particular site, rates of all the relevant items applicable to the site are added and the total rate thus arrived at is multiplied with the total area of the site.

32·7. PROFORMA FOR CALCULATION OF COST INDEX

As indicated above the various plinth area rates contained in the different tables referred above are based on the rates of material and labour in Delhi in 1976. Thus the estimated cost worked out by adopting the rates given in various tables will represent the preliminary estimated cost of the building/project in the year 1976.

Table 32·6 shows the proforma for calculation of cost index. The cost index in Delhi (for the purposes of working out the preliminary estimated-cost) was fixed as 100 as on 1.10 1976. Thus 1976 is considered as base year (with cost index taken as 100)

To workout the present day *preliminary estimated* cost of *building*, it is necessary to workout the *present day cost index* of the place where the building is to be constructed. The method of working out the present day cost index for any place is very simple. To work out the present day cost index for a particular station/site of work, first of all we have to fill up the current rates at site of work for 9 different items *viz.* brick, sand, cement, aggregate, timber, mild steel, mason, carpenter, cooly/beldar in column No. 5 of table. Thereafter divide figures in coluumn No. 5 by figures in column No. 4 and multiply the results with figure of weightage given in column No. 7 of the table in respect of each item. The sum of the result in column No. 8 of the Table 32·6 will give the value of present day cost index.

The total preliminary estimated cost of the building based on 1976 rates is multiplied by the present day cost index to arrive at the present day estimated cost of the building.

TABLE 321

PLINTH AREA RATES FOR BUILDINGS (CIVIL WORKS) AS ON 1.10.1976

S. No.	*Description of item*	*Plinth area rates in Rupees/sq.m.*
1.	**Offices, Colleges, Hospitals and Dispensary buildings**	
1.1	R.C.C. framed structures :	
	R.C.C. framed structures upto 6 storeys (floor height 3.4 m)	400.00
1.2	Load bearing structures : (floor height 3.4 m)	
	(*a*) Single storeyed and three storeyed	355.00
	(*b*) Double storeyed	340.00
	(*c*) Four storeyed	375.00
2.	**Schools, Community centres and Shopping Centres**	
2.1	R.C.C. framed structures upto 6 storeyed (floor to floor height 3.4)	365.00
2.2	Load bearing structures (floor height 3.4 m) :	
	(*a*) Single storeyed and three storeyed	310.00
	(*b*) Double storeyed	290.00
	(*c*) Four storeyed	325.00
3.	**Residential buildings**	
3.1	R.C.C. framed construction building upto 6 storeys (floor height 2.9 m)	385.00
3.2	Load bearing construction (floor height 2.9 m):	
	(*i*) Quarters of type A (36 sq.m plinth area); type B (56 sq. m. plinth area) and type C (84 sq. m. plinth area) category and servant quarters (19 sq. m. plinth area)	
	(*a*) Double storeyed	260.00
	(*b*) One and three storeyed	275.00
	(*c*) Four storeyed	290.00
	(*ii*) Type D-quarters (139 sq. m. plinth area) and Officer's hostels	
	(*a*) Double storeyed	310.00
	(*b*) One and three storeyed	325.00
	(*c*) Four storeyed	340.00

Contd....

TABLE 32·1 (*Contd...*)

PLINTH AREA RATES FOR BUILDINGS (CIVIL WORKS) AS ON 1.10.1976

S. No.	*Description of item*	*Plinth area rates in Rupees/sq.m.*
	(*iii*) Scooter and cycle sheds	250.00
	(iv) Garages	235.00
4.	**Extra provisions for building portion where ever required**	
4.1	*Extra provision applicable in case of R.C.C.* frame structures	
4.1.1	Every additional storey over 6 storeys and upto 9 storeys	7.00
4.1.2	Every additional storey over 9 storeys and upto 12 storeys	10.00
4.1.3	Every 0.30 m additional height of floor above normal floor height 3.4m/2.9m	17.00
4.1.4	Every 0. 30 m height of plinth over normal plinth height of 0.60 m (on ground floor areas only)	17.00
4.1.5	Every 0.30 m deeper foundations over normal depth of 1.20 m (on ground floor area only)	17.00
4.1.6	Making stronger foundations to take load of one additional floor at a later date (on area of additional floor at a later date).	50.00
4.1.7	Strip foundations in poor soil having bearing capacity less than 10 T/sq. m.	15.00
4.1.8	Resisting earthquake forces	34.00
4.1.9	R.C.C. Raft foundations	60.00
4.1.10	Pile foundations upto a depth of 15 m	84.00
4.1.11	Stronger structural members to take heavy loads above 500 kg/sq. m. upto 1000 kg/sq. m.	26.00
4.1.12	Larger modules over 35 sq.m.	30.00
4. 1.13	Termite proof treatment (on ground floor area only ,	9.00
4.1.14	Fire fighting	25.00

Contd...

TABLE 3·21 (*Contd...*)

PLINTH AREA RATES FOR BUILDINGS (CIVIL WORKS) AS ON 1.10.1976

S. No.	*Description of item*	*Plinth area rates in Rupees/sq.m.*
4.1.15	Basement floor, floor height 3.35 m, with normal provision of water proofing treatment with bitumenous felt.	550.00
4.1.16	Extra for basements with mastic asphalt water proofing treatment.	60.00
4.2	*Extra provisions applicable in case of load bearing type of construction.*	
4.2.1	Every 0.30 m additional height above normal height of 4/2.9 m	10.00
4.2.2	Every 0.30 m higher plinth over normal plinth height of 0.60 m (on ground floor area only)	10.00
4.2.3	Every 0.30 m deeper foundations over normal depth of 1.20 m (on ground floor area only)	12.00
4.2.4	Foundation on poor soil having bearing capacity less than 10 T/sq.m.	15.00
4.2.5	Foundations requiring under-reamed piles 6m long	68.00
42.6	Resisting earthquake forces	
	(*i*) In zone V	25.00
	(ii) More than two storeyed building in zone III & IV	—
	(iii) With a design seismic coefficient greater than 0.06	12.50
	(iv) Design seismic co-efficient equal to or less than 0.60.	12.50
4.2.7	R.C.C. Raft Foundations	60.00
4.2.8	Pile foundations upto a depth of 15 m	84.00
4.2.9	Termite proof treatment (on ground floor area only)	9.00
4.2.10	Fire fighting	25.00

Note—Rate for items are applicable on entire plinth area except for items 4.1.4, 4.1.5, 4.1.6, 4.1.13, 4.2.2, 4.2.3, & 4.29.

TABLE 32·2

INTERNAL SERVICES—PLUMBING AND ELECTRICAL

(Expressed as percentage of building cost for non-residential buildings and Lump Sum in case of residential buildings)

Rates as on 1.10.76

S. No.	Description of item	*Non-residential buildings*				*Residential Buildings*			
		Office & Colleges	*Hospital*	*School*	*Hostel*	*Type A and Servant Quarter*	*Type B Qtr*	*Type C Qtr.*	*Type D Qtr.*
1.	(*a*) Internal electric installations	12 ¼%	12 ¼%	12 ¼%	12 ¼%	Rs. 1250/-	Rs. 1350/-	Rs. 2000/-	Rs. 4000/-
	(*b*) Internal water Supply and sanitary installations	4%	10%	5%	15%	Rs. 2000/-	Rs. 2100/-	Rs.2500/-	Rs. 3500/-
	(*c*) External Service connections	5%	5%	5%	5%	5%	5%	5%	5%
	(*d*) Extra for power wiring and plugs	4%	4%	—	—	—	—	—	—
	(*e*) Extra for Telephone conducts	0.5%	0.5%	0.5%	—	—	—	—	—
	(*f*) Extra for central call bell system	1%	—	—	—	—	—	—	—

TABLE 32·3 : RATE PER LITRE OF WATER TANK AS ON 1.10.1976

S.No.	Description of item	Rate in Rupees per litre
1.	Overhead tank without independent staging	Rs. 0.65 per litre
2.	Overhead tank upto staging height 20 meters	Rs. 1.10 per litre
3.	Overhead tank with staging height between 20 metres & upto 30 metres	Rs. 1.25 per litre
4.	Overhead tank with staging height between 30 metres & 40 metres	Rs. 1.50 per litre
5.	Underground sump	Rs. 0.65 per litre

TABLE 32·4 : LUMPSUM COST PER LIFT AS ON 1.10.1976

S.No.	Description of item	Unit	Lump sum cost per lift			
1.	Passenger lifts		8 passenger capacity	13 passenger capacity	16 passenger capacity	20 passenger capacity
	Lifts for 8 floors					
	(a) Speed 0.7 m/Second	Each	Rs. 1,25,000/-	Rs. 1,40,000/-		
	(b) Speed 1.0 m/Second	Each	Rs. 1,50,000/-	Rs. 1,60,000/-	Rs. 1,90,000/-	Rs. 2,20,000/-
	(c) Speed 1.5 m/Second	Each	Rs. 1,80,000/-	Rs. 2,10,000/-	Rs. 2,30,000/-	Rs. 2,50,000/-
	(d) Speed 2.5 m/Second	Each	Rs. 2,70,000/-	Rs. 3,00,000/-	Rs. 3,50,000/-	Rs. 4,00,000/-
	(e) Add/deduct for each floor more/less than 8 floor	Each	Rs. 10,000/-	Rs. 12,000/-	Rs. 15,000/-	Rs. 20,000/-
2.	Goods lift		Upto 4 floors	Extra for every additional floor above 4 floors		
	(a) Upto two tonnes	Each	Rs. 1,40,000/-	Rs. 8,000/-		
	(b) More than two tonnes but upto 4 tonnes	Each	Rs. 2,20,000/-	Rs. 12,000/-		

TABLE 32·5

RATES IN RUPEES PER SQ. M. FOR SITE DEVELOPMENT INCLUDING EXTERNAL SERVICES AS ON 1.10.76

	Development of Sites	
1.	Levelling	Rs. 1.50 per sq. m.
2.	Internal roads & paths	Rs. 4.65 per sq. m.
3.	Sewers	Rs. 3.35 per sq. m.
4.	**Filtered Water Supply**	
	(*a*) Distribution lines 100 mm dia	Rs. 2.45 per sq. m.
	(*b*) Peripherial grid 150 mm to 300 m dia pipes	Rs. 1.85 per sq. m.
	(*c*) Unifiltered water supply distribution lines	Rs. 1.40 per sq. m.
5.	Storm water drains	Rs. 4.00 per sq. m.
6.	Horticulture operations	Rs. 2.50 per sq. m.
7.	Street Lighting	
	(*a*) With incandescent lamps	Rs. 0.35 per sq. m.
	(*b*) With incandescent lamps and with high pressure mercury vapour/lamps/flourscent lamps in important places	Rs. 0.80 per sq. m.
	(*c*) Complete with HPMV lamps or fluorescent lamps.	Rs. 1.30 per sq. m.
8.	H.T. Sub-station and L.T. distribution (to be provided in areas where expected electricity load is such that electric company will not give L.T supply).	Rs. 2.60 per sq. m.

Note — *1. The rates are per sq. m. and are to be applied on the entire area of the plot to be developed.*

2. Cost of bulk services i.e. tubewells, pumps, treatment plants, out fall sewer etc. are not covered in the above rates and will be extra.

TABLE 32·6

PROFORMA FOR CALCULATION OF PRESENT DAY BUILDING COST INDEX

Sl. No.	*Description*	*Unit*	*Rate at New Delhi corresponding to base 100 as on 1.10.76*	*Current rate at station /site of work*	*Ratio between Col. 5 & 4 ie. Col. 5 / col.4*	*Weightage*	*Cost index (col. 6 X Col. 7)*	*Remarks*
1	**2**	**3**	**4**	**5**	**6**	**7**	**8**	**9**
	Materials							
1.	Brick	1000	106.15	...	—	16	—	
2.	Sand	Cum.	21.91	...	—	5	—	
3.	Cement	Qtl.	35.28	...	—	21	—	
4.	Aggregate	Cum.	27.10	...	—	6.5	—	
5.	Mild Steel	Qtl.	183.20	...	—	10	—	
6.	Timber	Cum.	2021.00	...	—	18	—	
	Labour							
7.	Mason	Each	9.89	...	—	8.5	—	
8.	Carpenter	Each	9.89	...	—	4	—	
9.	Coolie/Beldar	Each	4.41	...	—	11	—	
					Total :	100	—	Cost index

TABLE : 32·7

SPECIFICATIONS APPLICABLE TO PLINTH AREA RATE ESTIMATES

A specifications applicable to plinth area rate estimate of Non-Residential buildings and Residential buildings are summarised as under :-

A. *Specifications for non-Residential Bldgs.*

(Office : Hospital, Schools etc.)

I. FOUNDATIONS — Spread foundation or isolated/combined footings.

II. SUPER STRUCTURE — RCC frame construction with filler wall in brick or load bearing construction with intermediate RCC columns, whichever considered necessary.

Internal partitions in brick masonry.

III. DOOR / WINDOWS : *Doors*

(*a*) Frames IInd class Indian teak wood or equivalent.

(*b*) Shutters — Panelled (in 2nd class Indian teak wood) or flush door shutters with commercial ply.

Windows frames & shutters

(*a*) 2nd class Indian teak wood or equivalent. In case of hospitals-Fly proof shutters for all doors & windows and iron grill for windows at ground floor.

Alternatively — Steel windows

IV. FITTINGS — Inodised Aluminium or equivalent.

V. FLOORING — Main Entrance — Terrazo tile.

Kota stone flooring in areas like lavatory block, corridors etc.

In other areas, flooring will depend on the type of building such as :

(*a*) Office — Moasic flooring upto 25% of total area and in the remaining area cement concrete flooring.

(*b*) In case of Hospital - 2·32 m high dado in mosaic in corridors and upto sill level in other rooms. In remaining area, ordinary cement concrete flooring.

(*c*) In case of school - cement concrete flooring.

VI. TERRACING : Lime concrete or mud phuska finished with tiles with suitable treatment of water proofing depending upon the area where buildings are to be constructed. In case of buildings to be airconditioned-additional layer of foam concrete to be provided in between roof slab & terracing.

VII. FINISHING :

(*i*) *External* : Water proofing cement paint.

(*ii*) *Internal* :

(*a*) In important rooms like ; Officer's rooms, Committee room, Doctor's room, Operation theatre's Principal's room etc. Dry distemper limited to 25% of total area.

(*b*) Remaining area-either colour wash or white wash.

B. *Specifications for Residential Bldgs.*

I. FOUNDATIONS : Spread foundation in brick or RCC with lean concrete underneath. RCC columns to have isolated/combined footing.

II. SUPER-STRUCTURE : Load bearing walls in brick/stone masonry with intermediate columns where ever necessary. In case of RCC frame construction, brick filler walls with internal partitions of half brick masonry.

III. DOOR & WINDOWS : *Frames* : Ist class kail wood or IInd class deodar wood or mild steel.

Shutters : Ist class kail wood or 2nd class deodar wood.

IV. FITTING : Oxidised iron.

V. FLOORING : 30 mm thick cement concrete flooring. Bath & W.C. to have mosaic flooring. In case of D type qtrs., mosaic flooring is also to be provided in Living room and Drawing/Dining room besides Bath & W.C and remaining area to have cement concrete flooring.

VI. TERRACING : Lime concrete or mud phuska finished with tiles.

VII. FINISHING : *External* : Colour wash.

Internal : White wash in A B & C type qtrs. In case of D type qtrs. - distemper in Drawing, Dining, Study & Bed rooms.

32.8 METHOD OF PREPARATION OF PRELIMINARY ESTIMATES

The method of preparation of preliminary estimates is explained with the help of following examples.

CASE I : PRELIMINARY ESTIMATE FOR A MULTISTOREYED OFFICE BUILDING

Example 32.1. *Prepare preliminary estimate for a six storeyed office building. The building is to be of a RCC framed construction type. The other data for preparation of the preliminary estimate is as under :-*

(*i*) *Plinth area on each floor* = 200 sq. m.

Hence the total plinth area of the building = 1200 sq. m.

(*ii*) *The floor to floor height* = 3.7 m

(*iii*) *Plinth height of the building* = 0.75 m

(*iv*) *The column spacing is such that the structural module is more than 35 sq. m. in area.*

(*v*) *The building is to be located in earthquake zone (zone v)*

(*vi*) *Provision of 2 lifts of 8 passenger capacity having speed of 1.5m/second is to be made.*

(*vii*) *Provision of water tank to be made as under :-*

(*a*) *An overhead tank of 10,000 litre capacity to be located on terrace.*

(*b*) *An underground sump of 1,00,000 litre capacity.*

(*viii*) *Provision of fire fighting, power wiring and telephone conduits are to be made.*

(*ix*) *Provision of water supply pumps to be made on lumpsum basis.*

(*x*) *For the purposes of calculation of cost index, the current rate at site of work for the different materials and labour may be taken as under*

(*a*) *Current rate of materials*

Bricks	= Rs. 400/*thousand*	*Sand*	= Rs. 120/cum
Cement	= Rs. 100/qtl.	Aggregate	= Rs. 170/cum
Mild steel	= Rs. 700/qtl.	*Timber*	= Rs. 8000/cum.

(*b*) *Current* rate of labour

Mason	= Rs. 50 each
Carpenter	= Rs. 50 each
Coolie/Beldar	= Rs. 15 each.

Solution : Preliminary cost estimate based on CPWD plinth area rates 1976 for 6 storeyed office building

S. No.	Description of item	Area in sq. m.	Rate/ sq.m.	Amount in Rupees
1	2	3	4	5
	Refer table 32.1			
1.	RCC Framed structures			
1.1	RCC Framed structure upto 6 storeyed Floor height 3.4m (Refer item 1.1)	1200 sq. m.	400.00	4,80,000
1.2	*Extra for* :			
1.2.1	Every additional 0.3 m height of floor above normal floor height of 3.4m. Floor to floor height being 3.7 m (Refer item 4.1.3)	1200 sq.m.	17.00	20,400
1.2.2	Every 0.30 m height plinth over normal plinth height of 0.60m (on ground floor area only). The height of plinth being 0.75 m. (Refer item 4.14)	200 sq. m. (G.F area only) Proportionate rate for 15cm additional height = $\frac{0.15}{0.00} \times 17$ = Rs. 8.50/sq.m.	8.50	17,00
1.2.3	Larger modules over 35 sq. m. (Refer item 4.1.12)	1200 sq.m.	30.00	36,000
1.2.4	Resisting earthquake forces (Refer item 4.1.8)	1200 sq. m.	34.00	40,800
1.2.5	RCC raft foundation (Refer item 4.1.9)	1200 sq.m.	60.00	72,000

(*Contd...*)

1	2	3	4	5
1.2.6	Fire Fighting (Refer item 4.1.14)	1200 sq.m.	25.00	30,000
2.	**Services**			
	Refer Table 32.2 Building cost to be considered for services - item no. 1.1+1.2.1+1.2.2 = Rs. 4,80,000 + Rs. 20,400 + Rs. 1,700 =Rs. 5,02,100			
2.1	Internal water supply and sanitary installation @ 4% of Rs. 5,02,100			20,085
2.2	External services connection @ 5% of Rs. 5,02,100			25,105
2.3	Internal electrical installation @ 12 ½ % of Rs. 5,02,100			62,765
2.4	Extra for power wiring & plugs @ 4% of Rs. 5,02,100			20,085
2.5	Extra for telephone conduits @ 0.5% of Rs. 5,02,100			2,510
2.6	Passenger lift of speed of 1.5m (Refer table 32.4)(rate of lift for 8 floors)	2 nos. @ Rs. 1,80,000 each		3,60,000
2.7	Deduct for each floor less than 8 floors for lifts. The building being six stories deduction in per lift is for 2 floor = 4 nos. @ Rs. 10,000 each - per floor Hence Number of floors for deduction = 4			(—) 40,000

(Contd...)

1	2	Amount in Rupees
2.8	Overhead tank without independent staging of capacity 10,000 litres @ Rs. 0.65/litre	= Rs. 6,500
2.9	Underground sump of 1,00,000 litres capacity @ Rs. 0.65/litres	= Rs. 65,000
	Total cost based on CPWD Plinth area rates of 1976	Rs. 12,02,950
	Based on the current rates of material and labour at site of work, the value, of present day cost index as worked out in table 35.8 = 398	
	Enhancing to cost index 398 i.e. Rs. 12,02,950 x 2.98 = Rs. 35,84,790	Rs. 35,84,790
	Add cost of pumps, L.S. of Rs. 1,00,000	Rs. 1,00,000
		Rs. 48,87,740
	Contingencies @ 3% on Rs. 48,87,740	Rs. 1,46,630
	Total	Rs. 50,34,370
	Say	Rs. 50,34,400
	∴ Present day preliminary estimated cost of office building = Rs. 50,34,400	

TABLE 32.8

Solution : Calculation of present day building cost index for example no. 32·1

S.No.	Description	Unit	Rate at New Delhi corres-ponding to base 100 as on 1.10.76	Current rate at station/ site of work	Ratio between col. 5 & 4 i.e. col. 5 / col.4	Weightage	Cost index (col.6 x col.7)	Remarks
1	2	3	4	5	6	7	8	9
1.	Bricks	1000	106.16	400.00	3.77	16	60.32	
2.	Sand	Cum	21.92	120.00	5.47	5	27.35	
3.	Cement	Qtl.	35.28	100.00	2.83	21	59.43	
4.	Aggregate	Cum	27.10	170.00	6.27	6.5	40.76	
5.	Mild steel	Qtl.	183.20	700.00	3.82	10	38.20	
6.	Timber	Cum	2021.00	8000.00	3.96	18	71.28	
	Labour							
7.	Mason	Each	9.89	50.00	5.05	8.5	42.93	
8.	Carpenter	Each	9.89	50.00	5.05	4	20.20	
9.	Coolie/Belder	Each	4.41	15.00	3.40	11	37.40	
							397.87	= Say 398

∴ Cost Index = 398

CASE II. PRELIMINARY ESTIMATE FOR A RESIDENTIAL BUILDING

Example 32·2. *Prepare preliminary estimate for a block of four 'C' type quarters. Double storey construction having two 'C' type quarters on ground floor and two on upper floor. The construction is to be of load bearing type. The other data for preparation of preliminary estimate is as under :-*

(*i*) *The calculations of plinth area of the block of 4--c type quarters is as under :-*

(*a*) *Plinth area of one C type quarter* = 80 sq. m.

(*b*) *Plinth area on account of covered verandah/balcony* = 3 sq. m (*assumed*)

(*C*) *Proportionate area per quarter for staircase* = 5 sq. m. (*assumed*)

Total plinth area per C type quarter = 88 sq. m.

Plinth area of the block on ground floor (having 2 C type quarters) = 2 x 88 = 176 sq. m.

Plinth are of the block on first floor = 2 x 88 = 176 sq.m.

Total Plinth area of the block of 4-C type quarters including area of staircase & verandah/balcony = 352 sq. m.

(*ii*) *Floor to floor height of the quarters* =3.2 m.

(*iii*) *Plinth height* = 0.6 m.

(*iv*) Depth of foundation = 1.5 m.

(*v*) *One water tank per quarter of capacity 1080 litres to be provided at terrace level.*

(*vi*) *For the purposes of calculation of cost index, the current rate of work may be assumed same as in case of office building in example 32.1*

Solution : Preliminary cost estimate based on CPWD plinth area rates 1976 for one block of 4—c type quarters (two storeyed load bearing construction)

Sl.No.	Description of item	Area in sq. m.	Rate/Sq. m.	Amount
(1)	(2)	(3)	(4)	(5)
	Refer table No. 35.1			
1.	Load bearing construction			
1.1	Floor height 2.90 m. (Double storeyed (refer item 3.2)	352 sq. m.	260.00	91,520
2.	*Extra for*			
2.1	Every 0.30m. additional height above normal height of 2.90 m. The floor to floor height being 3.2 m (Refer item 4.2.1.)	352 sq. m.	10.00	3,520
2.2	Every 0.30 m. deeper foundations over normal depth of 1.20 m Depth of foundation being 1.5 m (refer item 4.2.3 rate applicable on ground floor area only)	176 sq. m.	12.00	2,112
2.3	Termite proof treatment (on ground floor area only)	176 sq. m.	9.00	1,584
3.	*Services* *Building cost to be considered for services = item No. 1.1 + 2.2 *i.e.,* Rs. 91,520 + Rs. 3520 = Rs. 95,040 Refer table No. 35.2			
3.1	Internal electrical installation @ Rs. 2000/qtr. for 4 qtr @ Rs. 2000 per Qtr.			8,000
3.2	Internal water supply and sanitary installation @ Rs. 2500/qtr for 4 qtr.			10,000
3.3	External service connection @ 5% of Rs. 95,040 (refer Table 35.2)			4,752

1	2	3	4	5
3.4	Water tanks without independent staging of 1080 litres capacity per qtr for 4 qtrs @ Rs. 0.65/litre			2,808
	Preliminary cost based on CPWD plinth area rates 1976 for one block of 4 quarters			Rs. 124,296
	Enhancing to cost index 398 *i.e.* Rs. 124296 x 2.98 =			Rs. 370402
			Total :	Rs. 494,698
	Add contingencies @ 3% on Rs. 4,94,698			Rs. 14,841
			Total :	Rs. 509,539
	Preliminary estimated cost of one block of 4 'C' type quarters = Rs. 5,09,539 Say Rs. 5,09,540			

CASE III PRELIMINARY ESTIMATE FOR LAND DEVELOPMENT WORK

Example 32.3. *Prepare a preliminary estimate for land development.*

The data for the preparation of preliminary estimate is as under :-

(*i*) *Area of land to be developed = 1.5 hectare = 15,000 sq.m.*

(*ii*) *For the purposes of calculation of cost index the current rate of material & labour at site of work may be assumed same as in case of office building in example 32.1.*

Solution : Preliminary cost estimate based on CPWD plinth area rates 1976 for Land Development :

Area of land = 1.5 hectares = 15000 sq. m.

The entire area of land is to be considered without making any deduction in respect of area of land occupied by building located/to be located in the plot of land.

S.No.	*Description of item*	*Rate per sq. m.*		*Amount*
Refer Table No. 32.5				
1.	levelling	@ Rs. 1.50 per sq. m.		
2.	Internal paths & Roads	@ Rs. 4.65 per sq. m.		
3.	Sewers	@ Rs. 3.35 per sq. m.		
4.	Water supply distribution lines	@ Rs.2.45 per sq. m.		
5.	Unfiltered water supply distribution lines	@ Rs. 1.40 per sq. m.		
6.	Storm water drains	@ Rs. 4.00 per sq. m.		
7.	Horticulture operation	@ Rs. 2.50 per sq. m.		
8.	Street lighting with incandescent lamps	@ Rs. 0.35 per sq. m.		
9.	HT substation & LT distribution system	@ Rs. 2.60 per sq. m.		
	Total	Rs. 22.8 per sq. m.		
	Cost of land development based on CPWD plinth area rates 1976			Rs. 3,42,000.00
	for 15000 sq. m. @ Rs. 22.8 Per sq. m = 3,42,000,00			
	Enhancing to cost index 398 i.e. Rs. 3,42,000 x 2.98			Rs. 10,19.160.00
			Total	Rs. 13,61,160.00
	Add contingencies @ 3% on Rs. 13,61,160.00			Rs. 40,835.00
			Total	Rs. 14,01,995.00

Preliminary estimated cost for land development = say Rs. 14,02,000.00

CASE IV. PRELIMINARY ESTIMATE FOR A PROJECT

Example 32.4. *Prepare a preliminary estimate for a project comprising of a six storeyed office building and a two storeyed block of 4 C-type residential quarters. The buildings are to be located in a plot of land having an area of 1.5 Hectare. The estimate shall also include the cost of land development. The other data required for preparation of estimate shall be same as given in Example 32.1, 32.2 and 32.3 respectively.*

Solution : Preliminary estimated cost of the project will comprise of sum of the estimated cost of (i) Office building (*ii*) 4-C-Type Quarters and (*iii*) Land development for 1.5 Hectare of land.

Hence

	(i) Preliminary estimated cost of office buildings as per example 32.1		= Rs. 50,34,400
	(ii) Preliminary estimated cost of 4-C type quarters as per example 32.2		= Rs. 5,09,540
and	(iii) Preliminary estimated cost of land development for 1.5 Hectare of land as per example 32.3.		= Rs. 14,02,000
	Thus preliminary estimated cost of the project	.Total	= Rs. 69,45,940
		Say	Rs. 69,46,000

Questions for Review

1. Prepare preliminary estimate for a block of four B type quarters--two quarter being on ground floor and two on upper floor. Plinth area of each quarter is 56 sq.m. Make suitable assumptions in respect of provision of area for balcony and stair cases. Adopt cost index of 650 to arrive at present day cost of work.
2. Prepare preliminary cost estimate for a 12 storeyed office building.

   Following data may be adopted

   (*i*) Plinth area on each floor = 600 sq m.

   (*ii*) Floor to floor height = 3.6 m

   (*iii*) Plinth height = 1 m.

   (*iv*) C/C spacing of column in the Structural grid in either direction = 6 m

   (*v*) The building is to be located in earthquake zone (zone V)

   (*vi*) Provision of 4 No. 12 passenger lifts has to be made

   (*vii*) One over head tank of 40,000 litre capacity.

   (*viii*) One under ground sump of 1,40,000 litre capacity.

   (*ix*) Make other suitable provisions applicable to multistoreyed buildings

   (*x*) Adopt cost index of 680 to arrive at present day cost.

33

Project Management Through PERT/CPM

33·1. INTRODUCTION

Planning is essential for achieving any pre-determined objective. It is however seen that inspite of planning, majority of the projects do not keep up their original time schedule and their completion cost is also higher as compared with the estimated cost.

The main reason for overrun in time and cost is that the planning at the outset of the project was not adequate or the method of planning was not a refined one.

It is observed that the conventional methods of planning and scheduling have not been able to meet requirements of effective planning and control of the projects.

With the growing advanced technology and increasing size and complexity of modern projects, sound and effective method of planning and control assume great importance. We have therefore to device suitable methods of planning which can reduce the possibilities of time and cost overruns through quicker, more responsive and integrated control.

To meet this requirement, it becomes necessary to plan the project in detail taking into account the vast number of activities and the agencies involved in executing the project, paying due regard to the logical sequence of occurance of the activities as well as inter-dependency and inter-relationship between them.

To have proper control, we need not only timely and accurate information on the status of all constituents and significant activities and the effect of these on the total project but we also require quicker and simpler methods of rescheduling the activities to minimise the effect of delay in some activities on the overall completion of the project.

These requirements are satisfied by adopting the method of *network techniques* for planning, scheduling and controlling of projects. The network techniques which include PERT and CPM have now become almost an indispensable tool for modern management, to prevent unnecessary delays in the execution of projects and to add overall economy in the project cost by effective utilisation of the available resources.

Before explaining the principles of network technique, it is considered necessary to review all the conventional methods of planning and control with a view to recognise rumerous advantages of network techniques and the aid it can render for better management.

33·2. METHOD OF PLANNING AND SCHEDULING

The various planning and scheduling methods or techniques used for management of a project, can be divided into following two groups.

(*a*) *Conventional methods*

(*b*) *Network methods*

33·3. CONVENTIONAL METHODS

Conventional methods used for planning and scheduling broadly include the following :

1. Gantt or Bar chart
2. Milestone chart
3. Flow chart

1. **Gantt or Bar chart.** Gantt chart was introduced by a consultant named as Henry L. Gantt prior to world War-1. This method is also commonly known as bar chart method. It is a simple method which can display time schedule of different activities in the project as well as make a comparison between actual performance and the schedule. In this case a project is divided into different jobs or activities which are represented by horizontal bars on a time scale. The beginning and end of the bars represent the time of start and completion of the activity whereas its length represents the total time required to complete the activity. On the date of reviewing progress, the bars are shaded to indicate the portion of the work completed. Hence with the help of this chart, it is possible to know the scheduled target as well as the actual progress for different activities at the time of review. This inturn enables one to identify whether a particular activity is behind scheduled, ahead of schedule or on time. A typical Gantt or bar chart is shown in Fig. 33·1.

From Fig 33·1, it is seen that on the date of review of progress *i.e.* 15th of May, the activity no. 1 pertaining to construction of foundation (which had a total time allocation of 2 months) has beencompleted as per schedule. Activity no. 2 pertaining to construction of superstructure is behind schedule by 15 days. Activity no. 4 pertaining to fabrication of doors and windows frames (which was to be completed by end of May) is ahead of schedule

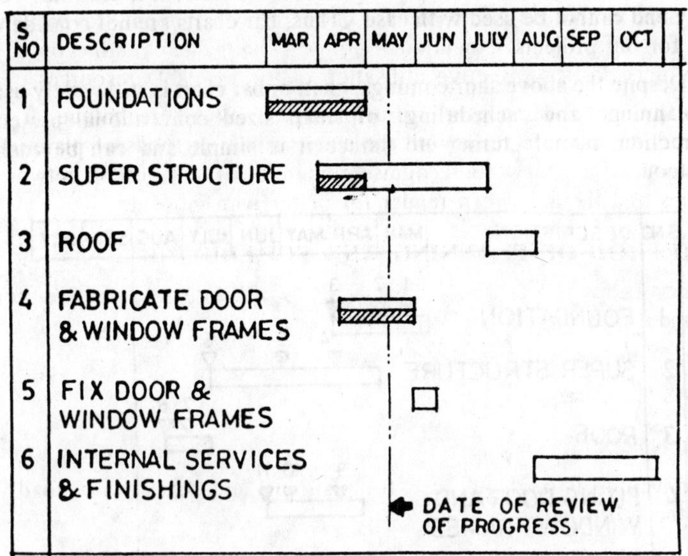

Fig. 33·1

Gantt or Bar chart

since this activity which was planned to be completed by end May was already complete on 15th of May. The remaining jobs are planned to start beyond end May and hence no progress has been reflected on the date of review.

Shortcomings of Gantt or bar chart. This method of planning suffers from the following shortcomings.

(*a*) It does not indicate the inter-relationship between various activities.

(*b*) It does not indicate the effect of the activity behind schedule or ahead of schedule on the overall completion of the project.

(*c*) It does not indicate which activity is critical and which has spare time.

(*d*) It does not give any idea of what exactly is completed on the date of review, and one has to only guess very broadly as to how much of the work (in terms of percentage *i.e.* 50% or 75% etc.) is completed.

(*e*) In case of bar chart, it is difficult to optimise the use of resources.

(*f*) Bar charts do not prove useful in research and development projects where number of uncertainties are involved and it is difficult to allocate definite time for different activities.

(*g*) In case of bar charts only major activities in the project are exhibited. If all activities of the project are to be shown, it becomes very lengthy and cannot be used with ease. Thus, bar charts are not considered useful for big projects.

Despite the above shortcomings, gnatt or bar chart is still widely used for planning and scheduling of small sized conventional projects (construction, manufacturing etc.) since it is simple and can be easily understood.

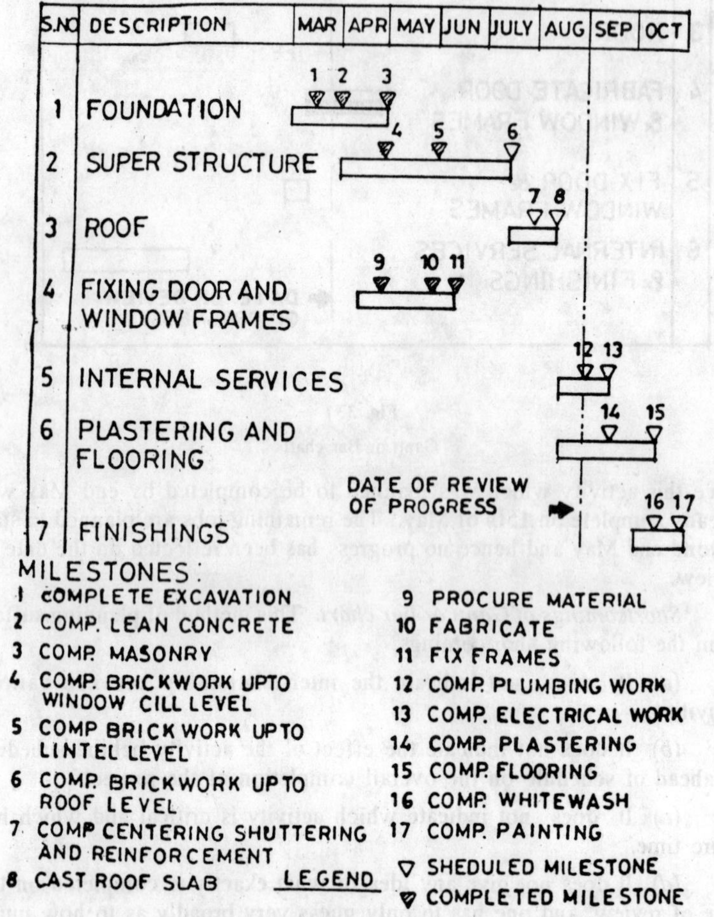

Fig. 33·2

Milestone chart for a building project

2. Milestone chart. This chart was developed around 1940. This chart is basically similar to Gantt chart except that it shows important events or milestones (in the form of vertical arrows) on the horizontal bars representing the activity. As a result thereof, the system of reporting progress becomes definite in this chart as it is possible to identify the progress in the form of completed milestone. It is also possible to draw a milestone chart without horizontal bars. A milestone chart is shown in Fig. 33·2.

Shortcoming of milestone chart. This chart suffers from the following limitations :

(*i*) It does not show inter-relationship of milestones and dependencies between different jobs.

(*ii*) It does not show the effect of one milestone being behind schedule on the overall completion of the project.

3. **Flow chart.** This is a brief and simple graphic representation of a time schedule. This chart shows relationship between activities and at the same time contains events and milestones. A typical flow chart of construction of a house is shown in Fig. 33·3.

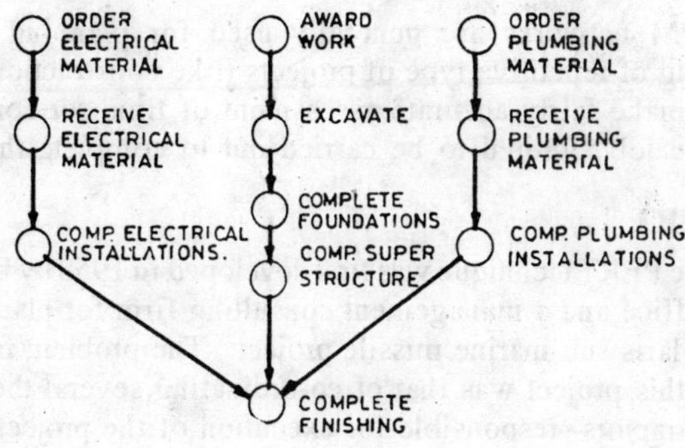

Fig. 33·3.
Flow chart

Shortcoming of flow chart. This chart has the following limitations:-

(*i*) It does not provide information regarding time of completion of activities.

(*ii*) It does not show the effect of slippage in one activity over the total period of completion of the project.

33·4. NETWORK METHOD

In view of the inherent draw-back and limitations of the conventional methods of planning and scheduling explained above, it was felt necessary to evolve effective and more efficient method of planning and scheduling of projects. Moreover, conventional methods were found to be very inadequate for planning, scheduling and monitoring of large projects involving thousands of activities. To tackle complex projects, it becomes necessary to plan them in greater detail paying due regard to the increased number of activities and increased number of executing agencies. Continued search for improved method for planning the projects, has resulted in the evolution of a new technique which is termed as *Network Technique.* The programme Evolution and Review Technique (PERT) and the Critical Path Method (CPM) are the two methods which get covered under the nomenclature of network techniques.

33·5. CRITICAL PATH METHOD (CPM)

This method was developed in USA in 1957 by Morgan E. Walker (of DU Pont) and James Kelley (of Remington Rand USA). This method was first used for planning and scheduling the construction of a new chemical plant by DU Pont. Later this method was used for scheduling the overhaul and maintenance for shut down of equipment in their works. By adoption of this method, DU Pont were able to reduce heavy losses of production sustained during the shut-downs.

CPM networks are generally used for planning, scheduling and controlling of repetitive type of projects (like construction projects) where one can make fairly accurate assessment of time for completion of each and every job required to be carried out to complete the project.

33·6. PERT

The PERT technique was first developed in 1958 by USA Navy special project office and a management consulting firm for planning and control of the Polaris sub-marine missile project. The problem faced by the USA Navy in this project was that of co-ordinating several thousands agencies and contractors responsible for execution of the project. The technique of PERT was specially developed to meet this situation and application of the same helped in reducing the overall project time by as much as two years.

PERT networks are generally used in research and development type of projects or projects which are non-repetitive in nature or for projects where it is difficult to allot definite time for performing each and every activity for want of information regarding past experience or knowledge to handle such projects.

However, with the passage of time numerous techniques and modifications developed in the PERT and CPM to meet widely and varying requirements of projects and distinction between two methods have now got blurred

Both PERT as well as CPM techniques describe the plan of a project in the form of arrow diagram, also known as network diagram.

33·7 ADVANTAGES OF PERT/CPM

The various advantages of PERT/CPM method are summarised below.

1. In the network technique of planning and scheduling, one is compelled to examine the complete project in advance and interact with everyone who is to be involved in the project thereby evolving a workable plan of the project. One is also forced to decide upon the extent of spliting of the project into smaller activities and to establish logical relationship between different activities.

2. It helps in dividing activities funtionwise and responsibility-wise which makes it possible to co-ordinate the works of different agencies involved in the completion of the project.

3. In enables one to determine reasonably correct schedule for the completion of different events and the project as a whole based on the availability of resources (Men, materials, money and machines etc.)

4. It permits one to take advance actions and timely decisions to reduce delays in completion of different events.

5. It permits greater flexibility in having optimum utilisation of resources thereby effecting economy.

6. It identifies activities critical to different stages of projects completion which in turn enables the management to pay greater attention to fewer activities instead of concentrating on all activities at all times with equal emphasis.

7. It enables one to exercise effective control on the project by way of periodic review and to adopt timely corrective measures to minimise slippages in the completion of the project.

33·8. ELEMENTS OF NETWORK

Before describing the method of developing a network, it is necessary to have an idea about the important terms and symbols.

Some of the important terms used in network techniques are described below.

1. Network. The network is a graphic plan showing jobs or activities that must be completed to achieve completion of a project. A network is also defined as an arrow diagram obtained by connecting all the activities of project in logical sequence. Thus the components which make up a network are ; activities and events.

2. Activity. Jobs that are required to be performed in executing a project are known as Activities. An activity is represented on a network

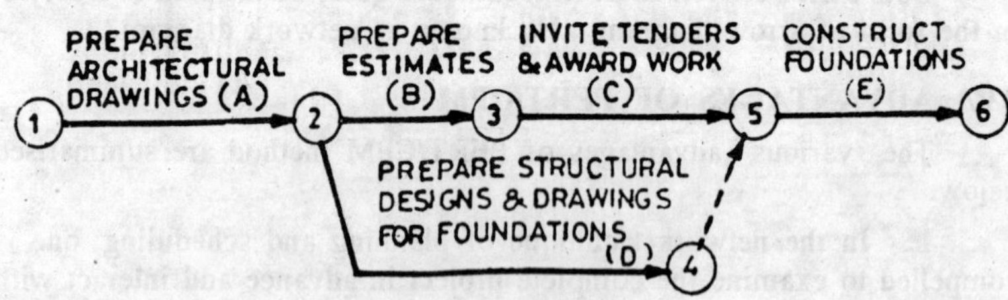

Fig. 33·4

Simple network showing events and activities

by an arrow. The tail of the arrow represents start of the activity and head of the arrow represents the completion of the Activity. The arrows are not to the scale and their length has no physical significance. The head of the arrow shows direction of time flow from one event to another. In a network the activities are arranged in a logical sequence of their occurrence. Thus it may be been that activity represents performance of work and it consumes time and resources.

3. Event. An event represents specific definable accomplishment in a work plan. In other words, this is a point in time representing start or completion of one or more activities. An event is generally represented on a network by geometrical symbol such as ; circle or rectangle and is identified by code nos. like 1, 2, 3 etc. An event consumes neither time nor resources. It is a point in time and not passage of time. The concept of event and activities can be further clarified by referring fig. 33·4. In the Fig. (33·4) the activities are designated by 1-2 or (A) ; 2-3 or (B) ; 3-5 or (C) etc. The activities are arranged in their logical sequence. From the review of diagram, it can be seen that activity- A controls activity B, C, D, and E. Alternatively it can be said that Activity B, C, D, and E depend on activity-A. The examples of events in the above Fig. are event no. 1 shows start of the work of preparation of architectural drawings or activity-A. Event no. 2 represents the completion of the work of preparation of architectural drawings and also represents start of activity B and D.

4. Dummy activity. It is a notional activity which neither consumes time nor resources but indicates the logical sequential relationship between other activities. Dummy activities are therefore introduced to represent a connection or inter dependency between events. Dummy activities are represented on a network by a dotted arrow. From Fig. (33·4) above, it can be seen that activity *E* can start only after activities C and D are completed. It is also seen that activity C and D are quite independent to each other. A dummy activity 4-5 is therefore introduced to maintain the logic.

Let us consider a case of a network having two separate sequence of activities as shown in Fig. (33·5). From the Fig. it can be seen that activity

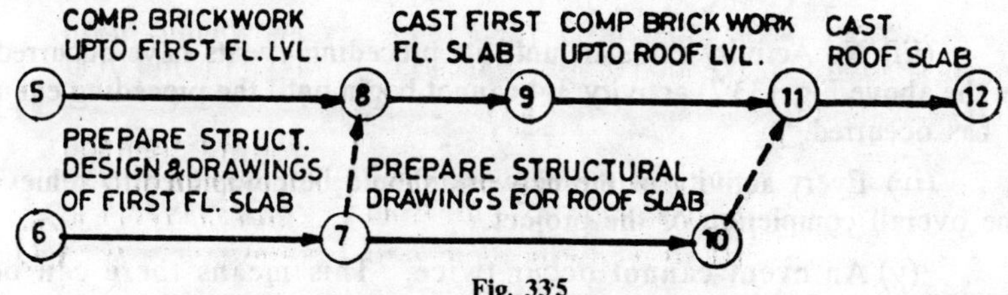

Fig. 33·5

Network diagram showing Dummy activities

no. 8-9 can start only when activity 6-7 is completed. Similarly activity 11-12 can start only when activity 7-10 is completed. This logic is reflected in the network by introducing dummy activity 7-8 and 10-11 as shown in the figure. Sometimes dummy activities are introduced for convenience or to simplify the drawing of the network diagram.

33·9. RULES TO BE FOLLOWED FOR DRAWING NETWORK

Important rules to be observed in drawing a network are as under :

(*i*) Each activity is represented by one and only one arrow Ref. Fig (33·6)

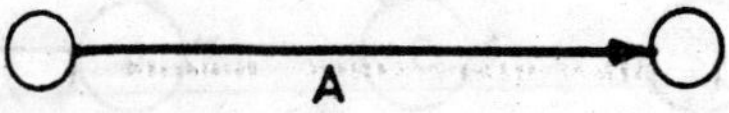

Fig. 33·6

(*ii*) An event cannot occur until all activities leading to it are completed. Ref. Fig. (33·7) In accordance with this rule, even no. 3 cannot occur until activities 1-3, and 2-3 are completed.

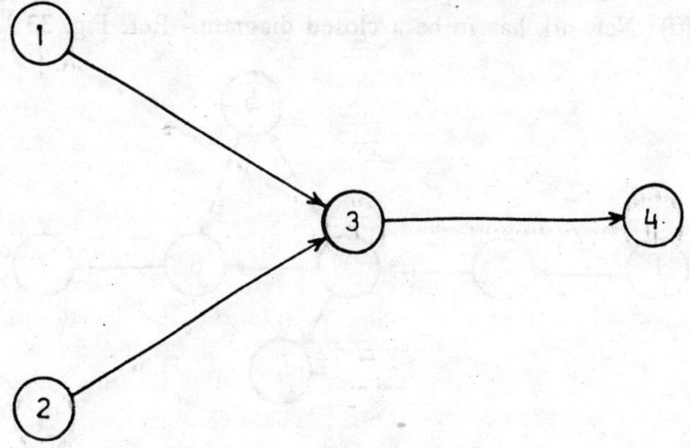

Fig. 33·7

(*iii*) No Activity can start until its preceding events have occurred. In the above Fig. (33·7) activity 3-4 cannot begin until the preceding event 3 has occurred.

(*iv*) Every activity of the network should be completed to achieve the overall completion of the project.

(*v*) An event cannot occur twice. This means there can be no network path looping back to a previously accomplished event.

This rule is explained with the help of Fig. 33·8. In this fig. activity no. 5-3 is forming a loop and hence the network is *incorrect.* This is on account of the fact, that if we want to interpret such a diagram, we will have to say that event-5 can occur only after events 2-3 and 4 have occurred therefore looping back event-3 which has already occured is not possible as it is not a logical sequence. Thus no event can depend for its completion upon the completion of a succeeding event .

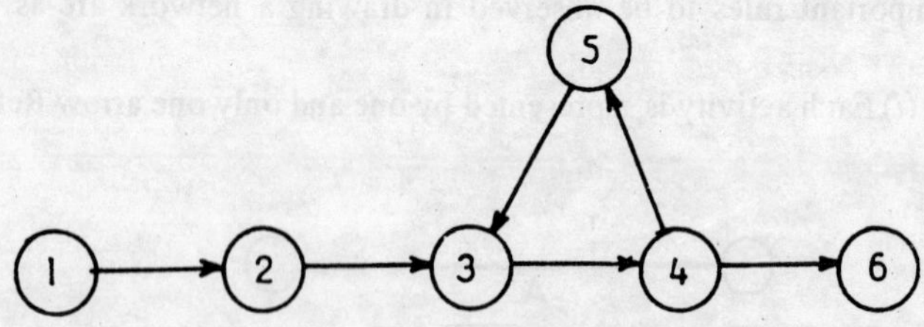

Fig. 33·8

(*vi*) In the network, time is assumed to flow from left to right.

(*vii*) Length of the arrow is chosen to suit drafting convenience. It has no other significance.

(*viii*) Network has to be a closed diagram. Ref. Fig. 33·9.

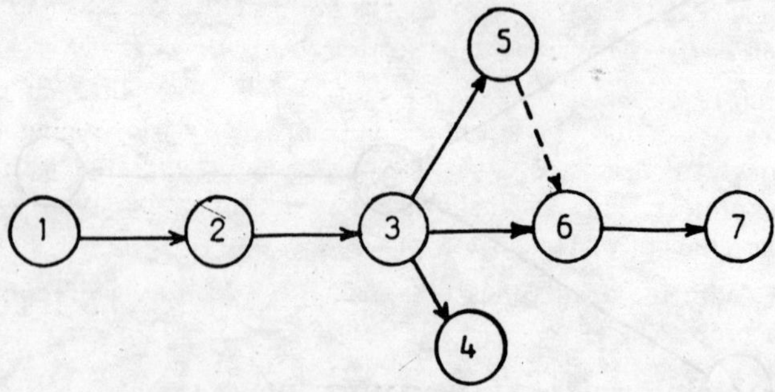

Fig. 33·9

In this fig. event no. 4 is shown unconnected. This situation is n *permissible* in a network. Event no. 4 has to be connected to some oth event depending upon inter-relationship/inter dependency.

(*ix*) All constraints and inter-dependencies must be shown in the network.

Examples of the constraints are ; delivery of materials, equipment, funds etc.

33·10. DEVELOPMENT OF NETWORK

Steps to be followed for developing a network are as under :-

(*a*) Decide and write down all the activities which are required to be performed for completing the project.

(*b*) Arrange these activities in their correct logical sequence.

To obtain correct sequence of activities in a work following 3 basic questions have to be posed and answered while drawing each arrow representing an activity.

(*i*) Which Activity must be completed before this particular Activity can be started ?

(*ii*) Which other activities can be done concurrently with this Activity ?

(*iii*) Which other activity must immediately follow this activity ?

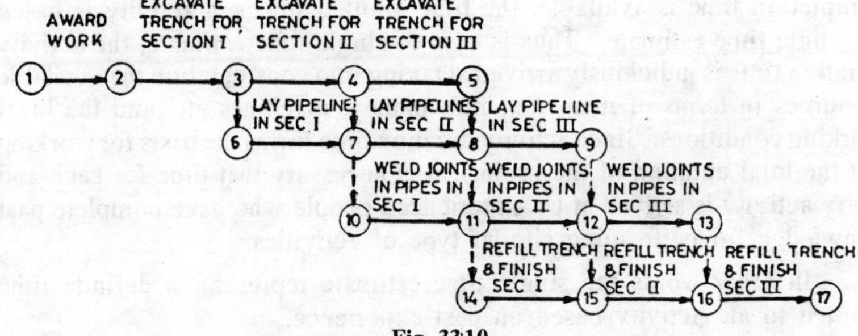

Fig. 33·10

Network diagram for a pipe laying project drawn based on concept of concurrent Activities

(*c*) Consider possibilities of sub-dividing different activities in phases or sections to introduce the concept of performing jobs simultaneously or concerrently. This helps in reducing overall time of completion of the project. Fig. 33·10 shows a network for a pipe laying work planned on the concept or concurrent performance of activities.

With the details thus obtained, it is possible to develop a network for a project.

33·11. TO ESTIMATE PROJECT DURATION

After the network for a project is developed based on the methods stipulated above next step is to estimate the completion time of the project, based on the time required to perform different activities comprising the project. To assess the project duration, following 4 steps have to be flowed in their sequential order.

1. Estimation of activity duration time (t_e)
2. Estimation of earliest expected event time (T_E).
3. Estimation of latest allowable event time.(T_L)
4. Calculation of event slack

These steps are elaborated below.

1. Estimation of activity duration time. (t_e) Activity duration time is a time that is required to perform an activity. This is designated by a symbol (t_e). On the network diagram the value of (t_e) is recorded below the relevant Activity arrow. The unit of time may be in days, weeks or months etc. In case of dummy activity value of 't_e' is zero. The Activity duration time can be assessed based on two methods.

(*a*) Single time estimate ;

(*b*) Three time estimate

(*a*) *Single time estimate.* In case of projects involving repetitive nature of activities, where adequate knowledge and experience of their completion time is available, the time allotted for each activity is based on single time estimate. Thus is case of single time estimate, the activity duration time is judiciously arrived at taking into consideration the available resources in terms of men, material, money, machines etc. and the local working conditions. Since activity duration time forms the basis for working out the total duration of the project, it is necessary that time for each and every activity is arrived at by experienced people who have complete past knowledge for performing similar type of activities.

In other words the single time estimate represents a definite time allotted to an Activity based on past experience.

(*b*) *Three time estimate.* In case of projects involving research and development or projects involving activities never handled before, it is not possible to allocate single time estimate to the activities for want of adequate information or knowledge. In such cases, the activity time estimate could be highly uncertain. To deal with such projects, three time estimates are made initially and the value of t_e is thereafter worked out by an empirical formula.

The three time estimates to be considered in such a case are :

(*i*) *Optimistic time* :

Time required to complete the Activity assuming every thingsgoes alright. This is designated by a symbol '*a*' (*ii*) *Passimistic time*

Time required to complete the Activity assuming everything goes wrong. This is designated by a symbol 'b'

(*iii*) *Most likely time*

Time required to complete the activity assuming that only such things go wrong which are normally expected to be wrong. This is represented by a symbol '*m*'

The three time estimates are then converted into a single time estimates or 't_e' for each Activity by the following formula :

$$t_e = \frac{a + 4m + b}{6}$$

The working of three times estimates is based on the theory of probability.

The main difference between the Critical Path Method (CPM) and the PERT is that in case of CPM the values of 't_e' are based on single time estimates and in case of PERT values of 't_e' are based on three time estimates.

(2) Estimation of earliest expected event time (T_E). After writing down the values of 't_e' for all activities on the network, next step is to calculate the minimum time required for the accomplishment of all the events in the network. The earliest possible time at which an event can occur assuming that preceding activities take the time estimated for them is known as earliest event time. All activities shown terminating at the event, must be completed before the event occurs. The earliest expected time for each event is designated by symbol (T_E).

Thus the earliest expected event time (T_E) for any given event is obtained by adding the activity time (t_e) for all preceding activities along the longest path from the beginning to the given event.

The Fig. (33·11) shows a simple network in which the activity duration in weeks have been written below each arrow. Since event no. 1 marks the start of the project, T_E for this event will be zero.

Let us consider event-5 in Fig. (33·11) Event 5 cannot occur until all the activities terminating at this event have been completed. We have therefore to examine two paths *i.e.* event path 1-2-5 and event path 1-4-5. Time at which event 5 can occur based on event path 1-2-5 works out to 5 W + 7 W = 12 weeks and that based on event path 1-4-5 works out to 6W + 5W = 11 weeks. Since the activity along event path 1-2-5 would not have completed before 12 weeks, hence T_E value for event no. 5 will be 12 weeks being derived from event path 1-2-5 which is longer than event path 1-4-5.

Based on the above working it can be seen that value of T_E for the end event (event no. 7) of the project works out to 23 weeks. Thus the earliest time at which the project can be completed is 23 weeks.

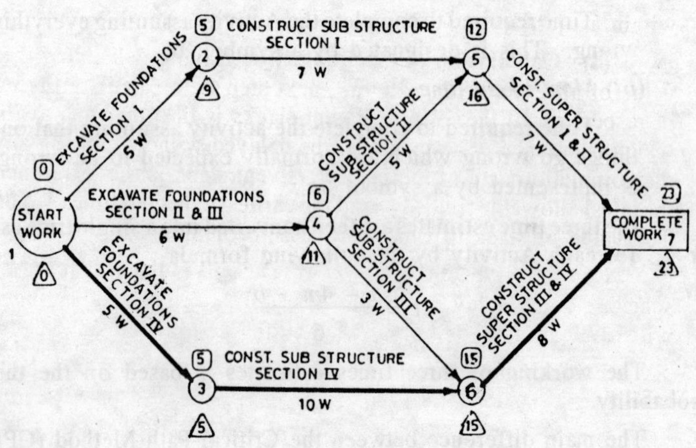

Fig. 33·11

Network showing calculations of T_E and T_L and critical path

(3) Estimation of latest allowable event time (T_L). The latest allowable event time is the latest time by which an event must take place if the total project time is not to be exceeded. The latest allowable time for each event is designated by symbol 'T_L'.

After fixing 'T_L' Value for the end event, the next step is to work out 'T_L' inrespect of all the events in the network. The method of calculating T_L is just reverse of that of calculating T_E *i.e* by subtracting the duration of succeeding activities along the logest path starting at the final event.

Ref. Fig. (33·11). If the project is required to be completed in 23 weeks, the T_L value of the end event has to be equal to 23 W. Thus T_L value for event 6 and event 5 in fig. (33·11) would be = 23 W–8 W = 15 W and 23 W–7 W = 16 W respectively.

To work out T_L value for event no. 4, two paths leading to it have to be considered. One path from event no. 6 and other from event no. 5.

Since event 6 can occur not later than 15 weeks and the activity 4–6 take 3 weeks, the latest event 4 can be allowed to occur is 15 W–3 W = 12 weeks. 'Similarly since event 5 cannot occur later than 16 weeks and activity 4-5 takes 5 weeks, the latest event 4 can be allowed to occur is 16 W - 5 W = 11 Weeks. Thus one path leading to event no. 4 gives the latest allowable time of 12 weeks and the other path leading to event no. 4 gives the latest allowable time of 11 weeks. Hence the T_L value of event no. 4 would be 11 weeks being the lower of the two values.

The TL value is normally written below the event by enclosing it within a triangle.

(4) Calculation of Event slack : Having calculated the T_E & T_L value of all events in a network, next step is to calculate the event slack in respect of all the events. Event slacks is the free time available upto which occurance of an event can be delayed without causing any delay in the completion date of the project. Event slack is the difference between the latest allowable time (T_L) and the earliest expected time (T_E) of an event. The value of slack can be positive zero or minus.

Zero slack occurring in any event will indicate that the event is on schedule. Any delay caused in the occurance of events having zero slack will result in corresponding delay in the completion of the project.

Positive slack occurring at any event will indicate that the event is ahead of schedule. Thus, in case there is delay in occurrence of an event to the extent of the positive slack available, it will not cause delay in the overall completion of the project. This information allows the management to temporarily divert the resources (men,materials, money & machines etc.) to other activities while rescheduling.

Negative slack occurring at the event shows that the event is behind schedule. Negative slack thus indicates that there will be delay in the completion of the project to the extent of the negative slack of the end event. The information allows the management to take advance action to speed up the concerned activities to minimise chances of delay.

The event slack for different events are worked out and indicated below the each event in Fig. 33·11.

33·12. IDENTIFICATION OF THE CRITICAL PATH

From the review of the network diagram in Fig. 33·11 it can be seen that there are a number of alternative paths along the network from start of the project to its completion. The critical path is path along the network which requires the longest time for the project from start to its completion and thus, decides the time of completion of the project as a whole. In Fig. 33·11 the path along events '1-3-6-7 is the critical path. The events on the critical path have the least algebraic slack value almost cases the critical path passes through all events whose T_E an T_L value are the same.

Any event on critical path is known as critical event. Similarly any activity on critical path is known as critical activity. There can be more than one critical paths in a network. Any delay in the accomplishment of events on the critical path, would necessarily result in corresponding delay in the completion of the project.

Identification of critical path has the following main advantages.

(*i*) The project management need not spend time and effort in reviewing all the activities in a project and instead concentrate only on the critical activities for ensuring timely completion of project.

(*ii*) In case the management decides to shorten the time of the completion of the project, attention can be concentrated on speeding up only critical activities.

33·13. COMPRESSING THE CRITICAL PATH

From the above, it is clear that critical path governs both the project duration as well as the minimum time required to complete a project. Very often the project duration worked out based on network analysis has to be reduced to meet the directive from top management to complete project in shorter time. Thus 'T_E' of the end events becomes greater than T_L (the directed time of completion of project) and there is negative slack along the critical path. In such situation it becomes necessary to compress the critical path, to avoid negative slack.

Compressing of critical path can be achieved by following one or more steps given below :

(*i*) By closer examination of the activity duration time (t_e) for different activities along critical path and reducing them wherever feasible.

(*ii*) By re-planning the network by introducing number of parallel or concurrent activities on critical path.

(*iii*) By transfer of resources (men, materials, money and machines etc.) from the non-critical activities to the critical activities (within permissible limits) to reduce activity duration time along the critical path.

(*iv*) By adding more resources to the critical activity to effect reduction in activity duration.

(*v*) By reducing the scope of work or by changing or lowering performance requirements of some activities, which are relatively few as compared to numerous activities in the project.

33·14. START AND FINISH TIME OF ACTIVITY

We have already explained the method of working out event slack. The event slack gives an indication as to what extent an event can be delayed without effecting the completion date of the project. In a similar manner the concept of activity slack gives us an idea of the free time available to perform the activity without causing any delay in the completion date of the project.

Activity slack thus permits the manager to make certain adjustments in the start and completion time of the activity having slack.

Similar to the case of events each activity has four different types of times.

These are

(*i*) Earliest activity start time (E.S.)

(*ii*) Earliest activity finish time (E.F.)

(*iii*) Latest activity finish time (L.F.)

(*iv*) Latest activity start time. (L.S.)

(*i*) **Earliest activity start time (E.S.) :** Earliest start time for an activity is the earliest time by which it can commence. Hence this will essentially be the same as the earliest expected time (Ts) of the preceding event or event at the tail of the activity arrow. Ref. Fig. 33·11.

T_E for event-2 is 5 weeks. The earliest activity 2-5 can start is the earliest expected time event 2 can occur *i.e.* 5 weeks.

$$E.S. = T_E \text{ for preceding event.}$$

(*ii*) **Earliest activity finish time (E.F).** An activity can finish at its earliest only when it starts at its earliest time and takes estimated duration (t_e) for its completion In Fig. 33·11 earliest activity finish time (E.F.) for activity 2-5 is equal to 5 W + 7 W = 12 weeks.

$$E.F = E.S + t_e$$

(*iii*) **Latest activity finish time (L.F).** The latest finish time for an activity is the latest time by which an activity can be allowed to be finished without delaying the completion of the project.

Hence the latest finish time for an activity will be = the latest allowable occurrence time for the succeeding event or event at the head of activity arrow.

$$\therefore \quad L.F \text{ is } = T_L \text{ for the succeeding event.}$$

(*iv*) **Latest activity start time (L.S.).** This is obtained by deducting activity time duration (t_e) from the latest activity finish time.

$$\therefore \quad L.S. \text{ is } = L.F - t_e.$$

33·15. ACTIVITY SLACK OR FLOAT

This is the difference between the time required for performance of an activity and the time available for it. Activity floats are of the following types:

(*i*) Total activity float

(*ii*) Free float

(*iii*) Independent float.

(*i*) **Total activity float.** The total time available to perform an activity minus the duration of activity is termed the total activity slack or float. It can also be defined as the difference between the latest activity start time and earliest activity start time.

∴ Total activity float = L.S. – E.S

(*ii*) **Free float.** This is the time available for a particular activity between its earliest finish and earliest start of next activity.

If the earliest finish time of an activity is delayed by duration greater than its free float, the start of the subsequent activity will be delayed.

∴ Free float = T_E of succeeding event (*i.e.* at the head of the arrow)-T_E of preceding event (*i.e.* at the tail of the activity arrow)-te.

Free slack can be shared by preceding activities.

(*iii*) **Independent float.** This is the excess time available if the preceding activity ends as late as possible and the succeeding activity starts as early as possible.

Independent float = T_E of succeeding event (at the head of arrow)
– T_L of preceding event (at the tail of arrow)
$-t_e$.

The independent float is available only to the concerned activity and can not be shared be proceeding or succeeding activity.

33·16. ANALYSIS AND SCHEDULING OF NETWORK

After drawing the arrow diagram and writing down the duration of each activity (t_e) a network is analysed by following the steps given below :

(*i*) Calculate the earliest expected event time and latest allowable event time for each event (*i.e.* T_E and T_L), assuming that time estimates for each activity (t_e) are correct.

(*ii*) Determine the critical path which governs the minimum project completion time.

(*iii*) Determine the event slack and the activity floats for jobs which are not on critical path.

(*iv*) Prepare activity schedule with the help of the above information and list out the dates on which each activity is to be started and completed.

(*v*) Examine possibility of diverting the un-utilised resources from non-critical activities to the critical activities with a view to reducing the project completion time. This can be achieved by taking advantage of float or slack available.

(*vi*) Analyse the network again after considering the re-distribution of resources and the float and workout revised starting and completion dates of the activities.

Questions for Review

1. What do you understand by the term PERT or CPM. What are the main advantages the network method have over the conventional methods of planning and scheduling.

2. Write short notes on the following :

(i) Critical path

(ii) Dummy activity

(iii) Rule to be observed while drawing network

(iv) Independent float.

3. How do you process to estimate project duration ? How do you identify the critical path ? What steps are followed to confirm a critical path.

4. Write a short note on Project management by adopting PERT.

34

Planning of Buildings

34.1. INTRODUCTION

All buildings exert direct and indirect influence on the people who use the buildings as well as the one's who see the buildings. The direct influence is judged from the feed back as to how far the building help in making its occupants comfortable, healthy and cheerful. The indirect influence is far reaching as it not only affects the occupants of the building who are influenced by what they see of the outside from the inside, but also the people who see the building from outside as it forms a part of the overall development and landscape. The relation of site with its environment and the site itself would influence the moulding of the architect's scheme. The topographical features of the site with natural and artificial surroundings are to be taken into account while planning and designing a building. The architect is faced with either of the two possible situations while planning.

(a) When the site for the proposed building is already earmarked.

(b) When the site for the building is not decided and the choice is left to the planner.

In situation under (a) above the architect has to evolve the scheme paying due regard to the existing buildings and the given environment.

Whereas in situtation under (b) the architect enjoys freedom in making choice of the site best suited for the proposed building.

34.2 SELECTION OF SITE :

Following factors should be kept in view while making the selection of site for a building.

(1) The site should preferably be situated on an elevated and levelled ground. It should not be located in a flood prone area.

(2) The soil at site should not be of black cotton or made up type and should have good value of bearing capacity.

(3) The water table of ground at the site should not be high.

(4) The site should not be irregular in shape or have sharp corners. The site should preferably be rectangular on square in shape.

(5) The site should be in a developed area having facilities like shopping, educational institutions, recreation, medical, post and telegraph, telephone, police station, fire station, transport and utility service like water supply, drainage and sanitation, gas supply, electricity etc.

(6) The site should be located away from quarries, kilns, industrial plants/buildings emitting smoke, steam, fumes of obnoxious odour, noise or other similar environmental pollutants.

(7) The site should have unobstructed natural light and air and the building on the proposed site should not get overshadowed from adjacent buildings.

(8) The site snould have clear status of the present ownership or the title of the property.

34.3 PRINCIPLES OF PLANNING :

The main objective of planning a building is to ensure that the different components of a building are so arranged that the occupants can perform desired function with ease and comfort. Good planning also requires that the entire area available within the building is gainfully utilised, with minimum area allocated to circulation. Maximum percentage of our buildings comprise of dwellings and as such the various principles of planning which are given below are more relavant to houses.

The various principles which should be kept in view while planning of buildings can be broadly summarised as under.

(1) Aspact

(2) Prospect

(3) Grouping

(4) Privacy

(5) Furniture requirements

(6) Roominess

(7) Circulation

(8) Sanitation

(9) Elegance

(10) Flexibility

(11) Economy

(1) Aspect : Aspect means the peculiarity of the arrangement of doors and windows in the external walls of a building which permits the occupants to enjoy the gifts of nature viz sun, breeze, outside scenery etc. Aspect gains special significance in case of residential buildings. This provision is necessary to ensure proper comfort conditions in the room and it also helps in providing hygienic conditions in the room as the sun rays destroy the insects and also impart cheerful living conditions in the room. A room which receives light and air from particular side is termed to have aspect of that direction. Needless to emphasize that different rooms/areas in the dwelling need particular aspect.

A kitchen in a dwelling should as far as possible have eastern aspect so that rising sun would purify and refresh the air and the kitchen would remain cool during the remaining part of the day. The living room may have southern or south-eastern aspact. The sun is towards south during winter and as such the living room with southern aspect would have the advantage of sunshine during winter and obviate sun during summer (since the sun would be on northern side or at high attitude towards south in the summer).

The bed-room may have west or south-west aspect as the breeze required during summer will be available from that side. However, a verandah or a gallery etc. should be provided on this side to protect the structure from hot afternoon sun and also to minimise heat radiation during night.

The reading room, store, studios, stair, class-room etc. should have north aspect since there will be no direct sun from north side for most part of the year. Besides this the light from north will be diffused and evenly distributed which result in achieving desired comfort conditions in such rooms.

(2) Prospect : Prospect is the term used to highlight the architectural treatment given to a building so as to make it aesthetically pleasing from outside and arranging external doors and windows in such a manner that the occupants are able to enjoy the desired outside views from certain rooms. Prospect is basically governed by the pecularities of the selected site. Hence like aspect, prospect of a building also require the disposition of external doors and windows in a building at particular places and in particular manner so as to expose the notable and pleasent features of the openings in the external facade of the building and concealing the undesirable views in a given site. Hence both aspect as well as prospect demand proper disposition of doors and windows in the external walls at particular places and in particular manner.

(3) Grouping : We know that every apartment in a building has got a definite function and there is some inter-relationship of sequence in between them. Grouping consists in arranging various rooms in the layout plan of the building in such a manner that all the rooms are placed in proper co-relation to their functions and in proximity with each other. The basic aim of grouping of the apartments is, to maintain the sequence of their function according to their inter-relationship with least interference.

For instance in a residential building dinning room should be close to the kitchen. The kitchen, on the other hand, should be kept away from drawing room or living room to avoid smoke or smell from kitchen spreading in these rooms. In India most of the houses do not have servants and the house wife has to be in the kitchen for major part of the day. She has also to see the visitor as well as keep an eye on the playing childern in the front open space. For the sake of convenience of the house-wife, kitchen should be so located in relation to the main enterance that it is easy to approach the entrance from kitchen, without crossing other rooms, in the shortest possible time.

The W.C. should be located away from the kitchen. Main bedrooms should be so located that there is independent and seprate access from each room towards the W.C. directly or through other un-important rooms.

In case of office buildings, hospital etc. administrative department should be located centrally for convenience and economy in the cost of providing services. Thus the concept of grouping plays a very important role in planning of buildings of all types.

(4) Privacy : Privacy is considered to be one of the most important principle of planning in all buildings specially in residential buildings. Privacy may be from one part to another part of the same building or it may be the privacy of all parts of the building from neighbouring buildings, public streets or bye ways etc.

The extent of privacy of a building from the street, bye-ways or neighbouring buildings depends mainly upon the functions performed in the building. Many-a-time privacy of only a part of building is necessary from exterior whereas the remaining building as a whole may be required to be exposed to view. This is achieved by proper layout of streets, approach roads, entrances, provision of trees, creepers etc.

The privacy within the building means screening interior of one room from other rooms or parts. Screening of all the apartments or some of them from the entrance, corridors etc. gets covered under the term privacy of part of building from exterior.

In case of residential buildings, privacy can be achieved by judicious planning of the building with respect to grouping, disposition of doors and windows, mode of hanging of doors, location of entrance pathways, drives etc. Some times, provision of lobbies, corridors, screeen, curtains etc. is also made to achieve internal privacy. Importance of privacy requires special consideration in case of bedrooms, toilets, lavatories, W.C., urinals etc. All these services should have an independent access from every bedroom without disturbing the others. Doors with single shutter are desirable for such rooms.

(5) Furniture Requirements : The furniture requirements of a room or an apartment depends upon the functions required to be performed therein. The furniture requirements of a living room in a dwelling will be different from that of a class room in a school or an operation theatre in a nursing home/

hospital. There are no rigid rules which govern the furniture requirements of a particular room in a dwelling. . It should however be sufficient to accommodate the normal needs of maximum number of persons who can use the room without overcrowing. In case of buildings, other than residential, it should be adequate to meet the requirements of the particular functions. The space requirements of non-residential building is planned paying regard to the furniture, equipment and other fittings or fixtures which are essential to meet the need of the particular functions required to be performed in the building.

In case of residential buildings, normally not much thought is given to the furniture requirements. It is however, desirable to prepare a sketch plan indicating required furniture as well as its location in different rooms (viz drawing room, bed room, kitchen etc.) so as to ensure that doors, windows cupboards and circulation spaces do not prevent the placement of required number of furniture items in the room.

(6) Roominess : The effect produced by deriving the maximum benefit from the minium dimensions of a room is termed as roominess. Roominess is hence the accomplishment of economy of space without cramping of the plan. Particularly in case of residential buildings where considerable storage space is needed for various purpose, adequate provision of wall cupboards, lofts wooden/R.C.C. shelves etc. should be made to make maximum use of every nook and corner of the building. Following points should be kept in view for creating desirable impression regarding roominess.

(i) A room squre in plan appears relatively smaller than a rectangular room of same area. It is also considered relatively smaller from utitlity point of view as compared with rectangular room of the same area. Length to breath proportion for a good room is taken as 1.2 to 1.5. If the ratio of length to breadth exceed 1.5 it creates an undesirable effect. A small room having its length more than 2 times its width is objectionable, as it creates tunnel effect.

(ii) A small room with high walls appears relatively smaller than its actual size and as such small rooms should have the minimum permissible height as per bye-laws.

(iii) The location of doors, windows and built-in-cupboard etc. should be such that they permit easy approachability and do not obstruct the placement of furniture etc.

(iv) It requires skill and serious thinking in making best use of the accommodation provided by suitable, arrangements of rooms, by locating doors and passages in such a way that the livability, utility, privacy and exterior appearance are not adversely affected.

(v) The design of the building should be evolved in such a manner that its floors, walls and ceiling creats a sense of un-interrupted surfaces carried consistantly through.

(7) Circulation : Circulation means internal thoroughfares or access provided in a room or between rooms on the same floor. Passage, halls and lobbies perform the function of circulation on the same floor, such provisions are termed as *horizontal circulation.* On the other hand, stairs, lifts, ramps etc, which serve the purpose of providing means of access between different floors get covered under the category of the term *vertical circulation.*

Following aspects should be kept in view to achieve good circulation

(i) For comfort and convenience, all passages, corridors, halls etc., on each floor should be short, straight, well ventilated and sufficiently lighted.

(ii) The location of enterance passages and staricase which serve as link between various rooms and floors, need careful consideration right at the intial stage of planning.

(iii) In a multistoreyed building, the staircase, which perhaps serve the only *unfailing* means of vertical circulation, should be planned paying due regard to the size of tread and riser, width of stair and landing, light and ventilation etc. Staricases should be so located that they do not introdue upon privacy of any room or cause distrubances in the horizontal circulation.

(iv) Toilets, should be planned near the staircase block for easy accessiblity.

(8) Sanitation : The term sanitation covers not only sanitary conveniences like W.C., urinals, bath rooms, wash basins etc. but also proper and adequte lighting ventilation and facilities for general cleaning of the building. From hygienic considerations, all parts of the building should be well-ventilated and lighted. The lighting of the interior of the building may be done by natural lighting, assisted natural lighting or by artificial lighting. Uniform distribution of light in necessary, specially in offices, schools, factories and other similar buildings where number of persons work in the same premises and each individual has to work at a specified place. For ensuring sun light for greater length of time it is desirable to provide vertical windows. For proper lighting the area of windows in a room should not be less than 1/10th of the floor area which may be increased to 1/5th for buildings like schools, offices, workshops, factories etc.

Good ventilation is important to achieve desired comfort conditions in a building. Ventilation is the change of air in a room. Ventilation is essential as it removes the air polluted by respiration and other objectionable unpleasent odour. Ventilation may be achieved by natural or artificial means. In case of natural ventilation the circulation of air takes place through doors, windows and ventilators.

In case of artificial ventilation the change of air is effected by mechnical means.

General cleaning of the building has to be done on daily basis to prevent accumulation of dust which is turn promotes the growth of bacteria and spread of disease. As far as possible, the floor should be smooth and of non-absorbent type. The floors should be given proper slope to faciliate washing and the floor traps should be suitably located for drainage of water. The interior of the rooms should have a plain surface finish to minimise deposition of dust.

(9) Elegance : Elegance is the term used to express the effect produced by the elevation and general layout of the building. Hence for a building to be elegant, it is necessary that its elevation should be so evolved that it should be aesthetically pleasing and its layout should fit in well in relation to the site and its environment.

(10) Flexibility : Flexibility means designing certain rooms required for specific purpose in such a manner that they may be used for overlapping functions as and when desired. This concept is particularly important for designing houses where area's can not be increased from consideration of cost yet the provision of additional facilities is desired during functions or other occasions of social gatherings. It is, therefore, desirable to plan drawing room and dining room with a removal partition wall or screen in between them so that a large room can be obtained by removing the partition/screen to accommodate large gathering.

(11) Economy : Economy is one of the very important factors which is required to be kept in view while envolving any scheme. Every unit of the built up area is a function of cost and as such the architect has to make sure that the building planned by him can be completed within the funds available for the project. Many-a-times it becomes necessary to carry out number of alterations in the plans to keep the proposal within the limitation of funds. Some of the factors which can be considered to achieve economy without sacrificing the basic principles of planning are :

(i) Conceive simple elevation without ornamental work

(ii) Standardise the size of various components of the building

(iii) Do not use rich specifications for internal and external finishes

(iv) Specify use of locally available materials to the extent possible

(v) Do not use timber for door/windows frames. Use R.C.C. frames on L-iron steel frames instead

(vi) Adopt single stack system for plumbing

34.4. ORIENTATION OF BUILDINGS :

Orientation of a building is the term used to define the setting or fixing the direction of the layout plan of a building with respect to the direction of north. In case of a building non-square in plan the orientation is indicated by the direction of the normal to the long axis of the building. Thus if a building plan is sited in such a manner that its length is parallel or along the East-West axis, such a building is termed to have North-South orientation.

The orientation of a building is termed as optimium when the building is designed and laid out in such a manner that it is able to achieve indoor comfort conditions by gainfully utilising the benficial effects of the elements of nature like sun, wind and rain.

The best orientation of a building from solar heat consideration requires that the plan of the building is so laid that the building as a whole receives maximum sun during winter months, and avoids excessive heat during summer months. At the same time the setting of the building should be such that the full advantage of the prevalent wind direction is taken to achieve desirable air motion indoor. To achieve natural ventilation it is not necessary to orient the building in a direction perpendicular to the prevalent outdoor wind. It is observed that the building could be oriented at any angle between 0° to 30° without loosing any beneficial effect of breeze.

It is, however, seen that in most of the cases the location of available site, local building bye-laws and other factors do not permit one to adopt the optimum orientation for the building. In such situations, a compromise orientation between desired orientation for sun and wind may have to be arrived at depending upon the dominent design criteria. Orientation in such cases should be based on need of summer or winter comfort. In other words the orientation choosen should be such that in places where summer causes greater thermal discomfort, the building should be so laid out that it leads the minimum solar heat into the building during summer and maximum solar radiation in winter. Similarly in case of building to be located in cold climate, the orientation should b such that it permits maximum exposure to sun or heating from sun in winter.

Types of climatic zone : From consideration of design, India can be divided into four types of climatic zone viz *(i) Hot and arid zone, (ii) Hot and humid zone, (iii) Warm and humid zone, (iv) Cold zone.* However there are certain regions in the country which alternately experience two or even three types of climates during the year with varying degree of intensity and duration. Climate of such regions is termed as composite.

The aspects which need special consideration in deciding the orientation of buildings in different zones are summarised below

(1) Orientation of buildings in Hot and arid zone: The characteristic feature of the climate in such a zone is high summer day temperature (upto 45° C and above), cool to very cold during winter and warm and humid during monsoon season (humidity as low as 20%), Agra, Allahabad, Ajmer, Bikaner,

Jaipur, Kanpur, Nagpur, Delhi and Varanasi and some of the towns which fall under this climatic zone

Orientation of buildings in this zone require exclusion of summer sun during day time, maximum exposure to sun during winter and adequate provision of air change and comfort ventiliation during monsoon. Provision of ceiling fans is necessary. Desert coolers give lot of comfort in summer.

Based on investiation carried out, C.B.R.I. (Roorkee) has suggested following orientation for building in different parts of the country falling in Hot and arid zones.

(i) ***Northern India (Punjab)*** **:** The buildings should be oriented along East and West, facing North.

(ii) ***Central India*** **:** The building should be oriented along East-South East and West-North, West facing North-North East.

(iii) ***Delhi*** **:** Delhi has a mixed climate varying from cool to temperate and finally to hot and dry. Based on recommendations of Delhi Development Authority, optimum orientation for a building is 20° East of South. Range of desirable orientation for a building should however be 30° West and East of South and 30° West and East of North.

(2) Orientation of buildings in Hot and Humid zone : The characteristic feature of hot and humid climate is that the mean daily maximum temperature is above 32°C and the relative humidity is above 40% during the hottest month of the year. Thus buildings in this zone are subject to low summer day-time temperature and high relative humidity mostly due to agressive monsoon, Such zones occur mostly along coastal belt. Bombay, Calcutta, Madras etc. are some of the towns which fall under this zone.

(3) Orientation of buildings in Warm and Humid zone : The characterstic feature of climate in this zone is more or less identical to *Hot and Humid zone* except that the relative humidity of climate in Warm and Humid zone range between 70% to 90%. Cochin, Dwarka, Guwahati and Puri and some of the towns which fall in this zone.

Based on C.B.R.I. suggestions, the orientation for buildings in Hot and Humid zone and Warm and Humid zone could be as under

(i) ***West Coast Regions (like Bombay) :*** The building should be oriented South-East and North-West facing South-West.

(ii) ***East Coast Regions (like Madras)*** **:** The building shall be oriented South-East and North-West.

(iii) ***Bengal :*** The building should be oriented along East and West facing South.

The design of buildings in both Hot & Humid zone as well as Warm

& Humid zone call for optimum ventilation of different rooms (by ensuring free movement of air through doors and windows) and adequate protection from heavy monsoon which extend for a period of around 4 months in a year. Desert coolers are not considered suitable for such zones. Ceiling fans are, however, necessary.

(4) Orientation of buildings in Cold zone : The characteristic feature of cold zone is that its temperature is around 6°C or less during the months of December and January and the place is located at a hill having Attitude of more than 1200 m above mean sea level. Darjeeling, Gulmarg, Leh, Mussoorie, Shimla, Srinagar are some of the towns which fall under this zone.

Since winter causes maximum discomfort, orientation of buildings in cold zone should be such as to ensure maximum heating due to sun in winter months. The walls and the roof of the building should be protected against heavy rain and snowfall.

34.5 DESIGN OF BUILDINGS FOR DIFFERENT CLIMATES :

Improved indoor comfort conditions can be achieved by paying due regard to, (i) Orientation of the building, (ii) by judicious layout of the rooms and by providing required area of openings to ensure proper light and ventilation. Based on the investigations carried out by various building research organisations in the country, following suggestions are made which should be kept in view while designing buildings in different climates.

For buildings in *hot climate*, living rooms on the South and West sides should be protected by verandahas, bath, W.C., store room etc. Sun shade may be provided to cut off the sun on the South and West sides. Eastern or North-Eastern corner in the best for kitchen. The kitchen should have cross-ventilation.

For bed rooms at least one wall must be on the out side for good ventilation and the room should be placed in the direction of the prevailing wind. Water closets should be located in such a manner that the foul smell should exit in direction away from the house. Provision of outdoor sleeping areas (open balconies, terraces etc) for summer nights is essential.

Rooms which remain normally occupied during day time should be located on the North and East side to secure maximum glarefree natural light.

In case of Public buildings like schools, hospitals etc. one of the long sides should preferably face North and South to receive more solar heat during winter months. The West side of the building should be protected by verandah or by other means.

Drawing-office, dark room etc. should be located on the North side. Trees, creepers etc. when planted on the sunny side help to reduce the effect of heat on the building by shading the wall surface. Tree planted suitably, (keeping the direction of summer winds in view) also serve as barriers to obstruct the hot winds striking the building.

Spraying of water on roof and applying coat of white wash on roof surfaces during summer months give cooling effect to the building.

Penetration of sunlight into the building can also be controlled by providing louvers/venetion blinds for glazed areas, tinted glass and curtains etc.

For designing building in cold climate/hilly areas, the living room should generally open on the South and West sides to achieve maximum solar heat gain. For windows double glazing is desirable to avoid heat loss during winter nights.

34.6 BUILDING BYE-LAWS :

Building bye-laws consists of certain rules and regulations framed by a municipal or town planning or urban development board to control the development of area under its jurisdiction. The aim of framing the bye-laws is to ensure provision of reasonable minimum requirements and standards (based on current scientific and engineering knowledge, experience and techniques and modern methods of construction) in the planning, designing and construction activities of buildings in the zone. The bye-laws are framed paying due regard to the weather conditions, local construction practice, availiablity of materials, labour and other similar factors.

34.7 OBJECTIVE OF BYE-LAWS

The buidling bye-laws are essential to achieve the following objectives:

(i) It prevents construction of buildings in a haphazard manner

(ii) It lays down guidelines to be followed by architect/engineer in envolving the building layout and plans to ensure planned development of the area as a whole

(iii) The buldings which are planned based on bye-laws are comfortable to live in, have proper light and ventilation and are safe

(iv) Bye-laws serve as a standard document for the local bodies to lawfully enforce the prescribed norms in the planning of buildings by the architect.

The various aspects of building activities covered by the bye-laws is summarised as under :

(1) Distance from electric lines

(2) Lines of building frontages

(3) Open space within a plot

(4) Built up area limitations

(5) Norms relating to size and height of rooms and other requirements of part of buildings

(6) Lighting and ventilation of rooms

(7) Parking spaces

(8) Exit requirements

(9) Fire protection requirements

(10) Structural safety requirements

(11) Building services

(1) Distance From Exectric Lines : As per National Building Code, no verandah, balcony or the like shall be allowed to be erected or re-erected or any additions or alterations made to a building within the distance quoted below in accordance with the current Indian Electricity Rules between the building and any overhead electric supply line.

Table 34.1

Distance of Buildings from Electric lines

S.NO.	Type of overhead line	Vertically	Horizontally
1.	Low and medium voltage line and service lines	2.5 m	1.2 m
2.	High voltage lines upto and including 11000 volts	3.7 m	1.2 m
3.	High voltage lines above 11000 volts and upto and including 33000 volts	3.7 m	2.0 m
4.	Extra high voltage line beyond 33000 volts	3.7 m (Plus 0.3 m for every additional 33000 V and part thereof)	2.0 m (Plus 0.3 m for every additional 33000 v and part thereof)

(2) Lines of building frontages : A line usually parallel to the plot boundaries or centre line of a road, beyond which nothing can be constructed towards the plot boundaries. This line is normally known as *setback or front building line*. The front building line is laid down in each case by the local authority as per the recommendations of Master/Zonal plan. Buildings like cinemas, factories, business centres etc. where large number of vehicles are anticipated require increased set back. Such increased set back line is termed as control line. Normally the distance of building control line is taken as :-

1/2 times the distance of building line. As a thumb rule the distance of building line and control line for building facing a National State Highway in an urban area is taken as 30 m and 45 m respectively from the centre line of the highway. The corresponding distance for building facing major roads in urban area is taken as 15 m and 24 m respectively.

As per National Building Code there should be a minimum frontage of 6 m for any building on any street.

The advantage of such building lines are :

(i) It permits acquisition of land for future widening of the street.

(ii) It prevents formation of blind corners at the intersection of the streets and ensures improved visibility for the user of the road.

(iii) It reduces impact of street dust and noise, smoke etc. produced by the vehicular traffic on the occupants of the building.

(iv) The portion of set back land can be conveniently used for parking, gardening etc.

(3) Open spaces within a plot : Every building should have provision of open spaces, inside and around the building to cater to the requirement of natural light and ventilation of the rooms abutting such open spaces. In case of buildings abutting on streets in the front, rear or sides, the extent of open spaces provided should be sufficient to permit future widening of such streets.

(a) ***Front open space :*** The width of the open space which is required to be provided foı a building depends upto the width of the street fronting the plot. As per National Building Code, every building fronting a street should have a front space forming on integral part of site as below :

Front Open Space (Minimum)	*Width of Street Fronting the plot*
1.5 m	upto 7.5 m*
3.0 m	7.5 to 18 m
4.5 m	18 to 30 m
6.0 m	Above 30 n.

* For building upto a maximum height of 7 m

For streets less than 7.5 m in width the distance of building shall be at least 5 m from the centre line of the street.

(b) ***Rear open space :*** Every residential building should have a rear open space forming an integral part of site of an average width of 3 m. For plots of depth less than 9 m, for building upto 7 m in height, the rear open space may be reduced to 1.5 m

(c) ***Side open space :*** Every semi-detached and detached building should have open air spaces as under :

(*i*) For detached buildings ....3 m (minimum) on both sides

(*ii*) For semi-detached buildings...3 m (minimum) on one side.

For buildings of a height above 10 m : The open sapçe (side and rear) shall be as given in table 34.2

Table 34.2

Side And Rear Open Spaces For Different Heights Of Buildings

Height of Building (m)	Side and Rear open space to be left around the building (m)
10	3
15	5
18	6
21	7
24	8
27	9
30	10
35	11
40	12
45	13
50	14
55 and above	16

For buildings above 24 m in height there should be a minimum front open space of 6 m.

(4) Built up area limitation : Built up area or covered area of a building on ground is equal to.

Plot area – Area of open spaces around the buiding

The ratio oftotal covered area of all floors to the area of plot on which the building is to be constructed in known as floor area ratio or F.A.R.

$$\text{F.A.R.} = \frac{\text{Covered area of all floors}}{\text{Plot area}}$$

The floor area ratio is also termed as *floor space index or F.S.I.* The maximum area which can be allowed to be built on a plot of land and the maximum height upto which construction can be carried out on a plot is governed by the F.A.R. prescribed for the site by the local authority. The various aspects that are considered in specifying the value of F.A.R. for a site are given below.

(i) Type of occupancy i.e. residential or non-residential

(ii) Type of construction

(iii) Width of street fronting the proposed building and the traffic load.

(iv) Locality where the building is proposed and the density of that area.

(v) Parking facilities

(vi) Water supply, drainage and fire fighting facilities

National Building Code (N. B. C.) has laid down certain value of F.A.R. for different occupancies and type of construction.

The local authority arrives at the values of F.A.R. for the area under its juridistion by adopting the basic F.A.R. value as give in N.B.C. and modifying the same based on local factors or consideration.

Hence if the area of a given plot is 400 sq.m and if the F.A.R. prescribed by the local authority for the site is 1.5 than the maximum total area of building that can be constructed on such a plot would be = 1.5 x 400 = 600 sq.m.

Depending upon the width of the road fronting the plot one can have a 4 storeyed building each floor having an area of 150 sq.m or 3 storeyed building each floor having an area of 200 sq.m.

Plot coverage : Besides F.A.R., the local authority also specify the maximum permissible ground coverage (covered area of the building on ground) for different type of occupanies (Residential or non-residential) which are required to be respected by the architect in evolvin the design of the building and deciding upon the number of stories/height of the building.

(5) Norms Relating to Size and Height of Rooms and Other Requirement of Part Buildings : Norms relating to size and height of rooms viz (a) Habitable rooms, (b) Kitchen, (c) Store Room, (d) Garage, (e) Mezzanine floor, (f) bathrooms and W.C. etc. and other requirement of part of buildings are given in art 1.5 of chapter I.

(6) Lighting and ventailation of rooms : All habitable rooms shall have for the admission of light and air one or more openings such as windows, glazed doors and ventialtors or fan lights, opening directly to the external air or into an open verandah not more than 2.4 m in width. In case it is necessary to have light and ventilation to habitable rooms through an internal courtyard, the minimum dimensions of such court-yard shall not be less than 3 m x 3 m for buildings upto 10 m in height. For higher buildings the minimum dimensions of the internal court-yard shall be governed by provisions in N.B.C.

The minimum aggregate area of openings of habitable rooms and kitchens excluding doors, shall be as under :

(a) Not less than 1/10th of the floor area of such rooms for buildings for dry hot climate

(b) Not less than 1/6th of the floor area of such rooms for buildings for wet, hot climate

(c) Not less than 1/8th of the floor area of such rooms for intermediate climate

(d) Not less than 1/12th of the floor area of such rooms for cold climate

However the aggregate area of door and windows shall not be less than 1/7th of the area of room.

No portion of a room shall be assumed to be lighted if it is more than 7.5 m from the opening assumed for lighting that portion

Ventilation shaft : If the design of the building is such that it is not possible to locate the water closets (W.C's) and bathrooms in the front, side or rear wall or towards interior open spaces, in such situations it is necessary to provide a ventilation shaft for ventilating the spaces for W.C. and bathrooms. The windows and ventilator of the W.C. & bathroom, open ou. in the shaft. The size of the ventilation shaft shall not be less than the value given below.

Height of building (m)	*Size of Ventilation Shaft (m^2)*	*Minimum side of shaft (m)*
Upto 10	1.5	1
12	3	1.2
18	4	1.5
24	5.4	1.8
30	8	2.4
above 30	9	3

For buildings of height above 30 m, a mechanical ventilation system shall be installed besides the provision of minimum ventilation shaft.

(7) Parking space : The parking spaces to be provided in a building shall be as per the recommendations contained in the Master Plan/Zonal plans and the regulations of the local authority. For areas not covered by the Master plan / Zonal plan and for occupancies where specific provision are not made, the provision of parking space shall be made as given below :

(i) One car space per 92.93 Sq.m of covered area. This parking can be provided in the form of open parking or covered parking.

Provision of area to be made for each car space in different location shall be as under :

(a)	Basement	35 sq.m
(b)	Stilts	30 sq.m
(c)	Open	25 sq.m

(ii) Parking space shall be provided with adequate vehicular access to a street and the area of drive, aisles and such other provisions required for adequate manoeuring of vehicle shall be inclusive of the parking space stipulated in these rules.

(iii) In addition to the parking space provided, for buildings of mercantile (commercial), industrial and storage type, one parking space of 3.5 x 7.5 m shall be provided for loading and unloading activities for each 1000 sq.m of floor area.

(iv) Parking spaces shall be paved and clearly marked for different types of vehicles.

(v) In the case of parking spaces provided in basements, at least two ramps of adequate width and slope shall be provided, located preferably at opposite ends.

(8) Exit requirements :

(i) Every building meant for human occupancy shall be provided with exits sufficient to permit safe escape of occupants in case of fire or other emergency.

(ii) All exits shall be free of obstructions.

(iii) Exists shall be clearly visible and the routes to reach the exit shall be clearly marked and sign posted to guide the population of floor concerned

(iv) All exit ways shall be properly illuminated.

(v) Fire fighting equipment where provided along exits shall be suitably located and marked in such a way as not to obstruct the exit way.

(vi) Alarm devices shall be installed to ensure prompt evacuation of the population concerned through the exits.

(vii) All exits shall provide continuous means of egress to the exterior of a building or to an exterior open space leading to a street.

(viii) From fire safety/ protection considerations all staircases must be segregated from the lift lobby/remainder of the floor/basement at each level by erecting fire resisting wall and (or) fire/smoke check door of not less than 2 hours and 1 hour fire rating respectively.

— —

Types of Exits : An exit may be doorway, corridor, passage way to an internal staircase or external staircase, ramps, or to a verandah and/or terraces which have access to the street or to roof of a building.

Lifts and escalators shall not be considered as exits.

(9) Fire protection requirements : Buildings shall be planned, designed and constructed in such a manner as to ensure that it incorporates bulit in fire protection (safety) provisions in accordance with part of National Building Code pertaining to Fire Protection of buildings. For multi storeyed buildings which are more than 15 m in height and for special buildings like assembly, institutional, industrial, storage and hazardous occupancies, following aspects shall need special considerations in the design of the building.

(i) Access to fire engines/vehicles through clear motrable access way all around the building.

(ii) Safe area for firemen to assemble and equip themselves for fire fighting operation.

(iii) Compartmentation of floor space for limiting the spread of fire.

(iv) Provision of required number of safe (from heat and smoke) escape routes like protected staircase etc. for safe evacuation of occupant's.

(v) All enclosed staircases entered through fire resisting automatically closing type doors.

(vi) Provision of fire lift for speedy means of access for firemen for fire fighting/rescue operations.

(vii) Pressurisation of stairwell, lifts shafts and lift lobbies when situated in the core block of building.

(viii) Service shafts to be sealed at each floor. Inspection doors for shafts to be fire resisting.

(ix) Requisite fire protection (sprinklers etc.) and ventilation for basements and sub-basements.

(x) Provision of wet risers, sprinklers, m. ilsifyre system/fire alarm system, fire detection system and protable fire fighting equipments (like extinguishers & hose reel) where ever necessary.

(xi) Provision of public address system.

(xii) Provision of curtailment of fire spread through air conditioning ducts, service shafts, elevator shafts.

(xiii) Provision of generator of adequate capacity as alternative source to supply of power to specified equipment, lift and areas of buildings (escape routes etc.) in emergency.

(xiv) Adequate water storage facilities.

(xv) Provision of safe ventilation of smoke, gases, fumes and heat.

(xvi) The underground water fire tank should be approachable by the fire tenders.

(xvii) Water mains, telephone lines, fire detection installation line and intercom line etc should not be laid in the same duct with the electric cables.

(xviii) Long corridors to be provided with self closing fire/smoke check doors.

(10) Structural saferty requirements : The structural design of various components of the bulding viz, foundation, masonry wall, plain concrete work, reinforced concrete work, pre-stressed concrete work, timber as well as structural steel work etc. shall be carried out in accordance with the relevant I.S. code of practice applicable to the respective component. The bulding shall be design to be strong enough to resist safely and effectively all types of loads and other forces to which it can resonably be expected to be subjected to during its life time. All materials and workmanship shall be of good quality and conformining to the relevant I.S. codes, standard specifications of local P.W.D. etc.

(11) Building services : This could be broadly divided into

(a) ***Electrical installation, air conditioning and heating, installation of lifts and escalators :***

The planning designing and installation of these services shall be carried out in accordance with provision contained in Part VIII of National Building Code pertaining to such service.

(b) ***Plumbing Services :*** Planning, designing, construction and installation of plumbing services which interlia include water supply and drainage for different types of buildings (residential and non-residential buildings) shall be carried in accordance with the provision contained in Part IX of National Building Code pertaining to Plumbing Services.

The water requirements for residences and buildings other than residences as given in N.C.B. are reproduced in Art. 29.5 of chapter 29 pertaining to water supply and drainage system.

(i) ***Plumbing Services for residences :*** The drainage and sanitation requirement for residences shall be as under :

(1) Dewellings with individual convenience should have at least the following fitments.

(a) One bathroom provided with tap

(b) One Water Closet (W.C.) and

(c) One nahani or sink, either in the floor or raised from the floor with a tap.

When only one W.C. is provided in a dewelling the bath and W.C. should be separately accommodated.

(2) Dwellings without individual conveniences should have the following fitments.

(a) One water tap with draining arrangement in each tenement.

(b) One water closet and one bath for every two tenements and

(c) Water taps in common bath rooms

(2) ***Plumbing Service for buildings other than residences :*** Drainage and Sanitation requirements for building other than residences :

The type and number of drainage and sanitation fitments to be provided in different types of public buildings like *(i)* Office building, *(ii)* Factories, *(iii)* Cinemas, concrete halls and theatres, *(iv)* Art galleries, libraries and museums, *(v)* Hospitals (indoor and outdoor patient ward, administrative buildings, medical staff quarters and nurse houses), *(vi)* Hostels, *(vii)* Restaurants, *(viii)* Schools and educational institutions, (ix) Hostels etc. have been given in detail in the N.B.C.

The requirment for fitments for drainage and sanitation for an office building as per N.B.C. should be as under :

Table 34.3

Fitments For Office Building

S.No.	*Fitments*	*For Male Personnel*	*For Female Personnel*
1.	Water closet	1 for every 25 persons	1 for every 15 persons
2.	Ablution tap	1 in each toilet one water tap shall be provided for every 50 persons in the vicinity of W.C. & urinals	1 in each toilet
3.	Wash basins	1 for every 25 persons	
4.	Drinking water fountain	1 for every 100 persons with a minimum one on each floor	

5.	Cleaners sink	1 per floor
6.	Urinals	Nil upto 6 persons 1 for 7 to 20 persons 2 for 21 to 45 persons 3 for 46 to 70 persons 4 for 71 to 100 persons From 101 to 200 persons Add at the rate of 3%. For over 200 persons add @ 2.5%

34.8 SUBMISSION OF DRAWINGS TO LOCAL AUTHORITIES :

Every proposal for construction of a building should be submitted to the local authorities with 4 sets of the following drawings :

(1) *Site plant*

(2) *Building plan*

(3) *Service plan*

(4) *Specifications*

(1) Site plan : Site plan also known as plot plan is prepared to established the location of the site or plot with respect to the surroundings. The site plan is drawn to a scale of not less than 1:1000, and it should reflect the following information

(i) The boundary of the site showing the plot number of property

(ii) Setback to be kept at the front, sides and back as per building bye-laws.

(iii) The position of the site in relation to the neighbouring streets

(iv) Name of the existing roads and streets including the names of the street, if any facing which the building it proposed to be situated

(v) Name of existing buildings, if any, standing on or adjacent to the site, identifying the location of the proposed building with respect to the adjoining properties, if any.

(vi) The means of access from the existing street to the proposed building

(vii) The width of streets (if any) in front, side or rear of the plot

(viii) The direction of North line and pervailing wind

(ix) Location of exisiting physical features like, wells, open nallahs, ponds, trees etc, if any

(x) Size, location and levels of existing underground water lines, fire hydrants, sewer lines, manholes, strom water drains, electric lines, gas lines, etc. if any.

(xi) The ground area of the property and the break-up of covered area of proposed building on each floor with a calculations for percentage covered in each floor in terms of total area of plot as required under the bye-laws governing the coverage of area on the plot.

(2) Building plan : The plans of the building, elevations and sections etc. should be drawn to a scale of 1:50 for plots measuring upto 250 Sq.m and for plots measuring more than 250 Sq.m the drawing should be drawn to a scale of 1:100. The drawing shall:

(i) Include floor plans of all floors together with the covered area clearly indicating the size and spacing of all framing members and sizes of rooms and the position and width of staircase, ramps and other exit ways, lifts wells, lifts machine room and lift pit details.

(ii) Show the use of all parts of the building.

(iii) Show exact location of essential services such as W.C., sink, bath and the like.

(iv) Include sectional drawings showing clearly the size of the footings, thickness of basement walls, wall construction, size and spacing of framing members, floor slabs and roof slabs with their material. The section shall indicate the heights of the parapet, and the drainage and the slope of the roof. At least one section should be taken through the staircase, kitchen and toilet, bath and W.C.

(v) Show all the elevations.

(vi) Give dimensions of the projected portions beyond the permissible building line.

(vii) Include terrace plan indicating the drainage and the slope of the roof.

(viii) Give indication of the north point relative to the plan.

(ix) Details of parking spaces provided.

(x) Give indiciation of all doors, windows and other openings including ventilators with sizes in proper schedule form.

(3) Service Plan : Besides building plan a service plan is also required to be submitted in which the lay out of water supply pipes and drainage lines are shown. This plan is prepared to a scale of not less than 1:100. The service plan shall show.

(i) Every floor of the building in which the proposed drains are to be laid.

(ii) All new drains as proposed with their corresponding ground levels.

(iii) Invert level of the proposed drains with corresponding ground levels.

(iv) The position of every manhole, gully, soil and water pipe, ventilating pipe, rain water pipe, W.C., urinal, bath, lavatory, sink, trap or other appliances in the premises proposed to be connected to any drain.

(v) Serwers are shown in red, waste water pipes and rain water pipes in blue and existing work in black

(4) Specification : General specifications of the proposed construction giving type and grade of material to be used are also required to be provided. All plans and document of specification alongwith prescribed form (duely filled in) and application are required to be signed by the registered/licensed architect/engineer/supervisor and submitted to the local authority for approval of scheme.

34.9 PROCEDURE FOR OBTAINING SANCTION OF BULDING PLANS/BUILDING PERMIT FROM LOCAL AUTHORITY

Any person intending to constructed a building or to carry out addition/alternation to an existing building has to get the building plan sanctioned (obtain building permit) from the local authority. The type and scale of different drawings and the prescribed forms together with the other informations which are required to be submitted to the local authority for this purpose are highlighted in the book pertaining to building bye-laws.

The building department of local authority scrutinises the plans in detail to ensure that the scheme contained in the drawings complies with the various requirements of the bye-laws. In case, during scrutiny it is observed that there is some voilation of bye-laws, such voilations are recorded in the form of objections and conveyed to the applicant. The plans are sanctioned only after the objections have been removed and drawings modified. A complete set of approved drawings alongwith the sanction letter (building permit) is handed over to the applicant for taking up the construction of the building at site.

The construction of building is to be carried out in accordance with the sanctioned building drawings. If during the construction of building, susbstantial departure from sanctioned plans is intended to be made (by way of internal on external additions/alternations) it is obligatory on the part of the applicants to obtain sanction of plan showing such deviations from the local authority before taking up such construction.

After the construction of the building is over, the applicant is required to submit a notice of completion of the building to the authority. On receipt of the notice of completion, the authority conducts inspection of the work and issues the occupancy certificate after the objections, if any, observed during inspection are removed physically.

The grant of building permit (sanction) normally remains valid for two years (from the date of sanction) for residential, industrial and commercial buildings upto 4 storeyed construction and for three years for larger complexes and multi-storyed buildings. In case the construction is not complete within the validity period it will be necessary for the owner to get the building permit relvalidated from the authority.

Contravention or disregard of any provision made by the local authority can render the offender liable to procution and imposition of penalities.

APPENDICES

APPENDIX I

USEFUL DATA FOR R.C.C. DESIGNS

Properties of Round Bars

Diameter in mm	*Area in sq. cm.*	*Perimeter in cm.*	*Weight in kg/metre*
5	0.20	1.57	0.15
6	0.28	1.89	0.22
8	0.50	2.51	0.39
10	0.79	3.14	0.62
12	1.13	3.77	0.89
14	1.54	4.40	1.21
16	2.01	5.03	1.58
18	2.54	5.65	2.00
20	3.14	6.28	2.46
22	3.80	6.91	2.98
24	4.52	7.54	3.55
25	4.91	7.85	3.85
28	6.16	8.80	4.83
32	8.04	10.05	6.31
36	10.18	11.33	7.99
40	12.37	12.57	9.87

Sectional Area of Groups of Steel Round Bars

Bar diameter in (mm)	*Area in cm² for various number of bars*											
	1	2	3	*4*	5	6	7	8	9	*10*	*11*	*12*
5	0.20	0.39	0.59	0.78	0.98	1.18	1.37	1.57	1.77	1.96	2.26	2.35
6	0.28	0.56	0.85	1.13	1.41	1.70	1.98	2.26	2.54	2.83	3.11	3.40
8	0.50	1.00	1.51	2.01	2.51	3.01	3.52	4.02	4.52	5.03	5.53	6.04
10	0.79	1.57	2.36	3.14	3.53	4.71	5.50	6.28	7.07	7.85	8.64	9.42
12	1.13	2.26	3.39	4.52	5.65	6.78	7.91	9.05	10.18	11.31	12.44	13.57
14	1.54	3.08	4.62	6.16	7.70	9.24	10.77	12.32	13.86	15.39	16.94	18.48
16	2.01	4.02	6.03	8.04	10.05	12.05	14.07	16.08	18.09	20.11	22.12	24.13
18	2.54	5.09	7.63	10.18	12.72	15.26	17.81	20.36	22.90	25.45	27.99	30.54
20	3.14	6.28	9.42	12.57	15.71	18.84	21.99	25.14	28.28	31.42	34.56	37.70
22	3.80	7.60	11.40	15.21	19.01	22.81	26.61	30.41	34.21	38.01	41.81	45.62
24	4.52	9.05	13.57	18.10	22.62	27.14	31.67	36.39	40.71	45.24	49.66	54.29
25	4.91	9.82	14.73	19.63	24.54	29.45	34.36	39.27	44.18	49.09	54.00	58.91
28	6.16	12.31	18.47	24.63	36.79	36.94	43.10	49.26	55.42	61.58	67.73	73.89
32	8.04	16.08	24.13	32.17	40.21	48.26	56.30	64.34	72.38	80.42	88.47	96.51
36	10.18	20.36	30.54	40.72	50.90	61.07	71.26	81.83	91.61	111.79	111.79	122.15
40	12.57	25.13	37.70	50.26	63.83	75.40	87.96	100.53	113.09	125.66	138.23	150.80

Area of Reinforcement Per Metre Width of Slab for Various Spacings

Spacing of bars in (cm)	*Area in cm^2 per metre width of slab*										
	5 mm	*6 mm*	*8 mm*	*10 mm*	*12 mm*	*14 mm*	*16 mm*	*18 mm*	*20 mm*	*22 mm*	*24 mm*
6	3.27	4.71	8.38	13.09	18.85	25.66	33.52	42.41	52.36	63.36	75.40
7	2.81	4.04	7.18	11.22	16.16	21.99	28.73	36.36	44.87	54.30	64.63
8	2.46	3.53	6.28	9.82	14.14	19.24	25.14	31.18	35.26	17.51	56.55
9	2.18	3.14	5.59	8.73	12.57	17.10	22.34	28.28	34.90	42.23	50.27
10	1.96	2.83	5.03	7.83	11.31	15.39	20.11	25.45	31.41	38.01	45.24
11	1.78	2.57	4.57	7.14	10.28	13.99	18.28	23.14	28.55	34.55	41.13
12	1.63	2.36	4.19	6.54	9.42	12.83	16.76	21.21	26.17	31.67	37.70
13	1.51	2.17	3.87	6.04	8.70	11.84	15.47	19.58	24.16	29.24	34.80
14	1.40	2.02	3.59	5.61	8.08	11.00	14.36	18.18	22.44	27.15	32.31
15	1.31	1.89	3.35	5.24	7.54	10.26	13.41	16.97	20.94	25.34	30.16
16	1.23	1.77	3.15	4.91	7.07	9.62	12.57	15.90	19.64	23.76	28.28
17	1.16	1.66	2.96	4.62	6.65	9.05	11.83	14.97	18.48	22.36	26.61
18	1.09	1.57	2.79	4.36	6.28	8.55	11.17	14.14	17.46	21.12	25.13
19	1.03	1.49	2.65	4.13	5.95	8.10	10.58	13.39	16.54	20.01	23.81
20	0.98	1.41	2.51	3.93	5.65	7.69	10.05	12.72	15.72	10.01	22.62

APPENDIX II

Useful Data for Retaining Walls

(Weights and angles of repose of various materials)

Material	*Weight in kg/cm^2*	*Angle of repose in degrees*
Dry sand	1600	30 to 35
Damp sand	1840	35
Wet sand	1920	25 to 26
Sand dry and consolidated	1920	35
Vegetable earth (dry)	1640	20 to 30
Vegetable earth (moist)	1600	45 to 49
Vegetable earth (wet)	1680	15 to 17
Vegetable earth (consolidated and dry)	1600	49
Clean gravel	1790	48
Gravel and sand	1920	26 to 30
Rubble stone	1920	45
Dry clay	1760	30
Damp clay	1840	45
Wet clay	1920	15
Mud	1600 to 1920	0
Ashes	640	40

Co-efficient of friction between concrete and different surfaces

Surfaces			*Co-efficient of friction*
Dry clay, gravel or sand	...	...	0.5 to 0.6
Moist clay or sand	...	...	0.3 to 0.35
Wet clay	...	...	0.1 to 0.3
Rock	...	...	0.5 to 0.6

Functions of Soil Angles

Degree ϕ	*cos* ϕ	*sin* ϕ	$\frac{1-\sin\phi}{1+\sin\phi}$	$\left[\frac{1-\sin\phi}{1+\sin\phi}\right]^2$	$\cos^2\phi$
15	0.966	0.259	0.589	0.348	0.933
17	0.956	0.292	0.548	0.300	0.914
20	0.940	0.342	0.490	0.240	0.883
22	0.927	0.375	0.455	0.207	0.859
25	0.906	0.423	0.408	0.166	0.821
26	0.899	0.438	0.390	0.150	0.810
27	0.891	0.454	0.376	0.141	0.794
30	0.860	0.500	0.333	0.111	0.740
31	0.857	0.515	0.322	0.104	0.735
32	0.848	0.530	0.308	0.094	0.720
33	0.839	0.545	0.295	0.087	0.704
34	0.829	0.559	0.283	0.080	0.688
35	0.819	0.574	0.271	0.073	0.671
36	0.809	0.588	0.260	0.068	0.655
37	0.798	0.602	0.250	0.060	0.638
38	0.788	0.616	0.238	0.057	0.621
39	0.777	0.625	0.230	0.050	0.603
40	0.766	0.643	0.218	0.048	0.588
41	0.755	0.656	0.210	0.040	0.570
42	0.743	0.670	0.199	0.040	0.522
43	0.731	0.681	0.190	0.040	0.552
44	0.719	0.695	0.180	0.032	0.509
45	0.707	0.707	0.174	0.030	0.500

APPENDIX III

Mensuration

Areas

Parallelogram = base × perpendicularheight

Triangle = base × $\frac{1}{2}$ perpendicularheight

Trapezium = $\frac{1}{2}$ sum of parallel sides × perpendicularht.

Circle = 0.7854 × square of diameter

Sector of circle = length of arc × $\frac{1}{2}$ radius

Parabola = base × $\frac{2}{3}$ height

Ellipse = longerdiameter × shortdiameter × 0.7854

Regular polygon = sum of sides × $\frac{1}{2}$ perpendiculardistance from centre of sides

Surface of sphere = π × square of diameter

Surface of cone = area of base + (circumference of base × $\frac{1}{2}$ slantheight)

Volume

Prism = area of base × height

Pyramid or cone = area of base × $\frac{1}{3}$ height

Sphere = 4.181 × radius^2

Position of centre of gravity

Triangle = $\frac{1}{3}$ perpendicular height from base

Parabola = $\frac{2}{3}$ height from base

Semi-parabola = $\frac{3}{5}$ length of the axis from the vertex, and $\frac{3}{8}$ of the semi–base from the axis

Parallelogram = At the intersection of the diagonals

Prism or cylinder = At the middle point of the line joining the centre of gravity of the bases

APPENDIX IV

Properties of Indian Standard Beams

Size in mm	Weight per metre (kg)	Sectional area	Thickness of web	Thickness of flange	Moment of inertia		Radii of Gyration		Section modulli	
					I_{xx}	I_{yy}	V_{xx}	V_{yy}	Z_{xx}	Z_{yy}
75 × 50	6.1	7.71	3.7	6.0	72.7	10.1	3.07	1.14	19.4	4.0
100 × 50	8.0	.10.21	4.0	6.4	168.0	12.7	4.06	1.21	33.6	5.1
125 × 75	11.9	15.12	4.4	6.5	408.8	43.4	5.19	1.69	65.1	11.6
150 × 80	14.2	18.08	4.8	6.8	688.8	55.2	6.17	1.75	91.8	13.8
175 × 90	16.7	21.30	5.1	6.9	1096.2	79.6	7.17	1.93	125.3	17.7
200 × 100	19.8	25.27	5.4	7.3	1096.6	115.4	8.19	2.13	169.7	23.1
225 × 100	23.5	29.92	5.8	8.6	2501.9	112.7	9.15	1.14	222.4	22.5
250 × 125	27.9	35.53	6.1	8.2	3717.8	193.4	10.23	2.33	297.4	30.9
275 × 140	33.0	42.02	6.4	8.8	5375.3	287.0	11.31	2.61	392.4	41.0
300 × 150	37.7	48.08	6.7	9.4	7332.9	376.2	12.35	2.80	488.9	50.2
325 × 165	43.1	54.90	7.0	9.8	9874.5	510.8	13.41	3.05	607.7	61.9
350 × 165	49.5	63.01	7.4	11.4	13158.3	631.6	14.45	3.07	751.0	76.6
400 × 165	56.9	72.43	8.0	12.5	14306.3	716.4	16.33	3.15	965.3	86.8
450 × 170	65.3	83.14	8.6	13.4	27536.1	853.0	18.20	3.20	1223.8	100.0
500 × 180	75.0	95.50	9.2	14.1	38579.0	1063.9	20.10	3.34	543.2	118.2
550 × 190	86.3	109.9	9.9	15.0	53161.6	1335.1	21.99	3.48	1933.1	140.5
600 × 210	99.5	126.69	10.5	15.5	72867.0	1821.9	23.98	3.79	2428.9	173.5

APPENDIX V

The Metric System

Length

10 millimetres	= 1 centimetre
10 centimetres	= 1 decimetre
10 decimetres	= 1 metre
10 metres	= 1 dekametre
10 dekametres	= 1 hectometre
10 hectometres	= 1 kilometre

Weight

10 milligrams	= 1 centigram
10 centigrams	= 1 decigram
10 decigrams	= 1 gram
10 grams	= 1 dekagram
10 dekagrams	= 1 hectogram
10 hectograms	= 1 kilogram

100 kilograms	= 1 quintal
1000 kilograms	= 1 metric tonne
200 kilograms	= 1 caret

Area

100 sq. millimetres	= 1 sq. centimetre
100 sq, centimetres	= 1 sq. decimetre
100 sq. decimetres	= 1 sq. metre
100 sq. metres	= 1 sq. dekametre
100 sq. dekametres	= 1 sq. hectometre
100 sq. hectometres	= 1 sq. kilometre

Capacity

10 millilitres	= 1 centilitre
10 centilitres	= 1 decilitre
10 decilitres	= 1 litre
10 litres	= 1 dekalitre
10 dekalitres	= 1 hectolitre
10 hectolitres	= 1 kilolitre

Volume

1000 cu. millimetres	= 1 cu. centimetre
1000 cu. centimetres	= 1 cu. decimetre
1000 cu. decimetres	= 1 cu. metre

APPENDIX VI
Common Conversion Factors

Linear Measure

Multiply by	*To convert*	*To*
25.4	inch	millimetres
0.305	foot	metres
0.9144	yard	metres
201.168	furlong	metres
1.609	mile	kilometres

Square Measure

Multiply by	*To convert*	*To*
6.45	sq. inch	sq, centimetres
0.093	sq. foot	sq. metres
0.838	sq. yard	sq. metres
40.447	acre	sq. decimetres
2.588	sq. mile	sq. kilometres

Cubic Measure

Multiply by	*To convert*	*To*
16.39	cubic inch	cubic centimetres
28.32	cubic ft.	cubic decimetres
0.756	cubic yards	cubic metres

Liquid Measure

Multiply by	*To convert*	*To*
0.142	gill	litres
0.568	pint	litres
1.136	Quart	litres
4.546	gallons	litres

General Conversion

Multiply by	*To convert*	*To*
70.3	lb./sq. inch	gram./sq. cm.
4.883	lb./sq. ft.	kgm./sq. metre
0.593	lb./cubic yard	kgm./cubic metre
16.02	lb./cubic ft.	kgm./cubic metre
0.0998	lb./gallon	kgm./litre
0.1383	foot pounds	kgm. metres
0.33	foot tons	tonne metres
746	horse power	watts
33000	horse power	ft. lb./minute
76	horse power	kgm./seconc
44	watts	ft. lb./minute
0.1	watts	kgm. m./second
88	miles/hour	ft./minute
197	metres/sec.	ft./minute
1.488	lb./ft.	kgm./metre
0.496	lb./yard	kgm./metre
3333.33	tons/ft.	kgm./metre
1111.11	tons/yard	kgm./metre
0.2818	lbs./mile	kgm./k. metre
10.916	tonnes/sq. ft.	tons/sq. metre
1.215	tons/sq. yard	tons/sq. metre
1.329	tons/cubic yard	tons/cubic metre
48.905	gallons/sq. ft.	litres/sq. metre
50.80	cwt.	kilograms
1016.3	tonne	kilograms

Terminology

Absorption: the assimilation of sound energy by a body without echo.
Accent lighting: direct lighting to emphasize a particular object or draw attention to a part of the field of view.
Accent: to enhance decoration by using a lesser quantity of a contrasting colour or finish.
Achromatic: a neutral colour, lacking in chroma or intensity: black, white, or gray.
Acoustic tile: modular sound absorption panels.
Acrylic: a synthetic fiber of moderate tenacity with resistance to scratch and light, and having dimensional stability.
Aesthetic: pertaining to an object's quality of beauty.
Aggregate: the sum of a collection of particulars.
Ambient light: measurement of the overall illumination of a room from reflected sunlight or Overhead lighting applications.
Ampere: a unit of electric current equal to the current of one Volt acting through a resistance of one Ohm.
Amplification: An increase in the volume of sound. The term is also used for the effect of colour intensification due to the interreflection of light from one coloured surface to another.
Angle of incidence: the angle (measured from the surface on the side of the approach) at which an element strikes an object.
Anneal: the process of strengthening through heating and gradual cooling.
Anodize: chemical process where a coating is etched onto metal
Antique: all items, including accessories, art, buildings, furnishings, lighting, sculpture, carvings and other personal possessions, which are over 100 years old.
Anthropometrics: the measurement of the size and proportions of the human body, and its applications to the design of furnishings, equipment, and spaces.
Antimicrobial treatment: a treatment that prevents the growth of mold, mildew, and bacteria on carpet.
Appurtenance: a subordinate part or accessory.
Arcade: an open or closed series of arches supported by columns used along passageways, courtyards or porches.
Awing: lightweight roof structure typically constructed of fabric or metal

installed over an exterior opening such as a doorway, porch or balcony to provide shelter.

Baluster: small, heavily proportioned pillar used for vertical support in a railing, typically made of wood or stone.

Balustrade: railing composed of series of balusters, bottom rail and top handrail, also used in smaller scale to embellish furniture.

Batik: ancient hand-printing technique using wax as a resist, which is applied, removed and re-applied for each colour layer to create various patterns including a crackled appearance.

Barrier-free design: which presents no physical barriers to access and free movement within an environment.

Bay window: a window set in a frame projecting outward from a wall to create an interior recess, usually of fixed glass.

Bay: a volumetric or structural unit defined by columns or walls.

Binder: a substance, which brings cohesion to a collection of elements.

Biodegradable: capable of being broken down into innocuous products through biological processes.

Bonded fabric: a fabric formed by combining an outer face fabric with a backing fabric using an adhesive or laminate.

Bow window: a bay window in the form of an unbroken curve.

Bubble diagram: an initial planning sketch in which clustered circles represent activity zones placed in proximity relationships.

Building byelaws: rules and regulations imposed by local municipality or Municipal Corporation for the construction of buildings.

Cantilever: any part of a structural member (commonly a beam), that projects beyond its support.

Carpet: a soft-surface floor covering used to blanket a floor from wall to wall.

Casement: a loosely woven or knit sheer drapery fabric.

Casement window: a window hinged at the side, and which swings open like a door.

Cathode: the terminal from which electrons emanate in a fluorescent or other discharge-type lamp.

Ceramic: made from clay and heated (fired) in a kiln. A hard and durable material.

Chandelier: ceiling mounted light fixture with arms branching out from a central support.

Chromaticity: the aspect of a colour that describes its dominant wavelength and purity

Clerestory: a window or bank of windows inserted between two roof levels to bring light into a room.

Closed plan: an architectural plan that divides the internal space of a structure into separate, discrete rooms.

Coefficient of utilization (cu): the ratio of light (lumens) from a luminaire received on a work plane to that emitted by the lamp.

Coffered ceiling: one with ornamental raised panels between closely spaced beams.

Colour fastness: the ability of a material to retain its colour when exposed to light or cleaning agents.

Colour rendering index: the general expression for the effect of a light source on the colour appearance of objects in comparison with their colour appearance under a referenced light source.

Colour retention: the ability of a material to retain its colour when exposed to light or cleaning agents.

Colour spectrum: the range of colour comprising visible light.

Compact fluorescent: an energy-saving replacement for incandescent sources, with greater efficiency and life.

Concrete: a material consisting of cement mixed in varying proportions with sand and gravel or other aggregates. With the addition of water, the mixture becomes mouldable, capable of assuming almost any shape. Concrete dries to a heavy, stone like mass of great strength.

Concrete blocks: large, generally hollow, bricklike blocks composed of concrete; widely used in building, especially for walls.

Conduction: the direct transfer of heat through a solid material, such as glass or metal.

Conduit: a tube used to protect electrical wires.

Conservation area: protection of an area with characteristic features typical to the area and defined in the local zoning regu- lations with historic and architectural importance.

Construction documents: the full volume of drawings, specifi- cations and contracts, which comprise the complete description of a built project.

Contract: a legally binding, written agreement between two parties outlining services to be rendered, responsibilities, and compensation.

Convection: the transfer of heat through natural air circulation.

Cotton: a natural staple fiber harvested from the flower of the cotton plant.

Couch: upholstered furniture used for sitting or reclining.

Courtyard: space permanently open to sky enclosed fully or partially by buildings and may be at ground level or any other level within or adjacent to a building.

Cove lighting: provided by light sources shielded by a ledge and distributing light upward over the ceiling.

Coved ceiling: a ceiling that curves into the supporting walls rather than meeting them at an angle.

Covered Area or Coverage: the Ground area covered immediately above the plinth level covered by the building but does not include the space covered by:

Critical path: the time frame and overlapping order of steps in the building and finishing process which determines the minimum time or necessary progression required for an event.

Decibel (dB, db): a unit for expressing the relative intensity of sounds on a scale from zero for the least perceptible sound to about 130 for the average pain level.

Delaminate: to come apart in layers.

Design concept: In the present context, it means the idea for the solution to the process of planning a building, furnishings, or composite interiors; the organization of a work of interior design, architecture or art problem.

Detail drawing: a drawing on a scale greater than that of the general drawings, showing specific composition of an object to be built.

Density: Concentration of population expressed in terms of number of persons per hectare in a particular area. Density is often the most commonly accepted indicator and provides a reference for the amount of land needed for supporting facilities with regard to a given population and is also an indicator of the type of physical development.

Diffusion: the random dispersion or scattering of a sound or light wave when it strikes a surface that is irregular, concave, or convex in form; or when the wave form is not in line with, or not equal in size to, the object struck, the wavelength will tend to bend around the object and continue on its path.

Dimensional stability: the property of a material, which indicates its resistance to change in size due to climatic conditions like temperature or humidity.

Dimmer: a device used to control the intensity of light emitted by a luminaire through the control of the voltage or current available to it.

Direct glare: the result of areas of excessive contrast (such as insufficiently shielded light sources) in the direct field of view.

Direct lighting: Lighting provided on the work plane or object from a prominent and identifiable source.

Double-glazing: a process of providing windows with two sheets of glass sealed together trapping air or a vacuum between them to provide superior thermal insulation.

Dowel: round wooden or metal pin used to reinforce furniture construction.

Down light: a luminaire from which light emanates downward; may be recessed, surface or pendant mounted.

Drapery: a loosely hung, often heavy, fabric curtain.

Drapery panel: a drapery length made of one or more widths of fabric that travel together on a rod.

Draw curtain: a fabric panel that can be opened and closed by means of a draw cord on a traverse rod.

Eave: lower edge of a roof extending beyond the exterior wall.

Ecletic: utilizing elements from multiple sources and styles

Ecology: the study of the relationship between organisms and their environment.

Efficacy: the measure of the effectiveness of a solution; the ratio of light

emitted by a lamp to the electrical power input, usually expressed in lumens per watt.
Effluent: something that flows out.
Egress: a means of exiting.
Elevation: a drawing, which represents a vertical image of one side of an object, room or building.
Embellishment: additional ornamentation.
Emulsion: a combination of two non-compatible liquids, such as oil and water.
Engineered wood product: any recombination of wood by-products to form a material of greater performance capacity than the original.
Ergonomics: the science that seeks to adapt the environment to its users.
Etch: to remove through corrosive action.
Fabric: cloth; more specifically, a construction of fibers, not necessarily woven.
Faade: the face or front view of a building or article.
False ceiling: that portion of a ceiling below the actual functional level or below other sections of the ceiling within the same space. Often, a false ceiling serves to articulate specific segments of a room such as a dining area, or to accommodate mechanical equipment.
Fire retardant: durable or non-durable applications to decrease flammability.
Felting: the joining of fibers into a fabric through the application of heat, agitation and moisture, or by mechanical treatment.
Fenestration: the arrangement and design of windows and other openings of a building.
Fiber: a material of natural or synthetic derivation capable of forming a continuous filament, such as yarn or thread.
Fiberglass: any number of plastic resins, such as polyesters, polypropylene, or nylon, reinforced with segments of glass fibers.
Fiber optics: the transmission of signals as pulses of light through an optical conductor, such as glass cord.
Filament: a single continuous strand of fiber.
Finish floor: the final visible flooring material or surface.
Firebrick: a very hard brick capable of withstanding the intense heat of a kiln or fireplace interior.
Fire proof: fire retardant barrier added to any object to inhibit the spread of fire
Flame-resistant fabric: a fabric whose fiber content or topical finish makes it difficult to ignite, slow to burn, and often self-extinguishing.
Flame-retardant finish: a chemical finish that renders a fabric made of a flammable fiber resistant to fire ignition and spread.
Flat slice: a method of cutting a log parallel to a plane through its center, which produces a vaulted, or cathedral like grain.
Flood lamp: a lamp providing a relatively wide light distribution pattern.

Floor Area Ratio (FAR): the quotient obtained by dividing the total covered area of all floors by the area of the plot. Floor area includes the mezzanine floor also.

F.A.R. = Total covered area of all floors / Plot area

The Floor Area Ratio can be used in zoning to limit the amount of construction in a certain area. A builder can plan for either a single-story building using the entire allowable area in one floor, or a multi-storey building that rises higher above the plane of the land.

Floor plan: a drawing that represents the horizontal arrangement of building parts and furnishings.

Fluorescent lamp: a low-pressure mercury electric discharge lamp in which a fluorescent coating (phosphor) transforms ultraviolet energy into visible light.

Flue: channel above a fireplace used to carry smoke and heat to the exterior of a structure.

Flux: the flow of light or the measure of the flow of electric charge in a wire.

Foot-candle (fc): a unit of illuminance on a surface equal to that produced by the light of one candle at a distance of one foot.

Fresco: painted mural or decoration done on a wet plaster wall or ceiling.

Galvanized iron: iron coated with zinc as a means of retarding rust.

Gallery: narrow exterior porch or balcony; succession of rooms to display art; rail with small wood or metal spindles attached to the outer edge of cabinetry.

Glare: any brightness that annoys, distracts or reduces visibility.

Glass: a mixture of silicates, alkalis, and lime that is extremely mouldable when heated to high temperatures, permitting blown, moulded, pressed, and stretched forms, and which cools to a rigid, nonabsorbent, transparent or translucent substance.

Glass blocks: brick-like forms of glass available in a variety of shapes and sizes. They can be set together or joined to other materials with mortar.

Glaze: protective or decorative glass-like coating formed on the surface of a ceramic piece by firing.

Grout: thin mortar used to fill in spaces between tiles, bricks or other surfaces

Hammered: a metal or stone finish produced by direct blows of a hammer on the material surface, or a finish giving the appearance.

Hardboard: a mat-formed flat panel consisting of particles of wood bonded together with a synthetic resin or other suitable binder.

Hardware: various materials, such as iron, brass, bronze, wood, chrome, plastic added to mechanise the operation of cabinetry drawers and doors, drapery, etc. For example, cremone bolt, doork-nocker, drapery rod, handle, hinge, kick plate, knob, lever, pull, etc.

Header: a horizontal structural member, which serves to tie vertical members together at the top of an element.

Honed: a medium to coarse stone finish produced by abrasion of a cut face.

HVAC: Heating, Ventilation and Air Conditioning.

Hydrophobic fibers: those, which resist water-borne soiling.

Incandescent filament lamp (bulb): a lamp in which light is produced by a filament heated to incandescence by an electric current.

Indirect lighting: luminaries, which distribute 90-100 percent of their emitted light first onto a reflective surface such as a ceiling or wall.

Industrial board: engineered wood products commonly used in the manufacturing of furniture and case goods.

Ingress: the way in or entrance.

Insulation: the prevention, by using certain materials, of an excessive transfer of electricity, heat, cold or sound between the inside and the outside of a structure or between portions of a structure.

Jacquard weave: a fabric of complex pattern, such as tapestry or brocade, produced on the Jacquard loom.

Jamb: interior sides of a door or window frame.

Joist: floor or ceiling beam support.

Junction box: a protective housing for the joining of electrical connections.

Kelvin (K): relating to the thermometric scale in which the unit of measurement equals the centigrade degree and according to which absolute zero, the temperature at which all atomic activity ceases, and the equivalent of '–273.16°' Celsius is '0' Kelvin.

Kelvin temperature (K): represents the relative whiteness of a light source; whether the source appears warm, cool, or neutral. The higher the temperature, the cooler the appearance of the source. The Kelvin temperature assigned is that of a unit black body radiator heated to the same whiteness.

Lacquer: a family of fast drying resin or synthetic finishes capable of being polished to high gloss sheen.

Laminate: the process of bonding together, thin sheets or small pieces of material to create a substance having properties the material would not otherwise possess, such as strength, durability or intricate form.

Land use: the purpose to which the site or part of the site or the building or part of the building is in use or permitted to be used by the Authority. Land use includes zoning of land use as stipulated in the development plan and the Zoning Regulations

Latex: milky, fluid, synthetic adhesive used to seal the back of carpet.

Lath: a framework of thin wood or metal ribs integral with a building skeleton for the support of tile, plaster, reinforced concrete, plastic forms etc.,

Letter of agreement: a legal contractual agreement between two parties that describes the obligations and responsibilities of each.

Level loop: a carpet style having all tufts in a loop form and of identical height; may be woven or tufted.

Life-cycle cost: the initial cost of a product (including installation) plus the cost of maintenance or operation over its useful life.

Lumen: a unit of light flow equal to the light emitted into a unit solid angle by a uniform point source of one candle.

Luminance: the density of light falling upon a surface, the units of which are the lux and footcandle.

Luminaire: a complete lighting unit consisting of a lamp, or lamps, together with parts designed to distribute the light, to position and protect the lamps, and to connect the lamps to the power supply; a ceiling or wall mounted light fixture, or a portable lamp.

Luminance: the luminous intensity of any surface in a given direction per unit area of that surface as viewed from that direction. All things visible have some luminance. Luminance is measured in candelas per square meter or candelas per square foot.

Luminance ratio: an expression of the relative brightness of any two areas in the visual field.

Marquetry: decorative wood veneer inlay.

Masonry: architectural construction of stone, brick, tile, concrete block, or glass block joined together with mortar; in broader usage, construction of plaster or concrete.

Mastic: originally a resin used in the production of adhesives, now commonly the generic term for flooring adhesives.

Matte: a flat, non-shiny finish, which thoroughly diffuses incident light.

Medium Density Fiberboard (MDF): a fine-grained smooth faced industrial board product, composed of wood fiber and adhesives. Easily cut, worked and finished.

Medullary rays: capillary-like structures within the body of a tree that run radially from the core to the bark and their expression in quarter-sawn lumber.

Metamerism: the shift in visual appearance of colours under varied illumination.

Metamers: lights of the same apparent colour but of different spectral energy.

Mezzanine Floor: an intermediate floor between the floor and ceiling of any storey.

Microbial: relating to micro-organisms such as bacteria.

Microblind: horizontal blinds with 13 mm (1/2") slats; also called micro-miniblinds.

Miniblinds: horizontal blinds with 25 mm (1") slats.

Modacrylic: synthetic fiber that is highly flame resistant and bulky. It has moderate resiliency, moderate dimensional stability, and high elastic recovery. It also has a low melting point and low abrasion resistance.

Modular: built of unit pieces or according to standardised sets of measurements.

Module: one of a series of units designed and scaled to integrate with others in different combinations to form, for example, a set of furnishings, a system of construction, or whole buildings. In current usage, the term is most often applied to mass-produced prefabricated units.

Mohair: a specialty wool fiber. The most resilient of all natural fibers, strong, and possessing a good affinity for dyes.
Moulding: an ornamental strip of wood or plaster that protrudes from a ceiling or wall surface.
Monolithic construction: a system in which the major part of a structure consists of a single, self-supporting mass, usually of reinforced concrete, plastic or fiberglass.
Mortar: cement, lime or plaster combined with sand and water. When wet, the substance is mouldable; it hardens to form the binding agent of masonry construction.
Motif: dominant or recurring theme throughout a project
Mullion: slender strips of wood or metal dividing panes of glass in door or window units.
Mural: scene or image painted or wallpapered onto a wall.
Muslin: plain woven cotton fabric
NBC (National Building Code): a set of specifications prepared to regulate materials and methods used in construction and establishing consistent standards to assure healthy, safe and sanitary conditions.
Niche: recessed opening, often arched at the top, set in a wall with a shelf, often lit from the top to display sculpture or objects of art.
Nonwoven fabric: a fabric which is either fibrous, but not made in a loom, or non-fibrous, made directly from a solution.
Nylon: the generic term (as well as trade name) for a family of plastics exhibiting high tensile strength in fiber or sheet form. It has excellent abrasion resistance, and resiliency, retains appearance well. Fiber modifications are antistatic, anti-soil, sunlight resistant, and flame retardant.
Occupancy or use: the main occupancy or use for which a building or a part of it is used or intended to be used *i.e.* contingent/subsidiary occupancies. Mixed occupancy buildings being those in which more than one occupancy is present in different portions of the buildings.
Olefin: a plastic used for fiber and container manufacturing. Advantages of olefin fibers include good abrasion resistance, tenacity, excellent resiliency, good dimensional stability, static resistant, and excellent resistance to most chemicals. It resists mildew and stains. Fiber modifications contribute to heat and light stabilization. Often referred to as polypropylene.
Open plan: an architectural plan organized with few fixed partitions; provides maximum flexibility in the use of interior space.
Open space: an area forming an integral part of a site left open to the sky.
Orientation: arrangement, alignment, or position in relation to other factors or elements.
Oriented strand board: an industrial board product composed of wood chips created and bound specifically to take advantage of their non-homogeneous nature to produce a product with specific structural properties.

Oxidation: a chemical process by which one material combines with oxygen to form a second (iron oxidizes to form rust).
Padding: fabric or materials used to provide additional cushioning.
Paneling: thin, flat wooden boards or other similarly rectangular pieces of construction material joined side by side to form an interior decorative surface for walls or ceilings.
Parquetry: inlay of wood that takes the form of geometric patterns; used primarily for floors and sometimes for tabletops.
Particleboard: a mat-formed flat panel consisting of particles of wood bonded together with a synthetic resin or other suitable binder.
Partition: a non-load-bearing subordinate wall between two piers or other supporting members.
Passive solar heating system: a technique of solar heating that uses parts of the building structure to collect, store and distribute solar heat without pumps or fans.
Patina: the sheen, colour and texture of an object, produced by age, use, waxing, or design. On metal, patina is the film that develops from long exposure to the atmosphere.
Pelmet: framework to conceal drapery hardware at the top of a window.
Pendant: hanging ornamentation or electrical fixture mounted from the ceiling.
Perspective: a system of realistic pictorial representation of objects and spaces in relative distance or depth. The optical effects perceived in viewing objects over a distance.
Pigment: substance imparting a colour to other materials.
Pile: the visible wear surface of carpet, consisting of yarn tufts in loop or cut configuration; also called face or nap.
Pile weave: a fabric construction in which cut or uncut loops protrude from the ground cloth.
Pile yarn: a yarn, which forms the tuft of carpet; also called face yarn.
Pill: the tendency of staple fibers to be shed from some fabrics and to form small tufts or balls on the face of the fabric.
Pitch: degree of roof slope; angle of inclination.
Plain slice: a method of cutting a log parallel to a plane through its center, which produces a vaulted or cathedral-like grain.
Plan: the configuration of spaces and rooms, walls and openings in an architectural structure; also, the graphic representation of such an arrangement.
Plane: a two-dimensional expanse; a flat surface.
Plaster: a paste, usually of lime, sand, and water, which hardens as it dries. Often used as a finish for interior wall and ceiling surfaces.
Plastic: a soft, ductile material, which is a member of any of the several families of synthetic polymer substances.
Pleating: to tuck and bind fabric for additional fullness.
Plenum: space between ceiling and structure where mechanical and electrical

components are typically run. In some instances the plenum itself may serve as a sort of super duct for air returning to the mechanical system from the volume beneath the ceiling.

Plinth Area: the built up covered area measured at the floor level of the basement or of any storey.

Plumbing: the system of pipes, which carries water and sewage within a building.

Plush carpet: a smooth carpet surface texture in which individual tufts are minimally visible and the overall effect is that of a single level of fiber ends.

Ply: a single thickness, layer or strand.

Plywood: a composite sheet of laminated veneers, some or all of which are made of wood, with the grain of adjacent strata arranged at perpendicular angles for increased strength.

Polished: smooth and glossy.

Polyester: a synthetic fiber with excellent strength, good abrasion resistance, and which resists wrinkling. Characterized by permanent body. Fiber modifications contribute to pill resistance, antistatic, and flame retardancy.

Polyethylene: a group of lightweight, flexible plastics characterized by a waxy surface and resistance to chemicals and moisture but not high temperatures.

Polypropylene: a family of plastics which are resistant to water-borne but not oil-borne staining.

Polystyrenes: a family of rigid, transparent-to-opaque plastics that are durable, capable of accepting varied finishes, and possessed of good insulation properties.

Porcelain: high-grade, translucent white ceramic ware fixed at extremely high temperatures; most common in fine dishes and ornaments, but having many industrial applications, such as plumbing fixtures electrical insulators, and tiles.

Portable luminaire: table or floor lamp, or wall unit, which is not permanently connected to the electrical power supply.

Prefabricate: to mass-produce standardized construction parts or modules for later assembly and/or combination.

Primary backing: a component of tufted carpet consisting of woven or non-woven fabric into which pile weave tufts are inserted by the tufting needles.

Primer: a coating that is applied before any other finish treatment to seal the pores in or otherwise prepare a surface to be treated.

Printing: as applied to textiles (fabric, carpet, etc.), the application of dyes according to a selective pattern to create a design by such methods as woodcut, silk screen, tie-dye, or injection.

Programming: the research phase of design, which determines the objectives and requirements of a design project.

Proportion: the relation of parts to each other or to the whole in terms of magnitude, quantity, or degree.

Proxemics: the study of human interaction with space and of personal and cultural spatial needs.

PVC (Poly Vinyl chloride): a tough, mouldable thermoplastic used in the manufacture of plumbing pipes and furniture.

Quality of lighting: distribution of luminance in a visual environment with regard to visual performance, visual comfort, ease of seeing, safety and aesthetics for the specific visual tasks involved.

Rag-rug: a plain weave rug woven with strips of fabric, historically rags or recycled clothing remnants.

Ravel: to untwist or unweave.

Rayon: a natural fiber made from cellulose, which has been chemically prepared. Soft hand, high moisture regain, smooth and soft fibers, good conductor of heat and static. Accepts durable press and dimensional stability finishes. Disadvantages include poor resiliency, progressive shrinkage, and weakness, especially when wet.

Recessed light: lighting fixture installed within a ceiling.

Rectilinear: straight lined.

Reflectance: the ratio of the light reflected by a surface to the light falling on it.

Reflected glare: results from specular reflections of high contrast from polished or glossy surfaces in the field of view.

Reflection: the process by which light or sound leaves a surface from the incident side.

Refraction: the change in direction or propagation of a sound or light wave due to a change in the velocity occurring when a change in materials, a thermal gradient or a wind velocity gradient is encountered.

Regain: a measure of the recovery of moisture lost to the atmosphere through drying.

Reinforced concrete: concrete embedded before hardening with steel rods that lend the material a tensile strength far beyond its original capacity.

Rendering: a pictorial representation of a proposed design, usual in perspective and full color.

Renovate: to restore condition by rebuilding, repairing or cleaning.

Restore: to renew or return to an original state or condition.

Retrofit: to fit with new parts or equipment not available at the time of initial construction or production.

Reverberation: a sound effect similar to that of a continuous echo.

Rheostat: a device used to control the intensity of light emitted by a luminaire through control of the voltage or current available to it.

Riser: the individual vertical faces of a stair. The term is also used for the collection of ducts or pipes, which provide vertical transport of air or water through a building.

Room Height: the vertical distance measured from the finished floor surface to the finished ceiling.

r-value: the thermal resistance of a material. A higher r-value indicates greater resistance.
Sandblast: abrasive cleaning with sand moved by an air jet.
Sash curtain: a semi-sheer fabric gathered or shirred onto rods at top and bottom and hung onto a window sash.
Schedule: a table that indicates the finish material used on floors, walls, and ceilings, or lists types of doors and windows.
Schematic design drawings: quick initial drawings used to generate or show ideas.
Seasoning: to bring to the proper state for use, as in seasoning wood by drying.
Set-back: a line usually parallel to the plot boundaries or center line of a road and laid down in each case by the Authority or as per recommendations of Master/Zonal Plan, beyond which nothing can be constructed towards the plot boundaries excepting with the permission of the Authority.
Shade: a low-value or dark color produced by adding black to a hue. A cover for window or lamp to control light distribution.
Shingles: narrow, overlapping pieces of wood or slate installed as roofing or siding.
Stucco: fine plaster or cement made of Portland cement, sand and lime applied to a structure used as ornamentation.
Silk: the only natural continuous filament fiber. Silk fiber is harvested from the cocoon of the silk worm. Silk fabric is highly valued for its strength, hand and luster.
Sisal: a natural plant fiber; smooth, straight, and yellow. May be used by itself or in blends with wool and acrylic for a softer bond. Sisal is used in wall coverings, especially in heavyduty commercial applications because of its durability and ease of application to a variety of surfaces.
Soffit: a dropped or underside horizontal surface.
Space plan: a drawing, which shows the arrangement of functional elements within an area.
Space planning: the functional planning of interior space; a design specialty, which concentrates on establishing space needs and utilization in the early stages of design.
Spall: the breaking away of a surface due to mechanical action within the material itself such as thermal expansion.
Specification: the criteria of minimum durability, cost and safety requirements of finished materials; all of the information necessary for the construction of custom-made items.
Specular angle: that angle between the perpendicular surface and the reflected ray. It is numerically equal to the angle of incidence.
Specular reflection: that process by which incident flux is redirected without diffusion.
Specular surface: shiny or glossy surface (including mirror and polished

metals) from which the reflection is predominantly specular.

Spun-bonded: fabric manufactured by a non-woven method in which fibers are deployed in a near-random fashion and bound by chemical or thermal means.

Standpipe: station pipe for use by a fire department to allow connection of a pumper truck to the outside of a building for delivery of water to hose stations on the interior of the building.

Stipple: to decorate by repeated touching as in dots.

Structural plan: a drawing, the primary purpose is to show the means of supporting a thing.

Substrate: the underlying or supporting layer.

Surface-mounted unit: a luminaire mounted directly on the ceiling.

Temper: to harden a material (metal or glass) through a specific process of heating and cooling, thereby releasing internal stresses created during the forming process.

Tenacity: resistance to tearing.

Tensile strength: the capacity to resist breaking or tearing apart under longitudinal stress.

Terrazzo: a polished concrete flooring traditionally made of crushed marble and cement.

Textile: a fiber construction; technically, a woven fabric.

Texture: tactile surface quality, perceived directly through touch or indirectly through vision.

Thermoplastic: a material that softens with application of heat and hardens again when cooled.

Thermoset: an irreversible property of a substance that is attained by application of heat to change the chemistry of the substance, thereby making it firm.

Tile: stone, concrete, or ceramic pieces, flattened or curved, used for roofing as well as wall, ceiling and floor covering. Also, thin modules of cork, vinyl, or other resilient material used primarily to protect and enhance interior walls, floors, and ceilings.

Tip-shear carpet: a carpet texture created by shearing either level loop or high-low loop carpet lightly so that only the higher filaments within the loops are sheared.

Torchere: an indirect floor lamp directing all, or nearly all, of the luminous flux upward.

Track: an electrical raceway permitting the flexible use and placement of various types of luminaries along its length.

Transformer: an electrical device for converting electrical potential.

Translucent: semi-opaque surface, which diffuses light while partially obscuring a view, *i.e.* privacy window, etched or frosted glass, window film

Traverse rod: a rod that uses carriers, pulleys, and cords to draw draperies open and closed.

Tread: the individual run, or horizontal distance from the face of one riser to the next, in a flight of stairs.

Trevira: polyester in which resistance to matting and crushing has been enhanced by autoclaving or heat setting the fibers.

Tuft bind: a measure of the effort required to remove a carpet tuft from its primary backing.

Tufting: a method of carpet construction which utilizes hundreds of needles to push pile yarns through a previously constructed primary backing sheet, forming loops or tufts of yarn which may be left uncut, cut, or selectively sheared to form a variety of surface textures, or by which high-low loops and sculptured effects can be created.

Twisting: the winding of two or more strands of fiber or yarn together to make a single multiple-ply yarn.

Universal design: appropriate to people of all ages, abilities, and sizes.

Upholstery: a soft covering of fabric on seating units, sometimes but not necessarily over padding, stuffing or springs.

Urethane: lightweight, cellular plastic capable of assuming nearly any density and thus any hardness from resilient to rigid. Urethane can be sprayed as surface coating, preformed as cushioning and insulation, or cast to form rigid components.

Utility core: a central space or unit, which contains all service elements, including bathrooms, heating, air conditioning etc.,

Valance: a decorative fabric heading at the top of a window.

Value: the attribute of perceived colour by which it seems to transmit or reflect a greater or lesser fraction of the light incident upon it.

Vaulted ceiling: a ceiling constructed as an extended arch, often semi-cylindrical in form (a barrel vault). Intersecting arches produce a groin vault; a ribbed vault reveals the framework of arched ribs.

Veiling reflection: partially or totally obscures the details to be seen by reducing the contrast.

Velvet: a fabric covered with a close, short, fine, dense nap or pile.

Veneer: paper-thin wood sheets applied to the outer surface of furniture, which are often inlaid or matched to create intricate, decorative patterns.

Vinyl: a versatile family of strong, lightweight plastics available in flexible, rigid, molded, film, foam, and cellular forms.

Viscose: a fiber manufactured from chemically treated cellulose.

Visual comfort probability (VCP): the rating of a lighting system expressed as a percentage of people who, when viewing from a specified location and in a specified direction, will be expected to find it acceptable in terms of discomfort glare.

Visual field: the sum of objects or points in space, which can be perceived when the head and eyes are kept fixed.

Visual surround: all portions of the visual field except the visual task.

Visual task: details and objects, which must be seen for the performance of a given activity, including the immediate background of details or objects.
Visual weight: the effect of visual impact regardless of actual weight, determined in part by color, texture and pattern.
Vitrify: to convert into glass by fusion; typically by heat.
VOC (Volatile Organic Compounds): a gaseous, sometimes harmful by-product of the curing process of some paints or adhesives.
Wafer board: a course-grained, rough-faced industrial board product, composed of wood flakes and adhesive; easily cut, but not intended for finish work.
Warp: the lengthwise yarns of a fabric, running through the loom, parallel to the edges of the run of fabric. The twist of an element away from its intended position caused by internal stresses due to changes in temperature or moisture.
Wardrobe: freestanding or built-in cabinet designed specifically to house clothing with hooks or rods.
Weaving: the process of interlacing two or more sets of yarns, usually at right angles to each other, to make textiles.
Weft: woven fabric. Also woof, the threads on a woven fabric which cross the warp.
Wilton: a specific weave of carpet in which yarns of different colors are raised in loops to form a pattern, after which the loops are cut to form a diverse plus pile.
Woof: the threads on a woven fabric, which cross the warp.
Working drawings: the final drawings that are used to obtain bids for and construct a design.
Work plane: the plane at which work is done and on which illuminance is specified and measured. Unless otherwise indicated, this is assumed to be a horizontal plane 0.76 m (30 in.) above the floor.
Wrought iron: iron, which is heated, bent and hammered into shape by hand, *i.e.* handforged iron.

INDEX

A

B

C

D

E

F

G

H

I

J

K

L

M

N

O

P

Q

R

S

T

Z